Student Solutions Manual

for use with

Chemistry

The Molecular Nature of Matter and Change

Third Edition

Martin S. Silberberg

Prepared by

Deborah Wiegand

University of Washington

and

Tristine Samberg

McGraw Hill

Boston Burr Ridge, IL Dubuque, IA Madison, WI New York San Francisco St. Louis
Bangkok Bogotá Caracas Lisbon London Madrid
Mexico City Milan New Delhi Seoul Singapore Sydney Taipei Toronto

McGraw-Hill Higher Education
A Division of The McGraw-Hill Companies

Student Solutions Manual for use with
CHEMISTRY: THE MOLECULAR NATURE OF MATTER AND CHANGE, THIRD EDITION.
MARTIN S. SILBERBERG.

Published by McGraw-Hill Higher Education, an imprint of The McGraw-Hill Companies, Inc., 1221 Avenue of the Americas, New York, NY 10020.

3 4 5 6 7 8 9 0 QPD QPD 0 9 8 7 6 5 4

ISBN 0-07-241052-3

www.mhhe.com

Contents

PREFACE

WELCOME TO YOUR STUDENT SOLUTIONS MANUAL

Your Student Solutions Manual (SSM) includes detailed solutions for the Follow-up problems and highlighted End-of-Chapter problems in the third edition of Martin Silberberg's *Chemistry: The Molecular Nature of Matter and Change.*

You should use the SSM in your study of chemistry as a study tool to:

- *better understand the reasoning behind problem solutions.* The plan-solution-check format illustrates the problem-solving thought process for all Follow-up problems and for selected End-of-Chapter problems.
- *better understand the concepts through their application in problems.* Explanations and hints for problems are in response to questions from students in general chemistry courses.
- *check your problem solutions.* Solutions provide comments on the solution process as well as the answer. In addition, alternate solutions are given for a few problems.

To succeed in general chemistry you must develop skill in problem-solving. Not only does this mean being able to follow a solution path and reproduce it on your own, but also to analyze problems you have never seen before and develop a solution strategy. Chemistry problems are story problems that bring together chemistry concepts and mathematical reasoning. In our years of teaching general chemistry we have found that the analysis of new problems is the most difficult step in problem-solving for students. You may face this in the initial few chapters or not until later in the year when the material is less familiar. When you find that you are having difficulty starting problems, don't be discouraged. This is an opportunity to learn new skills that will benefit you in future courses and your future career. The following two strategies, tested and found successful by our students, may help you develop the skills you need.

The first strategy is designed to make you more aware of your thought process as you solve problems. Draw a line down the right margin of your paper. As you solve a problem write your thought process in the margin – why you are doing each step and any questions that you ask during the solution. After completing the problem go back and make an outline of the process you used in the solution while reviewing the reasoning behind the solution. It may be useful to write a paragraph describing the solution process you used.

The second strategy helps you develop the ability to transfer a solution process to a new problem. After solving a problem, rewrite the problem to ask a different question. One way to rewrite the question is to ask the question backwards – find what has been given in the problem from the answer to the problem. Another approach is change the conditions – for instance, ask yourself what if the temperature is higher or there is twice as much carbon dioxide present? A third method is to change the reaction or process taking place – what if the substance is melting instead of boiling?

At times you may find slight differences between your answer and the one in the SSM. Two reasons may account for the differences. First, SSM calculations do not round answers until the final step so this may impact the exact numerical answer. Note that in preliminary calculations, one or two extra significant figures are retained and shown as subscripts in the answer. The second reason for discrepancies may be that your solution route was different than the one given in the SSM. Valid alternate solution paths exist for many problems, but space allows only a few alternate solutions to be included. So trust your solution as long as the discrepancy with the SSM answer is small and use the different calculation route to better understand the concepts used in the problem.

You may also find slight differences between the answer in the textbook appendix and the one in the SSM. These differences occur when questions are open-ended, such as, "Give an example of an acid." The example given in the SSM may differ from the one in the appendix, but both answers are correct.

Names and formulae of compounds are used interchangeably in the SSM to help familiarize you with the nomenclature of chemistry. Use this to review your ability to name compounds.

We thank Joan Weber, our editor at McGraw-Hill, for her assistance and both of our families for their support, editing efforts and technical expertise.

Tristine Samberg and Deborah Wiegand

March 2002

CHAPTER 1

KEYS TO THE STUDY OF CHEMISTRY

FOLLOW-UP PROBLEMS

1.1 a) Changing iodine solid to iodine vapor is a physical change since the composition of the iodine does not change.

b) Combustion of gasoline fumes is a chemical change since the gasoline is converted into other substances.

c) A scab forms as the result of a chemical change since the composition of the wound changes.

1.2 Plan: We know the area of each bolt and the area of fabric needed for each chair. Converting the area in ft^2 to m^2 using conversion factor: $1m^2 = (3.281\ ft)^2 = 10.76\ ft^2$

Solution:

$$3\,bolts\left(\frac{200.\,m^2}{bolt}\right)\left(\frac{10.76\,ft^2}{1\,m^2}\right)\left(\frac{1\,chair}{31.5\,ft^2}\right) = \mathbf{205\ chairs}$$

Check: 205 chairs would require 205 x 31.5 ft^2 = 6460 ft^2 of fabric. Three bolts contain 3 x 200. = 600. m^2 of fabric that converts to 6460 ft^2 of fabric.

1.3 Plan: The volume of the cylinder contents equals the volume of the water (31.8 mL) plus the volume of the cube. The volume of a cube equals (length of side)^3. The length of the side of the cube is in mm, so first we will convert this to cm, then find volume in units of cm^3. Using the conversion factor 1 cm^3 = 1mL, we convert the cube volume to mL and add it to the water volume also in mL. Then convert the mL to L: 1000m L = 1L.

Solution: Note that additional significant figures are retained in the calculation until the final step. The addition step limits the digits beyond the decimal point to one.

$$\left[15.0\,mm\left(\frac{1\,cm}{10\,mm}\right)\right]^3\left(\frac{1\,mL}{1\,cm^3}\right) = 3.375\,mL$$

$$(3.375\,mL + 31.8\,mL)\left(\frac{1\,L}{1000\,mL}\right) = \mathbf{0.0352\ L}$$

Check: Since the length of a side of the cube is greater than 1 cm its volume should be greater than 1 cm^3 or 1 mL. 3.375 mL is greater, so this result agrees with our expectations. Units of L are correct.

1.4 Plan: From the total time (1.5 h) and the number of raindrops per minute we can use the conversion factor 60 min = 1 h to determine the total number of raindrops. Multiplying the total number of raindrops times the mass of one raindrop gives the mass of the rain that falls in 1.5 h. Use conversion factors 1 g = 1000mg and 1 kg = 1000g to convert mg raindrops to kg raindrops.

Solution:

$$1.5\,h\left(\frac{60\ \text{min}}{1\ h}\right)\left(\frac{5.1\,x\,10^5\ raindrops}{\text{min}}\right)\left(\frac{65\ mg}{raindrop}\right)\left(\frac{1\ g}{1000\ mg}\right)\left(\frac{1\ kg}{1000\ g}\right) = \mathbf{3.0\,x\,10^3\ kg}$$

Check: The magnitude of the answer can be estimated: ~100min x 500000 raindrops/min x 0.0001kg/raindrop = 5000. The answer of 3000 kg is within this order of magnitude. Units are correct as kg of raindrops.

1.5 Plan: Mass can be calculated by multiplying the density by the volume. Then convert the mass in grams to kg using conversion factor 1000g = 1 kg.

Solution: $4.6\ cm^3\left(\frac{7.5\ g}{cm^3}\right)\left(\frac{1\ kg}{1000\ g}\right) = \mathbf{0.034\ kg}$

Check: The units are correct and the estimated answer is 5 x 7 = 35 g or 0.035 kg.

1.6 Plan: First convert the temperature in K to °C using formula T (in °C)= -273.15 + T (in K). Then convert the temperature in °C to °F using the formula T(in °F) = 9/5 T(in °C) +32.

Solution: T (in °C)= -273.15 + 234K = **-39 °C**

T(in °F) = 9/5(-39.15 °C) +32 = **-38 °F**

Check: The Celsius temperature is below zero as expected from a Kelvin temperature below 273. The temperature at which the Celsius and Fahrenheit scales have the same value is –40 degrees. Thus, with a Celsius temperature of –39 we'd expect the Fahrenheit temperature to be close, and it is.

1.7 Zeros between significant digits are significant. Zeros that end a number and lie either before or after the decimal point are significant.

a) 31.<u>0</u>7<u>0</u> mg; five significant figures

b) 0.06<u>0</u>6<u>0</u> g; four significant figures

c) 85<u>0</u>. °C; three significant figures – note terminal decimal point that makes zero significant.

d) 2.<u>000</u> x 10^2 mL; 4 significant figures

e) 3.9 x 10^{-6} m; 2 significant figures

f) 4.<u>0</u>1 x 10^{-4} L; 3 significant figures

1.8 Adding 25.65 and 37.4 will give an answer significant to the tenths digit. Dividing this answer with 3 significant figures by 73.55 with 4 significant figures and an exact number will give an answer with 3 significant figures.

$$\frac{25.65\ mL + 37.4\ mL}{73.55\ s\left(\frac{1\ min}{60\ s}\right)} = \mathbf{51.4\ mL / min}$$

END-OF-CHAPTER PROBLEMS

1.2 Gas molecules are distributed throughout the entire volume of a container while liquids and solids have a definite volume, so if the volume of the container is more than the volume of the liquid or solid the molecules will only fill part of the volume of the container.

a) Helium is a <u>gas</u> that expands to fill the entire volume of the balloon. The size of the balloon is determined by the balance of pressure exerted by the He gas atoms on the inner walls of the balloon and the pressure exerted by air molecules on the outside of the balloon.

b) Mercury forms a column of liquid inside the thin tube of the thermometer. The top of the mercury column has a surface. Mercury gas exists in the space between the liquid surface and sealed top of the thermometer.

c) Soup is a liquid that is contained by the sides of the soup bowl and forms a surface.

1.4 Physical property: characteristic of a substance that does not involve the production of another substance.

Chemical property: characteristic of a substance that results in the formation of another substance.

a) Color and physical state (gas, crystals) are examples of *physical* properties. The interaction between chlorine gas and sodium metal to produce another substance is an example of a *chemical* property.

b) Magnetism and color are examples of *physical* properties. Because no new substance is produced (iron remains iron after interaction with magnet), this is not a chemical property.

1.6 a) physical change – boiled canned soup can be cooled to obtain the original soup.

b) chemical change – a toasted slice of bread has changed in its composition such that it cannot change back into an untoasted slice of bread

c) physical change – although the log is in smaller pieces, the small pieces have the same chemical composition as the log.

d) chemical change – the wood is converted into other substances, such as ashes, soot and carbon dioxide gas.

1.8 a) Fuel has higher energy because energy is released from fuel to run car.

b) Wood has higher energy because heat energy is released as wood burns.

1.13 Central to Lavoisier's theory of combustion was his measurement of the masses of reactants and products. He found that the mass of reactants equaled the mass of the products when the masses of gases involved in the reaction were included. He discovered that the gas involved in combustion was part of the air and when this gas was absent from the air a candle would not burn. He named the gas he discovered oxygen.

1.16 A well-designed experiment

1) involves at least two variables that are believed to be related.
2) can be controlled so that only one variable is changed at a time.
3) is easily reproducible by other experimenters.
4) is able to connect what is changing to the cause of the change.

1.19 a) To convert from in^2 to cm^2, use $(2.54\ cm/1\ in)^2$ or $(6.45\ cm^2/1\ in^2)$.

b) To convert from km^2 to m^2, use $(1000\ m/1\ km)^2$ or $(1 \times 10^6\ m^2/1\ km^2)$

c) This problem requires two conversion factors: one for distance and one for time. To convert distance, mi to cm, use:

$(1.609\ km/1\ mi)x(1000\ m/1\ km)x(100\ cm/1\ m) = 1.609 \times 10^5\ cm/mi$.

To convert time, h to s, use:

(1 hr/60 min)(1 min/60 s) = 1 hr/3600 s.

Therefore, the complete conversion factor is (1.609 x 10^5 cm/mi)(1 hr/3600 s) = 5.792 x $10^8 \frac{cm \cdot hr}{mi \cdot s}$. Do the units cancel when you start with a measurement of mi/hr?

d) To convert from pounds (lb) to grams (g), use (1000 g/2.205 lb). To convert volume from ft^3 to cm^3 use conversion factor $\left[\left(\frac{0.3048\,m}{1\,ft}\right)\left(\frac{100\,cm}{1\,m}\right)\right]^3$ = 3.048 x 10^5 cm^3/ft^3.

1.21 The value of an extensive property depends on the amount of the substance, while the value of an intensive property does not depend on the amount of the substance. Density (b) and melting point (d) are intensive properties.

1.23 a) Density increases. Compressing the chlorine gas samples decreases the volume of the gas while the mass of chlorine gas molecules remains the same. The chlorine gas molecules are closer together in the smaller volume so the sample becomes more dense.

b) Density remains the same. The volume of a solid is unaffected by changes in atmospheric pressure.

c) Density decreases. Frozen water is less dense than liquid water because ice floats in water.

d) Density increases. Volume decreases with decrease in temperature so density increases.

e) Density remains the same. Submersion in water does not change the density of the diamond.

1.26 Use conversion factors: 1 m = 1 x 10^{12} pm; 1 m = 1 x 10^9 nm

$$radius = (128\ pm)\left(\frac{1\,m}{1x10^{12}\ pm}\right)\left(\frac{1x10^9\ nm}{1\,m}\right) = \mathbf{0.128\ nm}$$

1.28 Use conversion factors: 1 m = 100 cm; 1 inch = 2.54 cm.

$$length = 100.\ m\left(\frac{100\,cm}{1\,m}\right)\left(\frac{1\ inch}{2.54\,cm}\right) = \mathbf{3.94x10^3\ in}$$

1.30 a) Use conversion factors: $(1\ m)^2 = (100\ cm)^2$; $(1\ km)^2 = (1000\ m)^2$

($17.7\ cm^2$)($1\ m^2/1 \times 10^4\ cm^2$)($1\ km^2/1 \times 10^6\ m^2$) = **1.77 x 10^{-9} km^2**.

b) Use conversion factor: $(1\ inch)^2 = (2.54\ cm)^2$

($17.7\ cm^2$)($1\ in^2/6.45\ cm^2$)($\$3.25/\ in^2$) = **\$8.92**.

1.32 Use conversion factor: 1 kg = 2.205 lb. Assuming a body weight of 155 lbs,

$$155\,lb\left(\frac{1\,kg}{2.205\,lb}\right) = \mathbf{70.3\ kg}$$

1.34 a) $\left(\frac{5.52\ g}{cm^3}\right)\left(\frac{100\,cm}{m}\right)^3\left(\frac{kg}{1000\ g}\right) = \mathbf{5.52x10^3\ kg/m^3}$

b) $\left(\frac{5.52\ g}{cm^3}\right)\left(\frac{2.54\ cm}{in}\right)^3\left(\frac{12\ in}{ft}\right)^3\left(\frac{kg}{1000\ g}\right)\left(\frac{2.205\ lb}{kg}\right) = \mathbf{345\ lb/ft^3}$

1.36 a) $volume = \left(\frac{1.72\ \mu m^3}{cell}\right)\left(\frac{1\ mm}{1000\ \mu m}\right)^3 = \mathbf{1.72x10^{-9}\ mm^3/cell}$

b) $volume = \left(1x10^5\ cells\right)\left(\frac{1.72\ \mu m^3}{cell}\right)\left(\frac{1\ cm}{10000\ \mu m}\right)^3\left(\frac{1\ mL}{1\ cm^3}\right)\left(\frac{1\ L}{1000\ mL}\right) = \mathbf{2x10^{-10}\ L}$

1.38 a) mass of mercury = 185.56 g – 55.32 g = 130.24 g
volume of mercury = volume of vial = 130.24 g/(13.53 g/cm^3) = **9.626 cm^3**.
b) mass of water = 9.626 cm^3 x 0.997 g/cm^3 = 9.60 g
mass of vial filled with water = 55.32 g + 9.60 g = **64.92 g**

1.40 Volume of a cube = (length of side)^3
Use conversion factor: 1 cm = 10 mm

$$15.6\ mm\left(\frac{1\ cm}{10\ mm}\right) = 1.56\ cm;\ Al\ cube\ volume = (1.56\ cm)^3 = 3.79_{64}\ cm^3$$

Note: Keep extra significant figures until the end of problem, so use 3.79_{64} cm^3 as the cube volume with the subscripted 64 as the additional significant figures.
Density = mass/volume = 10.25 g/3.79_{64} cm^3 = **2.70 g/cm^3**

1.42 Equations 1.2 – 1.5 on pp. 31-32 show the conversions between the three temperature scales.

a) T (in °C) = [T(in °F) –32] $5/9$ = (72-32) $5/9$ = **22 °C**
T (in K) = T(in °C) + 273.15 = 22 + 273.15 = **295 K**
b) T (in K) = -164 + 273.15 = **109 K**
T (in °F) = $9/5$ T (in °C) + 32 = $9/5$ (-164) + 32 = **-263 °F**
c) T(in °C) = T(in K) – 273.15 = 0 – 273.15 = **-273 °C**
T (in °F) = $9/5$ T (in °C) + 32 = $9/5$ (-273) + 32 = **-459 °F**

1.44 $length = 0.025\ inch\left(\frac{2.54\ cm}{1\ inch}\right)\left(\frac{1\ m}{100\ cm}\right) = \mathbf{6.4x10^{-4}\ m}$

1.45 a) Use 1 m = 1 x 10^9 nm to convert wavelength: $545\ nm\left(\frac{1\ m}{10^9\ nm}\right) = \mathbf{5.45\ x\ 10^{-7}\ m}$

b) Use 1 A = 10^{-10} m and 1 m = 1 x 10^9 nm: $683\ nm\left(\frac{1\ m}{10^9\ nm}\right)\left(\frac{1\ \text{Å}}{10^{-10}\ m}\right) = \mathbf{6830\ \text{Å}}$

1.46 If we convert the mass of gold into its equivalent volume, we can determine the sheet thickness because the other two dimensions of volume (length x width = 40.0 ft^2) are given.

Volume of gold = mass ÷ density = 1.10 g/(19.32 g/cm^3) = 0.0569 cm^3

Volume = length x width x depth (or thickness), so 0.0569 cm^3 = 40.0 ft^2 x (depth)

$$0.0569\ cm^3 = 40.0\ ft^2\left(\frac{12\,in}{ft}\right)^2\left(\frac{2.54\,cm}{in}\right)^2(depth)$$

0.0569 cm^3 = (3.72 x 10^4 cm^2)(depth)

depth or thickness = **1.53 x 10^{-6} cm**

1.50 Rule: Zeros between significant digits are significant. Zeros that end a number containing a decimal point are significant.

1.52 a) no significant zeros; b) no significant zeros; c) 0.039<u>0</u>; d) 3.<u>0</u>9<u>00</u> x 10^4.

1.54 a) 0.00036; b) 35.83; c) 22.5

1.56 19 rounds to 20; 155 rounds to 160; 8.3 rounds to 3; 2.9 rounds to 3; 4.7 rounds to 5.

$$\left(\frac{20\times160\times8}{3\times3\times5}\right) = 600$$ (vs. 560 with original number of significant digits)

1.58 a) 1.34 m; b) 3.350x10^3 cm^3; c) 4.43x10^2 cm

1.60 a) 1.310000 x 10^5 (Note that all zeros are significant)

b) 4.7 x 10^{-4} (No zeros are significant)

c) 2.10006 x 10^5

d) 2.1605 x 10^3

1.62 a) 5550 Do not use terminal decimal point since zero is not significant.

b) 10070. Use terminal decimal point since final zero is significant.

c) 0.000000885

d) 0.003004

1.64 a) 8.025 x 10^4 Add exponents when multiplying powers of ten: 8.02 x 10^2 x 10^2

b) 1.0098 x 10^{-3} 1.0098 x 10^3 x 10^{-6}

c) 7.7 x 10^{-11} 7.7 x 10^{-2} x 10^{-9}

1.66 a) 4.06 x 10^{-19} J (489 x 10^{-9} m limits the answer to 3 significant figures)

b) 1.56 x 10^{24} molecules (1.19 x 10^2 g limits answer to 3 significant figures)

c) 1.82 x 10^5 J/mol (2.18 x 10^{-18} J/atom limits answer to 3 significant figures)

1.68 Exact numbers are those which have no uncertainty. Unit definitions and number counts of items in a group are examples of exact numbers.

a) The height of Angel Falls is a measured quantity. This is <u>not</u> an exact number.

b) The number of planets in the solar system is a number count. This <u>is</u> an exact number.

c) The number of grams in a pound is not a unit definition. This is <u>not</u> an exact number.

d) The number of millimeters in a meter is a definition of the prefix "milli-". This <u>is</u> an exact number.

1.70 Scale markings are 0.2 cm apart. The end of the metal strip falls between the mark for 7.4 cm and 7.6 cm. It is assumed that one can divide the space between markings into fourths. Thus, since the end of the metal strip falls between 7.45 and 7.55 we can report its length as 7.50 ± 0.05 cm. (Note: If the assumption is that one can divide the space between markings into halves only then the result is 7.5 ± 0.1 cm.)

1.73 a)

$$I_{avg} = \frac{8.72\,g + 8.74\,g + 8.70\,g}{3} = 8.72\,g$$

$$II_{avg} = \frac{8.56\,g + 8.77\,g + 8.83\,g}{3} = 8.72\,g$$

$$III_{avg} = \frac{8.50\,g + 8.48\,g + 8.51\,g}{3} = 8.50\,g$$

$$IV_{avg} = \frac{8.41\,g + 8.72\,g + 8.55\,g}{3} = 8.56\,g$$

Sets **I and II** are most accurate since their average value, 8.72 g, is closest to true value, 8.72 g.

b) Deviations are calculated as $|\text{measured value} - \text{average value}|$. Average deviation is the average of the deviations. To distinguish between the deviations an additional significant figure is retained.

$$I_{avgdev} = \frac{|8.72 - 8.72| + |8.74 - 8.72| + |8.70 - 8.72|}{3} = 0.013\,g$$

$$II_{avgdev} = \frac{|8.56 - 8.72| + |8.77 - 8.72| + |8.83 - 8.72|}{3} = 0.11\,g$$

$$III_{avgdev} = \frac{|8.50 - 8.50| + |8.48 - 8.50| + |8.51 - 8.50|}{3} = 0.010\,g$$

$$IV_{avgdev} = \frac{|8.41 - 8.56| + |8.72 - 8.56| + |8.55 - 8.56|}{3} = 0.11\,g$$

Set **III** is the most precise, but is the least accurate.

c) Set **I** has the best combination of high accuracy and high precision.

d) Set **IV** has both low accuracy and low precision.

1.76 a)

b)

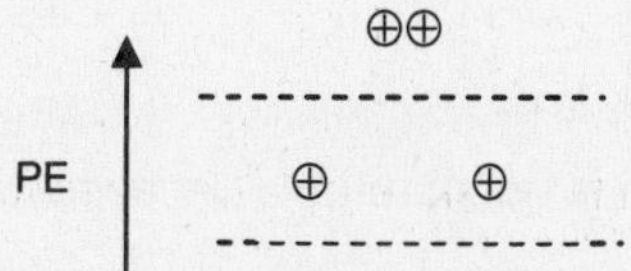

a) The balls on the relaxed spring have a lower potential energy and are more stable. The balls on the compressed spring have a higher potential energy, because the balls will move once the spring is released. This configuration is less stable.

b) The two ⊕ charges apart from each other have a lower potential energy and are more stable. The two ⊕ charges near each other have a higher potential energy, because they repel one another. This arrangement is less stable.

1.79 The volume of the room, and thus the volume of air in the room (assuming there is nothing else in the room to take up space) is calculated as follows:

$V_{air} = V_{room} = 11\ ft\ x\ 12\ ft\ x\ 8.5\ ft = 1.1_{22}\ x\ 10^3\ ft^3$

Conversion from ft^3 to L: $(1\ ft)^3 = (12\ inches)^3$; $(1\ inch)^3 = (2.54\ cm)^3$; $1 cm^3 = 1$ mL; 1000 mL = 1 L

$$volume = 1.1_{22}\ x\ 10^3\ ft^3 \left(\frac{1728\ in^3}{1\ ft^3}\right)\left(\frac{16.39\ cm^3}{1\ in^3}\right)\left(\frac{1\ mL}{1\ cm^3}\right)\left(\frac{1\ L}{1000\ mL}\right) = 3.1_{78}\ x\ 10^4\ L$$

At a rate of 1200L/min, how many minutes will it take to replace all the air in the room?

3.1_{78} x 10^4 L ÷ 1200 L/min = **26 minutes**

Note: 2 additional significant figures (subscripted numbers) were kept in the calculation until the final step.

1.82 a) Density of evacuated ball: $\left(\frac{0.12\ g}{560\ cm^3}\right)\left(\frac{1\ cm^3}{1\ mL}\right)\left(\frac{1000\ mL}{1\ L}\right) = 0.21\ g/L$

The evacuated ball will float because its density is less than that of air.

b) Because the density of CO_2 is greater than that of air, a ball filled with CO_2 will **sink**.

c) The added H_2 will contribute a mass of (0.0899 g/L)(0.560 L) = 0.0503 g. The H_2 filled ball will have a total mass of 0.0503 + 0.12 g = 0.17 g, and a resulting density of (0.17 g/0.560 L) = 0.30 g/L. The ball will **float** because density of the ball filled with hydrogen is less than density of air. *Note: The densities are additive because the volume of the ball and the volume of the H_2 gas are the same: 0.0899 + 0.21 g/L = 0.30 g/L.*

d) Because the density of O_2 is greater than that of air, a ball filled with O_2 will **sink**.

e) Density of ball filled with nitrogen: 0.21 g/L + 1.165 g/L = 1.38 g/L. Ball will **sink** because density of ball filled with nitrogen is greater than density of air.

f) To sink, the total mass of the ball and gas must weigh (1.189 g/L)(0.560 L) = 0.666 g. For ball filled with hydrogen: 0.666 – 0.17 g = 0.50 g. More than 0.50 g would have to be added to make the ball sink.

1.84 To decide if crown is made of pure gold the density of the crown must be calculated from its mass and volume.

$$\frac{\left(4\tfrac{13}{16}\ lb\right)\left(\frac{454\ g}{1\ lb}\right)}{(186\ mL)\left(\frac{1\ cm^3}{1\ mL}\right)} = 12\ g/cm^3$$

The crown is not pure gold because its density is not the same as the density of gold.

1.87 a) T(in °C) = T(in K) – 273.15 = 77.36 – 273.15 = **-195.79 °C**

b) T (in °F) = $\frac{9}{5}$ T (in °C) + 32 = $\frac{9}{5}$ (-195.79) + 32 = **-320.42 °F**

c) The mass of nitrogen is conserved, meaning that the mass of the nitrogen gas equals the mass of the liquid nitrogen. We first find the mass of liquid nitrogen and then convert that quantity to volume using the density of the <u>liquid</u>.

Mass of liquid nitrogen = mass of gaseous nitrogen = 895.0 L * (4.566 g/L) = 4087 g N_2

Volume of liquid N_2 = 4087 g liquid N_2 ÷ (809 g liquid N_2/L of liquid N_2) = **5.05 L**

1.89 a) $\left(\frac{5.9\,mi}{h}\right)\left(\frac{1000\ m}{0.6214\ mi}\right)\left(\frac{1\,h}{3600\,s}\right) = \mathbf{2.6\ m/s}$

b) $(98\,\text{min})\left(\frac{5.9\ mi}{h}\right)\left(\frac{1.609\,km}{1\ mi}\right)\left(\frac{1\,h}{60\,\text{min}}\right) = \mathbf{16\,km}$

c) $\dfrac{4.75\,x\,10^4\ ft}{\left(\frac{5.9\,mi}{h}\right)\left(\frac{5280\,ft}{1\,mi}\right)} = 1.5\,h$

If she starts running at 11:15, 1.5 hours later the time is **12:45 p.m.**

1.96 In visualizing the problem, the two scales can be set next to each other:

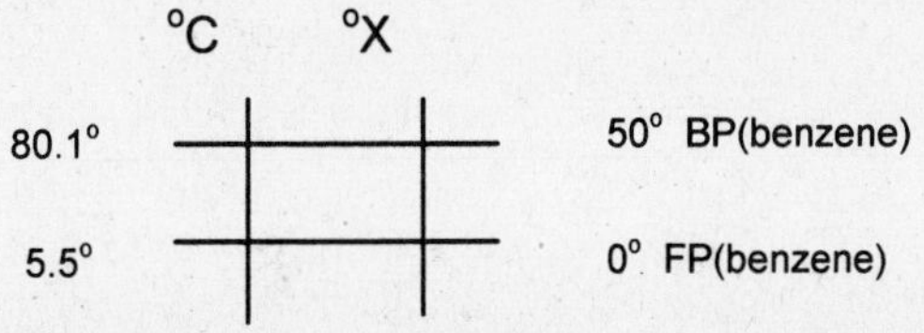

There are 50 divisions between the freezing point and boiling point of benzene on the °X scale and 74.6 divisions on the °C scale. So $^\circ X = \left(\frac{50\,^\circ X}{74.6\,^\circ C}\right)^\circ C$

This does not account for the offset of 5.5 divisions in the °C scale from the zero point on the °X scale.

So $^\circ X = \left(\frac{50\,^\circ X}{74.6\,^\circ C}\right)(^\circ C - 5.5^\circ C)$

<u>Check</u>: Plug in 80.1 °C and see if result agrees with expected value of 50 °X.

$$^\circ X = \left(\frac{50\,^\circ X}{74.6\,^\circ C}\right)(80.1^\circ C - 5.5^\circ C) = 50^\circ X$$

Use this formula to find the freezing and boiling points of water on the °X scale.

$$FP_{H_2O}\ in\ ^\circ X = \left(\frac{50\,^\circ X}{74.6\,^\circ C}\right)(0.00^\circ C - 5.5^\circ C) = -3.7^\circ X$$

$$BP_{H_2O}\ in\ ^\circ X = \left(\frac{50\,^\circ X}{74.6\,^\circ C}\right)(100.0^\circ C - 5.5^\circ C) = 63.3^\circ X$$

CHAPTER 2

THE COMPONENTS OF MATTER

FOLLOW-UP PROBLEMS

2.1 Plan: From the sample problem, the mass fraction of uranium in pitchblende is 71.4 g ÷ 84.2 g = 0.848. Because oxygen is the only other element present in pitchblende, the mass fraction of oxygen is (1.000 – 0.848) = 0.152. However, the problem specifies a mass of uranium, not mass of pitchblende. First, determine how much pitchblende is present. Then determine the mass of oxygen using the oxygen mass fraction.

Solution:

$$\text{mass pitchblende} = 2.3\,t\ \text{uranium} \times \frac{1.00\,t\ \text{pitchblende}}{0.848\,t\ \text{uranium}} = \mathbf{2.7\ t\ pitchblende}$$

$$\text{mass oxygen} = 2.7\,t\ \text{pitchblende} \times \frac{0.152\,t\ \text{oxygen}}{1.00\,t\ \text{pitchblende}} = \mathbf{0.41\ t\ oxygen}$$

Check: Because the majority of pitchblende is composed of uranium, it makes sense that oxygen has a smaller mass than uranium.

2.2 Plan: The superscript represents the *mass* number, A, (protons + neutrons) and subscript represents the *atomic* number, Z, (protons). Obtain the number of neutrons by calculating A – Z, recognize that the number of electrons = Z for an element, and identify element by comparing the Z value to the periodic table.

Solution:

a) A = 11 and Z = 5; therefore this element contains 5 p^+, 6 n^o, and 5 e^-. Q = Boron or **B**.

b) A = 41 and Z = 20; this element contains 20 p^+, 21 n^o, and 20 e^-. X = Calcium or **Ca**.

c) A = 131 and Z = 53; this element contains 53 p^+, 78 n^o, and 53 e^-. Y = Iodine or **I**.

2.3 Plan: Write an equation that expresses the atomic mass of boron as the sum of the two weighted isotopic abundances, and then substitute the known values.

Solution:

Atomic mass of B = (isotopic mass ^{10}B)(abundance of ^{10}B) + (isotopic mass ^{11}B)(abundance of ^{11}B)

$10.81 = 10.0129x + (11.0093)(1-x)$

$10.81 = 11.0093 - 0.9964x$

$-0.20 = -0.9964x$ (10.81 dictates that S.F.'s are restricted to the hundredths place in subtraction)

$x = 0.20$; $1 - x = 0.80$

% abundance of ^{10}B = **20.%**; % abundance of ^{11}B = **80.%**

Check: Boron-11 should have the larger % abundance because the average atomic mass (10.81 amu) is closer to the atomic mass of ^{11}B than the atomic mass of ^{10}B.

2.4 a) $\mathbf{S^{2-}}$. Sulfur is a nonmetal that gains 2 electrons to have the same number of electrons as $_{18}$Ar.

b) $\mathbf{Rb^+}$. Rubidium is an alkali metal that loses 1 electron to have the same number of electrons as $_{36}$Kr.

c) **Ba^{2+}**. Barium is an alkaline earth metal that loses 2 electrons to have the same number of electrons as $_{54}Xe$.

2.5 a) Zinc [Group 2B(12)] and oxygen [Group 6A(16)]
b) Silver [Group 1B(11)] and bromine [Group 7A(17)]
c) Lithium [Group 1A(1)] and chlorine [Group 7A(17)]
d) Aluminum [Group 3A(13)] and sulfur [Group 6A(16)]

2.6 a) Zn^{2+} and O^{2-} combine to form **ZnO**. The charge on one zinc ion, +2, is balanced by the charge on one oxide ion, -2, to give a compound with no net charge, ZnO.
b) Ag^{+} and Br^{-} combine to form **AgBr**. The charge on one silver ion, +1, is balanced by the charge on one bromide ion, -1, to give a compound with no net charge, AgBr.
c) Li^{+} and Cl^{-} combine to form **LiCl**. The charge on one lithium ion, +1, is balanced by the charge on one chloride ion, -1, to give a compound with no net charge, LiCl.
d) Al^{3+} and S^{2-} combine to form **Al_2S_3**. The charge on two aluminum ions, 2(+3) = +6, is balanced by the charge on three sulfide ions, 3(-2) = -6, to give a compound with no net charge, Al_2S_3.

2.7 a) Lead(IV) is Pb^{4+} and oxide is O^{2-}, so their combination yields **PbO_2**.
b) Sulfide has a –2 charge, so the copper has a +1 charge. Because copper can form more than one cation, the name needs to specifically identify the compound. The systematic name is **copper(I) sulfide** (common name = cuprous sulfide).
c) Bromide has a –1 charge, so the iron has a +2 charge. Iron can also form more than one cation, so its systematic name is **iron(II) bromide** (or ferrous bromide).
d) The mercuric ion is Hg^{2+}, so the correct formula is **$HgCl_2$**.

2.8 a) The cupric ion is Cu^{2+} and the sulfate ion is SO_4^{2-}; **$CuSO_4{\cdot}5H_2O$**.
b) The zinc ion is Zn^{2+} and the hydroxide ion is OH^-; **$Zn(OH)_2$**.
c) Lithium forms only one cation, so no numerals or suffixes are necessary; **lithium cyanide.**

2.9 a) The subscript for the ammonium ion and for the number of hydrogens in the ammonium ion have been switched. The ammonium ion is NH_4^+, so the correct formula is **$(NH_4)_3PO_4$**.
b) Parentheses are needed to around the polyatomic anion, OH^-; **$Al(OH)_3$**. The formula as written describes a compound with one Al atom, one oxygen atom, and three hydrogen atoms whereas the revised formula correctly reflects one Al ion combined with three hydroxide ions.
c) Mg is magnesium and can only have a +2 charge, therefore roman numerals are not needed. HCO_3^- is the hydrogen carbonate (or bicarbonate) ion. The correct name is **magnesium hydrogen carbonate**.
d) The suffix '-ic' is not used in combination with Roman numerals. NO_3^- is the nitr*ate* ion, not the nitr*ide* ion. The correct systematic name is **chromium(III) nitrate** (the common name is chromic nitrate).
e) Ca is the symbol for calcium, not cadmium. NO_2^- is the nitr*ite* ion, not nitr*ate*. The correct name is **calcium nitrite.**

2.10 a) Chloric acid is an oxoacid. "Chloric" corresponds to the chlorate ion (ClO_3^-) so the corresponding acid is **$HClO_3$**.
b) HF is a binary acid, **hydrofluoric acid.**

c) Acetic acid is an oxoacid and corresponds to the acetate ion, CH_3COO^-. The acid formula is **CH_3COOH** (also written as $HC_2H_3O_2$).

d) Sulfurous acid is an oxoacid and is derived from the sulfite ion, SO_3^{2-}. The acid formula is **H_2SO_3**.

e) HBrO is an oxoacid and has one less oxygen than the bromite ion (BrO_2^-). The "hypo-" prefix is used to reflect the correct name, **hyprobromous acid**.

2.11 a) sulfur trioxide b) silicon dioxide

c) N_2O d) SeF_6

2.12 a) The "-ous" suffix is not used in naming a binary covalent compound. The correct name is **disulfur dichloride**.

b) The first word in a binary compound has a prefix when more than one atom is present. The correct name is **dinitrogen monoxide**. Alternatively, the formula could also be corrected to NO so that it matches the nitrogen monoxide name.

c) Br is in a higher-numbered period and is named first. The correct name is **bromine trichloride**.

2.13 a) The formula is **H_2O_2** (see Table 2.5, peroxide ion = O_2^{2-}).

Molecular mass = (2 x 1.008 amu) + (2 x 16.00 amu) = **34.02 amu**.

b) **CsCl**; formula mass = 132.9 amu + 35.45 amu = **168.4 amu**.

c) **H_2SO_4**; molecular mass = (2 x 1.008 amu) + 32.07 amu + (4 x 16.00 amu) = **98.09 amu**.

d) **K_2SO_4**; formula mass = (2 x 39.10 amu) + 32.07 amu + (4 x 16.00 amu) = **174.27 amu**.

END-OF-CHAPTER PROBLEMS

2.1 An element is composed of only kind of atom whereas a compound is composed of more than one kind of atom.

2.4 a) Calcium chloride is a compound since more than one element is combined chemically.

b) Sulfur is an element since it consists of only one type of atom.

c) Baking powder is a mixture since it consists of more than one compound and the composition is not fixed.

d) Sand is a compound, silicon dioxide, but the sand found on beaches is a mixture of silicon dioxide and other substances such as rocks and shells.

2.9 The original sample was a mixture because it was composed of more than one component. The student separated a white, powdery substance that is distinct from the other substances in the original mineral sample in that it was soluble in water.

2.11 a) The law of mass conservation applies to all substances – elements, compounds and mixtures. Matter can neither be created nor destroyed, whether it is an element, compound or mixture.

b) The law of definite composition applies to compounds only, because it refers to a constant, or definite, composition of elements within a compound.

c) The law of multiple proportions applies to compounds only, because it refers to the combination of elements to form compounds.

2.13 a) Law of Definite Composition – The compound potassium chloride, KCl, is composed of the same elements and same fraction by mass, regardless of its source (Chile or Poland).

b) Law of Mass Conservation – The mass of the substance inside the flashbulb did not change during the chemical reaction (formation of magnesium oxide from magnesium and oxygen)

c) Law of Multiple Proportions – Two elements, O and As, can combine to form two different compounds that have different proportions of As present.

2.14 a) No. The mass percent of each element in a compound is fixed. The percentage of Na in the compound NaCl is 39.34% (22.99 amu / 58.44 amu), whether the sample is 0.5000 g or 50.00 g.

b) Yes. The mass of each element in a compound depends on the mass of the compound. A 0.5000 g sample of NaCl contains 0.1967 g of Na (39.34% of 0.5000 g), whereas a 50.00 g sample of NaCl contains 19.67 g of Na (39.34% of 50.00 g).

2.16 Experiments 1 and 2 together demonstrate the Law of Definite Composition. When 3.25 times the amount of reactants in experiment 1 is used in experiment 2, then 3.25 times the amount of products were made and the relative amounts of each product is the same in both experiments. In experiment 1 the ratio of white compound to colorless gas is 0.64:0.36 or 1.78:1 and in experiment 2 the ratio is 2.08:1.17 or 1.78:1. The two reactions also demonstrate the Law of Conservation of Mass since the total mass before reaction equals the total mass after reaction.

2.18 Fluorite is a mineral containing only calcium and fluorine.

a) mass of fluorine = mass of fluorite – mass of calcium = 2.76 g – 1.42 g = **1.34 g F**

b) To find the mass fraction of each element, divide the mass of each element by the mass of fluorite:

mass fraction of Ca = 1.42 g Ca / 2.76 g fluorite = **0.514**

mass fraction of F = 1.34 g F / 2.76 g fluorite = **0.486**

c) To find the mass percent of each element, multiply the mass fraction by 100:

mass % Ca = (0.514)(100) = **51.4%**

mass % F = (0.486)(100) = **48.6%**

2.20 a) If 1.25 g of MgO contains 0.754 g of Mg, then the mass ratio (or fraction) of magnesium in the oxide compound is 0.754 g/1.25 g = **0.603**.

b) $$435\,g\,MgO\left(\frac{0.603\,g\,Mg}{1\,g\,MgO}\right) = \mathbf{262\,g\,Mg}$$

Alternatively, the equation could be written as:

$$435\,g\,MgO\left(\frac{0.754\,g\,Mg}{1.25\,g\,MgO}\right) = \mathbf{262\,g\,Mg}$$

2.22 Since copper is a metal and sulfur is a nonmetal, the sample contains 88.39 g Cu and 44.61 g S. Calculate the mass fraction of each element in the sample:

$$mass\ fraction\ of\ Cu = \left(\frac{mass\ of\ element\ in\ cmpd}{mass\ of\ compound}\right) = \left(\frac{88.39\,g}{88.39\,g + 44.61\,g}\right) = 0.6645$$

$$mass\ fraction\ of\ S = \left(\frac{44.61\ g}{88.39\ g + 44.61\ g}\right) = 0.3354$$

Intermediate check: Do the mass fractions add to 1.000?

To find the mass of copper in the sample, multiply the sample mass by the mass fraction of Cu:

$$mass\ of\ Cu\ in\ sample = (5264\ kg\ compound)(1000\ g\ /\ kg)\left(\frac{0.6645\ g\ Cu}{1\ g\ compound}\right) = \mathbf{3.498x10^6\ g\ Cu}$$

Because the compound contains only Cu and S:

mass (g) of S = mass of compound – mass of Cu = 5264 kg – 3498 kg = **$1.766x10^6$ g S**

Alternatively, you could multiply the sample mass by the mass fraction of S to find the mass of S: (5264 kg)(1000 g/kg)(0.3354) = $1.766x10^6$ g S.

2.24 The law of multiple proportions states that if two elements form two different compounds, the relative amounts of the elements in the two compounds form a whole number ratio. To illustrate the law we must calculate the mass of one element to one gram of the other element for each compound and then compare this mass for the two compounds. The law states that the ratio of the two masses should be a small whole number ratio such as 1:2, 3:2, 4:3, etc.

$$Compound\ I: \frac{47.5\ g\ sulfur}{52.5\ g\ chlorine} = 0.905\ g\ S\ /\ g\ Cl$$

$$Compound\ II: \frac{31.1\ g\ sulfur}{68.9\ g\ chlorine} = 0.451\ g\ S\ /\ g\ Cl$$

$$Ratio = \frac{0.905}{0.451} = \frac{2}{1}$$

Thus, the ratio of the mass of sulfur per gram of chlorine in the two compounds is a small whole number ratio of 2 to 1, which agrees with the law of multiple proportions.

2.27 The coal type with the smallest mass percent of sulfur has the smallest environmental impact.

Mass % in Coal A = (11.3 g S / 378 g sample)(100) = 2.99% S (by mass)

Mass % in Coal B = (11.0 g S / 286 g sample)(100) = 3.85% S (by mass)

Mass % in Coal C = (20.6 g S / 675 g sample)(100) = 3.05% S (by mass)

Coal A has the smallest environmental impact.

2.29 Potassium nitrate, or KNO_3, is a compound composed of three elements – potassium, nitrogen and oxygen – in a specific ratio. If the ratio of elements changed, say to KNO_2, then the compound would be different (potassium nitrite) with different physical and chemical properties. Dalton postulated that atoms of an element are identical, regardless of whether that element is found in India or Italy. Dalton also postulated that compounds result from the chemical combination of specific ratios of different elements. Thus, Dalton's theory explains why KNO_3, a compound comprised of three different elements in a specific ratio, has the same chemical composition regardless of where it is mined or how it is synthesized.

2.30 Milikan determined the minimum *charge* on an oil drop and that the minimum charge was equal to the charge on one electron. Using Thomson's value for the *mass-to-charge ratio* of the electron and the determined value for the charge on one electron, Milikan calculated the mass of an electron (charge/(charge/mass)) to be 9.109×10^{-28} g.

2.34 Atomic number is the number of protons in the nucleus. Mass number is the sum of the number of protons and neutrons in the nucleus. The identity of an element is based on the number of protons and not the number of neutrons. The mass number can vary (by a change in number of neutrons) without changing the identity of the element.

2.37

Isotope	mass number	# of protons	# of neutrons	# of electrons
^{36}Ar	36	18	18	18
^{38}Ar	38	18	20	18
^{40}Ar	40	18	22	18

2.39 a) $^{16}_{8}O$ and $^{17}_{8}O$ have the same number of protons and electrons, but different numbers of neutrons. $^{16}_{8}O$ and $^{17}_{8}O$ are isotopes of oxygen, and $^{16}_{8}O$ has 8 neutrons whereas $^{17}_{8}O$ has 9 neutrons. **Same Z value.**

b) $^{40}_{18}Ar$ and $^{41}_{19}K$ have the same number of neutrons (22) but different numbers of protons and electrons. **Same N value.**

c) $^{60}_{27}Co$ and $^{60}_{28}Ni$ have different numbers of protons, neutrons, and electrons. **Same A value.**

2.41 a) $^{38}_{18}Ar$ b) $^{55}_{25}Mn$ c) $^{109}_{47}Ag$

2.43 a) $^{48}_{22}Ti$ b) $^{79}_{34}Se$ c) $^{11}_{5}B$

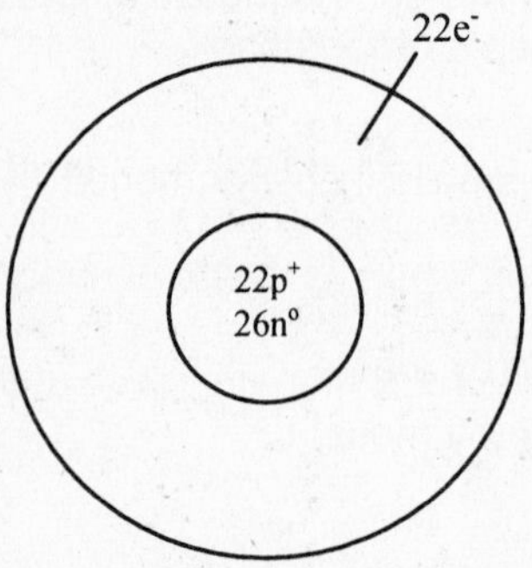

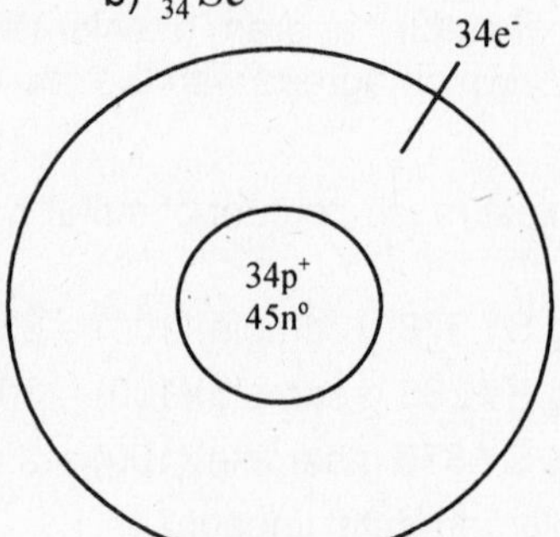

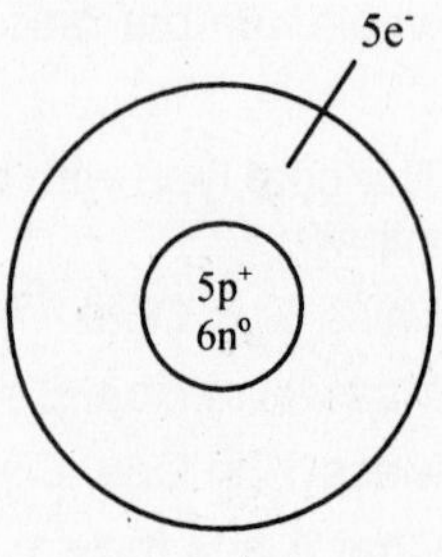

2.45 To calculate the atomic mass of an element take a weighted average based on the natural abundance of the isotopes.

atomic mass of gallium = (0.6011 x 68.9256 amu) + (0.3989 x 70.9247 amu) = **69.72 amu**

2.47 To find the percent abundance of each Cl isotope, let *x* equal the fractional abundance of ^{35}Cl and *(1-x)* equal the fractional abundance of ^{37}Cl.

Atomic mass of Cl = 35.4527

= 34.9689*x* + 36.9659*(1-x)*

$= 34.9689x + 36.9659 - 36.9659x$

$= 36.9659 - 1.9970x$

$1.9970x = 1.5132$

$x = 0.75774$ and $1\text{-}x = 0.24226$

% abundance ^{35}**Cl = 75.774%** % abundance ^{37}**Cl = 24.226%**

2.50 a) In the modern periodic table the elements are arranged in order of increasing atomic number.

b) Elements in a group have similar chemical properties.

c) Elements can be classified as metals, metalloids, or nonmetals.

2.53 The alkali metals (Group 1A(1)) readily lose one electron to form cations whereas the halogens (Group 7A(17)) readily gain one electron to form anions.

2.54

	Z	name	Symbol	group #	classification
a)	32	germanium	Ge	4A(14)	metalloid
b)	16	sulfur	S	6A(16)	nonmetal
c)	2	helium	He	8A(18)	nonmetal
d)	3	lithium	Li	1A(1)	metal
e)	42	molybdenum	Mo	6B(6)	metal

2.56 a) The symbol and atomic number of the heaviest alkaline earth metal are **Ra** and **88**.

b) The symbol and atomic number of the lightest metalloid in Group 5A(15) are **As** and **33.**

c) The symbol and atomic mass of the coinage metal whose atoms have the fewest electrons are **Cu** and **63.55 amu.**

d) The symbol and atomic mass of the halogen in Period 4 are **Br** and **79.90 amu**.

2.58 Ionic bonding occurs between an active metal and a nonmetal with the metal forming a cation and the nonmetal forming an anion. The opposite charges of the ions attract forming the ionic bond.

2.61 Coulomb's law states the energy of attraction in an ionic bond is directly proportional to the *product of charges* and inversely proportional to the *distance between charges.* Assuming that the ions in LiF and MgO are of similar size, the *distance between charges* in both ions is similar. However, the *product of charges* in MgO (+2 x -2) is greater than the *product of charges* in LiF (+1 x -1). Thus, MgO has stronger ionic bonding.

2.64 The monatomic ions of Group 1A(1) have a +1 charge (e.g. Li^+, Na^+, and K^+) whereas the monatomic ions of Group 7A(17) have a –1 charge (e.g. F^-, Cl^-, and Br^-). Elements gain or lose electrons to form ions with the same number of electrons as the nearest noble gas. For example, the alkali metal Rb loses one electron to form an ion with the same number of electrons as Kr. The halogen Br gains one electron to form an ion with the same number of electrons as Kr.

2.66 Potassium sulfide (K_2S) is an ionic compound formed from a metal (potassium) and a nonmetal (sulfur). Potassium atoms transfer electrons to sulfur atoms. Each potassium atom loses one electron to form an ion with +1 charge and the same number of electrons (18) as the noble gas argon. Each sulfur atom gains two electrons to form an ion with –2 charge and the same number of electrons (18) as the noble gas argon. The oppositely charge ions, K^+ and S^{2-}, attract each other to form an ionic compound with the ratio of two K^+ ions to one S^{2-} ion.

2.68 Potassium (K) is in Group 1A(1) and forms the $\mathbf{K^+}$ ion. Iodine (I) is in Group 7A(17) and forms the $\mathbf{I^-}$ ion.

2.70 Number of protons (atomic number) identifies the atom. Mass number = # protons + # neutrons.

	Name	Mass number	Group(column)	Period(row)
a)	Oxygen	17	6A(16)	2
b)	Fluorine	19	7A(17)	2
c)	Calcium	40	2A(2)	4

2.72 Lithium forms the Li^+ ion; oxygen forms the O^{2-} ion. The ionic compound that forms from the combination of these two ions must be electrically neutral, so two Li^+ ions combine with one O^{2-} ion to form the compound Li_2O. There are twice as many Li^+ ions as O^{2-} ions in a sample of Li_2O.
number of O^{2-} ions = (5.3 x 10^{20} Li^+ ions)(1 O^{2-} ion / 2 Li^+ ions) = $\mathbf{2.6x10^{20}\ O^{2-}}$ **ions** (see "Rules for Rounding Off", Section 1.6)

2.74 Coulomb's law states the energy of attraction in an ionic bond is directly proportional to the *product of charges* and inversely proportional to the *distance between charges.* (see also problem 2.61). The *product of the charges* is the same in both compounds because both sodium and potassium ions have a +1 charge. Attraction increases as distance decreases, so the ion with the smaller radius, Na^+, will form a stronger ionic interaction (**NaCl**).

2.76 An empirical formula describes the type and *ratio* of elements present in a compound whereas a molecular formula describes the type and *actual number* of elements in a molecule of the compound. The empirical formula and the molecular formula can be the same. For example, the compound formaldehyde has the molecular formula, CH_2O. The carbon, hydrogen and oxygen atoms are present in the ratio of 1:2:1. The ratio of elements cannot be further reduced, so formaldehyde's empirical formula and molecular formula are the same. Acetic acid has the molecular formula, $C_2H_4O_2$. The carbon, hydrogen and oxygen atoms are present in the ratio of 2:4:2, which can be reduced to 1:2:1. Therefore, acetic acid's empirical formula is CH_2O, which is different from its molecular formula. Note that the empirical formula does not *uniquely* identify a compound, because acetic acid and formaldehyde share the same empirical formula but are not the same compound.

2.78 The mixture is similar to the sample of hydrogen peroxide in that both contain 20 billion oxygen atoms and 20 billion hydrogen atoms. They differ in that they contain different types of molecules: H_2O_2 molecules in the hydrogen peroxide sample and H_2 and O_2 molecules in the mixture. In addition, the mixture contains 20 billion molecules (10 billion H_2 and 10 billion O_2) while the hydrogen peroxide sample contains 10 billion molecules.

2.82 a) To find the empirical formula for N_2H_4, divide the subscripts by the highest common multiple, 2:

Empirical formula of N_2H_4 = $N_{\frac{2}{2}}H_{\frac{4}{2}} = NH_2$

b) Empirical formula of glucose = $C_{\frac{6}{6}}H_{\frac{12}{6}}O_{\frac{6}{6}} = CH_2O$

2.84 a) Lithium is a metal that forms a +1 (group 1A) ion and nitrogen is a nonmetal that forms a –3 ion (group 5A). Compound is **Li_3N**, lithium nitride.

b) Oxygen is a nonmetal that forms a –2 ion (group 6A) and strontium is a metal that forms a +2 ion (group 2A). Compound is **SrO**, strontium oxide.

c) Aluminum is a metal that forms a +3 ion (group 3A) and chlorine is a nonmetal that forms a –1 ion (group 7A). Compound is **$AlCl_3$**, aluminum chloride.

2.86 a) $_{12}L$ is the element Mg (Z=12). Magnesium forms the Mg^{2+} ion. $_9M$ is the element F (Z=9). Fluorine forms the F^- ion. The compound formed by the combination of these two elements is **MgF_2,** magnesium fluoride.

b) $_{11}L$ is the element Na (Z=11). Sodium forms the Na^+ ion. $_{16}M$ is the element S (Z=16). Sulfur forms the S^{2-} ion. The compound formed by the combination of these two elements is **Na_2S**, sodium sulfide.

c) $_{17}L$ is the element Cl (Z=17). Chlorine forms the Cl^- ion. $_{38}M$ is the element Sr (Z=38). Strontium forms the Sr^{2+} ion. The compound formed by the combination of these two elements is **$SrCl_2$**, strontium chloride.

2.88 a) $SnCl_4$

b) iron(III) bromide (common name is ferric bromide)

c) CuBr (cuprous is +1 copper ion, cupric is +2 copper ion)

d) manganese(III) oxide

2.90 a) cobalt(II) oxide (cobalt forms more than one monatomic ion and must be specified)

b) Hg_2Cl_2 (mercury is an unusual case in which the +1 ion form is Hg_2^{2+}, not Hg^+)

c) lead(II) acetate trihydrate

d) Cr_2O_3 ('chromic' denotes a +3 charge, oxygen has a –2 charge)

2.92 a) Barium forms Ba^{2+} and oxygen forms O^{2-} so the neutral compound forms from one barium ion and one oxygen ion. Correct formula is **BaO**.

b) Iron(II) indicates Fe^{2+} and nitrate is NO_3^- so the neutral compound forms from one iron(II) ion and two nitrate ions. Correct formula is **$Fe(NO_3)_2$**.

c) Mn is the symbol for manganese. Mg is the correct symbol for magnesium. Correct formula is **MgS**. Sulfide is the S^{2-} ion and sulfite is the SO_3^{2-} ion.

2.94 a) Acids donate H^+ ion to solution, so the acid is a combination of H^+ and a negatively charged ion. Hydrogen sulfate is HSO_4^-, so its source acid is *(H^+ + HSO_4^-)* or H_2SO_4. Name of acid is sulfuric acid.

b) HIO_3, iodic acid.

c) Cyanide is CN^-; its source acid is HCN hydrocyanic acid.

d) H_2S, hydrosulfuric acid.

2.96 a) dinitrogen pentaoxide

b) ClF_3

c) silicon disulfide

2.98 Disulfur tetrafluoride, S_2F_4. Rule 1 on page 70 ("Names and Formulas of Binary Covalent Compounds) indicates that the element with the lower group number is named first.

2.100 a) Correct name is carbon monoxide.
b) Correct name is sulfur dioxide.
c) Correct name is dichlorine monoxide.

2.102 a) There are 12 atoms of oxygen in $Al_2(SO_4)_3$. The molecular mass is:

Al	=	2(26.98 amu)	=	53.96 amu
S	=	3(32.07 amu)	=	96.21 amu
O	=	12(16.00 amu)	=	192.0 amu
			=	**342.2 amu**

b) There are 9 atoms of hydrogen in $(NH_4)_2HPO_4$. The molecular mass is:

N	=	2(14.01 amu)	=	28.02 amu
H	=	9(1.008 amu)	=	9.072 amu
P	=	1(30.97 amu)	=	30.97 amu
O	=	4(16.00 amu)	=	64.00 amu
			=	**132.06 amu**

c) There are 8 atoms of oxygen in $Cu_3(OH)_2(CO_3)_2$. The molecular mass is:

Cu	=	3(63.55 amu)	=	190.6 amu
O	=	8(16.00 amu)	=	128.0 amu
H	=	2(1.008 amu)	=	2.016 amu
C	=	2(12.01 amu)	=	24.02 amu
			=	**344.6 amu**

2.104 a) $(NH_4)_2SO_4$

N	=	2(14.01 amu)	=	28.02 amu
H	=	8(1.008 amu)	=	8.064 amu
S	=	1(32.07 amu)	=	32.07 amu
O	=	4(16.00 amu)	=	64.00 amu
			=	**132.15 amu**

b) NaH_2PO_4

Na	=	1(22.99 amu)	=	22.99 amu
H	=	2(1.008 amu)	=	2.016 amu
P	=	1(30.97 amu)	=	30.97 amu
O	=	4(16.00 amu)	=	64.00 amu
			=	**119.98 amu**

c) $KHCO_3$

K	=	1(39.10 amu)	=	39.10 amu
H	=	1(1.008 amu)	=	1.008 amu
C	=	1(12.01 amu)	=	12.01 amu
O	=	3(16.00 amu)	=	48.00 amu
			=	**100.12 amu**

2.106 a) N_2O_5 = 2(14.01 amu) + 5(16.00 amu) = **108.02 amu**

b) $Pb(NO_3)_2$ = (207.2 amu) + 2(14.01 amu) + 6(16.00 amu) = **331.2 amu**

c) CaO_2 = (40.08) + 2(16.00 amu) = **72.08 amu**

2.108 a) Formula is SO_3. Name is sulfur trioxide.

Molecular mass = (32.07 amu) + 3(16.00 amu) = **80.07 amu**

b) Formula is C_3H_8. Since it contains only carbon and hydrogen it is a hydrocarbon and with three carbons its name is propane.

Molecular mass = 3(12.01 amu) + 8(1.008 amu) = **44.09 amu**

2.110 The compound's name is **disulfur dichloride**. (Note: Are you unsure when and when not to use a prefix? If you leave off a prefix, can you definitively identify the compound? The name *sulfur* dichloride would not exclusively identify the molecule in the diagram because sulfur dichloride could be any combination of sulfur atoms with two chlorine atoms. Use prefixes when the name may not uniquely identify a compound.) The empirical formula is $S_{\frac{2}{2}}Cl_{\frac{2}{2}}$ or **SCl**. The molecular mass is

2(32.07 amu) + 2(35.45 amu) = **135.04 amu**.

2.112 The components of a mixture can be separated by physical processes, such as boiling, whereas the components of a compound can only be separated by chemical processes.

2.115 a) Distilled water is a <u>compound</u> that consists of H_2O molecules only. How would you classify tap water?

b) Gasoline is a <u>homogeneous mixture</u> of hydrocarbon compounds of uniform composition that can be separated by physical means (distillation).

c) Beach sand is a <u>heterogeneous mixture</u> of different size particles of minerals and broken bits of shells.

d) Wine is a <u>homogenous mixture</u> of water, alcohol, tannins, and phenolic compounds that can be separated by physical means (distillation).

e) Air is a <u>homogeneous mixture</u> of different gases, mainly N_2, O_2, and Ar.

2.117 a) Salt dissolves in water while pepper does not. Procedure: add water to mixture, filter to remove solid pepper. Evaporate water to recover solid salt.

b) Sugar dissolves in water, sand does not. Procedure: add water to mixture, filter to remove sand. Evaporate water to recover sugar.

c) Oil is less volatile (lower tendency to vaporize to a gas) than water. Procedure: separate by distillation (boiling) collecting the water condensed from vapor and leaving the oil in the distillation flask.

d) Vegetable oil and vinegar (solution of acetic acid in water) are not soluble in each other. Procedure: Allow mixture to separate into two layers in a separatory funnel and drain off the lower (more dense) layer.

2.119 a) Filtration.

b) Extraction. The colored impurities are extracted into a solvent that is rinsed away from the raw sugar.

Or Chromatography. A sugar solution is passed through a column in which the impurities stick to the stationary phase and the sugar moves through the column in the mobile phase.

c) Extraction and filtration. The hot water extracts the soluble tannins, flavoring and caffeine from the coffee bean. The grounds are excluded from the coffee pot by filtering the mixture through a coffee filter.

2.121 a) $$fraction\ of\ volume = \frac{volume\ of\ nucleus}{volume\ of\ atom} = \frac{\frac{4}{3}\pi(2.5 x 10^{-15}\ m)^3}{\frac{4}{3}\pi(3.1 x 10^{-11}\ m)^3} = \mathbf{5.2 x 10^{-13}}$$

b) mass of nucleus = mass of atom – mass of electrons

= 6.64648 x 10^{-24} g – 2(9.10939 x 10^{-28} g)

= 6.64466 x 10^{-24}

$$fraction\ of\ mass = \frac{6.64466 x 10^{-24}\ g}{6.64648 x 10^{-24}\ g} = \mathbf{0.999726}$$

As expected the volume of the nucleus relative to the volume of the atom is small while its relative mass is large.

2.124 The key to solving these problems is carefully converting subscripts into multiplicative coefficients.

a) $[Co(NH_3)_6]Cl_3$ = (58.93 amu) + 6(14.01 amu) + 18(1.008 amu) + 3(35.45 amu)
= **267.48 amu**

b) $[Pt(NH_3)_4BrCl]Cl_2$ = (195.1 amu) + 4(14.01 amu) + 12(1.008 amu) + (79.90 amu) + 3(35.45 amu) = **449.5 amu**

c) $K_4[V(CN)_6]$ = 4(39.10 amu) + (50.94 amu) + 6(12.01 amu) + 6(14.01 amu) = **363.46 amu**

d) $[Ce(NH_3)_6][FeCl_4]_3$ = (140.1 amu) + 6(14.01 amu) + 18(1.008 amu) + 3(55.85 amu) + 12(35.45 amu) = **835.2 amu**

Check your answers by rechecking the multiplicative coefficients and adding the numbers in a different order on your calculator.

2.126 The rock is composed of 5.0% Fe_2SiO_4, 7.0% Mg_2SiO_4 and 88.0% SiO_2. It is comprised of four elements – iron, magnesium, silicon and oxygen. At first glance, one would expect that silicon and oxygen would have the highest mass percentages because the rock is mostly silicon dioxide.

1) First determine the fraction of each element in each mineral.

Molecular mass of Fe_2SiO_4 = 2(55.85 amu) + 28.09 amu + 4(16.00 amu) = 203.79 amu

Mass fraction of Fe = 2(55.85 amu) ÷ 203.79 amu = 0.5481

Mass fraction of Si = 28.09 amu ÷ 203.79 amu = 0.1378

Mass fraction of O = 4(16.00 amu) ÷ 203.79 amu = 0.3140

Molecular mass of Mg_2SiO_4 = 2(24.31 amu) + 28.09 amu + 4(16.00 amu) = 140.71 amu

Mass fraction of Mg = 2(24.31 amu) ÷ 140.71 amu = 0.3455

Mass fraction of Si = 28.09 amu ÷ 140.71 amu = 0.1996

Mass fraction of O = 4(16.00 amu) ÷ 140.71 amu = 0.4548

Molecular mass of SiO_2 = 28.09 amu + 2(16.00 amu) = 60.09 amu

Mass fraction of Si = 28.09 amu ÷ 60.09 amu = 0.4675

Mass fraction of O = 2(16.00 amu) ÷ 60.09 amu = 0.5325

2) The percent mass of each element in the rock can be found by multiplying the percent of each element in each mineral by the percent of that mineral in the rock.

% Fe = (0.050)(0.5481) * 100% = 2.7%

% Mg = (0.070)(0.3455) * 100% = 2.4%

% Si = [(0.050)(0.1378) + (0.070)(0.1996) + (0.880)(0.4675)] * 100% = 43.2%

% O = [(0.050)(0.3140) + (0.070)(0.4548) + (0.880)(0.5325)] *100% = 51.6%

3) The percentages add up to ~100% (rounding accounts for the small error) and the majority of the rock is silicon and oxygen as expected.

2.130 In each case, the mass of the starting materials (reactants) equals the mass of the ending materials (products), so the Law of Mass Conservation is observed. Each reaction also yields the product NaCl, not Na_2Cl or $NaCl_2$ or some other variation, so the Law of Definite Composition is observed. In this case, Na and Cl will always combine in a 1:1 ratio. In Case 2 and 3, the product NaCl forms and excess starting material is unused in the reaction.

2.131 a) For chloride ions:

$$mass\%\ Cl^- = \left(\frac{18980.\ mg\ Cl^-}{1\ kg\ seawater}\right)\left(\frac{1\ g}{1000\ mg}\right)\left(\frac{1\ kg}{1000 g}\right)(100\%) = 1.8980\%$$

Ion	Amount (mg/kg)	Mass %
Cl^-	18,980.	1.8980
Na^+	10,560.	1.0560
SO_4^{2-}	2,650.	0.2650
Mg^{2+}	1,270.	0.1270
Ca^{2+}	400.	0.0400
K^+	380.	0.0380
HCO_3^-	140.	0.0140

Comment: Should the mass % add up to 100? No, because the majority of seawater is H_2O.

b) Total mass of ions in 1 kg of seawater

= 18980. mg + 10560. mg + 2650. mg + 1270. mg + 400. mg + 380. mg + 140. mg

= 34380. mg

% Na^+ = (10560. mg Na^+/34380. mg total ions)(100) = **30.716%**

c) Alkaline earth metal ions are Mg^{2+} and Ca^{2+}. Total mass % = 0.1270 + 0.0400 = **0.1670%**

Alkali metal ions are K^+ and Na^+. Total mass % = 1.0560 + 0.0380 = **1.0940%**

Total mass percent for alkali metal ions is 6.6 times greater than the total mass percent for alkaline earth metal ions. Sodium ions (alkali metal ions) are dominant in seawater.

d) Anions are Cl^-, SO_4^{2-}, and HCO_3^-. Total mass % = 1.8980 + 0.2650 + 0.0140 = **2.1770%**

Cations are Na^+, Mg^{2+}, Ca^{2+}, and K^+. Total mass % = 1.0560 + 0.1270 + 0.0400 + 0.0380 = **1.2610%**

The mass fraction of anions is larger than the mass fraction of cations. Is the solution neutral since the mass of anions exceeds the mass of cations? Yes, although the mass is larger the number of positive charges equals the number of negative charges.

2.132 The molecular formula of succinic acid is $\mathbf{C_4H_6O_4}$. Dividing the subscripts by 2 yields the empirical formula $\mathbf{C_2H_3O_2}$. The molecular mass of succinic acid is

4(12.01 amu) + 6(1.008 amu) + 4(16.00 amu) = **118.09 amu.**

2.134 atomic mass of chromium = (0.0435)(49.946046 amu) + (0.8379)(51.940509 amu)

+ (0.0950)(52.940651 amu) + (0.0236)(53.938882 amu)

= **52.00 amu**

2.136 To find the molecular mass of potassium fluoride, add the atomic masses of potassium and fluorine. Fluorine has only one naturally occurring isotope, so the mass of this isotope equals the atomic mass of fluorine. The atomic mass of potassium is the weighted average of the two isotopic masses:

atomic mass of K = (0.93258)(38.9637 amu) + (0.06730)(40.9618 amu)

= 36.337 amu + 2.757 amu = 39.093 amu

The formula for potassium fluoride is KF, so its molecular mass is (18.9984 + 39.093) = **58.091 amu.**

2.138 a) If equal volumes of gases contain equal numbers of particles, the number of chlorine atoms in 1 L of chlorine gas must be the same as the number of chlorine atoms in 2 L of hydrogen chloride gas. Therefore, there must be two chlorine atoms per chlorine molecule, so chlorine gas is Cl_2.

b) Because equal volumes of hydrogen and chlorine react, there must be one atom of hydrogen per atom of chlorine in a molecule of hydrogen chloride.

c) Assuming a 1:1 atom ratio, a chlorine atom has a mass that is 1.000 g/0.0284 g, or 35.2 times the mass of a hydrogen atom. Using the value of 1.008 amu for the atomic mass of hydrogen, the Cl atom has an atomic mass of 35.5 amu (to 3 s.f.).

2.141 Calculate the mass percent of vanadium in each compound by dividing vanadium's contributing mass by the compound's molecular mass. The compound with the highest mass percent has the most vanadium per gram.

$$\%\,V\ in\ carnotite = \frac{2(50.94\,amu)}{2(39.10\,amu) + 2(238.0\,amu) + 15(16.00\,amu) + 2(50.94\,amu) + 6(1.008\,amu)}$$

$$= \frac{101.88}{902.1_3} = 0.1129 = 11.29\%$$

$$\%\,V\ in\ tyuyamunite = \frac{2(50.94\,amu)}{40.08\,amu + 2(238.0\,amu) + 18(16.00\,amu) + 2(50.94\,amu) + 12(1.008\,amu)}$$

$$= \frac{101.88}{918.0_6} = 0.1110 = 11.10\%$$

$$\%\,V\ in\ patronite = \frac{50.94\,amu}{50.94\,amu + 4(32.07\,amu)} = \frac{50.94}{179.2_2} = 0.2842 = 28.42\%$$

$$\% V \text{ in vanadinite} = \frac{3(50.94\ amu)}{5(207.2\ amu) + 3(50.94\ amu) + 12(16.00\ amu) + 35.45\ amu}$$

$$= \frac{152.82}{1416._3} = 0.1079 = 10.79\%$$

$$\% V \text{ in roscoelite} = \frac{2(50.94\ amu)}{39.10\ amu + 2(50.94\ amu) + 26.98\ amu + 3(28.09\ amu) + 12(16.00\ amu) + 2(1.008\ amu)}$$

$$= \frac{101.88}{446.25} = 0.22830 = 22.830\%$$

Patronite is the mineral source richest in vanadium; **vanadinite** has the least vanadium per gram. Without calculating the actual percentage, it is easy to see that patronite has the highest percentage of vanadium because the molecular mass is much smaller compared to the other minerals.

2.143 The molecular formula for TNT is $C_7H_5O_6N_3$. (What is its empirical formula?) The molecular mass of TNT is:

C =	7(12.01 amu)	=	84.08 amu
H =	5(1.008 amu)	=	5.040 amu
O =	6(16.00 amu)	=	96.00 amu
N =	3(14.01 amu)	=	42.03 amu
		=	227.14 amu

The mass percent of each element is:

C = (84.08 amu / 227.14 amu)(100) = **37.01% C**

H = (5.040 amu / 227.14 amu)(100) = **2.219% H**

O = (96.00 amu / 227.14 amu)(100) = **42.26% O**

N = (42.03 amu / 227.14 amu)(100) = **18.50% N**

Check: Do the percentages add up to 100%?

CHAPTER 3

STOICHIOMETRY: MOLE-MASS-NUMBER RELATIONSHIPS IN CHEMICAL SYSTEMS

FOLLOW-UP PROBLEMS

3.1 a) Plan: Given the mass of carbon in mg so first convert from mg to g. Then divide the grams of carbon by the molar mass of carbon to calculate the number of moles of carbon.

Solution: $$(315\,mg\,C)\left(\frac{1\,g}{1000\,mg}\right)\left(\frac{1\,mol\,C}{12.01\,g\,C}\right) = \mathbf{2.62 \times 10^{-2}\ mol\ C}$$

Check: Since 0.3 g is two orders of magnitude smaller than the mass of 1 mol of carbon it is expected that the number of moles to be about two orders of magnitude less than 1.

b) Plan: In this case the number of atoms of Mn are given so use Avogadro's number to get the equivalent number of moles of Mn. Then multiply the number of moles by the molar mass of manganese to determine the mass in grams.

Solution: $$(3.22 \times 10^{20}\ Mn\,atoms)\left(\frac{1\,mol}{6.022 \times 10^{23}\ atoms}\right)\left(\frac{54.94\,g\,Mn}{mol\,Mn}\right) = \mathbf{2.94 \times 10^{-2}\ g\ Mn}$$

Check: The number of atoms is less than Avogadro's number, the number of atoms in one mole, so it is expected that the mass of Mn is less than its molar mass.

3.2 a) Plan: First write the formula for tetraphosphorus decaoxide and then calculate its molar mass. Use Avogadro's number to convert the number of molecules to moles. Multiply by the molar mass of the compound to determine the mass in grams.

Solution: Formula indicates 4 phosphorus atoms and ten oxygen atoms: P_4O_{10}.

The molar mass, $\mathcal{M}$, is the sum of the atomic weights, expressed in g/mol:

P = 4(30.97) = 123.88 g/mol
O = 10(16.00) = 160.00 g/mol
= 283.88 g/mol of P_4O_{10}

$$(4.65 \times 10^{22}\ molecules\ P_4O_{10})\left(\frac{1\,mol}{6.022 \times 10^{23}\ molecules}\right)\left(\frac{283.88\,g\,P_4O_{10}}{mol\,P_4O_{10}}\right) = \mathbf{21.9\ g\ P_4O_{10}}$$

Check: The number of molecules is a little less than one tenth of Avogadro's number, so the number of grams should be a little less than one tenth of the molar mass. 21.9 g is a little less than 1/10(283.88) or 28.3 g.

b) Plan: The number of phosphorus atoms in one molecule is known from the formula. Multiply this number by the number of molecules will give the number of phosphorus atoms.

Solution: The formula P_4O_{10} indicates there are 4 P atoms per molecule of P_4O_{10}.

$$(4.65 \times 10^{22}\ molecules\ P_4O_{10})\left(\frac{4\ atoms\ P}{1\ molecule\ P_4O_{10}}\right) = \mathbf{1.86 \times 10^{23}\ atoms\ P}$$

Check: Since there are more phosphorus atoms than molecules the number of atoms is expected to be larger than the number of molecules. The calculated value of P atoms is larger than the given number of molecules.

3.3 Plan: First write the formula for ammonium nitrate and find its molar mass. Then multiply the number of nitrogen atoms in one molecule by the molar mass for nitrogen to find the mass of nitrogen in one mole of ammonium nitrate. Divide the mass of nitrogen in one mole of ammonium nitrate by the molar mass of ammonium nitrate to calculate the mass percent of nitrogen in ammonium nitrate. To find the mass of nitrogen in the given mass of ammonium nitrate, multiply the mass fraction (mass percent ÷ 100) by the mass of ammonium nitrate.

Solution: Formula for ammonium nitrate is NH_4NO_3. Thus, there are 2 atoms of N in each molecule.

a) Molar mass NH_4NO_3 = (2 x 14.01) + (4 x 1.008) + (3 x 16.00) = 80.05 g/mol

$$\left(\frac{2\ mol\ N}{mol\ NH_4NO_3}\right)\left(\frac{14.01\ g\ N}{mol\ N}\right)\left(\frac{1\ mol\ NH_4NO_3}{80.05\ g\ NH_4NO_3}\right)(100) = \mathbf{35.00\ \%N}$$

b) Mass of N = (0.3500)(35.8 kg)(1000 g/kg) = **1.25×10^4 g nitrogen**

Check: Ignoring the hydrogen in the formula for ammonium nitrate leaves 2N:3O. Since the molar masses of nitrogen and oxygen are similar the mass percent of nitrogen should be approximately 2/5 or 40%. It is close at 35%.

3.4 Plan: From the mass of sulfur calculate the moles of sulfur using the molar mass of sulfur. From the formula M_2S_3 the number of moles of M is known to be 2/3 of the number of moles of S. Also given is the mass of M which can be used to calculate the molar mass of M as mass of M divided by moles of M. Then use the periodic table to identify M by its molar mass.

Solution:

$$(2.88\ g\ S)\left(\frac{1\ mol\ S}{32.07\ g\ S}\right)\left(\frac{2\ mol\ M}{3\ mol\ S}\right) = 0.05987\ mol\ M$$

$$\frac{3.12\ g\ M}{0.05987\ mol\ M} = 52.1\ g\ M\ /\ mol\ M$$

The periodic table shows that chromium, Cr, with a molar mass of 52.00. M is **Cr** and M_2S_3 is **chromium(III) sulfide**.

Check: Given that the mass of the two elements in the compound are similar and since there are 2/3 as many M atoms as S atoms, the molar mass of M should be approximately 3/2 that of sulfur. The molar mass of sulfur, 32, times 3/2 is 48, which is close to the answer of 52.

3.5 Plan: The only two elements present in benzo[*a*]pyrene are carbon and hydrogen, so first assume some total mass of the substance, say 1.00 g, and calculate the number of moles of each element in that mass. Then find the lowest whole number mole ratio of the two elements and use that to write the empirical formula. To find the molecular formula, first calculate the molar mass of the empirical formula. Dividing the molar mass given for benzo[*a*]pyrene by the molar mass of the empirical formula should give an integer. Then multiply the subscripts in the empirical formula by this integer to get the molecular formula.

Solution: Assuming 1.00 g of benzo[*a*]pyrene gives 0.9521 g C and 0.0479 g H, find the moles of each element using their molar masses.

$$\frac{0.9521\ g\ C}{12.01\ g\ C\ /\ mol\ C} = 0.07928\ mol\ C\ in\ 1.00\ g\ compound$$

$$\frac{0.0479\ g\ H}{1.008\ g\ H\ /\ mol\ H} = 0.0475\ mol\ H\ in\ 1.00\ g\ compound$$

To find the smallest whole number ratio divide by the smaller of the two values for moles, 0.0475.

$$\frac{0.07928\ mol\ C}{0.0475} = 1.67\ mol\ C; \quad \frac{0.0475\ mol\ H}{0.0475} = 1.00\ mol\ H$$

The ratio 1.67:1 is not a whole number ratio since 1.67 is not a whole number. If the number were close to an integer, rounding it to the integer might work, but since 1.67 is not very close to 2 the ratio must be multiplied by an integer. Multiplying by 2 gives 3.34, still not close. Multiplying by 3 gives 5.01:3, which is 5:3, a whole number ratio. The conclusion is that the empirical formula is C_5H_3.

The molar mass for benzo[*a*]pyrene is given as 252.30. The molar mass for the empirical formula is

(5 x 12.01 g/mol) + (3 x 1.008 g/mol) = 63.07 g/mol

Dividing 252.30 g/mol by 63.07 g/mol gives 4.00. Now multiply the empirical formula by 4 to find the molecular formula of benzo[*a*]pyrene to be $\mathbf{C_{20}H_{12}}$.

Check: Check the molar mass for the molecular formula to make sure it matches the given molar mass. $\mathcal{M}$ = (20 x 12.01 g/mol) + (12 x 1.008 g/mol) = 252.296 g/mol which agrees with the given value.

3.6 Plan: All of the carbon in the solvent is converted to carbon dioxide during combustion. Calculate the mass of carbon in the solvent from the mass of carbon in the carbon dioxide produced. Similarly all the hydrogen from the solvent is converted to water during combustion so the mass of hydrogen in the water produced equals the mass of hydrogen in the solvent compound. The only element in the solvent that cannot be determined in this manner is chlorine since there is not a product containing chlorine. But, since the mass of carbon and hydrogen is known, subtracting their masses from the total mass of the solvent compound gives the mass of chlorine in the solvent. Once all masses are determined proceed to calculate moles of each and find the smallest whole number ratio of the three elements for the empirical formula. From the empirical formula calculate the empirical mass and divide that into the molar mass of the solvent compound to find the integer factor. This factor times the subscripts in the empirical formula gives the molecular formula of the compound.

Solution: The mass of carbon in the sample is calculated from the mass of carbon dioxide produced:

$$0.451\ g\ CO_2 \left(\frac{12.01\ g\ C}{44.01\ g\ CO_2}\right) = 0.123\ g\ C$$

The mass of hydrogen in the sample is calculated from the mass of water produced:

$$0.0617\ g\ H_2O \left(\frac{2\ x\ 1.008\ g\ H}{18.02\ g\ H_2O}\right) = 6.90\ x\ 10^{-3}\ g\ H$$

The mass of chlorine in the sample is the difference between the mass of the compound and the combined masses of the hydrogen and carbon:

0.250 g sample – 0.123 g C – 0.00690 g H = 0.120 g Cl

Next step is to calculate the moles of each element in the sample:

$$\frac{0.123\ g\ C}{12.01\ g\ C / mol\ C} = 0.0102\ mol\ C; \quad \frac{0.00690\ g\ H}{1.008\ g\ H / mol\ H} = 0.00684\ mol\ H; \quad \frac{0.120\ g\ Cl}{35.45\ g\ Cl / mol\ Cl} = 0.00338\ mol\ Cl$$

Dividing each mole value by the lowest value, 0.006875:

$$\frac{0.0102\ mol\ C}{0.00684} = 1.49\ mol\ C; \quad \frac{0.00684\ mol\ H}{0.00684} = 1.00\ mol\ H; \quad \frac{0.00338\ mol\ Cl}{0.00684} = 0.494\ mol\ Cl$$

Multiplying the ratio 1.49:1.00:0.494 by 2 gives 3:2:1, a whole number ratio. Thus, the empirical formula is C_3H_2Cl with empirical mass:

(3 x 12.01 g/mol) + (2 x 1.008 g/mol) + (35.45 g/mol) = 73.50 g/mol C_3H_2Cl

Dividing this empirical mass into the molar mass of the compound gives a factor of 146.99 g/mol ÷ 73.50 g/mol = 2. The molecular formula is twice the empirical formula, $\mathbf{C_6H_4Cl_2}$.

Check: The mass of carbon and chlorine in the sample are about equal. So since chlorine atoms weigh about three times as much as carbon atoms, the expected ratio of atoms in the molecule to be 3C:1Cl. This is indeed the ratio of carbon to chlorine that was calculated. The ratio of the sample masses of hydrogen and chlorine is 1:17 while their molar mass ratio is 1:35. Thus, twice as many hydrogen atoms as chlorine atoms are expected in the compound. The empirical formula does show 2 hydrogen atoms for each chlorine atom.

Also check that the molar mass of the molecular formula agrees with the given molar mass.

(6 x 12.01 g/mol) + (4 x 1.008 g/mol) + (2 x 35.45 g/mol) = 146.99 g/mol

3.7 Plan: Each of the questions requires that a description of a reaction be converted into a chemical equation. Steps to take are 1) deciding what substances are reacting and what substances are produced and writing their chemical formulas, 2) writing the equation, 3) balancing the atoms in the equation and 4) specifying the state of each reactant and product.

a) Solution: Sodium and water are reactants while hydrogen gas and sodium hydroxide are products. Sodium is element 11 with symbol Na and as a metal is a solid. Water is H_2O and exists in the liquid state at room temperature. Hydrogen occurs as a diatomic gas molecule, H_2. Sodium hydroxide consists of sodium ions, Na^+, and hydroxide ions, OH^-, and the problem states that it is in solution.

$Na(s) + H_2O(l) \rightarrow H_2(g) + NaOH(aq)$ is the equation.

As written, the reaction is not balanced because there are more hydrogen atoms as products than as reactants. There are 2 H atoms as reactant and 3 H atoms as product. To balance the hydrogens more needs to be added as reactants. The only option is to increase the number of water molecules to 2

$Na(s) + 2H_2O(l) \rightarrow H_2(g) + NaOH(aq)$

This now gives 4 H atoms as reactant and 3 as product, so increase the number of product H atoms. To increase the number of H atoms by 1 add a 2 in front of the sodium hydroxide.

$Na(s) + 2H_2O(l) \rightarrow H_2(g) + 2NaOH(aq)$

The hydrogens are now balanced with 4 H atoms on both sides of the equation. Now check the other atoms. There is one Na atom as reactant and two as product, so increase the number of Na atoms on the left by placing a coefficient of 2 in front of the Na metal.

$2Na(s) + 2H_2O(l) \rightarrow H_2(g) + 2NaOH(aq)$

Now the number of sodium and hydrogen atoms are balanced. For oxygen count 2 atoms as reactant and 2 as product, so all the atoms are balanced.

Check: Reactants (2Na, 4H, 2O) = Products 2Na, 4H, 2O)

b) Nitric acid and calcium carbonate are identified as reactants and carbon dioxide, water, and aqueous calcium nitrate as products. Formula of nitric acid is HNO_3. Calcium carbonate is an ionic compound of Ca^{2+} ions and CO_3^{2-} ions to give the formula $CaCO_3$.

One can interpret that calcium carbonate is a solid since marble is known to be a solid. Carbon dioxide is CO_2. Water is H_2O. Calcium nitrate combines of Ca^{2+} ions and NO_3^- ions in the formula $Ca(NO_3)_2$. Equation is

$$HNO_3(aq) + CaCO_3(s) \rightarrow CO_2(g) + H_2O(l) + Ca(NO_3)_2(aq)$$

Counting the atoms gives

Reactants: 1H, 1N, 6O, 1Ca, 1C

Products: 2H, 2N, 9O, 1Ca, 1C

which shows the hydrogen, nitrogen and oxygen atoms are not balanced. Nitrate and carbonate ions are polyatomic. When balancing an equation with a polyatomic ion it is best to first check if the ion occurs on both sides of the equation so that it can be balanced as the ion rather than the individual atoms. Carbonate appears only on the left side of the equation. Nitrate occurs both as a reactant in nitric acid and as a product in calcium nitrate. There are 1 NO_3^- as reactant and 2 NO_3^- as product. So, to balance the nitrate ions, add a coefficient of 2 for nitric acid.

$$2HNO_3(aq) + CaCO_3(s) \rightarrow CO_2(g) + H_2O(l) + Ca(NO_3)_2(aq)$$

Count atoms for this equation:

Reactants: 2H, 2N, 9O, 1Ca, 1C

Products: 2H, 2N, 9O, 1Ca, 1C

Now the reaction is balanced.

c) Phosphorus trifluoride, PF_3, is prepared, so it is a product. Phosphorus trichloride, PCl_3, and hydrogen fluoride, HF, are reactants. Hydrogen chloride, HCl, is a product. Equation is

$$PCl_3(g) + HF(g) \rightarrow PF_3(g) + HCl(g)$$

with 1P, 3Cl 1H and 1F on the left and 1P, 1Cl, 1H, and 3F on the left.

To equalize the number of chlorines on each side change the coefficient of HCl to 3:

$$PCl_3(g) + HF(g) \rightarrow PF_3(g) + 3HCl(g)$$

Now the chlorine atoms are balanced but the fluorine and hydrogen atoms are not. To equalize the number of fluorine atoms on each side change the coefficient of HF to 3:

$$PCl_3(g) + 3HF(g) \rightarrow PF_3(g) + 3HCl(g)$$

Check: Count atoms to find 1P, 3Cl, 3H, and 3F on both sides, so the equation is balanced.

d) Liquid nitroglycerine is the reactant that produces carbon dioxide, CO_2, water, H_2O, nitrogen, N_2, and oxygen, O_2, all as gases. The equation is

$$C_3H_5N_3O_9(l) \rightarrow CO_2(g) + H_2O(g) + N_2(g) + O_2(g)$$

Counting the atoms gives Reactants: 3C, 5H, 3N, 9O

Products: 1C, 2H, 2N, 5O

All the atoms except oxygen appear in only one of the products, so it will be easiest to balance carbon, hydrogen and nitrogen before balancing oxygen.

Starting with carbon (hydrogen and nitrogen would be just as good as starting points) place a coefficient of 3 for CO_2 to have 3C atoms as product.

$$C_3H_5N_3O_9(l) \rightarrow 3CO_2(g) + H_2O(g) + N_2(g) + O_2(g)$$

Now to balance the hydrogens by placing a coefficient of 2 for $C_3H_5N_3O_9$ and 5 for H_2O

$$2C_3H_5N_3O_9(l) \rightarrow 3CO_2(g) + 5H_2O(g) + N_2(g) + O_2(g)$$

The hydrogens are balanced, but now the carbons are no longer balanced (maybe starting with hydrogen would have been better?). The coefficient for CO_2 must be increased to 6 to equalize the number of C atoms.

$$2C_3H_5N_3O_9(l) \rightarrow 6CO_2(g) + 5H_2O(g) + N_2(g) + O_2(g)$$

Balance the nitrogen atoms as follows. There are 6N as reactant and 2 as product, so change the coefficient of N_2 to 3 to give 6N as product.

$2C_3H_5N_3O_9(l) \rightarrow 6CO_2(g) + 5H_2O(g) + 3N_2(g) + O_2(g)$

So, the carbon, hydrogen and nitrogen atoms are balanced, now check the oxygen atoms. There are 18 O atoms as reactant and 19 as product, so the number of nitroglycerin molecules need to be increased and that will change the number of carbon, hydrogen, and nitrogen atoms! Well, try it and this time start with hydrogen.

Balancing hydrogen: $4C_3H_5N_3O_9(l) \rightarrow 6CO_2(g) + 10H_2O(g) + 3N_2(g) + O_2(g)$

Balancing carbon: $4C_3H_5N_3O_9(l) \rightarrow 12CO_2(g) + 10H_2O(g) + 3N_2(g) + O_2(g)$

Balancing nitrogen: $4C_3H_5N_3O_9(l) \rightarrow 12CO_2(g) + 10H_2O(g) + 6N_2(g) + O_2(g)$

Now the totals are

Reactants: 12C, 20H, 12N, 36O

Products: 12C, 20H, 12N, 36O

And the reaction is balanced!

An alternative approach would be to insert a ½ for the coefficient of O_2 at the point that there were 18 O atoms as reactant and 19 as product. Then multiply all the coefficients by 2 to get rid of the fraction, although reactions are sometimes written with a fraction for a homonuclear diatomic molecule (molecule made up of two atoms of the same element).

3.8 a) Plan: First write the equation for the reaction and balance it. Convert the grams of aluminum to moles then use the stoichiometric ratio from the balanced equation to find the moles of iron formed from the given mass of aluminum. The final step is to convert the moles of iron to grams of iron.

Solution: Iron(III) oxide, Fe_2O_3, and aluminum, Al, are the reactants and are solids which is known either as a result of prior knowledge of the substances or of their description as powders in the problem. The products of the reaction are aluminum oxide, Al_2O_3, and iron, Fe. The problem states that aluminum oxide is a solid and the iron is molten, which means it is melted and exists in the liquid state. Now write the equation:

$Fe_2O_3(s) + Al(s) \rightarrow Al_2O_3(s) + Fe(l)$

Adding up atoms gives 2Fe, 3O, and 1Al as reactants and 1Fe, 3O, and 2Al as products.

Balancing the iron: $Fe_2O_3(s) + Al(s) \rightarrow Al_2O_3(s) + 2Fe(l)$

Balancing the aluminum: $Fe_2O_3(s) + 2Al(s) \rightarrow Al_2O_3(s) + 2Fe(l)$

Check the balancing: reactants: (2Fe, 3O, 2Al) = products (2Fe, 3O, 2Al)

Next calculate the grams of iron that are produced from 135 g aluminum.

$$135\, g\ Al \left(\frac{1\, mol\ Al}{26.98\, g\ Al}\right)\left(\frac{2\, mol\ Fe}{2\, mol\ Al}\right)\left(\frac{55.85\, g\ Fe}{1\, mol\ Fe}\right) = \mathbf{279\ g\ Fe}$$

Check: The same number of moles of aluminum will react as the number of moles of iron formed. Iron atoms are about twice as heavy as aluminum atoms so the mass of iron should be about twice the mass of aluminum, 2 x 135 g = 270 g. The calculated mass of iron, 279 g, is close to this value.

b) Plan: The answer to the number of aluminum atoms that react is equivalent to the number of aluminum atoms in 1.00 g of aluminum oxide. So if the number of moles of aluminum oxide in 1.00 g is multiplied by 2, since there are 2 moles of aluminum atoms in each mole of aluminum oxide, the product is the number of moles of aluminum atoms. Multiplying this number by Avogadro's number gives the number of aluminum atoms.

Solution:

$$1.00\ g\ Al_2O_3\left(\frac{1\ mol\ Al_2O_3}{101.96\ g\ Al_2O_3}\right)\left(\frac{2\ mol\ Al}{1\ mol\ Al_2O_3}\right)\left(\frac{6.022x10^{23}\ Al\ atoms}{1\ mol\ Al}\right) = \mathbf{1.18x10^{22}\ Al\ atoms}$$

Check: 1.00 g aluminum oxide is about 1/100 the mass of 1 mole of aluminum oxide so the number of aluminum oxide molecules should be 1/100 Avogadro's number, so 6 x 10^{21} molecules of Al_2O_3. There will be twice as many aluminum atoms as Al_2O_3 molecules, which is 1.2 x 10^{22} Al atoms, almost exactly the value calculated.

3.9 Plan: Write an equation for each step and balance it. Then add the two steps together making sure to adjust the coefficients for common substances to cancel them in the overall equation.

Solution:

First step: $SO_2(g) + O_2(g) \rightarrow SO_3(g)$

The sulfurs are balanced, but the oxygens are not. If a coefficient of 2 is added for SO_3 to increase the number of oxygens in the products, then a coefficient of 2 must be added for SO_2 to balance the sulfurs.

$$2SO_2(g) + O_2(g) \rightarrow 2SO_3(g)$$

Totaling the atoms shows this step is balanced.

Second step: $SO_3(g) + H_2O(l) \rightarrow H_2SO_4(aq)$

This step is balanced as written.

Now to add the two steps multiply the coefficients in the second step by 2 to cancel the sulfur trioxide in the overall equation.

First step: $2SO_2(g) + O_2(g) \rightarrow 2SO_3(g)$

Second step: $\underline{2SO_3(g) + 2H_2O(l) \rightarrow 2H_2SO_4(aq)}$

Overall: $2SO_2(g) + O_2(g) + \cancel{2SO_3(g)} + 2H_2O(l) \rightarrow \cancel{2SO_3(g)} + 2H_2SO_4(aq)$

$2SO_2(g) + O_2(g) + 2H_2O(l) \rightarrow 2H_2SO_4(aq)$

Check: Reactants (2S, 8O, 4H) = products (2S, 8O, 4H)

3.10 Plan: First write the equation and balance it. To find which reactant will limit the amount of product, calculate the moles of aluminum sulfide made from 10.0 g aluminum assuming the sulfur was present in excess. Then calculate the moles of aluminum sulfide made from 15.0 g sulfur assuming the aluminum is in excess. Whichever reactant will produce fewer moles of aluminum sulfide is identified as the limiting reactant and the number of moles of aluminum sulfide produced from the amount of the limiting reactant is used to calculate the grams of aluminum sulfide. To find the amount of the excess reactant calculate how much of it was used in the reaction from the amount of the limiting reactant. Subtract the amount used in the reaction from the starting amount to find the mass left after the reaction.

Solution: Balanced equation: $2Al(s) + 3S(s) \rightarrow Al_2S_3(s)$

Moles of aluminum sulfide from each reactant:

$$10.0\ g\ Al\left(\frac{1\ mol\ Al}{26.98\ g\ Al}\right)\left(\frac{1\ mol\ Al_2S_3}{2\ mol\ Al}\right) = 0.185\ mol\ Al_2S_3\ produced\ with\ excess\ S$$

$$15.0\ g\ S\left(\frac{1\ mol\ S}{32.07\ g\ S}\right)\left(\frac{1\ mol\ Al_2S_3}{3\ mol\ S}\right) = 0.156\ mol\ Al_2S_3\ produced\ with\ excess\ Al$$

Since fewer moles of aluminum sulfide are produced when sulfur is limiting, calculate the mass of aluminum sulfide that is equivalent to 0.156 mol Al_2S_3.

$$0.156\,mol\,Al_2S_3\left(\frac{150.17\,g\,Al_2S_3}{1\,mol\,Al_2S_3}\right) = \mathbf{23.4\,g\,Al_2S_3}$$

The mass of aluminum left is calculated from the mass of the limiting reactant.

$$15.0\,g\,S\left(\frac{1\,mol\,S}{32.07\,g\,S}\right)\left(\frac{2\,mol\,Al}{3\,mol\,S}\right)\left(\frac{26.98\,g\,Al}{1\,mol\,Al}\right) = 8.41\,g\,Al\,used\,in\,reaction$$

The mass of unreacted aluminum is calculated from the original mass and the mass that reacted.

Mass of aluminum unreacted = 10.0 g – 8.41 g = **1.6 g**

Check: From the balanced equation the stoichiometric ratio of aluminum to sulfur is 2:3. The ratio of their masses is also exactly 2:3. So, the limiting reactant should be the substance with the larger molar mass since there will be fewer atoms of it available to react. Sulfur's molar mass is 32.07 while aluminum's is 26.98 and by calculation sulfur is the limiting reactant. Another check is that when the mass of aluminum reacted was calculated from the mass of sulfur the result was a mass less than the starting mass of aluminum, so indeed some is not used in the reaction.

3.11 Plan: First write the equation and balance it. To find the theoretical yield calculate the mass of carbon dioxide that could form from the starting mass of marble. The percent yield is the ratio of the mass of carbon dioxide actually produced to the theoretical mass of carbon dioxide times 100.

Solution: Balanced equation: $CaCO_3(s) + 2HCl(aq) \rightarrow CaCl_2(aq) + H_2O(l) + CO_2(g)$

The theoretical mass of carbon dioxide:

Note: Additional significant figures are retained for the preliminary calculation and are shown as subscripts.

$$10.0\,g\,CaCO_3\left(\frac{1\,mol\,CaCO_3}{100.09\,g\,CaCO_3}\right)\left(\frac{1\,mol\,CO_2}{1\,mol\,CaCO_3}\right)\left(\frac{44.01\,g\,CO_2}{1\,mol\,CO_2}\right) = 4.39_{70}\,g\,CO_2$$

The percent yield:

$$\frac{3.65\,g}{4.39_{70}\,g}\,x\,100 = \mathbf{83.0\%}$$

Check: The starting mass of calcium carbonate is about 1/10 its molar mass, so the theoretical yield should be about 1/10 the molar mass of carbon dioxide, 0.1 x 44 = 4.4.

3.12 Plan: Molarity is moles of solute per liter of solution. Multiply the molarity of KI by the volume of solution in liters to find the moles of KI.

Solution: $(84\,mL\,solution)\left(\frac{1\,L}{1000\,mL}\right)\left(\frac{0.50\,mol\,KI}{L\,solution}\right) = \mathbf{0.042\,mol\,KI}$

Check: Since the molarity is 0.5 the ratio of mol KI to L solution should be 1:2. The ratio 0.042 mol : 0.084 L is 1:2.

3.13 Plan: Convert the given mass of sucrose to moles of sucrose then divide by the molarity to find the volume of solution.

Solution: $135\,g\,C_{12}H_{22}O_{11}\left(\frac{1\,mol\,C_{12}H_{22}O_{11}}{342.30\,g\,C_{12}H_{22}O_{11}}\right)\left(\frac{1\,L\,solution}{3.30\,mol\,C_{12}H_{22}O_{11}}\right) = \mathbf{0.120\,L}$

Check: Units cancel to give the answer in liters, which is what is requested in the problem. 135 g is about 1/3 of the mass of one mole. O.1 L of a 3M solution would also be about 1/3 mole. Calculation looks to be correct.

3.14 Plan: To calculate the molarity after a dilution use $M_{dil}V_{dil} = M_{conc}V_{conc}$. Then convert the molarity in moles per liter to grams per mL

Solution: $M_{dil} = \frac{M_{conc}V_{conc}}{V_{dil}} = \frac{7.50\ M \times 25.0\ mL}{500.\ mL} = 0.375\ M$

$$\left(\frac{0.375\ mol\ H_2SO_4}{L\ solution}\right)\left(\frac{98.09\ g\ H_2SO_4}{1\ mol\ H_2SO_4}\right)\left(\frac{1\ L}{1000\ mL}\right) = \mathbf{3.68x10^{-2}\ g\ H_2SO_4\ per\ mL\ solution}$$

Check: The first check is that the concentration after dilution is less than the concentration before dilution, 0.375 M < 7.50 M. Checking the calculation by doing a rough calculation gives 8 ÷ 20 x 100 ÷ 1000 = 0.04, which is close the answer of .0368 g/mL.

3.15 Plan: To compare effectiveness of antacids calculate the volume of 0.10 M HCl that is neutralized by a certain mass of the antacid. In Sample Problem 3.15 it was found that 3.4×10^{-2} L of 0.1 M HCl is neutralized by 0.10 g $Mg(OH)_2$. To calculate the volume of 0.10 M HCl neutralized by the same mass, 0.1 g, of $Al(OH)_3$ first write the balanced equation for the reaction between HCl and $Al(OH)_3$. To calculate the volume of acid use the stoichiometric ratio from the balanced equation, the molar mass of $Al(OH)_2$, and the molarity of the acid solution. Then compare the volume neutralized by the two antacids.

Solution: Effectiveness of aluminum hydroxide:

Balanced equation: $3HCl(aq) + Al(OH)_3(s) \rightarrow AlCl_3(aq) + 3H_2O(l)$

$$0.10\ g\ Al(OH)_3\left(\frac{1\ mol\ Al(OH)_3}{78.00\ g\ Al(OH)_3}\right)\left(\frac{3\ mol\ HCl}{1\ mol\ Al(OH)_3}\right)\left(\frac{1\ L\ sol'n}{0.10\ mol\ HCl}\right) = 0.038\ L\ HCl\ per\ gram\ Al(OH)_3$$

Aluminum hydroxide is more effective at neutralizing stomach acid than magnesium hydroxide based on an equivalent mass of each substance.

Check: Each mole of aluminum hydroxide neutralizes 3 moles of acid while each mole of magnesium hydroxide neutralizes only 2 moles of acid. So, unless the molar mass of aluminum hydroxide was more than 1.5 times that of magnesium hydroxide, 0.1 g of aluminum hydroxide would neutralize more acid than 0.1 g of magnesium hydroxide. The molar mass of aluminum hydroxide is only about 30% greater than the molar mass of magnesium hydroxide, so the finding that aluminum hydroxide is more effective appears correct.

3.16 a) Plan: To calculate the volume of $Pb(CH_3COO)_2$ solution containing 0.400 mol Pb^{2+} multiply the molarity of $Pb(CH_3COO)_2$ by the number of moles of lead ions per mole of lead acetate to get the molarity of lead ions. Then dividing 0.400 mol Pb^{2+} by the molarity of Pb^{2+} will give the volume.

(Note that the formula for acetate ions can be written as either $C_2H_3O_2^-$ or CH_3COO^-.)

Solution:

$$(1.50\ M\ Pb(CH_3COO)_2)\left(\frac{1\ mol\ Pb^{2+}}{1\ mol\ Pb(CH_3COO)_2}\right) = 1.50\ M\ Pb^{2+};\quad \frac{0.400\ mol\ Pb^{2+}}{1.50\ M} = \mathbf{0.267\ L\ soln}$$

Check: Calculating backwards gives

$$0.267\ L(1.50\ M\ Pb(CH_3COO)_2)\left(\frac{1\,mol\ Pb^{2+}}{1\,mol\ Pb(CH_3COO)_2}\right)=0.400\,molPb^{2+}$$

which gives the starting value, within rounding error, for moles of lead ions.

b) Plan: First write the balanced reaction for the precipitation of lead(II) chloride. Then calculate the amount of lead(II) chloride that would be formed from each reactant if the other reactant was in excess. The reactant that produces the smaller amount of product is the limiting reactant. The mass of lead(II) chloride that can form is that calculated from the limiting reactant.

Solution: Balanced reaction:

$Pb(CH_3COO)_2(aq) + 2NaCl(aq) \rightarrow PbCl_2(s) + 2NaCH_3COO(aq)$

Mass of product when lead(II) acetate is limiting reactant:

$$0.400\,mol\ Pb^{2+}\left(\frac{1\,mol\ PbCl_2}{1\,mol\ Pb^{2+}}\right)\left(\frac{278.1\,g\ PbCl_2}{1\,mol\ PbCl_2}\right)=111\,g\ PbCl_2$$

Mass of product when sodium chloride is limiting reactant:

$$(125\,mL\ sol'n)\left(\frac{1\,L}{1000\,mL}\right)\left(\frac{3.40\,mol\ NaCl}{L\ sol'n}\right)\left(\frac{1\,mol\ PbCl_2}{2\,mol\ NaCl}\right)\left(\frac{278.1\,g\ PbCl_2}{1\,mol\ PbCl_2}\right)=\mathbf{59.1\ g\ PbCl_2}$$

Since the amount of sodium chloride produces less product it is the limiting reactant and lead(II) acetate is present is excess. Thus, 59.1 g lead(II) chloride can form when 267 mL of 1.50 M lead(II) acetate is mixed with 125 mL of 3.40 M sodium chloride.

Check: The moles of each reactant are 0.400 of lead ions and 0.425 of chloride ions. Each mole of lead(II) chloride requires one mole of lead ions combining with two moles of chloride ions. For the sodium chloride to be limiting there would have to be ½ of 0.425 = 0.213 mol of lead ions. Since the amount of lead ions present is greater than this amount, chloride ions will limit the amount of product. One way to check the answer is to calculate the number of chloride ions in the lead(II) chloride product.

$$59.1\,g\ PbCl_2\left(\frac{1\,mol\ PbCl_2}{278.1\,g\ PbCl_2}\right)\left(\frac{2\,mol\ Cl^-}{1\,mol\ PbCl_2}\right)=0.425\,mol\ Cl^-$$

This calculation confirms the result of 59.1 g $PbCl_2$.

END-OF-CHAPTER PROBLEMS

3.2 a) There are **12 moles of C atoms** in 1 mole of sucrose, just as there are 12 atoms of carbon in 1 molecule of sucrose.

b) One mole of sucrose contains 6.022 x 10^{23} molecules of sucrose, and (12 x 6.022 x 10^{23}) or **7.226 x 10^{24} carbon atoms**.

$$(1\,mol\ C_{12}H_{22}O_{11})\left(\frac{12\,mol\ C}{1\,mol\ C_{12}H_{22}O_{11}}\right)\left(\frac{6.022\,x\,10^{23}\ C\ atoms}{1\,mol\ C\ atoms}\right)=7.226\,x\,10^{24}\ C\ atoms$$

3.6 a) The element on the **left** (green) has the higher molar mass because only 5 green balls are necessary to exactly counterbalance the mass of 6 yellow balls. Since the green ball is heavier, its atomic mass is larger, and therefore its molar mass is larger.

b) The element on the **left** (red) has more atoms per gram. This figure requires more thought because the number of red and blue balls is unequal and their masses are unequal. If each pan contained 3 balls, then the red balls would be lighter. The presence of six red balls means that they are that much lighter. Because the red ball is lighter, more red atoms are required to make 1 gram.

c) The element on the **left** (orange) has fewer atoms per gram. The orange balls are heavier, and it takes fewer orange balls to make 1 gram.

d) Both the left and right elements have the same number of atoms per mole. The number of atoms per mole (6.022×10^{23}) is constant and so is the same for every element.

3.8 a) The molar mass, $\mathcal{M}$, is the sum of the atomic weights, expressed in g/mol:

Sr = = 87.62 g Sr/mol $Sr(OH)_2$
O = (2 mol O)(16.00 g O/mol O) = 32.00 g O/mol $Sr(OH)_2$
H = (2 mol H)(1.008 g H/mol H) = 2.016 g H/mol $Sr(OH)_2$
= **121.64 g/mol of $Sr(OH)_2$**

b) $\mathcal{M}$ = (2 mol N)(14.01 g N/mol N) + (1 mol O)(16.00 g O/mol O) = **44.02 g/mol of N_2O**

c) $\mathcal{M}$ = (1 mol Na)(22.99 g Na/mol Na) + (1 mol Cl)(35.45 g Cl/mol Cl) + (3 mol O)(16.00 g O/mol O) = **106.44 g/mol of $NaClO_3$**

d) $\mathcal{M}$ = (2 mol Cr)(52.00 g Cr/mol Cr) + (3 mol O)(16.00 g O/mol O) = **152.00 g/mol of Cr_2O_3**

3.10 a) $\mathcal{M}$ = (1 mol Sn)(118.7 g Sn/mol Sn) + (2 mol O)(16.00 g O/mol O) = **150.7 g/mol of SnO_2**

b) $\mathcal{M}$ = (1 mol Ba)(137.3 g Ba/mol Ba) + (2 mol F)(19.00 g F/mol F) = **175.3 g/mol of BaF_2**

c) $\mathcal{M}$ = (2 mol Al)(26.98 g Al/mol Al) + (3 mol S)(32.07 g S/mol S) + (12 mol O)(16.00 g O/mol O) = **342.17g/mol of $Al_2(SO_4)_3$**

d) $\mathcal{M}$ = (1 mol Mn)(54.94g Mn/mol Mn) + (2 mol Cl)(35.45g Cl/mol Cl) = **125.84g/mol of $MnCl_2$**

3.12 The mass of a substance and its number of moles are related through the conversion factor of $\mathcal{M}$, the molar mass expressed in g/mol. The moles of a substance and the number of entities per mole are related by the conversion factor, Avogadro's number.

a) $\mathcal{M}$ of $KMnO_4$ = 39.10 + 54.94 + (4 x 16.00) = 158.04 g/mol of $KMnO_4$

$$mass\ KMnO_4 = (0.57\ mol\ KMnO_4)\left(\frac{158.04\ g\ KMnO_4}{1\ mol\ KMnO_4}\right) = \mathbf{9.0 \times 10^1\ g\ KMnO_4}$$

b) $\mathcal{M}$ of $Mg(NO_3)_2$ = 24.31 + (2 x 14.01) + (6 x 16.00) = 148.33 g/mol $Mg(NO_3)_2$

$$(8.18\ g\ Mg(NO_3)_2)\left(\frac{1\ mol\ Mg(NO_3)_2}{148.33\ g\ Mg(NO_3)_2}\right)\left(\frac{6\ mol\ O\ atoms}{1\ mol\ Mg(NO_3)_2}\right) = \mathbf{0.331\ mol\ O\ atoms}$$

c) $\mathcal{M}$ of $CuSO_4 \cdot 5H_2O$ = 63.55 + 32.07 + (4 x 16.00) + (5 x 18.016) = 249.70 g/mol (Note that the waters of hydration are included in the molar mass)

$$\left(\frac{8.1 \times 10^{-3}\ g\ CuSO_4 \cdot 5H_2O}{249.70\ g/mol\ CuSO_4 \cdot 5H_2O}\right)\left(\frac{9\ mol\ O\ atoms}{1\ mol\ CuSO_4 \cdot 5H_2O}\right)\left(\frac{6.022 \times 10^{23}\ O\ atoms}{1\ mol\ O\ atoms}\right) = \mathbf{1.8 \times 10^{20}\ O\ atoms}$$

3.14 a) $\mathcal{M}$ of $MnSO_4$ = (54.94 g Mn/mol Mn) + (32.07 g S/mol S) + (4 mol O x 16.00 g O/mol O) = 151.01 g/mol of $MnSO_4$

mass of $MnSO_4$ = 0.64 mol $MnSO_4$ x 151.01 g $MnSO_4$/mol $MnSO_4$ = **97 g $MnSO_4$**

b) $\mathcal{M}$ of $Fe(ClO_4)_3$ = (55.85g Fe/mol Fe) + (3 mol Cl x 35.45g Cl/mol Cl) + (12 mol O x 16.00g O/mol O) = 354.20 g/mol of $Fe(ClO_4)_3$

moles $Fe(ClO_4)_3$ = 15.8 g $Fe(ClO_4)_3$ ÷ 354.2 g $Fe(ClO_4)_3$/mol $Fe(ClO_4)_3$
= **4.46 x 10^{-2} mol $Fe(ClO_4)_3$**

c) $\mathcal{M}$ of NH_4NO_2 = (2 mol N x 14.01g N/mol N) + (4 mol H x 1.008g H/mol H)
+ (2 mol O x 16.00g O/mol O)
= 64.05 g/mol of NH_4NO_2

$$(92.6\, g\, NH_4NO_2)\left(\frac{1\, mol\, NH_4NO_2}{64.05\, g\, NH_4NO_2}\right)\left(\frac{2\, mol\, N}{1\, mol\, NH_4NO_2}\right)\left(\frac{6.022x10^{23}\, atoms}{1\, mol\, N}\right) = \mathbf{1.74x10^{24}\ N\ atoms}$$

3.16 a) Carbonate is a polyatomic anion with the formula, CO_3^{2-}. The correct formula for this ionic compound is Cu_2CO_3.
$\mathcal{M}$ of Cu_2CO_3 = (2 x 63.55) + 12.01 + (3 x 16.00) = 187.11 g/mol

$$mass\, Cu_2CO_3 = (8.41\, mol\, Cu_2CO_3)\left(\frac{187.11\, g\, Cu_2CO_3}{1\, mol\, Cu_2CO_3}\right) = \mathbf{1.57\, x\, 10^3\ g\, Cu_2CO_3}$$

b) $\mathcal{M}$ of N_2O_5 = (2 x 14.01) + (5 x 16.00) = 108.02 g/mol

$$mass\, of\, N_2O_5 = (2.04x10^{21}\, molecules\, N_2O_5)\left(\frac{1\, mol\, N_2O_5}{6.022x10^{23}\, molecules\, N_2O_5}\right)\left(\frac{108.02\, g\, N_2O_5}{1\, mol\, N_2O_5}\right) = \mathbf{0.366\, g\, N_2O_5}$$

c) The correct formula for this ionic compound is $NaClO_4$. There are Avogadro's number of entities (in this case, formula units) in a mole of this compound.
$\mathcal{M}$ of $NaClO_4$ = 22.99 + 35.45 + (4 x 16.00) = 122.44 g/mol

$$mol\, NaClO_4 = (57.9\, g\, NaClO_4)\left(\frac{1\, mol\, NaClO_4}{122.44\, g\, NaClO_4}\right) = \mathbf{0.473\, mol\, NaClO_4}$$

$$formula\, units\, NaClO_4 = (0.473\, mol\, NaClO_4)\left(\frac{6.022x10^{23}\, formula\, units\, NaClO_4}{1\, mol\, NaClO_4}\right) = \mathbf{2.85x10^{23}\ formula\ units}$$

d) The number of ions or atoms is calculated using Avogadro's number:

$$number\, of\, Na^+\, ions = (0.473\, mol\, NaClO_4)\left(\frac{1\, mol\, Na^+\, ions}{1\, mol\, NaClO_4}\right)\left(\frac{6.022x10^{23}\, Na^+\, ions}{1\, mol\, Na^+\, ions}\right) = \mathbf{2.85x10^{23}\ Na^+\ ions}$$

$number\ of\ ClO_4^-\ ions$ = **$2.85x10^{23}$ ions** *because the molar ratio is the same as above*

$$number\, of\, Cl\, atoms = (0.473\, mol\, NaClO_4)\left(\frac{1\, mol\, Cl\, atoms}{1\, mol\, NaClO_4}\right)\left(\frac{6.022x10^{23}\, Cl\, atoms}{1\, mol\, Cl\, atoms}\right) = \mathbf{2.85x10^{23}\ Cl\ atoms}$$

$$number\, of\, O\, atoms = (0.473\, mol\, NaClO_4)\left(\frac{4\, mol\, O\, atoms}{1\, mol\, NaClO_4}\right)\left(\frac{6.022x10^{23}\, O\, atoms}{1\, mol\, O\, atoms}\right) = \mathbf{1.14x10^{24}\ O\ atoms}$$

3.18 a) Ammonium bicarbonate is an ionic compound consisting of ammonium ions, NH_4^+ and bicarbonate ions, HCO_3^-. The formula of the compound is NH_4HCO_3.
$\mathcal{M}$ of NH_4HCO_3 = (14.01g/mol) + (5 x 1.008g/mol) + (12.01g/mol) + (3 x 16.00g/mol)
= 79.06 g/mol NH_4HCO_3

In 1 mole of ammonium bicarbonate, with a mass of 79.06 g, there are 5 H atoms with a mass of 5.040 g.

mass % of H in NH_4HCO_3 = 5.040 g H ÷ 79.06 g NH_4HCO_3 x 100% = **6.375%**

b) Sodium dihydrogen phosphate heptahydrate is a salt that consists of sodium ions, Na^+, dihydrogen phosphate ions, $H_2PO_4^-$, and seven waters of hydration. The formula is $NaH_2PO_4 \cdot 7H_2O$. Note that the waters of hydration are included in the molar mass.

$\mathcal{M}$ of $NaH_2PO_4 \cdot 7H_2O$ = (22.99 g/mol) + (16 x 1.008 g/mol) + (30.97g/mol) + (11 x 16.00 g/mol) = 246.09 g/mol $NaH_2PO_4 \cdot 7H_2O$

In each mole of $NaH_2PO_4 \cdot 7H_2O$ (with mass of 246.09g) there are 11 x 16.00g/mol or 176.00g of oxygen.

mass % O in $NaH_2PO_4 \cdot 7H_2O$ = 176.00 g O / 246.09 g $NaH_2PO_4 \cdot 7H_2O$ x 100 = **71.52%**

3.20 Mass fraction is identical to mass %, but is expressed in decimal rather than percentage form.

a) Cesium acetate is an ionic compound consisting of Cs^+ cations and $C_2H_3O_2^-$ anions. (Note that the formula for acetate ions can be written as either $C_2H_3O_2^-$ or CH_3COO^-.) The formula of the compound is $CsC_2H_3O_2$. One mole of $CsC_2H_3O_2$ weighs 191.9 g:

$\mathcal{M}$ of $CsC_2H_3O_2$ = 132.9 + (2 x 12.01) + (3 x 1.008) + (2 x 16.00) = 191.9 g/mol

mass fraction of C = 24.02 g/mol ÷ 191.9 g/mol = **0.1252**

b) The formula for this compound is $UO_2SO_4 \cdot 3H_2O$.

$\mathcal{M}$ of $UO_2SO_4 \cdot 3H_2O$ = 238.0 + (9 x 16.00) + 32.07 + (6 x 1.008) = 420.1 g/mol

mass fraction of O = 144.00 g/mol ÷ 420.1 g/mol = **0.3428**

3.23 The formula for cisplatin is $Pt(Cl)_2(NH_3)_2$

$\mathcal{M}$ of $Pt(Cl)_2(NH_3)_2$ = 195.1 + (2 x 35.45) + (2 x 14.01) + (6 x 1.008) = 300.1 g/mol

a) $$285.3\,g\,\mathrm{Pt(Cl)_2}(NH_3)_2\left(\frac{1\,mol\,\mathrm{Pt(Cl)_2}(NH_3)_2}{300.1\,g\,\mathrm{Pt(Cl)_2}(NH_3)_2}\right) = \mathbf{0.9507\ mol\ Pt(Cl)_2(NH_3)_2}$$

b) $$(0.98\ mol\ Pt(Cl)_2(NH_3)_2)\left(\frac{6\ mol\ H}{1\ mol\ Pt(Cl)_2(NH_3)_2}\right)\left(\frac{6.022x10^{23}\ atoms\ H}{1\ mol\ H}\right) = \mathbf{3.5 \times 10^{24}\ atoms\ H}$$

3.25 a) $\mathcal{M}$ of $Fe_2O_3 \cdot 4H_2O$ = (2 x 55.85) + (3 x 16.00) + (4 x 18.016) = 231.76 g/mol

$$(65.2\ kg\ Fe_2O_3 \cdot 4H_2O)\left(\frac{1000\,g}{1\,kg}\right)\left(\frac{1\,mol\ Fe_2O_3 \cdot 4H_2O}{231.76\,g\ Fe_2O_3 \cdot 4H_2O}\right) = \mathbf{281\ mol\ Fe_2O_3 \cdot 4H_2O}$$

b) There is one mole of Fe_2O_3 for every mole of rust, so there are also **281 moles of Fe_2O_3**.

c) Calculate grams of iron by determining the mass fraction of Fe (method 1), or determine the mole ratio and convert to grams (method 2).

Method 1: mass fraction of Fe in rust = 111.70 g Fe ÷ 231.76 g rust = 0.48196

Grams of Fe in rust = 0.48196 x (6.52 x 10^4 g) = **3.14 x 10^4 g Fe**

Method 2: $$(281\,mol\ Fe_2O_3 \cdot 4H_2O)\left(\frac{2\,mol\ Fe}{1\,mol\ Fe_2O_3 \cdot 4H_2O}\right)\left(\frac{55.85\,g\ Fe}{1\,mol\ Fe}\right) = \mathbf{3.14 \times 10^4\ g\,Fe}$$

3.27 Rank in order from highest effectiveness (highest %N) to lowest effectiveness (lowest %N):

Name	Formula	Molar Mass, g/mol	Mass %N
Potassium nitrate	KNO_3	101.11	14.01 g N ÷ 101.11 g KNO_3 = 13.86%
Ammonium nitrate	NH_4NO_3	80.05	28.02 g N ÷ 80.05 g NH_4NO_3 = 35.00%
Ammonium sulfate	$(NH_4)_2SO_4$	132.15	28.02 g N ÷ 132.15 g NH_4NO_3 = 21.20%
Urea	$CO(NH_2)_2$	60.06	28.02 g N ÷ 60.06 g $CO(NH_2)_2$ = 46.65%

Rank is $CO(NH_2)_2 > NH_4NO_3 > (NH_4)_2SO_4 > KNO_3$

3.29 If the molecular formula for hemoglobin was known, the number of Fe^{2+} ions in a molecule of hemoglobin could be calculated. Instead, use unit analysis to convert g/mol to ions/molecule:

Mass of Fe^{2+} ions in one mole of hemoglobin = $(0.0033)(6.8 \times 10^4$ g Fe/mol Fe)
= $2.2_{44} \times 10^2$ g Fe^{2+}

$$\left(\frac{2.2_{44} \times 10^2 \ g\ Fe^{2+}}{1\ mol\ hemoglobin}\right)\left(\frac{1\ mol\ Fe^{2+}}{55.85\ g\ Fe^{2+}}\right)\left(\frac{6.022 x 10^{23}\ Fe^{2+}\ ions}{1\ mol\ Fe^{2+}}\right)\left(\frac{1\ mol\ hemoglobin}{6.022 x 10^{23}\ molecules\ Hgn}\right) = 4.0$$

There are **4 Fe^{2+} ions** in a molecule of hemoglobin.

3.31 a) No, you can obtain the empirical formula from the number of moles of each type of atom in a compound, but not the molecular formula.

b) Yes you can obtain the molecular formula from the mass per cents and the total number of atoms. Solution plan:

1) Assume a 100.0 g sample and convert masses (from the mass % of each element) to moles using molar mass.

2) Identify the element with the lowest number of moles and use this number to divide into the number of moles for each element. You now have at least one elemental mole ratio (the one with the smallest number of moles) equal to 1.00 and the remaining mole ratios that are larger than one.

3) Examine the numbers to determine if they are whole numbers. If not, multiply each number by a whole number factor to get whole numbers for each element. You will have to use some judgement to decide when to round.

4) Write the empirical formula using the whole numbers from step 3.

5) Check the total number of atoms in the empirical formula. If it equals the total number of atoms given then the empirical formula is also the molecular formula. If not, then divide the total number of atoms given by the total number of atoms in the empirical formula. This should give a whole number. Multiply the number of atoms of each element in the empirical formula by this whole number to get the molecular formula. If you do not get a whole number when you divide return to step 3 and revise how you multiplied and rounded to get whole numbers for each element.

c) Yes, you can determine the molecular formula from the mass per cents and the number of atoms of one element in a compound. Solution plan:

1) Follow steps 1-4 in part b).

2) Compare the number of atoms given for the one element to the number in the empirical formula. Determine the factor the number in the empirical formula must be multiplied by to obtain the given number of atoms for that element. Multiply the empirical formula by this number to get the molecular formula.

d) No, the mass % will only lead to the empirical formula.

e) Yes, a structural formula shows all the atoms in the compound. Solution plan: Count the number of atoms of each type of element and record as the number for the molecular formula.

3.33 The empirical formula is the simplest formula that accurately reflects the *ratio* of elements in a compound.

a) C_2H_4 has a ratio of 2 carbon atoms to 4 hydrogen atoms, or 2:4. This ratio can be reduced to1:2, so that the empirical formula is $\mathbf{CH_2}$. The empirical formula mass is 12.01 + 2(1.008) = **14.03 g/mol**.

b) The ratio of atoms is 2:6:2, or 1:3:1. The empirical formula is $\mathbf{CH_3O}$ and its empirical formula mass is 12.01 + 3(1.008) + 16.00 = **31.03 g/mol**.

c) Since, the ratio of elements cannot be further reduced, the molecular formula and empirical formula are the same, $\mathbf{N_2O_5}$. The formula mass is 2(14.01) + 5(16.00) = **108.02 g/mol**.

d) The ratio of elements is 3 atoms of barium to 2 atoms of phosphorus to 8 atoms of oxygen, or 3:2:8. This ratio cannot be further reduced, so the empirical formula is also $\mathbf{Ba_3(PO_4)_2}$, with a formula mass of 3(137.3) + 2(30.97) + 8(16.00) = **601.8 g/mol**.

e) The empirical formula is $\mathbf{TeI_4}$, and the formula mass is (127.6 + 4*126.9) = **635.2 g/mol**.

3.35 a) CH_2 has empirical mass equal to 14.03 g/mol;

$$\frac{42.08g/mol}{14.03g/mol} \cong 3$$

multiplying the subscripts in CH_2 by 3 gives $\mathbf{C_3H_6}$

b) NH_2 has empirical mass equal to 16.03 g/mol;

$$\frac{32.05g/mol}{16.03g/mol} \cong 2$$

multiplying the subscripts in NH_2 by 2 gives $\mathbf{N_2H_4}$

c) NO_2 has empirical mass equal to 46.01 g/mol;

$$\frac{92.02g/mol}{46.01g/mol} \cong 2$$

multiplying the subscripts in NO_2 by 2 gives $\mathbf{N_2O_4}$

d) CHN has empirical mass equal to 27.03 g/mol;

$$\frac{135.14g/mol}{27.03g/mol} \cong 5$$

multiplying the subscripts in CHN by 5 gives $\mathbf{C_5H_5N_5}$

3.37 a) A compound's formula represents the ratio of elements present in 1) numbers of atoms or 2) numbers of moles. Therefore, the formula for this compound could be represented as $Cl_{0.063}O_{0.22}$, but a previous rule dictates that the subscripts are written as whole numbers. To convert to whole numbers, divide the subscripts by the smallest fraction, 0.063. The formula then becomes $ClO_{3.5}$. The ratio of elements is 1:3.5, or 2:7. The empirical formula is $\mathbf{Cl_2O_7}$.

b) The masses have to be converted to moles in order to determine the empirical formula.

$$(2.45\,g\,Si)\left(\frac{1\,mol\,Si}{28.09\,g\,Si}\right) = 0.0872\,mol\,Si; \frac{0.0872}{0.0872} = 1$$

$$(12.4\,g\,Cl)\left(\frac{1\,mol\,Cl}{35.45\,g\,Cl}\right) = 0.350\,mol\,Cl; \frac{0.350}{0.0872} \approx 4$$

The empirical formula is **$SiCl_4$**.

c) Assume a 100 g sample and convert the masses to moles.

$$(27.3\,g\,C)\left(\frac{1\,mol\,C}{12.01\,g\,C}\right) = 2.27\,mol\,C; \frac{2.27}{2.27} = 1$$

$$(72.7\,g\,O)\left(\frac{1\,mol\,O}{16.00\,g\,O}\right) = 4.54\,mol\,O; \frac{4.54}{2.27} \approx 2$$

The empirical formula is **CO_2**.

3.39 An oxide of nitrogen would have the general formula N_xO_y

a) To find the empirical formula, determine the lowest whole number combination for x and y. Assuming a 100.0 g sample of the oxide, it would contain 30.45 g of nitrogen and 100.0 g - 30.45 g = 69.55 g oxygen.

How many moles of nitrogen are in 30.45 g and how many moles of oxygen are in 69.55g?

$$mol\,N = (30.45\,g\,N)\left(\frac{1\,mol\,N}{14.01\,g\,N}\right) = 2.173\,mol\,N$$

$$mol\,O = (69.55\,g\,O)\left(\frac{1\,mol\,O}{16.00\,g\,O}\right) = 4.347\,mol\,O$$

Dividing both by the smallest value, 2.173, gives 1.000 mol N and 2.000 mol O. The empirical formula is **NO_2**.

b) The molar mass of the empirical formula, NO_2, is 46.03 g/mol.

$$\frac{90\,g/mol}{46.03\,g/mol} \approx 2$$

Multiply the subscripts in empirical formula by 2 to get the molecular formula: **N_2O_4**

3.41 The balanced equation for this reaction is: $M(s) + F_2(g) \rightarrow MF_2(s)$ since fluorine, like other halogens, exists as a diatomic molecule.

a) Using the molar relationship in the equation,

$$(0.600\,mol\,M)\left(\frac{1\,mol\,MF_2}{1\,mol\,M}\right)\left(\frac{2\,mol\,F}{1\,mol\,MF_2}\right) = \mathbf{1.20\ mol\ F}$$

b) If there are 1.20 moles of F in the MF_2, then the mass of fluorine in the sample is:

$$(1.20\,mol\,F)\left(\frac{19.00\,g\,F}{1\,mol\,F}\right) = 22.8\text{ g F}$$

The difference in mass, (46.8 g – 22.8 g) = **24.0 g**, is the mass of M.

c) 24.0 g of M is equal to 0.600 mol M, so molar mass of M = 24.0 g/0.600 mol = 40.0 g/mol. The element, M, is probably **calcium**.

3.44 Are carbon, hydrogen and oxygen the only elements in cortisol? Check by adding the per cents of each element. Since they total close to 100%, then the formula will include only these three elements. First find the empirical formula from the mass per cents and then use the empirical formula and the molar mass of cortisol to find the molecular formula.

Assuming a 100.0g sample of cortisol would contain 69.6g C, 8.34g H and 22.1g O, calculate the number of moles of each element and divide the total moles of each element by the lowest number:

$$mol\ C = (69.6\ g\ C)\left(\frac{1\ mol\ C}{12.01\ g\ C}\right) = 5.80\ mol\ C;\ \frac{5.80}{1.38} = 4.20$$

$$mol\ H = (8.34\ g\ H)\left(\frac{1\ mol\ H}{1.008\ g\ H}\right) = 8.27\ mol\ H;\ \frac{8.27}{1.38} = 5.99$$

$$mol\ O = (22.1\ g\ O)\left(\frac{1\ mol\ O}{16.00\ g\ O}\right) = 1.38\ mol\ O;\ \frac{1.38}{1.38} = 1.00$$

Both the values for hydrogen and oxygen give whole numbers, but for carbon the value is 4.20, which requires a decision whether to round the number or not. There are no hard and fast rules for when to round and when not to round, so it requires trial and error.

Trial 1: With rounding the 4.20 to 4, which is very tempting when two of the three elements are so close to a whole number, the empirical formula is C_4H_6O, molar mass of 70.09g/mol. Dividing the given molar mass for cortisol (362.47) by the molar mass of the empirical formula gives 5.18. Rounding this to 5 gives a molecular formula of $C_{20}H_{30}O_5$ and molar mass 350.44g/mol.

Trial 2: The first multiple of 4.20 that is a whole number is 5: 5 x 4.20 = 21. Multiplying all the element values by 5 gives the formula $C_{21}H_{30}O_5$, molar mass of 362.45g/mol.

The two formulas are similar - just one more carbon in the formula from trial 2. The best formula is the one with molar mass closest to the given value. Thus, $\mathbf{C_{21}H_{30}O_5}$ from trial 2 is the molecular formula based on the given data.

3.46 In combustion analysis, finding the moles of carbon and hydrogen is relatively simple because all of the carbon present in the sample is found in the carbon of CO_2, and all of the hydrogen present in the sample is found in the hydrogen of H_2O. The moles of oxygen are more difficult to find, because additional O_2 was added to cause the combustion reaction.

Step 1: Find the grams of C and H present in the sample, and subtract those masses from the original sample mass (0.1595 g) to find the grams of oxygen. Since the masses are so small, keep additional significant figures at the beginning of the calculation. It is acceptable to keep more significant figures during the intermediate steps and then round the final answer to the correct number of significant figures.

$$mass\ of\ C = (0.449\ g\ CO_2)\left(\frac{1\ mol\ CO_2}{44.01\ g\ CO_2}\right)\left(\frac{1\ mol\ C}{1\ mol\ CO_2}\right)\left(\frac{12.01\ g\ C}{1\ mol\ C}\right) = 0.1225\ g\ C$$

$$mass\ of\ H = (0.184\ g\ H_2O)\left(\frac{1\ mol\ H_2O}{18.016\ g\ H_2O}\right)\left(\frac{2\ mol\ H}{1\ mol\ H_2O}\right)\left(\frac{1.008\ g\ H}{1\ mol\ H}\right) = 0.02059\ g\ H$$

$$mass\ of\ O = 0.1595\ g - (0.1225 + 0.02059) = 0.0164\ g\ O$$

Step 2: Convert sample masses to moles and construct a preliminary empirical formula.

$$mol\ C = (0.1225\ g\ C)\left(\frac{1\ mol\ C}{12.01\ g\ C}\right) = 0.01020\ mol\ C$$

$$mol\ H = (0.02060\ g\ H)\left(\frac{1\ mol\ H}{1.008\ g\ H}\right) = 0.02044\ mol\ H$$

$$mol\ O = (0.0164\ g\ O)\left(\frac{1\ mol\ O}{16.00\ g\ O}\right) = 0.001025\ mol\ O$$

The preliminary empirical formula is $C_{0.01020}H_{0.02044}O_{0.001025}$.

Step 3: Determine the final empirical formula by expressing the ratio of elements in the smallest whole numbers.

$$C_{\frac{0.01020}{0.001025}}H_{\frac{0.02044}{0.001025}}O_{\frac{0.0010205}{0.0010205}} = C_{9.95}H_{19.94}O_1 \cong C_{10}H_{20}O$$

Step 4: Determine the molecular formula by dividing the molecular mass by the empirical mass. Multiply the subscripts in the empirical formula by that factor. Since (molecular mass/empirical mass) = 1, the molecular formula is also $\mathbf{C_{10}H_{20}O}$.

3.47 A balanced chemical equation describes:

a) which substances are present before and after a reaction

b) the molar (and molecular) ratios by which reactants react to form products

c) the physical states of all substances in the reaction

3.50 The correct answer is **b**). Both A and B are diatomic, so their formulas are expressed as A_2 and B_2. The answer d) gives the same reaction but it is incorrect because the coefficients are not in their reduced form.

3.51 a) 16 $Cu(s)$ + __ $S_8(s) \rightarrow$ 8 $Cu_2S(s)$

b) __ $P_4O_{10}(s)$ + 6 $H_2O(l) \rightarrow$ 4 $H_3PO_4(l)$

Hint: Balance the P first, because there is an obvious deficiency of P on the right side of the equation. Balance the H next, because H is present in only one reactant and only one product. Balance the O last, because it appears in both reactants and is harder to balance.

c) __ $B_2O_3(s)$ + 6 $NaOH(aq) \rightarrow$ 2 $Na_3BO_3(aq)$ + 3 $H_2O(l)$

Hint: Oxygen is again the hardest element to balance because it is present in more than one place on each side of the reaction. If you balance the easier elements first (B, Na, H), the oxygen will automatically be balanced.

d) 2 $CH_3NH_2(g)$ + 9/2 $O_2(g) \rightarrow$ 2 $CO_2(g)$ + 5 $H_2O(g)$ + __ $N_2(g)$

4 $CH_3NH_2(g)$ + 9 $O_2(g) \rightarrow$ 4 $CO_2(g)$ + 10 $H_2O(g)$ + 2 $N_2(g)$

Hint: Balance odd/even numbers of oxygen using the "half" method, and then multiply all coefficients by 2.

3.53 a) 2 $SO_2(g)$ + __ $O_2(g) \rightarrow$ 2 $SO_3(g)$

b) __ $Sc_2O_3(s)$ + 3 $H_2O(l) \rightarrow$ 2 $Sc(OH)_3(s)$

c) __ $H_3PO_4(aq)$ + 2 $NaOH(aq) \rightarrow$ __ $Na_2HPO_4(aq)$ + 2 $H_2O(l)$

d) __ $C_6H_{10}O_5(s)$ + 6 $O_2(g) \rightarrow$ 6 $CO_2(g)$ + 5 $H_2O(g)$

3.55 a) 4 $Ga(s)$ + 3 $O_2(g) \rightarrow$ 2 $Ga_2O_3(s)$

b) 2 $C_6H_{14}(l)$ + 19 $O_2(g) \rightarrow$ 12 $CO_2(g)$ + 14 $H_2O(g)$

c) 3 $CaCl_2(aq)$ + 2 $Na_3PO_4(aq) \rightarrow Ca_3(PO_4)_2\ (s)$ + 6 $NaCl(aq)$

3.59 Assuming you knew the identity of reactants D and E, you would first write a balanced equation for the reaction. Second, you would convert the reactants to mole quantities and determine the limiting reactant. Finally, you would use the stoichiometric relationship found in the balanced equation to find the mass of F, based on the mass of the limiting reactant.

3.61 a) $mol\ Cl_2 = (1.82\ mol\ HCl)\left(\frac{1\ mol\ Cl_2}{4\ mol\ HCl}\right) = \mathbf{0.455\ mol\ Cl_2}$

b) $grams\ Cl_2 = (0.455\ mol\ Cl_2)\left(\frac{70.90\ g\ Cl_2}{1\ mol\ Cl_2}\right) = \mathbf{32.3\ g\ Cl_2}$

3.63 a) In the balanced equation the ratio of moles of KNO_3, potassium nitrate, to moles of oxygen is 4:5.

Convert the kg of oxygen to moles of oxygen and then use the ratio to determine the number of moles of potassium nitrate needed:

$$88.6\ kg\ O_2\left(\frac{1000\ g}{1\ kg}\right)\left(\frac{1\ mol\ O_2}{32.00\ g\ O_2}\right)\left(\frac{4\ mol\ KNO_3}{5\ mol\ O_2}\right) = \mathbf{2.22 x 10^3\ mol\ KNO_3}$$

b) Convert moles of KNO_3 to grams of KNO_3:

$$2.22x10^3\ mol\ KNO_3\left(\frac{101.11\ g\ KNO_3}{1\ mol\ KNO_3}\right) = \mathbf{2.24x10^5\ g\ KNO_3}$$

3.65 The balanced equation is: $B_2H_6(g) + 6H_2O(l) \rightarrow 2H_3BO_3(s) + 6H_2(g)$.

$$mass\ H_3BO_3 = \left(\frac{33.61\ g\ B_2H_6}{27.67\ g/mol\ B_2H_6}\right)\left(\frac{2\ mol\ H_3BO_3}{1\ mol\ B_2H_6}\right)\left(\frac{61.83\ g\ H_3BO_3}{1\ mol\ H_3BO_3}\right) = \mathbf{150.2\ g\ H_3BO_3}$$

$$mass\ H_2 = \left(\frac{33.61\ g\ B_2H_6}{27.67\ g/mol\ B_2H_6}\right)\left(\frac{6\ mol\ H_2}{1\ mol\ B_2H_6}\right)\left(\frac{2.016\ g\ H_2}{1\ mol\ H_2}\right) = \mathbf{14.69\ g\ H_2}$$

3.67 Write the balanced equation by first writing the formulas for the reactants and products:

reactants: formula for phosphorus is given as P_4

formula for chlorine gas is Cl_2 (chlorine occurs as a diatomic molecule)

products: formula for phosphorus pentachloride - the name indicates one phosphorus atom and five chlorine atoms to give the formula PCl_5.

Formulas give the equation: $P_4 + Cl_2 \rightarrow PCl_5$

Balancing the equation: $P_4 + 10Cl_2 \rightarrow 4PCl_5$

The balanced equation gives a ratio of 1 mol of P_4 reacting with 10 mol of Cl_2.

$$(355\ g\ P_4)\left(\frac{1\ mol\ P_4}{123.88\ g\ P_4}\right)\left(\frac{10\ mol\ Cl_2}{1\ mol\ P_4}\right)\left(\frac{70.90\ g\ Cl_2}{1\ mol\ Cl_2}\right) = 2.03x10^3\ g\ Cl_2$$

2.03×10^3 g of chlorine gas is needed to react with 355 g of phosphorus.

3.69 a) $I_2(s) + Cl_2(g) \rightarrow 2\ ICl(s)$

$ICl(s) + Cl_2(g) \rightarrow ICl_3(s)$

b) Multiply the coefficients of the second equation by 2, so that ICl(s), an intermediate product can be eliminated from the overall equation.

$I_2(s) + Cl_2(g) \rightarrow 2\ ICl(s)$

$\underline{2ICl(s) + 2Cl_2(g) \rightarrow 2ICl_3(s)}$

$I_2(s) + Cl_2(g) + \cancel{2ICl(s)} + 2Cl_2(g) \rightarrow \cancel{2\ ICl(s)} + 2ICl_3(s)$

Overall equation: $I_2(s) + 3Cl_2(g) \rightarrow 2ICl_3(s)$

c) $$(31.4\ kg\ ICl_3)\left(\frac{1000\ g}{1\ kg}\right)\left(\frac{1\ mol\ ICl_3}{233.2\ g\ ICl_3}\right)\left(\frac{1\ mol\ I_2}{2\ mol\ ICl_3}\right)\left(\frac{253.8\ g\ I_2}{1\ mol\ I_2}\right) = \mathbf{1.71 \times 10^4\ g I_2}$$

3.71 a) $$mol\ CaO(from\ Ca) = (4.20\ g\ Ca)\left(\frac{1\ mol\ Ca}{40.08\ g}\right)\left(\frac{2\ mol\ CaO}{2\ mol\ Ca}\right) = \mathbf{0.105\ mol\ CaO}$$

b) $$mol\ CaO(from\ O_2) = (2.80\ g\ O_2)\left(\frac{1\ mol\ O_2}{32.00\ g\ O_2}\right)\left(\frac{2\ mol\ CaO}{1\ mol\ O_2}\right) = \mathbf{0.175\ mol\ CaO}$$

c) **Calcium** is the limiting reagent because less CaO can be made from the amount of calcium present than can be made from the amount of oxygen present.

d) $$mass\ CaO\ produced = (0.105\ mol\ CaO)\left(\frac{56.08\ g\ CaO}{1\ mol\ CaO}\right) = \mathbf{5.89\ g\ CaO}$$

3.73 The balanced chemical equation for this reaction is:

$2ICl_3 + 3H_2O \rightarrow ICl + HIO_3 + 5HCl$

Hint: Balance the equation by starting with oxygen. The other elements are in multiple reactants and/or products and are harder to balance initially.

Next find the limiting reactant by using the molar ratio to find the smaller number of moles of HIO_3 produced from each reactant given and excess of the other:

$$moles\ HIO_3\ (from\ ICl_3) = (685\ g\ ICl_3)\left(\frac{1\ mol\ ICl_3}{233.2\ g\ ICl_3}\right)\left(\frac{1\ mol\ HIO_3}{2\ mol\ ICl_3}\right) = \mathbf{1.47\ mol\ HIO_3}$$

$$moles\ HIO_3\ (from\ H_2O) = (117.4\ g\ H_2O)\left(\frac{1\ mol\ H_2O}{18.016\ g\ H_2O}\right)\left(\frac{1\ mol\ HIO_3}{3\ mol\ H_2O}\right) = 2.17\ mol\ HIO_3$$

Therefore, ICl_3 is the limiting reactant.

$$mass\ HIO_3 = (1.46_{86}\ mol\ HIO_3)\left(\frac{175.9\ g\ HIO_3}{1\ mol\ HIO_3}\right) = \mathbf{258\ g\ HIO_3}$$

The remaining mass of the excess reagent can be calculated from the amount of H_2O used to make 1.47 moles of HIO_3:

$$mass\ H_2O\ in\ excess = 117.4\ g\ H_2O - \left[1.46_{86}\ mol\ HIO_3\left(\frac{3\ mol\ H_2O}{1\ mol\ HIO_3}\right)\left(\frac{18.02\ g\ H_2O}{1\ mol\ H_2O}\right)\right] = \mathbf{38.0\ g\ H_2O}$$

3.75 Write the balanced equation: formula for carbon is C, formula for oxygen is O_2 and formula for carbon dioxide is CO_2: $C(s) + O_2(g) \rightarrow CO_2(g)$

The equation is balanced as written.

Which reactant, carbon or oxygen will limit the amount of carbon dioxide produced? Since one mole of carbon reacts with one mole of oxygen, the reactant present in a lower quantity of moles will be the limiting reactant. Carbon is present in a 0.100 mol amount and oxygen

$$mol\ O_2 = (8.00\ g\ O_2)\left(\frac{1\ mol\ O_2}{32.00\ g\ O_2}\right) = 0.250\ mol\ O_2$$

is present in 0.250 mol amount. Thus, carbon is the limiting reactant and it will be used to calculate the amount of carbon dioxide produced:

$$mass\ CO_2\ produced = (0.100\ mol\ C)\left(\frac{1\ mol\ CO_2}{1\ mol\ C}\right)\left(\frac{44.01\ g\ CO_2}{1\ mol\ CO_2}\right) = \mathbf{4.40\ g\ CO_2}$$

Some O_2, the reactant that is not limiting, will remain after the reaction. How much is left is the difference between the amount present before the reaction and the amount that reacted:

$$mass\ of\ O_2\ left = (0.250\ mol\ O_2 - 0.100\ mol\ O_2)\left(\frac{32.00\ g\ O_2}{1\ mol\ O_2}\right) = \mathbf{4.80\ g\ O_2}$$

3.77 The question asks for the mass of each <u>substance</u> present at the <u>end</u> of the reaction. "Substance" refers to both reactants and products. Solve this problem using multiple steps.

<u>Step 1:</u> Recognizing that this is a limiting reactant problem, first write a balanced chemical equation.

$Al(NO_2)_3(aq) + 3NH_4Cl(aq) \rightarrow AlCl_3(aq) + 3N_2(g) + 6H_2O(l)$

All ionic compounds in this reaction are soluble and have the (aq) designation. Nitrogen exists in its diatomic form and is a gas at normal pressures and temperatures.

<u>Step 2:</u> Using the molar relationships from the balanced equation, determine which reactant is limiting. Any product can be used to predict the limiting reactant; in this case, water is used. Additional significant figures are retained until the last step.

$$mol\ H_2O(from\ Al(NO_2)_3) = (62.5\ g\ Al(NO_2)_3)\left(\frac{1\ mol\ Al(NO_2)_3}{165.01\ g}\right)\left(\frac{6\ mol\ H_2O}{1\ mol\ Al(NO_2)_3}\right) = 2.27_{26}\ mol$$

$$mol\ H_2O(from\ NH_4Cl) = (54.6\ g\ NH_4Cl)\left(\frac{1\ mol\ NH_4Cl}{53.49\ g}\right)\left(\frac{6\ mol\ H_2O}{3\ mol\ NH_4Cl}\right) = 2.04_{15}\ mol$$

Therefore NH_4Cl is the limiting reactant. None of it remains at the end of the reaction.

<u>Step 3:</u> Use the limiting reactant to calculate the amounts of products formed.

$$mol\ NH_4Cl = (54.6\ g\ NH_4Cl)\left(\frac{1\ mol\ NH_4Cl}{53.49\ g}\right) = 1.02_{08}\ mol\ NH_4Cl$$

$$mass\ AlCl_3 = (1.02_{08}\ mol\ NH_4Cl)\left(\frac{1\ mol\ AlCl_3}{3\ mol\ NH_4Cl}\right)\left(\frac{133.3\ g\ AlCl_3}{1\ mol\ AlCl_3}\right) = 45.4\ g\ AlCl_3$$

$$mass\ N_2 = (1.02_{08}\ mol\ NH_4Cl)\left(\frac{3\ mol\ N_2}{3\ mol\ NH_4Cl}\right)\left(\frac{28.02\ g\ N_2}{1\ mol\ N_2}\right) = 28.6\ g\ N_2$$

$$mass\ H_2O = (2.04_{15}\ mol\ H_2O)\left(\frac{18.016\ g\ H_2O}{1\ mol\ H_2O}\right) = 36.8\ g\ H_2O$$

Step 4: Determine the amount of excess reactant remaining. The excess reagent is $Al(NO_2)_3$. First determine the amount of moles (and grams) of $Al(NO_2)_3$ that reacted with NH_4Cl:

$$(1.02_{08}\ mol\ NH_4Cl)\left(\frac{1\,mol\ Al(NO_2)_3}{3\,mol\ NH_4Cl}\right)\left(\frac{165.01\,g}{1\,mol\ Al(NO_2)_3}\right) = 56.1 g\ Al(NO_2)_3\ reacted$$

The remaining amount is the starting amount (62.5 g) minus the amount used in the reaction (56.1 g), or 6.4 g.

The mass of each substance present at the end of the reaction is:

$Al(NO_2)_3$ = 6.4 g $AlCl_3$ = 45.4 g

NH_4Cl = 0 g N_2 = 28.6 g

H_2O = 36.8 g

3.79 Overall yield = 0.82 x 0.65 = 0.53. Overall percent conversion of A to C is 53%.

3.81 The balanced equation for this reaction is: $WO_3(s) + 3H_2(g) \rightarrow W(s) + 3H_2O(l)$

Calculate the mass of H_2O *expected* from the reaction of 41.5 g WO_3:

$$theoretical\ mass = (41.5\,g\,WO_3)\left(\frac{1\,mol\,WO_3}{231.9\,g}\right)\left(\frac{3\,mol\,H_2O}{1\,mol\,WO_3}\right)\left(\frac{18.016\,g}{1\,mol\,H_2O}\right) = 9.67\,g\,H_2O$$

The reaction actually yielded (9.50 mL H_2O x 1.00 g H_2O/mL H_2O) = 9.50 g H_2O. The percent yield is (actual yield/theoretical yield) x 100%, or (9.50/9.67) x 100% = **98.2%**.

3.83 Write and balance equation: $CH_4(g) + Cl_2(g) \rightarrow CH_3Cl(g) + HCl(g)$

Which reactant is limiting?

$$mol\,CH_3Cl\,(from\,CH_4) = (18.5\,g\,CH_4)\left(\frac{1\,mol\,CH_4}{16.04\,g\,CH_4}\right)\left(\frac{1\,mol\,CH_3Cl}{1\,mol\,CH_4}\right) = 1.15_{34}\ mol\,CH_3Cl$$

$$mol\,CH_3Cl\,(from\,Cl_2) = (43.0\,g\,Cl_2)\left(\frac{1\,mol\,Cl_2}{70.90\,g\,Cl_2}\right)\left(\frac{1\,mol\,CH_3Cl}{1\,mol\,Cl_2}\right) = 0.606_{49}\ mol\,CH_3Cl$$

Chlorine is the limiting reactant, so the yield of chloromethane is

$$mass\,CH_3Cl = 0.606_{49}\ mol\,CH_3Cl\left(\frac{50.48\,g\,CH_3Cl}{1\,mol\,CH_3Cl}\right)(0.800) = \mathbf{24.5\,g\,CH_3Cl}$$

3.85 The balanced equation for this reaction is: $(CN)_2(g) + 7F_2(g) \rightarrow 2CF_4(g) + 2NF_3(g)$

The limiting reactant is found by calculating which reactant yields the least amount of CF_4:

$$mass\,CF_4\,(from\,(CN)_2) = \left(\frac{80.0\,g\,(CN)_2}{52.04\,g/mol\,(CN)_2}\right)\left(\frac{2\,mol\,CF_4}{1\,mol\,(CN)_2}\right) = 3.07_{46}\ mol\,CF_4$$

$$mass\,CF_4\,(from\,F_2) = \left(\frac{80.0\,g\,F_2}{38.00\,g/mol\,F_2}\right)\left(\frac{2\,mol\,CF_4}{7\,mol\,F_2}\right) = 0.601_{50}\ mol\,CF_4$$

F_2 is the limiting reagent:

$$(0.601_{50}\ mol\,CF_4)\left(\frac{88.01\,g\,CF_4}{1\,mol\,CF_4}\right) = \mathbf{52.9\,g\,CF_4}$$

3.88 If the reaction with a 75.0% yield produces 25.0 g of diborane, then the theoretical yield can be calculated by dividing the 25.0 g by 0.750 to give 33.3_{33} g of diborane.

Balance the equation: $3NaBH_4(s) + 4BF_3(g) \rightarrow 2B_2H_6(g) + 3NaBF_4(s)$

$$33.3_{33}\ g\ B_2H_6\left(\frac{1\ mol\ B_2H_6}{27.67\ g\ B_2H_6}\right)\left(\frac{3\ mol\ NaBH_4}{2\ mol\ B_2H_6}\right)\left(\frac{37.83\ g\ NaBH_4}{1\ mol\ NaBH_4}\right) = 68.4\ g\ NaBH_4$$

To make 25.0 g of B_2H_6 with a 75.0% yield for the reaction would require starting with **68.4 g of $NaBH_4$**.

3.89 The spheres represent particles of solute, and the amount of *solute* per given volume of *solution* determines its concentration.

a) Box C has more solute added because it contains 2 more spheres than Box A contains.

b) Box B has more solvent because solvent molecules have displaced two solute molecules.

c) Box C has a higher molarity, because it has more moles of solute per volume of solution.

d) Box B has a lower concentration (and molarity), because it has fewer moles of solute per volume of solution.

3.91 Volumes are **not** additive when two different solutions are mixed, so the final volume will be slightly different from 1000.0 mL. The correct method would state: "Take 100.0 mL of the 10.0 M solution and add water until the total volume is 1.00 L."

3.92 The key to solving these types of problems is to multiply the given quantities and use conversion factors (typically g $\leftrightarrow$ mol and mL $\leftrightarrow$ L) in such a way that yields the quantity (and correct units) you are solving for.

a) The formula for calcium acetate is $Ca(C_2H_3O_2)_2$.

$$mass = (175.8\ mL\ sol'n)\left(\frac{1\ L}{1000\ mL}\right)\left(\frac{0.207\ mol\ Ca(C_2H_3O_2)_2}{L\ sol'n}\right)\left(\frac{158.17\ g\ Ca(C_2H_3O_2)_2}{1\ mol\ Ca(C_2H_3O_2)_2}\right)$$

$$= \mathbf{5.76\ g\ Ca(C_2H_3O_2)_2}$$

b) The formula of potassium iodide is KI.

$$M = \left(\frac{21.1\ g\ KI}{500.\ mL\ sol'n}\right)\left(\frac{1\ mol\ KI}{166.0\ g\ KI}\right)\left(\frac{1000 mL}{1 L}\right) = \mathbf{0.254\ M}$$

c) The formula of sodium cyanide is NaCN.

$$mol = (145.6\ L\ sol'n)\left(\frac{0.850\ mol\ NaCN}{1\ L\ sol'n}\right) = \mathbf{124\ mol\ NaCN}$$

3.94 a) The formula for potassium sulfate is K_2SO_4.

$$(0.475\ L\ soln)\left(\frac{5.62x10^{-2}\ mol\ K_2SO_4}{L\ soln}\right)\left(\frac{174.27\ g}{1\ mol\ K_2SO_4}\right) = \mathbf{4.65\ g\ K_2SO_4}$$

b) The formula for calcium chloride is $CaCl_2$.

$$M = \left(\frac{6.55\ mg\ CaCl_2}{1\ mL\ sol'n}\right)\left(\frac{1\ g}{1000\ mg}\right)\left(\frac{1000\ mL}{1\ L}\right)\left(\frac{1\ mol\ CaCl_2}{110.98\ g\ CaCl_2}\right) = \mathbf{0.0590\ M\ CaCl_2}$$

c) The formula for magnesium bromide is $MgBr_2$.

$$\left(\frac{0.184\ mol\ MgBr_2}{L\ soln}\right)\left(\frac{1\ L}{1000\ mL}\right)\left(\frac{1\ mol\ Mg^{2+}}{1\ mol\ MgBr_2}\right)\left(\frac{6.022x10^{23}\ Mg^{2+}\ ions}{1\ mol\ Mg^{2+}}\right)$$

$$= \mathbf{1.11x10^{20}\ Mg^{2+}\ ions / mL\ soln}$$

3.96 a) $0.250\ M\ KCl\left(\frac{37.00\ mL}{150.00\ mL}\right) = \mathbf{0.0617\ M\ KCl}$

b) $0.0706\ M\ (NH_4)_2SO_4\left(\frac{25.71\ mL}{500.00\ mL}\right) = \mathbf{3.63x10^{-3}\ M\ (NH_4)_2SO_4}$

c) total volume after mixing = 3.58 mL + 500. mL = 503.58 mL

$$Na^+\ (from\ NaCl\ sol'n) = (0.288\ M)(3.58\ mL\ sol'n)\left(\frac{1\ L}{1000\ mL}\right)$$

$$= 1.03\ x\ 10^{-3}\ mol\ Na^+$$

$$Na^+\ (from\ NaSO_4\ sol'n) = (6.51\ x\ 10^{-3}\ M)(500.\ mL\ sol'n)\left(\frac{1\ L}{1000\ mL}\right)\left(\frac{2\ mol\ Na^+}{1\ mol\ Na_2SO_4}\right)$$

$$= 6.51\ x\ 10^{-3}\ mol\ Na^+$$

$$total\ mol\ Na^+\ after\ mixing = 1.03\ x\ 10^{-3}\ mol + 6.51\ x\ 10^{-3}\ mol = 7.54\ x\ 10^{-3}\ mol$$

$$M\ Na^+ = \left(\frac{7.54\ x\ 10^{-3}\ mol\ Na^+}{503.58\ mL\ sol'n}\right)\left(\frac{1000\ mL}{1\ L}\right) = \mathbf{0.0150\ M}$$

3.98 a) Since the problem does not state the molarity of the nitric acid, only that it is concentrated, you must calculate the mass using the information given. The term "70.0% by mass" means that you have 70.0 g of HNO_3 in 100.0 g of solution. This is not 70.0 g of HNO_3 in 100.0 mL of solution, because the density is 1.41 g/mL, not 1.00 g/mL (the density of water at 0°C). Using the mass percent and density, calculate the desired quantity in g HNO_3/L:

$$mass\ HNO_3\ per\ liter = \left(\frac{70.0\ g\ HNO_3}{100.0\ g\ soln}\right)\left(\frac{1.41\ g\ soln}{mL\ soln}\right)\left(\frac{1000\ mL}{L}\right) = \mathbf{987\ g\ HNO_3\ /\ L}$$

b) $M = \left(\frac{987\ g\ HNO_3}{L}\right)\left(\frac{1\ mol\ HNO_3}{63.02\ g\ HNO_3}\right) = \mathbf{15.7\ M\ HNO_3}$

3.100 From balanced equation given, the mole ratio is 2 mol HCl/1mol $CaCO_3$.

$$16.2\ g\ CaCO_3\left(\frac{1\ mol\ CaCO_3}{100.09\ g\ CaCO_3}\right)\left(\frac{2\ mol\ HCl}{1\ mol\ CaCO_3}\right)\left(\frac{1\ L\ sol'n}{0.383\ mol\ HCl}\right) = 0.845\ L\ or\ 845\ mL\ of\ solution$$

845 mL of 0.383 M HCl is required to react with 16.2 g of $CaCO_3$.

3.102 The balanced equation for this reaction is:

$BaCl_2(aq) + Na_2SO_4(aq) \rightarrow BaSO_4(s) + 2NaCl(aq)$

This is a limiting reactant problem because quantities of both reactants are given:

$$mol\ BaCl_2 = (0.0250\ L\ sol'n)\left(\frac{0.160\ mol\ BaCl_2}{L\ sol'n}\right) = 4.00x10^{-3}\ mol\ BaCl_2$$

$$mol\ Na_2SO_4 = (0.0680\ L\ sol'n)\left(\frac{0.055\ mol\ Na_2SO_4}{L\ sol'n}\right) = 3.74x10^{-3}\ mol\ Na_2SO_4$$

Since the stoichiometry is 1:1 for both reactants, Na_2SO_4 is limiting. An additional significant figure (3.7<u>4</u> x 10^{-3}) is kept during the intermediate steps, but the final answer is reported to two significant figures, as determined by the quantity 0.055 M.

$$mass\ BaSO_4 = (3.74x10^{-3}\ mol\ Na_2SO_4)\left(\frac{1\ mol\ BaSO_4}{1\ mol\ Na_2SO_4}\right)\left(\frac{233.4\ g\ BaSO_4}{1\ mol\ BaSO_4}\right) = \mathbf{0.87\ g\ BaSO_4}$$

3.105 a) How many mol HCl are in 5.0 gallons of 3.5 M HCl?

$$(5.0\ \text{gal sol'n})\left(\frac{4\ \text{qt}}{1\ \text{gal}}\right)\left(\frac{0.946\ L}{1\ qt}\right)\left(\frac{3.5\ mol\ HCl}{L\ sol'n}\right) = 66._{22}\ mol\ HCl$$

What volume of concentrated HCl contains $66._{22}$ mol HCl?

$$V_{conc\ HCl} = 66._{22}\ mol\ HCl\left(\frac{1\ L\ sol'n}{11.7\ mol\ HCl}\right) = 5.6\ L\ solution$$

Instructions: Be sure to wear goggles to protect your eyes! Pour approximately 3.0 gallons of water into the container. Add slowly and with mixing 5.6 L (or 1.5 gal) of concentrated HCl into the water. Dilute to 5.0 gallons with water.

b) Volume of concentrated solution that contains 9.55 g HCl:

$$V = 9.55\ g\ HCl\left(\frac{1\ mol\ HCl}{36.46\ g\ HCl}\right)\left(\frac{1\ L\ sol'n}{11.7\ mol\ HCl}\right)\left(\frac{1000\ mL}{1\ L}\right) = \mathbf{22.4\ mL\ solution}$$

3.109 Rank in order from highest fuel value (highest %H) to lowest fuel value (lowest %H):

Name	Formula	Molar Mass, g/mol	Mass %H
ethane	C_2H_6	30.07	6.048 g H ÷ 30.07 g C_2H_6 = 20.11%
propane	C_3H_8	44.09	8.064 g H ÷ 44.09 g C_3H_8 = 18.29%
benzene	C_6H_6	78.11	6.048 g H ÷ 78.11 g C_6H_6 = 7.743%
ethanol	C_2H_6O	46.07	6.048 g H ÷ 46.07 g C_2H_6O = 13.13%
cetyl palmitate	$C_{32}H_{64}O_2$	480.8	64.51 g H ÷ 480.8 g $C_{32}H_{64}O_2$ = 13.42%

Rank is $C_2H_6 > C_3H_8 > C_{32}H_{64}O_2 > C_2H_6O > C_6H_6$

3.110 $$x = \left(\frac{10.8\ g\ H_2O}{100.0\ g\ narceine}\right)\left(\frac{499.52\ g\ narceine}{1\ mol\ narceine}\right)\left(\frac{1\ mol\ H_2O}{18.02\ g\ H_2O}\right) = 2.99\ mol\ H_2O$$

The formula for narceine hydrate is narceine·**3**H_2O.

3.115 a) The key to solving this problem is determining the overall balanced equation. In the first step, <u>ferric</u> oxide (ferric denotes Fe^{3+}) reacts with carbon monoxide to form Fe_3O_4 and carbon dioxide:

$3Fe_2O_3(s) + CO(g) \rightarrow 2Fe_3O_4(s) + CO_2(g)$ (1)

In the second step, Fe_3O_4 reacts with more carbon monoxide to form ferrous oxide:

$Fe_3O_4(s) + CO(g) \rightarrow 3FeO(s) + CO_2(g)$ (2)

In the third step, ferrous oxide reacts with more carbon monoxide to form molten iron:

$FeO(s) + CO(g) \rightarrow Fe(s) + CO_2(g)$ (3)

The intermediate products are Fe_3O_4 and FeO, so multiply the equation (2) by 2 to cancel Fe_3O_4 and equation (3) by 6 to cancel FeO:

$3Fe_2O_3(s) + CO(g) \rightarrow \cancel{2Fe_3O_4(s)} + CO_2(g)$

$\cancel{2Fe_3O_4(s)} + 2CO(g) \rightarrow \cancel{6FeO(s)} + 2CO_2(g)$

$\cancel{6FeO(s)} + 6CO(g) \rightarrow 6Fe(s) + 6CO_2(g)$

$3Fe_2O_3(s) + 9CO(g) \rightarrow 6Fe(s) + 9CO_2(g)$

Then divide by 3 to obtain the smallest integer coefficients:

$Fe_2O_3(s) + 3CO(g) \rightarrow 2Fe(s) + 3CO_2(g)$

b) To find the mass of carbon monoxide needed, convert tons of Fe (1 metric ton = 1000 kg = 1 x 10^6 g) to moles and use the stoichiometric relationship from the balanced equation.

$$mass\ CO = (40.0\ ton\ Fe)\left(\frac{1x10^6\ g}{1\ ton}\right)\left(\frac{1\ mol\ Fe}{55.85\ g\ Fe}\right)\left(\frac{3\ mol\ CO}{2\ mol\ Fe}\right)\left(\frac{28.01\ g\ CO}{1\ mol\ CO}\right) = \mathbf{3.01x10^7\ g\,CO}$$

3.116 If 100.0 g of dinitrogen tetroxide reacts with 100.0 g of hydrazine (N_2H_4), what is the theoretical yield of nitrogen if no side reaction takes place? First, identify the limiting reactant:

If N_2O_4 is limiting, how much N_2 would be produced?

$$100.0\ g\ N_2O_4\left(\frac{1\ mol\ N_2O_4}{92.02\ g\ N_2O_4}\right)\left(\frac{3\ mol\ N_2}{1\ mol\ N_2O_4}\right) = 3.26\ mol\ N_2$$

If N_2H_4 is limiting, how much N_2 would be produced?

$$100.0\ g\ N_2H_4\left(\frac{1\ mol\ N_2H_4}{32.05\ g\ N_2H_4}\right)\left(\frac{3\ mol\ N_2}{2\ mol\ N_2H_4}\right) = 4.68\ mol\ N_2$$

The dinitrogen tetroxide is limiting since it limits the amount of N_2 produced. Thus, the theoretical yield is the 3.26 mol of N_2 produced from the 100.0 g of N_2O_4 and part of the 100.0 g of N_2H_4.

The actual yield is decreased by the N_2O_4 used in the side reaction. To produce 10.0g of NO in the side reaction, how much N_2O_4 would be used?

$$10.0\ g\ NO\left(\frac{1\ mol\ NO}{30.01\ g\ NO}\right)\left(\frac{2\ mol\ N_2O_4}{6\ mol\ NO}\right) = 0.111\ mol\ N_2O_4$$

How much N_2 would have been produced from 0.111 mol N_2O_4?

$$0.111\ mol\ N_2O_4\left(\frac{3\ mol\ N_2}{1\ mol\ N_2O_4}\right) = 0.333\ mol\ N_2$$

Actual yield = theoretical yield of N_2 - amount of N_2 not made due to side reaction

= 3.26mol - 0.333 mol = 2.92_7 mol N_2

$$\%\ yield = \left(\frac{2.92_7\ mol}{3.26\ mol}\right)(100\%) = \mathbf{89.8\%}$$

3.118 a) $2AB_2 + B_2 \rightarrow 2AB_3$

The contents of the boxes can be represented in their entirety, and then the equations can be simplified:

$6AB_2 + 5B_2 \rightarrow 6AB_3 + 2B_2$ Subtract the excess B_2 from both sides of the equation

$6AB_2 + 3B_2 \rightarrow 6AB_3$ Divide coefficients by 3 to obtain smallest coefficients

$2AB_2 + B_2 \rightarrow 2AB_3$

b) AB_2 is the limiting reactant since none of it is left at the end of the reaction.

c) From 3.0 mol of B_2 6.0 mol of AB_3 could be made. From 5.0 mol of AB_2 5.0 mol of AB_3 could be made. Thus, the maximum amount of AB_3 that will result from mixing 3.0 mol of B_2 and 5.0 mol of AB_2 is 5.0 mol.

d) If 5.0 mol AB_3 is made then 2.5 mol of B_2 would have been used. The unreacted B_2 is 3.0 mol – 2.5 mol = 0.5 mol.

3.120 To calculate molarity (moles of 6-benzylaminopurine in one liter of solution) need to convert 0.030mg 6-benzylaminopurine to moles 6-benzylaminopurine (6-bap) and to convert the moles per volume from a volume of 150.mL to 1L.

$$\left(\frac{0.030\, mg\ 6\cdot bap}{150.\, mL\ sol'n}\right)\left(\frac{1\, g}{1000\, mg}\right)\left(\frac{1\, mol\ 6\cdot bap}{225.26\, g\ 6\cdot bap}\right)\left(\frac{1000\, mL}{1\, L}\right) = \mathbf{8.9 \times 10^{-7}\ M}$$

3.122 Plan: Subtract the excess I_2 from the total amount of I_2 to obtain the amount of I_2 that reacted with the ascorbic acid in the tablet. Use the stoichiometric relationship between I_2 and $C_6H_8O_6$ in the first equation to determine the mass of $C_6H_8O_6$ present in the tablet.

Solution: The total amount of I_2 can be calculated using the volume and molarity of I_2:

$$starting\ mol\ I_2 = (0.05320\ L\ sol'n)\left(\frac{0.1030\, mol\ I_2}{1\, L\ sol'n}\right) = 5.480x10^{-3}\ mol\ I_2$$

The amount of excess I_2 can be calculated by determining the number of moles of $Na_2S_2O_3$ that reacted with it:

$$mol\ Na_2S_2O_3 = (0.02754\ L\ sol'n)\left(\frac{0.1153\, mol\ Na_2S_2O_3}{1\, L\ sol'n}\right) = 3.175x10^{-3}\ mol\ Na_2S_2O_3$$

$$excess\ I_2 = (3.175x10^{-3}\ mol\ Na_2S_2O_3)\left(\frac{1\, mol\ I_2}{2\, mol\ Na_2S_2O_3}\right) = 1.588x10^{-3}\ mol\ excess\ I_2$$

The amount of I_2 that reacted with the vitamin C in the tablet is the difference between the starting amount of I_2 and the excess amount of I_2, or (5.480×10^{-3} mol – 1.588×10^{-3} mol) = 3.892×10^{-3} mol I_2. The amount of $C_6H_8O_6$ in the tablet can now be calculated using the stoichiometric relationship in the first equation:

$$mass\ C_6H_8O_6 = (3.892x10^{-3}\ mol\ I_2)\left(\frac{1\, mol\ C_6H_8O_6}{1\, mol\ I_2}\right)\left(\frac{176.12\, g\ C_6H_8O_6}{1\, mol\ C_6H_8O_6}\right) = \mathbf{0.6855\ g\ C_6H_8O_6}$$

Check: Does this amount seem correct? Considering that a typical vitamin C tablet weighs 500 mg, this answer (685.5 mg) might seem a little high, but is within the right order of magnitude.

3.124 a) **C**. Figure A represents 12 spheres in one unit volume. After dilution by tripling the volume the 12 spheres would occupy three unit volumes giving four spheres per one unit volume.

b) **B**. Figure A represents 12 spheres in one unit volume. After dilution by doubling the volume the 12 spheres would occupy two unit volumes giving six spheres per one unit volume.

c) **D**. Figure A represents 12 spheres in one unit volume. After dilution by quadrupling the volume the 12 spheres would occupy four unit volumes giving three spheres per one unit volume.

3.128 After mixing the two solutions the concentration of KBr will be between the two initial concentrations of 0.053 M and 0.078 M. Since a larger volume of the 0.078 M solution is added to the mixture, the final concentration of KBr should be closer to 0.078 M.

total volume = 0.200 L + 0.550 L = 0.750 L

total mol KBr = (0.200 L x 0.053 M) + (0.550 L x 0.078 M) = 0.053_5 mol

molarity of mixture = 0.053_5 mol KBr/0.750 L soln = **0.071 M KBr**

3.129 To find the mass % of B, calculate the mass of B and divide by the given mass of the boron compound and multiply by 100%. To find the mass of B, use the amount of the NaOH and the 1:1 molar relationship of H_3BO_3 to NaOH to calculate moles of H_3BO_3:

$$moles\ H_3BO_3 = (0.02347\ L\ sol'n)\left(\frac{0.205\ mol\ NaOH}{1\ L\ sol'n}\right)\left(\frac{1\ mol\ H_3BO_3}{1\ mol\ NaOH}\right) = 4.81x10^{-3}\ mol\ H_3BO_3$$

Calculate the amount of B found in $4.81x10^{-3}$ mol H_3BO_3:

$$mass\ B = \left(4.81x10^{-3}\ mol\ H_3BO_3\right)\left(\frac{1\ mol\ B}{1\ mol\ H_3BO_3}\right)\left(\frac{10.81\ g\ B}{1\ mol\ B}\right) = 5.20x10^{-2}\ g\ B$$

The mass % of B in the sample is the mass of B from the boric acid (all of the boron in the sample is converted to boric acid) divided by the starting sample mass:

$$mass\%\ B = \frac{5.20x10^{-2}\ g\ B}{0.1352\ g\ compound}\ x\,100\% = \mathbf{38.5\ mass\ \%B}$$

3.132 Treat the combustion of methane, CH_4, separate from the combustion of propane, C_3H_8.
Combustion of methane: $CH_4 + 2O_2 \rightarrow CO_2 + 2H_2O$

$$(200.\ g\ mixture)\left(\frac{25.0\ g\ CH_4}{100\ g\ mixture}\right)\left(\frac{1\ mol\ CH_4}{16.04\ g\ CH_4}\right)\left(\frac{1\ mol\ CO_2}{1\ mol\ CH_4}\right)\left(\frac{44.01\ g\ CO_2}{1\ mol\ CO_2}\right) = 137._{19}\ g\ CO_2$$

Combustion of propane: $C_3H_8 + 5O_2 \rightarrow 3CO_2 + 4H_2O$

$$(200.\ g\ mixture)\left(\frac{75.0\ g\ C_3H_8}{100\ g\ mixture}\right)\left(\frac{1\ mol\ C_3H_8}{44.09\ g\ C_3H_8}\right)\left(\frac{3\ mol\ CO_2}{1\ mol\ C_3H_8}\right)\left(\frac{44.01\ g\ CO_2}{1\ mol\ CO_2}\right) = 449._{18}\ g\ CO_2$$

Total mass = $137._{19}$ g + $449._{18}$ g = **586 g CO_2** produced by mixture.

3.134 30:10:10 fertilizer means 30%N, 10%P_2O_5 and 10%K_2O. Assume the total sample of fertilizer is 100 g (other assumptions also valid), the sample would contain 30 g N, 10 g P_2O_5, and 10 g K_2O.

To find the ratio of N:P:K find the moles of each present in the sample. Note that an additional significant figure is kept until the final ratio is determined. A slightly different ratio (10:0.67:1.0) results if the additional significant figure is not carried through the calculation.

$$30\ g\ N\left(\frac{1\ mol\ N}{14.01\ g\ N}\right)=2.1_4\ mol\ N$$

$$10\ g\ P_2O_5\left(\frac{1\ mol\ P_2O_5}{141.94\ g\ P_2O_5}\right)\left(\frac{2\ mol\ P}{1\ mol\ P_2O_5}\right)=0.14_1\ mol\ P$$

$$10\ g\ K_2O\left(\frac{1\ mol\ K_2O}{94.2\ g\ K_2O}\right)\left(\frac{2\ mol\ K}{1\ mol\ K_2O}\right)=0.21_2\ mol\ K$$

The ratio asked for sets mol K = 1.0. To do this 0.21_2 mol must be multiplied by a factor to give 1.0. The factor is 1.0/0.212 = 4.7_1. Use this factor to find the mol of N and P that relate to 1.0 mol K

$$mol\ N = 2.1_4\ x\ 4.7_1 = 10._1\ mol\ N;\quad mol\ P = 0.14_1\ x\ 4.7_1 = 0.66_4\ mol\ P$$

Ratio of **N:P:K is 10:0.66:1.0**.

3.137 Assuming 100 g of mixture, there are 35.0 g CO and 65.0 g CO_2. All of the carbon in the mixture will be converted to the carbon of CO and CO_2, so calculate the grams of C in the CO and CO_2:

$$(35.0\ g\ CO)\left(\frac{1\ mol\ CO}{28.01\ g\ CO}\right)\left(\frac{1\ mol\ C}{1\ mol\ CO}\right)\left(\frac{12.01\ g}{1\ mol\ C}\right) = 15.0\ g\ C\ from\ CO$$

$$(65.0\ g\ CO_2)\left(\frac{1\ mol\ CO_2}{44.01\ g\ CO_2}\right)\left(\frac{1\ mol\ C}{1\ mol\ CO_2}\right)\left(\frac{12.01\ g}{1\ mol\ C}\right) = 17.7\ g\ C\ from\ CO_2$$

The mass % of C in the mixture is (15.0 g + 17.7 g)/100 g x 100% = **32.7 %C**.

3.139 a) The formula of citric acid obtained by counting the number of carbon atoms, oxygen atoms, and hydrogen atoms is **$C_6H_8O_7$**.

Molar mass = (6 x 12.01) + (8 x 1.008) + (7 x 16.00) = **192.12 g/mol**

b) First determine the mass of lemon juice from the volume and density:

$$(1.50\ qt\ lemon\ juice)\left(\frac{1\ L}{1.057\ qt}\right)\left(\frac{1000\ mL}{1\ L}\right)\left(\frac{1.09\ g\ lemon\ juice}{mL\ lemon\ juice}\right) = 1.54_{68}\ x\ 10^3\ g\ lemon\ juice$$

Next find the mass of citric acid in the calculated mass of lemon juice and convert the mass to moles:

$$(1.54_{68}\ x\ 10^3\ g\ lemon\ juice)\left(\frac{6.82\ g\ citric\ acid}{100\ g\ lemon\ juice}\right)\left(\frac{1\ mol\ citric\ acid}{192.12\ g\ citric\ acid}\right) = \mathbf{0.549\ mol\ citric\ acid}$$

3.140 a) nitrogen and oxygen combine to form nitrogen monoxide: $N_2 + O_2 \rightarrow 2NO$

nitrogen monoxide reacts with oxygen to form nitrogen dioxide: $2NO + O_2 \rightarrow 2NO_2$

nitrogen dioxide combines with water to form nitric acid and nitrogen monoxide:

$3NO_2 + H_2O \rightarrow 2HNO_3 + NO$

b)

$$\begin{array}{rl} N_2 + O_2 & \rightarrow 2NO \\ 2NO + O_2 & \rightarrow 2NO_2 \\ \underline{3NO_2 + H_2O} & \underline{\rightarrow 2HNO_3 + NO} \\ N_2 + 2O_2 + 2NO + 3NO_2 + H_2O & \rightarrow 3NO + 2NO_2 + 2HNO_3 \end{array}$$

To cancel the NO and NO_2 there needs to be one more NO as reactant and one more NO_2 as product. Multiplying the second reaction by 3/2 would give one more of each molecule:

$$N_2 + O_2 \rightarrow 2NO$$
$$3NO + {}^{3}/_{2}O_2 \rightarrow 3NO_2$$
$$\underline{3NO_2 + H_2O \rightarrow 2HNO_3 + NO}$$
$$N_2 + {}^{5}/_{2}O_2 + 3NO + 3NO_2 + H_2O \rightarrow 3NO + 3NO_2 + 2HNO_3$$

canceling the nitrogen monoxide and the nitrogen dioxide gives

$$N_2 + {}^{5}/_{2}O_2 + H_2O \rightarrow 2HNO_3$$

Eliminating the fraction gives $2N_2 + 5O_2 + 2H_2O \rightarrow 4HNO_3$

c) $$1.25x10^3\ t\ N_2\left(\frac{1x10^6\ g}{1\ t}\right)\left(\frac{1\ mol\ N_2}{28.02\ g\ N_2}\right)\left(\frac{2\ mol\ HNO_3}{1\ mol\ N_2}\right)\left(\frac{63.02\ g\ HNO_3}{1\ mol\ HNO_3}\right)\left(\frac{1\ t}{1x10^6\ g}\right)$$

$$= \mathbf{5.62x10^3\ t\ HNO_3}$$

3.143 This problem can be solved as a system of two equations and two unknowns.

The two equations are:
$Xe(g) + 2F_2(g) \rightarrow XeF_4(s)$
$Xe(g) + 3F_2(g) \rightarrow XeF_6(s)$

The two unknowns are:
x = mole XeF_4 produced
y = mole XeF_6 produced

Moles of Xe consumed = 1.85×10^{-4} moles present – 9.00×10^{-6} moles excess = 1.76×10^{-4}

Then x + y = 1.76×10^{-4} mol Xe consumed

2x + 3y = 5.00×10^{-4} mol F_2 consumed

Solve for x using the first equation and substitute the value of x into the second equation:

$x = 1.76 \times 10^{-4} - y$

$2(1.76 \times 10^{-4} - y) + 3y = 5.00 \times 10^{-4}$

$(3.52 \times 10^{-4}) - 2y + 3y = 5.00 \times 10^{-4}$

$y = (5.00 \times 10^{-4}) - (3.52 \times 10^{-4}) = 1.48 \times 10^{-4}$ mol XeF_6

$x = (1.76 \times 10^{-4}) - (1.48 \times 10^{-4}) = 0.28 \times 10^{-4}$ mol XeF_4

Convert moles of each product to grams using the molar masses:

Mass XeF_4 = (0.28×10^{-4} mol XeF_4) x (207.3 g/mol) = 5.8×10^{-3} g XeF_4

Mass XeF_6 = (1.48×10^{-4} mol XeF_6) x (245.3 g/mol) = 3.63×10^{-2} g XeF_6

Calculate % of each compound using the total weight of the product (0.0058 + 0.0363) = 0.0421 g:

$$\%\ XeF_4 = \frac{5.8x10^{-3}\ g\ XeF_4}{0.0421\ g\ total}\ x\ 100\% = \mathbf{14\%};\ \%\ XeF_6 = \frac{3.63x10^{-2}\ g\ XeF_6}{0.0421\ g\ total}\ x\ 100\% = \mathbf{86.2\%}$$

3.144 a) $$0.45\ g\ hemoglobin\left(\frac{6.0\ g\ heme}{100.0\ g\ hemoglobin}\right) = \mathbf{0.027\ g\ heme}$$

b) $$0.027\ g\ heme\left(\frac{1\ mol\ heme}{616.49\ g\ heme}\right) = \mathbf{4.4x10^{-5}\ mol\ heme}$$

c) $$(4.3_{80}\ x10^{-5}\ mol\ heme)\left(\frac{1\ mol\ Fe}{1\ mol\ heme}\right)\left(\frac{55.85\ g\ Fe}{mol\ Fe}\right) = \mathbf{2.4x10^{-3}\ g\ Fe}$$

d) Assume that heme forms hemin in a 1:1 ratio.

$$\left(4.3_{80}\, x10^{-5}\, mol\right)\left(\frac{1\, mol\ hemin}{1\, mol\ heme}\right)\left(\frac{651.94\, g\ hemin}{mol\ hemin}\right) = \mathbf{0.028\, g\, hemin}$$

3.146 a) To find mass percent of nitrogen first determine molecular formula then molar mass of each compound. Mass percent is then calculated from the mass of nitrogen in the compound divided by the molar mass of the compound.

Urea: CH_4N_2O, $\mathcal{M}$ = 60.06 g/mol

%N = 28.02 g/mol N ÷ 60.06 g/mol CH_4N_2O x 100% = **46.65%N in urea**

Arginine: $C_6H_{15}N_4O_2$, $\mathcal{M}$ = 175.22 g/mol

%N = 56.04 g/mol N ÷ 175.22 g/mol $C_6H_{15}N_4O_2$ x 100% = **31.98% N in arginine**

Ornithine: $C_5H_{13}N_2O_2$, $\mathcal{M}$ = 133.17 g/mol

%N = 28.02 g/mol N ÷ 133.17 g/mol $C_5H_{13}N_2O_2$ x 100% = **21.04% N in ornithine**

b)

$$\left(143.2\, g\ ornithine\right)\left(\frac{mol\ ornithine}{132.17\, g\ ornithine}\right)\left(\frac{1\, mol\ urea}{1\, mol\ orn}\right)\left(\frac{60.06\, g\ urea}{mol\ urea}\right)\left(\frac{46.65\, g\ N}{100.0\, g\ urea}\right) = \mathbf{30.36\ g\ N}$$

3.148 a) The theoretical yield of diethyl ether is

$$\left(50.0\, g\, CH_3CH_2OH\right)\left(\frac{mol\, CH_3CH_2OH}{46.07\, g\, CH_3CH_2OH}\right)\left(\frac{1\, mol\ diethylether}{1\, mol\, CH_3CH_2OH}\right)\left(\frac{74.12\, g\ diethylether}{1\, mol\ diethylether}\right) = 40.2_{21}\, g$$

The percent yield of diethyl ether is 33.9 g ÷ 40.2$_{21}$ g * 100% = **84.3%**

b) Only 84.3% of the ethanol reacts to form diethyl ether, leaving 15.7% to react to form ethylene. According to the problem, only half of this percentage actually reacts to form ethylene.

$$\left(0.500\right)\left(0.157\right)\left(50.0\, g\, CH_3CH_2OH\right)\left(\frac{1\, mol\, CH_3CH_2OH}{46.07\, g}\right)\left(\frac{1\, mol\, C_2H_4}{1\, mol\, CH_3CH_2OH}\right)\left(\frac{28.05\, g\, C_2H_4}{1\, mol\, C_2H_4}\right) = \mathbf{2.39\, g\, C_2H_4}$$

3.150 a) $dissolved\ cocaine \cdot HCl = \left(0.0500\, L\right)\left(\frac{2.50\, kg\, C_{17}H_{21}NO_4HCl}{1\, L}\right)\left(\frac{1000\, g}{1\, kg}\right) = \mathbf{125\, g\, C_{17}H_{21}NO_4HCl}$

b) Convert the cocaine·HCl from part a) into moles. One mole of cocaine forms from one mole of the hydrochloride salt when treated with NaOH. Convert moles of cocaine to grams, and use the smaller solubility of cocaine to determine how much water is needed to dissolve it.

$$125\, g\, C_{17}H_{21}O_4NHCl\left(\frac{1\, mol\, C_{17}H_{21}O_4NHCl}{339.81\, g\ C_{17}H_{21}O_4NHCl}\right) = 0.367_{85}\ mol\ C_{17}H_{21}O_4NHCl$$

$$\left(0.367_{85}\ mol\ C_{17}H_{21}O_4NHCl\right)\left(\frac{1\, mol\, C_{17}H_{21}O_4N}{1\, mol\, C_{17}H_{21}O_4NHCl}\right)\left(\frac{303.35\, g}{1\, mol\, C_{17}H_{21}O_4N}\right) = 111._{59}\ g\, C_{17}H_{21}O_4N$$

$$\frac{111._{59}\ g\ C_{17}H_{21}NO_4}{1.70\ g\ C_{17}H_{21}NO_4/1\, L} = 65.6\, L$$

The additional volume is the difference 65.6 L – 0.0500 L = **65.6 L** to 3 significant figures.

CHAPTER 4

THE MAJOR CLASSES OF CHEMICAL REACTIONS

FOLLOW-UP PROBLEMS

4.1 a) One mole of $KClO_4$ dissociates to form one mole of potassium ions and one mole of perchlorate ions.

$$KClO_4(s) \rightarrow K^+(aq) + ClO_4^-(aq)$$

Therefore 2 moles of solid $KClO_4$ produce **2 moles of K^+** ions and **2 moles of ClO_4^-** ions.

b) $$Mg(C_2H_3O_2)_2(s) \rightarrow Mg^{2+}(aq) + 2C_2H_3O_2^-(aq)$$

First convert grams of $Mg(C_2H_3O_2)_2$ to moles of $Mg(C_2H_3O_2)_2$ and then use molar ratios to determine the moles of each ion produced.

$$\left(354\ g\ Mg(C_2H_3O_2)_2\right)\left(\frac{1\ mol\ Mg(C_2H_3O_2)_2}{142.40\ g\ Mg(C_2H_3O_2)_2}\right) = 2.48_{60}\ mol\ Mg(C_2H_3O_2)_2$$

The dissolution of 2.49 mol $Mg(C_2H_3O_2)_2(s)$ produces **2.49 mol Mg^{2+}** and **4.97 mol $C_2H_3O_2^-$**.

c) $$(NH_4)_2CrO_4(s) \rightarrow 2\ NH_4^+(aq) + CrO_4^{2-}(aq)$$

First convert formula units to moles and then use molar ratios as in part b).

$$\left(1.88x10^{24}\ formula\ units\right)\left(\frac{1\ mol\ (NH_4)_2CrO_4}{6.022x10^{23}\ formula\ units}\right) = 3.12\ mol\ (NH_4)_2CrO_4$$

The dissolution of 3.12 mol $(NH_4)_2CrO_4(s)$ produces **6.24 mol NH_4^+** and **3.12 mol CrO_4^{2-}**

d) $$NaHSO_4(s) \rightarrow Na^+(aq) + HSO_4^-(aq)$$

The solution contains (1.32 L soln)(0.55 mol $NaHSO_4$/L soln) = 0.73 mol of $NaHSO_4$ and therefore **0.73 mol of Na^+** and **0.73 mol HSO_4^-**.

4.2 $$HBr(l) \rightarrow H^+(aq) + Br^-(aq)$$

$$moles\ of\ H^+ = \left(0.451\ L\ soln\right)\left(\frac{3.20\ mol\ HBr}{L\ soln}\right)\left(\frac{1\ mol\ H^+}{1\ mol\ HBr}\right) = \mathbf{1.44\ mol\ H^+}$$

4.3 a) The resulting ion combinations that are possible are iron(III) phosphate and cesium chloride. According to Table 4.1, iron(III) phosphate is a common phosphate and insoluble (rule 2), so a reaction occurs. Applying rule 3 (soluble), we see that cesium chloride is soluble.

Total ionic equation:

$Fe^{3+}(aq) + 3Cl^-(aq) + 3Cs^+(aq) + PO_4^{3-}(aq) \rightarrow FePO_4(s) + 3Cl^-(aq) + 3Cs^+(aq)$

Net ionic equation:

$Fe^{3+}(aq) + PO_4^{3-}(aq) \rightarrow FePO_4(s)$

b) The resulting ion combinations that are possible are sodium nitrate (soluble) and cadmium hydroxide (insoluble). A reaction occurs.

Total ionic equation:

$2Na^+(aq) + 2OH^-(aq) + Cd^{2+}(aq) + 2NO_3^-(aq) \rightarrow Cd(OH)_2(s) + 2Na^+(aq) + 2NO_3^-(aq)$

Note: The coefficients for Na^+ and OH^- are necessary to balance the reaction and must be included.

Net ionic equation:

$Cd^{2+}(aq) + 2OH^-(aq) \rightarrow Cd(OH)_2(s)$

c) The resulting ion combinations that are possible are magnesium acetate (soluble – Rule 2) and potassium bromide (soluble – Rule 1). No reaction occurs.

d) The resulting ion combinations that are possible are silver chloride (insoluble due to exception in Rule 3) and barium sulfate (insoluble due to exception in Rule 4). A reaction occurs.

Total ionic equation:

$2Ag^+(aq) + SO_4^{2-}(aq) + Ba^{2+}(aq) + 2Cl^-(aq) \rightarrow 2AgCl(s) + BaSO_4(s)$

The total and net ionic equations are identical because there are no spectator ions in this reaction.

4.4 Molecular equation: $\mathbf{2}HNO_3(aq) + Ca(\mathbf{OH})_\mathbf{2}(aq) \rightarrow Ca(NO_3)_2(aq) + \mathbf{2H_2O}(l)$

Total ionic equation:

$2H^+(aq) + 2NO_3^-(aq) + Ca^{2+}(aq) + 2OH^-(aq) \rightarrow Ca^{2+}(aq) + 2NO_3^-(aq) + 2H_2O(l)$

Net ionic equation: $2H^+(aq) + 2OH^-(aq) \rightarrow 2H_2O(l)$ or $H^+(aq) + OH^-(aq) \rightarrow H_2O(l)$

4.5 The molarity of the HCl solution is 0.1016 M. However, the molar ratio is not 1:1 as in the example problem. According to the balanced equation, the ratio is 2 moles of acid per 1 mole of base:

$$2HCl(aq) + Ba(OH)_2(aq) \rightarrow BaCl_2(aq) + 2H_2O(l)$$

Instead of trying to remember to divide by 2 or multiply by 2, perform the calculation by keeping track of units.

$$vol = (0.05000\ L\ HCl\ soln)\left(\frac{0.1016\ mol\ HCl}{L\ soln}\right)\left(\frac{1\ mol\ Ba(OH)_2}{2\ mol\ HCl}\right)$$

$$\times\left(\frac{L\ Ba(OH)_2\ soln}{0.1292\ mol\ Ba(OH)_2}\right) = \mathbf{0.01966\,L\ Ba(OH)_2\ soln}$$

4.6 a) In most compounds, oxygen has a –2 O.N. so oxygen is often a good starting point. If each oxygen atom has a –2 O.N., then each scandium must have a +3 oxidation state so that the sum of O.N.'s equal zero: 2(+3) + 3(-2) = 0.

b) O.N. of Ga = +3; O.N. of Cl = -1

c) The hydrogen phosphate ion is HPO_4^{2-}. Again, oxygen has a –2 O.N. Hydrogen has a +1 O.N. because it is combined with nonmetals. The sum of the O.N.'s must equal the ionic charge, so the following algebraic equation can be solved for P:

1(+1) + 1(P) + 4(-2) = -2; O.N. for P = +5

d) The formula of iodine trifluoride is IF_3. O.N. of F = -1 (in all compounds); O.N. of I = +3.

4.7 If an atom gains electrons, then it reduces and the reactant containing that atom is the oxidizing agent.

a) $2\overset{0}{Fe}(s) + 3\overset{0}{Cl_2}(g) \rightarrow 2\overset{+3}{Fe}\overset{-1}{Cl_3}(s)$

Fe is oxidized from 0 to +3; Fe is the reducing agent.

Cl is reduced from 0 to –1; Cl_2 is the oxidizing agent.

b) $2\overset{-3}{C}_2\overset{+1}{H}_6(g)+7\overset{0}{O}_2(g)\rightarrow 4\overset{+4}{C}\overset{-2}{O}_2(g)+6\overset{+1}{H}_2\overset{-2}{O}(l)$

C is oxidized from –3 to +4; C_2H_6 is the reducing agent.

O is reduced from 0 to –2; O_2 is the oxidizing agent.

c) $5\overset{+2}{C}\overset{-2}{O}(g)+\overset{+5}{I}_2\overset{-2}{O}_5\rightarrow \overset{0}{I}_2(s)+5\overset{+4}{C}\overset{-2}{O}_2(g)$

C is oxidized from +2 to +4; CO is the reducing agent.

I is reduced from +5 to 0; I_2O_5 is the oxidizing agent.

4.8

$-2e^-$

$$\overset{+1}{K}_2\overset{+6}{Cr}_2\overset{-2}{O}_7(aq)+2\overset{+1}{H}\overset{-1}{I}(aq)\rightarrow \overset{+1}{K}\overset{-1}{I}(aq)+2\overset{+3}{Cr}\overset{-1}{I}_3(aq)+\overset{0}{I}_2(s)+\overset{+1}{H}_2\overset{-2}{O}(l)$$

$+6e^-$

Add a coefficient of 2 to CrI_3 to initially balance Cr and add a coefficient of 2 to HI to initially balance I. Multiply the oxidation half-reaction by 3, so that electrons lost (6) equal electrons gained.

$$K_2Cr_2O_7(aq) + 6HI(aq) \rightarrow KI(aq) + 2CrI_3(aq) + 3I_2(s) + H_2O(l)$$

The equation is still unbalanced at this point. Add a coefficient of 2 to KI to balance K and add a coefficient of 7 to H_2O to balance O. Note that this does not affect the exchange of electrons because the oxidation states of the products remain the same as in the reactants. To balance H, the coefficient of HI must be changed to 14. This also balances the I. However, this change does effect the exchange of electrons, so double check that electrons lost equals electrons gained. Although there are 14HI molecules, only 6 of the I^- ions undergo oxidation to form 3 I_2 molecules.

$-6e^-$

$$\overset{+1}{K}_2\overset{+6}{Cr}_2\overset{-2}{O}_7(aq)+14\overset{+1}{H}\overset{-1}{I}(aq)\rightarrow 2\overset{+1}{K}\overset{-1}{I}(aq)+2\overset{+3}{Cr}\overset{-1}{I}_3(aq)+3\overset{0}{I}_2(s)+7\overset{+1}{H}_2\overset{-2}{O}(l)$$

$+6e^-$

4.9 a) The balanced equation derived in Example 4.9 also applies for this problem. All of the Ca^{2+} ion in the milk sample is precipitated as CaC_2O_4. We can use the stoichiometric relationships in the balanced equation to arrive at moles of Ca^{2+}.

$$mol\ Ca^{2+} = (0.00653\ L\ KMnO_4\ soln)\left(\frac{4.56x10^{-3}\ mol\ KMnO_4}{L\ soln}\right)$$

$$\times\left(\frac{5\ mol\ CaC_2O_4}{2\ mol\ KMnO_4}\right)\left(\frac{1\ mol\ Ca^{2+}}{1\ mol\ CaC_2O_4}\right) = 7.44x10^{-5}\ mol\ Ca^{2+}$$

$$M\ of\ Ca^{2+} = \left(\frac{7.44x10^{-5}\ mol\ Ca^{2+}}{0.00250\ L\ milk}\right) = \mathbf{2.98x10^{-2}\ M\ Ca^{2+}}$$

b) $conc\ of\ Ca^{2+}\ (g/L) = \left(\frac{2.98x10^{-2}\ mol\ Ca^{2+}}{L\ milk}\right)\left(\frac{40.08\ g\ Ca^{2+}}{mol\ Ca^{2+}}\right) = \mathbf{1.19\ g\ Ca^{2+}\ /L\ milk}$

4.10 a) combination; $S_8(s) + 16F_2(g) \rightarrow 8SF_4(g)$

Sulfur changes from 0 to +4 oxidation state; it is oxidized and S_8 is the reducing agent.

Fluorine changes from 0 to –1 oxidation state; it is reduced and F_2 is the oxidizing agent.

b) displacement; $2CsI(aq) + Cl_2(aq) \rightarrow 2CsCl(aq) + I_2(aq)$

Total ionic eqn: $2Cs^+(aq) + 2I^-(aq) + Cl_2(aq) \rightarrow 2Cs^+(aq) + 2Cl^-(aq) + I_2(aq)$

Net ionic eqn: $2I^-(aq) + Cl_2(aq) \rightarrow 2Cl^-(aq) + I_2(aq)$

Iodine changes from -1 to 0 oxidation state; it is oxidized and CsI is the reducing agent. Chlorine changes from 0 to -1 oxidation state; it is reduced and Cl_2 is the oxidizing agent.

c) displacement; $3Ni(NO_3)_2 + 2Cr(s) \rightarrow 2Cr(NO_3)_3(aq) + 3Ni(s)$

Total ionic eqn: $3Ni^{+2}(aq) + 6NO_3^-(aq) + 2Cr(s) \rightarrow 2Cr^{+3}(aq) + 6NO_3^-(aq) + 3Ni(s)$

Net ionic eqn: $3Ni^{+2}(aq) + 2Cr(s) \rightarrow 2Cr^{+3}(aq) + 3Ni(s)$

Nickel changes from +2 to 0 oxidation state; it is reduced and $Ni(NO_3)_2$ is the oxidizing agent. Chromium changes from 0 to +3 oxidation state; it is oxidized and Cr is the reducing agent.

END-OF-CHAPTER PROBLEMS

4.2 Ionic and polar covalent compounds are most likely to be soluble in water. Because water is polar, the partial charges in water molecules are able to interact with the charges, either ionic or dipole-induced, in other substances.

4.3 Ions must be present in an aqueous solution for it to conduct an electric current. Ions come from ionic compounds or from electrolytes such as acids and bases.

4.6 The box in (2) best represents a portion of magnesium nitrate solution. Upon dissolving the salt in water, magnesium nitrate or $Mg(NO_3)_2$ would dissociate to form one Mg^{2+} ion for every two NO_3^- ions, thus forming twice as many nitrate ions. Only box (2) has twice as many nitrate ions (red circles) and magnesium ions (blue circles)

4.10 a) Benzene is likely to be insoluble in water since benzene is non-polar and water is polar.

b) Sodium hydroxide, an ionic compound, is likely soluble in water since the ions from sodium hydroxide will be held in solution through ion-dipole attraction with water.

c) Ethanol will likely be soluble in water since the alcohol group (-OH) will hydrogen bond with the water.

d) Potassium acetate, an ionic compound, will likely dissolve in water to form sodium ions and acetate ions that are held in solution through ion-dipole attraction to water.

4.12 a) An aqueous solution that contains ions conducts electricity. CsI is a soluble ionic compound, and a solution of this salt in water contains Cs^+ and I^- ions. Its solution conducts electricity.

b) HBr is a strong acid that dissociates completely in water. Its aqueous solution contains H^+ and Br^- ions, so it conducts electricity.

4.14 a) Each mole of NH_4Cl dissolves in water to form 1 mole of NH_4^+ ions and 1 mole of Cl^- ions.

$$0.25\,mol\ NH_4Cl\left(\frac{2\,mol\ ions}{1\,mol\ NH_4Cl}\right) = \mathbf{0.50\,mol\ ions}$$

b) Each mole of $Ba(OH)_2 \cdot 8H_2O$ forms 1 mole of barium ions (Ba^{2+}) and 2 moles of hydroxide ions (OH^-).

$$(26.4\ g\ Ba(OH)_2 \cdot 8H_2O)\left(\frac{1\,mol\ Ba(OH)_2 \cdot 8H_2O}{315.4\ g\ Ba(OH)_2 \cdot 8H_2O}\right)\left(\frac{3\ mol\ ions}{1\,mol\ Ba(OH)_2 \cdot 8H_2O}\right) = \mathbf{0.251\ mol\ ions}$$

c) Each mol of LiCl produces 2 moles of ions.

$$\left(1.78\,x\,10^{20}\ F.U.\,LiCl\right)\left(\frac{1\,mol\ LiCl}{6.022x10^{23}\ F.U.\,LiCl}\right)\left(\frac{2\ mol\ ions}{1\,mol\ LiCl}\right) = \mathbf{5.91x10^{-4}\ mol\ ions}$$

4.16 To determine the total moles of ions released, write a dissolution equation showing correct molar ratios, and convert the given amounts to moles if necessary.

a) Recall that phosphate, PO_4^{3-}, is a polyatomic anion and does not dissociate further in water.

$$Na_3PO_4(s) \rightarrow 3Na^+(aq) + PO_4^{3-}(aq)$$

4 moles of ions are released when one mole of Na_3PO_4 dissolves, so the total number of moles released is (0.15 mol Na_3PO_4)(4 mol ions/mol Na_3PO_4) = **0.60 mol**.

b) $$NiBr_2 \cdot 3H_2O(s) \rightarrow Ni^{2+}(aq) + 2Br^-(aq)$$

The waters of hydration become part of the larger bulk of water. Convert the grams of $NiBr_2 \cdot 3H_2O$ to moles using the molar mass (be sure to include the mass of the water):

$$\left(47.9\ g\ NiBr_2 \cdot 3H_2O\right)\left(\frac{1\,mol\ NiBr_2 \cdot 3H_2O}{272.54_8\ g\ NiBr_2 \cdot 3H_2O}\right) = 0.175_{75}\ mol\ NiBr_2 \cdot 3H_2O$$

3 moles of ions are released when one mole of $NiBr_2 \cdot 3H_2O$ dissolves, so the total moles released is (0.175$_{75}$ mol $NiBr_2 \cdot 3H_2O$)(3 mol ions/ mol $NiBr_2 \cdot 3H_2O$) = **0.527 mol.**

c) $$FeCl_3(s) \rightarrow Fe^{3+}(aq) + Cl^-(aq)$$

Recall that a mole contains 6.022 x 10^{23} entities, so a mole of $FeCl_3$ contains 6.022 x 10^{23} units of $FeCl_3$, or more easily expressed as formula units. Since the problem specifies only 4.23 x 10^{22} formula units, we know that the amount is some fraction of a mole.

$$\left(4.23x10^{22}\ formula\ units\ FeCl_3\right)\left(\frac{1\ mol\ FeCl_3}{6.022x10^{23}\ formula\ units\ FeCl_3}\right)\left(\frac{4\ mol\ ions}{1\ mol\ FeCl_3}\right) = \mathbf{0.281\ mol\ ions}$$

4.18 a) One mole of aluminum chloride, $AlCl_3$, forms 1 mole of Al^{3+} and 3 moles of Cl^-.

$$\frac{2.45\,mol\ AlCl_3}{L\ soln}\left(\frac{1\,L}{1000\,mL}\right)(95.5\,mL\ soln)\left(\frac{1\,mol\ Al^{3+}}{1\,mol\ AlCl_3}\right) = \mathbf{0.234\,mol\ Al^{3+}}$$

$$\left(0.233_{975}\ mol\ Al^{3+}\right)\left(\frac{6.022x10^{23}\ Al^{3+}\ ions}{1\,mol\ Al^{3+}}\right) = \mathbf{1.41x10^{23}\ Al^{3+}\ ions}$$

$$\frac{2.45\,mol\ AlCl_3}{L\ soln}\left(\frac{1\,L}{1000\,mL}\right)(95.5\,mL\ soln)\left(\frac{3\,mol\ Cl^-}{1\,mol\ AlCl_3}\right) = \mathbf{0.702\,mol\ Cl^-}$$

$$\left(0.701_{925}\ mol\ Cl^-\right)\left(\frac{6.022x10^{23}\ Cl^-\ ions}{1\,mol\ Cl^-}\right) = \mathbf{4.23x10^{23}\ Cl^-\ ions}$$

b) One mole of sodium sulfate, Na_2SO_4 forms 2 moles of Na^+ and 1 mole of SO_4^{2-}.

$$\left(2.50\ L\right)\left(\frac{4.59g\ Na_2SO_4}{L}\right)\left(\frac{1\,mol\ Na_2SO_4}{142.05\,g}\right)\left(\frac{2\,mol\ Na^+}{1\,mol\ Na_2SO_4}\right) = \mathbf{0.162\,mol\,Na^+}$$

$$\left(0.161_{56}\ mol\ Na^+\right)\left(\frac{6.022x10^{23}\ Na^+\ ions}{1\,mol\ Na^+}\right) = \mathbf{9.73x10^{22}\ Na^+\ ions}$$

$$\left(2.50\ L\right)\left(\frac{4.59g\ Na_2SO_4}{L}\right)\left(\frac{1\,mol\ Na_2SO_4}{142.05\,g}\right)\left(\frac{1\,mol\ SO_4^{2-}}{1\,mol\ Na_2SO_4}\right) = \mathbf{0.0808\,mol\,SO_4^{2-}}$$

$$\left(0.0807_{81}\ mol\ SO_4^{2-}\right)\left(\frac{6.022x10^{23}\ SO_4^{2-}\ ions}{1\,mol\ SO_4^{2-}}\right) = \mathbf{4.86x10^{22}\ SO_4^{2-}\ ions}$$

c) 1 formula unit of magnesium bromide, $MgBr_2$, forms 3 ions: 1 Mg^{2+} and 2 Br^-.

$$\frac{2.68\,x\,10^{22}\ F.U.\ MgBr_2}{L\ soln}\left(\frac{1L}{1000mL}\right)\left(80.5mL\ soln\right)\left(\frac{1\ Mg^{2+}\ ion}{1\,F.U.\ MgBr_2}\right) = \mathbf{2.16\,x10^{21}\ Mg^{2+}\ ions}$$

$$\left(2.15_{74}\,x10^{21}\ Mg^{2+}\ ions\right)\left(\frac{1\,mol\ Mg^{2+}}{6.022x10^{23}\ Mg^{2+}\ ions}\right) = \mathbf{3.58x10^{-3}\ mol\ Mg^{2+}}$$

$$\frac{2.68\,x\,10^{22}\ F.U.\ MgBr_2}{L\ soln}\left(\frac{1L}{1000mL}\right)\left(80.5mL\ soln\right)\left(\frac{2\ Br^-\ ions}{1\,F.U.\ MgBr_2}\right) = \mathbf{4.31\,x10^{21}\ Br^-\ ions}$$

$$\left(4.31_{48}\,x10^{21}\ Br^-\ ions\right)\left(\frac{1\,mol\ Br^-}{6.022x10^{23}\ Br^-\ ions}\right) = \mathbf{7.16x10^{-3}\ mol\ Br^-}$$

4.20 The acids in this problem are all strong acids, so you can assume that all acid molecules dissociate completely to yield H^+ ions and associated anions. One mole of $HClO_4$, HNO_3 and HCl each produce one mole of H^+ upon dissociation, so mol H^+ = mol acid.

a) mol H^+ = mol $HClO_4$ = (0.140 L)(2.5 M) = **0.35 mol H^+**

b) mol H^+ = mol HNO_3 = (0.0068 L)(0.52 M) = **3.5 x 10^{-3} mol H^+**

c) mol H^+ = mol HCl = (2.5 L)(0.056 M) = **0.14 mol H^+**

4.24 Ions in solution that do not participate in the reaction do not appear in a net ionic equation. These spectator ions remain as dissolved ions throughout the reaction.

4.28 The precipitated ions, Ag^+ and Cl^-, are represented by silver and green balls while the ions that remain in solution, Na^+ and NO_3^-, are represented by brown and blue balls, respectively.

Molecular equation: $AgNO_3(aq) + NaCl(aq) \rightarrow AgCl(s) + NaNO_3(aq)$

Total ionic equat'n: $Ag^+(aq) + NO_3^-(aq) + Na^+(aq) + Cl^-(aq) \rightarrow AgCl(s) + Na^+(aq) + NO_3^-(aq)$

Net ionic equation: $Ag^+(aq) + Cl^-(aq) \rightarrow AgCl(s)$

4.29 A precipitate forms if reactant ions can form combinations that are insoluble, as determined by solubility rules in Table 4.1. Create cation-anion combinations other than the reactants and determine if they are insoluble.

a) No precipitate will form. The ions Na^+ and SO_4^{2-} will not form an insoluble salt according to the solubility rule #1: *All common compounds of Group 1A ions are soluble.* The ions Cu^{2+} and NO_3^- will not form an insoluble salt according to the solubility rule #2: *All common nitrates are soluble.*

b) A precipitate will form because silver ions, Ag^+, and iodide ions, I^-, will combine to form a solid salt, silver iodide, AgI.

4.31 a) New cation-anion combinations are potassium nitrate and iron(II) chloride. Rules 2 & 3 state that all common nitrates and chlorides (with some exceptions) are soluble, so no precipitate forms.

b) New cation-anion combinations are ammonium chloride and barium sulfate. Again, rule 3 states most chlorides are soluble, but rule 4 states that sulfate compounds containing barium are insoluble. Barium sulfate is a precipitate and its formula is $BaSO_4$.

4.33 a) A precipitate, Hg_2I_2, will form according to solubility rule #3 for soluble compounds.

<u>Molecular equation</u>: $Hg_2(NO_3)_2(aq) + 2KI(aq) \rightarrow Hg_2I_2(s) + 2KNO_3(aq)$

Total ionic equation:

$Hg_2^{2+}(aq) + 2NO_3^-(aq) + 2K^+(aq) + 2I^-(aq) \rightarrow Hg_2I_2(s) + 2NO_3^-(aq) + 2K^+(aq)$

<u>Net ionic equation</u>: $Hg_2^{2+}(aq) + 2I^-(aq) \rightarrow Hg_2I_2(s)$

The spectator ions are $K^+(aq)$ and $NO_3^-(aq)$

b) Precipitates, $BaSO_4$ and $Fe(OH)_2$, will form according to solubility rule #4 for soluble compounds and rule #1 for insoluble compounds, respectively

<u>Molecular equation</u>: $FeSO_4(aq) + Ba(OH)_2(aq) \rightarrow BaSO_4(s) + Fe(OH)_2(s)$

Total ionic equation:

$Fe^{2+}(aq) + SO_4^{2-}(aq) + Ba^{2+}(aq) + 2OH^-(aq) \rightarrow BaSO_4(s) + Fe(OH)_2(s)$

<u>Net ionic equation</u>: same as total ionic equation because there are no spectator ions

4.35 Write a balanced equation for the chemical reaction described in the problem. By applying your solubility rules to the two possible products ($NaNO_3$ and PbI_2), you determine that PbI_2 is the precipitate. By using molar relationships, you can determine how many moles of $Pb(NO_3)_2$ (and thus Pb^{2+} ion) are required to produce 0.628 g of PbI_2.

$$Pb(NO_3)_2(aq) + 2NaI(aq) \rightarrow PbI_2(s) + 2NaNO_3(aq)$$

$$(0.628\ g\ PbI_2)\left(\frac{1\ mol\ PbI_2}{461.0\ g\ PbI_2}\right)\left(\frac{1\ mol\ Pb(NO_3)_2}{1\ mol\ PbI_2}\right)\left(\frac{1\ mol\ Pb^{2+}}{1\ mol\ Pb(NO_3)_2}\right) = 1.36x10^{-3}\ mol\ Pb^{2+}$$

$$molarity\ of\ Pb^{2+}\ ion = \left(\frac{1.36x10^{-3}\ mol\ Pb^{2+}}{0.0350\ L}\right) = \mathbf{0.0389\ M}$$

4.37 The balanced equation for this reaction is $AgNO_3(aq) + Cl^-(aq) \rightarrow AgCl(s) + NO_3^-(aq)$. First determine the moles of Cl^- present in the 25.00 mL sample. Second, convert moles of Cl^- into grams, and convert the sample volume into grams using the given density. Dividing one result by the other yields the mass percent of Cl^-.

$$moles\ Cl^{-}\ present = \left(0.04363\ L\ AgNO_3\ soln\right)\left(\frac{0.3020\ mol\ AgNO_3}{L\ soln}\right)\left(\frac{1\ mol\ Cl^{-}}{1\ mol\ AgNO_3}\right) = 0.01317_{63}\ mol$$

$$\left(\frac{0.01317_{63}\ mol\ Cl^{-}}{}\right)\left(\frac{35.45\ g\ Cl^{-}}{1\ mol\ Cl^{-}}\right) = 0.4670_{98}\ g\ Cl^{-}\ in\ a\ 25.00\ mL\ sample\ of\ seawater$$

$$mass\ of\ seawater = 25.00\ mL\left(\frac{1.04\ g}{mL}\right) = 26.0\ g$$

$$mass\ percent\ of\ Cl^{-} = \left(\frac{0.4670_{98}\ g}{26.0\ g}\right)(100\%) = \mathbf{1.80\ \%}$$

4.43 a) The formation of a gas, $SO_2(g)$, and formation of water drive this reaction to completion, because both products remove reactants from solution.

b) The formation of precipitate, $Ba_3(PO_4)_2(s)$, drives this reaction to completion. This reaction is one between an acid and a base, so the formation of water molecules through the combination of H^+ and OH^- ions also drives the reaction.

4.45 a) Molecular equation: $KOH(aq) + HI(aq) \rightarrow KI(aq) + H_2O(l)$

Total ionic equation: $K^+(aq) + OH^-(aq) + H^+(aq) + I^-(aq) \rightarrow K^+(aq) + I^-(aq) + H_2O(l)$

Net ionic equation: $OH^-(aq) + H^+(aq) \rightarrow H_2O(l)$

The spectator ions are $K^+(aq)$ and $I^-(aq)$

b) Molecular equation: $NH_3(aq) + HCl(aq) \rightarrow NH_4Cl(aq)$

Total ionic equation: $NH_3(aq) + H^+(aq) + Cl^-(aq) \rightarrow NH_4^+(aq) + Cl^-(aq)$

NH_3 is a weak base and is written in the molecular, undissociated form. HCl is a strong acid and is written the dissociated form. NH_4Cl is a soluble compound, because all ammonium compounds are soluble.

Net ionic equation: $NH_3(aq) + H^+(aq) \rightarrow NH_4^+(aq)$

Cl^- is the only spectator ion.

4.47 Calcium carbonate dissolves in HCl(aq) because the carbonate ion, a base, reacts with the acid to form $CO_2(g)$.

Total ionic equation:

$CaCO_3(s) + 2H_3O^+(aq) + 2Cl^-(aq) \rightarrow Ca^{2+}(aq) + 3H_2O(l) + CO_2(g) + 2Cl^-(aq)$

Net ionic equation: $CaCO_3(s) + 2H_3O^+(aq) \rightarrow Ca^{2+}(aq) + 3H_2O(l) + CO_2(g)$

or $CaCO_3(s) + 2H^+(aq) \rightarrow Ca^{2+}(aq) + H_2O(l) + CO_2(g)$

4.49 Neutralization reaction: $KOH(aq) + CH_3COOH(aq) \rightarrow KCH_3COO(aq) + H_2O(l)$

$$M_{KOH}V_{KOH} = M_{CH_3COOH}V_{CH_3COOH}$$

$$(0.1080\ M)(15.98\ mL) = M_{CH_3COOH}(52.00\ mL)$$

$$M_{CH_3COOH} = \mathbf{0.03319M}$$

4.58 a) Sulfuric acid acts as an oxidizing agent since the oxidation number of sulfur changes in the reaction form +6 in H_2SO_4 to +4 in SO_2. The sulfur is gaining electrons, thus being reduced and acting as the oxidizing agent.

b) Sulfuric acid acts as an acid since the proton from sulfuric acid is donated to the base F^- forming hydrofluoric acid, HF. Oxidation numbers remain constant throughout.

4.60 Refer to the rules in Table 4.3. Remember that oxidation number (O.N.) is <u>not</u> the same thing as ionic charge. The O.N. is the charge an atom would have **if** electrons were transferred completely. The compounds in this problem contain covalent bonds, or a combination of ionic and covalent bonds ($Na_2C_2O_4$). Covalent bonds do not involve the complete transfer of electrons.

a) CF_2Cl_2. Rules 4 and 6 dictate that F and Cl each have an O.N. of –1; two F and two Cl yield a sum of –4, so C must have a **+4** O.N.

b) $Na_2C_2O_4$. Rule 1 dictates that Na has a +1 O.N.; rule 5 dictates that O has a –2 O.N.; two Na and four O yield a sum of –6 [(+2) + (-8)]. Therefore, the total of the O.N.'s on the two C atoms is +6 and each C is **+3**.

c) HCO_3^-. H is combined with nonmetals and has an O.N. of +1 (Rule 3); O has an O.N. of –2. To have an overall oxidation state equal to –1, C must be **+4** because (+1) + (+4) + (-6) = -1.

d) C_2H_6. Each H has an O.N. of +1; six H gives +6. The sum of O.N.'s for the two C atoms must be –6, so each C is **–3**.

4.62 a) NH_2OH:	(O.N. for N) + 3(+1 for H) + 1(-2 for O) = 0	O.N. for N = **-1**
b) N_2H_4:	{2(O.N. for N)} + 4(+1 for H) = 0	O.N. for N = **-2**
c) NH_4^+:	(O.N. for N) + 4(+1 for H) = +1	O.N. for N = **-3**
d) HNO_2:	(O.N. for N) + 1(+1 for H) + 2(-2 for O) = 0	O.N. for N = **+3**

4.64 a) AsH_3. H is combined with a nonmetal, so its O.N. is +1 (Rule 3). The O.N. for As is **-3**.

b) H_3AsO_4. The H's in this formula are acidic hydrogens. When an acid dissociates in water, it forms H_3O^+ or H^+ ion, therefore the O.N. of H in this compound is +1, or +3 for 3 H's. Oxygen's O.N. is –2, with total O.N. of –8 (4 times –2) so As needs to have an O.N. of **+5**.

c) $AsCl_3$. Cl has an O.N. of –1, total of –3, so As must have an O.N. of **+3**.

4.66 a) MnO_4^{2-}:	(O.N. for Mn) + 4(-2 for O) = -2	O.N. for Mn = **+6**
b) Mn_2O_3:	{2(O.N. for Mn)} + 3(-2 for O) = 0	O.N. for Mn = **+3**
c) $KMnO_4$:	1(+1 for K) + (ON for Mn) + 4(-2 for O) = 0	O.N. for Mn = **+7**

4.68 <u>Oxidizing agent</u>: substance that accepts the electrons released by the substance that is oxidized.

<u>Reducing agent</u>: substance that provides the electrons accepted by the substance that is reduced.

First, assign oxidation numbers to all atoms. Second, recognize that the agent is the compound that contains the atom that is gaining or losing electron, not just the atom itself.

a) $$5\overset{+1}{H_2}\overset{+3}{C_2}\overset{-2}{O_4}(aq) + 2\overset{+7}{Mn}\overset{-2}{O_4}(aq) + 6H^+(aq) \rightarrow 2Mn^{2+}(aq) + 10\overset{+4}{C}\overset{-2}{O_2}(g) + 8\overset{+1}{H_2}\overset{-2}{O}(l)$$

Hydrogen and oxygen do not change oxidation state. Mn changes from +7 to +2 (reduction) so MnO_4^- is the oxidizing agent. C changes from +3 to +4 (oxidation), so $H_2C_2O_4$ is the reducing agent.

b) Cu changes from 0 to +2 (is oxidized) and is the reducing agent. N changes from +5 (in NO_3^-) to +2 (in NO) and is reduced, so NO_3^- is the oxidizing agent.

4.70 a) Oxidizing agent is NO_3^- because nitrogen changes from +5 O.N. in NO_3^- to +4 O.N. in NO_2.
Reducing agent is Sn because its O.N. changes from 0 as the element to +4 in $SnCl_6^{2-}$.

b) Oxidizing agent is MnO_4^- because manganese changes from +7 O.N. in MnO_4^- to +2 O.N. in Mn^{2+}.
Reducing agent is Cl^- because its O.N. changes from -1 in Cl^- to 0 as the element to Cl_2.

4.72 a) The lowest O.N. for S [(Group 6A(16)] is 6 - 8 = -2, which occurs in S^{2-}. Therefore, when S^{2-} reacts in an oxidation-reduction reaction, S can only increase its O.N. (oxidize), so S^{2-} can only function as a reducing agent.

b) The highest O.N. for S is +6, which occurs in SO_4^{2-}. Therefore, when SO_4^{2-} reacts in an oxidation-reduction reaction, the S can only decrease its O.N. (reduce), so SO_4^{2-} can only function as an oxidizing agent.

c) The O.N. of S in SO_2 is +4, so it can increase or reduce its O.N. Therefore, SO_2 can function as either an oxidizing or reducing agent.

4.74 a) Step 1: Assign oxidation numbers to all elements
Step 2: Identify oxidized and reduced species. The O.N. of Cr decreases from +6 in K_2CrO_4 to +3 in $Cr(NO_3)_3$ so chromium is gaining electrons and is reduced. The O.N. of Fe increases from +2 in $Fe(NO_3)_2$ to +3 in $Fe(NO_3)_3$ so iron is losing electrons and is oxidized.
Step 3: Compute number of electrons lost in oxidation and gained in reduction.
Step 4: Multiply by factors to make e^- lost equal e^- gained.
Cr gained 3 electrons so Fe should be multiplied by 3 to lose 3 elctrons
Step 5: Finish balancing by inspection.

$$8HNO_3(aq) + K_2CrO_4(aq) + 3Fe(NO_3)_2(aq) \rightarrow$$
$$2KNO_3(aq) + 3Fe(NO_3)_3(aq) + Cr(NO_3)_3(aq) + 4H_2O(l)$$

Oxidizing agent: K_2CrO_4 Reducing agent: $Fe(NO_3)_2$

b) $$8HNO_3(aq) + 3C_2H_6O(l) + K_2Cr_2O_7(aq) \rightarrow$$
$$2KNO_3(aq) + 3C_2H_4O(l) + 7H_2O(l) + 2Cr(NO_3)_3(aq)$$

Oxidizing agent: $K_2Cr_2O_7$ Reducing agent: C_2H_6O

c) $$6HCl(aq) + 2NH_4Cl(aq) + K_2Cr_2O_7(aq) \rightarrow$$
$$2KCl(aq) + 2CrCl_3(aq) + N_2(g) + 7H_2O(l)$$

Oxidizing agent: $K_2Cr_2O_7$ Reducing agent: NH_4Cl

d) $$KClO_3(aq) + 6HBr(aq) \rightarrow 3Br_2(l) + 3H_2O(l) + KCl(aq)$$

Oxidizing agent: $KClO_3$ Reducing agent: HBr

4.76 a) $$mol\ MnO_4^- = (0.0432\ L\ KMnO_4\ soln)\left(\frac{0.105\ mol\ KMnO_4}{L\ KMnO_4\ soln}\right)\left(\frac{1\ mol\ MnO_4^-}{1\ mol\ KMnO_4}\right) = \mathbf{4.54x10^{-3}\ mol}$$

b) At the endpoint, the MnO_4^- consumes all of the H_2O_2. Two moles of MnO_4^- are required to react with 5 moles of H_2O_2.

$$mol\ H_2O_2 = (4.53_6 x10^{-3}\ mol\ MnO_4^-)\left(\frac{5\ mol\ H_2O_2}{2\ mol\ MnO_4^-}\right) = \mathbf{0.0113\ mol}$$

c) Convert moles to grams using the molar mass of H_2O_2 = 34.02 g/mol.

$$mass\ H_2O_2 = \left(0.0113_4\ mol\ H_2O_2\right)\left(\frac{34.02\ g\ H_2O_2}{1\ mol\ H_2O_2}\right) = \mathbf{0.386\ g\,H_2O_2}$$

d) Mass percent of H_2O_2 = (0.386 g / 13.8 g)(100%) = **2.80%**

e) The reducing agent is H_2O_2 because oxygen's O.N. changes from –1 to 0 (in O_2).

4.81   A common example of a combination/redox reaction is the combination of a metal and nonmetal to form an ionic salt, such as $2Na(s) + Cl_2(g) \rightarrow 2NaCl(s)$. A common example of a combination/<u>non</u>-redox reaction is the combination of a nonmetal oxide and water to form an acid, such as: $NO_2(g) + H_2O(l) \rightarrow HNO_3(aq)$.

4.83   a) $Ca(s) + 2H_2O(l) \rightarrow Ca(OH)_2(aq) + H_2(g)$

Displacement: one Ca atom displaces 2 H atoms.

b) $2NaNO_3(s) \rightarrow 2NaNO_2(s) + O_2(g)$

Decomposition: one reactant breaks into two products.

c) $C_2H_2(g) + 2H_2(g) \rightarrow C_2H_6(g)$

Combination: reactants combine to form one product.

4.85   Recall the definitions of each type of reaction:

combination: X+Y $\rightarrow$ Z;   decomposition: Z $\rightarrow$ X +Y

single displacement: X + YZ $\rightarrow$ XZ + Y   double displacement: WX + YZ $\rightarrow$ WZ + YX

a) combination

$$\overset{\text{oxidation}}{2\,\overset{0}{Sb}(s) + 3\,\overset{0}{Cl_2}(g) \rightarrow 2\,\overset{+3}{Sb}\overset{-1}{Cl_3}(s)}$$
$$\underset{\text{reduction}}{}$$

b) decomposition

$$2\,\overset{+3}{As}\overset{-1}{H_3}(g) \rightarrow 2\,\overset{0}{As}(s) + 3\,\overset{0}{H_2}(g)$$

c) single displacement

$$3\,\overset{0}{Mn}(s) + 2\,\overset{+3}{Fe}(NO_3)_3(aq) \rightarrow 3\,\overset{+2}{Mn}(NO_3)_2(aq) + 2\,\overset{0}{Fe}(s)$$

4.87   a) The combination between a metal and a nonmetal gives a binary ionic compound.

$$\overset{0}{Ca}(s) + \overset{0}{Br_2}(l) \rightarrow \overset{+2}{Ca}\overset{-1}{Br_2}(s)$$

b) Many metal oxides release oxygen gas upon thermal decomposition.

$$2\,\overset{+1}{Ag_2}\overset{-2}{O}(s) \overset{\Delta}{\longrightarrow} 4\,\overset{0}{Ag}(s) + \overset{0}{O_2}(g)$$

c) This is a single displacement reaction. Mn is a more reactive metal and displaces Cu^{2+} from solution (Figure 4.20 shows that Mn more easily oxidizes [0 to +2] than Cu).

$$\overset{0}{Mn}(s) + \overset{+2}{Cu}(NO_3)_2(aq) \rightarrow \overset{+2}{Mn}(NO_3)_2(aq) + \overset{0}{Cu}(s)$$

4.89   a) $N_2(g) + 3H_2(g) \rightarrow 2NH_3(g)$

b) $2NaClO_3(s) \overset{\Delta}{\longrightarrow} 2NaCl(s) + 3O_2(g)$

c) $Ba(s) + 2H_2O(l) \rightarrow Ba(OH)_2(aq) + H_2(g)$

4.91 a) Cs, a metal, and I_2, a nonmetal, react to form the binary ionic compound, CsI.

$$2Cs(s) + I_2(s) \rightarrow 2CsI(s)$$

b) Al is a stronger reducing agent than Mn and is able to displace Mn from solution, i.e. cause the reduction from $Mn^{2+}(aq)$ to $Mn^0(s)$.

$$2Al(s) + 3MnSO_4(aq) \rightarrow Al_2(SO_4)_3(aq) + 3Mn(s)$$

c) Sulfur dioxide, SO_2, is a nonmetal oxide that reacts with oxygen, O_2, to form the higher oxide, SO_3.

$$2SO_2(g) + O_2(g) \rightarrow 2SO_3(g)$$

d) Propane is a three-carbon hydrocarbon with the formula C_3H_8. It burns in the presence of oxygen, O_2, to form carbon dioxide gas and water vapor (see Section 4.6, *Combustion Reactions*, for the reaction involving butane, C_4H_{10}). Although this is a redox reaction that could be balanced using the oxidation number method, it is easier to balance by considering only atoms on either side of the equation. First balance carbon and hydrogen (because they only appear in one species on each side of the equation), and then balance oxygen.

$$C_3H_8(g) + 5O_2(g) \rightarrow 3CO_2(g) + 4H_2O(g)$$

e) Total ionic equation:

$2Al(s) + 3Mn^{2+}(aq) + 3SO_4^{2-}(aq) \rightarrow 2Al^{3+}(aq) + 3SO_4^{2-}(aq) + 3Mn(s)$

Net ionic equation:

$2Al(s) + 3Mn^{2+}(aq) \rightarrow 2Al^{3+}(aq) + 3Mn(s)$

Note that the molar coefficients are not further simplified because the number of electrons lost ($6e^-$) must equal the electrons gained ($6e^-$).

4.93

$$2HgO(s) \rightarrow 2Hg(s) + O_2(g)$$

$$mass\ O_2 = 4.27\ kg\ HgO\left(\frac{1000\ g}{1\ kg}\right)\left(\frac{1\ mol\ HgO}{216.6\ g\ HgO}\right)\left(\frac{1\ mol\ O_2}{2\ mol\ HgO}\right)\left(\frac{32.00\ g\ O_2}{1\ mol\ O_2}\right) = \mathbf{315\ g\ O_2}$$

Other product is mercury, Hg: 4.27 kg total – 0.315_{42} kg O_2 = **3.95 kg Hg**

4.95 To determine the reactant in excess, write the balanced equation (metal + $O_2 \rightarrow$ metal oxide), convert reactant masses to moles, and use molar ratios to see which reactant makes the smaller ('limiting') amount of product.

$$4Li(s) + O_2(g) \rightarrow 2Li_2O(s)$$

a) $$mol\ Li_2O\ (from\ Li) = (1.62\ g\ Li)\left(\frac{1\ mol\ Li}{6.94\ g\ Li}\right)\left(\frac{2\ mol\ Li_2O}{4\ mol\ Li}\right) = 0.116_{71}\ mol\ Li_2O$$

$$mol\ Li_2O\ (from\ O_2) = (6.00\ g\ O_2)\left(\frac{1\ mol\ O_2}{32.00\ g\ O_2}\right)\left(\frac{2\ mol\ Li_2O}{1\ mol\ O_2}\right) = 0.375\ mol\ Li_2O$$

Therefore, Li is the limiting reagent and **O_2 is in excess.**

b) 0.117 mol Li_2O

c) Since Li is limiting, none remains.

Mass of Li = **0 g Li**

Mass of O_2 remaining = starting O_2 mass – oxygen consumed in reaction with Li to make Li_2O

$$Mass\ of\ O_2 = (6.00\ g\ O_2) - \left[(0.116_{71}\ mol\ Li_2O)\left(\frac{1\ mol\ O_2}{2\ mol\ Li_2O}\right)\left(\frac{32.00\ g\ O_2}{1\ mol\ O_2}\right)\right]$$

Mass of O_2 = 6.00 g O_2 – 1.86_{74} g O_2 = **4.13 g O_2**

Alternatively:

$$Mass\ of\ O_2 = 6.00\ g\ O_2 - \left[(1.62\ g\ Li)\left(\frac{1\ mol\ Li}{6.94\ g\ Li}\right)\left(\frac{1\ mol\ O_2}{4\ mol\ Li}\right)\left(\frac{32.00\ g\ O_2}{1\ mol\ O_2}\right)\right]$$

Mass of O_2 = 6.00 g O_2 – 1.86_{74} g O_2 = **4.13 g O_2**

Mass of Li_2O produced = $(0.116_{71}\ mol\ Li_2O)\left(\frac{29.88\ g\ Li_2O}{1\ mol\ Li_2)}\right)$ = **3.50 g Li_2O**

4.97 Mass of oxygen in sample = 0.900 g – 0.700 g = 0.200 g

$$(0.200\ g\ O)\left(\frac{1\ mol\ O}{16.00\ g\ O}\right)\left(\frac{1\ mol\ KClO_3}{3\ mol\ O}\right)\left(\frac{122.55\ g\ KClO_3}{mol\ KClO_3}\right) = 0.510_{63}\ g\ KClO_3$$

$$mass\ percent = \left(\frac{0.510_{63}\ g\ KClO_3}{0.900\ g\ total}\right)(100\%) = \mathbf{56.7\%}$$

4.99 To find the mass of Fe, write a balanced equation for the reaction, determine whether Al or Fe_2O_3 is the limiting reactant, and convert to mass.

$$2Al(s) + Fe_2O_3(s) \rightarrow 2Fe(s) + Al_2O_3(s)$$

When the masses of both reactants are given, you must determine which reactant is limiting.

$$Moles\ of\ Fe\ (from\ Al) = (1.00x10^3\ g\ Al)\left(\frac{1\ mol\ Al}{26.98\ g\ Al}\right)\left(\frac{2\ mol\ Fe}{2\ mol\ Al}\right) = 37.1\ mol\ Fe$$

$$Moles\ of\ Fe\ (from\ Fe_2O_3) = (2.00\ mol\ Fe_2O_3)\left(\frac{2\ mol\ Fe}{1\ mol\ Fe_2O_3}\right) = 4.00\ mol\ Fe$$

Therefore, Fe_2O_3 is the limiting reactant.

Mass of Fe = 4.00 mol Fe x (55.85 g Fe/mol Fe) = **223 g Fe**

4.100 In ionic compounds, iron has two common oxidation states, +2 and +3. First write the balanced equations for the formation and decomposition of compound A. Then determine which reactant is limiting and from the amount of the limiting reactant calculate how much compound B will form.

Compound A is iron chloride with iron in the higher, +3, oxidation state. Thus the formula for compound A is $FeCl_3$ and the correct name is iron(III) chloride. Balanced equation for formation of $FeCl_3$:

$$2Fe(s) + 3Cl_2(g) \rightarrow 2FeCl_3(s)$$

The $FeCl_3$ decomposes to $FeCl_2$ with iron in the +2 oxidation state. Balanced equation for the decomposition of $FeCl_3$:

$$2FeCl_3(s) \rightarrow 2FeCl_2(s) + Cl_2(g)$$

To find out whether 50.6 g Fe or 83.8 g Cl_2 limits the amount of product, calculate the number of moles of iron(III) chloride that could form based on each reactant.

$$moles\ FeCl_3\ (from\ moles\ of\ Fe) = 50.6\ g\ Fe\left(\frac{1\ mol}{55.85\ g}\right)\left(\frac{2\ mol\ FeCl_3}{2\ mol\ Fe}\right) = 0.906_{00}\ mol\ FeCl_3$$

$$moles\ FeCl_3\ (from\ moles\ of\ Cl_2) = 83.8\ g\ Cl_2\left(\frac{1\ mol}{70.90\ g}\right)\left(\frac{2\ mol\ FeCl_3}{3\ mol\ Cl_2}\right) = 0.787_{96}\ mol\ FeCl_3$$

Since fewer moles of $FeCl_3$ are produced from the available amount of chlorine, then chlorine is limiting and we calculate the amount of compound B from the amount of chlorine gas.

$$0.787_{96}\ mol\ FeCl_3\left(\frac{2\ mol\ FeCl_2}{2\ mol\ FeCl_3}\right)\left(\frac{126.75\ g}{mol\ FeCl_2}\right) = \mathbf{99.9\,g\,FeCl_2}$$

4.104 On a molecular scale, chemical reactions are dynamic. If NO and Br_2 are placed in a container, molecules of NO and molecules of Br_2 will react to form NOBr. Some of the NOBr molecules will decompose and the resulting NO and Br_2 molecules will recombine with different NO and Br_2 molecules. In this sense, the reaction is dynamic because the original NO and Br_2 pairings do not remain permanently attached to each other. Eventually, the rate of the forward reaction (combination of NO and Br_2) will equal the rate of the reverse reaction (decomposition of NOBr) at which point the reaction is said to have reached equilibrium. If you could take a "snapshot" picture of the molecules at equilibrium, you would see a constant number of reactant (NO, Br_2) and product molecules (NOBr), but the pairings would not stay the same.

4.106 a) $\overset{0}{Fe}(s) + 2\,\overset{+1}{H^+}(aq) \rightarrow \overset{+2}{Fe^{2+}}(aq) + \overset{0}{H_2}(g)$

b) $737\ g\ sauce\left(\frac{49\ mg\ Fe^{2+}}{125\ g\ sauce}\right)\left(\frac{1\ g}{1000\ mg}\right)\left(\frac{1\ mol\ Fe^{2+}}{55.85\ g\ Fe^{2+}}\right)\left(\frac{6.022x10^{23}\ Fe^{2+}\ ions}{1\ mol\ Fe^{2+}}\right) = \mathbf{3.1x10^{21}\ ions}$

4.110 a) A solution of $Na_2C_2O_4$ (soluble salt) is added to a solution containing 1.9348 g of road salt (NaCl and $CaCl_2$, both soluble salts) and an insoluble precipitate is formed (CaC_2O_4). The Na^+ and Cl^- ions are spectators in this reaction because all Na^+ salts are soluble and most Cl^- salts are soluble.

$$Ca^{2+}(aq) + C_2O_4^{2-}(aq) \rightarrow CaC_2O_4(s)$$

b) You may recognize this reaction as a redox titration, because the permanganate ion, MnO_4^-, is a common oxidizing agent. The MnO_4^- oxidizes the oxalate ion, $C_2O_4^{2-}$ to CO_2. Mn changes from +7 to +2 (reduction) and C changes from +3 to +4 (oxidation). The equation that we can write that describes this process is:

$$H_2C_2O_4(aq) + MnO_4^-(aq) \rightarrow Mn^{2+}(aq) + CO_2(g)$$

$H_2C_2O_4$ is a weak acid, so it can not be written in a fully dissociated form ($2H^+(aq) + C_2O_4^{2-}(aq)$). $KMnO_4$ is a soluble salt, so it can be written in its dissociated form. $K^+(aq)$ is omitted because it is a spectator ion. We will balance the equation using the oxidation number method (you can verify by doing the half reaction method) and first assign oxidation numbers to all elements in the reaction.

$-2\,e^-$

$$\overset{+1}{H_2}\overset{+3}{C_2}\overset{-2}{O_4}(aq) + \overset{+7}{Mn}O_4^-(aq) \rightarrow Mn^{2+}(aq) + 2\overset{+4}{C}O_2(g)$$

$+5\,e^-$

Identify the oxidized and reduced species and multiply one or both species by the appropriate factors to make the electrons lost equal the electrons gained.

$$5H_2C_2O_4(aq) + 2MnO_4^-(aq) \rightarrow 2Mn^{2+}(aq) + 10CO_2(g)$$

Adding water and $H^+(aq)$ to finish balancing the equation is appropriate since the reaction takes place in acidic medium. Add $8H_2O(l)$ to right side of equation to balance the oxygen and then add $6H^+(aq)$ to the left to balance hydrogen.

$$5H_2C_2O_4(aq) + 2MnO_4^-(aq) + 6H^+(aq) \rightarrow 2Mn^{2+}(aq) + 10CO_2(g) + 8H_2O(l)$$

c) oxidizing agent = $KMnO_4$

d) reducing agent = $H_2C_2O_4$ (remember that the *agent* refers to the whole compound, not a particular element within a compound)

e) The balanced equations provides the accurate molar ratios between species. Working backwards, we know that

(moles of MnO_4^-) is related to moles of $C_2O_4^{2-}$ at the endpoint

(moles of $C_2O_4^{2-}$) is related to moles of CaC_2O_4

(moles of CaC_2O_4) is related to moles of $CaCl_2$

moles of $KMnO_4$= moles of MnO_4^- = (0.03768 L)(0.1019 M) = 3.839_{59} x 10^{-3} moles MnO_4^-

$$mol\ of\ C_2O_4^{2-} = \left(3.839_{59}\ x10^{-3}\ mol\ MnO_4^-\right)\left(\frac{5\ mol\ H_2C_2O_4}{2\ mol\ MnO_4^-}\right)\left(\frac{1\ mol\ C_2O_4^{2-}}{1\ mol\ H_2C_2O_4}\right) = 9.598_{98}\ x10^{-3}\ mol$$

$$mole\ of\ CaCl_2 = \left(9.598_{98}\ x10^{-3}\ mol\ C_2O_4^{2-}\right)\left(\frac{1\ mol\ CaC_2O_4}{1\ mol\ C_2O_4^{2-}}\right)\left(\frac{1\ mol\ Ca^{2+}}{1\ mol\ CaC_2O_4}\right)$$

$$\times\left(\frac{1\ mol\ CaCl_2}{1\ mol\ Ca^{2+}}\right) = 9.598_{98}\ x10^{-3}\ mol$$

Mass of $CaCl_2$ = (9.598_{98}x10^{-3} mol $CaCl_2$)(110.98 g $CaCl_2$/ 1mol $CaCl_2$) = 1.065_{29} g $CaCl_2$

Mass percent of $CaCl_2$ = (1.065_{29} g $CaCl_2$ / 1.9348 g sample)(100%) = **55.06% $CaCl_2$**

4.111 The possible combinations of ions that *could* be precipitates are Ag_2S, Ag_2SO_4, $Ca(NO_3)_2$, $CaSO_4$, $NaNO_3$, and Na_2S. Solubility rules dictate that only **Ag_2S** and **$CaSO_4$** form precipitates. Because the three solutions were mixed, the initial concentrations of the ions are 1/3 of their original value:

0.03_{33} M Ag^+	0.006_{67} M Ca^{+2}	0.03_{33} M Na^+
0.03_{33} M NO_3^-	0.006_{67} M S^{2-}	0.01_{67} M SO_4^{2-}

Because NO_3^- and Na^+ always form soluble salts, their concentrations remain unchanged.

In one precipitation reaction, silver sulfide is formed:

	$2Ag^+(aq)$ +	$S^{2-}(aq)$ →	$Ag_2S(s)$
Initial conc	0.0333 M	0.00667 M	0 M
Final conc	0.020 M	0 M	0.00667M

Because 2(0.00667M) = 0.01334 M of Ag^+ precipitated as Ag_2S, the final concentration of Ag^+ is (0.0333 - 0.0133 M) = 0.020 M.

In the other precipitation reaction, calcium sulfate is formed:

	$Ca^{2+}(aq)$ +	$SO_4^{2-}(aq)$ →	$CaSO_4(s)$
Initial conc	0.00667 M	0.0167 M	0 M
Final conc	0 M	0.010 M	0.00667M

In summary (and reporting to one significant figure, the final concentrations of each ion are:

0.02 M Ag^+ **0 M Ca^{+2}** **0.03 M Na^+**

0.03 M NO_3^- **0 M S^{2-}** **0.01 M SO_4^{2-}**

4.114 The balanced equation for this reaction is:

$CaMg(CO_3)_2(s) + 4HCl(aq) \rightarrow Ca^{2+}(aq) + Mg^{2+}(aq) + 2H_2O(l) + 2CO_2(g) + 4Cl^-(aq)$

Refer to problem 4.47 for a simpler, analogous equation.

To find mass %, convert the solution to moles of HCl, use the molar ratio to find moles of dolomite, convert to mass and divide by the mass of soil.

$$mol\ CaMg(CO_3)_2 = (0.03356\ L)\left(\frac{0.2516\ mol\ HCl}{L}\right)\left(\frac{1\ mol\ CaMg(CO_3)_2}{4\ mol\ HCl}\right) = 2.111x10^{-3}\ mol$$

$$mass\%\ CaMg(CO_3)_2 = \frac{(2.111x10^{-3}\ mol)\left(\frac{184.41\ g\ CaMg(CO_3)_2}{1\ mol\ CaMg(CO_3)_2}\right)}{12.86\ g\ soil} x\ 100\% = \mathbf{3.027\%}$$

4.116 a) To figure out the amount of HNO_3 formed, sum the three equations by multiplying by appropriate coefficients to cancel out the intermediate product, NO_2.

3 × $\{4NH_3(g) + 5O_2(g) \rightarrow 4NO(g) + 6H_2O(l)\}$

6 × $\{2NO(g) + O_2(g) \rightarrow 2NO_2(g)\}$

4 × $\{3NO_2(g) + H_2O(l) \rightarrow 2HNO_3(l) + NO(g)\}$

$12NH_3(g) + 21O_2(g) \rightarrow 8HNO_3(l) + 14H_2O(l) + 4NO(g)$

Stoichiometry can now be used to solve the remainder of the problems.

$$mass\ HNO_3 = (72500g\ NH_3)\left(\frac{1\ mol\ NH_3}{17.034\ g}\right)\left(\frac{8\ mol\ HNO_3}{12\ mol\ NH_3}\right)\left(\frac{63.018\ g}{1\ mol\ HNO_3}\right) = \mathbf{1.79x10^5\ g\ HNO_3}$$

b) $$mass\ O_2 = (72500\ g\ NH_3)\left(\frac{1\ mol\ NH_3}{17.034\ g}\right)\left(\frac{21\ mol\ O_2}{12\ mol\ NH_3}\right)\left(\frac{32.00\ g}{mol\ O_2}\right)\left(\frac{1\ kg}{1000\ g}\right) = \mathbf{238\ kg\ O_2}$$

c) $$mass\ H_2O = (72500\ g\ NH_3)\left(\frac{1\ mol\ NH_3}{17.034\ g}\right)\left(\frac{14\ mol\ H_2O}{12\ mol\ NH_3}\right)\left(\frac{18.016\ g}{mol\ H_2O}\right)\left(\frac{1\ kg}{1000\ g}\right) = \mathbf{89.4\ kg\ H_2O}$$

d) In Step 1, the O.N. of nitrogen changes from -3 to +2; nitrogen is oxidized and NH_3 is the reducing agent. The O.N. of oxygen changes from 0 to -2; oxygen is reduced and O_2 is the oxidizing agent. A summary table is provided below.

STEP	Element	Change in O.N.		ox or red?	Ox agent	Red agent
		from	to			
1	N	-3	+2	oxidized		NH_3
	O	0	-2	reduced	O_2	
2	N	+2	+4	oxidized		NO
	O	0	-2	reduced	O_2	
3	N	+4	+5 (HNO_3)	oxidized		NO_2
	N	+4	+2 (NO)	reduced	NO_2	

As you can see in Step 3, it is possible for the same substance to act as both the oxidizing and reducing agent in a reaction. This type of redox reaction is called a *disproportionation*.

4.120 a) $$\overset{+6}{Cr}O_4^{2-}(aq) + H\overset{+2}{Sn}O_2^-(aq) + H_2O(l) \rightarrow \overset{+3}{Cr}O_2^-(aq) + H\overset{+4}{Sn}O_3^-(aq) + OH^-(aq)$$

(+3 e^- gained by Cr; −2 e^- lost by Sn)

CrO_4^{2-} is reduced to CrO_2^-, therefore CrO_4^{2-} is the oxidizing agent.
$HSnO_2^-$ is oxidized to $HSnO_3^-$; therefore $HSnO_2^-$ is the reducing agent.
Multiply by appropriate factors so that the electrons gained equals electrons lost, and balance non-oxidizing species as necessary.

$$2CrO_4^{2-}(aq) + 3HSnO_2^-(aq) + H_2O(l) \rightarrow 2CrO_2^-(aq) + 3HSnO_3^-(aq) + 2OH^-(aq)$$

b) $$2K\overset{+7}{Mn}O_4(aq) + 3Na\overset{+3}{N}O_2(aq) + H_2O(l) \rightarrow 2\overset{+4}{Mn}O_2(s) + 3Na\overset{+5}{N}O_3(aq) + 2KOH(aq)$$

(+3 e^- gained by Mn; −2 e^- lost by N)

$KMnO_4$ is the oxidizing agent; $NaNO_2$ is the reducing agent.

c) $$4\overset{-1}{I^-}(aq) + \overset{0}{O_2}(g) + 2H_2O(l) \rightarrow 2\overset{0}{I_2}(s) + 4\overset{-2}{O}H^-(aq)$$

O_2 is the oxidizing agent; I^- is the reducing agent.

4.123 A 1.00 kg piece of glass of composition 75% SiO_2, 15% Na_2O, and 10% CaO would contain 0.75 kg SiO_2, 0.15 kg Na_2O, and 0.10 kg CaO. In this example the SiO_2 is added directly while the sodium oxide comes from decomposition of sodium carbonate and the calcium oxide from decomposition of calcium carbonate:

$Na_2CO_3(s) \rightarrow Na_2O(s) + CO_2(g)$

$CaCO_3(s) \rightarrow CaO(s) + CO_2(g)$

$$0.15\,kg\,Na_2O\left(\frac{1000\,g}{1\,kg}\right)\left(\frac{1\,mol\,Na_2O}{61.98\,g\,Na_2O}\right)\left(\frac{1\,mol\,Na_2CO_3}{1\,mol\,Na_2O}\right)\left(\frac{105.99\,g\,Na_2CO_3}{mol\,Na_2CO_3}\right)\left(\frac{1\,kg}{1000\,g}\right) = 0.26\,kg$$

$$0.10\,kg\,CaO\left(\frac{1000\,g}{1\,kg}\right)\left(\frac{1\,mol\,CaO}{56.08\,g\,CaO}\right)\left(\frac{1\,mol\,CaCO_3}{1\,mol\,CaO}\right)\left(\frac{100.09\,g\,CaCO_3}{mol\,CaCO_3}\right)\left(\frac{1\,kg}{1000\,g}\right) = 0.18\,kg$$

Combine **0.75 kg SiO_2, 0.26 kg Na_2CO_3, and 0.18 kg $CaCO_3$** and heat to make 1.00 kg glass.

4.126 a) There is not enough information to write complete chemical equations, but the following equations can be written:

$C_5H_{11}I_4NO_4(s) + Na_2CO_3(s) \rightarrow 4I^-(aq)$ + other products

$I^-(aq) + Br_2(l) + HCl(aq) \rightarrow IO_3^-(aq)$ + other products

For every mole of thyroxine reacted, 4 mol IO_3^- are produced.

$$\overset{+5e^- \text{ (red)}}{\overbrace{}}$$

b) $\overset{+5}{I}O_3^-(aq) + H^+(aq) + \overset{-1}{I^-}(aq) \rightarrow \overset{0}{I_2}(aq) + H_2O(l)$

-1e⁻ (ox)

This is a difficult equation to balance because the iodine species are both reducing and oxidizing. Start balancing the equation by placing a coefficient of 5 in front of $I^-(aq)$, so the electrons lost equal the electrons gained. Do not place a 5 in front of $I_2(aq)$, because not all of the $I_2(aq)$ comes from oxidation of $I^-(aq)$. Some of the $I_2(aq)$ comes from the reduction of $IO_3^-(aq)$. Place a coefficient of 3 in front of $I_2(aq)$ to correctly balance iodine. The reaction is now balanced from a redox standpoint, so finish balancing the reaction by balancing the oxygen and hydrogen.

$$IO_3^-(aq) + 6H^+(aq) + 5I^-(aq) \rightarrow 3I_2(aq) + 3H_2O(l)$$

IO_3^- is the oxidizing agent, and I^- is the reducing agent.

If 3 mol of I_2 are produced per mol of IO_3^-, and 4 mol of IO_3^- are produced per mole of thyroxine, then 12 mol of I_2 are produced per mole of thyroxine.

c) The balanced equation for this reaction is $I_2(aq) + 2S_2O_3^{2-}(aq) \rightarrow 2I^-(aq) + S_4O_6^{2-}(aq)$.

$$mass\ thyroxine = (0.01723\ L\ S_2O_3^{2-}\ soln)\left(\frac{0.1000\ mol\ S_2O_3^{2-}}{L\ S_2O_3^{2-}\ soln}\right)\left(\frac{1\ mol\ I_2}{2\ mol\ S_2O_3^{2-}}\right)$$

$$\times\left(\frac{1\ mol\ C_{15}H_{11}I_4NO_4}{12\ mol\ I_2}\right)\left(\frac{776.8\ g\ C_{15}H_{11}I_4NO_4}{mol\ C_{15}H_{11}I_4NO_4}\right) = 0.05576_{78}\ g\ C_{15}H_{11}I_4NO_4$$

$$mass\ \% = \frac{0.05576_{78}\ g\ C_{15}H_{11}I_4NO_4}{0.4332\ g\ extract} x100\% = \mathbf{12.87\%\ thyroxine\ in\ extract}$$

4.128 Balance the equation to obtain correct molar ratios. This is not a redox reaction as none of the O.N.'s change. Here is a suggested method for approaching balancing the equation.

-- Since PO_4^{2-} remains as a unit on both sides of the equation, treat it is a unit when balancing.

-- On first inspection, one can see that Na needs to be balanced by adding a "2" in front of $NaHCO_3$. This then affects the balance of C, so add a "2" in front of CO_2.

-- Hydrogen is not balanced, so change the coefficient of water to "2" as this will have the least impact on the other species.

-- Verify that the other species are balanced.

$Ca(H_2PO_4)_2(s)$ + **2**$NaHCO_3(s) \rightarrow$ **2**$CO_2(g)$ + **2**$H_2O(g)$ + $CaHPO_4(s)$ + $Na_2HPO_4(s)$

a) Determine whether $Ca(H_2PO_4)_2$ or $NaHCO_3$ limits the production of CO_2.

$$mol\ CO_2\ from\ NaHCO_3 = (0.31\ g\ NaHCO_3)\left(\frac{1\ mol\ NaHCO_3}{84.008\ g}\right)\left(\frac{2\ mol\ CO_2}{2\ mol\ NaHCO_3}\right) = 0.0037\ mol\ CO_2$$

$$mol\ CO_2\ from\ Ca(H_2PO_4)_2 = (0.35 g\ Ca(H_2PO_4)_2)\left(\frac{1\ mol\ Ca(H_2PO_4)_2}{234.052\ g}\right)\left(\frac{2\ mol\ CO_2}{1\ mol\ Ca(H_2PO_4)_2}\right)$$

$$= 0.0030\ mol\ CO_2$$

Since $Ca(H_2PO_4)_2$ is limiting, **$3.0\text{x}10^{-3}$ mol CO_2** will be produced.

b) Volume CO_2 = ($3.0\text{x}10^{-3}$ mol CO_2)(37.0 L/mol CO_2) = **0.11 L CO_2**

4.131 a) Complete combustion of hydrocarbons involves heating the hydrocarbon in the presence of oxygen to produce carbon dioxide and water.

Ethanol: $C_2H_5OH(l) + 3O_2(g) \rightarrow 2CO_2(g) + 3H_2O(l)$

Gasoline: $2C_8H_{18}(l) + 25O_2(g) \rightarrow 16CO_2(g) + 18H_2(g)$

b) The 1.00 L mixture contains 90.0% or 900 mL gasoline and 10% or 100 mL ethanol.

$$(900\ mL\ C_8H_{18})\left(\frac{0.742\ g\ C_8H_{18}}{mL\ C_8H_{18}}\right)\left(\frac{1\ mol\ C_8H_{18}}{114.22\ g}\right)\left(\frac{25\ mol\ O_2}{2\ mol\ C_8H_{18}}\right)\left(\frac{32.00\ g\ O_2}{mol\ O_2}\right) = 2.34x10^3\ g\ O_2$$

$$(100\ mL\ C_2H_5OH)\left(\frac{0.789\ g\ C_2H_5OH}{mL\ C_2H_5OH}\right)\left(\frac{1\ mol\ C_2H_5OH}{46.07\ g\ C_2H_5OH}\right)\left(\frac{3\ mol\ O_2}{1\ mol\ C_2H_5OH}\right)$$

$$\times\left(\frac{32.00\ g\ O_2}{mol\ O_2}\right) = 1.64x10^2\ g\ O_2$$

Total mass of oxygen required to completely burn the mixture is the sum of the two masses: 2.34×10^3 g + 1.64×10^2 g = **2.50×10^3 g**.

c) $$2.50\,x\,10^3\ g\ O_2\left(\frac{1\ mol\ O_2}{32.00\ g}\right)\left(\frac{22.4\ L}{mol\ O_2}\right) = \mathbf{1.75x10^3\ L}$$

d) $$1.75\,x\,10^3\ L\ O_2\left(\frac{1\ L\ air}{0.209\ L\ O_2}\right) = \mathbf{8.37x10^3\ L\ air}$$

4.134 Write balanced chemical equations for each process. Since CO is the reducing agent, it must oxidize (i.e. gain oxygen) to become CO_2.

a) $$\overset{+2}{Fe}O(s) + \overset{+2}{C}O(g) \rightarrow \overset{0}{Fe}(s) + \overset{+4}{C}O_2(g)$$

Electrons lost equal electrons gained, so equation is balanced as written.

$$mass\ CO = (156.8\ g\ FeO)\left(\frac{1\ mol\ FeO}{71.85\ g}\right)\left(\frac{1\ mol\ CO}{1\ mol\ FeO}\right)\left(\frac{28.01\ g}{1\ mol\ CO}\right) = \mathbf{61.13\,g\,CO}$$

b) $$\overset{+3}{Fe_2}O_3(s) + \overset{+2}{C}O(g) \rightarrow 2\overset{0}{Fe}(s) + \overset{+4}{C}O_2(g)$$

Two iron atoms gain a total of 6 e^- (reduction). The carbon loses 2 e^- (oxidation). Multiply the oxidation half-reaction by 3, so that electrons gained equal electrons lost.

$Fe_2O_3(s) + 3CO(g) \rightarrow 2Fe(s) + 3CO_2(g)$

$$mass\ CO = (156.8\ g\ Fe_2O_3)\left(\frac{1\ mol\ Fe_2O_3}{159.70\ g}\right)\left(\frac{3\ mol\ CO}{1\ mol\ Fe_2O_3}\right)\left(\frac{28.01\ g}{1\ mol\ CO}\right) = \mathbf{82.50\,g\,CO}$$

c) 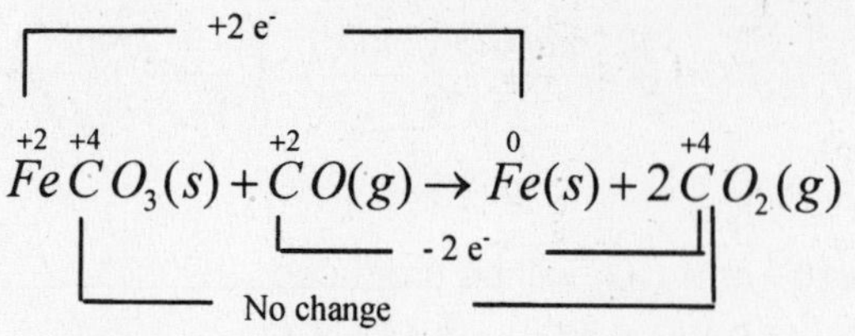 $\overset{+2\ +4}{FeCO_3}(s) + \overset{+2}{CO}(g) \rightarrow \overset{0}{Fe}(s) + 2\overset{+4}{CO_2}(g)$

(+2 e⁻ : Fe, +2 → 0; −2 e⁻ : C in CO, +2 → +4; No change : carbonate C, +4 → +4)

The electron transfer balances because the carbonate carbon does not gain or lose e⁻ when it converts to a carbon dioxide carbon.

$$mass\ CO = (156.8\ g\ FeCO_3)\left(\frac{1\ mol\ FeCO_3}{115.86\ g}\right)\left(\frac{1\ mol\ CO}{1\ mol\ FeCO_3}\right)\left(\frac{28.01\ g}{1\ mol\ CO}\right) = \mathbf{37.91\ g\ CO}$$

4.138 a) The second reaction is a redox process because the O.N. of iron changes from 0 to +2 (it oxidizes) while the O.N. of hydrogen changes from +1 to 0 (it reduces).

b) Use the molar ratios from the first balanced equation to calculate the desired quantities. First, calculate the moles of HCl present:

mol HCl = (3.00 M HCl)($2.50x10^3$ L) = $7.50x10^3$ mol

$$mass\ Fe_2O_3\ removed = (7.50x10^3\ mol\ HCl)\left(\frac{1\ mol\ Fe_2O_3}{6\ mol\ HCl}\right)\left(\frac{159.70\ g}{1\ mol\ Fe_2O_3}\right) = \mathbf{2.00x10^5\ g\ Fe_2O_3}$$

$$mass\ FeCl_3\ produced = (7.50x10^3\ mol\ HCl)\left(\frac{2\ mol\ FeCl_3}{6\ mol\ HCl}\right)\left(\frac{162.20\ g}{1\ mol\ FeCl_3}\right) = \mathbf{4.06x10^5\ g\ FeCl_3}$$

c) Use the molar ratios from the second balanced equation to calculate the desired quantities.

$$mass\ Fe\ lost = (7.50x10^3\ mol\ HCl)\left(\frac{1\ mol\ Fe}{2\ mol\ HCl}\right)\left(\frac{55.85\ g}{1\ mol\ Fe}\right) = \mathbf{2.09x10^5\ g\ Fe}$$

$$mass\ FeCl_2\ produced = (7.50x10^3\ mol\ HCl)\left(\frac{1\ mol\ FeCl_2}{2\ mol\ HCl}\right)\left(\frac{126.75\ g}{1\ mol\ FeCl_2}\right) = \mathbf{4.75x10^5\ g\ FeCl_2}$$

d) The mass of $FeCl_2$ is related to the mass of Fe through reaction 2.

$$mass\ FeCl_2 = (0.280\ g\ Fe)\left(\frac{1\ mol\ Fe}{55.85\ g}\right)\left(\frac{1\ mol\ FeCl_2}{1\ mol\ Fe}\right)\left(\frac{126.75\ g}{1\ mol\ FeCl_2}\right) = 0.635_{45}\ g\ FeCl_2$$

The mass of $FeCl_3$ is related to the mass of Fe_2O_3 through reaction 1.

$$mass\ FeCl_3 = (1.00\ g\ Fe_2O_3)\left(\frac{1\ mol\ Fe_2O_3}{159.70\ g}\right)\left(\frac{2\ mol\ FeCl_3}{1\ mol\ Fe_2O_3}\right)\left(\frac{162.20\ g}{1\ mol\ FeCl_3}\right) = 2.03_{13}\ g\ FeCl_3$$

Dividing the mass of $FeCl_2$ by $FeCl_3$ yields a mass ratio of (0.635_{45} g)/(2.03_{13} g) = **0.313 g $FeCl_2$/g $FeCl_3$**.

CHAPTER 5

GASES AND THE KINETIC-MOLECULAR THEORY

FOLLOW-UP PROBLEMS

5.1 Plan: Figure 5.4D shows that the pressure of the gas (CO_2 in this problem) in the flask is less than the atmospheric pressure. To calculate the pressure of the gas, subtract the change in height (Δh) from the atmospheric pressure (barometric reading). Use conversion factors in Table 5.2 to convert pressure in mmHg to units of torr, pascals and lb/in^2.

Solution: P_{CO_2} = 753.6 mmHg – 174.0 mmHg = 579.6 mmHg

Converting from mmHg to torr: $P_{CO_2} = 579.6\, mmHg\left(\frac{1\, torr}{1\, mmHg}\right) = \mathbf{579.6\ torr}$

Converting from mmHg to pascals: $P_{CO_2} = 579.6\, mmHg\left(\frac{1.01325x10^5\ Pa}{760\, mmHg}\right) = \mathbf{7.727x10^4\ Pa}$

Converting from mmHg to lb/in^2: $P_{CO_2} = 579.6\, mmHg\left(\frac{14.7\, lb / in^2}{760\, mmHg}\right) = \mathbf{11.2 lb / in^2}$

Check: The pressure of the carbon dioxide as calculated is less than the atmospheric pressure. For conversion to torr the value of the pressure should stay the same. The order of magnitude for each conversion corresponds to the calculated answer. For conversion to pascals the order of magnitude calculation is 10^2 x $10^5/10^2 = 10^5$. For conversion to lb/in^2 the order of magnitude calculation is 10^2 x $10^1/10^2 = 10^1$.

5.2 Plan: Given in the problem is an initial volume, initial pressure, and final pressure for the argon gas. The final volume can be calculated from the relationship $P_iV_i = P_fV_f$ where "i" stands for initial and "f" stands for final. Unit conversions for mL to L and atm to kPa must be included.

Solution: $V_f = \frac{(105\, mL)\left(\frac{1\, L}{1000\, mL}\right)(0.871\, atm)}{(26.3\, kPa)\left(\frac{1\, atm}{101.325\, kPa}\right)} = \mathbf{0.352 L}$

Check: As the pressure goes from 0.871 atm to 26.3 kPa is it increasing or decreasing? From Table 5.2, 1 atm corresponds to 101 kPa, so 26.3 kPa would be about 0.26 atm. Thus, the pressure has decreased by a factor of about 3 and the volume should increase by the same factor. The calculated volume of 0.352 L is approximately 3 times greater than the initial volume of 0.105 L.

5.3 Plan: The problem asked for a temperature with given initial temperature, initial volume, and final volume. The relationship to use is $V_i/T_i = V_f/T_f$. The units of volume in both cases are the same, but the initial temperature must be converted to Kelvin.

Solution: $T_f = \frac{(9.75\, cm^3)(273.15 + 0\, K)}{6.83\, cm^3} = \mathbf{390. K}$

Check: The volume increases by a factor of about 1.4 so the temperature must have increased by the same factor. The initial temperature of 273 K times 1.4 is 380 K, so the answer 390 K appears to be correct.

5.4 Plan: In this problem the amount of gas is decreasing. Since the container is rigid, the volume of the gas will not change with the decrease in moles of gas. The temperature is also constant. So, the only change will be that the pressure of the gas will decrease since fewer moles of gas will be present after removal of the 5.0 g of ethylene. To calculate the final pressure use the relationship $n_i/P_i = n_f/P_f$. Since the ratio of moles of ethylene is equal to the ratio of grams of ethylene there is no need to convert the grams to moles.

Solution: $P_f = \dfrac{(793\,torr)(35.0\,g - 5.0\,g)}{(35.0\,g)} = \mathbf{680.\,torr}$

Check: The amount of gas decreases by a factor of 6/7. Since pressure is proportional to the amount of gas, the pressure should decrease by the same factor, 6/7 x 793 = 680.

5.5 Plan: From Sample Problem 5.5 the temperature of 21°C and volume of 438 L are given. The pressure is 1.37 atm and the unknown is the moles of oxygen gas. Use the ideal gas equation to calculate the number of moles of gas.

Solution: $n = \dfrac{(1.37\,atm)(438\,L)}{\left(0.08206\dfrac{atm\cdot L}{mol\cdot K}\right)(273.15 + 21\,K)} = 24.9\,mol\,O_2$

Mass of O_2 = (24.9 mol O_2) x (32.00 g/mol) = **797 g O_2**

Check: The grams of oxygen can be checked by using the relationship $n_i/P_i = n_f/P_f$.

$$mass\,O_2 = \frac{(885\,g\,O_2)(1.37\,atm)}{1.53\,atm} = 792\,g\,O_2$$

5.6 Plan: Density of a gas can be calculated using a version of the ideal gas equation, $d = \dfrac{\mathcal{M}P}{RT}$

Two calculations are required, one with T = 273 K and P = 380 torr and the other at STP which is defined as T = 273 K and P = 1 atm at which every gas has a molar volume of 22.4 L.

Solution: Density at T = 273 K and P = 380 torr

$$d = \frac{(44.01\,g/mol)(380\,torr)(1\,atm/760\,torr)}{(0.08206\,atm\cdot L/mol\cdot K)(273\,K)} = 0.982\,g/L$$

Density at T = 273 K and P = 1 atm.

$$d = \frac{1\,mol}{22.4\,L}(44.01\,g/mol) = 1.96\,g/L$$

The density of a gas increases proportionally to the increase in pressure.

Check: In the two density calculations the temperature of the gas is the same, but the pressure differs by a factor of two. Comparing the density of CO_2 at STP to the gases in Figure 5.8 the value of 1.96 g/L looks reasonable since the molar mass of CO_2 is greater than the gases shown. The calculation of density shows that it is proportional to pressure. The pressure in the first case is half the pressure at STP, so the density at 380 torr should be half the density at STP and it is.

5.7 Plan: Use the ideal gas equation for density, $d = \frac{\mathcal{M}P}{RT}$, and solve for molar mass.

Solution: $$\mathcal{M} = \frac{dRT}{P} = \frac{(1.26\ g/L)(0.08206\ atm \cdot L/mol \cdot K)(283.2\ K)}{(102.5\ kPa)\left(\frac{1\ atm}{101.325\ kPa}\right)} = \mathbf{28.9_{46}\ g/mol}$$

Check: Dry air would consist of about 80% N_2 and 20% O_2. Estimating a molar mass for this mixture gives (0.80 x 28) + (0.20 x 32) = 28.8 g/mol, which is close to the calculated value.

5.8 Plan: Calculate the number of moles of each gas present and then the mole fraction of each gas. The partial pressure of each gas equals the mole fraction times the total pressure. Total pressure equals 1 atm since the problem specifies STP.

Solution: $$n_{He} = 5.50\ g\left(\frac{1\ mol}{4.003\ g}\right) = 1.37\ mol\ He$$

$$n_{Ne} = 15.0\ g\left(\frac{1\ mol}{20.18\ g}\right) = 0.743\ mol\ Ne$$

$$n_{Kr} = 35.0\ g\left(\frac{1\ mol}{83.80\ g}\right) = 0.418\ mol\ Kr$$

Total number of moles of gas = 1.37 + 0.743 + 0.418 = 2.53 mol

$$P_{He} = \left(\frac{1.37\ mol\ He}{2.53\ mol}\right)(1\ atm) = \mathbf{0.543\ atm}$$

$$P_{Ne} = \left(\frac{0.743\ mol\ Ne}{2.53\ mol}\right)(1\ atm) = \mathbf{0.294\ atm}$$

$$P_{Kr} = \left(\frac{0.418\ mol\ Kr}{2.53\ mol}\right)(1\ atm) = \mathbf{0.165\ atm}$$

Check: One way to check is that the partial pressures add to the total pressure, 0.542 + 0.294 + 0.165 = 1.001 atm which agrees with the total pressure of 1 atm at STP.

5.9 Plan: The gas collected over the water will consist of H_2 and H_2O gas molecules. The partial pressure of the water can be found from the vapor pressure of water at the given temperature (Table 5.4). Subtracting this partial pressure of water from total pressure gives the partial pressure of hydrogen gas collected over the water. Calculate the moles of hydrogen gas using the ideal gas equation. The mass of hydrogen can then be calculated by converting the moles of hydrogen from the ideal gas equation to grams.

Solution: From Table 5.4 the partial pressure of water is 13.6 torr at 16°C.

$$P_{H_2} = 752\ torr - 13.6\ torr = \mathbf{738\ torr}$$

$$mass\ of\ H_2 = \left(\frac{(738\ torr)\left(\frac{1\ atm}{760\ torr}\right)(1.495\ L)}{(0.08206\ atm \cdot L/mol \cdot K)(289\ K)}\right)\left(\frac{2.016\ g}{mol}\right) = \mathbf{0.123\ g\ H_2}$$

Check: Since the pressure and temperature are close to STP, estimate the moles of hydrogen from the molar volume at STP, 22.4 L/mol.

$\left(\frac{1\,mol}{22.4\,L}\right)(1.5\,L)(2\,g/mol) = 0.13g$ which is close to the calculated value.

5.10 Plan: Write a balanced equation for the reaction. Then calculate the moles of HCl(g) from the starting amount of sodium chloride. Find the volume of the HCl(g) from the molar volume at STP.

Solution: Balanced equation is $H_2SO_4(aq) + 2NaCl(aq) \rightarrow Na_2SO_4(aq) + 2HCl(g)$.

$$0.117\,kg\,NaCl\left(\frac{1000\,g}{kg}\right)\left(\frac{1\,mol\,NaCl}{58.44\,g}\right)\left(\frac{2\,mol\,HCl}{2\,mol\,NaCl}\right)\left(\frac{22.4\,L}{mol}\right)\left(\frac{1000\,mL}{1\,L}\right) = \mathbf{4.48x10^4\ mL\ HCl\ gas}$$

Check: 117 g NaCl is about 2 moles, which would form 2 moles of HCl(g). Twice the molar volume is 44.8 L, which is the answer as calculated.

5.11 Plan: Balance the equation for the reaction. Then determine the limiting reactant by finding the moles of each reactant from the ideal gas equation. Calculate the moles of remaining, excess reactant. This is the only gas left in the flask so it is used to calculate the pressure inside the flask.

Solution: Balanced equation is $NH_3(g) + HCl(g) \rightarrow NH_4Cl(s)$

The stoichiometric ratio of NH_3 to HCl is 1:1 so the reactant present in the lower quantity of moles is the limiting reactant.

$$n_{NH_3} = \frac{(0.452\,atm)(10.0\,L)}{(0.08206\,atm \cdot L/mol \cdot K)(295\,K)} = 0.186_{72}\,mol$$

$$n_{HCl} = \frac{(7.50\,atm)(0.155\,L)}{(0.08206\,atm \cdot L/mol \cdot K)(271\,K)} = 0.0522_{75}\,mol$$

The HCl is limiting so the moles of ammonia gas left after the reaction would be $0.186_{72} - 0.0522_{75} = 0.134_{45}$ mol.

$$P = \frac{(0.134_{45}\,mol)(0.08206\,atm \cdot L/mol \cdot K)(295\,K)}{(10.0\,L)} = \mathbf{0.325\ atm}$$

Check: Doing a rough calculation of moles gives for NH_3 (0.5 x 10)/(0.1 x 300) = 0.17 mol and for HCl (8 x 0.1)/(0.1 x 300) = 0.027 mol which means 0.14 mol NH_3 is left. Plugging this into a rough calculation of pressure gives (.14 x .1 x 300)/10 = 0.4 atm, which is close to the calculated answer.

5.12 Plan: Graham's Law can be used to solve for the effusion rate of the ethane since the rate and molar mass of helium is known along with the molar mass of ethane. In the same way that running slower increases the time to go from one point to another, so the rate of effusion decreases as the time increases. The rate can be expressed as 1/time.

Solution:

$$\frac{Rate_{He}}{Rate_{C_2H_6}} = \frac{time_{C_2H_6}}{time_{He}} = \frac{\sqrt{\mathcal{M}_{C_2H_6}}}{\sqrt{\mathcal{M}_{He}}}$$

$$time_{C_2H_6} = (1.25\,\text{min})\left(\sqrt{\frac{30.07\,g/mol}{4.003\,g/mol}}\right) = \mathbf{3.43\ min}$$

Check: The ethane should move slower than the helium. This is consistent with the calculation that the ethane molecule takes longer to effuse. The second check is an estimate. The square root of 30 is estimated as 5 and the square root of 4 is 2. The time 1.25 min x 5/2 = 3.1 min, which validates the calculated answer of 3.42 min.

END-OF-CHAPTER PROBLEMS

5.1 a) The liquid maintains its volume, whereas the gas expands its volume to fill the larger container.

b) As it is heated, the liquid occupies nearly the same volume. The increased kinetic energy of the gas molecules that results from the increase in temperature forces the container to expand and its volume to increase.

c) The volume of the liquid remains essentially constant, but the volume of the gas would be reduced by the pressure exerted by the external force.

5.6 The ratio of the heights of columns of mercury and water is inversely proportional to the ratio of the densities of the two liquids.

$$\frac{h_{H_2O}}{h_{Hg}} = \frac{d_{Hg}}{d_{H_2O}}$$

$$h_{H_2O} = 725\,mmHg\left(\frac{1\,cm}{10\,mm}\right)\left(\frac{13.5\,g\,Hg/mL}{1.00\,g\,H_2O/mL}\right) = \mathbf{979\,cm\,H_2O}$$

5.8 Since the height of the mercury column in contact with the gas is higher than the column in contact with the air, the gas is exerting less pressure on the mercury than the air.

$$P_{gas} = \left(738.5\,torr\left(\frac{1\,mmHg}{1\,torr}\right)\right) - \left(2.35\,cmHg\left(\frac{10\,mm}{1\,cm}\right)\right) = 715.0\,mmHg$$

$$715.0\,mmHg\left(\frac{1\,atm}{760\,mmHg}\right) = \mathbf{0.9408\,atm}$$

5.10 The difference in height of Hg is directly related to pressure in atmospheres.

$$P(atm) = (0.734\,m\,Hg)\left(\frac{1000\,mmHg}{1\,m\,Hg}\right)\left(\frac{1\,atm}{760\,mmHg}\right) = \mathbf{0.966\,atm}$$

5.12 a) $(0.745\,atm)\left(\frac{760\,mmHg}{1\,atm}\right) = \mathbf{566\,mmHg}$

b) $(992\,torr)\left(\frac{1.01325\,bar}{760\,torr}\right) = \mathbf{1.32\,bar}$

c) $(365\,kPa)\left(\frac{1\,atm}{101.325\,kPa}\right) = \mathbf{3.60\,atm}$

d) $(804\,mmHg)\left(\frac{1\,atm}{760\,mmHg}\right)\left(\frac{101.325\,kPa}{1\,atm}\right) = \mathbf{107\,kPa}$

5.18 At constant temperature and volume, the pressure of the gas is directly proportional to number of moles of gas. Verify this by examining the ideal gas law. At constant T & V, the ideal gas law equation becomes P = n(RT/V) or P = n x constant.

5.20 a) As the pressure on a gas increases, the molecules move closer together decreasing the volume. When the pressure is tripled the volume decreases to one third of the original volume at constant temperature (Boyle's Law).

b) As the temperature of a gas increases the gas molecules gain kinetic energy. With higher energy the gas molecules collide with the walls of the container with greater force which increases the size (volume) of the container. If the temperature increases by a factor of 2.5 (at constant pressure) then the volume increases by the same factor (Charles' Law).

c) As the number of molecules of gas increase the force they exert on the container increases. This results in an increase in the volume of the container. Adding two moles of gas to one mole increases the number of moles by a factor of three, which increases the volume by a factor of three (Avogadro's Law).

5.22 a) The temperature is decreased by a factor of 2, so the volume is decreased by a factor of 2 (Charles' Law).

b) The temperature increases by a factor of $\left(\frac{700+273}{350+273}\right)$ = 1.56, so the volume increases by a factor of 1.56 (Charles' Law).

c) The pressure is increased by a factor of 4, so the volume decreases by a factor of 4 (Boyle's Law).

5.24 The temperature must be lowered to reduce the volume of a gas. Charles' Law states that at constant pressure and with a fixed amount of gas the volume of a gas is directly proportional to the absolute temperature of the gas.

$$\frac{V_{initial}}{V_{final}} = \frac{T_{initial}}{T_{final}}$$

$$T_{final} = [(198+273.15)K]\left(\frac{2.50\ L}{5.10\ L}\right) = 230._{96}\ K$$

$$T_{final}\ in\ ^\circ C = 230._{96} - 273.15 = -\mathbf{42\ ^\circ C}$$

5.26 To find V_2, summarize and convert gas variables and apply the combined gas law equation.

$$\frac{P_1V_1}{T_1} = \frac{P_2V_2}{T_2};\quad V_2 = \frac{P_1V_1T_2}{T_1P_2}$$

P_1 = (153.3 kPa)(1 atm/101.325 kPa) = 1.513 atm	P_2 = 1.000 atm
V_1 = 25.5 L	V_2 = ?
T_1 = 298 K	T_2 = 273 K

$$V_2 = \frac{(1.513\ atm)(25.5\ L)(273\ K)}{(1.000\ atm)(298\ K)} = \mathbf{35.3\ L}$$

Instead of plugging in numbers, use your knowledge of gas behavior to set up ratios. What happens to the volume when the pressure decreases from 1.513 atm to 1.000 atm? The volume should increase, so multiply the volume by a ratio > 1, or (1.513/1.000). When the temperature decreases from 298 K to 273 K, the volume also decreases so multiply by a ratio < 1, or (273/298). The change in volume is represented as follows:

$$V_2 = (25.5\,L)\left(\frac{1.513}{1.000}\right)\left(\frac{273}{298}\right) = 35.3\,L$$

5.28 Given the volume, pressure and temperature of a gas the number of moles of the gas can be calculated using the ideal gas equation, $n = \frac{PV}{RT}$. The gas constant, R = 8.206 x 10^{-2} atm·L/mol·K, gives pressure in atmospheres and temperature in Kelvin. The given pressure in torr must be converted to atmospheres and the temperature converted to Kelvin.

$$n = \frac{\left(228\,torr\left(\frac{1\,atm}{760\,torr}\right)\right)5.0\,L}{(0.08206\,atm\cdot L/mol\cdot K)(27+273.15K)} = \mathbf{0.061\,mol}$$

5.30 Solve the ideal gas equation for moles and convert to mass using the molar mass of ClF_3.

$$n = \frac{\left(699\,mmHg\,x\frac{1.00\,atm}{760\,mmHg}\right)(0.207\,L)}{(0.08206\;atm\cdot L/mol\cdot K)((45+273)K)} = 7.29_{58}x10^{-3}\;mol\,ClF_3$$

$$mass\,ClF_3 = (7.29_{58}x10^{-3}\;mol\,ClF_3)\left(\frac{92.45\,g}{1\,mol\,ClF_3}\right) = \mathbf{0.675\,g\,ClF_3}$$

5.33 <u>Plan</u>: Assuming that while rising in the atmosphere the balloon will neither gain nor lose gas molecules, the number of moles of gas calculated at sea level will be the same as the number of moles of gas at the higher altitude. Using the ideal gas equation, (n x R) is a constant equal to (PV/T). Given the sea-level conditions of volume, pressure and temperature and the temperature and pressure at the higher altitude for the gas in the balloon, we can set up an equation to solve for the volume at the higher altitude. Comparing the calculated volume to the given maximum volume of 835 L will tell us if the balloon has reached its maximum volume at this altitude.

<u>Solution</u>: $\frac{P_1V_1}{T_1} = \frac{P_2V_2}{T_2}$

$$V_2 = \left(\frac{755\,torr\left(\frac{1\,atm}{760\,torr}\right)(55.0\,L)}{(23+273.15)K}\right)\left(\frac{(-5+273.15)K}{0.066\,atm}\right) = \mathbf{750\,L}$$

The calculated volume of the gas at the higher altitude is less than the maximum volume of the balloon. **No**, the balloon will not reach its maximum volume.

<u>Check</u>: Should we expect that the volume of the gas in the balloon should increase? At the higher altitude the pressure decreases which increases the volume of the gas. At the higher altitude the temperature decreases which decreases the volume of the gas. Which of these will dominate? The pressure decreases by a factor of 0.99/0.066 = 15. If we label

the initial volume V_1 then the resulting volume is $15V_1$. The temperature decreases by a factor of 296/268 = 1.1, so the resulting volume is $V_1/1.1$ or $0.91V_1$. The increase in volume due to the change in pressure is greater than the decrease in volume due to change in temperature, so the volume of gas at the higher altitude should be greater than the volume at sea level.

5.35 The molar mass of H_2 is less than the average molar mass of air (mostly N_2), so air is more dense. To collect a beaker of $H_2(g)$, invert the beaker so that the air will be replaced by the lighter H_2. The molar mass of CO_2 is greater than the average molar mass of air, so $CO_2(g)$ is more dense. Collect the CO_2 holding the beaker upright so the lighter air will be displaced out the top of the beaker.

5.38 Using the ideal gas equation and the molar mass of xenon, 131.3 g/mol, we can find the density of xenon gas at STP. Standard temperature is 0°C and standard pressure is 1 atm.

$$d = \frac{\mathcal{M} \times P}{RT}$$

$$d_{Xe} = \frac{(131.3\ g/mol)(1\ atm)}{(0.08206\ atm \cdot L/mol \cdot K)(273\ K)} = \mathbf{5.86\ g/L}$$

5.40 Apply the ideal gas law to determine the number of moles. Convert moles to mass and divide by the volume to obtain density in g/L.

$$mols\ AsH_3 = \frac{PV}{RT} = \frac{(1.00\ atm)(0.0400\ L)}{\left(0.08206\frac{atm \cdot L}{mol \cdot K}\right)(273\ K)} = 1.78_{55} x10^{-3}\ mol\ AsH_3 = \mathbf{1.79x10^{-3}\ mol\ AsH_3}$$

$$density\ of\ AsH_3 = \frac{n\mathcal{M}}{V} = \frac{(1.78_{55} x10^{-3}\ mol\ AsH_3)\left(\frac{77.94\ g\ AsH_3}{1\ mol\ AsH_3}\right)}{0.0400\ L} = \mathbf{3.48\ g/L}$$

5.42 Rearrange the formula d = $\mathcal{M}$P/RT to solve for molar mass: $\mathcal{M}$ = dRT/P.

$$\mathcal{M} = \frac{\left(\frac{206\ ng}{0.206\ \mu L}\right)\left(\frac{1\ g}{1 x 10^9\ ng}\right)\left(\frac{1 x 10^6\ \mu L}{1\ L}\right)(0.08206\ atm \cdot L/mol \cdot K)(45 + 273.15\ K)}{(388\ torr)(1\ atm/760\ torr)} = \mathbf{51.1\ g/mol}$$

5.44 Use the ideal gas law to determine the number of moles of Ar and O_2. The gases are combined ($n_{TOT} = n_{Ar} + n_{O2}$) into a 400 mL flask (V) at 27°C (T). Determine the total pressure from n_{TOT}, V, and T.

$$n_{Ar} = \frac{PV}{RT} = \frac{(1.20\ atm)(0.600\ L)}{\left(0.08206\frac{atm \cdot L}{mol \cdot K}\right)(227 + 273K)} = 0.0175_{48}\ mol\ Ar$$

$$n_{O_2} = \frac{PV}{RT} = \frac{\left(501\ torr\ x\left(\frac{1\ atm}{760\ torr}\right)\right)(0.200\ L)}{\left(0.08206\frac{atm \cdot L}{mol \cdot K}\right)(127 + 273K)} = 0.00401_{66}\ mol\ O_2$$

Therefore, $n_{TOT} = n_{Ar} + n_{O2} = 0.0175_{48} + 0.00401_{66} = 0.0215_{65}$ mol.

$$P_{flask} = \frac{n_{TOT}RT}{V} = \frac{(0.0215_{65}\ mol)\left(0.08206\frac{atm\cdot L}{mol\cdot K}\right)(27+273K)}{0.400\ L} = \mathbf{1.33\ atm}$$

5.48 The problem gives the mass, volume, temperature and pressure of a gas so we can solve for molar mass using $\mathcal{M}$ = mRT/PV. The problem also states that the gas is a hydrocarbon, which by, definition, contains only carbon and hydrogen atoms. We are also told that each molecule of the gas contains five carbon atoms so we can use this information and the calculated molar mass to find out how many hydrogen atoms are present and the formula of the compound.

$$\mathcal{M} = \frac{(0.482\ g)(0.08206\ atm\cdot L/mol\cdot K)(101+273.15\ K)}{(767\ torr)(1\ atm/760\ torr)(0.204\ L)} = 71.8_{81}\ g/mol$$

$$5C(12.01\ g/mol) + x\ H(1.008\ g/mol) = 71.8_{81}\ g/mol$$

$$x = 11.7_{39} \approx 12$$

So the formula is $\mathbf{C_5H_{12}}$.

5.50 a) $$n_{TOT} = \frac{PV}{RT} = \frac{(850.\ torr)\left(\frac{1\ atm}{760\ torr}\right)(21\ L)}{\left(0.08206\frac{atm\cdot L}{mol\cdot K}\right)(45+273K)} = \mathbf{0.90\ mol}$$

b) The information given in ppm is a way of expressing the proportion, or fraction, of SO_2 present in the mixture. Since n is directly proportional to V, the *volume* fraction can be used in place of the *mole* fraction used in equation 5.11. There are 7.95 x 10^3 parts SO_2 in a million parts of mixture, so $volume\ fraction = \frac{7.95x10^3}{1x10^6} = 7.95x10^{-3}$.

Therefore, p_{SO2} = volume fraction *x* p_{TOT} = (7.95 x 10^{-3})(850. torr) = **6.76 torr.**

5.51 We can find the moles of oxygen from the standard molar volume of gases and use the stoichiometric ratio from the balanced equation to determine the moles of phosphorus that will react with the oxygen.

$$\left(\frac{35.5\ L\ O_2}{22.4\ L/mol\ O_2}\right)\left(\frac{1\ mol\ P_4}{5\ mol\ O_2}\right)\left(\frac{123.88\ g}{mol\ P_4}\right) = \mathbf{39.3\ g\ P_4}$$

Alternatively, you could use PV = nRT to solve for moles of O_2, and then use the mole ratio to determine g of P_4.

$$n = \frac{(1.00\ atm)(35.5\ L)}{\left(0.08206\frac{atm\cdot L}{mol\cdot K}\right)(273\ K)} = 1.58\ mol\ O_2\ ;\ \ (1.58\ mol\ O_2)\left(\frac{1\ mol\ P_4}{5\ mol\ O_2}\right)\left(\frac{123.88\ g}{mol\ P_4}\right) = 39.3\ g\ P_4$$

5.53 To find the mass of PH_3, write the balanced equation, convert mass of P_4 to moles, solve for moles of H_2 using the standard molar volume (or use ideal gas law), and proceed with the limiting reactant problem.

$$P_4(s) + 6H_2(g) \rightarrow 4PH_3(g)$$

$$mol\ P_4 = (37.5\ g\ P_4)\left(\frac{1\ mol\ P_4}{123.88\ g\ P_4}\right) = 0.302_{71}\ mol\ P_4$$

$$mol\ H_2 = \frac{volume}{std\ molar\ volume} = \frac{83.0\ L}{22.414\ L/mol} = 3.70_{30}\ mol\ H_2$$

Determine the limiting reactant using the mole ratios from the balanced equation.

$$mol\ PH_3\ (from\ P_4) = (0.302_{71}\ mol\ P_4)\left(\frac{4\ mol\ PH_3}{1\ mol\ P_4}\right) = 1.21_{08}\ mol\ PH_3$$

$$mol\ PH_3\ (from\ H_2) = (3.70_{30}\ mol\ H_2)\left(\frac{4\ mol\ PH_3}{6\ mol\ H_2}\right) = 2.46_{87}\ mol\ PH_3$$

Therefore, P_4 is the limiting reactant.

$$mass\ PH_3 = (1.21_{08}\ mol\ PH_3)\left(\frac{33.99\ g\ PH_3}{1\ mol\ PH_3}\right) = \mathbf{41.2\,g\,PH_3}$$

5.55 First write the balanced equation:

$$2Al(s) + 6HCl(aq) \rightarrow 2AlCl_3(aq) + 3H_2(g)$$

The moles of hydrogen produced can be calculated from the ideal gas equation and then the stoichiometric ratio from the balanced equation is used to determine the moles of aluminum that reacted. The problem specifies "hydrogen gas collected over water", so the partial pressure of water must first be subtracted. Table 5.4 reports pressure at 26°C and 28°C, so take the average of the two values.

$p_{H2} = p_{TOT} - p_{H2O}$ = (751 torr – 26.8 torr)(1 atm/760 torr) = 0.953 atm

$$n_{H_2} = \frac{PV}{RT} = \frac{(0.953\ atm)(0.0358\ L)}{(0.08206\ atm \cdot L/mol \cdot K)(27 + 273\ K)} = 1.38_{59}x10^{-3}\ mol\ H_2$$

$$mass\ Al = (1.38_{59}x10^{-3}\ mol\ H_2)\left(\frac{2\ mol\ Al}{3\ mol\ H_2}\right)\left(\frac{26.98\ g}{mol\ Al}\right) = \mathbf{2.49\,x\,10^{-2}\ g\ Al}$$

5.57 To find mL of SO_2, write the balanced equation, convert the given mass of P_4S_3 to moles, use the molar ratio to find moles of SO_2, and use the ideal gas law to find volume.

$$P_4S_3(s) + 8O_2(g) \rightarrow P_4O_{10}(s) + 3SO_2(g)$$

$$mol\ SO_2 = (0.800\ g\ P_4S_3)\left(\frac{1\ mol\ P_4S_3}{220.09\ g\ P_4S_3}\right)\left(\frac{3\ mol\ SO_2}{1\ mol\ P_4S_3}\right) = 0.109\ mol\ SO_2$$

$$V_{SO_2} = \frac{nRT}{P} = \frac{(0.0109\ mol)\left(0.08206\frac{atm \cdot L}{mol \cdot K}\right)(32 + 273K)}{(725\ torr)\left(\frac{1\ atm}{760\ torr}\right)} = 0.286\ L\ or\ \mathbf{286\ mL}$$

5.59 First write the balanced equation

$$2XeF_6(s) + SiO_2(s) \rightarrow 2XeOF_4(l) + SiF_4(g)$$

Given the amount of xenon hexafluoride that reacts, we can find the number of moles of silicon tetrafluoride gas formed. Then using the ideal gas equation with the moles of gas, the temperature and the volume, we can calculate the pressure of the silicon tetrafluoride gas.

$$n_{SiF_4} = 2.00\,g\,XeF_6\left(\frac{1\,mol\,XeF_6}{245.3\,g}\right)\left(\frac{1\,mol\,SiF_4}{2\,mol\,XeF_6}\right) = 4.07_{66}\,x\,10^{-3}\,mol\,SiF_4$$

$$P_{SiF_4} = \frac{(4.07_{66}\,x\,10^{-3}\,mol\,SiF_4)(0.08206\,atm\cdot L/mol\cdot K)(298\,K)}{1.00\,L} = \mathbf{9.97\,x\,10^{-2}\,atm}$$

5.63 At STP (or any identical temperature and pressure) the volume occupied by a mole of any gas will be identical. This is due to the fact that at the same temperature, all gases have the same average kinetic energy, resulting in the same pressure.

5.66 The molar masses of the three gases are 2.016 for H_2 (Flask A), 4.003 for He (Flask B), and 16.04 for CH_4 (Flask C). Since hydrogen has the smallest molar mass of the three gases, 4 g of H_2 will contain more gas molecules than 4 g of He or 4 g of CH_4. Since helium has a smaller molar mass than methane, 4 g of He will contain more gas molecules than 4 g of CH_4.

a) $P_A > P_B > P_C$. The pressure of a gas is proportional to the number of gas molecules. So, the gas sample with more gas molecules will have a greater pressure.

b) $\overline{E_k}$ (A) = $\overline{E_k}$ (B) = $\overline{E_k}$ of (C). Average kinetic energy depends only on temperature. The temperature of each gas sample is 273 K, so they all have the same average kinetic energy.

c) $rate_A > rate_B > rate_C$. When comparing the speed of two gas molecules, the one with the lower mass travels faster.

d) total $\overline{E_k}$ (A) > total $\overline{E_k}$ (B) > total $\overline{E_k}$ (C). Since the average kinetic energy for each gas is the same (part b of this problem) then the total kinetic energy would equal the average times the number of molecules. Since the hydrogen flask contains the most molecules its total kinetic energy will be the greatest.

e) $d_A = d_B = d_C$. Under the conditions stated in this problem each sample has the same volume, 5 L, and the same mass, 4 g. Thus, the density of each is 4 g/5 L = 0.8 g/L.

f) collision frequency(A) > collision frequency(B) > collision frequency(C). The number of collisions depends on both the speed and the distance between gas molecules. Since hydrogen is the lightest molecule it has the greatest speed and the 5 L flask of hydrogen also contains the most molecules, so collisions will occur more frequently between hydrogen molecules than between helium molecules. By the same reasoning collisions will occur more frequently between helium molecules than between methane molecules.

5.67 To find the ratio of effusion rates, calculate the inverse of the ratio of the square roots of the molar masses (equation 5.13).

$$\frac{rate_{H_2}}{rate_{UF_6}} = \frac{\sqrt{\mathcal{M}_{UF_6}}}{\sqrt{\mathcal{M}_{H_2}}} = \frac{\sqrt{352.0\,g/mol}}{\sqrt{2.016\,g/mol}} = \mathbf{13.21}$$

5.69 a) The gases have the same $\overline{E_k}$ because they are at the same temperature. The heavier Ar atoms are moving slower than the lighter He atoms to maintain the same $\overline{E_k}$. Therefore, **Curve 1** better represents the behavior of Ar.

b) A gas that has a slower molecular speed would effuse more slowly, so **Curve 1** is the better choice.

c) Fluorine gas exists as a diatomic molecule, F_2, with **M** = 38.00 g/mol. Therefore, F_2 is much closer in size to Ar than He, so **Curve 1** more closely represents F_2's behavior.

5.71 $$\frac{rate_{He}}{rate_{F_2}} = \frac{time_{F_2}}{time_{He}} = \frac{\sqrt{\mathcal{M}_{F_2}}}{\sqrt{\mathcal{M}_{He}}}$$

$$time_{F_2} = \sqrt{\frac{38.00}{4.003}}(4.55\,\text{min}) = \mathbf{14.0\,min}$$

5.73 White phosphorus is a molecular form of the element phosphorus consisting of some number, x, of phosphorus atoms. Determine the number of phosphorus atoms, x, in one molecule of white phosphorus from the rate of effusion of the gaseous phosphorus molecules.

$$\mathcal{M}_{P_x} = \left(\left(\frac{1}{0.404}\right)\sqrt{20.18}\right)^2 = 124\ g\ P_x\ /\ mol$$

$$x = \frac{124\ g\ P_x\ /\ mol}{30.97\ g\ P\ /\ mol} = 4$$

There are **four atoms** of phosphorus in a molecule of white phosphorus, P_4.

5.75 Intermolecular attractions cause the real pressure to be *less than* ideal pressure, so it causes a *negative* deviation. The size of the intermolecular attraction is related to the constant *a*. According to Table 5.6, a_{N2} = 1.39, a_{Kr} = 2.32 and a_{CO2} = 3.59. Therefore, CO_2 experiences a greater negative deviation in pressure than the other two gases: N_2 < Kr < CO_2.

5.77 Nitrogen gas behaves more ideally at 1atm than at 500 atm because at lower pressures the gas molecules are farther apart. An ideal gas is defined as consisting of gas molecules that act independently of the other gas molecules. When gas molecules are far apart they act ideally, because intermolecular attractions are less important and the volume of the molecules is a smaller fraction of the container volume.

5.80 Molar mass has units of g/mol, so solve for the number of hemoglobin (Hb) moles combined with O_2. Moles of oxygen (from ideal gas law) combine with Hb in a 4:1 ratio.

$$mol\ O_2 = \frac{PV}{RT} = \frac{\left(743\ torr \times \frac{1\ atm}{760\ torr}\right)(0.00153\ L)}{\left(0.08206\frac{atm \cdot L}{mol \cdot K}\right)(37 + 273K)} = 5.87_{99}x10^{-5}\ mol$$

$$mol\ hemoglobin = \left(5.87_{99}x10^{-5}\ mol\ O_2\right)\left(\frac{1\ mol\ hemoglobin}{4\ mol\ O_2}\right) = 1.47_{00}x10^{-5}\ mol$$

Therefore, $\mathcal{M}$(Hb) = (1.00 g)/(1.47_{00} x 10^{-5} mol) = **6.80 x 10^4 g/mol.**

5.83 a) Pressure calculated from ideal gas equation

$$P_{IGL} = \frac{(0.5850\,kg\,Cl_2)\left(\frac{1\,mol\,Cl_2}{0.07090\,kg}\right)(0.08206\,atm \cdot L/mol \cdot K)(225 + 273.15\,K)}{15.00\,L} = \mathbf{22.5\,atm}$$

b) Pressure calculated from van der Waals equation

$$\left(P + \frac{n^2 a}{V^2}\right)(V - nb) = nRT$$

$$\left(P + \frac{(8.251\,mol)^2\left(\frac{6.49\,atm \cdot L^2}{mol^2}\right)}{(15.00\,L)^2}\right)\left(15.00\,L - \left(8.251\,mol\,x\,\frac{0.0562\,L}{mol}\right)\right) = (8.251\,mol)\left(\frac{0.08206\,atm \cdot L}{mol \cdot K}\right)(498\,K)$$

$$P = \left(\frac{337.2\,atm \cdot L}{14.54\,L}\right) - 1.964\,atm = \mathbf{21.2\,atm}$$

5.86 a) Partial pressures are calculated from Dalton's Law of Partial Pressures: $P_A = X_A(P_{total})$

$$P_{N_2} = (0.786)(1.00\,atm)\left(\frac{760\,torr}{1\,atm}\right) = \mathbf{597\,torr}$$

$$P_{O_2} = (0.209)(1.00\,atm)\left(\frac{760\,torr}{1\,atm}\right) = \mathbf{159\,torr}$$

$$P_{CO_2} = (0.0004)(1.00\,atm)\left(\frac{760\,torr}{1\,atm}\right) = \mathbf{0.3\,torr}$$

$$P_{H_2O} = (0.0046)(1.00\,atm)\left(\frac{760\,torr}{1\,atm}\right) = \mathbf{3.5\,torr}$$

To check calculate total pressure from partial pressures:

597 torr + 159 torr + 0.3 torr + 3.5 torr = 760 torr

The calculated total agrees with the given total pressure of 760 torr or 1 atm.

b) Mole fractions can be calculated by rearranging Dalton's Law of Partial Pressures: $X_A = P_A/P_{total}$ and multiply by 100 to express mole fraction as per cent.

$$X_{N_2} = \left(\frac{569\,torr}{760\,torr}\right)(100) = \mathbf{74.9\%}$$

$$X_{O_2} = \left(\frac{104\,torr}{760\,torr}\right)(100) = \mathbf{13.7\%}$$

$$X_{CO_2} = \left(\frac{40\,torr}{760\,torr}\right)(100) = \mathbf{5.3\%}$$

$$X_{H_2O} = \left(\frac{47\,torr}{760\,torr}\right)(100) = \mathbf{6.2\%}$$

Add percentages to check that the total is 100%: 74.9 + 13.7 + 5.3 + 6.2 = 100.1%.

c) Number of molecules of O_2 can be calculated using the Ideal Gas Equation.

$$molecules\ O_2 = \frac{N_A PV}{RT} = \frac{\left(\frac{6.022x10^{23}\ molecules}{1\ mol}\right)\left(104\ torr \times \frac{1\ atm}{760\ torr}\right)(0.50\ L)}{\left(0.08206\frac{atm \cdot L}{mol \cdot K}\right)(37+273K)} = \mathbf{1.6 x 10^{21}\ moleculesO_2}$$

5.88 a) Use $\frac{P_1V_1}{T_1} = \frac{P_2V_2}{T_2}$ with P_1 = 1400. mmHg, V_1 = 208 mL, T_1 = 286 K and P_2 = 1.00 atm, T_2 = 298 K.

$$V_2 = \left(\frac{1400.\ mmHg\left(\frac{1\ atm}{760\ mmHg}\right)(208\ mL)}{286\ K}\right)\left(\frac{298\ K}{1.00\ atm}\right) = \mathbf{399\ mL}$$

b) Partial pressure of nitrogen can be calculated from its fraction:

P_{N2} = 0.77(1400. mmHg) = 1078 mmHg (carry additional significant figures)

$$n_{N_2} = \frac{PV}{RT} = \frac{\left(1078\ mmHg \times \frac{1\ atm}{760\ mmHg}\right)(0.208\ L)}{\left(0.08206\frac{atm \cdot L}{mol \cdot K}\right)(286\ K)} = \mathbf{1.2 x 10^{-2}\ mole\ N_2}$$

5.90 The balanced equation and reactant amounts are given, so the first step is to identify the limiting reactant.

$$n_{Cu} = 4.95\,cm^3\left(8.95\,g/cm^3\right)\left(\frac{1\ mol\ Cu}{63.55\ g}\right) = 0.697_{13}\ mol\ Cu$$

$$n_{HNO_3} = 230.0\ mL\ sol'n\left(\frac{1\ cm^3}{1\ mL}\right)\left(1.42\ g/cm^3\right)\left(\frac{0.680\ g\ HNO_3}{1\ g\ sol'n}\right)\left(\frac{1\ mol\ HNO_3}{63.02\ g}\right) = 3.52_{41}\ mol\ HNO_3$$

Since the stoichiometric ratio of copper to nitrogen dioxide is 1:2, then 2 x 0.697_{13} = 1.39_{43} mol of nitrogen dioxide would be made from 0.697_{13} mol of copper. Since the stoichiometric ratio of nitric acid to nitrogen dioxide is 2:1, then 0.5 x 3.52_{41} = 1.76_{21} mol of nitrogen dioxide would be made from 3.52_{41} mol of nitric acid. Since less product can be made from the copper, it is the limiting reactant and excess nitric acid will be left after the reaction goes to completion. From the moles of limiting reactant a maximum of 1.39_{43} mol of nitrogen dioxide can be produced. Use the calculated number of moles and the given temperature and pressure in the ideal gas equation to find the volume of nitrogen dioxide produced. Note that nitrogen dioxide is the only gas involved in the reaction.

$$V_{NO_2} = \frac{(1.39_{43}\ mol)(0.08206\ atm \cdot L/mol \cdot K)(28.2+273.15\ K)}{(735\ torr)(1\ atm/760\ torr)} = \mathbf{35.7\ L}$$

5.95 The empirical formula for aluminum chloride is $AlCl_3$ (**M** = 133.33 g/mol). Calculate the molar mass of gaseous species from the ratio of effusion rate. This molar mass, divided by the empirical weight, should give a whole-number multiple that will yield the molecular formula.

$$\frac{rate_{unk}}{rate_{He}} = \frac{\sqrt{\mathcal{M}_{He}}}{\sqrt{\mathcal{M}_{unk}}}$$

$$0.122 = \frac{\sqrt{4.003}}{\sqrt{\mathcal{M}_{unk}}}; \; \mathcal{M}_{unk} = \left(\frac{\sqrt{4.003}}{0.122}\right)^2 = 269\ g/mol$$

The whole number multiple is 269/133.33 = 2.02 ≈ 2. Therefore, the molecular formula of the gaseous species is $\mathbf{Al_2Cl_6}$.

5.97 Plan: First write the balanced equation for the reaction: $2SO_2 + O_2 \rightarrow 2SO_3$. The total number of moles of gas will change as the reaction occurs since 3 moles of reactant gas forms 2 moles of product gas. From the volume, temperature and pressures given we can calculate the number of moles of gas before and after the reaction. For each mole of SO_3 formed, the total number of moles of gas decreases by ½ mole. Thus, twice the decrease in moles of gas equals the moles of SO_3 formed.

Solution:

Moles of gas before and after reaction

$$n_{before\ rxn} = \frac{(1.95\ atm)(2.00\ L)}{(0.08206\ atm \cdot L/mol \cdot K)(900\ K)} = 5.28_{07} \times 10^{-2}\ mol$$

$$n_{after\ rxn} = \frac{(1.65\ atm)(2.00\ L)}{(0.08206\ atm \cdot L/mol \cdot K)(900\ K)} = 4.46_{83} \times 10^{-2}\ mol$$

moles of SO_3 produced = 2 x decrease in the total number of moles

= 2 x ($0.0528_{07} - 0.0446_{83}$)

= $\mathbf{1.63 \times 10^{-2}}$ **mol**

Check: If the starting amount is 0.0528 total moles of SO_2 and O_2, then x + y = 0.0528 mol, where x = moles of SO_2 and y = moles of O_2. After the reaction:

$$(x - z) + (y - 0.5z) + z = 0.0447\ mol$$

where z = moles of SO_3 formed = moles of SO_2 reacted = 2(moles of O_2 reacted).

Subtracting the two equations gives:

$$x - (x - z) + y - (y - 0.5z) - z = 0.0528 - 0.0447$$

$$z = 0.0163$$

The approach of setting up two equations and solving them gives the same result as above.

5.101 a) The balanced equation for the reaction is $Ni(s) + 4CO(g) \rightarrow Ni(CO)_4(g)$. Although the problem states that Ni is impure, you can assume the impurity is unreactive and thus not include it in the reaction. The mass of Ni can be calculated from the stoichiometric relationship between moles of CO (using the ideal gas law) and Ni.

P = (100.7 kPa)(1 atm/101.325 kPa) = 0.9938_{32} atm

V = $(3.55\ m^3) \times [(100\ cm/1m)^3] \times (1\ L/1000\ cm^3) = 3550\ L$

$$n_{CO} = \frac{PV}{RT} = \frac{(0.9938_{32}\ atm)(3550\ L)}{\left(0.08206\frac{atm \cdot L}{mol \cdot K}\right)(50 + 273K)} = 133._{11}\ mol\ CO$$

$$mass\ Ni = (133._{11}\ mol\ CO)\left(\frac{1\ mol\ Ni}{4\ mol\ CO}\right)\left(\frac{58.70\ g\ Ni}{mol\ Ni}\right) = \mathbf{1.95x10^3\ g\,Ni}$$

b) First determine the *volume* of $Ni(CO)_4$ formed from the given P, T, and moles of $Ni(CO)_4$ formed. Convert the volume to m^3 and divide into the answer in (a).

$$n_{Ni(CO)_4} = (133._{11}\ mol\ CO)\left(\frac{1\ mol\ Ni(CO)_4}{4\ mol\ CO}\right) = 33.27_{75}\ mol\ Ni(CO)_4$$

$$V = \frac{nRT}{P} = \frac{(33.27_{75}\ mol)\left(0.08206\frac{atm \cdot L}{mol \cdot K}\right)(155 + 273K)}{(21\ atm)} = 55.6_{55}\ L\ or\ 0.0557\ m^3$$

Therefore, 1.95×10^3 g Ni/0.0556_{55} m^3 = **3.5×10^4 g of Ni** are obtained per cubic meter of the carbonyl (2 sig figs due to P = 21 atm).

<u>Alternate Method</u>:

Assume the volume is 1 m^3. Use the ideal gas law to solve for moles of $Ni(CO)_4$ and convert moles to grams using the molar mass.

P = 21 atm

T = 155°C = 428 K

V = 1.00 m^3 = (1.00 m^3)(100 cm/m)^3(1 mL/1 cm^3)(1 L/1000 mL) = 1.00×10^3 L

$$n_{Ni(CO)_4} = \frac{PV}{RT} = \frac{(21\ atm)(1.00x10^3\ L)}{\left(0.08206\frac{atm \cdot L}{mol \cdot K}\right)(428\ K)} = 5.98x10^2\ mol\ Ni(CO)_4$$

$$mass\ Ni = (5.98x10^2\ mol\ Ni(CO)_4)\left(\frac{1\ mol\ Ni}{1\ mol\ Ni(CO)_4}\right)\left(\frac{58.70\ g\ Ni}{mol\ Ni}\right) = 3.5x10^4\ g\ Ni$$

c) The amount of CO needed to form $Ni(CO)_4$ is the same amount that is released on decomposition. The vapor pressure of water at 35°C (42.2 torr) must be accounted for. Use the ideal gas law to calculate the volume of CO, and compare this to the volume of $Ni(CO)_4$.

$P_{CO} = P_{TOT} - P_{H2O}$ = 769 torr – 42.2 torr = 727 torr

$$V_{CO} = \frac{nRT}{P} = \frac{(133._{11}\ mol)\left(0.08206\frac{atm \cdot L}{mol \cdot K}\right)(35 + 273K)}{\left(\frac{727\ torr}{760\ torr\ /\ atm}\right)} = 3.51_{70}\ x10^3\ L\ or\ 3.52\ m^3$$

$$\frac{V_{CO}}{V_{Ni(CO)_4}} = \frac{3.51_{70}\ m^3}{0.0556_{55}\ m^3} = \mathbf{63.2 m^3\ CO\ per\ m^3\ of\ Ni(CO)_4}$$

<u>Alternate Method</u>:

Solve for moles for CO using the molar ratio from the balanced equation.

$$mol\ CO = (5.98x10^2\ Ni(CO)_4)\left(\frac{4\ mol\ CO}{1\ mol\ Ni(CO)_4}\right) = 2.39x10^3\ mol\ CO$$

$P_{CO} = P_{TOT} - P_{H2O} = 769 - 42.2 = 727$ torr
T = 35°C = 308 K

$$V_{CO} = \frac{nRT}{P} = \frac{\left(2.39x10^3\ mol\right)\left(0.08206\frac{atm\cdot L}{mol\cdot K}\right)\left(308\ K\right)}{\left(727/760\right)} = 6.32x10^4\ L = 63\ m^3$$

Answer is reported to 2 significant figures as in alternate method (b).

5.103 a) A preliminary equation for this reaction is $4C_xH_yN_z + nO_2 \rightarrow 4CO_2 + 2N_2 + 10H_2O$. Since the organic compound does not contain oxygen the only source of oxygen as a reactant is oxygen gas. To form 4 volumes of CO_2 would require 4 volumes of O_2 and to form 10 volumes of H_2O would require 5 volumes of O_2. Thus, a total of 9 volumes of O_2 was required.

b) Since the volume of a gas is proportional to the number of moles of the gas we can equate volume and moles. From a volume ratio of 4 CO_2 : 2 N_2 : 10 H_2O we deduce a mole ratio of 4C:4N:20H or 1C:1N:5H for an empirical formula of CH_5N.

5.107 To find the factor by which a diver's lungs would expand, find the factor by which P changes from 125 ft to the surface, and apply Boyle's Law. To find that factor, calculate $P_{seawater}$ at 125 ft by converting the given depth from ft-seawater to mm-Hg to atm and adding the surface pressure.

$$P_{sea\ water}(mm\ H_2O) = \left(125\ ft\ H_2O\right)\left(\frac{12\ in}{1\ ft}\right)\left(\frac{2.54\ cm}{1\ in}\right)\left(\frac{10\ mm}{1\ cm}\right) = 3.81x10^4\ mm\ H_2O$$

$$P_{sea\ water}(mmHg) = \left(3.81x10^4\ mm\right)\left(\frac{d_{sea\ water}}{d_{Hg}}\right) = \left(3.81x10^4\ mm\right)\left(\frac{1.04\ g/mL}{13.5\ g/mL}\right) = 2.94x10^3\ mmHg$$

$$P_{sea\ water}(atm) = \left(2.94x10^3\ mmHg\right)\left(\frac{1\ atm}{760\ mmHg}\right) = 3.87\ atm$$

Therefore, P_{TOT} = 3.87 atm + 1.00 atm (at surface) = 4.87 atm. Use Boyle's Law to find the volume change of the diver's lungs:

$$V_2 = V_1 \times \frac{P_1}{P_2} = V_1 \times \frac{4.87\ atm}{1.00\ atm} = \mathbf{4.87\ V_1}$$

To find the depth to which the diver could ascend safely, use the given safe expansion factor (1.5) and the pressure at 125 ft, P_{125}, to find the safest ascended pressure, P_{safe}.

$$\frac{P_{125}}{P_{safe}} = 1.5$$

$$P_{safe} = \frac{4.87\ atm}{1.5} = 3.25\ atm$$

Convert the pressure in atm to pressure in ft of seawater using the conversion factors above. Subtract this distance from the initial depth to find how far the diver could ascend.

$$P_{safe}(ft\ seawater) = (4.87\ atm - 3.25\ atm)\left(\frac{760\ mmHg}{1\ atm}\right)\left(\frac{13.5\ g/mL}{1.04\ g/mL}\right)$$

$$\times\left(\frac{1\ cm}{10\ mmHg}\right)\left(\frac{1\ in}{2.54\ cm}\right)\left(\frac{1\ ft}{12\ in}\right) = 52.4\ ft$$

Therefore, the safely ascend 52.4 ft to a depth of (125 – 52.4) = **73 ft**.

5.109 First write the balanced equation: $2H_2O_2(aq) \rightarrow 2H_2O(l) + O_2(g)$
According to the description in the problem, a given volume of peroxide solution (0.100 L) will release a certain number of "volumes of oxygen gas" (20). Assume that 20 is exact.

0.100 L sol'n will produce (20 x 0.100 L) = 2.00 L O_2 gas

You can convert volume of $O_2(g)$ to moles of gas using the ideal gas equation. But since the problem specifies conditions at STP, a quicker conversion uses the standard molar volume of 22.414 L/mol.

mol $O_2(g)$ = 2.00 L ÷ (22.414 L/mol) = 0.0892 mol $O_2(g)$

The grams of hydrogen peroxide can now be solved stoichiometrically using the balanced chemical equation.

$$mass\ H_2O_2 = (0.0892\ mol\ O_2)\left(\frac{2\ mol\ H_2O_2}{1\ mol\ O_2}\right)\left(\frac{34.016\ g\ H_2O_2}{1\ mol\ H_2O_2}\right) = \mathbf{6.07\ g\ H_2O_2}$$

5.112

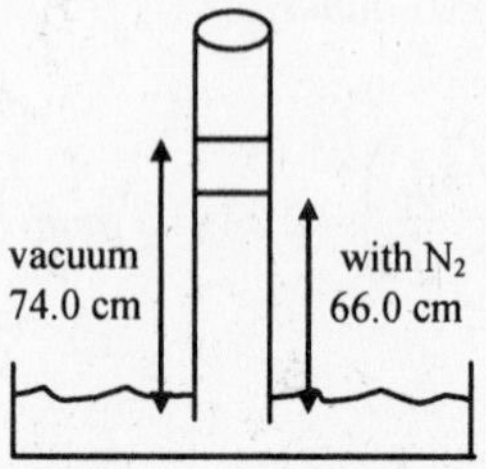

The diagram at left describes the two Hg height levels within the barometer. To find the mass of N_2, find P and V (T given) and use the ideal gas equation. The P_{N2} is directly related to the change in column height of Hg. The volume of the space occupied by the $N_2(g)$ is calculated from the dimensions of the barometer.

$$P_{N_2} = (74.0\ cmHg - 66.0\ cmHg)\left(\frac{10\ mm}{1\ cm}\right)\left(\frac{1\ atm}{760\ mm\ Hg}\right) = 0.105\ atm$$

$$V_{N_2} = (100.\ cm - 66.0\ cm)(1.20\ cm^2)\left(\frac{1\ L}{1000\ cm^3}\right) = 0.0408\ L$$

$$n_{N_2} = \frac{PV}{RT} = \frac{(0.105\ atm)(0.0408\ L)}{\left(0.08206\frac{atm\cdot L}{mol\cdot K}\right)(24 + 273K)} = 1.76x10^{-4}\ mol\ N_2$$

$$mass\ N_2 = (1.76x10^{-4}\ mol\ N_2)\left(\frac{28.02\ g\ N_2}{mol\ N_2}\right) = \mathbf{4.93x10^{-3}\ g\ N_2}$$

5.114 a) Xenon would show greater deviation from ideal behavior than argon since xenon is a larger atom than xenon. The electron cloud of Xe is more easily distorted so intermolecular attractions are greater. Xe's larger size also means that the volume the gas occupies becomes a greater proportion of the container's volume at high pressures.

b) Water gas would show greater deviation from ideal behavior than neon gas since the attractive forces between water molecules are greater than the attractive forces between neon atoms. We know the attractive forces are greater for water molecules because it forms a liquid at a higher temperature than neon (water is a liquid at room temperature while neon is a gas at room temperature).

c) Mercury vapor would show greater deviation from ideal behavior than radon gas since the attractive forces between mercury atoms is greater than that between radon atoms. We know that the attractive forces for mercury are greater because it is a liquid at room temperature while radon is a gas.

d) Water is a liquid at room temperature; methane is a gas at room temperature (think about where you have heard of methane gas before – bunsen burners in lab, cows' digestive system). So, water molecules have stronger attractive forces than methane molecules and should deviate from ideal behavior to a greater extent than methane moelcules.

5.117 V and T are not given, so the ideal gas equation cannot be used. The total mass of the mixture is given:

$$m_{TOT} = m_{Kr} + m_{CO2} = 35.0 \text{ g}$$

Since $\mathcal{M}$ (g/mol) * n (mol) = mass (g), the equation can be rewritten.

$$m_{TOT} = (\mathcal{M}_{Kr})(n_{Kr}) + (\mathcal{M}_{CO2})(n_{CO2}) = 35.0 \text{ g}$$

The number of moles of each species can be found by using the given pressure information to build mole fraction expressions.

$$X_{Kr} = \frac{P_{Kr}}{P_{TOT}} = \frac{0.250\ atm}{0.708\ atm} = 0.353;\ X_{CO_2} = 1 - 0.353 = 0.647$$

Use the definition of mole fraction to solve for moles of Kr in terms of moles CO_2.

$$X_{Kr} = \frac{n_{Kr}}{n_{Kr} + n_{CO_2}} = 0.353$$

$$n_{Kr} = 0.353\left(n_{Kr} + n_{CO_2}\right)$$

$$0.647 n_{Kr} = 0.353 n_{CO_2};\ \ n_{Kr} = 0.546 n_{CO_2}$$

Substitute $0.546n_{CO2}$ for n_{Kr} into the m_{TOT} expression given above. Use this equation to solve for moles of CO_2, convert to mass, and then subtract from the total mass to find amount of Kr.

$$m_{TOT} = (\mathcal{M}_{Kr})(n_{Kr}) + (\mathcal{M}_{CO2})(n_{CO2}) = 35.0 \text{ g}$$

$$m_{TOT} = (83.80 \text{ g/mol})(0.546 n_{CO2}) + (44.01 \text{ g/mol})(n_{CO2}) = 35.0 \text{ g}$$

$$89.8(\text{g/mol})n_{CO2} = 35.0 \text{ g}$$

$$n_{CO2} = 0.390 \text{ mol}$$

Therefore, mass of CO_2 = (0.390 mol)(44.01 g/mol) = **17.2 g CO_2**.

Mass of Kr = 35.0 g – 17.2 g = **17.8 g Kr**.

5.120 The root mean speed is related to temperature and molar mass as shown in equation 5.13.

$$u_{rms} = \sqrt{\frac{3RT}{\mathcal{M}}}$$

$$u_{rms}^{Ne} = \sqrt{\frac{3(8.314\ J/mol \cdot K)(370\ K)}{0.02018\ kg/mol}} = 6.76 x 10^2\ m/s$$

$$u_{rms}^{Ar} = \sqrt{\frac{3(8.314\ J/mol \cdot K)(370\ K)}{0.03995\ kg/mol}} = 4.81\ x\ 10^2\ m/s$$

$$u_{rms}^{He} = \sqrt{\frac{3(8.314\ J/mol \cdot K)(370\ K)}{0.004003\ kg/mol}} = 1.52\ x\ 10^3\ m/s$$

The units work out because a joule is equivalent to kg•m^{2}s^{-2}.

5.122 a) The number of moles of water gas can be found using the ideal gas equation.

$$n_{H_2O} = \frac{PV}{RT} = \frac{(9.0\ atm)(2.5\ x\ 10^{-4}\ L)(0.75)}{\left(0.08206 \frac{atm \cdot L}{mol \cdot K}\right)(175 + 273)K} = 4.5_{90}\ x\ 10^{-5}\ mol\ H_2O$$

The mass of water is (4.5$_{90}$ x 10^{-5} mol) x (18.016 g/mol) = 8.2$_{70}$ x 10^{-4} g H_2O in kernel. Since this mass is 1.5% of the mass of the kernel, dividing the mass by 0.15 will give the total mass of the kernel: 8.2$_{70}$ x 10^{-4} g H_2O ÷ 0.015 = **0.055 g**.

b) $$V_{H_2O} = \frac{nRT}{P} = \frac{(4.5_{90}\ x\ 10^{-5}\ mol)\left(0.08206 \frac{atm \cdot L}{mol \cdot K}\right)(298\ K)}{1.00\ atm} = 1.1x10^{-3}\ L = \mathbf{1.1 mL}$$

5.123 a) Derive $u_{rms} = \sqrt{\frac{3RT}{\mathcal{M}}}$

Set the given relationships equal to each other.

$$\frac{1}{2} m\bar{u}^2 = \frac{3}{2}\left(\frac{R}{N_A}\right)T$$

Solve for u^2; substitute molar mass, $\mathcal{M}$, for mN_A (mass of one molecule x Avogadro's number of molecules).

$$\bar{u}^2 = 3\left(\frac{R}{mN_A}\right)T$$

$$\bar{u} = \sqrt{\frac{3RT}{\mathcal{M}}}$$

b) Derive Graham's Law $\frac{rate_A}{rate_B} = \frac{\sqrt{\mathcal{M}_B}}{\sqrt{\mathcal{M}_A}}$

At a given T, the average kinetic energy for two substances, with molecular masses m_1 and m_2, is:

$$\bar{E}_k = \frac{1}{2} m_1 \bar{u}_1^2 = \frac{1}{2} m_2 \bar{u}_2^2$$

Rearranging and taking the square root of both sides gives:

$$\frac{m_1}{m_2} = \frac{\bar{u}_2^2}{\bar{u}_1^2} \Rightarrow \frac{\sqrt{m_1}}{\sqrt{m_2}} = \frac{\bar{u}_2}{\bar{u}_1}$$

The average molecular speed, u, is directly proportional to the rate of effusion. Therefore substitute "rate" for each "u". In addition, the molecular mass is directly proportional to the molar mass, so substitute "$\mathcal{M}$" for each "m" :

$$\frac{\sqrt{\mathcal{M}_1}}{\sqrt{\mathcal{M}_2}} = \frac{rate_2}{rate_1}$$

5.127 a) First find the number of moles of carbon dioxide and use the T and P given to calculate volume from the ideal gas equation.

$$T = 35 + 273 = 308\ K$$

$$P = 780\ torr\left(1\ atm/760\ torr\right) = 1.026\ atm$$

$$n_{CO_2} = 18.0\ g\ glucose\left(\frac{1\ mol\ glucose}{180.16\ g}\right)\left(\frac{6\ mol\ CO_2}{1\ mol\ glucose}\right) = 0.5995\ mol$$

$$V_{CO_2} = \frac{(0.5995\ mol)(0.08206\ atm\cdot L/mol\cdot K)(308\ K)}{1.026\ atm} = \mathbf{14.8\,L}$$

This solution assumes that partial pressure of O_2 does not interfere with the reaction conditions.

b) From the stoichiometric ratios calculate the moles of each gas and then use Dalton's law of partial pressures to determine the pressure of each gas.

$$n_{C_6H_{12}O_6} = 9.0\ g\left(1\ mol / 180.16\ g/mol\right) = 0.050\ mol$$

$$n_{O_2} = n_{CO_2} = n_{H_2O} = 0.050\ mol\ glucose\left(6\ mol / 1\ mol\ glucose\right) = 0.30\ mol$$

At 35°C, the vapor pressure of water is 42.2 torr. No matter how much water is produced, the partial pressure of H_2O will still be 42.2 torr. The remaining pressure, 780 torr – 42.2 torr = 738 torr is the sum of partial pressures for O_2 and CO_2.

$$P_{H_2O} = \mathbf{42.2\,torr}$$

$$P_{O_2} = P_{CO_2} = X_A P_{total} = \left(\frac{0.30}{2(0.30)}\right)(738\ torr) = \mathbf{370\,torr}$$

5.131 To find the number of steps through the membrane, calculate the molar masses to find the ratio of effusion rates. This ratio is the enrichment factor for each step.
From the margin note, $\mathcal{M}_{235}$ = 349.03 g/mol and $\mathcal{M}_{238}$ = 352.04 g/mol

$$enrichment\ factor = \frac{rate_{235}}{rate_{238}} = \frac{\sqrt{\mathcal{M}_{238UF6}}}{\sqrt{\mathcal{M}_{235UF6}}} = \frac{\sqrt{352.04\ g/mol}}{\sqrt{349.03\ g/mol}} = 1.0043$$

Therefore, the abundance of $^{235}UF_6$ after one membrane is 0.72% x 1.0043, and after "N" membranes:
Abundance of $^{235}UF_6$ after "N" membranes = 0.72% * $(1.0043)^N$
Desired abundance of $^{235}UF_6$ = 3.0% = 0.72% * $(1.0043)^N$
Solving for N:
4.2 = $(1.0043)^N$
ln 4.2 = ln $(1.0043)^N$
ln 4.2 = N * ln (1.0043)
N = (ln 4.2)/(ln 1.0043) = **334 steps**

5.133 The amount of each gas that leaks from the balloon is proportional to its effusion rate. Using 45% or 0.45 as the rate for H_2 the rate for O_2 can be determined from Graham's Law.

$$\frac{rate_{O_2}}{rate_{H_2}} = \frac{\sqrt{\mathcal{M}_{H_2}}}{\sqrt{\mathcal{M}_{O_2}}} \Rightarrow \frac{rate_{O_2}}{0.45} = \frac{\sqrt{2.016}}{\sqrt{32.00}} \Rightarrow rate_{O_2} = 0.11$$

The relative amount of H_2 left is 0.55 and of O_2 is 0.89, which gives an O_2/H_2 ratio of **1.6**.

5.135 The problem is to find the total pressure of the gases after reaction has occurred at 550. K. The pressure can be calculated from the ideal gas equation if volume, temperature and number of moles are known. The volume and temperature are given, 2.50 L and 550. K, but the total number of moles of gas must be calculated.

The number of moles of fluorine can be calculated from the initial conditions:

$$n_{F_2} = \frac{PV}{RT} = \frac{(350.\,torr)\left(\frac{1\,atm}{760\,torr}\right)(2.50\,L)}{\left(0.08206\frac{atm\cdot L}{mol\cdot K}\right)(250.\,K)} = 5.61_{21}\,x10^{-2}\,mol\,F_2$$

The initial number of moles of iodine is calculated from the mass of iodine:

$$n_{I_2} = (2.50\,g\,I_2)\left(\frac{1\,mol}{253.8\,g\,I_2}\right) = 9.85_{03}\,x\,10^{-3}\,mol\,I_2$$

Once the fluorine and iodine gases react, the container will contain the product gas, iodine heptafluoride, and whichever reactant gas is left in excess. From the balanced reaction,

$7F_2(g) + I_2(g) \rightarrow 2IF_7(g)$,

find which reactant limits the amount of product.

$$n_{IF_7}\,(from\,F_2) = (5.61_{21}\,x\,10^{-2}\,mol\,F_2)\left(\frac{2\,mol\,IF_7}{7\,mol\,F_2}\right) = 1.60_{34}\,x\,10^{-2}\,mol\,IF_7$$

$$n_{IF_7}\,(from\,I_2) = (9.85_{03}\,x\,10^{-3}\,mol\,I_2)\left(\frac{2\,mol\,IF_7}{1\,mol\,I_2}\right) = 1.97_{01}\,x\,10^{-2}\,mol\,IF_7$$

Less product is made from the amount of fluorine than from the amount of iodine so fluorine is the limiting reactant. Thus, there will be no fluorine in the container after reaction. And in the container there will be 1.60_{34} x 10^{-2} mol of iodine heptafluoride produced in the reaction and 1.83_{30} x 10^{-3} mol of iodine (calculated below) left after the reaction to give a total of 1.78_{67} x 10^{-2} mol of gas.

$$n_{I_2}\,(left\,after\,reaction) = (9.85_{03}\,x\,10^{-3}\,mol\,I_2) - \left(5.61_{21}\,x\,10^{-2}\,mol\,F_2\left(\frac{1\,mol\,I_2}{7\,mol\,F_2}\right)\right) = 1.83_{30}\,x\,10^{-3}\,mol\,I_2$$

The final pressure is calculated from the ideal gas equation.

$$P = \frac{(1.78_{67}\,x\,10^{-2}\,mol)(0.08206\,L\cdot atm/mol\cdot K)(550.\,K)}{2.50\,L} = \mathbf{0.323\,atm}$$

The partial pressure of iodine is calculated from Dalton's Law:

P_{I2} = (0.323 atm)(1.83 x 10^{-3} mol I_2/1.79 x 10^{-2} mol gas) = **0.0330 atm**

CHAPTER 6

THERMOCHEMISTRY: ENERGY FLOW AND CHEMICAL CHANGE

FOLLOW-UP PROBLEMS

6.1 Plan: The system is the reactant and products of the reaction. Since heat is absorbed by the surroundings, the system releases heat and q is negative. Because work is done on the system, w is positive. Use equation 6.2 to calculate ΔE.

Solution:

$$\Delta E = q + w = \left(-26.0\ kcal \times \frac{1\ kJ}{0.2390\ kcal}\right) + \left(15.0\ Btu \times \frac{1.055\ kJ}{1\ Btu}\right) = -109\ kJ + 15.8\ kJ = \mathbf{-93\ kJ}$$

Check: A negative ΔE seems reasonable since energy must be removed from the system to condense gaseous reactants into a liquid product.

6.2 Plan: Since heat is a "product" in this reaction, the reaction is exothermic (ΔH < 0) and reactants are above the products in an enthalpy diagram.

Solution:

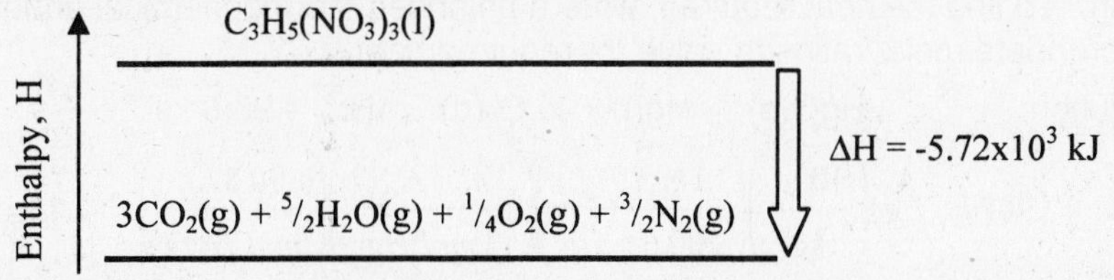

6.3 Plan: The heat released by the nail is calculated using equation 6.7. Convert the ΔT to Kelvin degrees and obtain the specific heat capacity for iron from Table 6.4.

Solution: ΔT = 25°C – 37°C = -12°C = -12 K.

$$q = m\,c\,\Delta T = (5.5\ g)(0.450\ J/g\text{-}K)(-12\ K) = \mathbf{-30.\ J}$$

Check: Heat is released by the nail, so q should be negative.

6.4 Plan: To find ΔT, find the T_{final} by applying equation 6.7. The diamond loses heat ($-q_{diamond}$) whereas the water gains heat (q_{H2O}). Although the Celsius degree and Kelvin degree are the same size, be careful when interchanging the units. To be safe, always convert the Celsius temperature to Kelvin temperatures. Use the initial temperatures, masses, and specific heat capacities to solve the expression below.

Solution:

$$-q_{diamond} = q_{H2O}$$

$$-(m\,c_{diamond}\,\Delta T) = m\,c_{H2O}\,\Delta T$$

$$\text{mass of diamond} = 10.25\ \text{carat} \times (0.2000\ g/\text{carat}) = 2.050\ g$$

$$T_{init}\ (\text{diamond}) = 74.21 + 273.15 = 347.36\ K$$

$$T_{init}\ (\text{water}) = 27.20 + 273.15 = 300.35\ K$$

$$-(2.050\ g)(0.519\ J/g\text{-}K)(T_{final} - 347.36) = (26.05\ g)(4.184\ J/g\text{-}K)(T_{final} - 300.35)$$

$$(-1.06_{40}\ J/K)(T_{final} - 347.36) = (108.9_{93}\ J/K)(T_{final} - 300.35)$$

$$-1.06_{40}\,T_{final} + 369._{57} = 108.9_{93}\,T_{final} - 32736._{11}$$

$$33105._{68} = 110.0_{57}\,T_{final}$$

$$T_{final} = 300.8_0\ K \text{ or } 27.6_5\ ^\circ C$$

$$\Delta T_{diamond} = 300.8_0 - 347.36 = \mathbf{-46.6\ K}$$
$$\Delta T_{H2O} = 300.8_0 - 300.35 = \mathbf{0.4_5\ K = 0.4\ K}$$

Check: Rounding errors account for the slight differences in the final answer when compared to the identical calculation using Celsius temperatures. The temperature of the diamond decreases as the temperature of the water increases. The temperature of the water does not increase significantly because its specific heat capacity is large in comparison to the diamond, so it takes more heat to raise 1 g of the substance by 1°C.

6.5 Plan: The bomb calorimeter gains heat from the combustion of graphite, so

$$-q_{graphite} = q_{calorimeter}$$

Convert the mass of graphite from grams to moles and use the given kJ/mol to find $q_{graphite}$. The heat lost by graphite equals the heat gained from the calorimeter, or ΔT multiplied by $C_{calorimeter}$.

Solution: -(mol graphite x kJ/mol graphite) = $C_{calorimeter}\,\Delta T_{calorimeter}$

[(-0.8650 g)(1 mol C/12.01 g)(-393.5 kJ/mol)] = $C_{calorimeter}$(2.613 K)

$$C_{calorimeter} = \mathbf{10.85\ kJ/K}$$

Check: A quick check shows the answer is reasonable: 1 ÷ 12 x 400 ÷ 3 = 11.

6.6 Plan: To find the heat required, write a balanced thermochemical equation and use appropriate molar ratios to solve for required heat.

Solution: $HgO(s) \rightarrow Hg(l) + \tfrac{1}{2} O_2(g) \quad \Delta H_{rxn} = 90.8\ kJ$

$$heat = (907\ kg\ Hg)\left(\frac{10^3\ g}{kg}\right)\left(\frac{1\ mol\ Hg}{200.6\ g\ Hg}\right)\left(\frac{1\ mol\ HgO}{1\ mol\ Hg}\right)\left(\frac{90.8\ kJ}{mol\ HgO}\right) = \mathbf{4.11x10^5\ kJ}$$

6.7 Plan: Manipulate the two equations so that their sum will result in the target equation. Reverse the first equation (and change the sign of ΔH); reverse the second equation and multiply the coefficients (and ΔH) two.

Solution:

$2NO(g) + {}^3/_2O_2(g) \rightarrow N_2O_5(s) \quad \Delta H = -223.7\ kJ$

$\underline{2NO_2(g) \rightarrow 2NO(g) + O_2(g) \quad \Delta H = 114.2\ kJ}$

$2NO_2(g) + {}^1/_2O_2(g) \rightarrow N_2O_5(s) \quad \Delta H = \mathbf{-109.5\ kJ}$

Check: The desired target equation has been obtained.

6.8 a) $C(graphite) + 2H_2(g) + {}^1/_2O_2(g) \rightarrow CH_3OH(l) \quad \Delta H_f^o = -238.6\ kJ$

b) $Ca(s) + {}^1/_2O_2(g) \rightarrow CaO(s) \quad \Delta H_f^o = -635.1\ kJ$

c) $C(graphite) + {}^1/_4S_8(rhombic) \rightarrow CS_2(l) \quad \Delta H_f^o = 87.9\ kJ$

6.9 Plan: Apply equation 6.8 to this reaction, substitute given values, and solve for the $\Delta H_f^o(CH_3OH)$.

Solution:

$\Delta H_{comb}^o = \Sigma\, \Delta H_f^o(products) - \Sigma\, \Delta H_f^o(reactants)$

$\Delta H_{comb}^o = [\Delta H_f^o(CO_2(g)) + 2\Delta H_f^o(H_2O(g))] - [\Delta H_f^o(CH_3OH(l)) + {}^3/_2\, \Delta H_f^o(O_2(g))]$

$-638.5\ kJ = [1\ mol(-393.5\ kJ/mol) + 2\ mol(-241.8\ kJ/mol)] - [\Delta H_f^o(CH_3OH(l)) + {}^3/_2(0)]$

$-638.5\ kJ = (-877.1\ kJ) - \Delta H_f^o(CH_3OH(l)$

$\Delta H_f^o(CH_3OH(l) = \mathbf{-238.6\ kJ}$

Check: You solved for this value in Follow-Up 6.8. Are they the same?

END-OF-CHAPTER PROBLEMS

6.4 The internal energy of the body is the sum of the cellular and molecular activities occurring from skin level inward. The body's internal energy can be increased by adding food, which adds energy to the body through the breaking of bonds in the food. The body's internal energy can also be increased through addition of work and heat, like the rubbing of another person's warm hands on the body's cold hands. The body can lose energy if it performs work, like pushing a lawnmower, and can lose energy by losing heat to a cold room.

6.6 The amount of the change in internal energy in the two cases is the same. By the law of energy conservation, the change in energy of the universe is zero. This requires that the change in energy of the system (heater or air conditioner) equals an opposite change in energy of the surroundings (room air). Since both systems consume the same amount of electrical energy, the change in energy of the heater equals that of the air conditioner.

6.8 The change in a system's energy is $\Delta E = q + w$. If the system receives heat, then its q_{final} is greater than $q_{initial}$ so q is positive. Since the system performs work, its $w_{final} < w_{initial}$ so w is negative. The change in energy is (+425 J) + (-425 J) = **0 J**.

6.10 A system that releases thermal energy has a negative value for q and a system that has work done on it has a positive value for work. So,

$$\Delta E = -675\text{ J} + (525\text{ cal} \times 4.184\text{ J/cal}) = -675\text{ J} + (2.20 \times 10^3\text{ J}) = \mathbf{1.52 \times 10^3\ J}$$

6.12 a) $$heat\ in\ kJ = \left(3.3x10^{10}\ J\right)\left(\frac{1\ kJ}{1x10^3\ J}\right) = \mathbf{3.3x10^7\ kJ}$$

b) $$heat\ in\ kcal = \left(3.3x10^{10}\ J\right)\left(\frac{1\ cal}{4.184\ J}\right)\left(\frac{1\ kcal}{1000\ cal}\right) = \mathbf{7.9x10^6\ kcal}$$

c) $$heat\ in\ Btu = \left(3.3x10^{10}\ J\right)\left(\frac{1\ Btu}{1055\ J}\right) = \mathbf{3.1x10^7\ Btu}$$

6.15 $$1.0\ lb\ fat\left(\frac{4.1\,x\,10^3\ kcal}{lb\ fat}\right)\left(\frac{1000\ cal}{1\ kcal}\right)\left(\frac{4.184\ J}{1\ cal}\right)\left(\frac{1\ kJ}{1000\ J}\right)\left(\frac{1\ h}{1850\ kJ}\right) = \mathbf{9.3\ h}$$

6.17 Since many reactions are performed in an open flask, the reaction proceeds at constant pressure. The determination of ΔH (constant pressure conditions) requires a measurement of heat only, whereas ΔE requires measurement of heat and PV work.

6.19 a) Exothermic. The system (water) is releasing heat in changing from liquid to solid.
b) Endothermic. The system (water) is absorbing heat in changing from liquid to gas.
c) Exothermic. The process of digestion breaks down food and releases energy.
d) Exothermic. Heat is released as a person runs and muscles perform work.
e) Endothermic. Heat is absorbed as food calories and converted to body tissue.
f) Endothermic. Heat (and work) is absorbed by the wood being chopped.

g) Exothermic. The furnace releases heat from fuel combustion. Alternatively, if the system is defined as the air in the house, the change is endothermic since the air's temperature in increasing by the input of heat energy from furnace.

6.22 An exothermic reaction releases heat, so the reactants have greater H ($H_{initial}$) than the products (H_{final}). $\Delta H = H_{final} - H_{initial} < 0$.

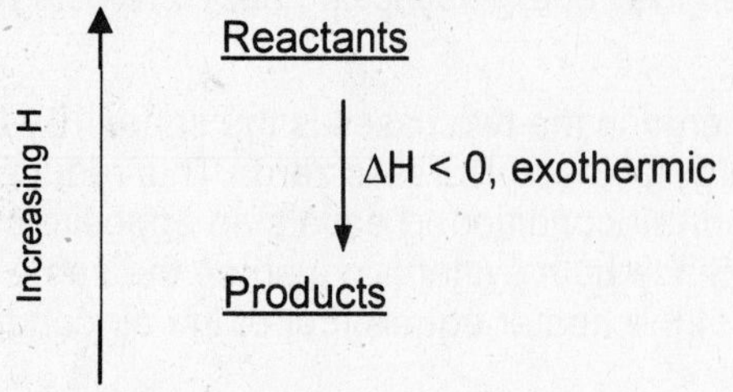

6.24 a) Combustion of methane: $CH_4(g) + 2O_2(g) \rightarrow CO_2(g) + 2H_2O(g)$ + heat

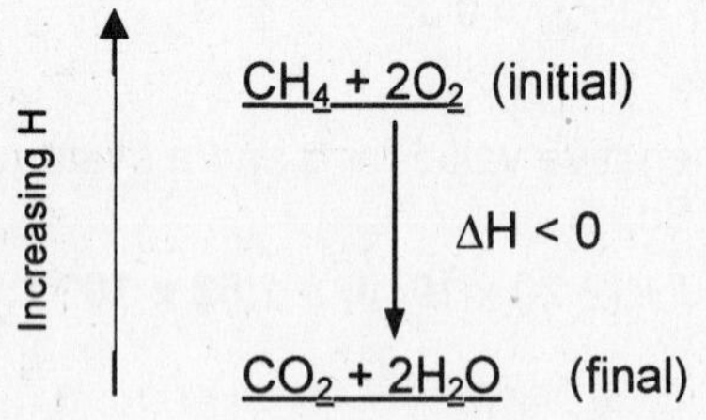

b) Freezing of water: $H_2O(l) \rightarrow H_2O(s)$ + heat

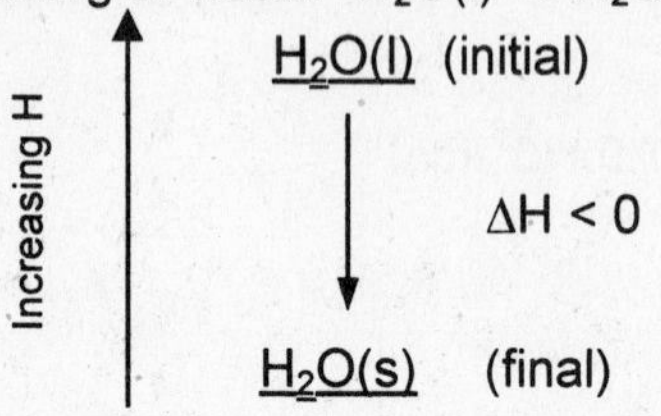

6.26 a) Combustion of hydrocarbons and related compounds requires oxygen (and a heat catalyst) to yield carbon dioxide gas, water vapor and heat.

$$C_2H_5OH(l) + 3O_2(g) \xrightarrow{\Delta} 2CO_2(g) + 3H_2O(g) + \text{heat}$$

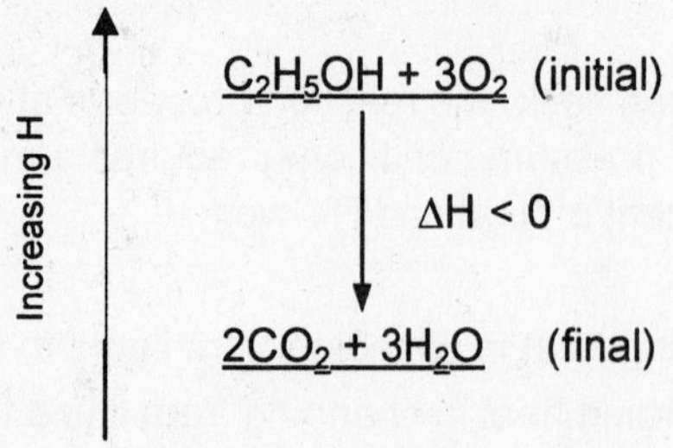

b) Nitrogen dioxide, NO_2, forms from N_2 and O_2.

$N_2(g) + 2O_2(g) + \text{heat} \rightarrow 2NO_2(g)$

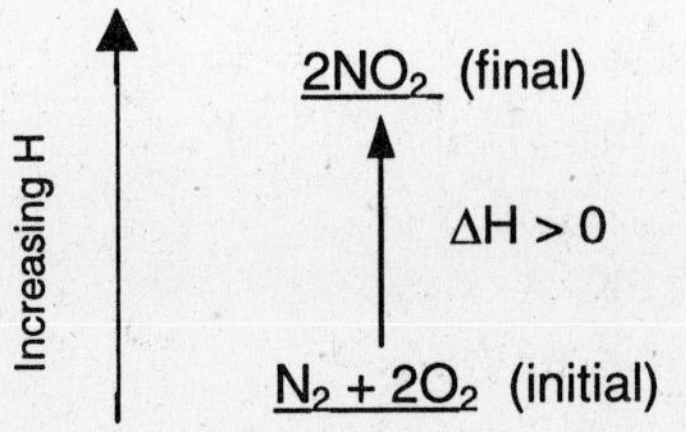

6.28 For methane: $CH_4(g) + 2O_2(g) \rightarrow CO_2(g) + 2H_2O(l)$ which requires that 4 C-H bonds and 2 O=O bonds be broken and 2 C=O bonds and 4 O-H bonds be formed.

For formaldehyde: $CH_2O(g) + O_2(g) \rightarrow CO_2(g) + H_2O(l)$ which requires that 2 C-H bonds, 1 C=O bond and 1 O=O bond be broken and 2 C=O bonds and 2 O-H bonds be formed.

The fact that methane contains more C-H bonds and fewer C-O bonds than formaldehyde suggests that more energy is released in the combustion of methane than of formaldehyde.

6.31 To determine the specific heat capacity of a substance, you need its mass, the heat added (or lost) and the change in temperature.

6.33 Specific heat capacity is the quantity of heat required to raise one gram of a substance by one kelvin. Heat capacity is also the quantity of heat required for a one kelvin temperature change, but is for an object instead of a specified amount of a substance. Thus, specific heat capacity is used when talking about an element or compound while heat capacity is used for a calorimeter or other object.

a) Use heat capacity since the fixture is a combination of substances.

b) and c) Use specific heat capacity since the copper wire and water are pure substances.

6.35 The heat required to raise the temperature of water by 80.°C is found by using equation 6.7, or q = c x mass x ΔT. The specific heat capacity, c_{H2O}, is found in Table 6.4. Because the celsius degree is the same size as the kelvin degree, ΔT = 80.°C = 80. K.

$$q = \left(4.184 \frac{J}{g \cdot K}\right)(12.0\,g)(80.\,K) = 4.0x10^3\,J = \mathbf{4.0 kJ}$$

6.37 Remember to use Kelvin temperatures.

$$q = c\ x\ mass\ x\ \Delta T$$

$$85.0\,kJ(1000\,J/kJ) = (0.900\,J/g \cdot K)(295\,g)(T_f - 276.15\,K)$$

$$320._{15}\,K = T_f - 276.15\,K$$

$$T_f = 596._{30}\,K\ or\ \mathbf{323\,^\circ C}$$

6.39 Since the bolts have the same mass, and one must cool as the other heats, the intuitive answer is $(T_1 + T_2)/2$ = 77.5 °C. This question can also be solved using an algebraic expression. To find T_{final}, let m_{Fe} represent the mass of the iron bolts, and use equation 6.7 to write expressions for both the hot and cold bolts. Convert °C temperatures (100.°C, 55°C) to Kelvin temperatures (373 K, 328 K).

$q_{hot} = c_{Fe}\, m_{Fe}\, (T_{final} - 373\text{ K})$ $\qquad$ $q_{cold} = c_{Fe}\, m_{Fe}\, (T_{final} - 328\text{ K})$

Adjust the signs of q to indicate heat is lost ($-q_{hot}$) and gained ($+q_{cold}$) and set them equal to each other.

$$-q_{hot} = +q_{cold}$$
$$-[\cancel{c_{Fe}}\,\cancel{m_{Fe}}\,(T_{final} - 373\text{ K})] = \cancel{c_{Fe}}\,\cancel{m_{Fe}}\,(T_{final} - 328\text{ K})$$
$$-T_{final} + 373\text{ K} = T_{final} - 328\text{ K}$$
$$2T_{final} = 701\text{ K}$$
$$T_{final} = \mathbf{351\text{ K or }78^\circ C}$$

6.41 Use the same procedure in 6.39, but mass is now a factor because the two water samples are not identical. Both volumes are converted to mass using the density (e.g. 85 mL x 1.00 g/mL = 85 g). Convert °C temperatures (22°C, 82°C) to Kelvin temperatures (295 K, 355 K).

$$-q_{hot} = +q_{cold}$$
$$-[4.184\text{ J/(g-K)} \times 85\text{ g} \times (T_{final} - 355\text{ K})] = 4.184\text{ J/(g-K)} \times 165\text{ g} \times (T_{final} - 295\text{ K})$$
$$-355.6\text{ J/K} \times (T_{final} - 355\text{ K}) = 690.4\text{ J/K} \times (T_{final} - 295\text{ K})$$
$$-355.6\, T_{final} + 126238\text{ K} = 690.4\, T_{final} - 203668\text{ K}$$
$$1046.0\, T_{final} = 329906 \text{ (this number only known to 3 sig figs)}$$
$$T_{final} = 315\text{ K or } \mathbf{42^\circ C}$$

6.43 Heat gained by water plus heat lost by copper tubing must equal zero, so $q_{water} = -q_{copper}$. However, some heat will be lost to the insulated container. Thus, the relationship is

$$q_{water} + q_{vessel} = -q_{copper}$$

$$\left(\frac{4.184\, J}{g\cdot K}\right)(59.8\, g)(T_f - 297.95\, K) + (10.0\, J/K)(T_f - 297.95\, K) = -\left(\frac{0.387\, J}{g\cdot K}\right)(505\, g)(T_f - 373.05\, K)$$

$$4.55638 \times 10^2\, T_f = 1.50434 \times 10^5\, K$$

$$T_f = 330.2\, K \text{ or } \mathbf{57.0^\circ C}$$

6.48 The reaction has a positive ΔH_{rxn}, because this reaction requires the input of energy to break the O – O bond: $O_2(g) + \text{energy} \rightarrow 2O(g)$.

6.49 As a substance changes from the gaseous state to the liquid state, energy is released so ΔH would be negative for the condensation of 1 mol of water. The value of ΔH for the vaporization of 2 mol of water would be twice the value of ΔH for the condensation of 1 mol of water vapor but would have an opposite sign (+ΔH).

6.50 a) This reaction is exothermic because ΔH is negative.

b) Because ΔH is a state function, the total energy required for the reverse reaction, regardless of how the change occurs, is the same <u>magnitude</u> but different <u>sign</u> of the forward reaction. Therefore, ΔH = **+20.2 kJ**.

c) The ΔH_{rxn} is specific for the reaction as written, meaning that 20.2 kJ is released when 1/8 of a mole of sulfur reacts. In this case, 3.2 moles of sulfur react and we therefore expect that much more energy will be released.

$$(3.2\,mol\,S_8)\left(\frac{-20.2\,kJ}{\frac{1}{8}\,mol\,S_8}\right) = -517\,kJ = \mathbf{-5.2x10^2\,kJ}$$

d) The mass of S_8 requires conversion to moles and then a calculation identical to c) can be performed.

$$(20.0g\,S_8)\left(\frac{1\,mol\,S_8}{256.6}\right)\left(\frac{-20.2\,kJ}{\frac{1}{8}\,mol\,S_8}\right) = \mathbf{-12.6kJ}$$

6.52 a) $\frac{1}{2}N_2(g) + \frac{1}{2}O_2(g) \rightarrow NO(g)$ $\Delta H_{rxn} = 90.29$ kJ

b) Decomposition: $NO(g) \rightarrow \frac{1}{2}N_2(g) + \frac{1}{2}O_2(g)$

$$1.50\,g\,NO\left(\frac{1\,mol\,NO}{30.01\,g}\right)\left(\frac{-90.29\,kJ}{mol\,NO}\right) = \mathbf{-4.51kJ}$$

6.54 For the reaction written, 2 moles of H_2O_2 release 196.1 kJ of energy upon decomposition. The problem specifies $(732\,kg\,H_2O_2)\left(\frac{1000\,g}{1\,kg}\right)\left(\frac{1\,mol\,H_2O_2}{34.016\,g}\right) = 2.15 \times 10^4$ mol H_2O_2.

Therefore, the heat released by 2.15×10^4 moles is $(2.15x10^4\,mol\,H_2O_2)\left(\frac{-196.1\,kJ}{2\,mol\,H_2O_2}\right) =$

$-2.11x10^6$ kJ.

6.58 a) $C_2H_4(g) + 3O_2(g) \rightarrow 2CO_2(g) + 2H_2O(g)$ $\Delta H_{rxn} = -1411$ kJ

b) $70.0\,kJ\left(\frac{1\,mol\,C_2H_4}{1411\,kJ}\right)\left(\frac{28.05\,g\,C_2H_4}{mol}\right) = \mathbf{1.39\,g\,C_2H_4}$

6.62 Two chemical equations can be written based on the description given:

$C(s) + O_2(g) \rightarrow CO_2(g)$ ΔH_1 (1)

$CO(g) + \frac{1}{2}O_2(g) \rightarrow CO_2(g)$ ΔH_2 (2)

The second reaction can be reversed, its ΔH sign changed, and appropriate coefficients can be multiplied to one or both of the equations to allow addition of the two reactions. In this case, no coefficients are necessary because the CO_2 cancels.

$C(s) + O_2(g) \rightarrow \cancel{CO_2(g)}$ ΔH_1

$\cancel{CO_2(g)} \rightarrow CO(g) + 1/2O_2(g)$ $-\Delta H_2$

$C(s) + \frac{1}{2}O_2(g) \rightarrow CO(g)$ $\Delta H_{rxn} = \Delta H_1 + (-\Delta H_2)$

How are the ΔH's for each reaction determined? Equation (2) was used in Sample Problem 6.7, with $\Delta H_2 = -283.0$ kJ. The ΔH_1 can be found by using the heats of formation in Table 6.5: $\Delta H_1 = \Delta H_f(CO_2) = -393.5$ kJ/mol – (0 + 0) = -393.5 kJ/mol. Assume one mole for equation (1) and $\Delta H_{rxn} = \Delta H_1 - \Delta H_2 = -393.5$ kJ – (-283.0 kJ) = **-110.5 kJ**. Does this compare with the Table 6.5 value for $\Delta H_f(CO)$?

6.63 To obtain the overall reaction, add the first reaction to the reverse of the second.

$Ca(s) + {}^{1}/_{2}O_2(g) \rightarrow CaO(s)$	ΔH = -635.1 kJ
$CaO(s) + CO_2(g) \rightarrow CaCO_3(s)$	ΔH = -178.3 kJ
$Ca(s) + {}^{1}/_{2}O_2(g) + CO_2(g) \rightarrow CaCO_3(s)$	ΔH_{rxn} = **-813.4 kJ**

6.65 Vaporization is the change in state from a liquid to a gas. The two equations describing these chemical reactions can be combined to yield the equation for vaporization.

$H_2(g) + {}^{1}/_{2}O_2(g) \rightarrow H_2O(g)$	ΔH = -241.8 kJ
$H_2O(l) \rightarrow H_2(g) + {}^{1}/_{2}O_2(g)$	ΔH = +285.8 kJ
$H_2O(l) \rightarrow H_2O(g)$	ΔH_{rxn} = **44.0 kJ**

6.67 Equation 3 is: $N_2(g) + 2O_2(g) \rightarrow 2NO_2(g)$ ΔH_{rxn} = 66.4 kJ
In figure P6.67 A represents reaction 1 with a larger amount of energy absorbed, B represents reaction 2 with a smaller amount of energy released and C represents reaction 3 as the sum of A and B.

6.70 The standard heat of reaction, ΔH°_{rxn}, is the enthalpy change for <u>any</u> reaction where all substances are in their standard states. The standard heat of formation, ΔH°_f, is the enthalpy change that accompanies the <u>formation of one mole</u> of a compound in its standard state from elements in their standard states. Standard state is 1 atm for gases, 1M for solutes, and pure state for liquids and solids. Standard state does not include a specific temperature, but a temperature must be specified in a table of standard values.

6.72 a) ${}^{1}/_{2}Cl_2(g) + Na(s) \rightarrow NaCl(s)$ The element chlorine occurs as Cl_2, not Cl.
b) $H_2(g) + {}^{1}/_{2}O_2(g) \rightarrow H_2O(l)$ The element hydrogen exists as H_2, not H and the formation of water is written with water as the product in the liquid state.
c) No changes.

6.73 a) $Ca(s) + Cl_2(g) \rightarrow CaCl_2(s)$
b) $Na(s) + {}^{1}/_{2}H_2(g) + C(\text{graphite}) + {}^{3}/_{2}O_2(g) \rightarrow NaHCO_3(s)$
c) $C(\text{graphite}) + 2Cl_2(g) \rightarrow CCl_4(l)$
d) ${}^{1}/_{2}H_2(g) + {}^{1}/_{2}N_2(g) + {}^{3}/_{2}O_2(g) \rightarrow HNO_3(l)$

6.75 The enthalpy change of a reaction is the sum of the ΔH_f of the products minus the sum of ΔH_f of the reactants. Since the ΔH_f values (Appendix B) are reported as energy per one mole, use the appropriate coefficient to reflect higher number of moles.

$$\Delta H^\circ_{rxn} = \Sigma[\Delta H^\circ_f(\text{products})] - \Sigma[\Delta H^\circ_f(\text{reactants})]$$

a) $\Delta H^\circ_{rxn} = 2\Delta H^\circ_f[SO_2(g)] + 2\Delta H^\circ_f[H_2O(g)] - 2\Delta H^\circ_f[H_2S(g)] - 3\Delta H^\circ_f[O_2(g)]$
$= 2\text{ mol}(-296.8\text{ kJ/mol}) + 2\text{ mol}(-241.8\text{ kJ/mol}) - 2\text{ mol}(-20.2\text{ kJ/mol}) - 3(0)$
= **-1036.8 kJ**

b) The balanced equation is $CH_4(g) + 4Cl_2(g) \rightarrow CCl_4(l) + 4HCl(g)$
$\Delta H^\circ_{rxn} = 1\text{ mol}(-139\text{ kJ/mol}) + 4\text{ mol}(-92.31\text{ kJ/mol}) - 1\text{ mol}(-74.87\text{ kJ/mol}) - 4\text{ mol}(0)$
= **-433 kJ**

6.77 Use Hess's Law to solve for heat of formation of copper(II) oxide.

$$\Delta H_{rxn} = \left(2\,x\,\Delta H_f^o\,CuO(s)\right) - \left(\Delta H_f^o\,Cu_2O(s)\right) - \left(\tfrac{1}{2}\,x\,\Delta H_f^o\,O_2(g)\right)$$

$$-146.0\,kJ = 2\left(\Delta H_f^o\,CuO(s)\right) - \left(-168.6\,kJ\right) - (0)$$

$$\Delta H_f^o = \mathbf{-157.3\ kJ/mol\ CuO(s)}$$

6.80 a) $\Delta H^\circ_{rxn} = \Delta H^\circ_f[Pb(s)] + \Delta H^\circ_f[PbO_2(s)] + 2\Delta H^\circ_f[H_2SO_4(l)] - 2\Delta H^\circ_f[PbSO_4(s)] - 2\Delta H^\circ_f[H_2O(l)]$

= 0 + 1 mol(-276.6 kJ/mol) + 2 mol(-814.0 kJ/mol) – 2 mol(-918.4 kJ/mol) –
2 mol(-285.85 kJ/mol)

= 503.9 kJ

b) Reverse the first equation (changing the sign of ΔH°_{rxn}) and multiply the coefficients (and ΔH°_{rxn}) of the second reaction by 2.

$2PbSO_4(s) \rightarrow Pb(s) + PbO_2(s) + \cancel{2SO_3(g)}$	$\Delta H^\circ_{rxn} = 768$ kJ
$\cancel{2SO_3(g)} + 2H_2O(l) \rightarrow 2H_2SO_4(l)$	$\Delta H^\circ_{rxn} = -264$ kJ
$2PbSO_4(s) + 2H_2O(l) \rightarrow Pb(s) + PbO_2(s) + 2H_2SO_4(l)$	ΔH°_{rxn} = **504 kJ**

6.81 a) $C_{18}H_{36}O_2(s) + 26O_2(g) \rightarrow 18CO_2(g) + 18H_2O(g)$

b) $\Delta H_{rxn} = 18[\Delta H_f^\circ\,CO_2(g)] + 18[\Delta H_f^\circ\,H_2O(g)] - [\Delta H_f^\circ\,C_{18}H_{36}O_2(s)] - 26[\Delta H_f^\circ\,O_2(g)]$

= 18(-393.5 kJ) + 18(-241.826 kJ) – (-948 kJ) – 26(0 kJ)

= -10488 kJ/mol

c) $$1.00\,g\,C_{18}H_{36}O_2\left(\frac{1\,mol\,C_{18}H_{36}O_2}{284.47\,g\,C_{18}H_{36}O_2}\right)\left(\frac{-10488\,kJ}{mol\,C_{18}H_{36}O_2}\right) = \mathbf{-36.9\,kJ}$$

$$-36.9\,kJ\left(\frac{1\,kcal}{4.184\,kJ}\right) = \mathbf{-8.81\,kcal}$$

In other words, stearic acid releases 36.9 kJ/g or 8.81 kcal/g of energy when burned.

d) Yes, according to the calculations, 11 grams of fat would produce (11.0 * 8.81) = 96.9 kcal.

6.83 a) A first read of this problem suggests there is insufficient information to solve the problem. Upon more careful reading, you find that the question asks volumes <u>for each mole of helium</u>. According to Avogadro's Law, one mole of any gas occupies 22.4 L at STP. Although the conditions are not at standard temperature, you can convert the volume to the conditions stated.

T_1 = 273 K (standard temperature) T_2 = 273 + 15 = 288 K
or T_2 = 273 + 30 = 303 K

$$volume\ at\ 15^\circ C = \left(\frac{22.4\,L}{mol}\right)\left(\frac{288\,K}{273\,K}\right) = \mathbf{23.6\,L\ per\ mol\ He}$$

$$volume\ at\ 30^\circ C = \left(\frac{22.4\,L}{mol}\right)\left(\frac{303\,K}{273\,K}\right) = \mathbf{24.9\,L\ per\ mol\ He}$$

b) Internal energy is the sum of the potential and kinetic energies of each He atom in the system (the balloon). The energy of one mole of helium atoms can be described as a function of temperature, E = 3/2nRT, where n = 1 mole. Therefore, the internal energy at 15°C and 30°C can be calculated. The inside back cover lists values of R with different units.

$$E(288\ K) = \frac{3}{2}(1\,mol)\left(8.314\frac{J}{mol \cdot K}\right)(288\ K) = 3.59x10^3\ J = 3.59\ kJ$$

$$E(303\ K) = \frac{3}{2}(1\,mol)\left(8.314\frac{J}{mol \cdot K}\right)(303\ K) = 3.78x10^3\ J = 3.78\ kJ$$

$$\Delta E = 3.78\ kJ - 3.59\ kJ = 0.19\ kJ = 190\ J$$

$$or\ \Delta E = \frac{3}{2}nR\Delta T = \frac{3}{2}(1\,mol)\left(8.314\frac{J}{mol \cdot K}\right)(303 - 288\,K) = \mathbf{187\ J}$$

c) When the balloon expands as temperature rises, the balloon performs PV work. However, the problem specifies that pressure remains constant, so work done <u>on</u> the surroundings by the balloon is defined by equation 6.4: w = -PΔV. When pressure and volume are multiplied together, the unit is L-atm. Since we would like to express work in joules, we can create a conversion factor between L-atm and J by comparing two gas constants.

$$0.08206\frac{atm \cdot L}{mol \cdot K} = 8.314\frac{J}{mol \cdot K}$$

$$w = -(1\,atm)(24.9 - 23.6\ L)\left(\frac{8.314\frac{J}{mol \cdot K}}{0.08206\frac{atm \cdot L}{mol \cdot K}}\right) = \mathbf{-1.3x10^2\ J}$$

d) See the derivation of equation 6.6

$$q_p = \Delta E + P\Delta V = 187\ J + 130\ J = \mathbf{320\ J}$$

e) ΔH = q_P = **320 J**.

f) When a process occurs at constant pressure, the change in heat energy of the system can be described by a state function called enthalpy. The change in enthalpy equals the heat (q) lost at constant pressure: ΔH = ΔE + PΔV = ΔE – w = (q + w) – w = q_p.

6.84 a) Write each equation as a total ionic equation.

$K^+(aq) + OH^-(aq) + H^+(aq) + NO_3^-(aq) \rightarrow K^+(aq) + NO_3^-(aq) + H_2O(l)$

The terms [ΔH_f^o $K^+(aq)$] and [ΔH_f^o $NO_3^-(aq)$] cancel out, to yield this equation:

ΔH° = [ΔH_f^o $H_2O(l)$] - [ΔH_f^o $OH^-(aq)$] - [ΔH_f^o $H^+(aq)$]

= (-285.840 kJ) – (-229.94 kJ) - 0

= **-55.90 kJ**

$Na^+(aq) + OH^-(aq) + H^+(aq) + Cl^-(aq) \rightarrow Na^+(aq) + Cl^-(aq) + H_2O(l)$

The terms [ΔH_f^o $Na^+(aq)$] and [ΔH_f^o $Cl^-(aq)$] cancel out, to yield the same equation as above, so ΔH° again equals **-55.90 kJ**.

You can confirm these results by calculating ΔH° using the ΔH_f^o's of the undissociated species.

b) The net ionic reaction in both cases is the same, thus the heat of reaction is the same. The heat of formation for the spectator ions is added to the total as a product and is subtracted as a reactant, thus spectator ions have no impact on the heat of the reaction.

6.86 a) From the information given, the reaction is written as follows:

$SiO_2(s) + 3C(s) \rightarrow SiC(s) + 2CO(g)$ ΔH_{rxn}^o = 624.7 kJ/mol SiC (sign is positive because reaction is endothermic)

Use Hess's Law and Appendix B to solve for [ΔH_f° SiC(s)].

ΔH_{rxn}° = [ΔH_f° SiC(s)] + 2[ΔH_f° CO(g)] - [ΔH_f° SiO_2(s)] - 3[ΔH_f° C(s)]

624.7 kJ/mol = [ΔH_f° SiC(s)] + 2(-110.5 kJ/mol) – (-910.0 kJ/mol) – 0

624.7 kJ/mol = [ΔH_f° SiC(s)] + 689.9 kJ/mol

[ΔH_f° SiC(s)] = **-65.2 kJ/mol**

b) The problem states that 624.7 kJ is absorbed *per mole* of SiC formed. Convert kJ/mol to kJ/kg using the molar mass of SiC (40.10 g/mol).

$$\left(\frac{kJ}{kg\ SiC}\right) = \left(\frac{624.7\ kJ}{mol\ SiC}\right)\left(\frac{1\ mol\ SiC}{40.10\ g}\right)\left(\frac{1000\ g}{kg}\right) = \mathbf{1.558x10^4\ kJ/kg\ SiC}$$

6.90 The ΔH°_{rxn}, which in this case is ΔH°_{comb}, is the combination of the heats of formation of the reactants and products. We can write a chemical equation that shows the combustion of oleic acid, assuming that the combustion is complete and yields only CO_2(g) and H_2O(g):

$$C_{18}H_{34}O_2(s) + {}^{53}/_2 O_2(g) \rightarrow 18CO_2(g) + 17H_2O(g)$$

ΔH°_{comb} = 18ΔH°_f[CO_2(g)] + 17ΔH°_f[H_2O(g)] - ΔH°_f[$C_{18}H_{34}O_2$(s)] - $^{53}/_2$ ΔH°_f[O_2(g)]

-1.11x10^4 kJ = 18 mol(–393.5 kJ/mol) + 17 mol(-241.8 kJ/mol) –(1 mol)ΔH°_f[$C_{18}H_{34}O_2$(s)] – 0

(1 mol)ΔH°_f[$C_{18}H_{34}O_2$(s)] = 1.11 x 10^4 kJ - 7.083 x 10^3 kJ - 4.111 x 10^3 kJ

ΔH°_f[$C_{18}H_{34}O_2$(s)] = **-94 kJ/mol**

6.93 a) $$\frac{J}{Btu} = \left(\frac{4.184\ J}{cal}\right)\left(\frac{1\ cal}{g\cdot{}^\circ C}\right)\left(\frac{453.6\ g}{1\ lb}\right)\left(\frac{1.0^\circ C}{1.8^\circ F}\right) = 1054\frac{J}{lb\cdot{}^\circ F} = \mathbf{1054\frac{J}{Btu}}$$

b) $$\left(\frac{100{,}000\ Btu}{1.00\ therm}\right)\left(\frac{1054\ J}{1.00\ Btu}\right) = \mathbf{1.05\times10^8\ J/therm}$$

c) CH_4(g) + 2O_2(g) → CO_2(g) + 2H_2O(g)

ΔH° = (-393.5 kJ) + 2(-241.8 kJ) – (-74.9 kJ) – 2(0 kJ) = -802.2 kJ

$$1.00\ therm\left(\frac{1.05x10^8\ J}{1.00\ therm}\right)\left(\frac{1\ kJ}{1000\ J}\right)\left(\frac{1\ mol\ CH_4}{802.2\ kJ}\right) = \mathbf{131\ mol\ CH_4}$$

d) $$cost = \left(\frac{\$0.46}{1\ therm}\right)\left(\frac{1.00\ therm}{131\ mol\ CH_4}\right) = \mathbf{\$0.0035}$$

e) $$308\ gal\left(\frac{3.78\ L}{1\ gal}\right)\left(\frac{1000\ mL}{1\ L}\right)\left(\frac{1.00\ g}{mL}\right)\left(\frac{4.184\ J}{g\cdot K}\right)(25\ K)\left(\frac{1.00\ therm}{1.05x10^8\ J}\right)\left(\frac{\$0.46}{therm}\right) = \mathbf{\$0.53}$$

6.95 Chemical equations can be written that describe the three processes. Assume one mole of each substance of interest so that units are expressed as kJ.

C(graphite) + 2H_2(g) → CH_4(g)	$\Delta H^\circ_f = \Delta H^\circ_{rxn}$ = -74.9 kJ	(1)
CH_4(g) → C(g) + 4H(g)	$\Delta H^\circ_{atom} = \Delta H^\circ_{rxn}$ = 1660 kJ	(2)
H_2(g) → 2H(g)	$\Delta H^\circ_{atom} = \Delta H^\circ_{rxn}$ = 432 kJ	(3)

The third equation is reversed and its coefficients are multiplied by 2 to add the three equations.

$C(graphite) + \sout{2H_2(g)} \rightarrow \sout{CH_4(g)}$ $\Delta H^\circ_{rxn} = -74.9$ kJ

$\sout{CH_4(g)} \rightarrow C(g) + \sout{4H(g)}$ $\Delta H^\circ_{rxn} = 1660$ kJ

$\sout{4H(g)} \rightarrow \sout{2H_2(g)}$ $\Delta H^\circ_{rxn} = -864$ kJ

$C(graphite) \rightarrow C(g)$ $\Delta H^\circ_{rxn} = \Delta H^\circ_{atom} =$ **721 kJ** per one mole C(graphite)

6.99 The ΔH°_{rxn} value is based on the stoichiometry of the equation as written. Therefore, the information can be interpreted as either meaning

1) 7.0×10^4 kJ are released when 2 moles of tristearin are consumed, or
2) 7.0×10^4 kJ are released when 163 moles of O_2 are consumed, or
3) 7.0×10^4 kJ are released when 114 mole of CO_2 are formed, etc.

a) $$heat\ released = \left(\frac{7.0x10^4\ kJ}{163\ mol\ O_2}\right) = \mathbf{4.3x10^2\ kJ / mol\ O_2}$$

The value can also be reported as $\Delta q = -4.3 \times 10^2$ kJ/mol O_2. Recall that the negative sign only indicates that heat is released.

b) $$heat\ released = \left(\frac{7.0x10^4\ kJ}{114\ mol\ CO_2}\right) = \mathbf{6.1x10^2\ kJ / mol\ CO_2}$$

c) $$heat\ released = \left(\frac{7.0x10^4\ kJ}{2\ mol\ C_{57}H_{110}O_6}\right)\left(\frac{1\ mol\ C_{57}H_{110}O_6}{891.45\ g}\right) = \mathbf{39\ kJ / g\ C_{57}H_{110}O_6}$$

d) Use the ideal gas law, PV = nRT, to solve for n, the number of moles of O_2. Then use the $O_2/C_{57}H_{110}O_6$ molar ratio and $C_{57}H_{110}O_6$ molar mass to find the mass of tristearin.

$$n = \frac{PV}{RT} = \frac{\left(\frac{755\ torr}{760\ torr / 1\ atm}\right)(325\ L)}{(0.08206\ L\cdot atm / mol\cdot K)(37 + 273K)} = 12.6_{92}\ mol\ O_2$$

$$mass\ tristearin = (12.6_{92}\ mol\ O_2)\left(\frac{2\ mol\ C_{57}H_{110}O_6}{163\ mol\ O_2}\right)\left(\frac{891.45\ g}{mol\ C_{57}H_{110}O_6}\right) = \mathbf{139\ g\ C_{57}H_{110}O_6}$$

6.102 Assume that coffee has the same specific heat capacity and density as water. The heat gained by the coffee (H_2O) equals the heat lost by the iron:

$$q_{H2O} = -q_{Fe}$$

$$0.50\ L\left(\frac{1000\ mL}{1\ L}\right)\left(\frac{1.00\ g}{mL}\right)\left(\frac{4.184\ J}{g\cdot K}\right)(T_f - 291\ K) = -502\ g\left(\frac{0.450\ J}{g\cdot K}\right)(T_f - 1073\ K)$$

$T_f = 367\ K$ or **94° C**

6.104 Combustion reactions can be written and ΔH°_{rxn} can be calculated for each hydrocarbon.

$CH_4(g) + 2O_2(g) \rightarrow CO_2(g) + 2H_2O(g)$

$\Delta H^\circ_{rxn} = (-393.5) + 2(-241.8) - (-74.8) = -802.3$ kJ per mol $CH_4(g)$

$C_2H_4(g) + 3O_2(g) \rightarrow 2CO_2(g) + 2H_2O(g)$

$\Delta H^\circ_{rxn} = 2(-393.5) + 2(-241.8) - (52.30) = -1322.9$ kJ per mol $C_2H_4(g)$

$C_2H_6(g) + {}^7/_2O_2(g) \rightarrow 2CO_2(g) + 3H_2O(g)$

$\Delta H^\circ_{rxn} = 2(-393.5) + 3(-241.8) - (-84.68) = -1427.7$ kJ per mol $C_2H_6(g)$

a) A negative enthalpy change denotes an exothermic reaction, so the combustion of $C_2H_6(g)$ yields the most heat, followed by $C_2H_4(g)$ and $CH_4(g)$; **III > II > I**.

b) Use molecular weights to convert kJ/mol to kJ/g.

$$kJ\ per\ gram\ CH_4 = (-802.3\ kJ/mol)\left(\frac{1\ mol\ CH_4}{16.04\ g}\right) = -50.02\ kJ/g$$

$$kJ\ per\ gram\ C_2H_4 = (-1322.9\ kJ/mol)\left(\frac{1\ mol\ C_2H_4}{28.05\ g}\right) = -47.16\ kJ/g$$

$$kJ\ per\ gram\ C_2H_6 = (-1427.7\ kJ/mol)\left(\frac{1\ mol\ C_2H_6}{30.07\ g}\right) = -47.48\ kJ/g$$

On a per mass basis, CH_4 yields more heat followed by $C_2H_6(g)$ and $C_2H_4(g)$; **I > III > II.**

6.106 a) The heat of reaction is calculated from the heats of formation found in Appendix B. The ΔH°_f's for all of the species, except $SiCl_4$, are found in Appendix B. Use reaction 3, with its given ΔH°_{rxn}, to find $\Delta H^\circ_f[SiCl_4(g)]$.

$\Delta H^\circ_{rxn} = \Delta H^\circ_f[SiO_2(s)] + 4\Delta H^\circ_f[HCl(g)] - \Delta H^\circ_f[SiCl_4(g)] - 2\Delta H^\circ_f[H_2O(g)]$

$-139.5\ kJ = (-910.9\ kJ) + 4(-92.31\ kJ) - \Delta H^\circ_f[SiCl_4(g)] - 2(-241.826\ kJ)$

$\Delta H^\circ_f[SiCl_4(g)] = -656.9_{88}$ kJ/mol

The heats of reaction for the first two steps can now be calculated.

$\Delta H^\circ_{rxn1} = \Delta H^\circ_f[SiCl_4(g)] - \Delta H^\circ_f[Si(s)] + 2\Delta H^\circ_f[Cl_2(g)]$

$\Delta H^\circ_{rxn1} = -656.9_{88}\ kJ - 0 - 2(0) = \mathbf{-657.0\ kJ}$

$\Delta H^\circ_{rxn2} = \Delta H^\circ_f[SiCl_4(g)] + 2\Delta H^\circ_f[CO(g)] - \Delta H^\circ_f[SiO_2(g)] - 2\Delta H^\circ_f[C(gr)] - 2\Delta H^\circ_f[Cl_2(g)]$

$\Delta H^\circ_{rxn2} = -656.9_{88}\ kJ + 2(-110.5\ kJ) - (-910.9\ kJ) - 2(0) - 2(0) = \mathbf{32.9\ kJ}$

b) Adding reactions 2 and 3 yield: $2C(gr) + 2Cl_2(g) + 2H_2O(g) \rightarrow 2CO(g) + 4HCl(g)$

$\Delta H^\circ_{rxn2+3} = 32.9\ kJ + (-139.5\ kJ) = \mathbf{-106.6\ kJ}$

Confirm this result by calculating ΔH°_{rxn} using Appendix B values.

$\Delta H^\circ_{rxn2+3} = 2\Delta H^\circ_f[CO(g)] + 4\Delta H^\circ_f[HCl(g)] - 2\Delta H^\circ_f[C(gr)] - 2\Delta H^\circ_f[Cl_2(g)] - 2\Delta H^\circ_f[H_2O(g)]$

$\Delta H^\circ_{rxn2+3} = 2(-110.5\ kJ) + 4(-92.31\ kJ) - 2(0) - 2(0) - 2(-241.826\ kJ) = \mathbf{-106.6\ kJ}$

6.108 a) $-P\Delta V = -nR\Delta T = (1\ mol) \times (8.31451\ J/mol \cdot K) \times (819\ K) = \mathbf{6.81 \times 10^3\ J}$

b) $\Delta T = (6.81 \times 10^3\ J/1.00\ J/g \cdot K) \times (1\ mol/28.02\ g) = 243\ K = \mathbf{243^\circ C}$

6.110 Only reaction 3 contains $N_2O_4(g)$, and only reaction 1 contains $N_2O_3(g)$, so we can use those reactions as a starting point. N_2O_5 appears in both reactions 2 and 5, but note the physical states present: solid and gas. As a rough start, adding reactions 1, 3 and 5 yield the desired reactants and products, with some undesired intermediates:

(reverse 1)	$N_2O_3(g) \rightarrow NO(g) + NO_2(g)$	$\Delta H^\circ_{rxn1} = 39.8$ kJ
(twice 3)	$4NO_2(g) \rightarrow 2N_2O_4(g)$	$\Delta H^\circ_{rxn3} = -114.4$ kJ
(reverse 5)	$N_2O_5(s) \rightarrow N_2O_5(g)$	$\Delta H^\circ_{rxn5} = 54.1$ kJ

To cancel out the $N_2O_5(g)$ intermediate, reverse equation 2. This also cancels out some of the undesired $NO_2(g)$ but adds $NO(g)$ and $O_2(g)$. Finally, add equation 4 to remove those intermediates:

(reverse 1)	$N_2O_3(g) \rightarrow \cancel{NO(g)} + \cancel{NO_2(g)}$	$\Delta H^\circ_{rxn1} = 39.8$ kJ
(twice 3)	$\cancel{4NO_2(g)} \rightarrow 2N_2O_4(g)$	$\Delta H^\circ_{rxn3} = -114.4$ kJ
(reverse 5)	$N_2O_5(s) \rightarrow \cancel{N_2O_5(g)}$	$\Delta H^\circ_{rxn5} = 54.1$ kJ
(reverse 2)	$\cancel{N_2O_5(g)} \rightarrow \cancel{NO(g)} + \cancel{NO_2(g)} + \cancel{O_2(g)}$	$\Delta H^\circ_{rxn2} = 112.5$ kJ
(4)	$\cancel{2NO(g)} + \cancel{O_2(g)} \rightarrow \cancel{2NO_2(g)}$	$\Delta H^\circ_{rxn4} = -114.2$ kJ
	$N_2O_3(g) + N_2O_5(s) \rightarrow 2N_2O_4(g)$	ΔH°_{rxn} = **-22.2 kJ**

6.112 a) $\Delta H^\circ_{rxn} = \Delta H^\circ_f[N_2H_4(aq)] + \Delta H^\circ_f[NaCl(aq)] + \Delta H^\circ_f[H_2O(l)] - 2\Delta H^\circ_f[NH_3(aq)] - \Delta H^\circ_f[NaOCl(aq)]$

Note that the Appendix B value for N_2H_4 is for the liquid state, so this term must be calculated. In addition, Appendix B does not list a value for NaCl(aq), so this term must be broken down into $\Delta H^\circ_f[Na^+(aq)]$ and $\Delta H^\circ_f[Cl^-(aq)]$.

$-151 \text{ kJ} = \Delta H^\circ_f[N_2H_4(aq)] + (-239.66 \text{ kJ}) + (-167.46 \text{ kJ}) + (-285.840 \text{ kJ}) - 2(-80.83 \text{ kJ}) - (-346 \text{ kJ})$

$\Delta H^\circ_f[N_2H_4(aq)]$ = **34 kJ/mol**

b) Determine the moles of O_2 present, then multiply by the ΔH°_{rxn} for the first reaction.

Moles of $O_2 = (5.00 \times 10^3 \text{ L})(2.50 \times 10^{-4} \text{ mol/L}) = 1.25 \text{ mol } O_2$

$\Delta H^\circ_{rxn} = \Delta H^\circ_f[N_2(g)] + 2\Delta H^\circ_f[H_2O(l)] - \Delta H^\circ_f[N_2H_4(aq)] - \Delta H^\circ_f[O_2(g)]$

$\Delta H^\circ_{rxn} = 0 + 2(-285.840 \text{ kJ}) - (34._{30} \text{ kJ}) - 0 = -605._{98} \text{ kJ}$

$$heat\ of\ reaction = (1.25\, mol\, O_2)\left(\frac{-605._{98}\ kJ}{1\, mol\, O_2}\right) = \mathbf{-757\ kJ}$$

6.114 a) The balanced chemical equation for this reaction is

$$CH_4(g) + 2O_2(g) \rightarrow CO_2(g) + 2H_2O(g)$$

Instead of burning one mole of methane, $(25.0\, g\, CH_4)\left(\frac{1\, mol\, CH_4}{16.04\, g}\right) = 1.56$ mol of methane are burned. Therefore, 1.56 mol of CO_2 and 3.12 mol of H_2O form upon combustion of 25.0 g of methane.

$\Delta H^\circ_{rxn} = 1.56 \text{ mol}(-393.5 \text{ kJ/mol}) + 3.12 \text{ mol}(-241.8 \text{ kJ/mol}) - 1.56 \text{ mol}(-74.8 \text{ kJ/mol})$

$\Delta H^\circ_{rxn} = -1.25 \times 10^3 \text{ kJ}$

Therefore, 25.0 g of CH_4 releases **1.25×10^3 kJ** when burned in excess O_2.

(In problem 6.101(a), you determined that the heat evolved upon combustion of one mole of methane was -802.3 kJ /mol $CH_4(g)$. Is this consistent with your calculation above?)

b) The heat released by the reaction is "stored" in the gaseous molecules by virtue of their specific heat capacities, c, using the equation $\Delta H = mc\Delta T$. The problem specifies heat capacities on a molar basis, so we modify the equation to use moles, instead of mass.

The gases that remain at the end of the reaction are CO_2 and H_2O. All of the methane and oxygen molecules were consumed. However, the oxygen was added as a component of air which is 79% N_2 and 21% O_2, and there is leftover N_2.

Moles of $CO_2(g) = 1.56$ mol

Moles of $H_2O(g) = 3.12$ mol

Moles of $N_2(g) = (3.12 \text{ mol } O_2 \text{ used})(0.79 \text{ mol } N_2 / 0.21 \text{ mol } O_2) = 11.7$ mol

The overall equation is

$\Delta H_{comb} = n_{CO2}c_{CO2}(T_f - 0)°C + n_{H2O}c_{H2O}(T_f - 0)°C + n_{N2}c_{N2}(T_f - 0)°C$

The specific heat capacities are given in units, J/mol-K, which is equivalent to J/mol-°C because the Kelvin degree and Celsius degree are the same size. For simplicity in calculation to allow T_{init} = 0 (instead of T_{init} = 273), use the J/mol-°C unit.

$$1.25x10^6\ J = (1.56\,mol)\left(57.2\frac{J}{mol\cdot {}^oC}\right)(T_f) + (3.12\,mol)\left(36.0\frac{J}{mol\cdot {}^oC}\right)(T_f) + (11.7\,mol)\left(30.5\frac{J}{mol\cdot {}^oC}\right)$$

$$1.25x10^6\ J = (89.232\ J/{}^\circ C)(T_f) + (112.32\ J/{}^\circ C)(T_f) + (356.85\ J/{}^\circ C)(T_f)$$

$$1.25x10^6\ J = (558.4\ J/{}^\circ C)(T_f)$$

T_f = **2.24x10³ °C**

CHAPTER 7

QUANTUM THEORY AND ATOMIC STRUCTURE

FOLLOW-UP PROBLEMS

7.1 Plan: Given the frequency of the light use the equation c = $\nu\lambda$ to solve for wavelength.

Solution: $\lambda = \frac{c}{\nu} = \frac{3.00x10^{8}\ m/s}{7.23x10^{14}\ s^{-1}} = 4.15x10^{-7}\ m$

In nanometers: $4.15\text{x}10^{-7}\ \text{m}\left(\frac{1\text{x}10^{9}\ nm}{1\,\text{m}}\right) =$ **415 nm**

In Angstroms: $4.15\text{x}10^{-7}\ \text{m}\left(\frac{0.01\ \text{Å}}{1\text{x}10^{-12}\ \text{m}}\right) =$ **4150 Å**

Check: The purple region of the visible light spectrum occurs between approximately 400 to 450 nm, so 415 nm is in the correct range for wavelength of purple light.

7.2 Plan: To calculate the energy for each wavelength we use the formula E = hc/λ.

Solution:

$$E_{uv} = \frac{hc}{\lambda} = \frac{(6.626x10^{-34}\ J\cdot s)(3.00x10^{8}\ m/s)}{1x10^{-8}\ m} = 2x10^{-17}\ J$$

$$E_{vis} = \frac{hc}{\lambda} = \frac{(6.626x10^{-34}\ J\cdot s)(3.00x10^{8}\ m/s)}{5x10^{-7}\ m} = 4x10^{-19}\ J$$

$$E_{ir} = \frac{hc}{\lambda} = \frac{(6.626x10^{-34}\ J\cdot s)(3.00x10^{8}\ m/s)}{1x10^{-4}\ m} = 2x10^{-21}\ J$$

As the wavelength of light increases from ultraviolet (uv) to visible (vis) to infrared (ir) the energy of the light decreases.

Check: The decrease in energy with increase in wavelength follows the inverse relationship between energy and wavelength in the equation E = hc/λ.

7.3 Plan: With the equation for the de Broglie wavelength, λ = h/mu and the given de Broglie wavelength, calculate the electron speed.

Solution: $u = \frac{h}{m\lambda} = \frac{6.626x10^{-34}\ J\cdot s}{(9.109x10^{-31}\ kg)(100x10^{-9}\ m)} =$ **$7.27x10^{3}$ m / s**

Check: Perform a rough calculation to check order of magnitude: $\frac{10^{-34}}{10^{-31}\ x\,10^{-7}} = 10^{38-34} = 10^{4}$

The units work since J = kg·m2/s.

7.4 Plan: Use Heisenberg's uncertainty principle to calculate the uncertainty in the position of the baseball given the uncertainty in its speed.

Solution: $\Delta x \geq \frac{h}{4\pi m \Delta u} \geq \frac{6.626x10^{-34} J \cdot s}{4\pi(0.142\ kg)(0.447\ m/s)} \geq$ **$8.31x10^{-34}$ m**

Any human eye, even one belonging to an umpire, could detect the position of the baseball with an uncertainty that is so small it is insignificant.

Check: An order of magnitude calculation ($10^{-34}/(10^{1}x10^{1}x10^{-1}x10^{-1}) = 10^{-34}$) agrees with the result.

7.5 Plan: Following the rules for l (integer from 0 to n-1) and m_l (integer from –l to +l) write quantum numbers for n = 4.

Solution: For n = 4, l = 0, 1, 2, 3

For l = 0, m_l = 0

For l = 1, m_l = -1, 0, 1

For l = 2, m_l = -2, -1, 0, 1, 2

For l = 3, m_l = -3, -2, -1, 0, 1, 2, 3

Check: The total number of orbitals with a given n is n^2. For n = 4, the total number of orbitals would be 16. Adding the number of orbitals identified in the solution, gives 1 (for l = 0) + 3 (for l = 1) + 5 (for l = 2) + 7 (for l = 3) = 16 orbitals.

7.6 Plan: Identify n and l from the subshell designation and knowing the value for l find the m_l values.

Solution:

Sublevel name	n value	l value	m_l values
2p	2	1	-1,0,1
5f	5	3	-3,-2,-1,0,1,2,3

Check: The number of orbitals for each sublevel equals 2l + 1. Sublevel 2p should have 3 orbitals and sublevel 5f should have 7 orbitals. Both of these agree with the number of m_l values for the sublevel.

7.7 Plan: Use the rules for designating quantum numbers to fill in the blanks.

For a given n, l can be any integer from 0 to n-1.

For a given l, m_l can be any integer from – l to + l.

The subshells are given a letter designation, in which s represents l = 0, p represents l = 1, d represents l = 2, f represents l = 3.

Solution: Values added are in bold type.

	n	l	m_l	Name
a)	**4**	**1**	0	4p
b)	2	1	0	**2p**
c)	3	2	-2	**3d**
d)	**2**	**0**	**0**	2s

Check: All the given values are correct designations.

END-OF-CHAPTER PROBLEMS

7.2 a) Figure 7.3 describes the electromagnetic spectrum by wavelength and frequency. Wavelength increases from left (10^{-2} nm) to right (10^{12} nm). The trend in increasing wavelength is: x-ray < UV < visible < infrared < microwave < radio waves.

b) Frequency is inversely proportional to wavelength according to equation 7.1, so frequency has the opposite trend: radio < microwave < infrared < visible < UV < x-ray.

c) Energy is directly proportional to frequency according to equation 7.2. Therefore, the trend in increasing energy matches the trend in increasing frequency: radio < microwave < infrared < visible < UV < x-ray. High-energy electromagnetic radiation disrupts cell function. It makes sense that you want to limit exposure to ultraviolet and x-ray radiation.

7.5 In order to explain the formula he developed for the energy vs. wavelength data of blackbody radiation, Max Planck assumed that only certain quantities of energy, called quanta, could be emitted.

7.7 Wavelength is related to frequency through the equation λ = c/ν. Recall that a Hz is a reciprocal second, or 1/s. Assume that the number "960" has three significant figures.

$$\lambda = \frac{3.00x10^8\ m/s}{(960\ kHz)\left(\frac{1000\ Hz}{kHz}\right)\left(\frac{1/s}{1\ Hz}\right)} = \mathbf{313\ m}$$

$$(313\,m)\left(\frac{1x10^9\ nm}{m}\right) = \mathbf{3.13x10^{11}\ nm};\quad (313\,m)\left(\frac{1x10^{10}\ \text{Å}}{m}\right) = \mathbf{3.13x10^{12}\ \text{Å}}$$

7.9 Frequency is related to energy through the equation E = hν. Note that 1 Hz = 1 s^{-1}.
E = (6.626x10^{-34} J-s)(3.6x10^{10} s^{-1}) = **2.4x10^{-23} J**

7.11 Since energy is directly proportional to frequency (E = hν) and frequency and wavelength are inversely related (ν = c/ λ), it follows that energy is inversely related to wavelength. As wavelength decreases, energy increases. In terms of increasing energy, **red < yellow < blue**. You can confirm this trend by calculating energy.

$$E_{660} = h\left(\frac{c}{\lambda}\right) = (6.626x10^{-34}\ J\cdot s)\left(\frac{3.00x10^8\ m/s}{(660\,nm)\left(\frac{1\,m}{10^9\ nm}\right)}\right) = 3.01x10^{-19}\ J$$

$$E_{595} = 3.34x10^{-19}\ J$$

$$E_{453} = 4.39x10^{-19}\ J$$

7.13 Frequency and wavelength can be calculated using the speed of light: c = λν.

$$\lambda = \frac{c}{\nu} = \frac{2.99792x10^8\ m/s}{22.235\,GHz}\left(\frac{1\,GHz}{10^9\ s^{-1}}\right)\left(\frac{10^9\ nm}{m}\right) = \mathbf{1.3483x10^7\ nm}$$

$$1.3483x10^7\ nm\left(\frac{1\,m}{10^9\,nm}\right)\left(\frac{1\,Å}{10^{-10}\,m}\right) = \mathbf{1.3483x10^8\ Å}$$

7.16 a) $$\nu = c/\lambda = \frac{(3.00x10^8\ m/s)}{242x10^{-9}\ m} = 1.23_{97}x10^{15} = \mathbf{1.24x10^{15}\ s^{-1}}$$

E_{242} = hν = (6.626 x 10^{-34} J·sec)(1.23_{97} x 10^{15} s^{-1}) = **8.21 x 10^{-19} J.**

b) The most energetic photon has the shortest wavelength, so calculate the frequency and energy for λ = 2200 Å. (1 Å = 10^{-10} m)

$$\nu = c/\lambda = \frac{(3.00x10^8\ m/s)}{2200x10^{-10}\ m} = 1.3_{64}x10^{15} = \mathbf{1.4x10^{15}\ s^{-1}}$$

$E_{2200\ A}$ = (6.626 x 10^{-34} J·sec)(1.3_{64} x 10^{15} s^{-1}) = **9.0 x 10^{-19} J.**

7.18 Bohr's model includes the assumption that electrons can move from one level to another by absorbing energy (move to a higher level) or by emitting energy (move to a lower level). A solar system model does not allow for the movement of electrons between levels. The theoretical basis Bohr used was that the energy of the lines in the spectra of the elements corresponds to the energy required for an electron to move to another level.

7.20 The quantum number *n* is related to the energy level of the electron. An electron *absorbs* energy to change from lower energy (low *n*) to higher energy (high *n*), so (a) and (d) correspond to absorption of energy. An electron *emits* energy as it drops from a higher energy level to a lower one, so (b) and (c) correspond to the emission of energy.

7.22 The Bohr model has successfully predicted the line spectra for the H atom and Be^{3+} ion since both are one electron species. The line spectra for H would not match the line spectra for Be^{3+} since the H nucleus contains one proton while the Be^{3+} nucleus contains 4 protons, thus the force of attraction of the nucleus for the electron would be greater in the beryllium ion than in the hydrogen atom.

7.23 Calculate wavelength by substituting the given values into equation 7.3, where n_1 = 2 and n_2 = 5 because $n_2 > n_1$. Although more significant figures could be used, five significant figures are adequate for this calculation.

$$\frac{1}{\lambda} = (1.096776x10^7\ m^{-1})\left(\frac{1}{2^2} - \frac{1}{5^2}\right) = 2.3032x10^6\ m^{-1}$$

$$\lambda = \left(\frac{1}{2.3032x10^6\ m^{-1}}\right)\left(\frac{10^9\ nm}{m}\right) = \mathbf{434.17\ nm}$$

7.25 For the infrared series of the H atom, n_1 equals 3. The least energetic spectral line in this series would represent an electron moving from the next highest energy level, n = 4.

$$\frac{1}{\lambda} = 1.096776x10^7\ m^{-1}\left(\frac{1}{3^2} - \frac{1}{4^2}\right) = 5.332x10^5\ m^{-1}$$

$$\lambda = \left(\frac{1}{5.332x10^5\ m^{-1}}\right)\left(\frac{10^9\ nm}{m}\right) = \mathbf{1875\ nm}$$

Checking this wavelength with Figure 7.9 we find the line in the infrared series with the greatest wavelength occurs at approximately 1850 nm.

7.27 To find the transition energy, apply equation 7.4 and multiply by Avogadro's number.

$$\Delta E_{atom} = -2.18x10^{-18}\,J\left(\frac{1}{2^2}-\frac{1}{5^2}\right) = -4.58x10^{-19}\ J\ per\ atom$$

$$\Delta E_{mol} = \left(\frac{-4.58x10^{-19}\ J}{atom}\right)\left(\frac{6.022x10^{23}\ atoms}{mol}\right) = -2.76x10^{5}\ J/mol\ \ or\ \ -\mathbf{276\,kJ/mol}$$

The negative sign denotes that light is emitted. Alternately, the wavelength of the emitted light was calculated in problem 7.23. This wavelength corresponds to a frequency and energy which, when multiplied by Avogadro's number, yields the same answer.

$$\Delta E_{mol} = N_A \times h\upsilon = \left(\frac{6.022x10^{23}\ atoms}{mol}\right)(6.626x10^{-34}\ J\cdot s)\left(\frac{3.00x10^{8}\ m/s}{434.2x10^{-9}\ m}\right) = 2.76x10^{5}\ J/mol$$

7.29 Looking at an energy chart will help answer this question.

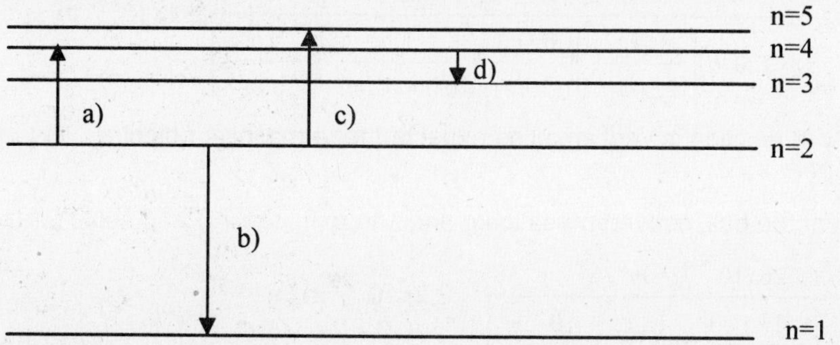

Frequency is proportional to energy so the smallest frequency will be d) n=4 to n=3 and the largest frequency b) n=2 to n=1. Transition a) n=2 to n=4 will be smaller than transition c) n=2 to n=5 since level 5 is a higher energy than level 4. In order of increasing frequency, d < a < c < b.

7.31 Either equation 7.3 or 7.4 is appropriate for solving this problem. To use equation 7.3, convert λ from nm to m so that the units cancel with Rydberg's constant. To use equation 7.4, convert λ from nm to m and then calculate ΔE using E = hc/λ. The calculation below uses equation 7.4.

$$\Delta E = \frac{hc}{\lambda} = \frac{(6.626x10^{-34}\ J\cdot s)(2.998x10^{8}\ m/s)}{97.20x10^{-9}\ m} = 2.044x10^{-18}\ J$$

$$\Delta E = -2.18x10^{-18}\,J\left(\frac{1}{n^2_{final}}-\frac{1}{n^2_{init}}\right);\ \ 2.044x10^{-18}\ J = -2.18x10^{-18}\,J\left(\frac{1}{n^2_{final}}-\frac{1}{1^2}\right)$$

$$-0.9375+1 = \frac{1}{n^2_{final}};\ \ n^2_{final} = 15.99;\ therefore\ n_{final} = \mathbf{4}$$

7.34 The energy can be calculated from ΔE = hc/λ.

$$\Delta E = \frac{(6.626x10^{-34}\ J \cdot s)(3.00x10^{8}\ m/s)}{589x10^{-9}\ m} = \mathbf{3.37x10^{-19}\ J\ per\ photon}$$

$$3.37x10^{-19}\ J/photon\left(\frac{6.022x10^{23}\ photons}{mole}\right)\left(\frac{1\ mol\ photons}{1\ einstein}\right)\left(\frac{1\ kJ}{1000\ J}\right) = \mathbf{203 kJ\ per\ einstein}$$

7.37 Macroscopic objects have significant mass. A large m in the denominator of $\lambda = h/mu$ results in a very small wavelength. Macroscopic objects do exhibit a wavelike motion, but the wavelength is too small for humans to see it.

7.39 a) $$\lambda = \frac{h}{mu} = \frac{6.626x10^{-34}\ kg \cdot m^2/s}{(220\ lb)\left(\frac{1\ kg}{2.205\ lb}\right)\left(\frac{19.6\ mi}{h}\right)\left(\frac{1\ m}{0.00062\ mi}\right)\left(\frac{1\ h}{3600\ s}\right)} = \mathbf{7.56x10^{-37}\ m}$$

b) Uncertainty in his velocity (speed) is 0.1mi/h, which converts to 0.04 m/s.

$$\Delta x \geq \frac{h}{4\pi m \Delta u} \geq \frac{6.626x10^{-34}\ kg \cdot m^2/s}{4\pi(220\ lb)\left(\frac{1\ kg}{2.205\ lb}\right)\left(\frac{0.1\ mi}{h}\right)\left(\frac{1\ m}{0.00062\ mi}\right)\left(\frac{1\ h}{3600\ s}\right)} \geq \mathbf{1x10^{-35}\ m}$$

The uncertainty in position is very small as expected for a massive object.

7.41 To find the speed of the ball, convert mass to kg and λ to m in eqn 7.5 and solve for u.

$$u = \frac{h}{m\lambda} = \frac{6.626x10^{-34}\ kg \cdot m^2/s}{\left(56.5\ g \times \frac{1\ kg}{1000\ g}\right)\left(5400\ Å \times \frac{10^{-10}\ m}{1\ Å}\right)} = \mathbf{2.2x10^{-26}\ m/s}$$

7.43 The de Broglie wavelength equation will give the mass-equivalent of a photon with known wavelength and velocity. The term "mass-equivalent" is used instead of "mass of photon" because photons are quanta of electromagnetic energy that have no mass. A light photon's velocity is the speed of light, $2.99792x10^8$ m/s.

$$m = \frac{h}{\lambda u} = \frac{6.626x10^{-34}\ kg \cdot m^2/s}{(589x10^{-9}\ m)(2.99792x10^{8}\ m/s)} = \mathbf{3.75x10^{-36}\ kg\ per\ photon}$$

7.47 A peak in the radial probability distribution at a certain distance means that the total probability of finding the electron is greatest within a thin spherical volume having a radius very close to that distance. Since principal quantum number (n) correlates with distance from the nucleus, the peak for n=2 would occur at a greater distance from the nucleus than 0.529 Å. Thus, the probability of finding an electron at 0.529 Å is much greater for the 1s orbital than for the 2s.

7.48 a) Principal quantum number, n, relates to the size of the orbital. More specifically it relates to the distance from the nucleus at which the probability of finding an electron is greatest. This distance is determined by the energy of the electron.

b) Angular momentum quantum number, l, relates to the shape of the orbital. It is also called the azimuthal quantum number.

c) Magnetic quantum number, m_l, relates to the orientation, or positioning, of the orbital in space.

7.49 Refer to Sample Problem 7.6 for additional examples.

	SUBLEVEL NAME	n	l	POSSIBLE m_l	NO. OF ORBITALS
a)	1s	1	0	0	1
b)	4d	4	2	-2, -1, 0, +1, +2	5
c)	3p	3	1	-1, 0, +1	3
d)	3s	3	0	0	1
	3p	3	1	-1, 0, +1	3
	3d	3	2	-2, -1, 0, +1, +2	5
					Total = **9**

Recall that l can only be 0 to n -1.

7.51 Magnetic quantum numbers can have integer values from $-l$ to $+ l$.

a) For $l = 2$, m_l can equal –2, -1, 0, 1, or 2.

b) For $n = 1$, l can equal only 0 and thus m_l can equal only 0.

c) For $n = 4$ and $l = 3$, m_l can equal –3, -2, -1, 0, 1, 2, 3.

7.53 a) s

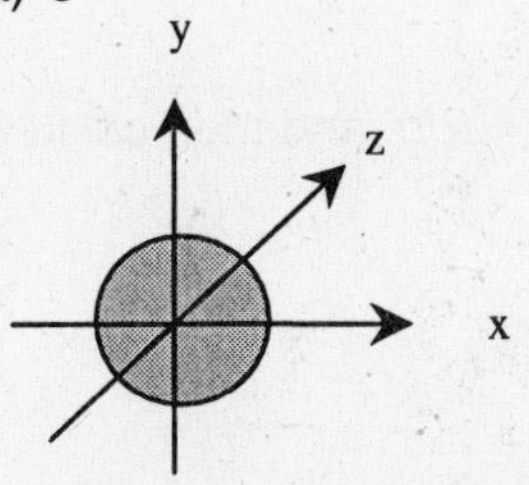

b) p_x

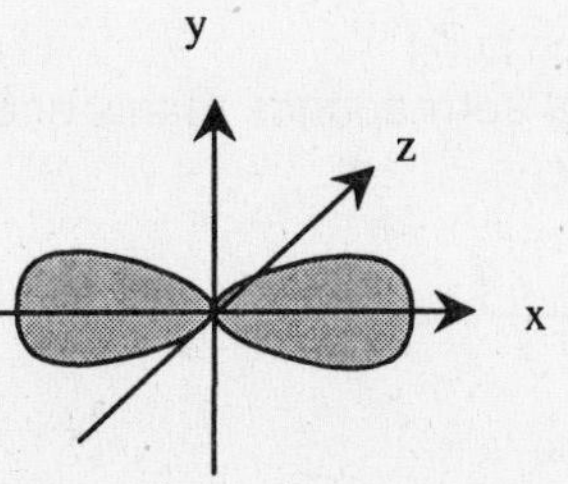

7.55

	n	l	SUBLEVEL DESIGNATION	POSSIBLE m_l	NO. OF ORBITALS
a)	4	2	d	-2,-1,0,1,2	5
b)	5	1	p	-1,0,1	3
c)	6	3	f	-3,-2,-1,0,1,2,3	7

7.57 a) For the 5s subshell, $n = 5$ and $l = 0$. Since $m_l = 0$, there is one orbital.

b) For the 3p subshell, $n = 3$ and $l = 1$. Since $m_l = -1, 0, +1$, there are three orbitals.

c) For the 4f subshell, $n = 4$ and $l = 3$. Since $m_l = -3, -2, -1, 0, +1, +2, +3$, there are seven orbitals.

7.59 a) With $l = 0$ the only allowable m_l value is 0. To correct, either change l or m_l value.

Correct: $n = 2$, $l = 1$, $m_l = -1$; $n = 2$, $l = 0$, $m_l = 0$.

b) Combination is allowed.

c) Combination is allowed.

d) With l = 2, +3 is not an allowable m_l value. To correct, either change l or m_l value. Correct: n = 5, l = 3, m_l = +3; n = 5, l = 2, m_l = 0.

7.62 a) The mass of the electron is 9.1094 x 10^{-31} kg. In order to correctly cancel units, report Planck's constant in units kg·m^2/s.

$$E = -\text{constant}\left(\frac{1}{n^2}\right) \quad where\ \text{constant} = \frac{h^2}{8\pi^2 m_e a_0^2}$$

$$\frac{h^2}{8\pi^2 m_e a_0^2} = \frac{\left(6.626x10^{-34}\,kg\cdot m^2/s\right)^2}{8\pi^2\left(9.1094x10^{-31}\,kg\right)\left(52.92x10^{-12}\,m\right)^2} = \mathbf{2.180x10^{-18}kg\cdot m^2/s^2}\ (J)$$

For the H atom, Z = 1 and Bohr's constant = -2.18 x 10^{-18} J. For the hydrogen atom, derivation using classical principles or quantum-mechanical principles yields the same constant.

b) The n = 3 energy level is higher in energy than the n = 2 level. Because the zero point of the atom's energy is defined as an electron's infinite distance from the nucleus, a lower energy level is described by a larger negative number. Although this may be confusing, it makes sense that an energy *change* would be a positive number.

$$\Delta E = -\frac{h^2}{8\pi m_e a_0^2 n^2}\left(\frac{1}{n_{final}} - \frac{1}{n_{initial}}\right)$$

$$\Delta E = -2.179_{63}x10^{-18}\,J\left(\frac{1}{3^2} - \frac{1}{2^2}\right) = \mathbf{3.027_{27}x10^{-19}\,J}$$

c) Find the frequency that corresponds to this energy using ν = E/h, and then calculate wavelength using λ = c/ν.

$$\nu = \frac{3.027_{27}x10^{-19}\,J}{6.626x10^{-34}\,J\cdot s} = 4.568_{77}x10^{14}\,s^{-1}$$

$$\lambda = \frac{2.998x10^8\,m/s}{4.568_{77}x10^{14}\,s^{-1}} = 6.561_{93}x10^{-7}\,m\ or\ \mathbf{656.2\ nm}$$

Yes, there is a red line at 656.3 nm on the hydrogen spectrum.

7.63 a) The lines do not begin at the origin because an electron must absorb a minimum amount of energy before it has enough energy to overcome the attraction of the nucleus and leave the atom. This minimum energy is the energy of photons of light at the threshold frequency.

b) The lines for K and Ag do not begin at the same point because the amount of energy that an electron must absorb to leave the K atom is less than the amount of energy that an electron must absorb to leave the Ag atom where the attraction between the nucleus and outer electron is stronger than in a K atom.

c) Wavelength is inversely proportional to energy. Thus, the metal that requires a larger amount of energy to be absorbed before electrons are emitted will require a shorter wavelength of light. Electrons in Ag atoms require more energy to leave, so Ag requires a shorter wavelength of light than K to eject an electron.

d) The slopes of the line show an increase in kinetic energy as the frequency (or energy) of light is increased. Since the slopes are the same this means that for an increase of one unit of frequency (or energy) of light the increase in kinetic energy of an electron ejected from K is the same as the increase in the kinetic energy of an electron ejected from Ag. After an electron is ejected, the energy that it absorbs above the threshold energy becomes kinetic energy of the electron. For the same increase in energy above the threshold energy for either K or Ag, the kinetic energy of the ejected electron will be the same.

7.66 The Bohr model has been successfully applied to predict the spectral lines for one-electron species other than H. Common one-electron species are small cations with all but one electron removed. Since the problem specifies a metal ion, assume that the possible choices are Li^{+2} or Be^{+3}, and solve Bohr's equation to verify if a whole number for n can be calculated. Recall that the negative sign is a convention based on the zero point of the atom's energy; it is deleted in this calculation to avoid taking the square root of a negative number.

$$E = h\nu = \left(6.626x10^{-34}\ J\cdot s\right)\left(2.961x10^{16}\ s^{-1}\right) = 1.9620x10^{-17}\ J$$

$$\Delta E = 1.9620x10^{-17} = 2.18x10^{-18}\ J\left(\frac{Z^2}{n_{final}^2} - \frac{Z^2}{n_{initial}^2}\right)$$

$$where\ n_{initial} = \infty\ and\ n_{final} = 1\ for\ the\ highest\ energy\ line$$

$$Z = \sqrt{\frac{\left(1.9620x10^{-17}\ J\right)}{2.18x10^{-18}\ J}} = 3$$

With Z, nuclear charge, equal to 3 the metal ion must be **Li^{+2}**.

7.69 The electromagnetic spectrum shows that the visible region goes from 400 to 750 nm. Thus, wavelengths b, c, and d are for the three transitions in the visible series with n_{final} = 2. Wavelength a is in the ultraviolet region of the spectrum and the ultraviolet series has n_{final} = 1. Wavelength e is in the infrared region of the spectrum and the infrared series has n_{final} = 3.

$$\frac{1}{(1212.7\ \text{Å})\left(\frac{10^{-10}\ m}{\text{Å}}\right)} = 1.096776x10^{7}\ m^{-1}\left(\frac{1}{1^2} - \frac{1}{n_{initial}^2}\right);\ n_{initial} = \mathbf{2}$$

$$\frac{1}{(4340.5\ \text{Å})\left(\frac{10^{-10}\ m}{\text{Å}}\right)} = 1.096776x10^{7}\ m^{-1}\left(\frac{1}{2^2} - \frac{1}{n_{initial}^2}\right);\ n_{initial} = \mathbf{5}$$

$$\frac{1}{(4861.3\ \text{Å})\left(\frac{10^{-10}\ m}{\text{Å}}\right)} = 1.096776x10^{7}\ m^{-1}\left(\frac{1}{2^2} - \frac{1}{n_{initial}^2}\right);\ n_{initial} = \mathbf{4}$$

$$\frac{1}{(6562.8\ \text{Å})\left(\frac{10^{-10}\ m}{\text{Å}}\right)} = 1.096776x10^{7}\ m^{-1}\left(\frac{1}{2^2} - \frac{1}{n_{initial}^2}\right);\ n_{initial} = \mathbf{3}$$

$$\frac{1}{(10938\,\text{Å})\left(\frac{10^{-10}\,m}{\text{Å}}\right)} = 1.096776x10^{7}\,m^{-1}\left(\frac{1}{3^2} - \frac{1}{n_{initial}^{\;2}}\right);\ n_{initial} = \mathbf{6}$$

7.71 a) The energy of visible light is lower than that of UV light. Thus, metal A must be barium because the attraction between barium's nucleus and outer electron is less than the attraction in tantalum or tungsten. In all three elements the outer electron is in the same energy level (6s) but the nuclear charge is less in barium than tantalum or tungsten. The longest wavelength corresponds to the lowest energy (work function).

$$\lambda_{Ba} = \frac{hc}{\phi_{Ba}} = \frac{(6.626x10^{-34}\,J\cdot s)(3.00x10^{8}\,m/s)}{4.30x10^{-19}\,J} = 4.62x10^{-7}\,m\,or\ \mathbf{462\,nm}$$

b) Tantalum and tungsten can be distinguished by the wavelengths between the longest for tantalum and the longest for tungsten.

$$\lambda_{Ta} = \frac{hc}{\phi_{Ta}} = \frac{(6.626x10^{-34}\,J\cdot s)(3.00x10^{8}\,m/s)}{6.81x10^{-19}\,J} = 2.92x10^{-7}\,m\,or\ 292\,nm$$

$$\lambda_{W} = \frac{hc}{\phi_{W}} = \frac{(6.626x10^{-34}\,J\cdot s)(3.00x10^{8}\,m/s)}{7.16x10^{-19}\,J} = 2.78x10^{-7}\,m\,or\ 278\,nm$$

The range of wavelengths that distinguish B and C (tungsten and tantalum) are between **278 nm and 292 nm**. Tungsten would emit an electron in this range of wavelengths while tantalum would not.

7.73 The frequency and wavelength of the least energetic photon that can break a carbon-carbon bond is calculated from the average bond energy:

$$\lambda = \frac{hc}{E} = \frac{(6.626x10^{-34}\,J\cdot s)(3.00x10^{8}\,m/s)}{\left(\frac{347\,kJ}{mol}\right)\left(\frac{mol}{6.022x10^{23}\,bonds}\right)\left(\frac{1000\,J}{1\,kJ}\right)} = 3.45x10^{-7}\,m\,or\ \mathbf{345\,nm}$$

$$\nu = \frac{c}{\lambda} = \frac{3.00x10^{8}\,m/s}{3.45x10^{-7}\,m} = \mathbf{8.70x10^{14}\,s^{-1}}$$

Photons of wavelength 345 nm fall in the **ultraviolet** region of the electromagnetic spectrum.

7.75 a) The l value must be at least 1 for m_l to be –1, but cannot be greater than n-1 = 2. Increase the l value to 1 or 2 to create an allowable combination.

b) The l value must be at least 1 for m_l to be +1, but cannot be greater than n -1 = 2. Decrease the l value to 1 or 2 to create an allowable combination.

c) The l value must be at least 3 for m_l to be +3, but cannot be greater than n -1 = 6. Increase the l value to 3, 4, 5, or 6 to create an allowable combination.

d) The l value must be at least 2 for m_l to be –2, but cannot be greater than n -1 = 3. Increase the l value to 2 or 3 to create an allowable combination.

7.77 a) Ionization occurs when the electron is completely removed from the atom, or when n = ∞. We can use equation 7.4 to find the quantity of energy needed to completely remove the electron, called the ionization energy (IE). The charge on nucleus must affect the IE because a larger nucleus would exert a greater pull on the escaping electron. The Bohr equation applies to H and other one-electron species. The general expression is:

$$\Delta E_{final} = -2.18x10^{-18}\frac{J}{atom}\left(\frac{Z^2}{n^2_{final}} - \frac{Z^2}{n^2_{initial}}\right) \quad where\ n^2_{final} = \infty\ and\ Z = \text{nuclear charge}$$

$$\Delta E_{final} = -2.18x10^{-18}\frac{J}{atom}\left(0 - \frac{Z^2}{n^2_{initial}}\right)$$

b) In the ground state n = 1, the initial energy level for the single electron in B^{4+}. Once ionized n = ∞, the final energy level. The ionization energy is calculated as follows (remember that 1/∞ = 0).

$$\Delta E_{final} = -2.18x10^{-18}\frac{J}{atom}\left(0 - \frac{5^2}{1^2}\right)\left(6.022x10^{23}\frac{atoms}{mol}\right)\left(\frac{1\,kJ}{10^3\,J}\right) = \mathbf{3.28x10^4\,kJ/mol}$$

c) Minimum wavelength corresponds to the ionization energy per atom.

$$\Delta E_{final} = -2.18x10^{-18}\frac{J}{atom}\left(0 - \frac{2^2}{3^2}\right) = 9.6889x10^{-19}\,J/atom$$

$$\lambda = \frac{hc}{E} = \frac{(6.626x10^{-34}\,J\cdot s)(3.00x10^8\,m/s)}{9.6889x10^{-19}\,J} = 2.05\,x\,10^{-7}\,m\,or\,\mathbf{205\,nm}$$

d) $$\Delta E_{final} = -2.18x10^{-18}\frac{J}{atom}\left(0 - \frac{4^2}{2^2}\right) = 8.72x10^{-18}\,J/atom$$

$$\lambda = \frac{hc}{E} = \frac{(6.626x10^{-34}\,J\cdot s)(3.00x10^8\,m/s)}{8.72x10^{-18}\,J} = 2.28\,x\,10^{-8}\,m\,or\,\mathbf{22.8\,nm}$$

7.81 <u>Plan</u>: Refer to Chapter 6 for the calculation of the amount of heat energy absorbed by a substance from its heat capacity and temperature change (q = C x mass x ΔT). Using this equation calculate the energy absorbed by the water. This energy equals the energy from the microwave photons. The energy of each photon can be calculated from its wavelength: E = hc/λ. Dividing the total energy by the energy of each photon gives the number of photons absorbed by the water.

<u>Solution</u>:

$$q = (4.184\,J/g\cdot°C)(252\,g)(98°C - 20.°C) = 8.2_{24}x10^4\,J$$

$$E_{photon} = \frac{(6.626x10^{-34}\,J\cdot s)(3.00x10^8\,m/s)}{1.55x10^{-2}\,m} = 1.28_{25}x10^{-23}\,J/photon$$

$$\#\,of\,photons = \frac{8.2_{24}x10^4\,J}{1.28x10^{-23}\,J/photon} = \mathbf{6.4x10^{27}\,photons}$$

<u>Check</u>: The order of magnitude appears to be correct for the calculation of total energy absorbed: 10^1 x 10^2 x 10^1 = 10^4. The order of magnitude of the energy for one photon also appears correct: 10^{-34} x $10^8/10^{-2}$ = 10^{-24}. And the order of magnitude of the number of photons is estimated as $10^4/10^{-24}$ = 10^{28}, which agrees with the calculated $6x10^{27}$.

7.83 a) The overlap between the n_1 = 1 series and the n_1 = 2 series would occur between the longest wavelengths for n_1 = 1 and the shortest wavelengths for n_1 = 2.

Longest wavelength in n_1 = 1 series has n_2 equal to 2.

$$\frac{1}{\lambda} = 1.097x10^7\ m^{-1}\left(\frac{1}{1^2} - \frac{1}{2^2}\right);\ \lambda = 122\ nm$$

Shortest wavelength in n_1 = 2 series has $n_2 = \infty$.

$$\frac{1}{\lambda} = 1.097x10^7\ m^{-1}\left(\frac{1}{2^2} - \frac{1}{\infty^2}\right);\ \lambda = 365\ nm$$

Since the longest wavelength for n_1 = 1 series is shorter than shortest wavelength for n_1 = 2 series, there is **no overlap** between the two series.

b) The overlap between the n_1 = 3 series and the n_1 = 4 series would occur between the longest wavelengths for n_1 = 3 and the shortest wavelengths for n_1 = 4.

Longest wavelength in n_1 = 3 series has n_2 equal to 4.

$$\frac{1}{\lambda} = 1.097x10^7\ m^{-1}\left(\frac{1}{3^2} - \frac{1}{4^2}\right);\ \lambda = 1875\ nm$$

Shortest wavelength in n_1 = 4 series has $n_2 = \infty$.

$$\frac{1}{\lambda} = 1.097x10^7\ m^{-1}\left(\frac{1}{4^2} - \frac{1}{\infty^2}\right);\ \lambda = 1459\ nm$$

Since the n_1 = 4 series shortest wavelength is shorter than the n_1 = 3 series longest wavelength, the **series do overlap**.

c) Shortest wavelength in n_1 = 5 series:

$$\frac{1}{\lambda} = 1.097x10^7\ m^{-1}\left(\frac{1}{5^2} - \frac{1}{\infty^2}\right);\ \lambda = 2279\ nm$$

Longest wavelengths in n_1 = 4 series, calculated as above:

n_2 = 5, l = 4051 nm

n_2 = 6, l = 2625 nm

n_2 = 7, l = 2165 nm

Two lines in the n_1 = 4 series, n_2 = 5 and n_2 = 6, **overlap** with the n_1 = 5 series

d) More overlap occurs at longer wavelengths. Eventually the spectrum becomes a continuous band.

7.88 a) The energy of the electron is a function of its speed leaving the surface of the metal. The mass of the electron is 9.109 x 10^{-31} kg.

$$E_k = \frac{1}{2}mu^2 = \frac{1}{2}(9.109x10^{-31}\ kg)(6.40x10^5\ m/s)^2 = \mathbf{1.87x10^{-19}\ J}$$

b) The minimum energy required to dislodge the electron (Φ) is a function of the incident light. In this example, the incident light is higher than the threshold frequency, so the kinetic energy of the electron, E_k, must be subtracted from the total energy of the incident light, hν, to yield the work function, Φ.

$$E_{photon} = h\nu = h\frac{c}{\lambda} = \frac{(6.626x10^{-34}\ J\cdot s)(2.998x10^8\ m/s)}{(358.1x10^{-9}\ m)} = 5.547_{26}x10^{-19}\ J$$

$$\Phi = h\nu - E_k = 5.547_{26}x10^{-19}\ J - 1.86_{55}x10^{-19}\ J = \mathbf{3.68x10^{-19}\ J}$$

7.90 a) Figure 7.3 indicates that the 641 nm wavelength of Sr falls in the **red** region and the 493 nm wavelength of Ba falls in the **green** region.

b) Use the formula E=hc/λ to calculate J/photon. Convert the 1.00 g amounts of $BaCl_2$ ($\mathcal{M}$ = 208.2 g/mol) and $SrCl_2$ ($\mathcal{M}$ = 158.52 g/mol) to moles, then to atoms, and assume each atom undergoes one electron transition (which produces the colored light) to find number of photons. Multiply J/photon by number of photons to find total energy.

$$E_{641} = h\frac{c}{\lambda} = \frac{(6.626x10^{-34}\ J\cdot s)(3.00x10^{8}\ m/s)}{641x10^{-9}\ m} = 3.10_{11}x10^{-19}\ J/photon$$

$$number\ of\ photons = (1.00\ g\ SrCl_2)\left(\frac{1\ mol\ SrCl_2}{158.52\ g}\right)\left(\frac{6.022x10^{23}\ atoms\ Sr}{mol\ SrCl_2}\right)\left(\frac{1\ photon}{atom\ Sr}\right) = 3.79_{89}x10^{21}\ photons$$

$$Total\ E = (3.10_{11}x10^{-19}\ J/photon)(3.79_{89}x10^{21}\ photons) = 1.18x10^{3}\ J = \mathbf{1.18kJ\ (Sr)}$$

$$E_{493} = h\frac{c}{\lambda} = \frac{(6.626x10^{-34}\ J\cdot s)(3.00x10^{8}\ m/s)}{493x10^{-9}\ m} = 4.03_{20}x10^{-19}\ J/photon$$

$$number\ of\ photons = (1.00\ g\ BaCl_2)\left(\frac{1\ mol\ BaCl_2}{208.2\ g}\right)\left(\frac{6.022x10^{23}\ atoms\ Ba}{mol\ BaCl_2}\right)\left(\frac{1\ photon}{atom\ Ba}\right) = 2.89_{24}x10^{21}\ photons$$

$$Total\ E = (4.03_{20}x10^{-19}\ J/photon)(2.89_{24}x10^{21}\ photons) = 1.17x10^{3}\ J = \mathbf{1.17kJ\ (Ba)}$$

7.92 a) At this wavelength the sensitivity to absorbance of light by Vitamin A is maximized while minimizing interference due to the absorbance of light by other substances in the fish-liver oil.

b) The wavelength 329 nm lies in the **ultraviolet region** of the electromagnetic spectrum.

c) A known quantity of vitamin A (1.67 x 10^{-3} g) is dissolved in a known volume of solvent (250. mL) to give a *standard* concentration with a known response (1.018 units). An equality can be made between the two concentration-to-absorbance ratios:

$$\frac{(0.1232\ g/500.\ mL)*(fraction\ of\ Vit.\ A)}{0.724\ units} = \frac{(1.67x10^{-3}\ g\ Vit.\ A/250.\ mL)}{1.018\ units}$$

$$fraction\ of\ Vit.\ A = \frac{(1.67x10^{-3}\ g\ Vit.\ A/250.\ mL)(0.724\ units)}{(1.018\ units)(0.1232\ g/500.\ mL)} = \mathbf{0.0193g\ Vit.A/1g\ oil}$$

Check this result by substituting it back into the equality:

$$\frac{(2.38x10^{-3}\ g/500.\ mL)}{0.724\ units} = \frac{(1.67x10^{-3}\ g/250.\ mL)}{1.018\ units}$$

$$6.57x10^{-6} = 6.56x10^{-6}\ (agreement\ would\ be\ better\ with\ more\ S.F.'s)$$

7.96 The amount of energy is calculated from the wavelength of light:

$$E_{photon} = \frac{hc}{\lambda} = \frac{(6.626x10^{-34}\ J\cdot s)(2.998x10^{8}\ m/s)}{550x10^{-9}\ m} = 3.6_{12}x10^{-19}\ J/photon$$

The amount of energy that hits the book:

$$E_{total} = 75W\left(\frac{1\,J/s}{1\,W}\right)(0.05)(0.10) = 0.37_5\ J/s$$

The number of photons that hits the book each second:

$$\left(\frac{0.37_5\ J}{s}\right)\left(\frac{1\ photon}{3.6_{12}\,x10^{-19}\ J}\right) = \mathbf{1.0x10^{18}\ photons / s}$$

7.98 In the visible series with $n_{final} = 2$, the transitions will end in either the 2s or 2p orbitals. With the restriction that the angular momentum quantum number can change by only ±1, the allowable transitions are from a p orbital to 2s, from an s orbital to 2p, and from a d orbital to 2p. The problem specifies a change in *energy level*, so n_{init} must be 3, 4, 5, etc. (Although a change from 2p to 2s would result in a +1 change in ℓ, this is not a change in energy level).

The first four transitions are as follows:

3s → 2p

3d → 2p

4s → 2p

3p → 2s

By looking ahead to Figure 8.7, you can confirm that these transitions are listed from lowest to highest energy.

CHAPTER 8

ELECTRON CONFIGURATION AND CHEMICAL PERIODICITY

FOLLOW-UP PROBLEMS

8.1 Plan: The superscripts can be added to indicate the number of electrons in the element, and hence its identity. A horizontal orbital diagram is written for simplicity, although it does not indicate the sublevels have different energies. Based on the orbital diagram, we identify the electron of interest and determine its four quantum numbers.

Solution: number of electrons = 2 + 2 + 4 = 8; element = oxygen.

Orbital diagram:

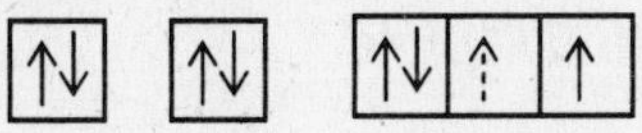

1s 2s 2p

The dashed arrow represents the sixth electron (review Hund's rule). This electron has the following quantum numbers: $n = 2$, $l = 1$ (for p orbital), $m_l = 0$, $m_s = +^1/_2$.

8.2 a) For Ni (Z = 28), the full electron configuration is $1s^22s^22p^63s^23p^64s^23d^8$. The condensed configuration is [Ar] $4s^23d^8$. The partial orbital diagram for the valence electrons is

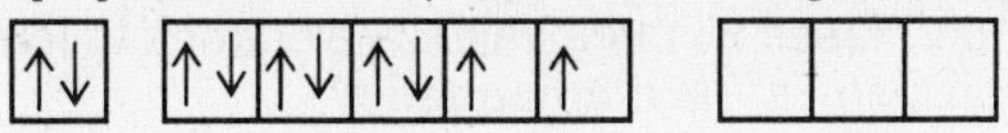

4s 3d 4p

There are 28 – 10(valence) = 18 inner electrons.

b) For Sr (Z = 38), the full electron configuration is $1s^22s^22p^63s^23p^64s^23d^{10}4p^65s^2$. The condensed configuration is [Kr] $5s^2$. The partial orbital diagram is

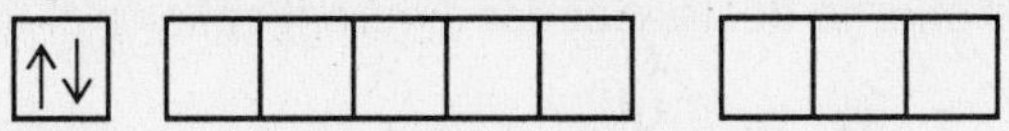

5s 4d 5p

There are 38 – 2(valence) = 36 inner electrons.

c) For Po, the full configuration is $1s^22s^22p^63s^23p^64s^23d^{10}4p^65s^24d^{10}5p^66s^24f^{14}5d^{10}6p^4$. Do the superscripts add up to 84? The condensed configuration is [Xe] $6s^24f^{14}5d^{10}6p^4$. The partial orbital diagram represents valence electrons only; the inner electrons are those in the previous noble gas (Xe or $1s^22s^22p^63s^23p^64s^23d^{10}4p^65s^24d^{10}5p^6$) and *filled* transition ($5d^{10}$) and inner transition ($4f^{14}$) levels.

6s 6p

There are 84 – 6(valence) = 78 inner electrons.

8.3 a) Cl < Br < Se. Cl has a smaller n than Br and Se (rule 1). Br experiences a higher Z_{eff} than Se and is smaller (rule 2.)

b) Xe < I < Ba. Xe and I have the same n, but Xe experiences a higher Z_{eff} and is smaller. Ba has the highest n and is the largest.

8.4 a) Sn < Sb < I. These elements have the same n, so the "period" rule applies. Iodine has the highest IE, because its outer electron is most tightly held and hardest to remove.

b) Ba < Sr < Ca. The "group" rule applies in this case. Barium's outer electron receives the most shielding, therefore it is easiest to remove and has the lowest IE.

8.5 The exceptionally large jump from IE_3 to IE_4 means that the fourth electron is an inner electron. Thus, Q has three valence electrons. Since Q is in period 3, it must be aluminum, Al: $1s^22s^22p^63s^23p^1$.

8.6 a) Barium loses two electrons to be isoelectronic with Xe: Ba ([Xe] $6s^2$) $\rightarrow$ Ba^{2+} ([Xe]) + $2e^-$

b) Oxygen gains two e^- to be isoelectronic with Ne: O ([He] $2s^22p^4$ + $2e^-$ $\rightarrow$ O^{2-} ([Ne])

c) Lead can lose two electrons to form an "inert pair" configuration:

Pb ([Xe] $6s^24f^{14}5d^{10}6p^2$) $\rightarrow$ Pb^{2+} ([Xe] $6s^24f^{14}5d^{10}$) + $2e^-$

or lead can lose four electrons form a "pseudo-noble gas" configuration:

Pb ([Xe] $6s^24f^{14}5d^{10}6p^2$) $\rightarrow$ Pb^{4+} ([Xe] $4f^{14}5d^{10}$) + $4e^-$

8.7 a) The V atom ([Ar] $4s^23d^3$) loses the two 4s electrons and one 3d electron to form V^{3+} ([Ar] $3d^2$). There are two unpaired d electrons, so V^{3+} is **paramagnetic**.

b) The Ni atom ([Ar] $4s^23d^8$) loses the two 4s electrons to form Ni^{2+} ([Ar] $3d^8$). There are two unpaired d electrons, so Ni^{2+} is **paramagnetic**.

3d [↑↓][↑↓][↑↓][↑][↑]

c) The La atom ([Xe] $6s^2$ $5d^1$) loses all three valence electrons to form La^{3+} ([Xe]). There are no unpaired electrons, so La^{3+} is **diamagnetic**.

8.8 a) Ionic size increases down a group, so F^- < Cl^- < Br^-.

b) These species are isoelectronic (all have 10 electrons), so size increases with increasing negative charge: Mg^{2+} < Na^+ < F^-.

c) Ionic size increases as charge decreases for different cations of the same element, so Cr^{3+} < Cr^{2+}.

END-OF-CHAPTER PROBLEMS

8.1 The periodic table is arranged in order of increasing atomic number, that is, number of protons in the atomic nucleus. Scientific opinion would be very skeptical of a new element between tin (Sn, Z = 50) and antimony (Sb, Z =51) because it would imply that the element had part of a proton.

8.3 a) Average of atomic masses of Na and Rb: (22.99 + 85.47)/2 = 54.23. With the actual atomic mass of K as 39.10, the averaged value is not a good predicted value for the atomic mass of K.

b) Average of melting points of Cl_2 and Br_2 = (-101.0 °C + 113.6 °C)/2 = 6.3 °C. This predicted value is not too far off from the actual value of –7.2 °C.

8.6 The m_s quantum number relates only to the electron and describes it spin. The n, l, and m_l quantum numbers describe the orbital's energy, shape, and orientation in 3-D space.

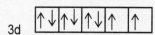

8.9 Shielding is the ability of inner electrons to decrease the full attractive force of the nucleus on outer electrons by blocking some of the positive charge of the nucleus. Effective nuclear charge is the nuclear charge that the electron actually experiences. The effective nuclear charge is the nuclear charge minus the shielding.

8.11 a) The l = 1 quantum number can only refer to a p orbital. These quantum numbers designate the 2p orbitals, which hold a maximum of **6** electrons.

b) There are five 3d orbitals, therefore a maximum of **10** electrons can have the 3d designation.

c) There is one 4s orbital which holds a maximum of **2** electrons.

8.13 a) **6** electrons can be found in the three 4p orbitals, 2 in each orbital.

b) **2** electrons can have the 3 quantum numbers given for one of the 3p orbitals with the difference between the two being in the m_s quantum number.

c) **14** electrons can be found in the 5f orbitals (l = 3 designates f orbitals).

8.16 Hund's rule states that electrons will fill empty orbitals in the same sublevel before filling half-filled orbitals. This lowest-energy arrangement has the maximum number of unpaired electrons with parallel spins. The correct electron configuration for nitrogen is shown in (a), which is contrasted to an incorrect configuration shown in (b). The arrows in the 2 p orbitals of configuration (a) could alternatively all point down.

(a) - correct

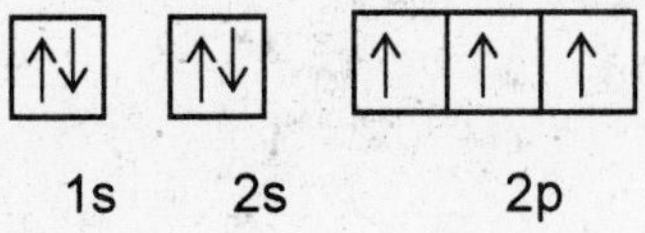

(b) - incorrect

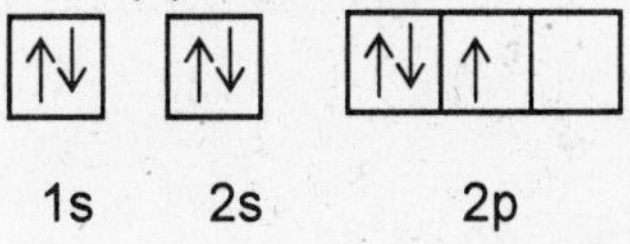

8.18 For elements in the same group (vertical column in periodic table), the electron configuration of the outer electrons is identical except for the n value. For elements in the same period (horizontal row in periodic table), their configurations vary because each succeeding element has one additional electron. The electron configurations are similar only in the fact that the same level (principal quantum number) is the outer level.

8.20 The total electron capacity for an energy level is $2n^2$, so the n = 4 energy level holds a maximum of $2(4^2)$ = 32 electrons. A filled n = 4 energy level would have the following configuration: $4s^24p^64d^{10}4f^{14}$.

8.21 Assume that the electron is in the ground state configuration and that electrons fill in a p_x–p_y–p_z order. By convention, we assign the first electron to fill an orbital an m_s = +½. Also by convention, m_l = -1 for the p_x orbital, m_l = 0 for the p_y orbital, and m_l = +1 for the p_z orbital.

a) The outermost electron in a rubidium atom would be in a 5s orbital. The quantum numbers for this electron are n = 5, l = 0, m_l = 0, and m_s = +½.

b) The S^- ion would have the configuration $[Ne]3s^23p^5$. The electron added would go into the $3p_z$ orbital. Quantum numbers are n = 3, l = 1, m_l = +1, and m_s = -½.

c) Ag atoms have the configuration $[Kr]5s^14d^{10}$. The electron lost would be from the 5s orbital with quantum numbers n = 5, l = 0, m_l = 0, and m_s = +½.

d) The F atom has the configuration $[He]2s^22p^5$. The electron gained would go into the $2p_z$ orbital. Quantum numbers are n = 2, l = 1, m_l = +1, and m_s = -½.

8.23 a) Rb (Z = 37); $1s^22s^22p^63s^23p^64s^23d^{10}4p^65s^1$

b) Ge (Z = 32); $1s^22s^22p^63s^23p^64s^23d^{10}4p^2$

c) Ar (Z = 18); $1s^2 2s^2 2p^6 3s^2 3p^6$

8.25 a) Cl (Z = 17); $1s^2 2s^2 2p^6 3s^2 3p^5$
b) Si (Z = 14); $1s^2 2s^2 2p^6 3s^2 3p^2$
c) Sr (Z = 38); $1s^2 2s^2 2p^6 3s^2 3p^6 4s^2 3d^{10} 4p^6 5s^2$

8.27 Valence electrons are those electrons beyond the previous noble gas configuration and unfilled d and f sublevels.
a) Ti (Z = 22); [Ar] $4s^2 3d^2$

[Ar] ⟦↑↓⟧ ⟦↑│↑│ │ │ ⟧ ⟦ │ │ ⟧
4s   3d   4p

b) Cl (Z = 17); [Ne] $3s^2 3p^5$

[Ne] ⟦↑↓⟧ ⟦↑↓│↑↓│↑⟧
3s   3p

c) V (Z = 23); [Ar] $4s^2 3d^3$

[Ar] ⟦↑↓⟧ ⟦↑│↑│↑│ │ ⟧ ⟦ │ │ ⟧
4s   3d   4p

8.29 a) Mn (Z = 25); [Ar] $4s^2 3d^5$

[Ar] ⟦↑↓⟧ ⟦↑│↑│↑│↑│↑⟧ ⟦ │ │ ⟧
4s   3d   4p

b) P (Z = 15); [Ne] $3s^2 3p^3$

[Ne] ⟦↑↓⟧ ⟦↑│↑│↑⟧
3s   3p

c) Fe (Z = 26); [Ar] $4s^2 3d^6$

[Ar] ⟦↑↓⟧ ⟦↑↓│↑│↑│↑│↑⟧ ⟦ │ │ ⟧
4s   3d   4p

8.31 a) Element = O, Group 6A(16), period 2   ⟦↑↓⟧ ⟦↑↓│↑│↑⟧

b) Element = P, Group 5A(15), period 3   ⟦↑↓⟧ ⟦↑│↑│↑⟧

8.33 a) ⟦↑↓⟧ ⟦↑↓│↑↓│↑⟧   Cl; Group 7A(17); period 3
3s   3p

b) ⟦↑↓⟧ ⟦↑↓│↑↓│↑↓│↑↓│↑↓⟧ ⟦↑│↑│↑⟧   As; Group 5A(15); period 4
4s   3d   4p

8.35 a) The orbital diagram shows the element is in period 4 (n=4 as outer level). The configuration is $1s^22s^22p^63s^23p^64s^23d^{10}4p^1$ or $[Ar]4s^23d^{10}4p^1$. One electron in the p level indicates the element is in group 3A(13). The element is Ga.

b) The orbital diagram shows the 2s and 2p orbitals filled which would represent the last element in period 2, Ne. The configuration is $1s^22s^22p^6$ or $[He]2s^22p^6$. Filled s and p orbitals indicate the group 8A(18).

8.37 It is easiest to determine the types of electrons by writing a condensed electron configuration.

a) O (Z = 8); $[He]\ 2s^22p^4$. There are 2 inner electrons (represented by [He]) and 6 outer electrons. The number of valence electrons equals the outer electrons in this case.

b) Sn (Z = 50); $[Kr]\ 5s^24d^{10}5p^2$. There are 36 (from [Kr]) + 10 (from the filled d level) = 46 inner electrons. There are 4 outer electrons (highest energy level is n=5) and 4 valence electrons.

c) Ca (Z = 20); $[Ar]\ 4s^2$. There are 2 outer electrons, 2 valence electrons, and 18 inner electrons.

d) Fe (Z = 26); $[Ar]\ 4s^23d^6$. There are 2 outer electrons (from n = 4 level), 8 valence electrons (the d orbital electrons count in this case because the sublevel is not full), and 18 inner electrons.

e) Se (Z = 34); $[Ar]\ 4s^23d^{10}4p^4$. There are 6 outer electrons (2 + 4 in the n = 4 level), 6 valence electrons (filled d sublevels count as inner electrons), and 28 ((18 + 10) or (34 - 6)) inner electrons.

8.39 a) The electron configuration $[He]2s^22p^1$ has a total of 5 electrons (3 + 2 from He configuration) which is element boron with symbol **B**. Boron is in group 3A(13). Other elements in this group are Al, Ga, In, and Tl.

b) The electrons in this element total 16, 10 from neon configuration plus 6 from the rest of the configuration. Element 16 is sulfur, **S**, in group 6A(16). Other elements in group 6A(16) are O, Se, Te, and Po.

c) Electrons total 3 + 54 (from xenon) = 57. Element 57 is lanthanum, **La**, in group 3B(3). Other elements in this group are Sc, Y, and Ac.

8.41 a) Carbon; other Group 4A(14) elements include Si, Ge, Sn, and Pb.

b) Vanadium; other Group 5B(5) elements include Nb, Ta, and Ha.

c) Phosphorus; other Group 5A(15) elements include N, As, Sb, and Bi.

8.43 The ground state configuration of Na is $1s^22s^22p^63s^1$. Upon excitation, the $3s^1$ electron is promoted to the 3p level, with configuration $1s^22s^22p^63p^1$.

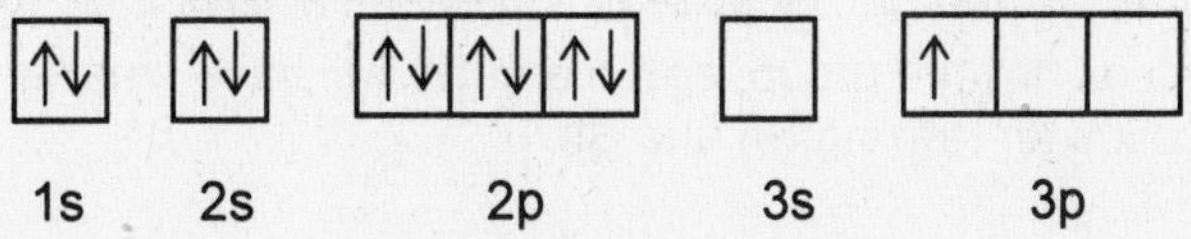

8.46 Atomic size increases down a main group and decreases across a period. Ionization energy decreases down a main group and increases across a period. These opposite trends result because as the atom gets larger, the outer electron is further from the attraction of the positive charge of the nucleus, which is what holds the electron in the atom. It thus takes less energy (lower IE) to remove the outer electron in a larger atom than to remove the outer electron in a smaller atom. As the atomic size decreases across a period due to higher Z_{eff}, it takes more energy (higher IE) to remove the outer electron.

8.48 For a given element, successive ionization energies always increase. As each successive electron is removed the positive charge on the ion increases which results in a stronger attraction between the leaving electron and the ion.

When a large jump between successive ionization energies is observed, the subsequent electron must come from another energy level. Thus, by looking at a series of successive ionization energies, we can determine the number of valence electrons. For instance the electron configuration for potassium is $[Ar]4s^1$. The first electron lost is the one from the 4s level. The second electron lost must come from the 3p level. Thus we see a significant jump in the amount of energy for the second ionization when compared to the first ionization.

8.50 A high, endothermic IE_1 means it is very difficult to remove the first outer electron. This value would exclude any metal, because metals lose an outer electron easily. A very negative, exothermic EA_1 suggests that this element easily gains one electron. These values indicate the element belongs to the halogens, Group **7A(17)**, which form –1 ions.

8.53 a) Increasing atomic size: **K < Rb < Cs**. These three elements are all part of the same group, the alkali metals. Atomic size increases down a main group (larger outer electron orbital), so potassium is the smallest and cesium is the largest.

b) Increasing atomic size: **O < C < Be**. These three elements are in the same period and atomic size decreases across a period (increasing effective nuclear charge), so beryllium is the largest and oxygen the smallest.

c) Increasing atomic size: **Cl < S < K.** Chlorine and sulfur are in the same period so chlorine is smaller since it is further to the right in the period. Potassium is the first element in the next period so it is larger than either Cl or S. K is larger than Na because it is further down the periodic table than K, but in the same group as Na. And Na is larger than Cl or S because it is further to the left in the same period as Cl and S. Thus, since K>Na>S>Cl then potassium is larger than chlorine and sulfur.

d) Increasing atomic size: **Mg < Ca < K**. Calcium is larger than magnesium because Ca is further down the alklaine earth metal group on the periodic table than Mg. Potassium is larger than calcium because K is further to the left than Ca in period 4 of the periodic table.

8.55 a) **Ba < Sr < Ca**. The "group" rule applies in this case. Barium's outer electron receives the most shielding, therefore it is easiest to remove and has the lowest IE.

b) **B < N < Ne**. These elements have the same n, so the "period" rule applies. B experiences the lowest Z_{eff} and has the lowest IE. Ne has the highest IE, because it's very difficult to remove an electron from the stable noble gas configuration.

c) **Rb < Se < Br**. IE decreases with increasing atomic size, so Rb (largest atom) has the smallest IE. Se has a lower IE than Br because IE increases across a period.

d) **Sn < Sb < As**. IE decreases down a group, so Sn and Sb will have smaller IE's than As. The "period" rule applies for ranking Sn and Sb.

8.57 The successive ionization energies show a significant jump between the first and second IE's and between the third and fourth IE's. This indicates that 1) the first electron removed occupies a different orbital than the second electron removed, 2) the second and third electrons occupy the same orbital, 3) the third and fourth electrons occupy different orbitals, and 4) the fourth and fifth electrons occupy the same orbital. The electron configurations for period 2 elements range from $1s^22s^1$ for lithium to $1s^22s^22p^6$ for neon. The three different orbitals are 1s, 2s and 2p. The first electron is removed from the 2p orbital and the second from the 2s orbital in order for there to be a significant jump in the ionization energy between the two. The third electron is removed from the 2s orbital while the fourth is removed from the 1s orbital since IE_4 is much greater than IE_3. The configuration is $\mathbf{1s^22s^22p^1}$ which represents the five electrons in boron, **B**.

8.59 a) **Na** would have the highest IE_2 because ionization of a second electron would require breaking the stable [Ne] configuration:

First ionization: Na ([Ne] $3s^1$) $\rightarrow$ Na^+ ([Ne]) + e^- (low IE)

Second ionization: Na^+ ([Ne]) $\rightarrow$ Na^{+2} ([He]$2s^22p^5$) + e- (high IE)

b) **Na** would have the highest IE_2 because it has one valence electron and is smaller than K.

c) You might think that Sc would have the highest IE_2, because removing a second electron would require breaking the stable, filled 4s shell. However, **Be** has the highest IE_2 because Be's small size makes it difficult to remove a second electron.

8.61 Three of the ways that metals and nonmetals differ are: 1) metals conduct electricity, nonmetals do not; 2) when they form stable ions metal ions tend to have a positive charge, nonmetal ions tend to have a negative charge; and 3) metal oxides are ionic and act as bases, nonmetal oxides are covalent and act as acids. How many other differences are there?

8.62 Metallic character increases down a group and decreases across a period. These trends are the same as those of atomic size and opposite those of ionization energy.

Larger elements in groups 4A, Sn and Pb, have atomic electron configurations that look like $ns^2(n-1)d^{10}np^2$. Both of these elements are metals so they will form positive ions. To reach the noble gas configuration of xenon the atoms would have to lose 14 electrons, which is not likely. Instead the atoms lose either 2 or 4 electrons to attain a stable configuration with either the ns and (n-1)d filled for the 2+ ion or the (n-1)d orbital filled for the 4+ ion. The Sn^{2+} and Pb^{2+} ions form by losing the two p electrons:

Sn ([Kr]$5s^24d^{10}5p^2$) $\rightarrow$ Sn^{2+} ([Kr]$5s^24d^{10}$) + $2e^-$

Pb ([Xe]$6s^25d^{10}6p^2$) $\rightarrow$ Pb^{2+} ([Xe]$6s^25d^{10}$) + $2e^-$

The Sn^{4+} and Pb^{4+} ions form by losing the two p and two s electrons:

Sn ([Kr]$5s^24d^{10}5p^2$) $\rightarrow$ Sn^{4+} ([Kr]$4d^{10}$) + $4e^-$

Pb ([Xe]$6s^25d^{10}6p^2$) $\rightarrow$ Pb^{4+} ([Xe]$5d^{10}$) + $4e^-$

8.64 Possible ions for tin and lead have +2 and +4 charges. The +2 ions form by loss of the outermost two p electrons, while the +4 ions form by loss of these and the outermost two s electrons.

8.68 Metallic behavior increases down a group and decreases across a period.

a) **Rb** is more metallic because it is to the left and below Ca.

b) **Ra** is more metallic because it lies below Mg.

c) **I** is more metallic because it lies below Br.

8.70 a) **As** should be less metallic than antimony because it lies above Sb in the same group of the periodic table.

b) **P** should be less metallic because it lies to the right of silicon in the same period of the periodic table.

c) **Be** should be less metallic since it lies above and to the right of sodium on the periodic table.

8.72 The reaction of a *nonmetal* oxide in water produces an *acidic* solution. An example of a Group 6A(16) nonmetal oxide is SO_3(g): $SO_3(g) + H_2O(g) \rightarrow H_2SO_4(aq)$

8.74 For main group elements the most stable ions have electron configurations identical to those noble gas atoms.

a) Cl^-, $1s^22s^22p^63s^23p^6$. Chlorine atoms are one electron short of the noble gas configuration, so a –1 ion will form by adding an electron to have the same electron configuration as an argon atom.

b) Na^+, $1s^22s^22p^6$. Sodium atoms contain one more electron than the noble gas configuration of neon. Thus, a sodium atom loses one electron to form a +1 ion.

c) Ca^{2+}, $1s^22s^22p^63s^23p^6$. Calcium atoms contain two more electrons than the noble gas configuration of argon. Thus, a calcium atom loses two electrons to form a +2 ion.

8.76 a) Elements in the main group form ions that attain a filled outer level, or noble gas configuration. In this case, Al achieves the noble gas configuration of Ne ($1s^22s^22p^6$) by losing its 3 outer shell electrons to form $\mathbf{Al^{3+}}$.

$Al\ (1s^22s^22p^63s^23p^1) \rightarrow Al^{3+}\ (\mathbf{1s^22s^22p^6}) + 3e^-$

b) Sulfur achieves the noble gas configuration of Ar by gaining 2 electrons to form $\mathbf{S^{2-}}$.

$S\ (1s^22s^22p^63s^23p^4) + 2e^- \rightarrow S^{2-}\ (\mathbf{1s^22s^22p^63s^23p^6})$

c) Strontium achieves the noble gas configuration of Kr by losing 2 electrons to form $\mathbf{Sr^{2+}}$.

$Sr\ (1s^22s^22p^63s^23p^64s^23d^{10}4p^65s^2) \rightarrow Sr^{2+}\ (\mathbf{1s^22s^22p^63s^23p^64s^23d^{10}4p^6}) + 2e^-$

8.78 To find the number of unpaired electrons look at the electron configuration expanded to include the different orientations of the orbitals, such as p_x and p_y and p_z. In the noble gas configurations all electrons are paired because all orbitals are filled.

a) Configuration of 2A(2) group elements: [noble gas]ns^2, no unpaired electrons.

b) Configuration of 5A(15) group elements: [noble gas]$ns^2np_x^1np_y^1np_z^1$. 3 unpaired electrons, one each in p_x, p_y, p_z.

c) Configuration of 8A(18) group elements: noble gas configuration with no half-filled orbitals, no unpaired electrons.

d) Configuration of 3A(13) group elements: [noble gas]ns^2np^1. One unpaired electron in one of the p orbitals.

8.80 Substances are paramagnetic if they have unpaired electrons. This problem is more challenging if you have difficulty picturing the orbital diagram from the electron configuration. Obviously an odd number of electrons will necessitate that at least one electron is unpaired, so odd electron species are paramagnetic. But a substance with an even number of electrons can also be paramagnetic, because even numbers of electrons do not guarantee all electrons are paired.

a) Ga (Z = 31) = [Ar] $4s^23d^{10}4p^1$. The s and d sublevels are filled, so all electrons are paired. The lone p electron is unpaired, so this element is **paramagnetic**.

b) Si (Z = 14) = [Ne] $3s^23p^2$. This element is **paramagnetic** with two unpaired electrons because the p subshell looks like this | ↑ | ↑ | | , not this | ↑↓ | | | .

c) Be (Z = 4) = [He] $2s^2$, so it is **not paramagnetic**.

d) Te (Z = 52) = [Kr] $5s^24d^{10}5p^4$ is **paramagnetic** with two unpaired electrons.

8.82 a) V^{3+}, $[Ar]3d^2$. Transition metals first lose the 4s electrons in forming ions, so to form the +3 ion a vanadium atom of configuration $[Ar]4s^23d^3$ loses two 4s electrons and one 3d electron.

b) Cd^{2+}, $[Kr]4d^{10}$. Cadmium atoms lose two electrons from the 4s orbital to form the +2 ion.

c) Co^{3+}, $[Ar]3d^7$. Cobalt atoms lose two 4s electrons and one 3d electron to form the +3 ion.

d) Ag^+, $[Kr]4d^{10}$. Silver atoms lose the one electron in the 5s orbital to form the +1 ion.

a) and c) are paramagnetic; b) and d) are diamagnetic.

8.84 For palladium to be diamagnetic, all of its electrons must be paired. You might first write the condensed electron configuration for Pd as $[Kr]\ 5s^24d^8$. However, the partial orbital diagram is not consistent with diamagnetism.

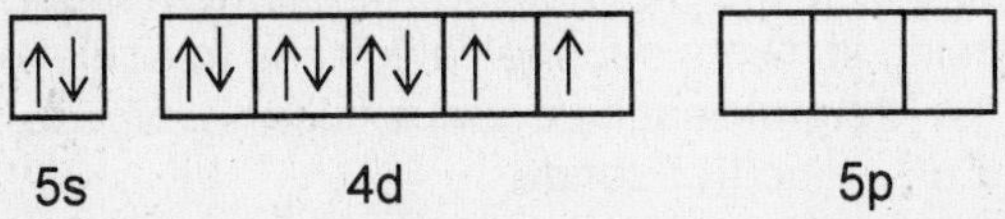

Promoting an s electron into the d sublevel (as in (c)) still leaves two electrons unpaired.

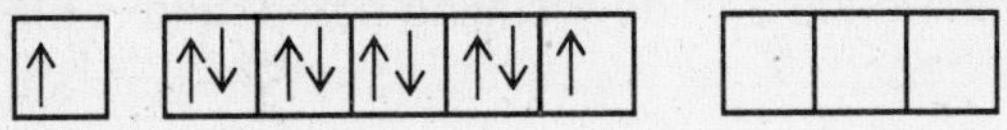

The only configuration that supports diamagnetism is **(b) [Kr] $4d^{10}$**.

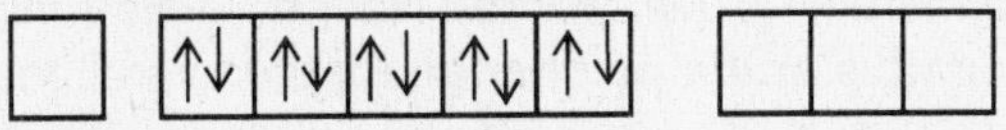

The size of ions increases down a group. For ions that are isoelectronic (have the same electron configuration) size decreases with increasing atomic number.

a) Increasing size: $Li^+ < Na^+ < K^+$. Size increases down group 1A.

b) Increasing size: $Rb^+ < Br^- < Se^{2-}$. These three ions are isoelectronic with the same electron configuration as krypton. Size decreases with atomic number in an isoelectronic series.

c) Increasing size: $F^- < O^{2-} < N^{3-}$. The three ions are isoelectronic with an electron configuration identical to neon. Size decreases with atomic number in an isoelectronic series.

8.86 a) **$Li^+ < Na^+ < K^+$**. Size increases down a group.

b) **$Rb^+ < Br^- < Se^{2-}$**. Size increases as atomic number decreases in an isoelectronic series.

c) **$F^- < O^{2-} < N^{3-}$**. Same as b).

8.90 a) K is further left (greater radius) while Sr is further down (also greater radius).

b) Z_{eff} remains fairly constant after the first 2 or 3 elements across a transition series.

c) Na is farther left (more metallic) but Ca is farther down (also more metallic).

d) Se is farther right (more acidic oxide) but P is farther up (also more acidic oxide).

8.92 a) The smallest metal will be the element to the furthest left on the table that is still a metal. Aluminum, **Al**, is this metal with configuration $1s^22s^22p^63s^23p^1$.

b) The heaviest lanthanide is lutetium, **Lu**, with configuration, $1s^22s^22p^63s^23p^64s^23d^{10}4p^65s^24d^{10}5p^66s^24f^{14}5d^1$.

c) The lightest transition metal in Period 5 is yttrium, **Y**, with configuration $1s^22s^22p^63s^23p^64s^23d^{10}4p^65s^24d^1$.

d) In Period 3 the element that must gain two electrons to have the same electron configuration as argon is sulfur, **S**, with electron configuration $1s^22s^22p^63s^23p^4$.

e) The strontium (**Sr**) 2+ ion would be isoelectronic with krypton. Electron configuration of Sr is $1s^22s^22p^63s^23p^64s^23d^{10}4p^65s^2$.

f) Going down a group metal oxides are more basic. So, of the two metalloids in Group 5A, arsenic and antimony, the one higher on the table, **As**, would form the more acidic oxide. Configuration of As: $1s^22s^22p^63s^23p^64s^23d^{10}4p^3$.

8.95 a) A chemically unreactive Period 4 element would be Kr. Both the Sr^{2+} ion and Br^- ion are isoelectronic with Kr. Their combination results in **$SrBr_2$,** strontium bromide.

b) Ar is the Period 3 noble gas. Ca^{2+} and S^{2-} are isoelectronic with Ar. The resulting compound is **CaS**, calcium sulfide.

c) The smallest filled d subshell is the 3d shell, so the element must be in Period 4. Zn forms the Zn^{+2} ion by losing its two s subshell electrons to achieve a *pseudo-noble gas* configuration ([Ar] $3d^{10}$). The smallest halogen is fluorine, whose anion is F^-. The resulting compound is **ZnF_2**, zinc(II) fluoride.

d) Ne is the smallest element in Period 2, but it is not ionizable. Li is the largest atom whereas F is the smallest atom in Period 2. The resulting compound is **LiF**, lithium fluoride.

8.97 a) Electron affinities decrease down Group 1A(1) elements, so one would predict that $Li + e^- \rightarrow Li^-$ will have a more negative ΔH then $Na + e^- \rightarrow Na^-$. When combined, $\Delta H_{Li} - \Delta H_{Na}$ yields a **negative enthalpy change** for the overall reaction.

b) This reaction involves ionization energies since the potassium is losing one electron to form the +1 ion and the opposite occurs for the sodium +1 ion. Thus, the enthalpy change would equal the ionization energy for potassium minus the ionization energy for sodium. Since sodium has a greater ionization energy than potassium the **enthalpy change is negative**.

8.101 Blue or purple clothing appears blue or purple because those wavelengths are being reflected to the observer's eye. Blue wavelengths have higher energy than the orange/red wavelengths, and so you would want to wear blue clothing to stay cooler because the higher energy would be reflected, rather than absorbed, by the clothing.

8.103 We know that the maximum number of orbitals per energy level, n, is n^2. The $n = 4$ level contains 16 orbitals (s=1, p=3, d=5, f=7; total = 16). The $n = 5$ level must contain 25 orbitals so an additional sublevel is needed to be consistent with the n^2 rule. The additional sublevel, called "g", contains 9 orbitals (s=1, p=3, d=5, f=7, g=9; total =25).

By substituting the 5g level into the periodic memory aid diagram, we can extend the orbital filling trend shown in Figure 8.10 as follows:

1s ⟶
2s ⟶ 2p
3s ⟶ 3p
4s ⟶ 3d ⟶ 4p
5s ⟶ 4d ⟶ 5p
6s ⟶ 4f ⟶ 5d ⟶ 6p
7s ⟶ 5f ⟶ 6d ⟶ **7p**
8s ⟶ **5g** ⟶ **6f** ⟶ **7d** ⟶ **8p**

An element with Z = 126 could have the configuration [Rn] $7s^25f^{14}6d^{10}7p^68s^25g^6$. The outermost electron is in the 8s shell, where n =8 and l = 0. There is one orbital in this sublevel.

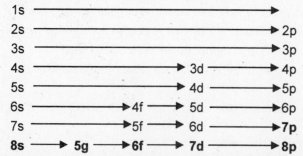

8.105 a) Recall from Chapter 7 that E=hν and c = λν, therefore E = h(c/λ).

$$E_{436nm} = h\left(\frac{c}{\lambda}\right) = \left(6.63x10^{-34}\ J \cdot s\right)\left(\frac{3.00x10^{8}\ m/s}{436x10^{-9}\ m}\right) = \mathbf{4.56x10^{-19}\ J}$$

$$E_{254\,nm} = h\left(\frac{c}{\lambda}\right) = \left(6.63x10^{-34}\ J \cdot s\right)\left(\frac{3.00x10^{8}\ m/s}{254x10^{-9}\ m}\right) = \mathbf{7.83x10^{-19}\ J}$$

b) The lower excited state has an energy of **7.83x10^{-19} J**, and the higher excited state has an energy of (7.83x10^{-19} J + 4.56x10^{-19} J) or **1.24x10^{-18} J**.

c) $$\lambda = \frac{hc}{E} = \frac{\left(6.63x10^{-34}\ J \cdot s\right)\left(3.00x10^{8}\ m/s\right)}{1.24x10^{-18}\ J} = 1.60x10^{-7}\ m = \mathbf{160.\ nm}$$

8.107 The electron configuration for cerium, Ce, is [Xe]$6s^2 4f^1 5d^1$. To form the +4 ion the cerium atom would lose four electrons (two from 6s, one from 5d, and one from 4f) and have the noble gas configuration for xenon. Thus, the ion would be expected to be stable. The electron configuration for europium, Eu, is [Xe]$6s^2 4f^7$. To form the +2 ion the europium atom would lose two electrons from the 6s level. The half-filled 4f subshell with one electron in each f orbital would be relatively stable.

CHAPTER 9

MODELS OF CHEMICAL BONDING

FOLLOW-UP PROBLEMS

9.1 Plan: First write out the condensed electron configuration and electron-dot structure for magnesium atoms and chlorine atoms. In the formation of the ions, each magnesium atom will lose two electrons to form the +2 ion and each chlorine atom will gain one electron to form the –1 ion. Write the condensed electron configuration and electron-dot structure for each of the ions. The formula of the compound is found by combining the ions in a ratio that gives a neutral compound.

Solution: Condensed electron configuration:

Mg ([Ne]$3s^2$) + Cl ([Ne]$3s^23p^5$) $\rightarrow$ Mg^{+2} ([Ne]) + Cl^- ([Ne]$3s^23p^6$)

To balance the charge (or the number of electrons lost and gained) two chlorine atoms:

Mg ([Ne]$3s^2$) + 2Cl ([Ne]$3s^23p^5$) $\rightarrow$ Mg^{+2} ([Ne]) + $2Cl^-$ ([Ne]$3s^23p^6$)

Lewis electron-dot symbols:

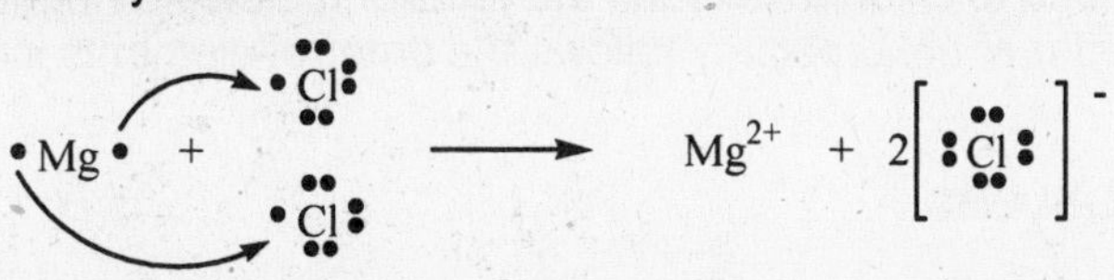

The formula of the compound would contain two chloride ions for each magnesium ion, $MgCl_2$.

9.2 Plan:

a) All bonds are single bonds from silicon to a second row element. The trend in bond length can be predicted from the fact that atomic radii decrease across a row, so fluorine will be smaller than oxygen, which is smaller than carbon. Bond length will therefore decrease across the row while bond energy increases.

b) Bonds are all between two nitrogen atoms, but differ in the number of electrons shared between the two atoms. The more electrons shared the shorter the bond length and the greater the bond energy.

Solution:

a) Bond length: Si–F< Si–O < Si–C

Bond strength: Si–C < Si–O < Si–F

b) Bond length: N≡N < N=N < N–N

Bond energy: N–N < N=N < N≡N

Check:

Bond lengths from Table 9.3: Si–F, 156pm < Si–O, 161 pm < Si–C, 186 pm

Bond energies from Table 9.2: Si–C, 301 kJ < Si–O, 368 kJ < Si–F, 565 kJ

Bond lengths from Table 9.3: N≡N, 110 pm < N=N, 122 pm < N–N, 146 pm

Bond energies from Table 9.2: N–N, 160 kJ < N=N, 418 kJ < N≡N, 945 kJ

The values from the tables agree with the order predicted.

9.3 Plan: Bond polarity can be determined from the difference in electronegativity of the two atoms. The more electronegative atom holds the δ- charge and the other atom holds the δ+ charge.

Solution: a) From Figure 9.16, EN of Cl = 3.0, EN of F = 4.0, EN of Br = 2.8. Cl–Cl will be the least polar since there is no difference between the electronegativity of the two chlorine atoms. Cl–F will be more polar than Cl–Br since the electronegativity difference between Cl and F is greater than the electronegativity difference between Cl and Br.

$$Cl-Cl < \overset{\delta+}{Br}-\overset{\delta-}{Cl} < \overset{\delta+}{Cl}-\overset{\delta-}{F}$$

b) From Figure 9.16, EN of Cl = 3.0, EN of Si = 1.8, EN of P = 2.1, EN of S = 2.5. Si–Si bond is nonpolar, therefore the least polar of the four bonds. The most polar is Si–Cl with an EN difference of 1.2, next is P–Cl with an EN difference of 0.9 and next is S–Cl with an EN of 0.5.

$$Si-Si < \overset{\delta+}{S}-\overset{\delta-}{Cl} < \overset{\delta+}{P}-\overset{\delta-}{Cl} < \overset{\delta+}{Si}-\overset{\delta-}{Cl}$$

Check: For a) the order follows the increase in electronegativity going up the group of halogens. For b) the order follows the increase in electronegativity across a row. The closer to the end of row 3 (chlorine being at the end of row 3 next to the noble gas argon) that the element that is bonded to chlorine is found the smaller the electronegativity difference of the bond. So the order of bond polarity follows the order of elements in row 3: Si, P, S.

END-OF-CHAPTER PROBLEMS

9.1 a) As ionization energy increases across a period, the atom tends to lose electrons less readily, so metallic character decreases.

b) As atomic radius decreases across a period, electrons are held more tightly, so metallic character decreases.

c) As the number of outer electrons increases, the nuclear charge increases as well, so atoms tend to lose electrons less readily and metallic character decreases.

d) As the effective nuclear charge increases, the outer electrons are held more tightly, so metallic character decreases.

9.4 Metallic behavior increases to the left and down on the periodic table.

a) **Cs** is more metallic since it is further down the alkali metal group than Na.

b) **Rb** is more metallic since it is both to the left and down from Mg.

c) **As** is more metallic since it is further down Group 5A than N.

9.6 Ionic bonding occurs between metals and nonmetals, covalent bonding between nonmetals, and metallic bonds between metals.

a) Bond in CsF is **ionic** because Cs is a metal and F is a nonmetal.

b) Bonding in N_2 is **covalent** N is a nonmetal.

c) Bonding in Na(s) is **metallic** because this is a monatomic, metal solid.

9.8 a) Bonding in O_3 would be **covalent** since O is a nonmetal.

b) Bonding in $MgCl_2$ would be **ionic** since Mg is a metal and Cl is a nonmetal.

c) Bonding in BrO_2 would be **covalent** since both Br and O are nonmetals.

9.10 Lewis electron-dot symbols show valence electrons as dots. Place one dot at a time on the four sides (this method explains the structure in b) and then pair up dots until all valence electrons are used.

9.12 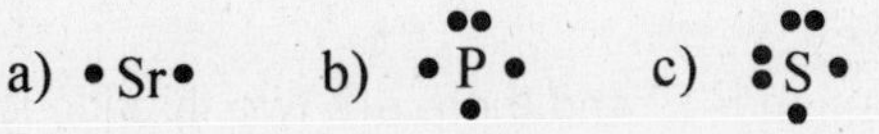

9.14 a) Assuming X is an A-group element, the number of dots (valence electrons) equals the group number. Therefore, X is a **6A(16)** element. It's general electron configuration is **[noble gas]ns^2np^4**, where n is the energy level.

b) X has three valence electrons and is a **3(A)13** element with general e^- configuration **[noble gas]ns^2np^1**.

9.17 Lattice energy refers to the strength of attractions between the ions in an ionic compound. As the charge on the ions increase, the strength of this attraction increases. The attraction also increases with decreasing distance between the two ions, so smaller ionic radii mean stronger lattice energy.

9.20 a) Barium is a metal and loses 2 electrons to achieve a noble gas configuration:

Ba ([Xe]$6s^2$) $\rightarrow$ Ba^{2+} ([Xe]) + $2e^-$

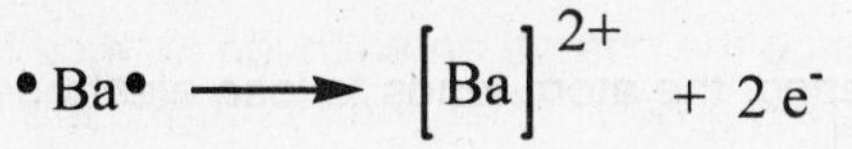

Chlorine is a nonmetal and gains 1 electron to achieve a noble gas configuration:

Cl ([Ne]$3s^23p^5$) + $1e^-$ $\rightarrow$ Cl^- ([Ne]$3s^23p^6$)

The two electrons lost by Ba are gained by two Cl atoms. The ionic compound formed is **$BaCl_2$**.

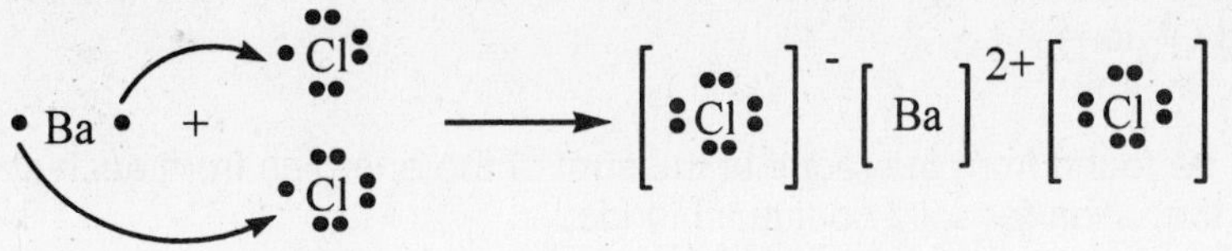

b) Sr ([Kr]$5s^2$) $\rightarrow$ Sr^{2+} ([Kr]) + $2e^-$ O ([He]$2s^22p^4$) + $2e^-$ $\rightarrow$ O^{2-} ([He]$2s^22p^6$)

The ionic compound formed is **SrO**.

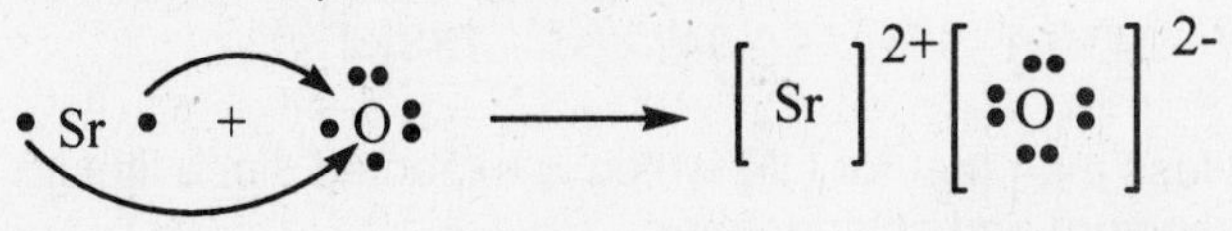

c) Al ([Ne]$3s^23p^1$) $\rightarrow$ Al^{3+} ([Ne]) + $3e^-$ F ([He]$2s^22p^5$) + $1e^-$ $\rightarrow$ F^- ([He]$2s^22p^6$)

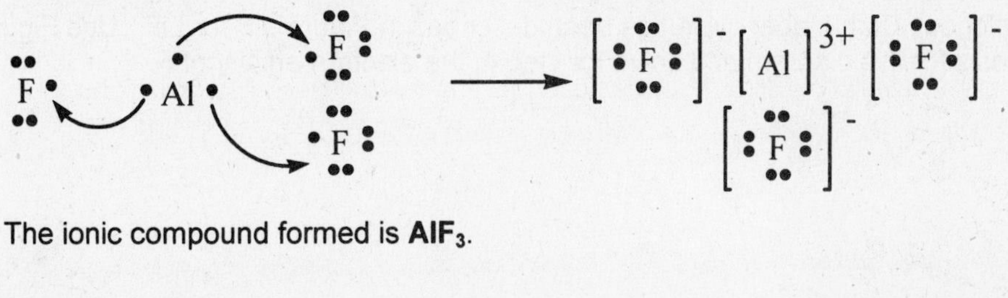

The ionic compound formed is **AlF_3**.

d) Rb ([Kr]$5s^1$) → Rb^+ ([Kr]) + 1e^- O ([He]$2s^22p^4$) + 2e^- → O^{2-} ([He]$2s^22p^6$)

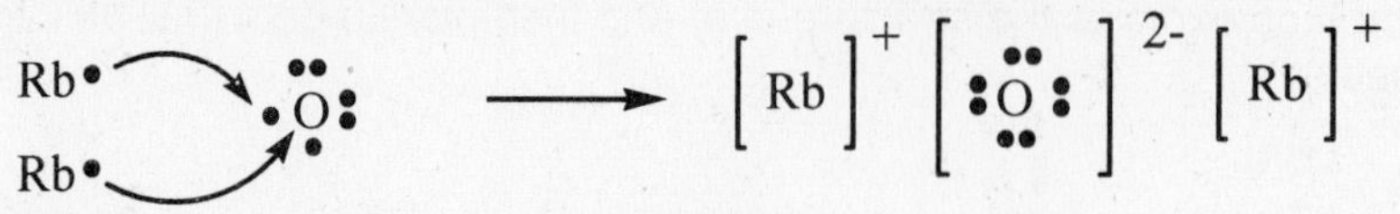

The ionic compound formed is **Rb_2O**.

9.22 a) X in XF_2 is a cation with +2 charge since the anion is F^- and there are two fluoride ions in the compound. Group **2A(2)** metals form +2 ions.

b) X in MgX is an anion with –2 charge since Mg^{2+} is the cation. Elements in Group **6A(16)** form –2 ions.

c) X in X_2SO_4 must be a cation with +1 charge since the polyatomic sulfate ion has a charge of –2. X comes from Group **1A(1)**.

9.24 a) X in X_2O_3 is a cation with +3 charge. The oxygen in this compound has a –2 charge. To produce an electrically neutral compound, 2 cations with +3 charge bond with 3 anions with –2 charge: 2(+3) + 3(-2) = 0. Elements in Group **3A(13)** form +3 ions.

b) The carbonate ion, CO_3^{2-}, has a –2 charge, so X has a +2 charge. Group **2A(2)** elements form +2 ions.

c) X in Na_2X has a –2 charge, balanced with the +2 overall charge from the two Na atoms. Group **6A(16)** elements gain 2 electrons to form –2 ions with a noble gas configuration.

9.26 a) **BaS** would have the higher lattice energy since the charge on each ion is twice the charge on the ions in CsCl and lattice energy is greater when ionic charges are larger.

b) **LiCl** would have the higher lattice energy since the ionic radius of Li^+ is smaller than that of Cs^+ and lattice energy is greater when the distance between ions is smaller.

9.28 a) **BaS** has the lower lattice energy because the ionic radius of Ba^{2+} is larger than Ca^{2+}. A larger ionic radius results in a greater distance between ions. The lattice energy decreases with increasing distance between ions.

b) **NaF** has the lower lattice energy since the charge on each ion (+1, -1) is half the charge on the Mg^{2+} and O^{2-} ions.

9.30 Lattice energy can be found from the fact that the sum of the energies from each step equals the heat of formation for solid sodium chloride.

$$\Delta H_f^o = \Delta H_{Na(s)\to Na(g)} + \frac{1}{2}\Delta H_{Cl_2(g)\to 2Cl(g)} + \Delta H_{Na(g)\to Na^+(g)+e^-} + \Delta H_{Cl(g)+e^-\to Cl^-(g)} + \Delta H_{lattice}$$

$$\Delta H_{lattice} = -411\,kJ - 109\,kJ - (\frac{1}{2}\cdot 243\,kJ) - 496\,kJ - (-349\,kJ) = -\mathbf{788\,kJ}$$

The lattice energy for NaCl is less than that for LiF, which is expected since lithium and fluoride ions are smaller than sodium and chloride ions.

9.33 An analogous Born-Haber cycle has been described in Figure 9.6 for LiF. Use Figure 9.6 as a basis for this diagram and solve for step 4, the electron affinity of F.

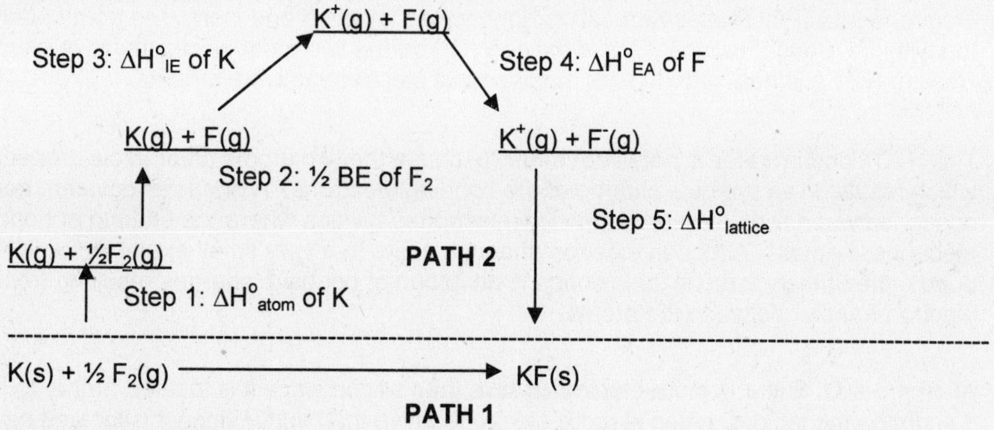

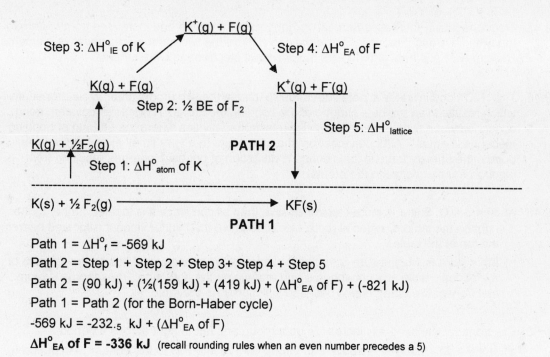

Path 1 = ΔH°_f = -569 kJ

Path 2 = Step 1 + Step 2 + Step 3+ Step 4 + Step 5

Path 2 = (90 kJ) + (½(159 kJ) + (419 kJ) + (ΔH°_{EA} of F) + (-821 kJ)

Path 1 = Path 2 (for the Born-Haber cycle)

-569 kJ = -232.$_5$ kJ + (ΔH°_{EA} of F)

ΔH°_{EA} of F = -336 kJ (recall rounding rules when an even number precedes a 5)

9.34 When two chlorine atoms are far apart there is no interaction between them. Once the two atoms move closer together, the nucleus of each atom attracts the electrons on the other atom. As the atoms move closer this attraction increases, but the repulsion of the two nuclei also increases. When the atoms are very close together the repulsion between nuclei is much stronger than the attraction between nuclei and electrons. The final internuclear distance for the chlorine molecule is the distance at which maximum attraction is achieved in spite of the repulsion. At this distance the energy of the molecule is at its lowest value.

9.35 The bond energy is the energy required to overcome the attraction between H atoms and Cl atoms in one mole of HCl molecules in the gaseous state. Energy input is needed to break the bond, so bond energy is always endothermic and $\Delta H^\circ_{\text{bond breaking}}$ is positive. The same amount of energy needed to break the bond is released upon its formation, so $\Delta H^\circ_{\text{bond forming}}$ is the same magnitude as $\Delta H^\circ_{\text{bond breaking}}$, but opposite in sign (always exothermic and negative).

9.39 a) **I—I < Br—Br < Cl—Cl.** Bond strength increases as the atomic radii of atoms in the bond decreases. Atomic radii decrease up a group in the periodic table, so I is the largest and Cl is the smallest of the three.

b) **S—Br < S—Cl < S—H.** Radii of H are the smallest and those of Br the largest, so the bond strength for S—H is the greatest and that for S—Br is the weakest.

c) **C—N < C=N < C≡N.** Bond strength increases as the number of electrons in the bond increases. The triple bond is the strongest and the single bond is the weakest.

9.41 For given pair of atoms, in this case carbon and oxygen, bond strength increases with increasing bond order. The C=O bond (bond order = 2) is stronger than the C–O bond (bond order =1). These covalent bonds occur between atoms in the formic acid molecule and are much greater than the intermolecular forces that occur between formic acid molecules.

9.43 Electronegativity increases from left to right across a period and increases from bottom to top within a group. Fluorine (F) and oxygen (O) are the two most electronegative elements. Cesium (Cs) and francium (Fr) are the two least electronegative elements.

9.45 The H–O bond in water is **polar covalent**, because the two atoms differ in electronegativity which results in an unequal sharing of the bonding electrons. A nonpolar covalent bond occurs between two atoms with identical electronegativities where the sharing of bonding electrons is equal. Although electron sharing occurs to a very small extent in some ionic bonds, the primary force in ionic bonds is attraction of opposite charges resulting from electron transfer between the atoms.

9.48 a) **Si < S < O**. Sulfur is more electronegative than silicon since it is located further to the right on the table. Oxygen is more electronegative than sulfur since it is located nearer the top of the table.

b) **Mg < As < P**. Magnesium is the least electronegative because it lies on the left side of the periodic table and phosphorus and arsenic on the right side. Phosphorus is more electronegative than arsenic because it is higher on the table.

9.50 Electronegativity generally increases up a group and left to right across a period.

a) **N > P > Si**. Nitrogen is above P in Group 5(A)15 and P is to the right of Si in period 3.

b) **As > Ga > Ca**. All three elements are in Period 4, with As the right-most element.

9.52 a) N—B ← b) N—O → c) C—S (no poles) d) S—O → e) N—H ← f) Cl—O ←

9.54 The more polar bond will have a greater difference in electronegativity, ΔEN.

a) $\Delta EN_a = 1.0$ b) $\Delta EN_b = 0.5$ c) $\Delta EN_c = 0$ d) $\Delta EN_d = 1.0$ e) $\Delta EN_e = 0.9$ f) $\Delta EN_f = 0.5$

(a), **(d)** and **(e)** have greater bond polarity.

9.56 a) Bonds in S_8 are **nonpolar covalent**. All the atoms are nonmetals so the substance is covalent and bonds are nonpolar because all the atoms are of the same element.

b) Bonds in RbCl are **ionic** because Rb is a metal and Cl is a nonmetal.

c) Bonds in PF_3 are **polar covalent**. All the atoms are nonmetals so the substance is covalent. The bonds between P and F are polar because their electronegativity differs. (By 1.9 units for P—F.)

d) Bonds in SCl_2 are **polar covalent**. S and Cl are nonmetals and differ in electronegativity. (By 0.5 unit for S—Cl.)

e) Bonds in F_2 are **nonpolar covalent.** F is a nonmetal. Bonds between two atoms of the same element are nonpolar.

f) Bonds in SF_2 are **polar covalent**. S and F are nonmetals that differ in electronegativity. (By 1.5 units for S—F.)

Increasing bond polarity: **SCl_2 < SF_2 < PF_3**.

9.58 Increasing ionic character occurs with increasing ΔEN.

a) $\Delta EN_{HBr} = 0.7$, $\Delta EN_{HCl} = 0.9$, $\Delta EN_{HI} = 0.4$. **H—I → < H—Br → < H—Cl →**

b) $\Delta EN_{HO} = 1.4$, $\Delta EN_{CH} = 0.4$, $\Delta EN_{HF} = 1.9$. **H—C → < H—O → < H—F →**

c) $\Delta EN_{SCl} = 0.5$, $\Delta EN_{PCl} = 0.9$, $\Delta EN_{SiCl} = 1.2$. **S—Cl < P—Cl < Si—Cl**

9.61 a) A solid metal is a shiny solid that conducts heat, is malleable, and melts at high temperatures. (Other answers include relatively high boiling point and good conductor of electricity.)

b) Metals lose electrons to form positive ions and metals form basic oxides.

9.65 a) Breaking bonds is an endothermic process and bond formation is exothermic process. The signs of the BE's calculated below reflect endothermic (+) and exothermic (-) processes.

Balanced reaction: $2C_2H_2(g) + 5O_2(g) \rightarrow 4CO_2(g) + 2H_2O(g)$

B.E. of reac = (4 x C-H) + (2 x C≡C) + (5 x O=O in O_2)

= 4 mol(413 kJ/mol) + 2 mol(839 kJ/mol) + 5 mol(498 kJ/mol)

= +5820. kJ (+ sign indicates energy required to break bonds)

B.E. of prod = (8 x C=O) + (4 x O-H)

= 8 mol(799 kJ/mol) + 4 mol(467 kJ/mol)

= -8260. kJ (- sign indicates energy is released when bonds are formed)

To find the heat of reaction, add the bond energies.

ΔH_{rxn} = 5820. kJ + (– 8260. kJ) = -2440. kJ/2 mol C_2H_2 = **-1220. kJ per mole of acetylene**

Note: Look ahead to section 10.2 to confirm that $\Delta H_{rxn} = \Delta H_{\text{bonds broken}} + \Delta H_{\text{bonds formed}}$

b) $$\Delta H = (500.0\, g\, C_2H_2)\left(\frac{-1220.\, kJ}{mol\, C_2H_2}\right)\left(\frac{1\, mol\, C_2H_2}{26.04\, g\, C_2H_2}\right) = -2.342x10^4\, kJ$$

2.342 x 10⁴ kJ of heat are given off.

c) $$mass\ of\ CO_2 = (500.0\, g\, C_2H_2)\left(\frac{1\, mol\, C_2H_2}{26.04\, g\, C_2H_2}\right)\left(\frac{4\, mol\, CO_2}{2\, mol\, C_2H_2}\right)\left(\frac{44.01\, g\, CO_2}{1\, mol\, CO_2}\right) = \mathbf{1.690x10^3\ g\ CO_2}$$

d) $$moles\ O_2 = (500.0\, g\, C_2H_2)\left(\frac{1\, mol\, C_2H_2}{26.04\, g\, C_2H_2}\right)\left(\frac{5\, mol\, O_2}{2\, mol\, C_2H_2}\right) = 48.00_{31}\, mol\, O_2$$

$$V = \frac{nRT}{P} = \frac{(48.00_{31}\, mol\, O_2)(0.08206\, atm\cdot L/mol\cdot K)(298\, K)}{18.0\, atm} = \mathbf{65.2\ L\ O_2}$$

9.67 a) Construct a Born-Haber cycle for the formation of MgCl (as in Figure 9.6). Let ΔH°_f equal Path 1 and sum the other steps to create Path 2. Assume energies given refer to a "per-mole of MgCl" amount.

Path 2 = (ΔH°_{atom} of Mg) + (½ ΔH°_{BE} of Cl) + (ΔH°_{IE1} of Mg) + (ΔH°_{EA} of Cl + $\Delta H^\circ_{\text{lattice MgCl}}$)

Path 2 = 148 kJ + ½(243 kJ) + 738 kJ + (-349 kJ) + (-783.5 kJ) = -125 kJ

ΔH°_f = Path 1 = Path 2 = **-125 kJ**

b) The sign of ΔH°_f will determine if MgCl is stable relative to its elements. MgCl is stable because energy is released ($\Delta H^\circ_f < 0$) upon its formation. MgCl would be unstable if energy were required to form it from its elements.

c) Write an expression for the formation of $MgCl_2$ from Mg and Cl_2 in their elemental form. Use the value of ΔH°_f obtained in part (a), reverse its sign and multiply by two. Sum the two equations to yield the desired reaction.

~~Mg(s)~~ + ~~$Cl_2(g)$~~ → $MgCl_2(s)$ $\Delta H^o_f = -641.6$ kJ
2MgCl(s) → ~~2Mg(s)~~ + ~~$Cl_2(g)$~~ $\Delta H^o_f = 2(+125$ kJ)
2MgCl(s) → $MgCl_2(s)$ + Mg(s) ΔH^o = **-392 kJ**

d) MgCl is not stable relative to the formation of $MgCl_2$ because the formation of $MgCl_2$ from MgCl releases energy (exothermic reaction). Therefore, MgCl has a higher energy and is less stable.

9.70 Plan: Find the bond energy for an H—I bond from Table 9.2. For part a) calculate wavelength from this energy using the relationship from chapter 7: E = hc/λ. For part b) calculate the energy for a wavelength of 254 nm and then subtract the energy from part a) to get the excess energy. Speed can be calculated from the excess energy since $E_K = \frac{1}{2}mu^2$.

Solution: Bond energy for H—I is 295 kJ/mol (Table 9.2).

a) $$\lambda = \frac{hc}{E} = \frac{(6.626x10^{-34}\ J \cdot s)(6.022x10^{23}\ atoms/mol)(3.00x10^8\ m/s)}{2.95x10^5\ J/mol}\left(\frac{1x10^9\ nm}{m}\right) = \mathbf{406\ nm}$$

b) $$E = \frac{hc}{\lambda} = \frac{(6.626x10^{-34}\ J \cdot s)(6.022x10^{23}\ atoms/mol)(3.00x10^8\ m/s)}{254\ nm\left(\frac{1\ m}{1x10^9\ nm}\right)} = 4.71x10^5\ J/mol$$

Excess energy = 471 kJ – 295 kJ = 176 kJ/mol or **$2.92x10^{-19}$ J/atom**.

c) $$u = \sqrt{\frac{2E}{m}} = \sqrt{\frac{2(2.92x10^{-19}\ kg \cdot m^2/s^2)}{(1.008\ g/mol)\left(\frac{1\ mol}{6.022x10^{23}\ atoms}\right)\left(\frac{1\ kg}{1000\ g}\right)}} = \mathbf{1.87x10^4\ m/s}$$

Check: The energy difference and speed are similar in magnitude to those calculated in chapter 7 for atoms, so the answers appear reasonable.

9.74 a) To determine the minimum photon energy and frequency, convert the bond energy of Cl–Cl (243 kJ/mol from Table 9.2) from kJ/mol to J/molecule, and then use Planck's relationship to find ν:

$$E\ (J/molecule) = \left(\frac{243\ kJ}{mol}\right)\left(\frac{1000\ J}{kJ}\right)\left(\frac{1\ mol}{6.022x10^{23}\ molecules}\right) = \mathbf{4.04x10^{-19}\ J/molecule}$$

$$\nu = \frac{E}{h} = \frac{4.03_{52}x10^{-19}\ J}{6.626x10^{-34}\ J \cdot s} = \mathbf{6.09x10^{14}\ s^{-1}}$$

b) To find the longest wavelength, convert the C–Cl bond energy (339 kJ/mol from Table 9.2) from kJ/mol to J/bond and use the c = λν relationship to find λ:

$$E\ (J/bond) = \left(\frac{339\ kJ}{mol}\right)\left(\frac{1000\ J}{kJ}\right)\left(\frac{1\ mol}{6.022x10^{23}\ bonds}\right) = 5.62_{94}x10^{-19}\ J/bond$$

$$\lambda = \frac{hc}{\Delta E} = \frac{(6.626x10^{-34}\ J \cdot s)(3.00x10^8\ m/s)(10^9\ nm/m)}{5.62_{94}x10^{-19}\ J} = \mathbf{353\ nm}$$

9.77 a) Bond energies in the table are defined as the energy needed to break the bond into the atoms in the gas phase. Thus, the bond breaks homolytically to give one electron to each atom to form the neutral atoms.

b) Ions will form when heterolytic breakage occurs. If a bond breaks to give two electrons to one atom that atom will have an excess of electrons beyond the number of valence electrons and be a negatively charged ion. The other atom will have fewer than the usual number of valence electrons, forming a positively charged ion.

c) The bond polarity will be greater for $H_3C—CCl_3$ than for $H_3C—CH_3$, so the energy to break the bond in $H_3C—CCl_3$ will be greater. In $H_3C—CH_3$ there is no difference in electronegativity between the two CH_3 groups, but in $H_3C—CCl_3$ there is a difference in the electronegativity of the CH_3 and CCl_3 groups. Chlorine is more electronegative than hydrogen making the CCl_3 group more electronegative than the CH_3 group. Thus the $H_3C—CCl_3$ bond is polar. A polar bond is stronger than a nonpolar bond and requires more energy to break.

d) Yes. When a bond breaks homolytically, energy is required to pull the atoms apart. In heterolytic breakage, energy is required to pull ions apart. The energy to pull ions apart is greater than the energy to pull atoms apart because the positively charged ion attracts the negatively charged ion.

9.79 a) To compare the two energies the ionization energy must be converted to the energy to remove an electron from an atom.

$$IE\ (in\ \frac{J}{atom}) = \left(\frac{731\ kJ}{mol\ Ag}\right)\left(\frac{1000\ J}{1\ kJ}\right)\left(\frac{1\ mol\ Ag}{6.022x10^{23}\ atoms\ Ag}\right) = 1.21x10^{-18}\ J$$

Because the energy to remove an electron from a gaseous atom of silver is greater than the energy to remove an electron from solid silver (IE > Φ), it is easier to remove an electron from a **solid silver surface**.

b) The electrons in solid silver are held less tightly that the electrons in gaseous silver because the electrons in metals are delocalized meaning they are shared among all the metal nuclei. The delocalized attraction of many nuclei to an electron (solid silver) is weaker than the localized attraction of one nucleus to an electron (gaseous silver).

9.81 The heat of reaction equals the sum of the bond energies of the bonds broken (reactant bonds, endothermic, ΔH is +) plus the bond energies of the bonds formed (product bonds, exothermic, ΔH is -). See 9.65 for additional explanation.

In each case, two methane molecules, or 8 moles of C-H bonds are formed:

$\Delta H^{\circ}_{bonds\ formed}$ = 8(C-H bonds) = (8 mol)(415 kJ/mol) = -3320. kJ (- sign indicates energy is released when bonds are formed)

For ethane:

ΔH°_{rxn} = [(1 mol)(BE_{C-C}) + (6 mol)(BE_{C-H}) + (1 mol)(BE_{H-H})] + (–3320.kJ)

(1mol)(-65.07 kJ/mol) = [(1mol)(347 kJ/mol) + (6mol)(BE_{C-H}) + (1mol)(432 kJ/mol)] - 3320. kJ

-65.07 kJ = 347 kJ + (6 mol)(BE_{C-H}) + 432 kJ – 3320. kJ

BE_{C-H} = **413 kJ/mol**

For ethene:

ΔH°_{rxn} = [(1 mol)($BE_{C=C}$) + (4 mol)(BE_{C-H}) + (2 mol)(BE_{H-H})] + (-3320.kJ)

(1mol)(-202.21 kJ/mol) = [(1mol)(614 kJ/mol) + (4mol)(BE_{C-H}) + (2mol)(432 kJ/mol)] - 3320. kJ

-202.21 kJ = 614 kJ + (4 mol)(BE_{C-H}) + 864 kJ – 3320. kJ

BE_{C-H} = **410. kJ/mol**

For ethyne:

ΔH°_{rxn} = [(1 mol)($BE_{C\equiv C}$) + (2 mol)(BE_{C-H}) + (3 mol)(BE_{H-H})] + (-3320.kJ)

(1mol)(-376.74 kJ/mol) = [(1mol)(839 kJ/mol) + (2mol)(BE_{C-H}) + (3mol)(432 kJ/mol)] - 3320.kJ

-376.74 kJ = 839 kJ + (2 mol)(BE_{C-H}) + 1296 kJ – 3320. kJ

BE_{C-H} = **404 kJ/mol**

Each of the C-H bond energies calculated are close to the table value of 413 kJ/mol, but slightly different. These results demonstrate that identical atom-to-atom bond energies can vary slightly within different molecules, and that the tabulated value is only an *average* bond energy.

CHAPTER 10

THE SHAPES OF MOLECULES

FOLLOW-UP PROBLEMS

10.1 a) Plan: Follow the four steps outlined in this section to draw the Lewis structure.

Solution: In H_2S, sulfur is the central atom because H can never be the central atom. The total number of valence electrons available is 1 + 1 + 6 = 8 e^-. Draw a single bond from each H atom to the central S atom, leaving 8 – 2(2) = 4 e^- to add to the structure. The remaining 4 e^- are distributed about the S atom, completing its octet.

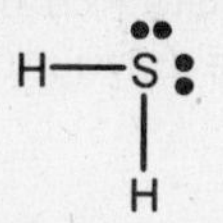

Check: Verify that each atom has an octet (except H which has 2 electrons). Add the number of electrons represented in the structure (4 in lone pairs and 4 in bonds) and verify that there are 8 valence electrons.

b) In OF_2, oxygen has the lower group number and is the central atom. The total number of valence electrons available is 6 + 7 + 7 = 20 e^-. Draw two bonds and distribute the remaining (20 – 4) 16 e^- around the atoms, starting with the two fluorines.

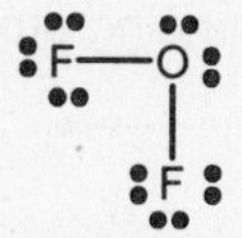

c) In $SOCl_2$, sulfur is the central atom because it has the lowest group number and highest period number. The total number of valence electrons available is 6 + 6 + 7 + 7 = 26 e^-. Draw three bonds and distribute the remaining (26 –6) 20 e^- on the structure, starting with the surrounding atoms. The surrounding atoms receive 18 e^-, leaving two e^- for placement on the central S atom.

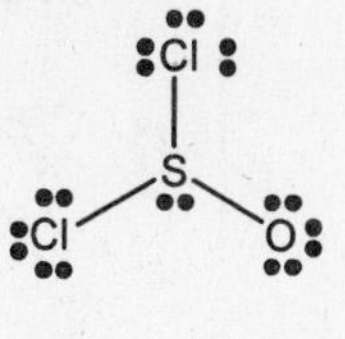

10.2 a) For NH_3O it is tempting to draw the Lewis structure with N as the central atom, surrounded by the three H's and one O. Although the dots and lines can be drawn giving each atom an octet, it is an incorrect structure for two reasons. First, the prefix *hydroxyl* indicates there is an O-H bond present. Second, oxygen typically forms two bonds – in this case one to N and one to H. The correct structure is shown on the right.

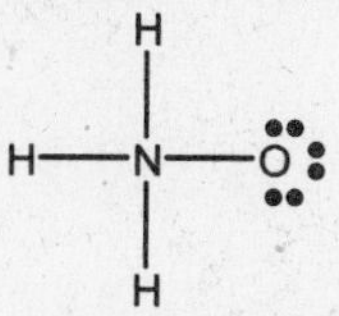

incorrect

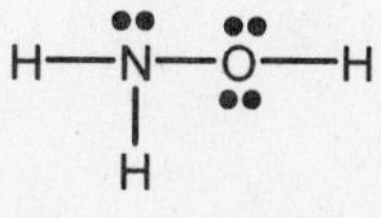

correct

b) Carbons are rarely surrounding atoms since they require 4 bonds, so both carbons are central atoms. Since there are no O-H bonds and H's must be on the end, the oxygen must be located between the two carbons. The name *ether* indicates a particular structure, in which an oxygen is single-bonded to two carbons. Organic compounds are covered in chapter 15.

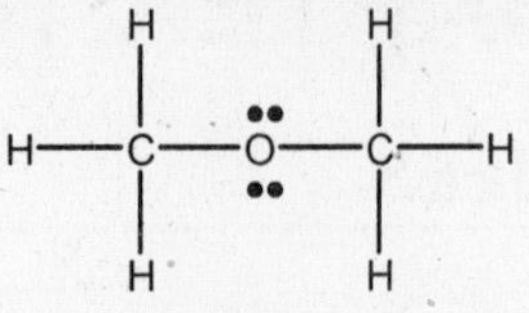

10.3 a) CO has a total of (4 + 6) = 10 valence e$^-$. Start by drawing CO with a single bond, and then attempt to distribute the remaining 8 e$^-$ so that both C and O receive an octet. Since this is impossible, you must create a triple bond to achieve the octet.

b) HCN has a total of (1 + 4 + 5) = 10 valence electrons. Carbon is the central atom because it has a lower group number. Start by drawing single bonds and distributing the remaining valence electrons around the surrounding atoms, similar to structure (i).

(i) (ii)

Nitrogen has an octet but carbon doesn't. By creating a C≡N triple bond using the two lone pairs around nitrogen, an octet is achieved around all atoms, as in structure (ii) which is the correct structure.

c) CO_2 has a total of (4 + 6 + 6) = 16 valence electrons. Carbon is the central atom because it has a lower group number. Start by drawing single bonds and filling in the remaining valence e$^-$ around the surrounding atoms:

Satisfy the octet of carbon by creating double bonds using lone e$^-$ pairs from each O.

10.4 The structure drawn in the text is shown below on the left. Arrows indicate the movement of electrons to achieve the other two resonance structures.

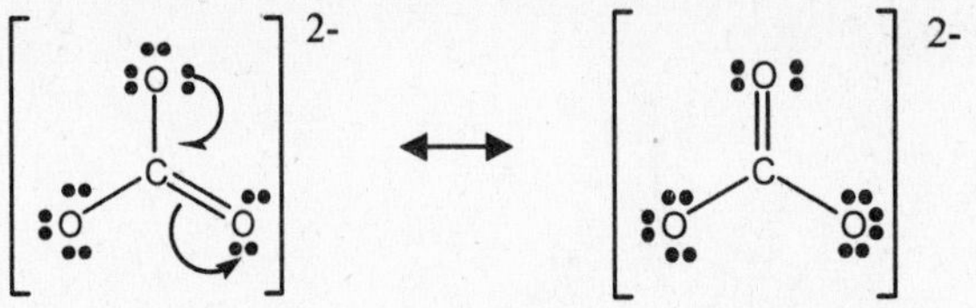

or

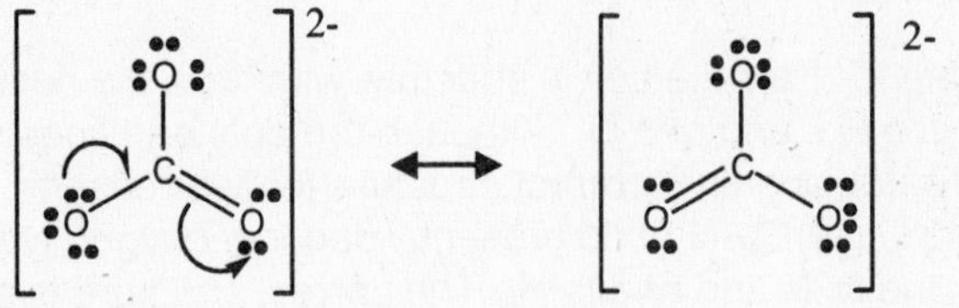

10.5 a) In $POCl_3$, P is the central atom (smallest group number, highest period number) and can have more than an octet. The structure contains 32 valence e$^-$. Upon examining the formal charges in the structures below, **II** has the lowest formal charges and is the dominant Lewis structure.

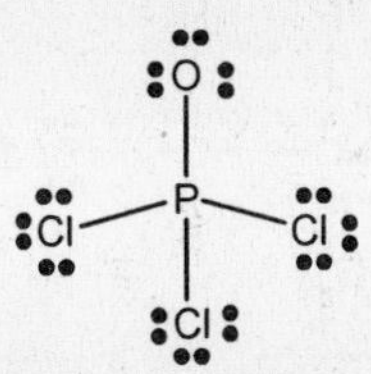

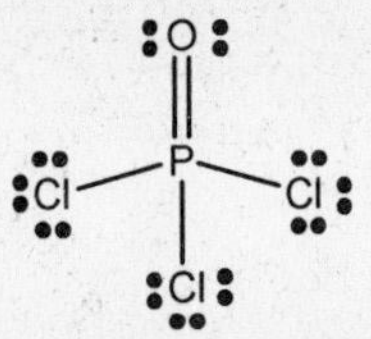

I **II**

$FC_P = 6 – (0 + ½(8)) = +2$ $FC_P = 6 – (0 + ½(10)) = +1$

$FC_O = 6 – (6 + ½(2)) = -1$ $FC_O = 6 – (4 + ½(4)) = 0$

In both structures the formal charge for Cl is 0.

b) In ClO_2, Cl is the central atom because it has the highest period number. This is an odd-electron species with 19 valence e^-. Following the rules for drawing Lewis structures, we start by creating single Cl-O bonds and then adding the remaining valence electrons to the surrounding atoms first.

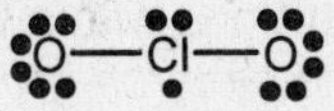

Because Cl is in period 3, it can access d orbitals and have more than an octet. The following resonance structures are possible, with structure **III** the most dominant because of its minimized formal charge.

I **II** **III**

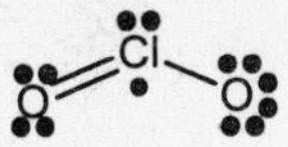

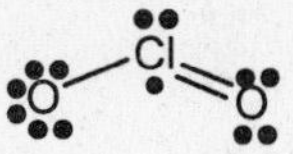

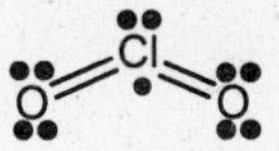

$FC_{Cl} = +1$ $FC_{Cl} = +1$ $FC_{Cl} = 0$

Verify that the formal charges on oxygen for the structure III are also minimized.

c) XeF_4 contains 36 valence e^-. Draw the structure by starting with single Xe-F bonds and then filling in remaining electrons. Xe can exceed the octet. The appropriate structure is given below, and Xe's formal charge is (8 – (4 + ½(8))=0. FC_{Cl} also equals 0.

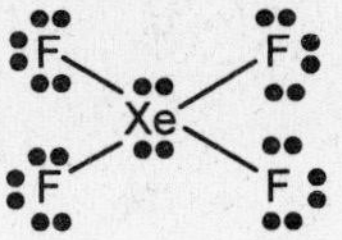

10.6 a) Plan: Draw Lewis structures for all substances and assume that all reactant bonds are broken and all product bonds are formed.

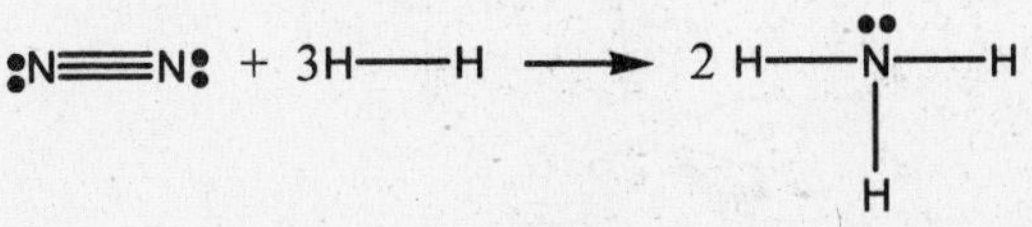

Solution:

Bonds broken: 1(N ≡ N) + 3(H—H) = (1mol)(945kJ/mol) + (3mol)(432kJ/mol) = +2241 kJ

Bonds formed: [2 x 3(N—H)] = (6mol)(-391 kJ/mol) = -2346 kJ

$\Delta H^\circ_{rxn} = \Delta H^\circ_{bonds\ broken} + \Delta H^\circ_{bonds\ formed}$ = 2241 kJ – 2346 kJ = **-105 kJ**

Check:

This value can be compared to ΔH°_{rxn} calculated from ΔH°_f values found in Appendix B.

$\Delta H^\circ_{rxn} = \Delta H^\circ_f$(products) - ΔH°_f(reactants) = (2 mol)(-45.9 kJ/mol) – (0 + 0) = -91.8 kJ.
The values are similar, but not identical, because bond energies are average values.

b) 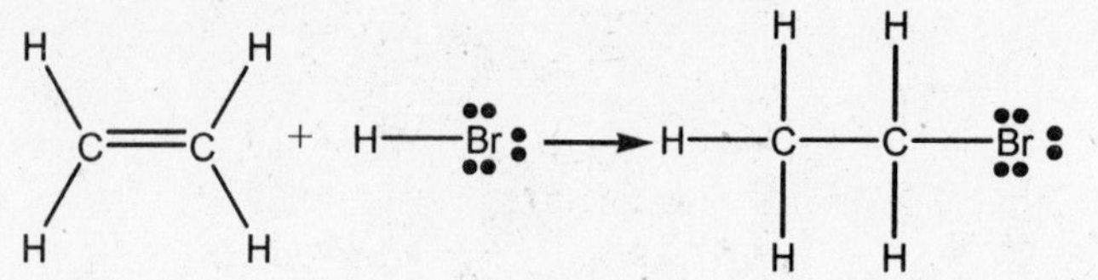

Bonds broken: 1(C=C) + 4(C—H) + 1(H—Br) = 614 + 4(413) + 363 = +2629 kJ

Bonds formed: 1(C—C) + 5(C—H) + 1(C—Br) = (-347) + 5(-413) + (-276) = -2688 kJ

$\Delta H^{o}_{rxn} = \Delta H^{o}_{bonds\ broken} + \Delta H^{o}_{bonds\ formed}$ = 2629 – 2688 = **-59 kJ**

10.7 a) The Lewis dot structure for CS_2 is given below. This molecule is linear and has the designation, AX_2, because each double bond acts like a bonding group. There is no deviation in this linear molecule's bond angle (180°) because there are no lone electron pairs and both bonding groups are identical.

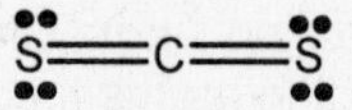

b) The $PbCl_2$ molecule has 18 valence electrons with a Lewis structure shown below. The molecule follows the AX_2E formula. The *electron-group arrangement* is trigonal planar, but the *shape* is bent or V-shaped. The lone pair compresses the ideal bond angle of 120°. Although this is a combination between a metal and nonmetal (typically an ionic compound), the difference in electronegativity is 1.1 and therefore is a polar covalent bond. Pb is electron deficient because the Cl only forms one bond when combined with a nonhalogen.

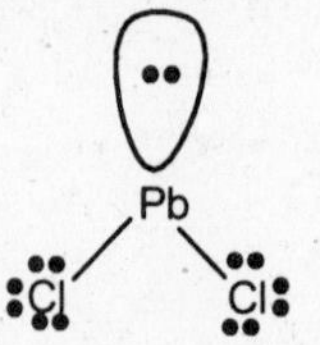

c) The CBr_4 molecule has 32 valence electrons and follows the AX_4 formula. This molecule is a perfect tetrahedron (with 109.5° bond angles) because all bonds are identical.

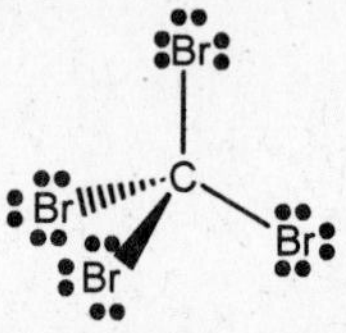

d) The SF_2 molecule (AX_2E_2) has a total of four electron groups about the central S atom, so the *electron group arrangement* is tetrahedral. The molecule is V-shaped because of the two unshared pairs of electrons. The ideal tetrahedral bond angle is compressed by the lone pair – lone pair interaction, so that F-S-F bond angle is less than 109.5°.

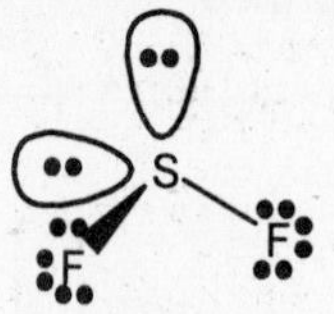

10.8 a) The ICl_2^- ion has 22 valence electrons, and the Lewis structure shows 5 electron groups around the central atom. This trigonal bypyramidal arrangement of electron groups results in a linear shape (180°) because the lone pairs occupy equatorial positions.

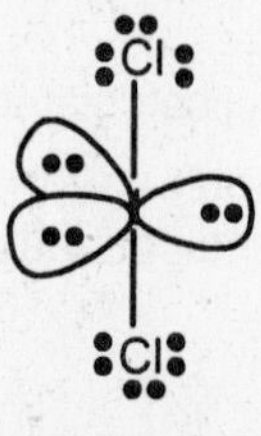

b) The ClF_3 molecule has 28 valence electrons, and the Lewis structure shows 5 electron groups around the central atom. Two lone pairs occupy equatorial positions, resulting in a <u>T-shape</u>. The lone pairs repel each other and the bonding electrons to compress the F-Cl-F bond angles to less than 90°.

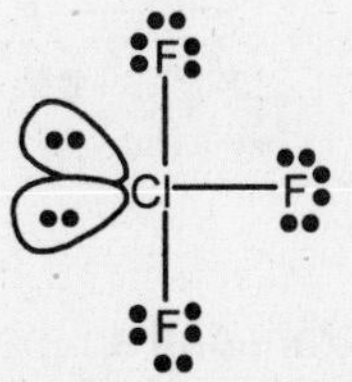

c) The SOF_4 molecule has 40 valence electrons. A first attempt for the Lewis structure is shown on the left; the correct structure is shown on the right. Why is the structure on the left incorrect? Recall that for central atoms with expanded shells, the atoms expand to form *more* bonds and *minimize* formal charge. The formal charge of S in the left structure is 6 – (0 + ½(10)) = +1, whereas the formal charge of S in the right structure is 6 – (0 + ½(12)) = 0. Is the location of the S-O double important? Yes, the double bond, like a lone pair, exerts a stronger repulsion than the single bonds and therefore occupies an equatorial position.

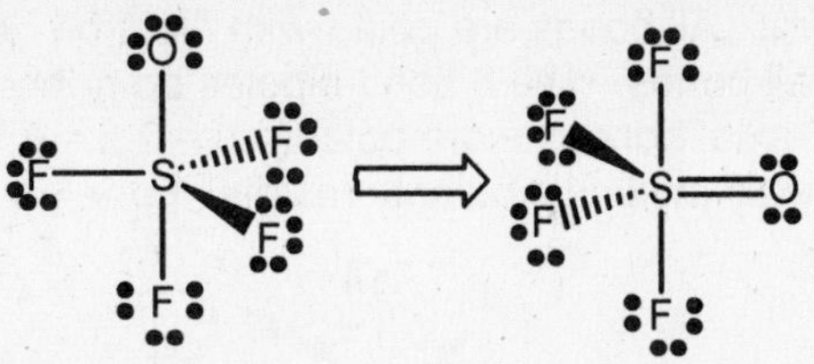

The molecule is <u>trigonal bipyramidal</u>, and the double bond causes deviation from ideal bond angles: the F_{eq}-S-F_{eq} angle is < 120°, F_{ax}-S-F_{eq} angle is < 90° and the F_{ax}-S-F_{ax} angle is < 180°.

10.9 a) The Lewis structure for sulfuric acid, H_2SO_4, was determined at the end of section 10.1. The central S atom has 4 electron groups around it, so the shape is tetrahedral about sulfur. The HO-S-OH bond angle is compressed by the double bonds to < 109.5° while the O=S=O angle is > 109.5°. Each central O also has 4 surrounding electron groups, but the presence of the lone pairs result in a V-shape about the O, with a compressed S-O-H bond angle that is < 109.5°.

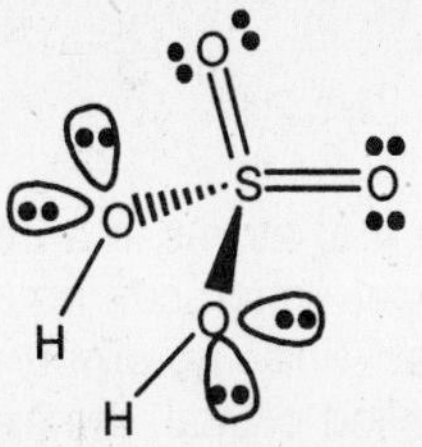

b) Carbon consistently forms 4 bonds, so the only possible structure containing a triple bond is shown below. The CH_3- group is called a *methyl* group and has a tetrahedral geometry with bond angles ~109.5° (no deviation). The other two carbon atoms have two *groups* surrounding them (remember a multiple bond counts as one group) so the shape around those carbon atoms is linear, with 180° bond angles.

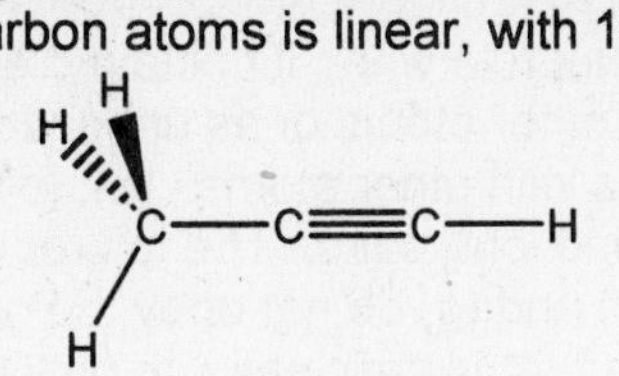

c) You may be tempted to draw a Lewis structure as S-F-S-F, however halogens typically form one bond to satisfy the octet when bonding with other nonhalogens. The lone pairs on each S result in a V-shape around each S atom, and compressed F-S-S bond angles (< 109.5°).

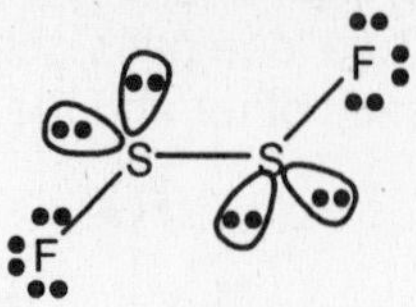

10.10 a) Dichloromethane has a tetrahedral shape. All bonds are polar, and there is a resulting overall dipole on the molecule.

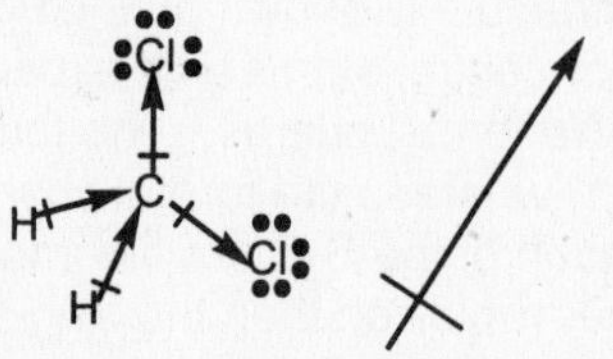

b) IOF_5 is octahedral. All bonds are polar, with electron density located away from the central atom in all bonds. The 6 bond dipoles do <u>not</u> counterbalance one another because the I-F axial bond is more polar (4.0 – 2.5 = 1.5) than the I-O axial bond (3.5 – 2.5 = 1). An overall molecule polarity results. Lone pairs on surrounding atoms are left out for simplicity.

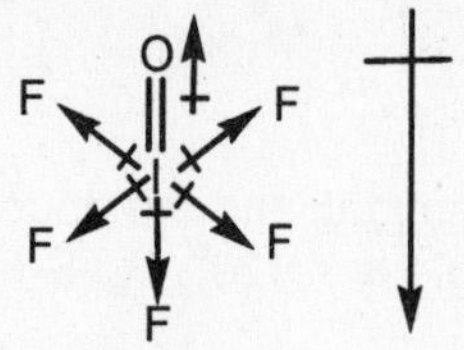

c) Nitrogen tribromide has a trigonal pyramidal shape. Due to its asymmetry, the molecule is polar.

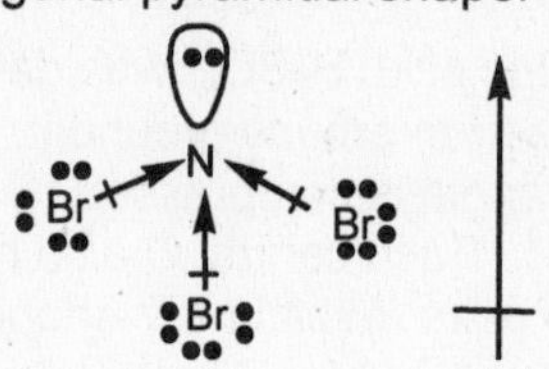

END-OF-CHAPTER PROBLEMS

10.1 He, F, and H cannot serve as central atoms in a Lewis structure. Helium and hydrogen cannot since their outer level is n=1 which can hold a maximum of two electrons. Helium already contains two electrons, so it has no empty or partially filled outer orbitals available for bonding. H has one partially filled orbital, the 1s orbital, so it can form one, but no more than one, bond. Fluorine has the electron configuration $1s^22s^22p^5$, so the n=2 outer level contains one partially filled orbital and fluorine can form one bond. Fluorine cannot have an expanded octet like the other halogens since its outer level, n=2, contains only s and p orbitals.

10.3 The central atom, X, obeys the octet rule when it contain 8 electrons in its valence shell, either through bonds formed with other atoms or as unbonded electron pairs. (a) and (e) achieve an octet by forming bonds with other atoms. (b), (d) and (f) achieve an octet through a combination of bonds and lone pairs. The ion represented by (h) achieves an octet through lone pairs alone. (c) and (g) do <u>not</u> obey the octet, because (c) exceeds the octet with 10 valence electrons and (c) is deficient with only 6 valence electrons.

10.5 Apply the steps in section 10.1 to arrive at a Lewis dot structure. First, determine relative placement of the atoms by identifying the central atom. Second, determine the total number of valence electrons in the structure. Third, create single bonds connecting each surrounding atom to the central atom, and distribute the remaining electrons around the surrounding atoms to form octets. If necessary, create multiple bonds to satisfy the octet on the central atom or recognize if the central atom can violate the octet rule.

a) SiF_4 (32 valence e^-)

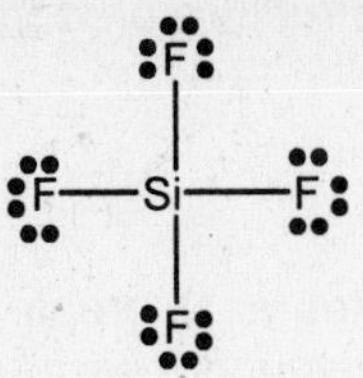

b) $SeCl_2$ (20 val e^-)

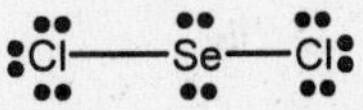

c) COF_2 (24 val e^-)

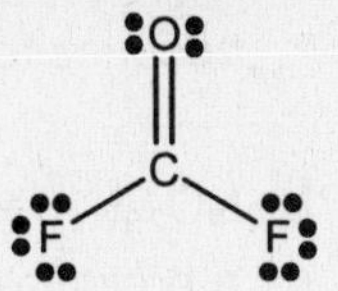

10.7 a) PF_3 (26 val e^-)

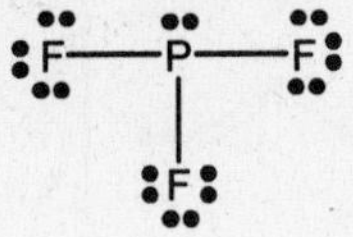

b) H_2CO_3 (24 val e^-)

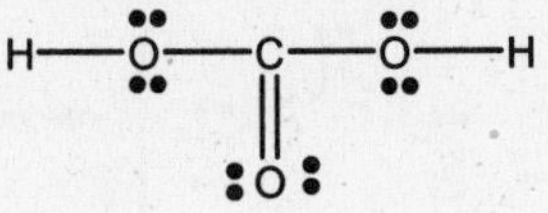

c) CS_2 (16 val e^-, isoelectronic with CO_2)

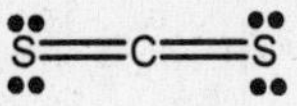

10.9 a) NO_2 is an electron deficient molecule with only 7 electrons around the nitrogen. The structure includes a double bond between the nitrogen and one of the oxygens. Both resonance forms have the same formal charges.

b) NO_2F has a Lewis structure with a double bond between N and one of the oxygens and a single bond from N to the other O. No lone pairs are on the nitrogen.

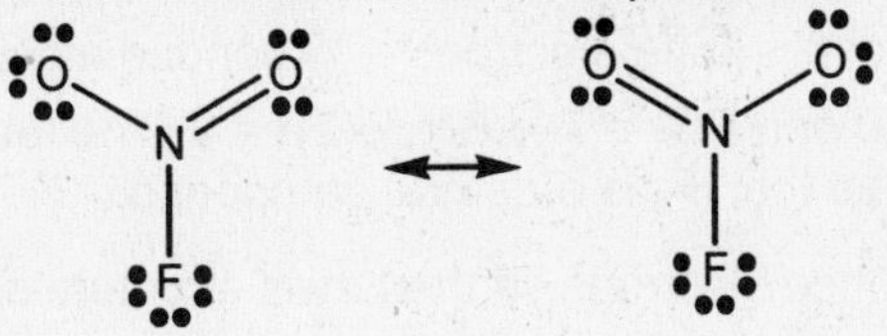

10.11 a) N_3^- (16 valence e^-)

Upon applying the rules for drawing Lewis dot structures, we arrive at the preliminary structure shown below:

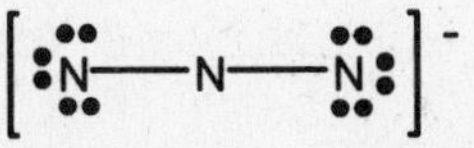

The central N atom does not have an octet, and there are two options to create N's octet. The first option involves forming 2 double bonds on the central N atom:

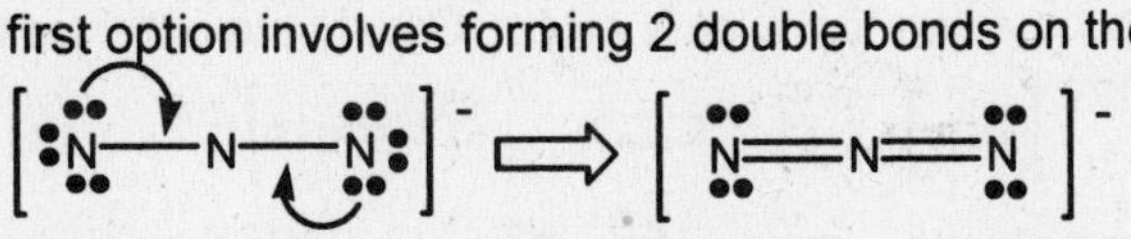

The second option involves forming a triple bond, using two lone pairs from either outer N atom:

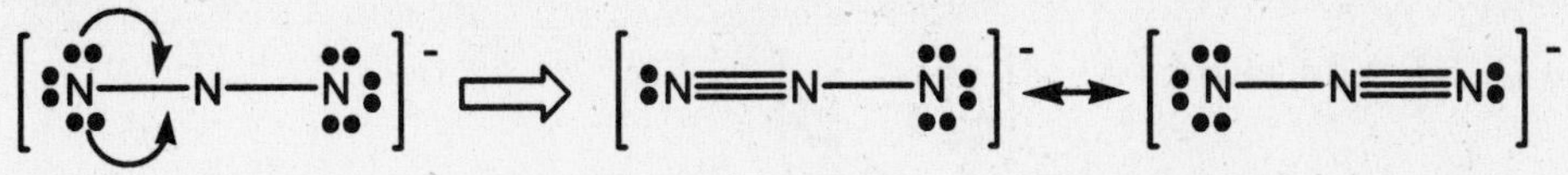

The resonance forms are written as follows:

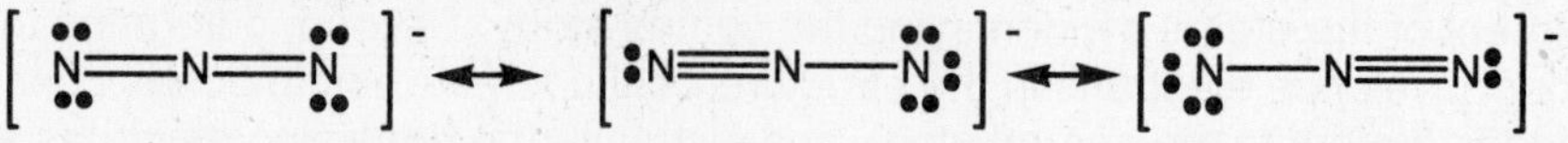

FC = -1 +1 -1 0 +1 -2 -2 +1 0

Which resonance form(s) are most stable? You could calculate the formal charge (FC) for each atom, verifying that the sum of all FC's equals the charge on the species, in this case, -1. An analysis of FC's suggests that the double bonded structure predominates, because formal charges are minimized.

b) NO_2^- (18 valence e^-).

A preliminary dot structure is shown on the left. To create an octet for N, create a double bond using a lone pair from either O. The resonance forms and their formal charges are shown on the right.

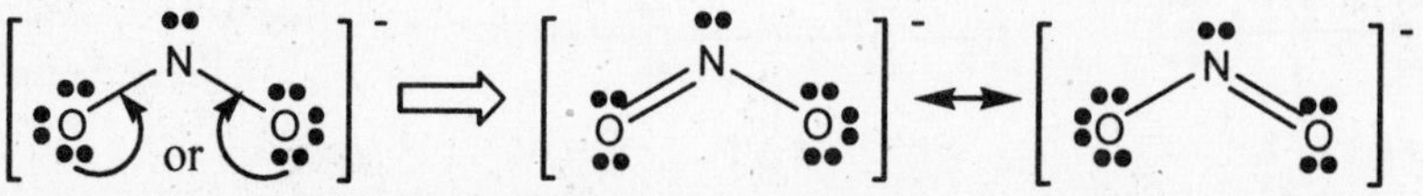

Both resonance forms have identical formal charges and therefore contribute equally to the structure. We would expect that both N-O bond lengths would be equal with a bond length midway between that found for a single and double N-O bond.

10.13 a) The Lewis structure for IF_5 (42 valence e^-) includes an expanded octet for I with 10 e^- in I—F bonds, six lone pairs on each F and one lone pair on central atom I.

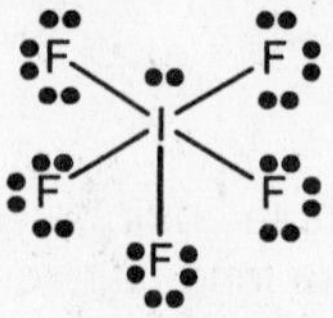

Formal charge (FC) = val e^- - (lone electrons – ½(bonded electrons))

$FC_I = 7 - (2 + ½(10)) = 0$ and $FC_F = 7 - (6 + ½(2)) = 0$. The total formal charge is 0, which agrees with the fact that IF_5 is a neutral compound.

b) AlH_4^- has a total of 8 valence electrons so the Lewis structure shows four Al—H bonds and no lone pairs.

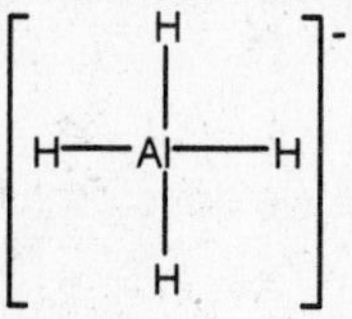

$FC_{Al} = 3 - (0 + ½(8)) = -1$ and $FC_H = 1 - (0 + ½(2)) = 0$. Total formal charge is –1 which agrees with the charge on the ion.

10.15 a) CN^- has 10 valence e. Start by drawing a single bond between C and N and distribute the remaining e^- around either the C or N atom. Satisfy the octet for both atoms by creating a triple bond.

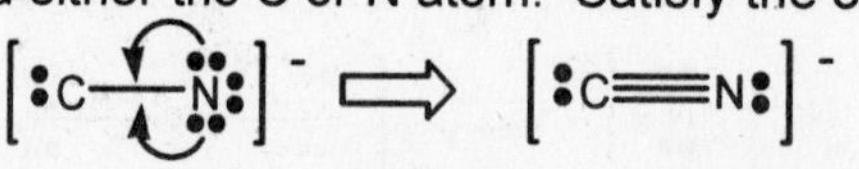

$FC_C = 4 - (2 + ½(6)) = -1$ $FC_N = 5 - (2 + ½(6)) = 0$

<u>Check</u> that the sum of the formal charges equals the charge on the species, in this case –1.

b) ClO^- has 14 valence e^-.

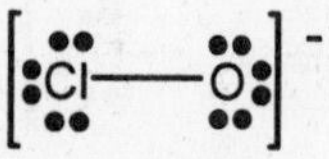

$FC_{Cl} = 7 - (6 + ½(2)) = 0$ $FC_O = 6 - (6 - ½(2)) = -1$

10.17 a) BrO_3^- with 26 valence e^- gives a Lewis structure (**I**) with all atoms having a full octet. The formal charges are –1 for each O and +2 for Br. Using lone pairs from two of the oxygens to form double bonds can minimize the formal charges. This will give Br an expanded octet. In structure (**II**) the formal charge on the double bonded oxygens and on the Br has been reduced to 0. The one O with a single bond still has a formal charge of –1. The total formal charge is –1, which agrees with the charge on the ion. Structure **II** is the most important resonance form of BrO_3^-. $O.N._{Br} = +5$ and $O.N._O = -2$.

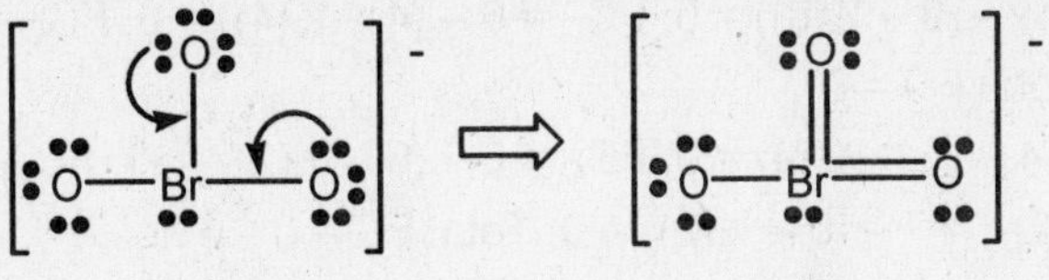

I II

b) For SO_3^{2-} structure **I** shows an octet for each atom with formal charges of +1 for S and –1 for each oxygen. The formal charges can be minimized if a double bond is formed with one oxygen as in structure **II** and S has an expanded octet. The formal charges are now 0 for S, 0 for the O in the double bond and –1 for each of the other two oxygens for a total formal charge of –2 equal to the charge on the ion. Structure **II** is the most important resonance form of SO_3^{2-}. $O.N._S = +4$ and $O.N._O = -2$.

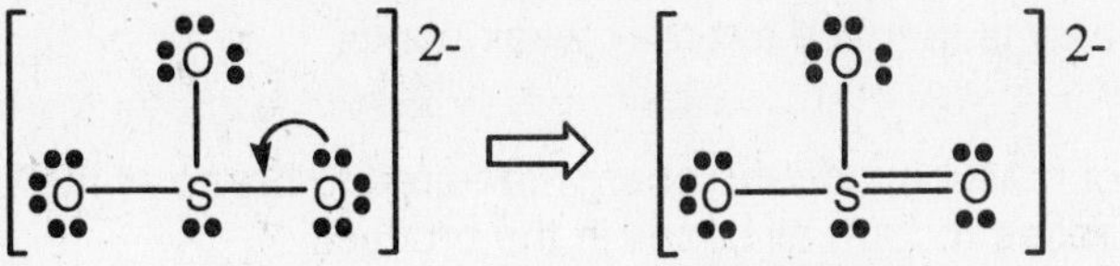

I II

10.19 a) BH_3 is electron deficient, with only six electrons around B.

b) AsF_4^- has an expanded valence shell with 10 electrons around As. Note that the extra electron pair exists as a lone pair on As. The lone pair will not form a As=F bond because fluorine cannot accommodate an expanded octet.

c) $SeCl_4$ also has an expanded valence shell with 10 electrons around Se. Because it is *isolectronic* (contains same number of e^-) with AsF_4^-, the Lewis structures are similar.

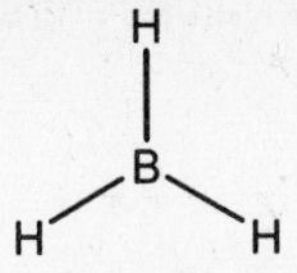

(a)

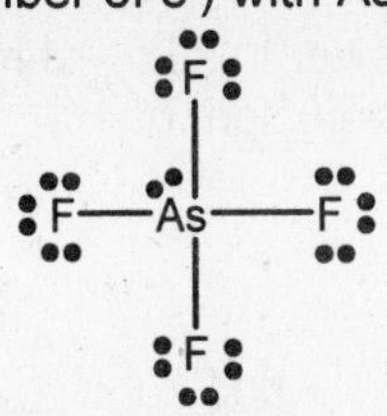

(b)

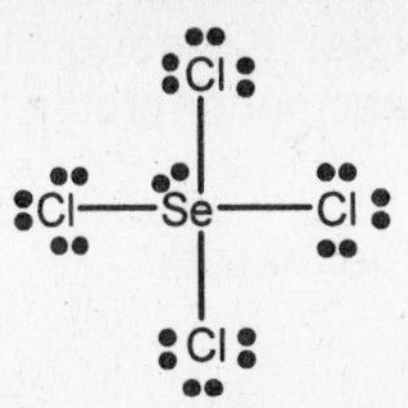

(c)

10.21 a) In BrF_3, Br has an expanded octet with 3 bonds and 2 lone pairs.

b) In ICl_2^-, I has an expanded octet with 2 bonds and 3 lone pairs.

c) In BeF_2, Be is deficient with only four electrons in the two bonds with F.

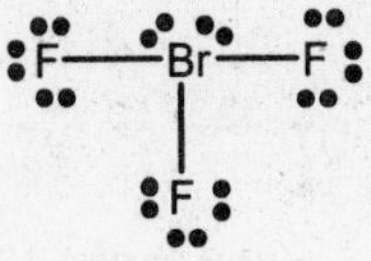

(a)

(b)

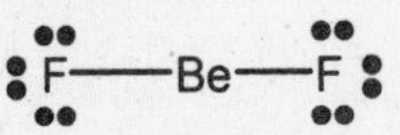

(c)

10.23 Beryllium chloride has the formula $BeCl_2$. Chloride ion has the formula Cl^-.

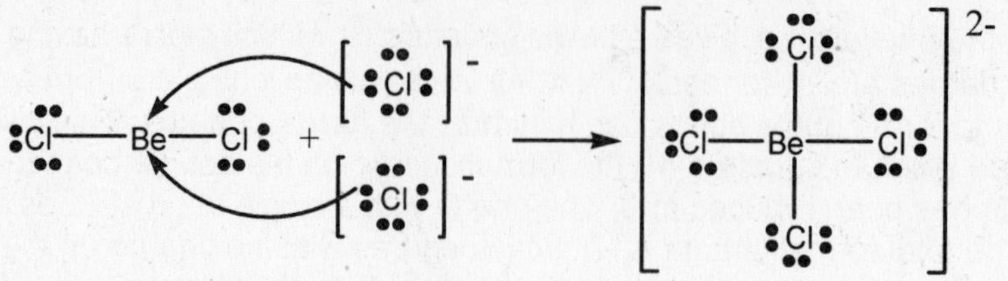

10.26 Structure **A**: $FC_C = 4 - (0 - ½(8)) = 0$; $FC_O = 6 - (4 - ½(4)) = 0$; $FC_{Cl} = 7 - (6 - ½(2)) = 0$
Total FC = 0

Structure **B**: $FC_C = 4 - (0 - ½(8)) = 0$; $FC_O = 6 - (6 - ½(2)) = -1$; $FC_{Cl=} = 7 - (4 - ½(4)) = +1$
$FC_{Cl-} = 7 - (6 - ½(2)) = 0$; Total FC = 0

Structure **C**: $FC_C = 4 - (0 - ½(8)) = 0$; $FC_O = 6 - (6 - ½(2)) = -1$; $FC_{Cl=} = 7 - (4 - ½(4)) = +1$
$FC_{Cl-} = 7 - (6 - ½(2)) = 0$; Total FC = 0

The most important resonance structure is **A** since the formal charges are minimized for all the atoms.

10.28 Reaction between molecules requires the breaking of existing bonds and the formation of new bonds. Substances with weak bonds are more reactive than those with strong bonds because less energy is required to break weak bonds.

10.30 To find the heat of reaction, add the energy required to break all the bonds in the reactants to the energy released to form all bonds in the product.

Reactant bonds broken: 1(C=C) + 4(C—H) + 1(Cl—Cl)
= (1 mol)(614 kJ/mol) + (4 mol)(413 kJ/mol) + (1mol)(243 kJ/mol)
= 2509 kJ

Product bonds formed: 1(C—C) + 4(C—H) + 2 (C—Cl)
= (1 mol)(-347 kJ/mol) + (4mol)(-413 kJ/mol) + (2mol)(-339 kJ/mol)
= -2677 kJ

$\Delta H^\circ_{rxn} = \Delta H^\circ_{\text{bonds broken}} + \Delta H^\circ_{\text{bonds formed}} = 2509\text{ kJ} + (-2677\text{ kJ}) = \mathbf{-168\text{ kJ}}$

<u>Note</u>: It is correct to report the answer in kJ or kJ/mol as long as the value refers to a reactant or product with a molar coefficient of 1.

10.32 The reaction:

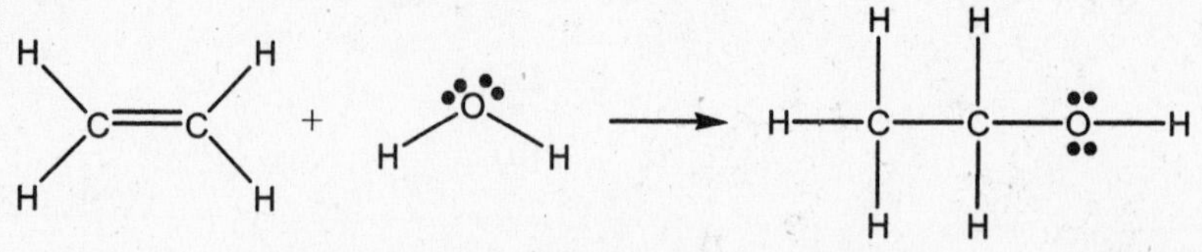

$\Delta H^\circ_{\text{bonds broken}}$ = 1 C=C = 1 mol(614 kJ/mol)
4 C-H = 4 mol(413 kJ/mol)
2 O-H = 2 mol(467 kJ/mol)
3200 kJ

$\Delta H^\circ_{\text{bonds formed}}$ = 5 C-H = 5 mol(-413 kJ/mol)
1 C-C = 1 mol(-347 kJ/mol)
1 C-O = 1 mol(-358 kJ/mol)
1 O-H = 1 mol(-467 kJ/mol)
-3237 kJ

$\Delta H^\circ_{rxn} = \Delta H^\circ_{\text{bonds broken}} + \Delta H^\circ_{\text{bonds formed}} = 3200\text{ kJ} + (-3237\text{ kJ}) = \mathbf{-37\text{ kJ}}$

<u>Note</u>: It is correct to report the answer in kJ or kJ/mol as long as the value refers to a reactant or product with a molar coefficient of 1.

10.34 The reaction:

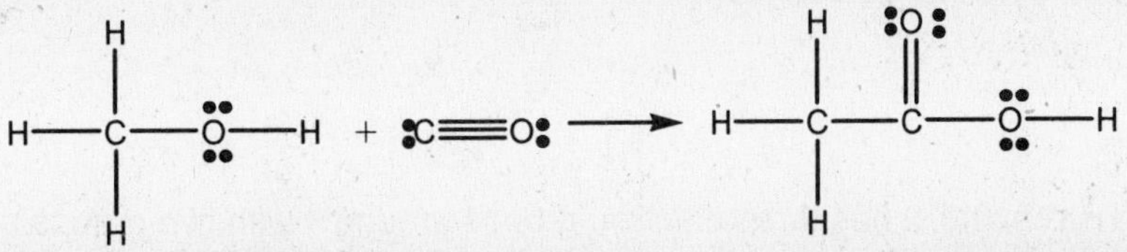

$\Delta H^\circ_{bonds\ broken}$ = 1 C—O = 1 mol(358 kJ/mol)
3 C—H = 3 mol(413 kJ/mol)
1 O—H = 1 mol(467 kJ/mol)
1 C≡O = 1 mol(1070 kJ/mol)
= 3134 kJ

$\Delta H^\circ_{bonds\ formed}$ = 3 C-H = 3 mol(-413 kJ/mol)
1 C-C = 1 mol(-347 kJ/mol)
1 C=O = 1 mol(-745 kJ/mol)
1 C-O = 1 mol(-358 kJ/mol)
1 O-H = 1 mol(-467 kJ/mol)
= -3156 kJ

$\Delta H^\circ_{rxn} = \Delta H^\circ_{bonds\ broken} + \Delta H^\circ_{bonds\ formed}$ = 3134 kJ-+ (-3156 kJ) = **-22 kJ**

10.37 When all electron pairs are involved in bonding (i.e. there are no lone electron pairs), the *molecular shape* name is the same as the *electron-group arrangement* name.

10.39 The molecular shapes with a tetrahedral electron-group arrangement are those with four electron groups: tetrahedral (AX_4), trigonal pyramidal (AX_3E), and bent or V-shaped (AX_2E_2).

10.41 a) A molecule that is V-shaped has two bonds and generally has either one (AX_2E) or two (AX_2E_2) lone electron pairs.

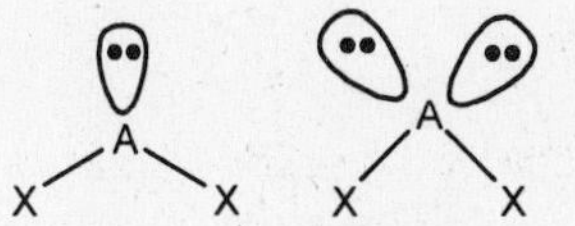

b) A trigonal planar molecule follows the formula AX_3 and has no lone electron pairs.

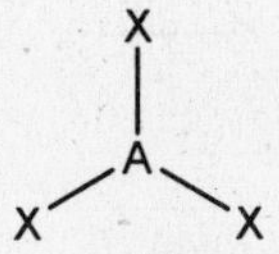

c) A trigonal bipyramidal molecule contains five electron groups (single bonds) and no lone pairs (AX_5).

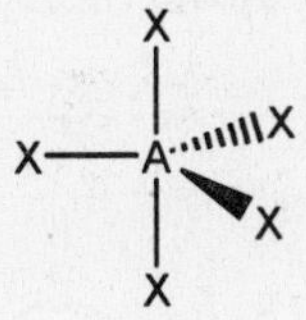

d) A T-shape molecule has three bonding groups and two lone pairs (AX_3E_2).

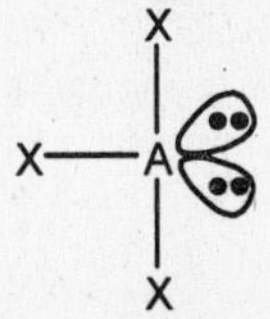

e) A trigonal pyramidal molecule follows the formula AX_3E.

f) A square pyramidal molecule shape follows the formula AX_5E.

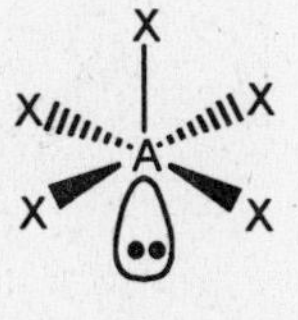

10.43 a) Lewis structure for O_3 is

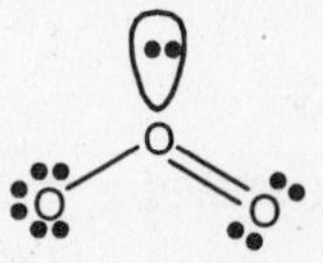

The central O has three electron groups around it with two groups as bonds and one as a lone pair. The electron group arrangement is <u>trigonal planar</u>, the molecular shape is <u>bent</u>, and the ideal bond angle is <u>120°</u>.

b) Lewis structure for H_3O^+ is

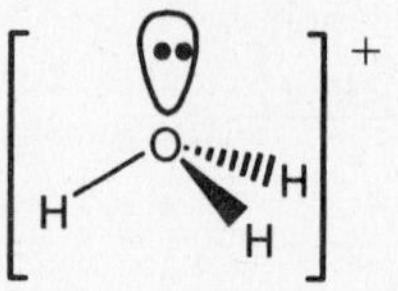

The central O atom has four electron groups around it with three bonds and one lone pair. The electron-group arrangement is <u>tetrahedral</u>, the molecular shape is <u>trigonal pyramidal</u>, and the ideal bond angle is <u>109.5°</u>.

c) Lewis structure for NF_3 is

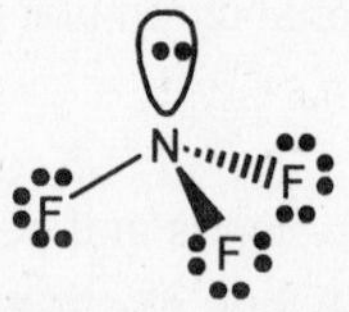

The electron-group arrangement is <u>tetrahedral</u>, the molecular shape is <u>trigonal pyramidal</u> and the ideal bond angle is <u>109.5°</u>.

10.45

Lewis Structure	Electron-group arrangement	Molecular Shape	Ideal Bond Angle
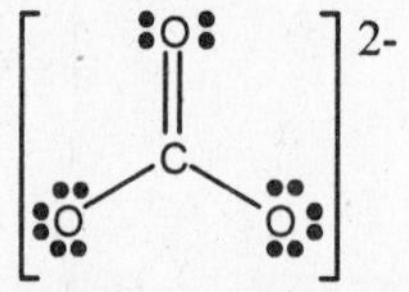 + 2 resonance forms	Trigonal planar (3 groups)	Trigonal planar	120°
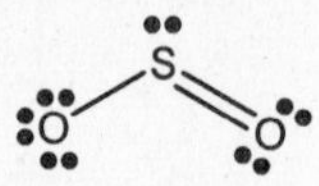 + 1 resonance form	Trigonal planar (3 groups)	V-shaped or bent	120°
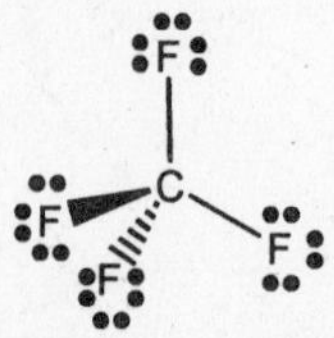	Tetrahedral (4 groups)	Tetrahedral	109.5°

10.47 a) Structure shows three electron groups and three bonds around central atom. Shape is trigonal planar, classification is AX_3, and ideal bond angle is 120°.

b) Structure shows four electron groups with three bonds and one lone pair. Shape is trigonal pyramidal, classification is AX_3E, and ideal bond angle is 109.5°.

c) Structure shows five electron groups with five bonds. Shape is trigonal bipyramidal, classification is AX_5, and ideal bond angles are 90° and 120°.

10.49 a) The chlorite ion, ClO_2^-, has 20 valence electrons. The central atom, Cl, has two lone pairs and two bonds (AX_2E_2) so the *shape* is bent (or V-shape). The ideal bond angle depends on the number of groups around the central atom, i.e. the electron-group arrangement. Because there are 4 groups around Cl, the *electron group arrangement* is tetrahedral, so the ideal bond angle is 109.5°. The two lone pairs take more space, compressing the ideal bond angle so that it is smaller than 109.5°.

b) Phosphorus pentafluoride, PF_5, has 40 valence electrons. Because there are no lone pairs to distort the bond angles, there is no deviation from the ideal bond angle of 120° (equatorial plane) and 90° (axial plane). The central atom, P, has 5 groups (AX_5) so the shape is trigonal bypyramidal.

c) Selenium tetrafluoride, SeF_4, has 34 valence electrons. The central atom, Se, has 5 electron pairs of which one is a lone pair (AX_4E). The lone pair occupies the equatorial position, resulting in a see-saw shape. The lone pair also causes a deviation of the ideal bond angles, resulting in angles smaller than 120° and 90°.

d) Krypton difluoride, KrF_2, has 22 valence electrons. The central atom, Kr, has 5 electron groups of which three are lone pairs (AX_2E_3). The electron-group arrangement is trigonal bypyramidal, and the three lone pairs occupy the equatorial positions. This arrangement results in a linear molecular shape. The ideal bond angle, 180°, is unaffected by the location of the lone pairs, so there is no deviation.

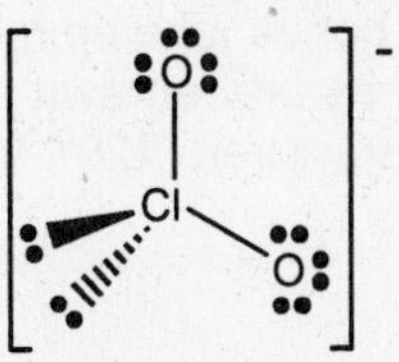

(a)

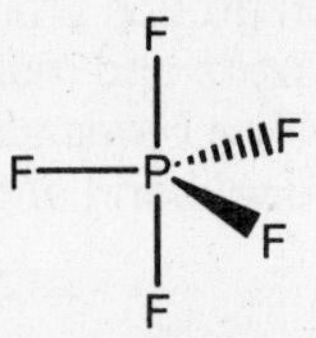

assume lone pairs on F

(b)

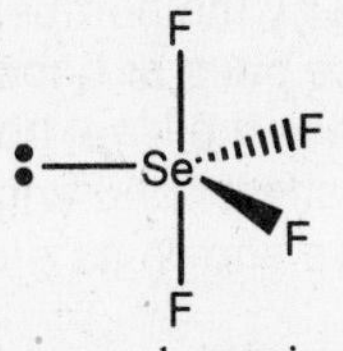

assume lone pairs on F

(c)

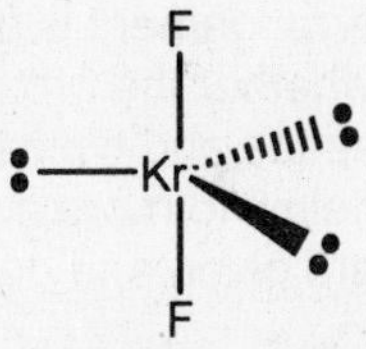

assume lone pairs on F

(d)

10.51 a) In CH_3OH carbon and oxygen are the two central atoms. The shape around the carbon is tetrahedral with the classification AX_4 and bond angles = 109.5°. The shape around the oxygen is bent with the classification AX_2E_2 and bond angles that are <109.5° due to the lone pairs.

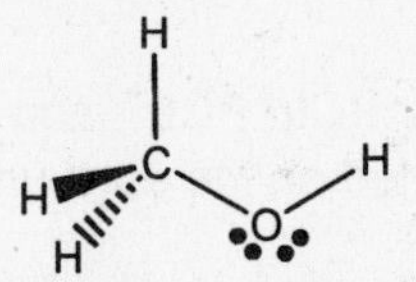

b) In N_2O_4 the two nitrogens are the central atoms. There are three electron groups around each nitrogen with the classification AX_3 so the shape is trigonal planar around each nitrogen. The bond angles are <120° because the greater electron density in double bonds repels the other single bonds slightly.

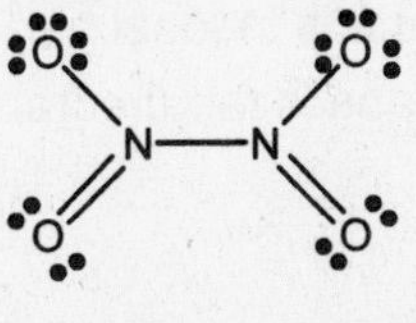

10.53 a)

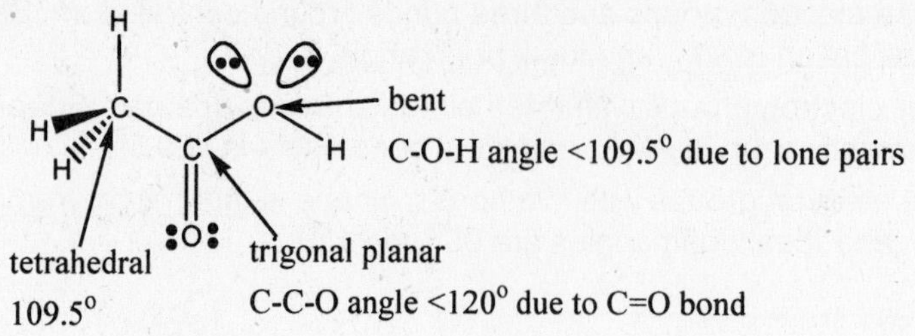

b)

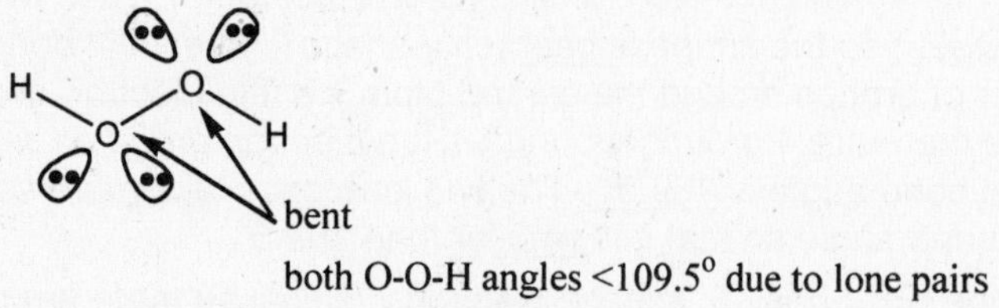

10.55 Increasing bond angle: $OF_2 < NF_3 < CF_4 < BF_3 < BeF_2$

OF_2, NF_3, and CF_4 all have electron-group arrangements that are tetrahedral with ideal bond angles of 109.5°. CF_4 will have 109.5° bond angles since all the electron groups are involved in equivalent bonds. The F—N—F bond angle in NF_3 will be less than 109.5° because the lone pair will compress the bond angles. This occurs in OF_2 as well but to a greater extent than in NF_3 since there are 2 lone pairs in OF_2 and only one in NF_3. Thus, the bond angle in OF_2 is smaller than the angle in NF_3. BF_3 is trigonal planar with bond angles of 120°. BeF_2 is linear with the largest bond angles at 180°.

10.57 Bond angles are determined by the *electron group arrangement*. The electron group arrangement is determined by the number of electron groups around the central atom, whether the groups are lone pairs or bonded. Lone pairs and multiple bonds take up more space and typically compress the ideal bond angle. The Lewis structures are redrawn below to more accurately reflect bond angles. The ideal bond angles are given (120°, 90°) are given and deviations are shown as < or >.

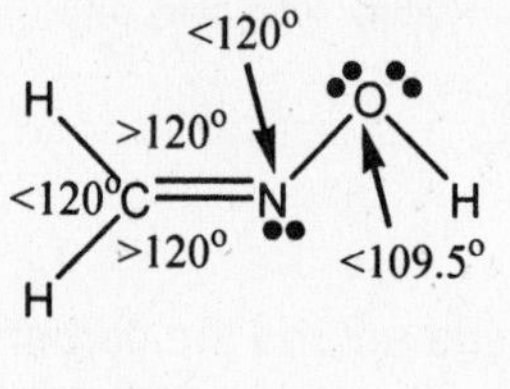

(a)

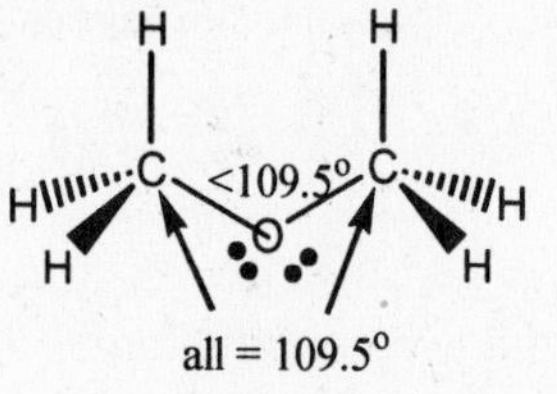

(b)

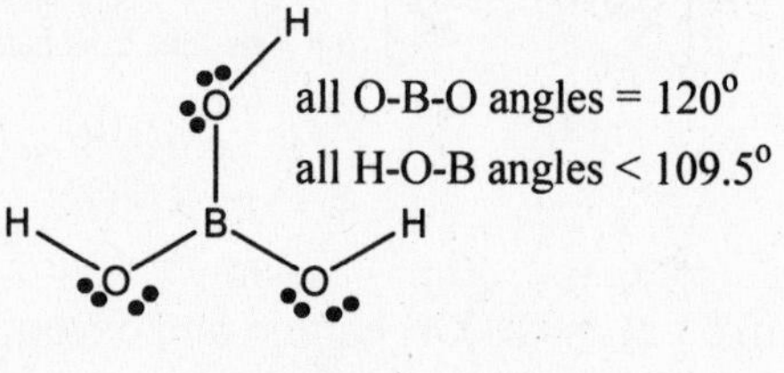

(c)

10.60 To determine the molecular shapes for PCl_5, PCl_4^+, and PCl_6^- the Lewis structures must be drawn. Lone electron pairs are assumed on the surrounding Cl atoms.

PCl_5:

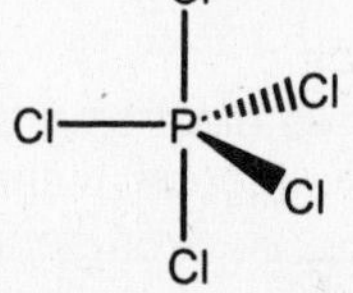

PCl_4^+:

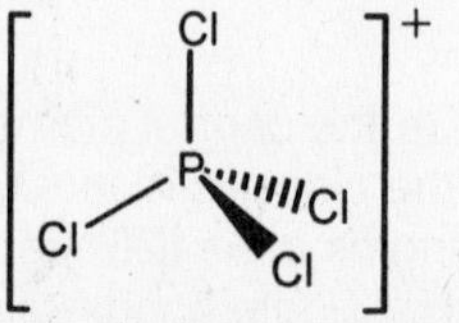

PCl_6^-:

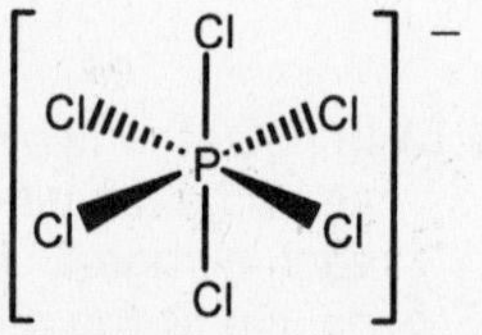

PCl_5 follows the classification AX_5; shape is trigonal bipyramidal.

PCl_4^+ follows the classification AX_4; shape is tetrahedral.

PCl_6^- follows the classification AX_6; shape is octahedral.

As PCl_5 solidifies half the molecules change from a trigonal bipyramidal shape to a tetrahedral shape and the other half change from the trigonal bipyramidal shape to an octahedral shape. The angles in the gas phase are 120° and 90° in the trigonal bipyramidal shape. These change to all 109.5° angles in PCl_4^+ and all 90° angles in PCl_6^-.

10.61 Bonds are polar if there is a difference in electronegativity of the atoms participating in the bond. Polar bonds have magnitude and direction. If the polar bonds are not balanced, i.e. the direction and the magnitude of the bond dipoles do not cancel each other, the molecule is polar.

10.64 a) The bonds with greatest polarity are those with the greatest difference in electronegativity between the two atoms in the bond. S—Cl ΔEN = 0.5; F—F ΔEN = 0; C—S, ΔEN = 0; C—F, ΔEN = 1.5; Br—Cl, ΔEN = 0.2. The greatest electronegativity difference is 1.5 for C—F bonds, so $\mathbf{CF_4}$ contains the bonds that are most polar.

b) Molecular dipole moments depend on the shape of the molecule. $\mathbf{SCl_2}$ and **BrCl** are the only two of the given molecules with dipole moments.

SCl_2 is bent so the polarity of the bonds will not cancel and the molecule has a dipole moment. F_2 has no dipole moment because its bonds are nonpolar. CS_2 also has nonpolar bonds. The polar bonds in CF_4 cancel because they are arranged in a symmetrical tetrahedral shape and all the bonds have the same polarity. Thus, CF_4 does not have a dipole moment. The BrCl bond is polar so the molecule has a dipole moment.

10.66 a) In SO_3, the bond dipoles cancel because the molecule is trigonal planar, so $\mathbf{SO_2}$ has the greater dipole moment.

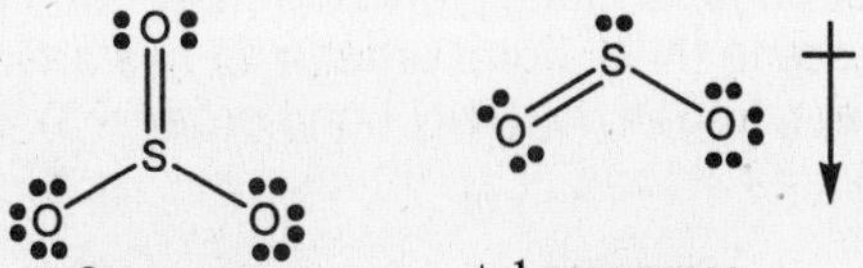

+ 2 resonance structures   + 1 resonance structure

b) Both ICl and IF are diatomic, linear and polar. However, **IF** has a greater ΔEN, so it has the greater dipole moment.

c) SiF_4 has polar bonds, but the bond dipoles cancel in the symmetrical tetrahedral arrangement. SF_4 has a see-saw shape, so the bond dipoles cannot cancel. $\mathbf{SF_4}$ has the greater dipole moment.

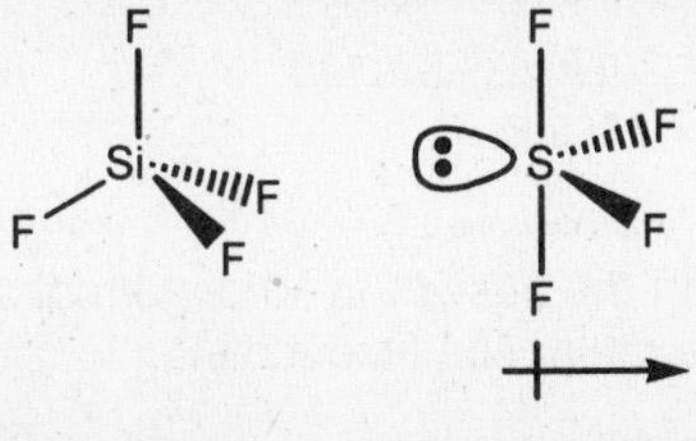

d) Both H_2O and H_2S have similar structures. Therefore, $\mathbf{H_2O}$ has a greater dipole because its bonds have a higher ΔEN.

10.68 The compound CH_2Cl_2 can be drawn as three possible structures:

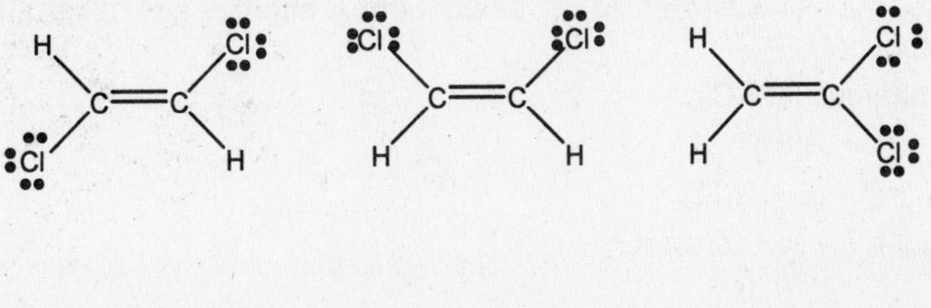

The double bond between the carbons prevents the rotation of one carbon relative to the other, so the first two structures are different molecules. The third structure is different from the first two structures because both chlorine atoms are bonded to the same carbon. Because the product is ClH_2–CH_2Cl, meaning one Cl atom is bonded to each C, the third structure must be Y because its reaction product would be CH_3—$CHCl_2$. The first two structures are X and Z, respectively, because the first structure has no dipole moment. Structures Y and Z have dipole moments.

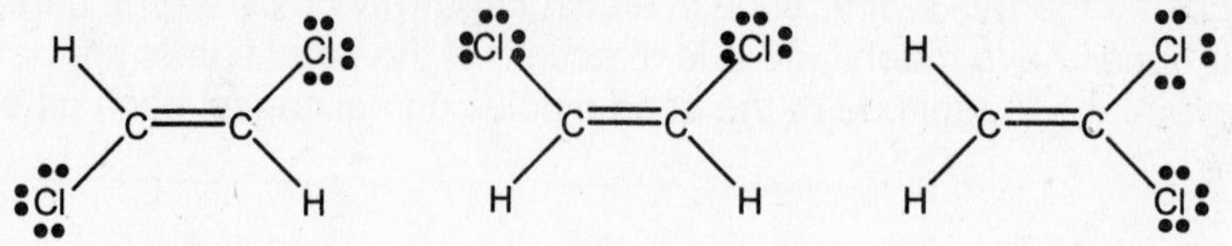

X **Z** **Y**

10.70 a) In drawing the Lewis structures, the H atoms must always be terminal atoms (i.e. on the end) because H can only form one bond and have 2 electrons in its valence shell. The N atoms must have exactly 8 valence electrons because they follow neither the electron-deficient or expanded valence shell exceptions.

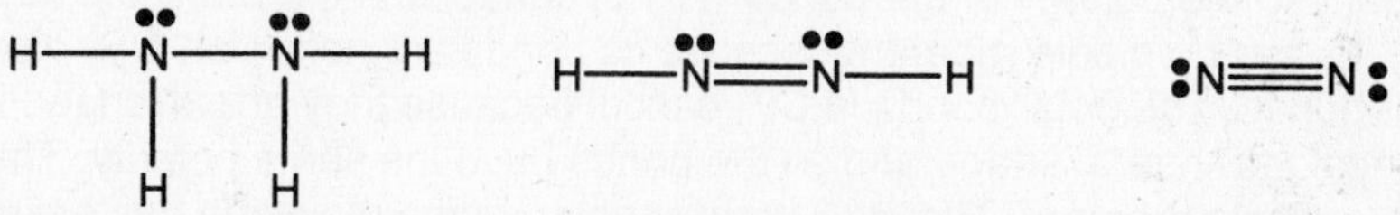

Hydrazine (14 val e$^-$) Diazene (12 val e$^-$) N_2 (10 val e$^-$)

Review bond characteristics described in section 9.3. For a given pair of atoms, *a shorter bond is a stronger bond.* Hydrazine (N-N bond order = 1) has the longest weakest bond. Diazene (N-N bond order = 2) has an intermediate bond length and strength. Molecular nitrogen, N_2 (N-N bond order = 3), has the shortest and strongest bond.

b)

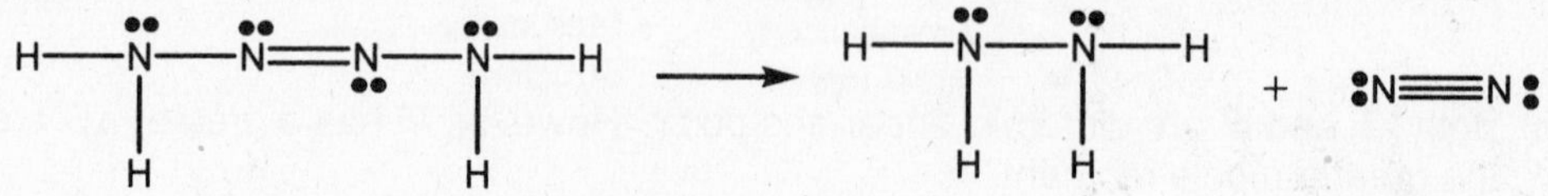

$\Delta H^\circ_{\text{bonds broken}}$ = 4 N-H = 4 mol(391 kJ/mol)
2 N-N = 2 mol(160 kJ/mol)
1 N=N = 1 mol(418 kJ/mol)
= 2302 kJ

$\Delta H^\circ_{\text{bonds formed}}$ = 4 N-H = 4 mol(-391 kJ/mol)
1 N-N = 1 mol(-160 kJ/mol)
1 N≡N = 1 mol(-945 kJ/mol)
= -2669 kJ

$\Delta H^\circ_{rxn} = \Delta H^\circ_{\text{bonds broken}} + \Delta H^\circ_{\text{bonds formed}}$ = 2302 kJ + (-2669 kJ) = **-367 kJ**

Note: It is correct to report the answer in kJ or kJ/mol as long as the value refers to a reactant or product with a molar coefficient of 1.

10.73 a) Formal charges for Al_2Cl_6:

$FC_{Al} = 3 - (0 + ½(8)) = -1$

$FC_{Cl,\ ends} = 7 - (6 + ½(2)) = 0$

$FC_{Cl,\ interior} = 7 - (4 + ½(4)) = +1$ (Check: Formal charges add to zero, the charge on the compound)

Formal charges for I_2Cl_6:

$FC_{I} = 7 - (4 + ½(8)) = -1$

$FC_{Cl,\ ends} = 7 - (6 + ½(2)) = 0$

$FC_{Cl,\ interior} = 7 - (4 + ½(4)) = +1$

b) The aluminum atoms in Al_2Cl_6 have four bonding pairs and no unbonded pairs, with a general formula AX_4 (see Figure 10.8), corresponding to a tetrahedral shape. The tetrahedron is not a planar shape. The molecular shape would look like two tetrahedral shapes that share one edge.

The iodine atoms in I_2Cl_6 have four bonding pairs and two unbonded pairs, with a general formula AX_4E_2 (see Figure 10.11), corresponding to a square planar shape. Therefore, **I_2Cl_6 is planar**.

10.76

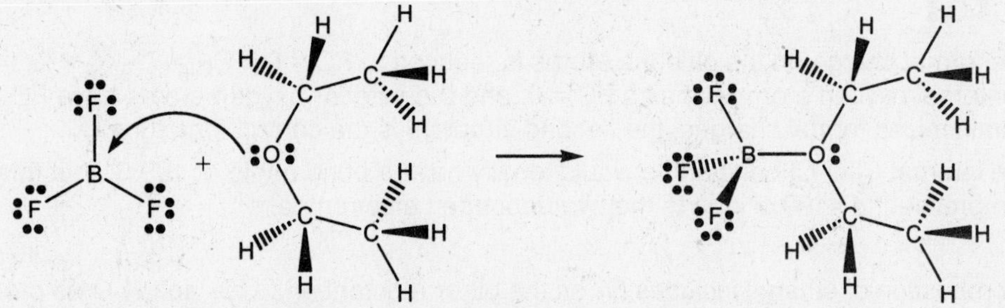

The reactant BF_3 is trigonal planar (bond angle = 120°). Upon reaction, the shape around the B atom changes to tetrahedral (bond angle = 109.5°). The shape around O in the reactant is bent, and the two lone pairs compress the bond angle to < 109.5°. Upon reaction, the shape around oxygen changes to trigonal pyramidal (AX_3E).

10.79 a)

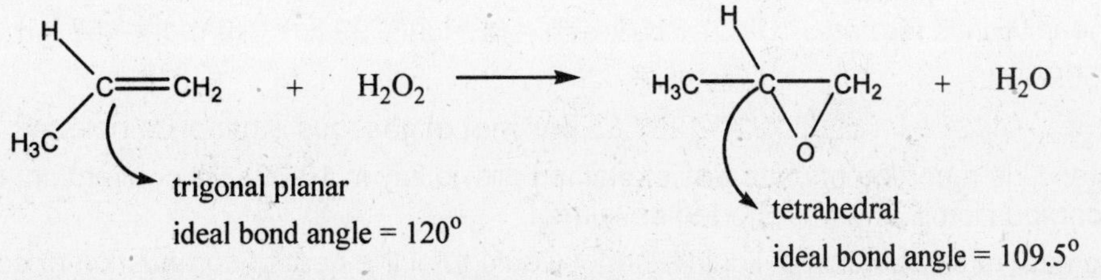

b) The C-C-H bond angle in propene is compressed (<120°) because the double bond takes up more room than the other bonds. The central carbon atom in propylene forms four bonds and therefore should have bond angles of 109.5°. However, the presence of the three-member ring compresses the C-C-O bond angle (and other interior angles) to 60°. The remaining bond angles have a little more space as a result, and their bond angles are >109.5°.

10.81 An initial attempt at drawing the Lewis structure for Cl_2O_7 would yield this structure:

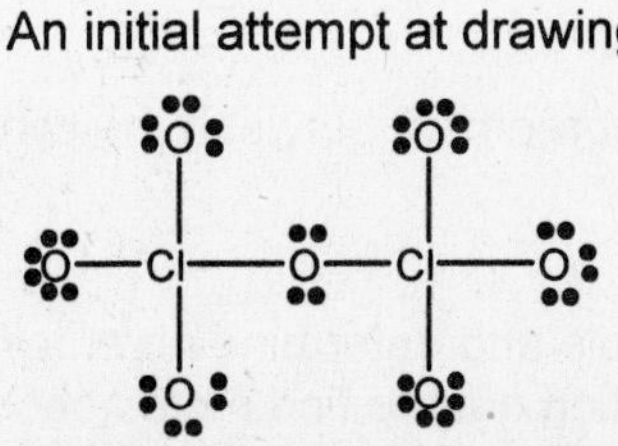

Check to insure that the number of electrons drawn equal the number of valence electrons (56 e⁻). An examination of formal charges shows that the Cl atoms have FC = +3 [$FC_{Cl} = 7 - (0 + ½(8))$], the internal oxygen atom has a FC = 0, and the exterior oxygen atoms have FC = -1. Although these formal charges sum to zero, the formal charge on chlorine is high so this is not the likely structure.

To reduce the formal charge on chlorine, we need to increase the number of bonding electrons to chlorine. Although this would violate chlorine's octet, chlorine can have an expanded valence shell because it can access its 3d orbitals.

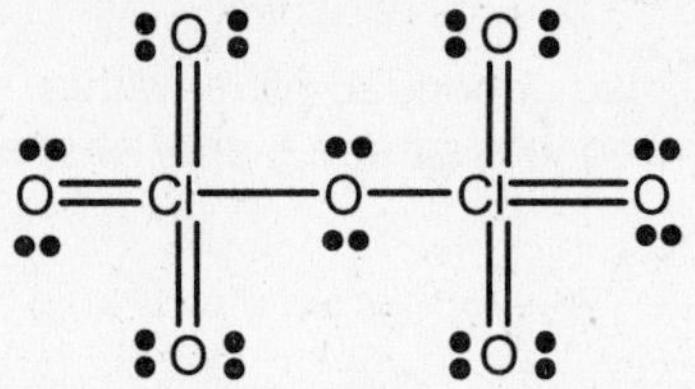

The formal charge on the chlorine atoms is reduced to FC = 0 [FC_{Cl} = 7 – (0 + ½(14))]. The internal oxygen atom still has a FC = 0, and the exterior oxygen atoms have FC = 0. With minimized formal charges, the second structure is the correct structure.

The internal Cl-O-Cl bond angle would ideally have a bond angle of 109.5°, but this angle is compressed to <109.5° due to the two unbonded electron pairs.

10.85 a) Combustion of ethanol includes O_2 as the other reactant and CO_2 and H_2O as products.

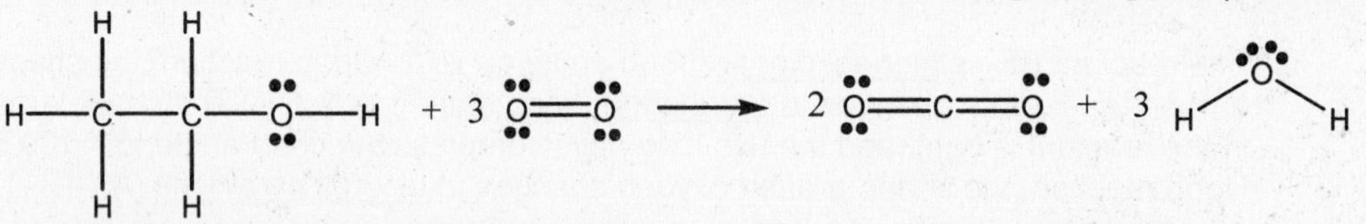

Reactant bonds broken: 1(C—C) + 5(C—H) + 1(C—O) + 1(O—H) + 3(O=O)
= (1 mol)(347 kJ) + (5 mol)(413 kJ) + (1 mol)(358 kJ) + (1 mol)(467 kJ) + (3 mol)(498 kJ)
= 4731 kJ

Product bonds formed: 4(C=O) + 6(O—H) = (4 mol)(-799 kJ) + (6 mol)(-467 kJ)
= -5998 kJ

ΔH_{comb} = 4731 kJ – 5998 kJ = **-1267 kJ per mol** of gaseous ethanol combusted

Note: The reporting of units was explained previously in 10.30. By convention, the heat of combustion is always reported as kJ/mol.

b) The heat of combustion of liquid ethanol would total the heat of combustion of gaseous ethanol as calculated in part a) plus the heat required to convert one mole of liquid ethanol to the gas state.

$\Delta H_{comb(liquid)}$= -1267 kJ + (1mol)(40.5 kJ/mol) = **-1226 kJ per mol** of liquid ethanol burned

c) $\Delta H_{comb(liquid)} = \sum n\Delta H_f^o(prod) - \sum m\Delta H_f^o(reac)$

$$\Delta H_{comb(liquid)} = 2(\Delta H_f^o(CO_2(g))) + 3(\Delta H_f^o(H_2O(g))) - (\Delta H_f^o(C_2H_5OH(l))) - 3(\Delta H_f^o(O_2(g)))$$

$$\Delta H_{comb(liquid)} = 2(-393.5\,kJ) + 3(-241.826\,kJ) - (-277.63\,kJ) - 3(0) = -1234.8\,kJ$$

The two values differ by less than 1%, so there is good agreement between the two methods of calculating heat of combustion for ethanol.

10.88 Plan: To learn the identity of the acid, assume a 100 g sample and determine an empirical formula from the elemental composition. Then use the titration data to find moles of NaOH/mole of acid, which tells the number of acidic protons per molecule. Use this factor to find the molecular formula and write the Lewis structure.

Solution:

$$(2.24\ g\ H)\left(\frac{1\ mol\ H}{1.008\ g\ H}\right) = 2.22\ mol\ H \div 2.22 = 1\ H$$

$$(26.7\ g\ C)\left(\frac{1\ mol\ C}{12.01\ g\ C}\right) = 2.22\ mol\ C \div 2.22 = 1\ C$$

$$(71.1\ g\ O)\left(\frac{1\ mol\ O}{16.00\ g\ O}\right) = 4.44\ mol\ O \div 2.22 = 2\ O$$

Therefore, the empirical formula is CHO_2.

The number of mmol of NaOH used in the titration is:

$$mmol\ of\ NaOH = \left(\frac{0.040\ mmol\ NaOH}{mL}\right)(50.0\ mL) = 2.0\ mmol\ NaOH$$

Because 1.00 mmol of acid react with 2.0 mmol of base, the acid must be diprotic, i.e. each mole of acid forms 2 H^+ ions in solution. Therefore, the molecular formula is 2 x CHO_2 or **$C_2H_2O_4$**.

In drawing the Lewis structure, the two H's must be on the end and attached to an O (hydrogens attached to C are not acidic). $C_2H_2O_4$ belongs to a family of compounds called carboxylic acids. Carboxylic acids have the functional group –COOH, shown below:

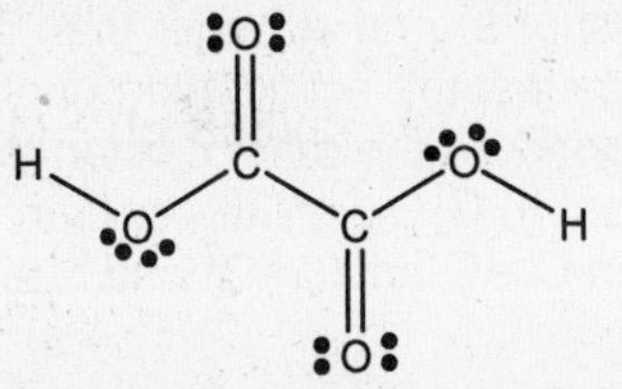

10.92 a) OH (called *hydroxyl*) has 7 valence electrons, so it is an octet deficient molecule. The Lewis structure is

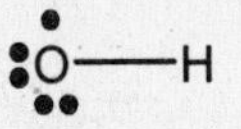

b) The heat of formation for OH(g) refers to the reaction ½O_2(g) + ½H_2(g) → OH(g).

$$\Delta H^o_{rxn} = \Delta H^o_{bonds\ broken} + \Delta H^o_{bonds\ formed}$$

The ΔH_{rxn}^o is the same as ΔH_f^o(OH) for this reaction. Recall from section 9.3 that bond breakage is always endothermic and bond formation is always exothermic. The above equation can then be rewritten as:

$$\Delta H_f^o(OH) = ½BE_{O=O} + ½BE_{H\text{-}H} - BE_{O\text{-}H}$$

$$(1\ mol)(39\ kJ/mol) = (½\ mol)(498\ kJ/mol) + (½\ mol)(432\ kJ/mol) - BE_{O\text{-}H}$$

$$BE_{O\text{-}H} = ½(498\ kJ) + ½(432\ kJ) - 39.0\ kJ = \mathbf{426\ kJ/mol}$$

Note: The reporting of units was explained previously in 10.30. By convention, the heat of combustion is always reported as kJ/mol.

c) To break the two O—H bonds in water would require (2 mol)(467 kJ/mol) = 934 kJ. Of this energy, 426 kJ is required to break the second bond (the bond energy for OH(g) calculated in part b), so the amount required to break the first O—H bond is 934 kJ/mol – 426 kJ/mol = **508 kJ/mol**.

10.96 $H_2C_2O_4$:

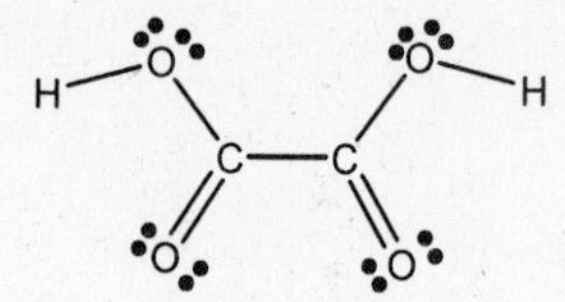

$HC_2O_4^-$:

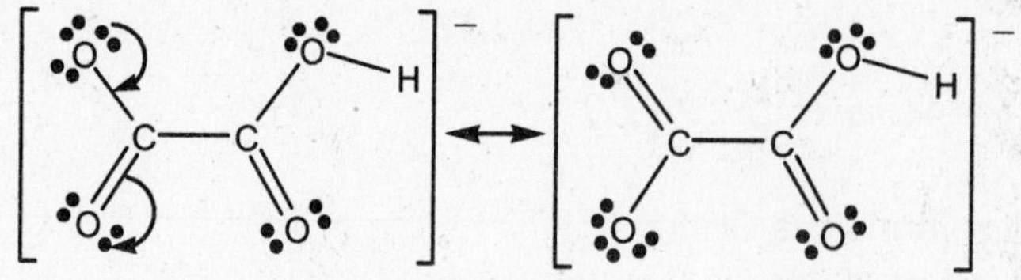

$C_2O_4^{2-}$:

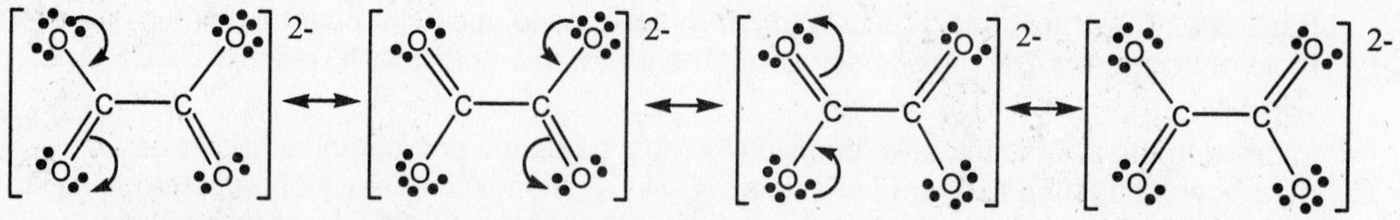

In $H_2C_2O_4$, the C=O bonds will be shorter and stronger than the C—O bonds. In $HC_2O_4^-$, the O which bonds to the C that has lost the acidic hydrogen will be equivalent and will be longer and weaker than the C=O bond on the other carbon, but shorter and stronger than the C—O bond on the other carbon. In $C_2O_4^{2-}$ all the carbon to oxygen bonds will be the same length (intermediate between C—O and C=O) and strength.

10.100

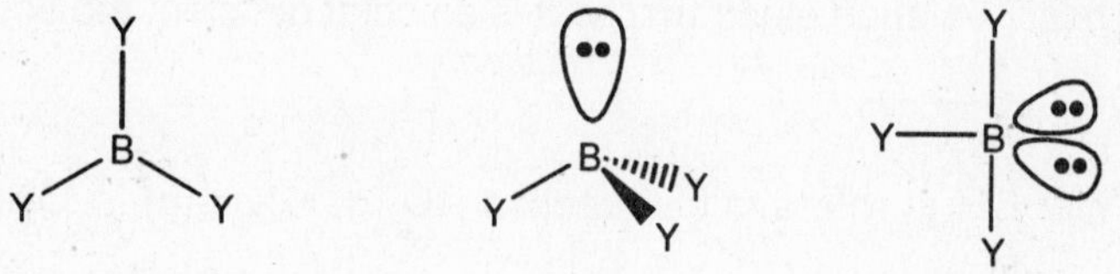

(a)
3 groups

(b)
4 groups

(c)
5 groups

The trigonal planar molecule in (a) is symmetrical and nonpolar; it cannot be the structure of BY_3. The trigonal pyramidal molecule in (b) and the T-shaped molecule in (c) can be polar.

CHAPTER 11

THEORIES OF COVALENT BONDING

FOLLOW-UP PROBLEMS

11.1 a) In BeF_2 the electron-group arrangement is linear so hybridization around Be is *sp*.

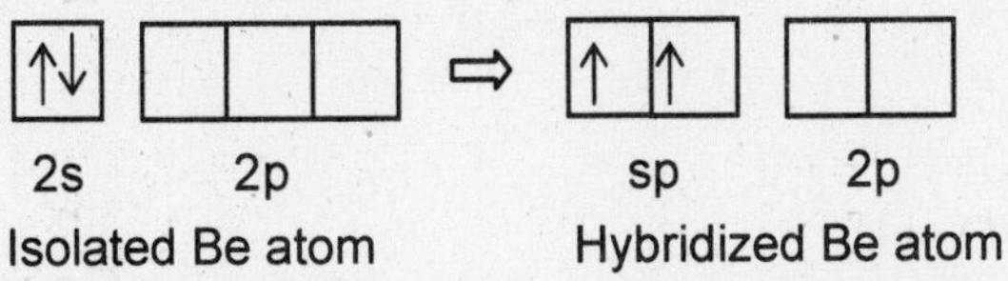

b) In $SiCl_4$ there are four electron groups around the silicon so its hybridization is sp^3.

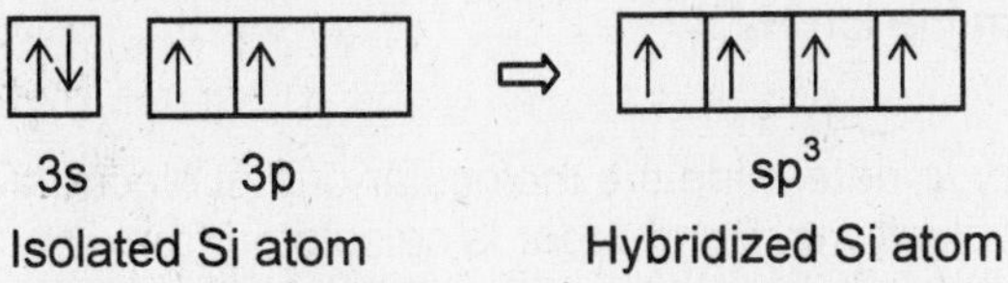

c) In XeF_4 the xenon is surrounded by 6 electron groups – 4 bonds and 2 lone pairs.

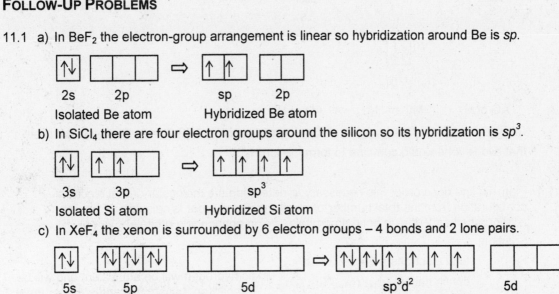

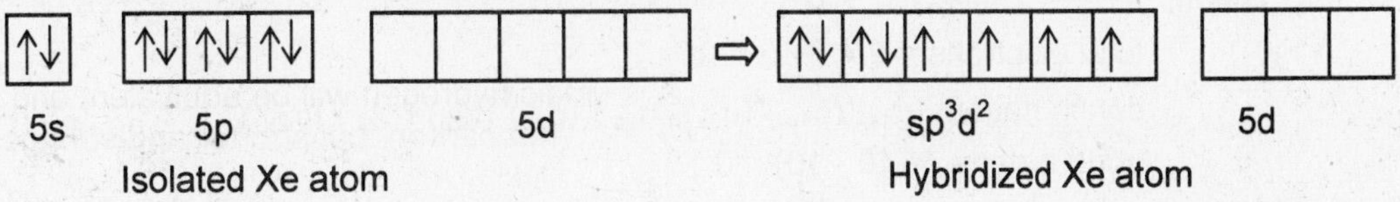

11.2 Plan: First determine the Lewis structure for the molecule. Then count the number of electron groups around each atom. Hybridization is sp if there are two groups, sp^2 if there are three groups and sp^3 if there are four groups. No hybridization occurs with only one group of electrons. The bonds are then designated as sigma or pi. A single bond is a sigma bond. A double bond consists of one sigma and one pi bond. A triple bond includes one sigma bond and two pi bonds. Hybridized orbitals overlap head on to form sigma bonds whereas pi bonds form through the sideways overlap of p or d orbitals.

Solution:

a) The Lewis structure of hydrogen cyanide is H—C≡N: . The single bond between carbon and hydrogen is a sigma bond formed by the overlap of a hybridized sp orbital on carbon with the 1s orbital from hydrogen. Between carbon and nitrogen are three bonds. One is a sigma bond formed by the overlap of a hybridized sp orbital on carbon with a hybridized sp orbital on nitrogen. The other two bonds between carbon and nitrogen are pi bonds formed by the overlap of p orbitals from carbon and nitrogen. One sp orbital on nitrogen is filled with a lone pair of electrons.

b) The Lewis structure of carbon dioxide, CO_2, is :Ö=C=Ö:

Both oxygen atoms are sp^2 hybridized and form a sigma bond and a pi bond with carbon. The sigma bonds are formed by the overlap of a hybridized sp orbital on carbon with a hybridized sp^2 orbital on oxygen. The pi bonds are formed by the overlap of an oxygen p orbital with a carbon p orbital. Two sp^2 orbitals on each oxygen are filled with a lone pair of electrons.

11.3 Plan: Draw the molecular orbital diagram. Determine the bond order from calculation:

BO = ½(# e^- in bonding orbitals - #e^- in antibonding orbitals). A bond order of zero indicates the molecule will not exist. A bond order greater than zero indicates that the molecule is at least somewhat stable and is likely to exist.

Solution:

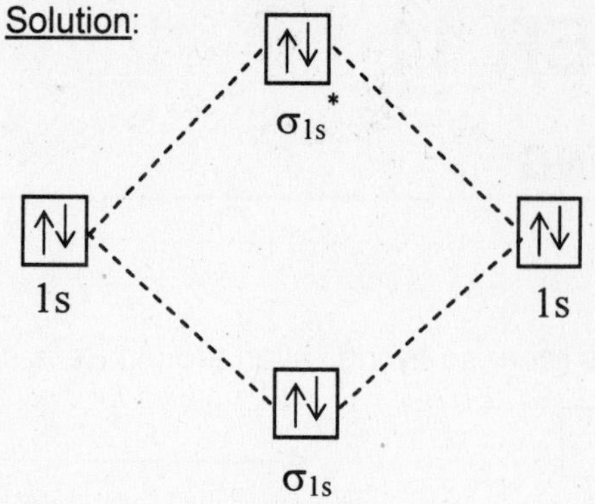

AO of H^- MO of H_2^{2-} AO of H^-

Configuration for H_2^{2-} is $(\sigma_{1s})^2(\sigma_{1s}^*)^2$. Bond order of H_2^{2-} is ½(2-2) = 0. Thus, it is not likely that two H^- ions would combine to form the ion H_2^{2-}.

11.4 Plan: To find bond order it is necessary to determine the molecular orbital electron configuration from the total number of electrons. Bond order is calculated from the configuration as ½(# e^- in bonding orbitals - #e^- in antibonding orbitals).

Solution:

F_2^{2-}: total electrons = 9 + 9+ 2 = 20.

configuration: $(\sigma_{1s})^2(\sigma_{1s}^*)^2(\sigma_{2s})^2(\sigma_{2s}^*)^2(\sigma_{2p})^2(\pi_{2p})^4(\pi_{2p}^*)^4(\sigma_{2p}^*)^2$

bond order = ½(10 - 10) = 0

F_2^-: total electrons = 9 + 9+ 1 = 19.

configuration: $(\sigma_{1s})^2(\sigma_{1s}^*)^2(\sigma_{2s})^2(\sigma_{2s}^*)^2(\sigma_{2p})^2(\pi_{2p})^4(\pi_{2p}^*)^4(\sigma_{2p}^*)^1$

bond order = ½(10 - 9) = ½

F_2: total electrons = 9 + 9 = 18.

configuration: $(\sigma_{1s})^2(\sigma_{1s}^*)^2(\sigma_{2s})^2(\sigma_{2s}^*)^2(\sigma_{2p})^2(\pi_{2p})^4(\pi_{2p}^*)^4$

bond order = ½(10 - 8) = 1

F_2^+: total electrons = 9 + 9 - 1 = 17.

configuration: $(\sigma_{1s})^2(\sigma_{1s}^*)^2(\sigma_{2s})^2(\sigma_{2s}^*)^2(\sigma_{2p})^2(\pi_{2p})^4(\pi_{2p}^*)^3$

bond order = ½(10 - 7) = 1½

F_2^{2+}: total electrons = 9 + 9 - 2 = 16.

configuration: $(\sigma_{1s})^2(\sigma_{1s}^*)^2(\sigma_{2s})^2(\sigma_{2s}^*)^2(\sigma_{2p})^2(\pi_{2p})^4(\pi_{2p}^*)^2$

bond order = ½(10 - 6) = 2

Bond energy increases as bond order increases: $F_2^{2-} < F_2^- < F_2 < F_2^+ < F_2^{2+}$

Bond length decreases as bond energy increases so the order of increasing bond length will be opposite that of increasing bond energy.

Increasing bond length: $F_2^{2+} < F_2^+ < F_2 < F_2^-$.

F_2^{2-} will not form a bond, so it has no bond length and is not included in the list.

Check: Since the highest energy orbitals are antibonding orbitals it makes sense that the bond order increases as electrons are removed from the antibonding orbitals.

END-OF-CHAPTER PROBLEMS

11.1 Table 11.1 describes the types of shapes that form from a given set of hybrid orbitals.
a) sp^2 b) sp^3d^2 c) sp d) sp^3 e) sp^3d

11.3 Carbon and silicon have the same number of valence electrons, but the outer level of electrons is $n = 2$ for carbon and $n = 3$ for silicon. Thus, silicon has d orbitals in addition to s and p orbitals available for bonding in its outer level to form up to 6 hybrid orbitals whereas carbon has only s and p orbitals available in its outer level to form 4 hybrid orbitals.

11.5 The *number* of orbitals remains the same as the number of orbitals before hybridization. The *type* depends on the orbitals mixed.
a) There are six unhybridized orbitals; therefore **six** hybrid orbitals result. The type is $\mathbf{sp^3d^2}$.
b) **Four** $\mathbf{sp^3}$ hybrid orbitals form from three *p* and one *s* atomic orbitals.

11.7 To determine hybridization draw the Lewis structure and count the number of electron groups. Hybridize that number of orbitals.
a) Lewis structure:

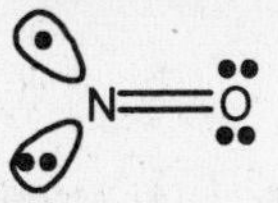

The three electron groups around nitrogen are $\mathbf{sp^2}$ hybridized.

b) Lewis structure:

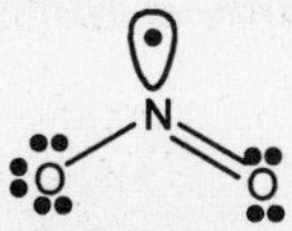

The nitrogen has three electron groups so hybridization is $\mathbf{sp^2}$.

c) Lewis structure:

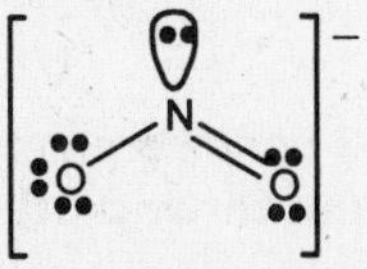

The nitrogen has three electron groups so hybridization is $\mathbf{sp^2}$.

11.9 a) Lewis structure:

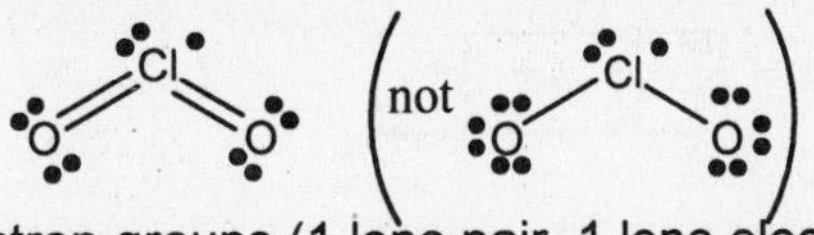

The Cl has four electron groups (1 lone pair, 1 lone electron, and 2 double bonds) so hybridization is $\mathbf{sp^3}$. The second structure is incorrect because formal charges are not minimized. Note that in ClO_2 the π bond is formed by the overlap of d-orbitals from chlorine with p-orbitals from oxygen.

b) Lewis structure:

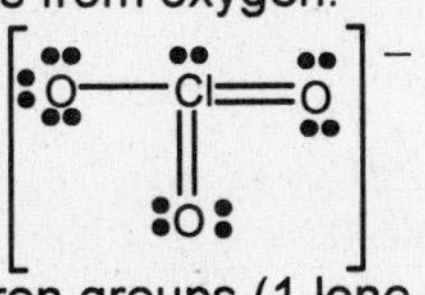

The Cl has four electron groups (1 lone pair and 3 bonds) so hybridization is $\mathbf{sp^3}$.

c) Lewis structure:

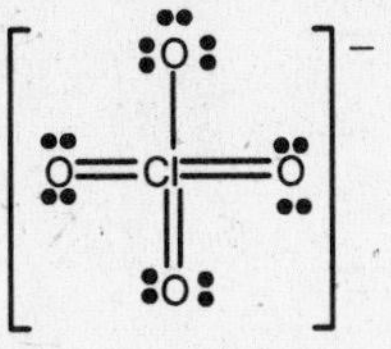

The Cl has four electron groups (4 bonds) so hybridization is **sp^3**.

11.11 a) Lewis structure of SiH_3Cl:

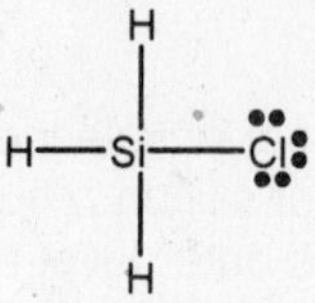

Around silicon are four electron groups with hybridization sp^3 made from **one *s* and three *p* atomic orbitals**.

b) Lewis structure of CS_2:

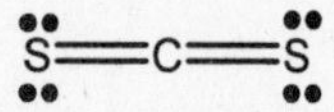

Around carbon are two electron groups with hybridization sp made from **one *s* and one *p* orbital**.

11.13 a) Lewis structure of SCl_3F:

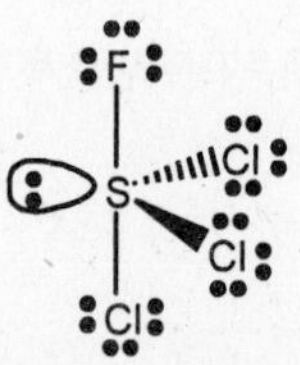

Sulfur is surrounded by 5 electron groups (4 bonding pairs, 1 lone pair) with hybridization sp^3d, formed from **one *s* orbital, three *p* orbitals, and one *d* orbitals**.

b) Lewis structure of NF_3:

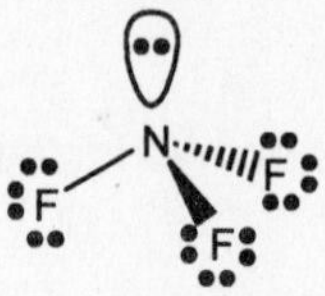

Nitrogen is surrounded by 4 electron groups (3 bonding pairs, 1 lone pair) with hybridization sp^3, formed from **one *s* orbital and three *p* orbitals**.

11.15 a) Germanium is the central atom in $GeCl_4$. Hybridization is sp^3 around Ge.

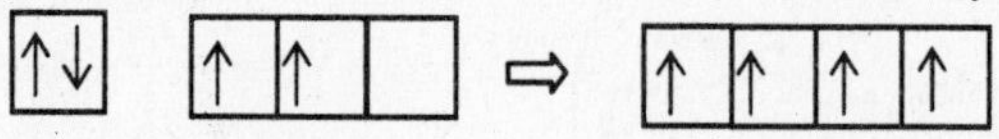

4s 4p sp^3

Isolated Ge atom Hybridized Ge atom

b) Boron is the central atom in BCl_3 and is sp^2 hybridized.

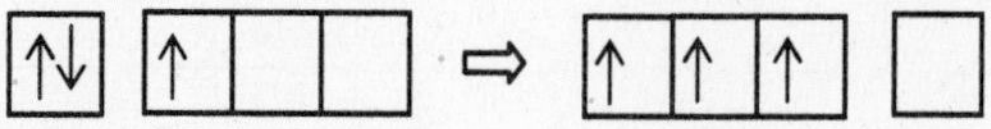

2s 2p sp^2 2p

Isolated B atom Hybridized B atom

c) Carbon is the central atom in CH_3^+ with sp^2 hybridization.

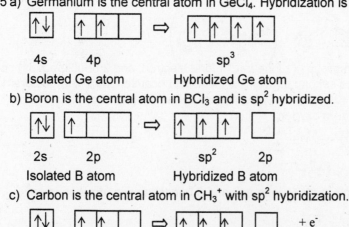

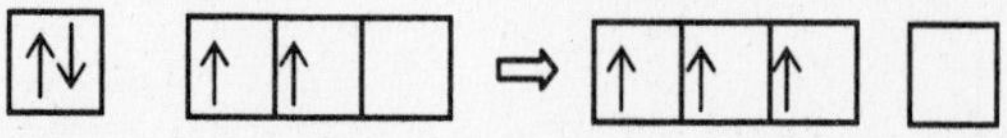

$+ e^-$

2s 2p sp^2 2p

Isolated C atom Hybridized C atom

11.17 a) In $SeCl_2$, Se is the central atom and is sp^3 hybridized. Two sp^3 orbitals are filled with lone electron pairs and two sp^3 orbitals bond with the chlorine atoms.

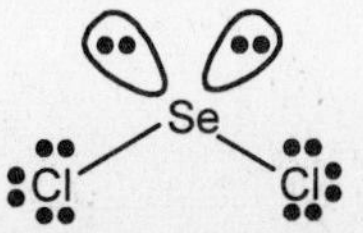

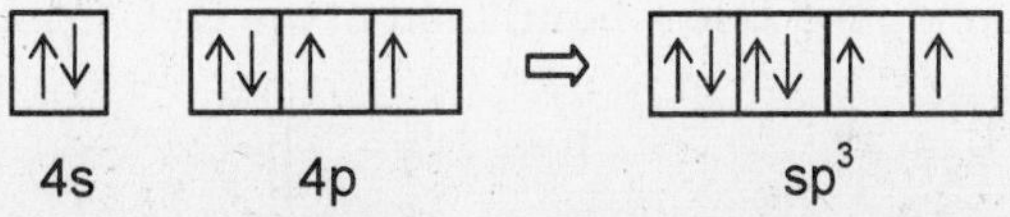

b) In H_3O^+, O is the central atom and is sp^3 hybridized. One sp^3 orbital is filled with a lone electron pair and three sp^3 orbitals bond with the hydrogens.

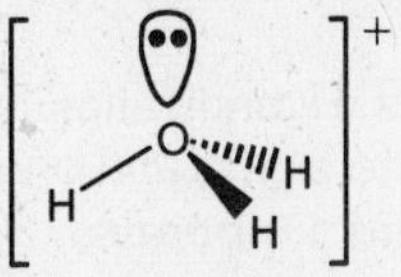

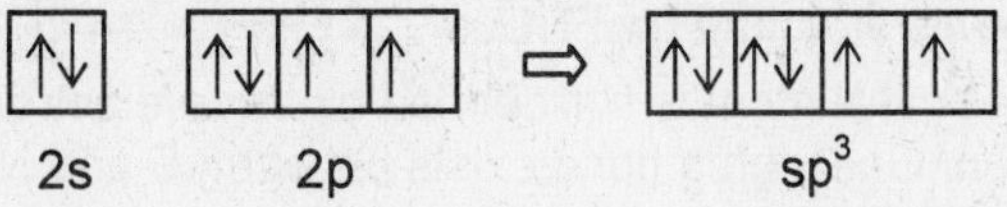

c) I is the central atom in IF_4^- with 6 electron groups surrounding it. Six electron groups give the octahedral geometry, with sp^3d^2 hybrid orbitals. The sp^3d^2 hybrid orbitals are composed of one *s* orbital, three *p* orbitals and two *d* orbitals.

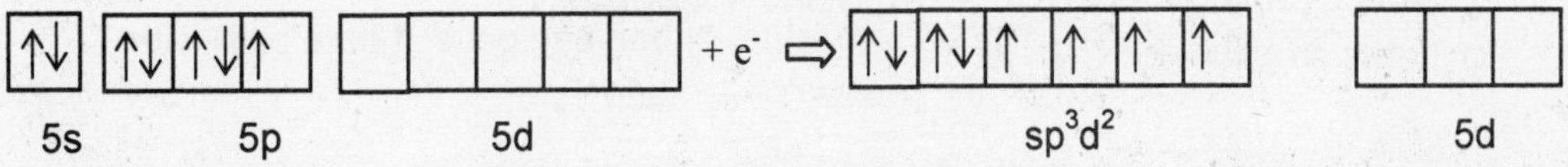

11.20 a) False. A σ bond and a π bond comprise a double bond.

b) False. A triple bond consists of one σ and two π bonds.

c) True

d) True

e) False. A π bond consists of one pair of electrons.

f) False. End to end overlap results in a bond with electron density along the bond axis between the two nuclei. (This is a sigma bond).

11.21 a) Nitrogen is the central atom in NO_3^-. Nitrogen has 3 surrounding electron groups so it is **sp^2** hybridized. Nitrogen forms **three σ bonds** (one each for the N–O bonds) and **one π bond** (part of the N=O double bond).

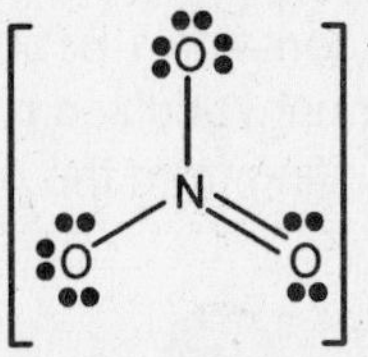

+ 2 resonance forms

b) Carbon is the central atom in CS_2. Carbon has 2 surrounding electron groups so it is **sp** hybridized. Carbon forms **two σ bonds** (one each for the C–S bonds) and **two π bonds** (part of the two C=S double bonds).

c) Carbon is the central atom in CH_2O. Carbon has 3 surrounding electron groups so it is **sp^2** hybridized. Carbon forms **three σ bonds** (one each for the two C–H bonds and one C–O bond) and **one π bond** (part of the C=O double bond).

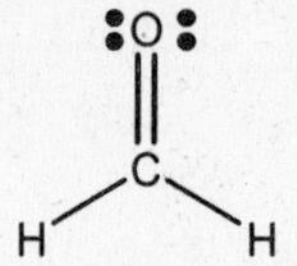

11.23 a) The Lewis structure for FNO is 

The central N atom is surrounded by three electron groups. Hybridization is **sp^2** around nitrogen. One sigma bond exists between F and N, and one sigma and one pi bond exist between N and O. Nitrogen participates in a total of **2 σ and 1 π bonds**.

b) The Lewis structure for C_2F_4 is shown below, with C as the two central atoms:

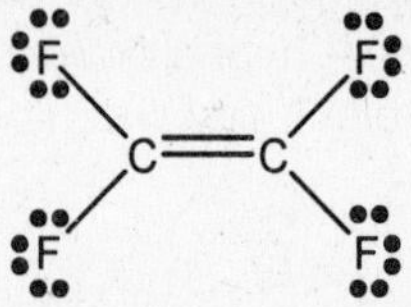

Each carbon has three electron groups with **sp^2** hybridization. The bonds between C and F are sigma bonds. The C—C bond consists of one sigma and one pi bond. Each carbon participates in a total of **3 σ and 1 π bonds**.

c) Lewis structure of $(CN)_2$ is

Each carbon is **sp** hybridized with a sigma bond between the two carbons and a sigma and two pi bonds comprising each C—N triple bond. Each carbon participates in a total of **2 σ and 2 π bonds**.

11.25 The double bond in 2-butene restricts rotation of the molecule, so that *cis* and *trans* structures result (see Figure 11.12). The two structures are shown below:

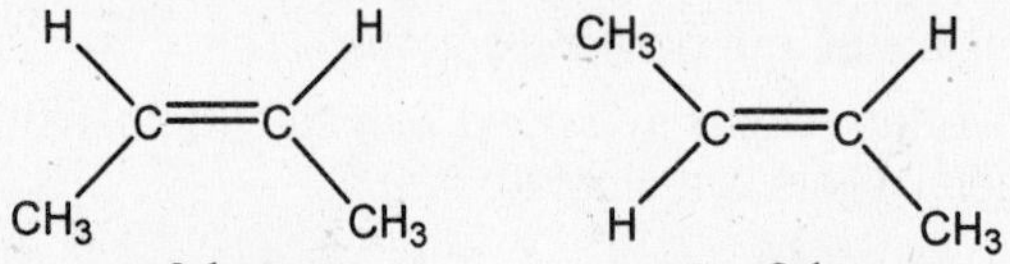

cis-2-butene *trans*-2-butene

The carbons participating in the double bond each have three surrounding groups, so they are sp^2 hybridized. The =C–H σ bonds result from the head-on overlap of a C sp^2 orbital and a H s orbital. The C–CH_3 bonds are also σ bonds, resulting from the head-on overlap of an sp^2 orbital and an sp^3 orbital. The C=C bond contains 1 σ bond (head on overlap of two sp^2 orbitals) and 1 π bond (sideways overlap of unhybridized p orbitals). Finally, C–H bonds in the methyl (-CH_3) groups are σ bonds resulting from the overlap of C's sp^3 orbital with H's s orbital.

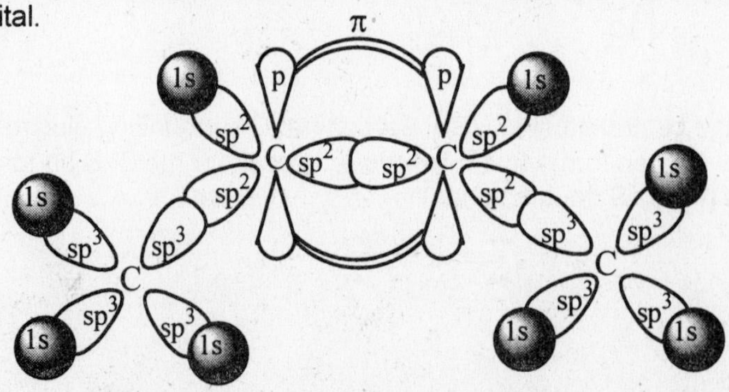

11.26 Four molecular orbitals form from the four p atomic orbitals. In forming molecular orbitals the total number of molecular orbitals must equal the number of atomic orbitals. Two of the four molecular orbitals formed are bonding orbitals and two are antibonding

11.28 a) Bonding MO's have lower energy than antibonding MO's. The bonding MO's lower energy, even lower than its constituent atomic orbitals, accounts for the stability of a molecule in relation to its individual atoms. However, the sum of energy of the MO's must equal the sum of energy of the AO's.

b) The node is the region of an orbital where the probability of finding the electron is zero, so the nodal plane is the plane that bisects the node perpendicular to the bond axis. According to Figure 11.13A, there is no node along the bond axis (probability is positive between the two nuclei) for the bonding MO. The antibonding MO does have a nodal plane, shown as "waves cancel" on Figure 11.13B.

c) The bonding MO has higher electron density between nuclei than the antibonding MO.

11.30 a) **Two** electrons are required to fill a σ-bonding molecular orbital. Each molecular orbital requires two electrons.

b) **Two** electrons are required to fill a π-antibonding molecular orbital. There are two π-antibonding orbitals, each holding a maximum of two electrons.

c) **Four** electrons are required to fill the two σ molecular orbitals (two electrons to fill the σ-bonding and two to fill the σ-antibonding) formed from two 1s atomic orbitals.

11.32 The horizontal line in all cases represents the bond axis.

a) bonding s + p_z

antibonding s – p_z

b) bonding p_x + p_x

antibonding p_x - p_x

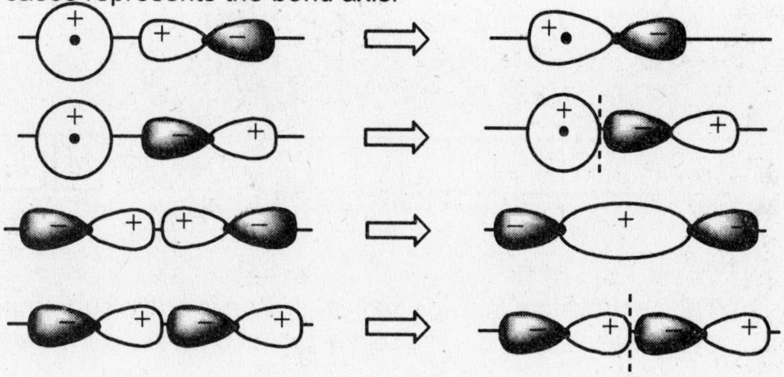

11.34 a) The molecular orbital configuration for Be_2^+ with a total of 7 electrons is $(\sigma_{1s})^2(\sigma_{1s}^*)^2(\sigma_{2s})^2(\sigma_{2s}^*)^1$ and bond order = ½(4-3) = ½. With a bond order of ½ the Be_2^+ ion will be **stable**.

b) No, the ion has one unpaired electron so it is **paramagnetic**, not diamagnetic.

c) Valence electrons would be those in the molecular orbitals at the $n = 2$ level so valence electron configuration is $\mathbf{(\sigma_{2s})^2(\sigma_{2s}^*)^1}$.

11.36 The sequence of MO's for C_2 is shown in Figure 11.20 (the $(\sigma_{1s})^2$ and $(\sigma_{1s}^*)^2$ MO's are not shown for convenience). For each species, determine the total number of electrons, the valence molecular orbital electron configuration and bond order.

C_2^- Total valence electrons = 4 + 4 + 1 = 9

Valence configuration: $(\sigma_{2s})^2(\sigma_{2s}^*)^2(\pi_{2p})^4(\sigma_{2p})^1$

Bond order = ½(7 – 2) = 2.5

C_2 Total valence electrons = 4 + 4 = 8

Valence configuration: $(\sigma_{2s})^2(\sigma_{2s}^*)^2(\pi_{2p})^4$

Bond order = ½(6 – 2) = 2

C_2^+ Total valence electrons = 4 + 4 –1 = 7

Valence configuration: $(\sigma_{2s})^2(\sigma_{2s}^*)^2(\pi_{2p})^3$

Bond order = ½(5 – 2) = 1.5

a) Bond energy increases as bond order increases: $\mathbf{C_2^+ < C_2 < C_2^-}$

b) Bond length decreases as bond energy increases so the order of increasing bond length will be opposite that of increasing bond energy. Increasing bond length: $\mathbf{C_2^- < C_2 < C_2^+}$.

11.39 a) There are **9** σ and **2** π bonds. Each of the six C—H bonds are sigma bonds. Each C—C bond contains a sigma bond. The double bonds between the carbons consist of a pi bond in addition to the sigma bond.

b) No, cis/trans structural arrangements are not possible because one of the carbons in each double bond has two hydrogens bonded to it. Cis/trans structural arrangements only occur when both carbons in the double bond are bonded to two groups that are not identical.

11.41 a) The central atom in IF_2^- is iodine. As shown by its Lewis dot structure, 5 electron groups surround the iodine so its hybridization is sp^3d. Write the partial orbital diagram for the central atom before and after the orbitals are hybridized and fill with electrons according to the Aufbau principle (unpaired spins fill first).

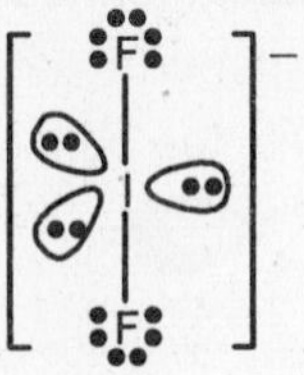

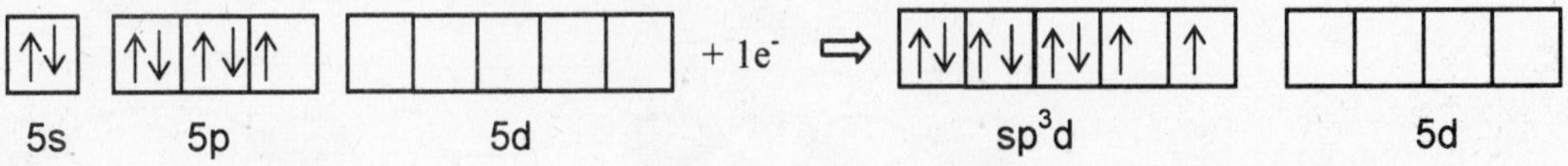

b) The central atom in ICl_3 is iodine. Iodine is again surrounded by 5 electron groups so its hybridization is sp^3d. The partial orbital diagram is the same as shown in (a).

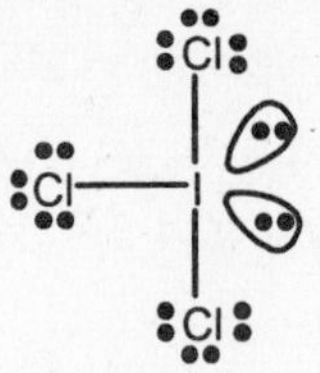

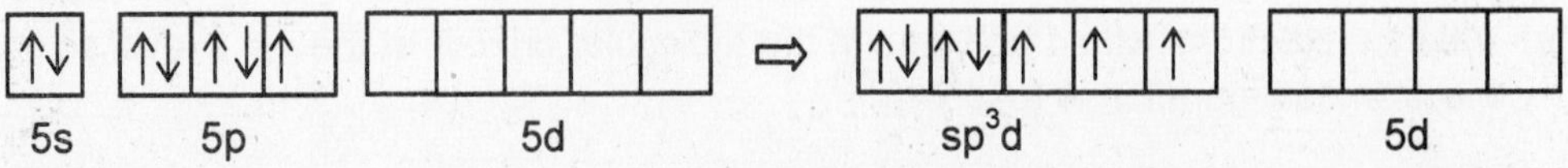

c) The central atom in $XeOF_4$ is Xe, with six surrounding electron groups and sp^3d^2 hybridization. The electron in the Xe 5d orbital overlaps with the O 2p orbital to form the π bond.

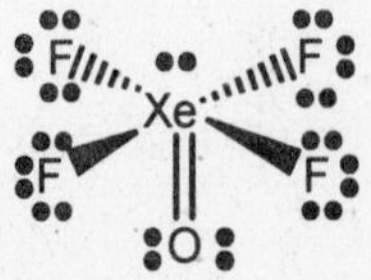

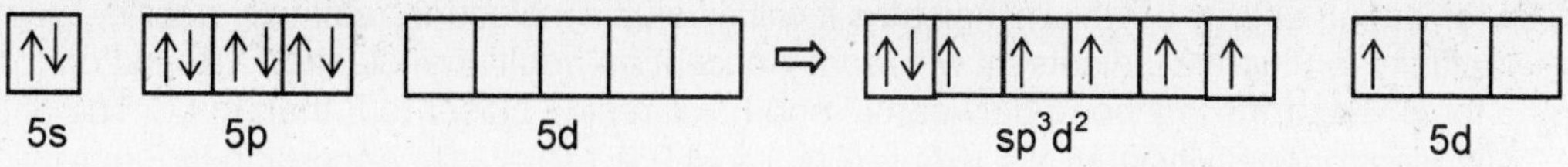

d) The central atom in BHF_2 is B, with three surrounding electron groups and sp^2 hybridization.

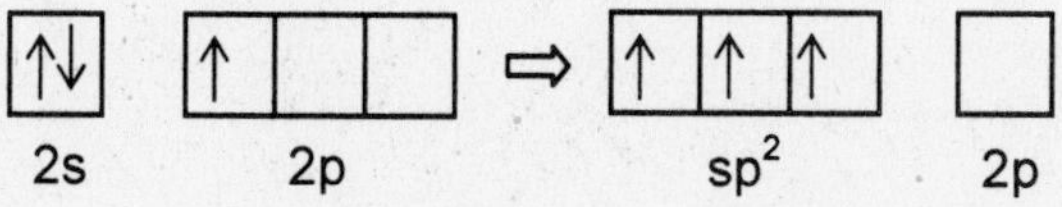

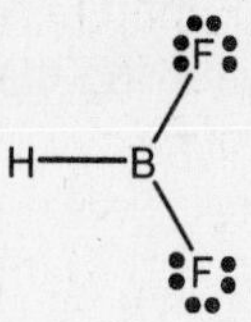

11.42 a) There are **17** σ bonds in isoniazid. Every atom-to-atom connection contains a σ bond.

b) All carbons have three surrounding electron groups, so their hybridization is $\mathbf{sp^2}$. The ring N also has three surrounding electron groups, so its hybridization is also $\mathbf{sp^2}$. The other two N's have four surrounding electron groups and are $\mathbf{sp^3}$ hybridized.

11.44 a) B changes from $\mathbf{sp^2 \rightarrow sp^3}$. Boron in BF_3 has three electron groups with sp^2 hybridization. In BF_4^-, the 4 electron groups surround B with sp^3 hybridization.

b) P changes from $\mathbf{sp^3 \rightarrow sp^3d}$. Phosphorus in PCl_3 is surrounded by 4 electron groups (3 bonds to Cl and 1 lone pair) for sp^3 hybridization. In PCl_5, phosphorus is surrounded by 5 electrons groups for sp^3d hybridization.

c) C changes from $\mathbf{sp \rightarrow sp^2}$. Two electrons groups surround C in C_2H_2 and 3 electron groups surround C in C_2H_4.

d) Si changes from $\mathbf{sp^3 \rightarrow sp^3d^2}$. Four electron groups surround Si in SiF_4 and 6 electron groups surround Si in SiF_6^{2-}.

e) **No change**. S in SO_2 and in SO_3 is surrounded by 3 electrons groups for sp^2 hybridization.

11.47 a) The representation with **two S=O double bonds** is better since it minimizes formal charges. For sulfur, the formal charge in the single bond representation is +2 while in the double bond representation it decreases to zero. The formal charge for the oxygen atoms double bonded to the sulfur increases from –1 to 0. The oxygens that are single bonded in both cases have the same formal charge in both representations, -1.

b) In both representations the sulfate ion is **tetrahedral** because 4 electron groups surround S in both cases. The double bonded representation would show some deviation from the ideal angle of 109.5° due to the double bonds. The single bond hybridization is $\mathbf{sp^3}$.

c) Since sulfur's valence p orbitals are used in the sigma bonds the π bonds are formed from the valence 3d **orbitals in sulfur overlapping with 2p orbitals in oxygen**.

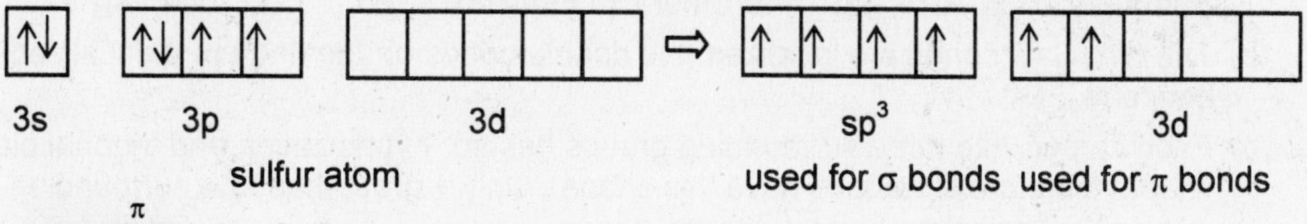

d)

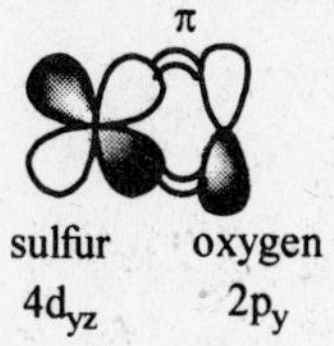

11.54 Ionization energy of lithium indicates it will give up an electron relatively easily. Electron affinity for fluorine indicates it will readily accept an additional electron. Based on these facts, electrons in a bond between Li and F will reside closer to F than to Li. The difference in electronegativity is $\Delta EN = (EN_F - EN_{Li}) = (4.0 - 1.0) = 3.0$. According to Figure 9.18, this difference corresponds to a mostly ionic bond. However, electron sharing occurs in *every* bond, even if to a very small extent. Therefore, a shared electron pair will tend to reside much closer to F than Li, which gives rise to the bond polarity. This point is shown in the MO diagram by having the bonding pair in the σ MO, which is much closer in energy to the atomic orbitals of F.

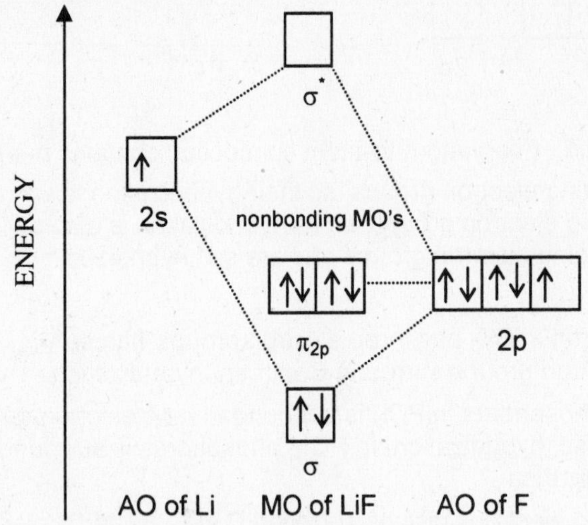

11.56 The resonance structures are

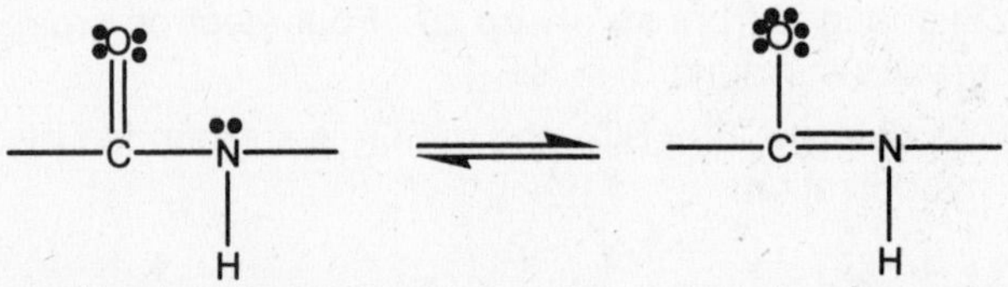

The resonance structures indicate that the electrons in the double bond are delocalized in the overlapping of p orbitals from oxygen, carbon and nitrogen. The overlap of p orbitals (π bonds) restricts rotation around the C--N bond.

11.58 a) The 6 carbons in the ring each have three surrounding electron groups with sp^2 hybrid orbitals. The two carbons participating in the C=O bond are also sp^2 hybridized. The single carbon in the $-CH_3$ group has 4 electron groups and is sp^3 hybridized.

There are only 2 central oxygen atoms, one in a C–O–H configuration and the other in a C–O–C configuration. Both of these atoms have 4 surrounding electron groups and are sp^3 hybridized.

Summary: C in $-CH_3$: $\mathbf{sp^3}$ all other C(8 total): $\mathbf{sp^2}$ O (2 total): $\mathbf{sp^3}$

b) The **two** C=O bonds are localized; the double bonds on the ring are delocalized as in benzene.

c) Each carbon with three surrounding groups has sp^2 hybridization and trigonal planar shape; therefore **8** carbons have this shape. Only **1** group has four surrounding groups with sp^3 hybridization and tetrahedral shape.

CHAPTER 12

INTERMOLECULAR FORCES: LIQUIDS, SOLIDS, AND PHASE CHANGES

FOLLOW-UP PROBLEMS

12.1 Plan: The variables ΔH_{vap}, P_1, T_1, and T_2 are given, so substitute them into Equation 12.1 and solve for P_2. Convert both temperatures from °C to K. Convert ΔH_{vap} to J so that the units cancel with R.

Solution: $T_1 = 34.1 + 273.15 = 307.2$ K

$T_2 = 85.5 + 273.15 = 358.6$ K

$$\ln\frac{P_2}{P_1} = \left(\frac{-40.7x10^3\, J/mol}{8.314\, J/mol \cdot K}\right)\left(\frac{1}{358.6\,K} - \frac{1}{307.2\,K}\right) = 2.28$$

$$\frac{P_2}{P_1} = 9.82;\ thus\ P_2 = 40.1\,torr \times 9.82 = \mathbf{394\ torr}$$

Note: Watch significant figures when subtracting numbers:

1/358.6 = 0.002789 and 1/307.2 = 0.003255

0.002789 – 0.003255 = 0.000466

In the subtraction the number of significant figures decreased from 4 to 3.

Check: The temperature increased, so the vapor pressure should be higher.

12.2 a) The —A:·····H—B— sequence is present in both structures below:

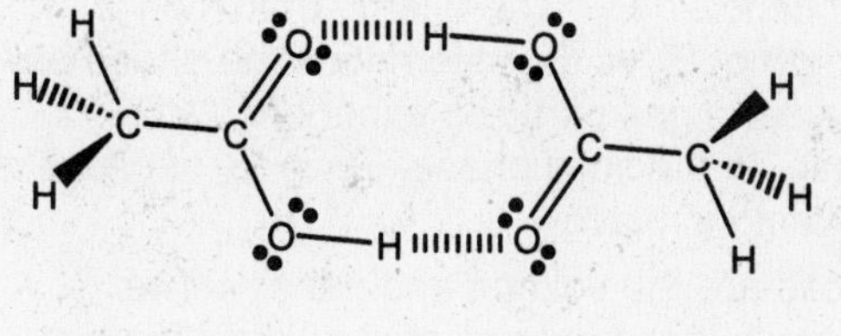

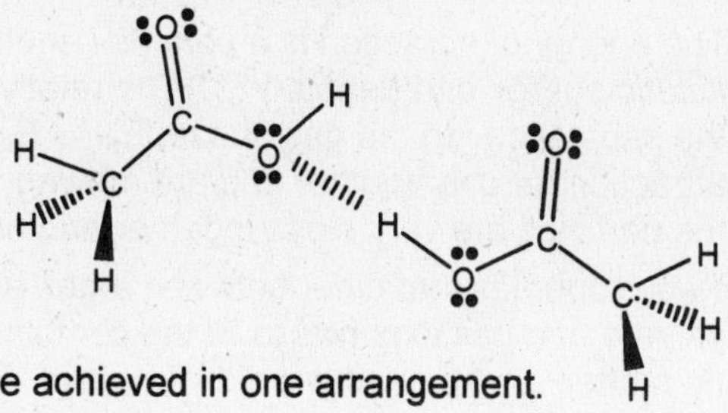

b) The —A:·····H—B— sequence can only be achieved in one arrangement.

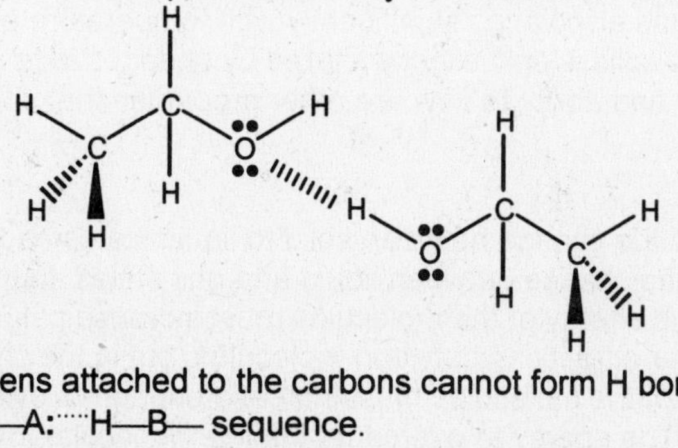

The hydrogens attached to the carbons cannot form H bonds because they do not satisfy the —A:·····H—B— sequence.

c) Hydrogen bonding is not possible because there are no O-H, N-H, or F-H bonds.

12.3 a) Both CH_3Br and CH_3F are polar molecules that experience dipole-dipole and dispersion intermolecular interactions. Because CH_3Br ($\mathcal{M}$ = 94.9 g/mol) is ~3 times larger than CH_3F ($\mathcal{M}$ = 34.0 g/mol), dispersion forces result in a higher boiling point (BP) for **CH_3Br**.

b) $CH_3CH_2CH_2OH$, n-propanol, forms H-bonds whereas $CH_3CH_2OCH_3$, ethyl methyl ether, forms only dipole-dipole and dispersion attractions. **$CH_3CH_2CH_2OH$** has the higher BP.

c) Both C_2H_6 and C_3H_8 are nonpolar and experience dispersion forces only. **C_3H_8**, with the higher molar mass, experiences greater dispersion forces and has the higher boiling point.

12.4 Plan: The density of Fe is 7.874 g/cm^3, but only 68% of the volume is occupied by Fe atoms. Calculate the volume/mole Fe ratio, and multiply by 0.68 to determine the volume/mol Fe atoms ratio. Dividing by the volume of a single Fe will yield the units of atoms/mol, which is Avogadro's number.

Solution: $$volume\ /\ mol\ Ba = \frac{55.85\ g\ /\ mol}{7.874\ g\ /\ cm^3} = 7.093\ cm^3\ /\ mol\ Ba$$

The volume of just the atoms (not including the empty spaces between atoms) is:

$$volume\ /\ mol\ Ba\ atoms = 7.093\ cm^3\ /\ mol \times 0.68 = 4.8\ cm^3\ /\ mol\ Ba\ atoms$$

The number of atoms in one mole of Ba is obtained by dividing by the volume of one Ba atom:

$$atoms\ /\ mol = \left(4.8\frac{cm^3}{mol\ Fe\ atoms}\right)\left(\frac{1\ Fe\ atom}{8.38x10^{-24}\ cm^3}\right) = \mathbf{5.8x10^{23}\ Fe\ atoms\ /\ mol}$$

Check: The calculated value is within 4.4% error of the true value, 6.02 x 10^{23} atoms/mol (%error = (true -measured) ÷ true).

END-OF-CHAPTER PROBLEMS

12.1 The energy of attraction is a *potential* energy and denoted E_p. The energy of motion is *kinetic* energy and denoted E_k. The relative strength of E_p vs. E_k determines the phase of the substance. In the gas phase, $E_p << E_k$ because the gas particles experience little attraction for one another and are moving very fast. In the solid phase, $E_p >> E_k$ because the particles are very close together and only vibrating in place.

Two properties that differ between a gas and a solid are the volume and density. The volume of a gas expands to fill the container it is in while the volume of a solid is constant no matter what container holds the solid. Density of a gas is much less than the density of a solid. The density of a gas also varies significantly with temperature and pressure changes. The density of a solid is only slightly altered by changes in temperature and pressure. Compressibility and ability to flow are other properties that differ between gases and solids.

12.4 a) Heat of fusion refers to the change between solid to liquid states and heat of vaporization refers to the change between liquid and gas states. In the change from solid to liquid the kinetic energy of the molecules must increase only enough to partially offset the intermolecular attractions between molecules, but in the change from liquid to gas the kinetic energy of the molecules must increase enough to overcome the intermolecular forces. The energy to overcome the intermolecular forces for the molecules to move freely in the gaseous state is much greater than the amount of energy to allow the molecules to move more easily past each other but still stay very close together.

b) The net force holding molecules together in the solid state is greater than that in the liquid state. Thus, to change solid molecules to gaseous molecules in sublimation requires more energy than to change liquid molecules to gaseous molecules in vaporization.

c) At a given temperature and pressure the magnitude of ΔH_{vap} is the same as the magnitude of ΔH_{cond}. The only difference is in the sign: $\Delta H_{vap} = -\Delta H_{cond}$.

12.5 a) Intermolecular – Oil evaporates when individual oil molecules can escape the attraction of other oil molecules in the liquid phase.

b) Intermolecular – The process of butter (fat) melting involves a breakdown in the rigid, solid structure of fat molecules to an amorphous, less ordered system. The attractions between the fat molecules are weakened, but the bonds within the fat molecules are not broken.

c) Intramolecular – A process called oxidation tarnishes pure silver. Oxidation involves the breaking of bonds and formation of new bonds.

d) Intramolecular – The decomposition of O_2 molecules into O atoms requires the breaking of chemical bonds, i.e. the force that holds the two O atoms together in an O_2 molecule .

Both a) and b) are physical changes, whereas c) and d) are chemical changes. In other words, <u>inter</u>molecular forces are involved in physical changes while <u>intra</u>molecular forces are involved in chemical changes.

12.7 a) Condensation. The water vapor in the air condenses to liquid when the temperature drops during the night.

b) Fusion (melting). Solid ice melts to liquid water.

c) Evaporation. Liquid water on clothes evaporates to water vapor.

12.9 The propane gas molecules slow down as the gas is compressed. Therefore much of the kinetic energy lost by the propane molecules is released to the surroundings upon liquefaction.

12.13 In closed containers two processes, evaporation and condensation, occur simultaneously. Initially there are few molecules in the vapor phase, so more liquid molecules evaporate than gas molecules condense. Thus, the number of molecules in the gas phase increases causing the vapor pressure of hexane to increase. Eventually, the number of molecules in the gas phase reaches a maximum where the number of liquid molecules evaporating equals the number of gas molecules condensing. In other words, the evaporation rate equals the condensation rate. At this point there is no further change in the vapor pressure.

12.14 a) At the critical temperature, the molecules are moving so fast that they can no longer be condensed. This temperature decreases with weaker intermolecular forces because the forces are not strong enough to overcome molecular motion. Alternatively, as intermolecular forces increase, the critical temperature increases.

b) As intermolecular forces increase, the boiling point increases because it becomes more difficult to separate molecules from the liquid phase.

c) As intermolecular forces increase, the vapor pressure decreases for the same reason given in b).

d) As intermolecular forces increase, the ΔH_{vap} increases because more energy is needed to separate molecules from the liquid phase.

12.18 When water at 100°C touches skin, the heat released is from the lowering of the temperature of the water. The specific heat of water is approximately 75 **J**/mol·K. When steam at 100°C touches skin, the heat released is from the condensation of the gas with a heat of condensation of approximately 41 **kJ**/mol. Thus, the amount of heat released from gaseous water condensing will be greater than the heat from hot liquid water cooling and burn from the steam will be worse than that from hot water.

12.19 The total heat required is the sum of three processes:

1) Warming the ice from –5.00°C to 0.00°C

$$q = m_{H_2O}c_{solid}\Delta T = (12.0\,g)\left(2.09\frac{J}{g\cdot{}^{\circ}C}\right)(0.00-(-5.00)^{\circ}C) = 125\text{ J}$$

2) Phase change of ice at 0.00°C to water at 0.00°C

$$q = n\Delta H^{\circ}_{fus} = (12.0\,g)\left(\frac{1\,mol\;H_2O}{18.016\,g}\right)(6.02\,kJ/mol) = 4.01\,kJ = 4010\text{ J}$$

3) Warming the liquid from 0.00°C to 0.500°C

$$q = m_{H_2O}c_{liq}\Delta T = (12.0\,g)\left(4.21\frac{J}{g\cdot{}^{\circ}C}\right)(0.500-0.00^{\circ}C) = 25.3\,J$$

The three heats are positive because each process takes heat from the surroundings (endothermic). The phase change requires much more energy than the two temperature change processes. The total heat is (125 J + 4010 J + 25.3 J) = 4160 J = **4.16 x 10³ J**.

12.21 The Clausius-Clapeyron equation gives the relationship between vapor pressure and temperature. Boiling point is defined as the temperature when vapor pressure of liquid equals atmospheric pressure, usually assumed to be exactly 1 atm. So, the number of significant figures in the pressure of 1 atm given is not one, but is unlimited since it is an exact number. In the calculation below 1.00 atm is used to emphasize the additional significant figures.

$$\ln\frac{P_2}{1.00\,atm} = \left(\frac{-35.5x10^3\;J/mol}{8.314\;J/mol\cdot K}\right)\left(\frac{1}{(273K+109)}-\frac{1}{(273K+122)}\right) = -0.367_{88}$$

$$P_2 = (1.00\,atm)\left(e^{-0.367_{88}}\right) = \mathbf{0.692\,atm\;or\;526\,torr}$$

12.23 At the boiling point, the vapor pressure equals the external pressure. Summarize the pressure and temperature variables and use the Clausius-Clapeyron equation (6.1) to find the ΔH_{vap}.

P_1 = 641 torr $\qquad$ T_1 = 85.2°C = 358.4 K

P_2 = 1 atm = 760. torr $\qquad$ T_2 = 95.7°C = 368.8 K

$$\ln\frac{P_2}{P_1} = \frac{-\Delta H_{vap}}{R}\left(\frac{1}{T_2}-\frac{1}{T_1}\right)$$

$$\ln\frac{760.\,torr}{641\,torr} = \frac{-\Delta H_{vap}}{8.314\;J/mol\cdot K}\left(\frac{1}{368.8\,K}-\frac{1}{358.4\,K}\right)$$

$$\ln 1.18 = \frac{-\Delta H_{vap}}{8.314\;J/mol\cdot K}\left(-7.8_{68}x10^{-5}\frac{1}{K}\right)$$

$$\frac{-\Delta H_{vap}}{8.314\ J/mol} = -21_{61};\ \Delta H_{vap} = 1.8x10^4\ J/mol = \mathbf{18 kJ/mol}$$

Note that the answer has only 2 significant figures because the difference in reciprocal temperatures has only 2 significant figures. The subscripted numbers are additional digits retained until the final step of the calculation.

12.25

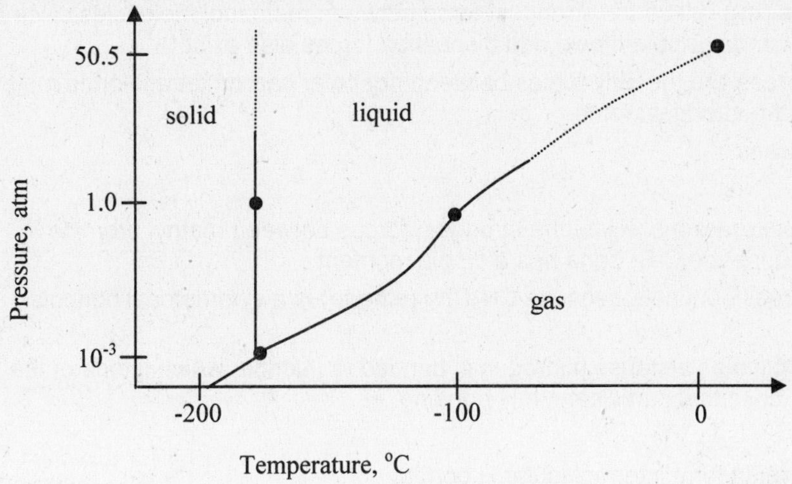

The pressure scale is distorted to represent the large range in pressures given in the problem, so the liquid-solid curve looks different than the one shown in Figures 12.9. The important features of the graph include the distinction between the gas, liquid and solid states and melting point T located directly above the critical T.

Solid ethylene is more dense than liquid ethylene.

12.28 This problem is also an application of the Clausius-Clapeyron equation. Convert the temperatures from °C to K and ΔH_{vap} from kJ/mol to J/mol to allow cancellation with the units in R.

P_1 = 2.3 atm $\qquad$ T_1 = 25.0°C = 298.2 K

P_2 = ? $\qquad$ T_2 = 150.°C = 423 K

ΔH_{vap} = 24.3 kJ/mol = 24.3 x 10^3 J/mol

$$\ln\frac{P_2}{2.3\ atm} = \frac{-24.3x10^3\ J/mol}{8.314\ J/mol\cdot K}\left(\frac{1}{423\ K} - \frac{1}{298.2\ K}\right)$$

$$\ln\frac{P_2}{2.3\ atm} = \frac{-24.3x10^3\ J/mol}{8.314\ J/mol\cdot K}\left(-9.8_{94}x10^{-4}\frac{1}{K}\right) = 2.8_{92}$$

$$\frac{P_2}{2.3\ atm} = e^{2.8_{92}}$$

P_2 = **41atm**

12.32 To form hydrogen bonds the atom bonded to hydrogen must have two characteristics, small size and high electronegativity, so that the atom has a very high electron density. With this high electron density, the attraction for a hydrogen on another molecule is very strong. Selenium is much larger than oxygen (atomic radius of 119 pm vs. 73 pm) and less electronegative than oxygen (2.4 for Se and 3.5 for O) resulting in an electron density on Se in H_2Se that is too small to form hydrogen bonds.

12.34 All particles (atoms and molecules) exhibit dispersion forces, but these are the weakest of intermolecular forces. The dipole-dipole forces in polar molecules dominate the dispersion forces.

12.37 a) **Hydrogen bonding** will be the strongest force between methanol molecules since they contain O—H bonds. Dipole-dipole and dispersion forces also exist.

b) **Dispersion forces** are the only forces between nonpolar carbon tetrachloride molecules and, thus, are the strongest forces.

c) **Dispersion forces**.

12.39 a) **Dipole-dipole** interactions will be the strongest forces between methyl bromide molecules because the C-Br bond has a dipole moment

b). **Dispersion** forces dominate because CH_3CH_3 (ethane) is a symmetrical nonpolar molecule

c) **H-bonding** dominates because hydrogen is bonded to fluorine, which is one of the 3 species (N, O, or F) that participate in H-bonding.

12.41 a) **Isopropanol** would form intermolecular H bonds.

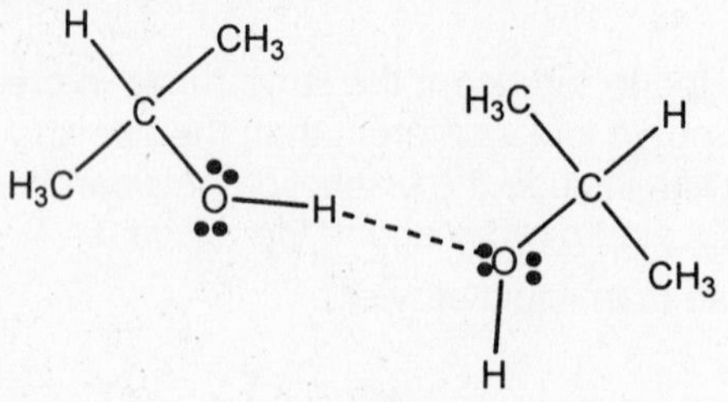

b) **Hydrogen fluoride** would form intermolecular H bonds.

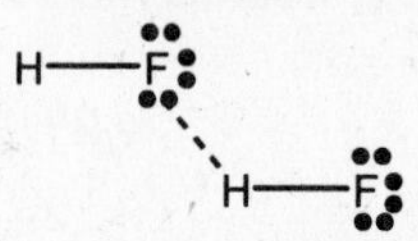

12.43 a) **Dispersion.** Hexane, C_6H_{14}, is a nonpolar molecule.

b) **H-bonding**. A single water molecule can engage in as many as four H-bonds.

c) **Dispersion.** Although the individual Si-Cl bonds are polar, the molecule has a symmetrical, tetrahedral shape and is therefore nonpolar.

12.45 a) **Iodide ion** has greater polarizability than the bromide ion because the iodide ion is larger. Electrons can be polarized over a larger area in a larger atom or ion.

b) **Ethene (CH_2=CH_2)** has greater polarizability than ethane (CH_3CH_3) because π electrons are more easily polarized than electrons in σ bonds.

c) **Selenium hydride** has greater polarizability than water because the selenium atom is larger than the oxygen atom.

12.47 Weaker attractive forces result in a higher vapor pressure because the molecules have a smaller barrier to escape the liquid and go to the gas phase.

a) **C_2H_6**. C_2H_6 is a smaller molecule with weaker dispersion forces than C_4H_{10}.

b) **CH_3CH_2F**. CH_3CH_2F has no H-F bonds, so it only exhibits dipole-dipole forces which are weaker than those of CH_3CH_2OH (H-bonding).

c) **PH_3**. PH_3 has weaker intermolecular forces (dipole-dipole) than NH_3 (H-bonding).

12.49 The stronger the intermolecular forces the higher the boiling point.

a) **Lithium chloride** would have a higher boiling point than hydrogen chloride because the ions in lithium chloride are held together by ionic forces which are stronger than the dipole-dipole intermolecular forces between hydrogen chloride molecules in the liquid phase.

b) **Ammonia, NH_3**, would have a higher boiling point than phosphine (PH_3) because the intermolecular forces in ammonia are stronger than those in phosphine. H-bonding exists between ammonia molecules but weaker dipole-dipole forces hold phosphine molecules together.

c) **Iodine** would have a higher boiling point than xenon. Both are nonpolar with dispersion forces, but the forces between iodine molecules would be stronger than those between xenon atoms since the iodine molecules are more polarizable as a result of their larger size.

12.51 The molecule in the pair with the weaker intermolecular attractive forces will have the lower boiling point.

a) **C_4H_8**. The cyclic molecule, cyclobutane, has less surface area exposed, so its dispersion forces are less than the straight chain molecule, C_4H_{10}.

b) **PBr_3**. The dipole-dipole forces of phosphorous tribromide are weaker than the ionic forces of sodium bromide.

c) **HBr**. The dipole-dipole forces of hydrogen bromide are weaker than the H-bonding forces of water.

12.53 Both the trend in atomic size and in electronegativity predicts that the trend in increasing strength of H-bonds is N—H < O—H < F—H. As the atomic size decreases and electronegativity increases the electron density of the atom increases. High electron density strengthens the attraction to a H atom on another molecule. Fluorine is the smallest of the three and also the most electronegative, so its H-bonds would be the strongest. Oxygen is smaller and more electronegative than nitrogen, so H-bonds for water would be stronger than H-bonds for ammonia.

12.57 The shape of the drop depends upon the competing cohesive forces (attraction of molecules within the drop itself) and adhesive forces (attraction between molecules in the drop and molecules of the waxed floor). If the cohesive forces are strong and outweigh the adhesive forces, the drop will be as spherical as gravity will allow. If, on the other hand, the adhesive forces are significant, the drop will spread out. Both water (H-bonding) and mercury (metallic bonds) have strong cohesive forces, whereas cohesive forces in oil (dispersion) are relatively weak (Table 12.2). Figure 12.20 explains the difference between these forces when *glass*, not *wax*, is used. Neither water nor mercury will have significant adhesive forces to the nonpolar wax molecules, so these drops will remain nearly spherical. The adhesive forces between the oil and wax can compete with the weak, cohesive forces of the oil (dispersion) and so the oil drop spreads out.

12.59 Surface tension is defined as the energy needed to increase the surface area by a given amount so units of energy per surface area describe this property.

12.61 The stronger the intermolecular force, the greater the surface tension. All three molecules exhibit H-bonding, but the extent of H-bonding increases with the number of O-H bonds present in each molecule. Molecule b) can form more H-bonds than molecule c), which in turn can form more H-bonds than molecule a). As the number of H-bonds increases, the intermolecular forces and surface tension increases so

$CH_3CH_2CH_2OH$ < $HOCH_2CH_2OH$ < $HOCH_2CH(OH)CH_2OH$.

12.63 Viscosity is greater for molecules with stronger intermolecular forces. The stronger the force attracting the molecules to each other the harder it is for one molecule to move past another. Thus, the substance will not flow easily if the intermolecular force is strong. Viscosity is a measure of the resistance of a liquid to flow. The ranking of increasing viscosity is the same as that of increasing surface tension (problem 12.61):

$CH_3CH_2CH_2OH$ < $HOCH_2CH_2OH$ < $HOCH_2CH(OH)CH_2OH$.

12.65 **No**, the forces involved when a paper towel absorbs apple juice are not the same as when it absorbs cooking oil. Apple juice is primarily water. The dominant intermolecular force in apple juice is hydrogen bonding. Cooking oil is nonpolar. Its dominant intermolecular force is dispersion. The paper-juice adhesion is stronger than the paper-oil adhesion because the juice molecules can hydrogen bond to the cellulose molecules in the paper.

12.68 Water is a good solvent for polar and ionic substances and a poor solvent for nonpolar substances. Water is a polar molecule and dissolves polar substances because their intermolecular forces are of similar strength. Water is also able to dissolve ionic compounds and keep ions separated in solution through ion-dipole interactions. Nonpolar substances will not be very soluble in water since their dispersion forces are much weaker than the hydrogen bonds in water. A solute whose intermolecular attraction to a solvent molecule is less than the attraction between two solvent molecules will not dissolve because its attraction cannot replace the attraction between solvent molecules.

12.69 Figure 12.21 shows the ability of a single water molecule to form 4 H-bonds. The two hydrogen atoms form an H-bond each to oxygen atoms on neighboring water molecules. The two lone pairs on the oxygen atom form H-bonds with hydrogen atoms on neighboring molecules.

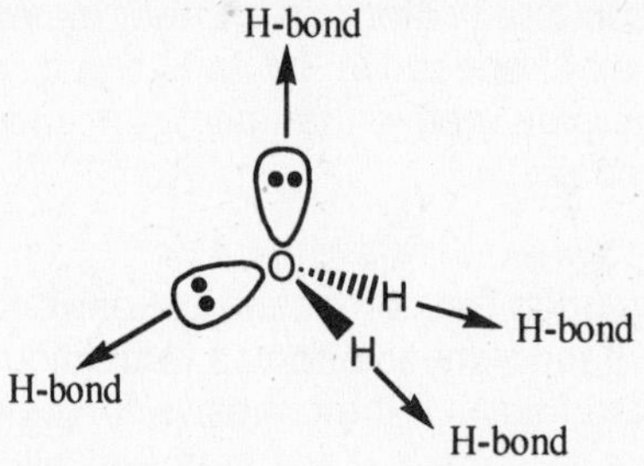

12.72 The high capillary action of water allows it to flow into the roots and up the plant from the ground.

12.78 The unit cell is a simple cubic cell. According to the bottom row in Figure 12.27, two atomic radii (or one atomic diameter) equal the width the cell.

12.81 The energy gap is the energy difference between the highest filled energy level (valence band) and the lowest unfilled energy level (the conduction band). In conductors and superconductors the energy gap is zero because the valence band overlaps with the conduction band. In semiconductors the energy gap is small but greater than zero. In insulators the energy gap is large and thus insulators do not conduct electricity.

12.83 The density of a solid depends on the atomic mass of the element (greater mass = greater density), the atomic radius (how many atoms can fit in a given volume) and the type of unit cell, which determines the packing efficiency (how much of the volume is occupied by empty space).

12.84 The simple cubic structure unit cell contains 1 atom; the body-centered cubic unit cell contains 2 atoms; the face-centered cubic cell contains 4 atoms.

a) Ni is **face-centered** cubic since there are 4 atoms/unit cell.

b) Cr is **body-centered** cubic since there are 2 atoms/unit cell.

c) Ca is **face-centered** cubic since there are 4 atoms/unit cell.

12.86 Table 12.5 summarizes the five major types of crystalline solids.

a) You may be familiar with tin, Sn, as a component of tin cans. Tin is a metal (malleable, excellent electrical conductor) that forms **metallic** bonds.

b) Silicon is in the same group as carbon, so it exhibits similar bonding properties. Since diamond and graphite are both **network covalent** solids, it makes sense that Si forms the same type of bonds.

c) Xenon, Xe, is a noble gas and is monatomic. Xe is an **atomic** solid.

12.88 a) Nickel forms a **metallic solid** since nickel is a metal whose atoms are held together by metallic bonds.

b) Fluorine forms a **molecular solid** since the F_2 molecules are held together by dispersion forces.

c) Methanol forms a **molecular solid** since the CH_3OH molecules are held together by H-bonds.

12.90 Figure 12.32 shows the face-centered cubic array of zinc blende, ZnS. Both ZnS and ZnO have a 1:1 ion ratio, so the ZnO unit cell will also contain **four** Zn^{2+} ions.

12.92 a) To determine the number of Zn^{2+} ions and Se^{2-} ions in each unit cell, picture the lattice (see Figure 12.32) and count the number of ions at the corners, faces, and center of unit cell. Looking at selenide ions there is one ion at each corner and one ion on each face. The total number of selenide ions is $^1/_8$ (8 corner ions) + ½ (6 face ions) = **4 Se^{2-} ions**. There are also **4 Zn^{2+} ions** due to the 1:1 ratio of Se ions to Zn ions.

b) Mass of unit cell = (4 x mass of Zn atom) + (4 x mass of Se atom)
= (4 x 65.39 amu) + (4 x 78.96 amu)
= **577.40 amu**

c) Given the mass of one unit cell and the ratio of mass to volume (density) divide the mass, converted to grams (conversion factor is 1 amu = $1.66054x10^{-24}$g), by the density to find the volume of the unit cell.

$$(577.40\,amu)\left(\frac{1.66054x10^{-24}\ g}{1\,amu}\right)\left(\frac{1\ cm^3}{5.42\ g}\right) = \mathbf{1.77x10^{-22}\ cm^3}$$

d) The volume of a cube equals (length of side)^3.

$$length = \sqrt[3]{1.77x10^{-22}\ cm^3} = \mathbf{5.61x10^{-8}\ cm\ or\ 5.61\,Å}$$

12.94 To classify a substance according to its electrical conductivity first locate it on the periodic table as a metal, metalloid or nonmetal. In general metals are conductors, metalloids are semiconductors and nonmetals are insulators.

a) Phosphorous is a nonmetal and an **insulator**.

b) Mercury is a metal and a **conductor**.

c) Germanium is a metalloid in Group 4A(14) and is beneath carbon and silicon the periodic table. Pure germanium crystals are **semiconductors** and are used to detect gamma-rays emitted by radioactive materials. Germanium can also be doped with phosphorous (similar to the doping of silicon) to form an n-type semiconductor or be doped with lithium to form a p-type semiconductor.

12.96 First classify the substance as an insulator, conductor or semiconductor (see problem 12.94). The electrical conductivity of conductors decreases with increasing temperature whereas that of semiconductors increases with temperature. Temperature increases have little impact on the electrical conductivity of insulators.

a) Antimony, Sb, is a metalloid so it is a semiconductor. Its electrical conductivity **increases** as the temperature increases.

b) Tellurium, Te, is a metalloid so it is a semiconductor. Its electrical conductivity **increases** as temperature increases.

c) Bismuth, Bi, is a metal so it is a conductor. Its electrical conductivity **decreases** as temperature increases.

12.98 To calculate an approximate radius for polonium, first multiply the molar mass of Po by the inverse of the density of Po to find the volume per mole Po *metal.* Use the packing efficiency for the simple cubic cell (52 %) to find the volume per mole of Po *atoms.* Find the volume of one Po atom by using Avogadro's number and calculate the radius of one atom.

$$volume\ /\ mol\ of\ Po\,metal = \frac{1}{d_{Po}} \times \mathbf{M} = \frac{1}{9.142\ g\ /\ cm^3} \times (209\ g\ /\ mol) = 22.9\ cm^3\ /\ mol$$

$$volume\ /\ mol\ of\ Po\,atoms = 22.9\ cm^3\ /\ mol \times (0.52) = 11.9\ cm^3\ /\ mol$$

$$volume\ /\ Po\,atoms = \left(\frac{11.9\ cm^3}{mol\ Po\,atoms}\right)\left(\frac{1\,mol\ Po\,atoms}{6.022x10^{23}\ Po\,atoms}\right) = 1.97x10^{-23}\ cm^3\ /\ Po\,atom$$

$$V\ of\ Po\,atom = \frac{4}{3}\pi r^3 \ and\ r = \sqrt[3]{\frac{3V}{4\pi}} = \sqrt[3]{\frac{3(1.97x10^{-23}\ cm^3\ /\ atom)}{4\pi}} = \mathbf{1.68x10^{-8}\ cm}$$

12.105 A substance whose physical properties are the same in all directions is called isotropic; otherwise the substance is anisotropic. Liquid crystals flow like liquids but have a degree of order that gives them the anisotropic properties of a crystal.

12.111 Germanium and silicon are elements in group 4A with 4 valence electrons. If germanium or silicon is doped with an atom with more than 4 valence electrons, an n-type semiconductor is produced. If it is doped with an atom with fewer than 4 valence electrons, a p-type semiconductor is produced.

a) Phosphorus has 5 valence electrons so an **n-type semiconductor** will form by doping Ge with P.

b) Indium has 3 valence electrons so a **p-type semiconductor** will form by doping Si with In.

12.113 The degree of polymerization is the number of monomer, or repeat, units that are bonded together in the polymer chain. Since the molar mass of one monomer unit is 104.14 g/mol, there are (3.5×10^5 g/mol ÷ 104.14 g/mol) 3361 monomer units. Reporting in correct significant figures, **n = 3400**.

12.115 Figure 12.47 describes the radius of gyration. The length of the repeat unit, ℓ_o, is given. Calculate the degree of polymerization, n, by dividing the molar mass of the polymer chain ($\mathcal{M} = 2.5 \times 10^5$ g/mol) by the molar mass of one monomer unit ($\mathcal{M} = 42.08$ g/mol). Substitute these values into equation below:

$$R_g = \sqrt{\frac{nl_o^2}{6}} = \sqrt{\frac{\left(\frac{2.5x10^5\ g/mol}{42.08\ g/mol}\right)(0.252\ nm)^2}{6}} = \mathbf{7.9\,nm}$$

12.118 At 22°C the vapor pressure of water is 19.8 torr.

a) Once compressed, the N_2 gas would still be saturated with water. The vapor pressure depends on the temperature, which has not changed. So the partial pressure of water in the compressed gas remains the same at **19.8 torr**.

b) In order to maintain the vapor pressure of water at 19.8 torr some of the water must condense to liquid. The moles of H_2O prior to compression can be calculated using the ideal gas law.

$$n_{H_2O(l)} = \frac{(19.8\ torr)(1\ atm/760\ torr)(4.00\ L)}{(0.08206\ atm \cdot L/mol \cdot K)(295\ K)} = 4.30_{48} x10^{-3}\ mol$$

The moles of H_2O remaining in the gas phase after compression can also be calculated using the ideal gas law.

$$n_{H_2O(l)} = \frac{(19.8\ torr)(1\ atm/760\ torr)(2.00\ L)}{(0.08206\ atm \cdot L/mol \cdot K)(295\ K)} = 2.15_{24} x10^{-3}\ mol$$

moles of liquid H_2O = ($4.30_{48} \times 10^{-3}$) – ($2.15_{24} \times 10^{-3}$) = $2.15_{24} \times 10^{-3}$ mol

mass of liquid H_2O = $2.15_{24} \times 10^{-3}$ mol x 18.02 g/mol = **0.0388 g**

12.120 a) Figure 12.3 shows the change of state with changing *temperature*, whereas the problem requests a diagram showing the effect of changing *pressure*. Therefore the new diagram will have pressure on the y-axis. On Figure 12.9B, a vertical line can be drawn at T = 2°C. Moving along the line from bottom to top, the pressure increases. The line only bisects the gas phase and liquid phase, so a solid is not possible at this temperature. The phase change from gas to liquid occurs at ~0.1 atm, which was "guesstimated" by the proximity to the triple point.

b) The triple point for CO_2 occurs at –57°C, so the –50°C constant temperature line is just to the right of the triple point. The –50°C constant temperature line goes through all three phases, so the P vs. time graph reflects that. The phase change from gas to liquid is estimated to occur at 8 atm (based on proximity to triple point) whereas the liquid-solid phase change occurs around 20 atm.

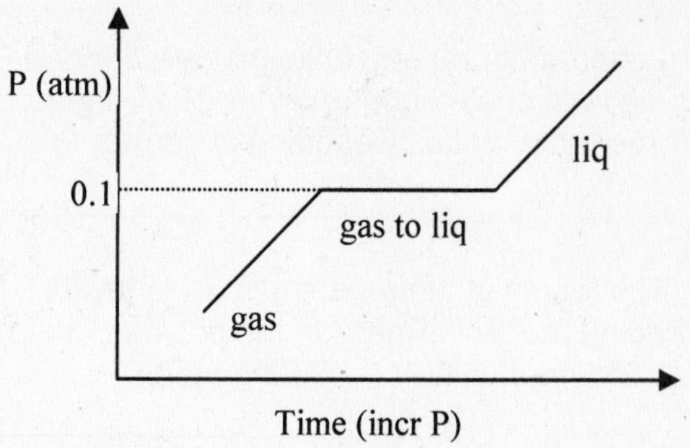

(a)

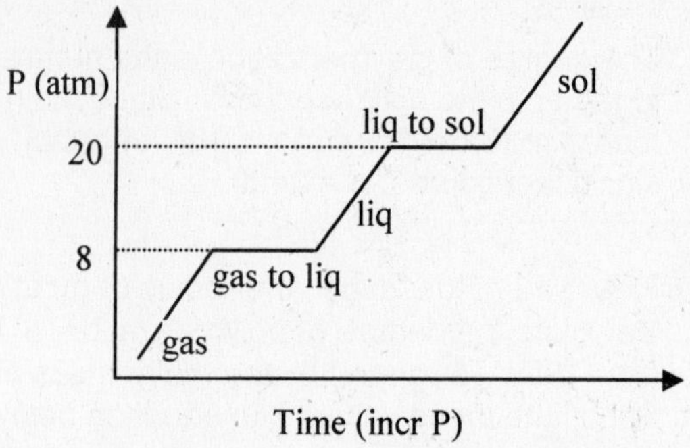

(b)

12.122 More heat is required to sublime ice than to vaporize liquid water. Vaporization would equal the sum of the heat to raise the temperature of the water to the boiling point plus the heat to convert the liquid water to gas. Sublimation would add to this the heat to change the solid ice to liquid water. The heat required for sublimation of carbon dioxide would also be greater than that required for vaporization.

12.123 a) The silicon ingot is a cylinder with r = ½ (diameter) = 2.625 in = 6.6675 cm. Use the density of Si (d = 2.34 g/cm^3) and mass of the ingot to determine its volume. Solve for the height of the cylinder using $V = \pi r^2 h$. Divide cylinder height by wafer thickness to determine the number of wafers possible.

$$volume\ of\ ingot = \frac{m}{d_{Si}} = \frac{5.00x10^3\ g}{2.34\ g/cm^3} = 2137\ cm^3$$

$$height\ of\ cylinder = \frac{V}{\pi r^2} = \frac{2137\ cm^3}{\pi(6.6675\ cm)^2} = 15.3\ cm\left(3\ sig\ figs\ from\ mass\right)$$

$$number\ of\ wafers = \frac{15.3\ cm}{\left(1.02x10^{-4}\ m/wafer\right)\left(100\ cm/m\right)} = \mathbf{1.50x10^3\ wafers}$$

b) A single wafer is also a cylinder with dimensions h = 1.02 x 10^{-4} m and r = 6.6675 cm. The volume of the cylinder can be converted to mass using the density.

$$wafer\ volume = \pi r^2 h = \pi(6.6675\ cm)^2\left(1.02x10^{-4}\ m \times \frac{100\ cm}{m}\right) = 1.42\ cm^3$$

$$wafer\ mass = d_{Si}V = \left(2.34\ g/cm^3\right)\left(1.42\ cm^3\right) = \mathbf{3.33\ g}$$

c) Silicon reacts with oxygen to form silicon dioxide, $SiO_2(s)$. This oxide coating interferes with the operation of the wafer and is removed by gaseous hydrogen fluoride, HF(g). In a double displacement reaction, silicon's 4 valence electrons combine with 4 fluorine atoms: $SiO_2(s) + 4HF(g) \rightarrow SiF_4(g) + 2H_2O(g)$

d) $$mol\ HF = \left(3.33\ g\ Si \times 0.00750\right)\left(\frac{1\ mol\ Si}{28.09\ g}\right)\left(\frac{1\ mol\ SiO_2}{1\ mol\ Si}\right)\left(\frac{4\ mol\ HF}{1\ mol\ SiO_2}\right) = \mathbf{3.56x10^{-3}\ mol\ HF}$$

12.126 The Clausius-Clapeyron equation can be solved for the temperature that will give a vapor pressure of 5.0x10^{-5} torr.

$$\ln\frac{5.0x10^{-5}\ torr}{1.20x10^{-3}\ torr}=\left(\frac{-59.1x10^{3}\ J/mol}{8.314\ J/mol\cdot K}\right)\left(\frac{1}{T_2}-\frac{1}{293\ K}\right)$$

T_2 = **259 K or –14°C**

12.135 a) Both furfuryl alcohol and 2-furoic acid can form hydrogen bonds.

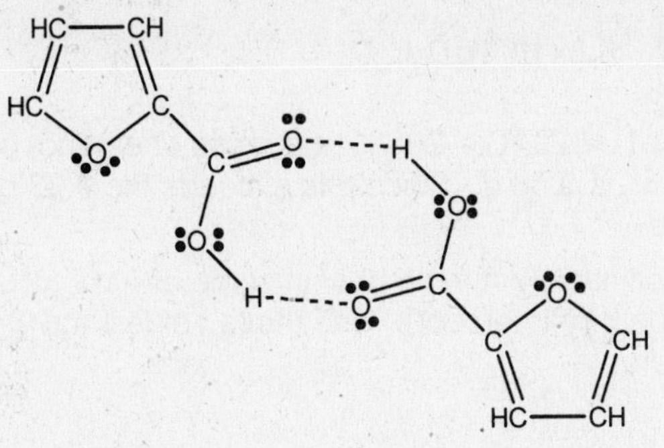

2-furoic acid

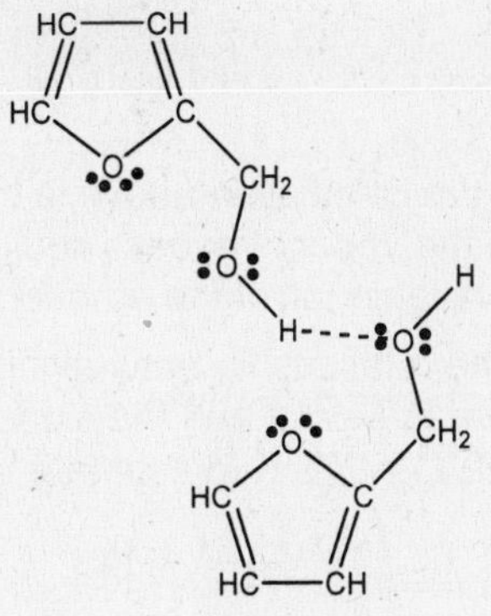

furfuryl alcohol

b) Both 2-furoic acid and furfuryl alcohol can form internal H bonds by forming a H-bond between the H attached to the O and the O in the ring.

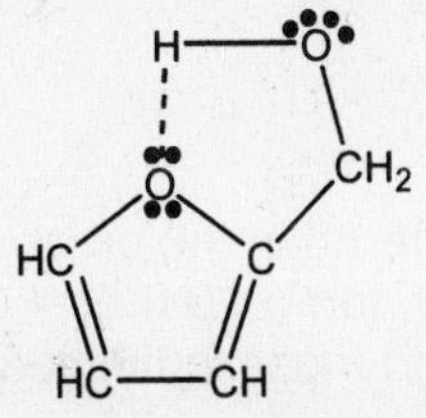

furfuryl alcohol

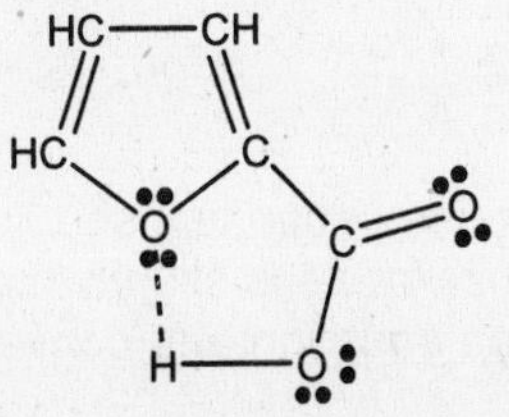

2-furoic acid

12.138 a) Determine the vapor pressure of ethanol in the bottle by applying the Clausius-Clapeyron equation (see Figure 12.6 – use P_2 = 760 torr and T_2 = 78.5°C). The ΔH_{vap} is given in Figure 12.1.

P_1 = ? T_1 = -11°C = 262 K

P_2 = 760. torr T_2 = 78.5°C = 351.6 K

ΔH_{vap} = 40.5 kJ/mol = 40.5 x 10^3 J/mol

$$\ln\frac{760.torr}{P_1}=\frac{\left(-40.5x10^{3}\ J/mol\right)}{8.314\ J/mol\cdot K}\left(\frac{1}{351.6\ K}-\frac{1}{262\ K}\right)$$

$$\ln\frac{760.torr}{P_1}=4.7_{38};\quad \frac{760.torr}{P_1}=e^{4.738};\ P_1=6.6_{54}\ torr$$

Note: The pressure should be small because not many ethanol molecules escape the liquid surface at such a cold temperature.

Determine the number of moles by substituting P, V, and T into the ideal gas equation. Assume that the volume the liquid takes up in the 4.7 L space is negligible.

$$mol\ C_2H_6O = \frac{PV}{RT} = \frac{\left(6.6_{54}\ torr \times \frac{1\ atm}{760\ torr}\right)(4.7\ L)}{\left(0.08206\frac{atm \cdot L}{mol \cdot K}\right)(262\ K)} = 1.9_{14}x10^{-3}\ mol\ C_2H_6O$$

Convert moles of ethanol to mass of ethanol using the molar mass (**M** = 46.07 g/mol).

$$mass\ C_2H_6O = \left(1.9_{14}x10^{-3}\ mol\right)\left(\frac{46.07\ g}{mol}\right) = \mathbf{8.8x10^{-2}\ g\ C_2H_6O}$$

b) If the calculation above is carried out at T_1 = 20.°C = 293 K, the mass of ethanol present in the vapor, if excess liquid was present, is 0.56 g. Since this exceeds the 0.33 g available, all of the ethanol will vaporize.

c) The relationship between lnP and 1/T is linear (y = mx + b) with slope of $-\Delta H_{vap}/R$. Use one P/T point and the slope to determine b (y-intercept). Use these values and T = 0.0°C = 273 K to estimate vapor pressure.

$$\ln P = \left(\frac{-\Delta H_{vap}}{R}\right)\left(\frac{1}{T}\right) + b$$

$$\ln(10.\ torr) = \left(\frac{-40.5x10^3\ J/mol}{8.314\ J/mol \cdot K}\right)\left(\frac{1}{270.8\ K}\right) + b \quad where\ b = 20.29$$

$$\ln P = \left(\frac{-40.5x10^3\ J/mol}{8.314\ J/mol \cdot K}\right)\left(\frac{1}{273}\right) + 20.29 \Rightarrow P = e^{2.44_{64}} = 11.5_{47}\ torr$$

This pressure makes sense because it should be greater than the pressure calculated in part a) because the temperature is slightly higher : 11.5 torr (at 0.0°C) > 6.65 torr (at 11°C). Use the ideal gas equation and convert moles to mass to find the amount of liquid present at 0.0°C.

$$mol\ C_2H_6O = \frac{PV}{RT} = \frac{\left(11.5_{47}\ torr \times \frac{1\ atm}{760\ torr}\right)(4.7\ L)}{\left(0.08206\frac{atm \cdot L}{mol \cdot K}\right)(273\ K)} = 3.1_{88}x10^{-3}\ mol\ C_2H_6O$$

$$mass\ C_2H_6O = \left(3.1_{88}x10^{-3}\ mol\right)\left(\frac{46.07\ g}{mol}\right) = \mathbf{1.5x10^{-1}\ g\ C_2H_6O}$$

12.140 At the boiling point the vapor pressure equals the atmospheric pressure, so the two boiling points can be used to find the heat of vaporization for amphetamine.

$$\ln\frac{760\ torr}{13\ torr} = \left(\frac{-\Delta H_{vap}}{8.314\ J/mol}\right)\left(\frac{1}{474\ K} - \frac{1}{356\ K}\right)$$

ΔH_{vap} = 48.$_{37}$ kJ/mol

Use the Clausius-Clapeyron equation to find the vapor pressure of amphetamine at 20.°C

$$\ln\frac{P_{293K}}{13\ torr} = \left(\frac{-4.8_{37}x10^3\ J/mol}{8.314\ J/mol \cdot K}\right)\left(\frac{1}{293\ K} - \frac{1}{356\ K}\right)$$

P_{293K} = 0.38$_{71}$ torr or 5.0$_{93}$ x10^{-4} atm

At this pressure and a temperature of 293 K use the ideal gas equation to calculate the concentration of amphetamine in the air.

$$\frac{n}{V} = \frac{5.0_{93} x10^{-4}\ atm}{(0.08206\, atm \cdot L / mol \cdot K)(293\, K)} = 2.1_{18} x10^{-5}\ mol / L$$

Converting moles to grams and liters to cubic meters:

$$(2.1_{18} x10^{-5}\ mol / L)(135.20\ g / mol)(1L / 1000\,cm^3)(1x10^6\ cm^3 / m^3) = \mathbf{2.9\,g/m^3}$$

12.143 a) To calculate the time required to heat the sample to its melting point, multiply the sample mass (g) by the specific heat (J/g-°C) and the temperature change (°C). Use the heating rate to convert heat (J) to time (min).

$$m_A c_{solid}(T_f - T_i) \times heating\ rate = (25\,g)(1.0\,J/g\cdot^\circ C)(-20.^\circ C - (-40.^\circ C))\left(\frac{1\min}{450\,J}\right) = \mathbf{1.1 min}$$

b) To calculate the time required for melting (phase change), multiply the sample mass (g) by the ΔH_{fus} (J/g) and use the heating rate to convert heat (J) to time (min).

$$time = (25\,g)(180\,J/g)\left(\frac{1\min}{450\,J}\right) = \mathbf{10. min}$$

c) Additional information needed to complete the curve includes:

1) Time to heat liquid from –20.°C to 85°C

$$time = m_A c_{liq}\Delta T(rate) = (25\,g)(2.5\,J/g\cdot^\circ C)(85^\circ C - (-20.^\circ C))\left(\frac{1\min}{450\,J}\right) = 15\min$$

2) Time to vaporize sample (phase change)

$$time = (25\,g)(500.\,J/g)\left(\frac{1\min}{450\,J}\right) = 28\min$$

3) Time to heat gas from 85°C to 100°C

$$time = m_A c_{gas}\Delta T(rate) = (25\,g)(0.5\,J/g\cdot^\circ C)(100.^\circ C - 85^\circ C)\left(\frac{1\min}{450\,J}\right) = 0.4\min$$

The following data points are obtained for the heating curve:

Process	Time [min] (x-axis)	Temp [°C] (y-axis)
Initial conditions	0	-40
Heating solid	1.1	-20
Solid-liquid phase change	1.1 + 10 = 11	-20
Heating liquid	11 + 15 = 25	85
Liquid-gas phase change	25 + 28 = 53	85
Heating gas	53 + 0.4 = 53.4	100

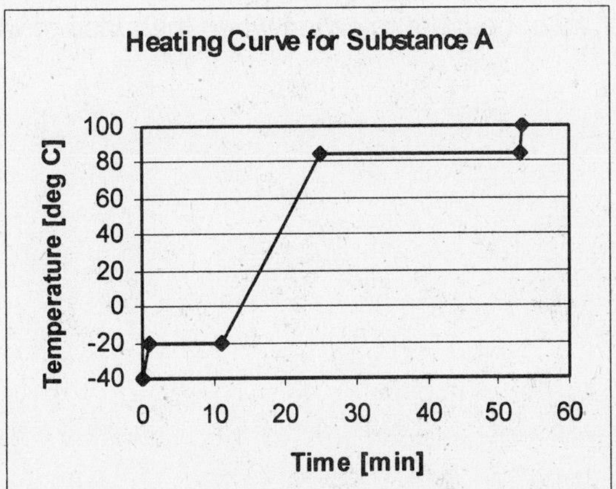

12.148 Body-centered cubic unit cells contain 2 atoms (Figure 12.27).

$$mass\ of\ Na\ unit\ cell = (2\ atoms)\left(\frac{22.99\ amu}{atom}\right)\left(\frac{1.66054x10^{-24}\ g}{1\ amu}\right) = \mathbf{7.635x10^{-23}\ g}$$

12.152 a) The number of moles of H_2O vapor can be calculated using your knowledge of partial pressures and the ideal gas law.

$P_{tot\ inside\ dome} = P_{H2O} + P_{air} = 1.0$ atm

Assume that the dome is filled with outside humid air and determine the mass of water.

P = 31.0 torr

V = $2.4 \times 10^6\ m^3 \times (1000\ L/m^3) = 2.4 \times 10^9$ L

T = 22.0°C = 295 K

$$mol\ H_2O = \frac{PV}{RT} = \frac{\left(\frac{31.0\ torr}{760\ torr / atm}\right)(2.4x10^9\ L)}{\left(0.08206\frac{atm \cdot L}{mol \cdot K}\right)(295\ K)} = 4.0_{44}x10^6\ mol\ H_2O$$

$$mass\ H_2O = (4.0x10^6\ mol)(18.02\ g / mol) = 7.2_{87}x10^7\ g\ H_2O$$

How much water is present when the humidity is lower, i.e. the partial pressure = 10.0 torr?

$$mol\ H_2O = \frac{PV}{RT} = \frac{\left(\frac{10.0\ torr}{760\ torr / atm}\right)(2.4x10^9\ L)}{\left(0.08206\frac{atm \cdot L}{mol \cdot K}\right)(295\ K)} = 1.3_{04}x10^6\ mol\ H_2O$$

$$mass\ H_2O = (1.3x10^6\ mol)(18.02\ g / mol) = 2.3_{51}x10^7\ g\ H_2O$$

The difference in mass between humid conditions and drier conditions is $(7.2_{87} \times 10^7\ g) - (2.3_{51} \times 10^7\ g) = 4.9_{36} \times 10^7$ g. Therefore, $(4.9_{36} \times 10^7\ g)$ (1 metric ton/10^6 g) = **49 metric tons** of water must be removed each time a dome-ful of inside air is replaced with outside of air.

b) Condensation (gas to liquid) is the reverse process to vaporization (liquid to gas), so the heat released is the product of the amount (n) and the heat of vaporization ($-\Delta H_{vap}$). Assuming the vapor has been cooled from 22.0°C to 0.0°C, the heat required for the phase change is:

$$q = n(-\Delta H_{vap}) = (4.9x10^7\ g)\left(\frac{1\ mol}{18.02\ g / mol}\right)(-40.7\ kJ / mol) = -1.1x10^8\ kJ$$

Therefore, $\mathbf{1.1 \times 10^8}$ **kJ** of heat are released when the mass of water condenses.

CHAPTER 13

THE PROPERTIES OF MIXTURES: SOLUTIONS AND COLLOIDS

FOLLOW-UP PROBLEMS

13.1 Plan: Compare the intermolecular forces in the solutes with the intermolecular forces in the solvent. The intermolecular forces for the more soluble solute will be more similar to the intermolecular forces in the solvent than the forces in the less soluble solute.

Solution:

a) 1,4-Butanediol is more soluble in water than butanol. Intermolecular forces in water are primarily hydrogen bonding. The intermolecular forces in both solutes also involve hydrogen bonding with the hydroxyl groups. Compared to butanol, each 1,4-butanediol molecule will form more hydrogen bonds with water because 1,4-butanediol contains two hydroxyl groups in each molecule whereas butanol contains only one –OH group. Since 1,4-butanediol has more hydrogen bonds to water than butanol, it will be more soluble than butanol.

b) Chloroform is more soluble in water than carbon tetrachloride because chloroform is a polar molecule and carbon tetrachloride is nonpolar. Polar molecules are more soluble in water, a polar solvent.

13.2 Plan: Solubility of a gas can be found from Henry's law: $S_{gas} = k_H \times P_{gas}$. The problem gives k_H for N_2 but not its partial pressure. To calculate partial pressure use the relationship from chapter 5: $P_{gas} = X_{gas} \times P_{total}$ where X represents the mole fraction of the gas.

Solution: To find partial pressure use the 78% N_2 given for the mole fraction:

$P_{N2} = 0.78 \times 1$ atm = 0.78 atm

Use Henry's law to find solubility at this partial pressure:

$S_{N2} = 7 \times 10^{-4}$ mol/L·atm x 0.78 atm = 5.46×10^{-4} mol/L = $\mathbf{5 \times 10^{-4}}$ **mol/L**

Check: The solubility should be close the value of k_H since the partial pressure is close to 1 atm.

13.3 Plan: Molality (m) is defined as amount (mol) of solute per kg of solvent. Use the molality and the mass of solvent given to calculate the amount of glucose. Then convert amount (mol) of glucose to mass (g) of glucose.

Solution: Mass of solvent must be converted to kg: 563 g x (1kg/1000g) = 0.563 kg

$$(0.563\,kg)(2.40x10^{-2}\,m)\left(\frac{180.16\,g\,C_6H_{12}O_6}{mol}\right) = \mathbf{2.43\,g\,C_6H_{12}O_6}$$

Check: To check estimate the molality from the calculated mass of glucose and given mass of solvent. The number of moles of glucose is approximately 2.4 g ÷ 180 g/mol = 0.013 mol glucose. Using the calculated value for amount of glucose and the given mass of solvent calculate the molality: 0.013 ÷ 0.56 kg = 0.023 m, which is close to the given value of 0.0240 m.

13.4 Plan: Mass percent is the mass (g) of solute per 100 g of solution. For each alcohol, divide the mass of the alcohol by the total mass. Multiply this number by 100 to obtain mass percent. To find mole fraction, first find the amount (mol) of each alcohol then divide by the total moles.

Solution: Mass percent:

$$mass\,\%\,propanol = \frac{35.0\ g\ propanol}{35.0\ g + 150.\ g}(100) = 18.9\,\%$$

$$mass\,\%\,ethanol = \frac{150.\ g\ ethanol}{35.0\ g + 150.\ g}(100) = 81.1\,\%$$

Mole fraction:

$$mol\ ethanol = (150.\ g\ ethanol)\left(\frac{1\ mol\ ethanol}{46.07\ g\ ethanol}\right) = 3.25_{59}\ mol;$$

$$mol\ propanol = (35.0\ g\ propanol)\left(\frac{1\ mol\ propanol}{60.09\ g\ propanol}\right) = 0.582_{46}\ mol$$

$$X_{ethanol} = \frac{3.25_{59}\ mol\ ethanol}{3.25_{59}\ mol + 0.582_{46}\ mol} = \mathbf{0.848};\quad X_{propanol} = \frac{0.582_{46}\ mol\ propanol}{3.25_{59}\ mol + 0.582_{46}\ mol} = \mathbf{0.152}$$

Check: Since the mass of ethanol is 4 times that of propanol, the mass percent of ethanol should be about 4 times greater. 81% is about 4 times 19%. The mole fraction should be similar, but the fraction that is ethanol should be slightly greater than the mass fraction of 0.81 – and it is at 0.85.

13.5 Plan: To find the mass percent, molality and mole fraction of HCl, the following is needed. 1) mole of HCl in 1 L solution from molarity; 2) mass of HCl in 1L solution from molarity times molar mass of HCl; 3) mass of 1 L solution from volume times density; 4) mass of solvent in 1L solution by subtracting mass of solute from mass of solution; 5) mole of solvent by multiplying mass of solvent by molar mass of water.

Mass percent is calculated by dividing mass of HCl by mass of solution.

Molality is calculated by dividing moles of HCl by mass of solvent in kg.

Mole fraction is calculated by dividing mol of HCl by the sum of mol of HCl plus mol of solvent.

Solution: Note that extra significant figures are kept in initial calculations.

1) mol HCl in 1L solution = 11.8 mol

2) mass HCl in 1 L solution = 11.8 mol x 36.46 g/mol = 430.23 g

3) mass of 1 L solution = 1 L x 1000 mL/L x 1.190 g/mL = 1190. g or 1.190 kg

4) mass of solvent in 1 L solution = 1190. kg – 430.23 g = 759.77 g

5) mol of solvent in 1 L solution = 759.77 g ÷ 18.02 g/mol = 42.162 mol

Mass percent of HCl = 430.23 g HCl ÷ 1190. g x 100 = **36.2%**

Molality of HCl = 11.8 mol HCl ÷ 0.75977 kg = **15.5 *m***

Mole fraction of HCl = 11.8 mol HCl ÷ (11.8 mol + 42.162 mol) = **0.219**

Check: The mass of HCl is about 1/3 of the mass of the solution, so 36% is in the ballpark for the mass % HCl. Since the moles of HCl is about ¼ the moles of water, the mole fraction of HCl should come out around 0.2.

13.6 Plan: Raoult's law states that the vapor pressure of a solvent is proportional to the mole fraction of the solvent: $P_{solvent} = X_{solvent} \times P^0_{solvent}$. To calculate the drop in vapor pressure, a similar relationship is used with the mole fraction of the solute substituted for that of the solvent.

Solution: Mole fraction of aspirin in methanol:

$$X_{aspirin} = \frac{2.00\,g\left(\frac{1\,mol\ aspirin}{180.16\,g}\right)}{2.00\,g\left(\frac{1\,mol\ aspirin}{180.16\,g}\right) + 50.0\,g\left(\frac{1\,mol\ methanol}{32.04\,g}\right)} = 7.06x10^{-3}$$

$$\Delta P = X_{aspirin} \times P^0_{methanol} = (7.06x10^{-3})(101\,torr) = \mathbf{0.713\ torr}$$

Check: The drop in vapor pressure is expected to be fairly insignificant since the amount of aspirin in solution is so small – only 0.01 mol aspirin in 1.5 mol solvent.

13.7 Plan: The question asks for the concentration of ethylene glycol that would prevent freezing at 0.00°F. First, convert 0.00°F to °C. The change in freezing point of water will then be this temperature subtracted from the freezing point of pure water, 0.00°C. Use this value for ΔT in ΔT_f = 1.86°C/*m* × molality of solution and solve for molality.

Solution: Temperature conversion

$$T(^\circ C) = \left(T(^\circ F) - 32.0^\circ F\right)\left(\frac{5^\circ C}{9^\circ F}\right) = \left(0.00^\circ F - 32.0^\circ F\right)\left(\frac{5^\circ C}{9^\circ F}\right) = -17.8^\circ C$$

$$molality = \frac{\Delta T_f}{1.86^\circ C/m} = \frac{0.00 - 17.8^\circ C}{1.86^\circ C/m} = \mathbf{9.56\ m}$$

The minimum concentration of ethylene glycol would have to be **9.56 *m*** in order to prevent the water from freezing at 0.00°F.

Check: In Sample Problem 13.8 a 3.6 *m* solution lowered the freezing point by about 7 degrees. Thus to lower the temperature by 18 degrees the concentration must be 18/7 times 3.6 or 9.3 *m*. Calculated value of 9.56 *m* is close to estimated value.

13.8 Plan: The osmotic pressure is calculated as the product of molarity, temperature and gas constant.

Solution:

$$\pi = (0.30\,M)(0.08206\,atm \cdot L/mol \cdot K)(273.15K + 37) = \mathbf{7.6\ atm}$$

Check: The answer, 7.6 atm, is a significant osmotic pressure, which is expected from a relatively concentrated solution of 0.3 M.

END-OF-CHAPTER PROBLEMS

13.2 Sodium chloride dissolves in water because the ion-dipole interactions between the ion and oppositely charged end of the water dipole replace the ion-ion interactions in the solid. Ion-dipole interactions form the first "layer" of hydration, followed by H-bonding between water molecules in subsequent layers.

13.4 **Sodium stearate** would be a more effective soap because the hydrocarbon chain in the stearate ion is longer than the chain in the acetate ion. A soap forms suspended particles called micelles with the polar end of the soap interacting with the water solvent molecules and the nonpolar ends forming a nonpolar environment inside the micelle. Oils dissolve in the nonpolar portion of the micelle. Thus a better solvent for the oils in dirt is a more nonpolar substance. The long hydrocarbon chain in the stearate ion is better at dissolving oils in the micelle than the shorter hydrocarbon chain in the acetate ion.

13.7 A more concentrated solution will have more solute dissolved in the solvent. Potassium nitrate, KNO_3, is an ionic compound and therefore soluble in a polar solvent like water. Potassium nitrate is not soluble in the nonpolar solvent CCl_4. Because potassium nitrate dissolves to a greater extent in water, **KNO_3 in H_2O** will result in the more concentrated solution.

13.9 To identify the strongest type of intermolecular force check the formula of the solute and identify the forces that could occur. Then look at the formula for the solvent and determine if the forces identified for the solute would occur with the solvent. The strongest force is ion-dipole followed by dipole-dipole (including H bonds). Next in strength is ion-induced dipole force and then dipole-induced dipole force. The weakest intermolecular interactions are dispersion forces.

a) **Ion-dipole** forces are the strongest intermolecular forces in the solution of the ionic substance cesium chloride in polar water.

b) **Hydrogen bonding** (type of dipole-dipole force) is the strongest intermolecular force in the solution of polar propanone (or acetone) in polar water.

c) **Dipole-induced dipole** forces are the strongest forces between the polar methanol and nonpolar carbon tetrachloride.

13.11 a) **Hydrogen bonding** occurs between the H atom on water and the lone electron pair on the O atom in dimethyl ether (CH_3OCH_3). However, none of the hydrogen atoms on dimethyl ether participate in H-bonding because the C–H bond does not have sufficient polarity.

b) The dipole in water induces a dipole on the Ne(g) atom, so **dipole-induced dipole** interactions are the strongest intermolecular forces in this solution.

c) Nitrogen gas and butane are both nonpolar substances, so **dispersion forces** are the principal attractive forces.

13.13 Diethyl ether has the formula:

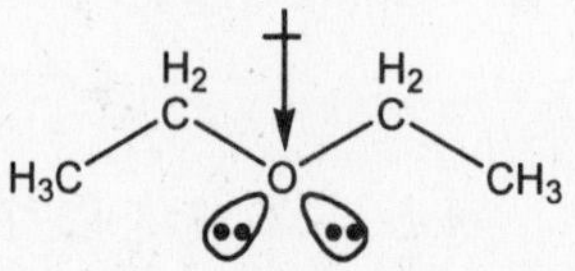

$CH_3CH_2OCH_2CH_3$ is polar with dipole-dipole interactions as the dominant intermolecular forces. Examine the solutes to determine which has intermolecular forces more similar to those for the diethyl ether. This solute is the one that would be more soluble.

a) **HCl** would be more soluble since it is a covalent compound with dipole-dipole forces whereas NaCl is an ionic solid. Dipole-dipole forces between HCl and diethyl ether are more similar to the dipole forces in diethylether than the ion-dipole forces between NaCl and diethylether.

b) **CH_3CHO** (acetaldehyde) would be more soluble. The dominant interactions in H_2O are H-bonding, a stronger type of dipole-dipole force. The dominant interactions in CH_3CHO are dipole-dipole. The solute-solvent interactions between CH_3CHO and diethylether are more similar to the solvent intermolecular forces than the forces between H_2O and diethylether.

c) **CH_3CH_2MgBr** would be more soluble. CH_3CH_2MgBr has a polar end (-MgBr) and a nonpolar end (CH_3CH_2-) whereas $MgBr_2$ is an ionic compound. The nonpolar end of CH_3CH_2MgBr and diethylether would interact with dispersion forces, while the polar end of CH_3CH_2MgBr and the dipole in diethylether would interact with dipole-dipole forces.

13.16 Gluconic acid is a very polar molecule because it has –OH groups attached to every carbon. The abundance of –OH bonds allows gluconic acid to participate in extensive H-bonding with water, hence its great solubility in water. On the other hand, caproic acid has 5-carbon, non-polar, hydrophobic ("water hating") tail that does not easily dissolve in water. The intermolecular dispersion forces in the nonpolar tail are more similar to the intermolecular dispersion forces in hexane, hence its greater solubility in hexane.

13.18 $\Delta H_{solvation}$ would include breaking the intermolecular force between solvent particles ($\Delta H_{solvent}$) and formation of the intermolecular force between solvent and solute particles (ΔH_{mix}).

13.23 This compound would be very soluble in water. A large exothermic value in $\Delta H_{solution}$ (enthalpy of solution) means that the solution has a much lower energy state than the isolated solute and solvent particles, so the system tends to the formation of the solution. Entropy that accompanies dissolution always favors solution formation. Entropy becomes important when explaining why solids with an endothermic $\Delta H_{solution}$ (and higher energy state) are still soluble in water.

13.24

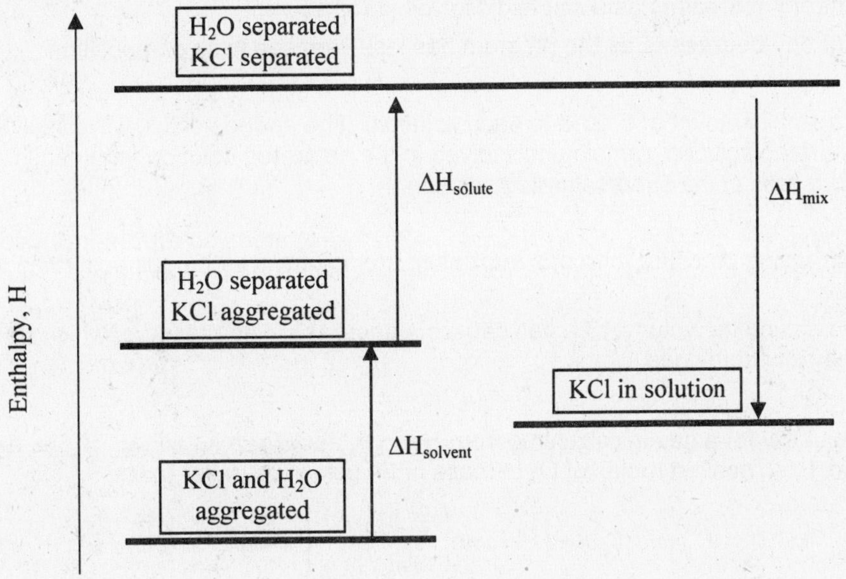

13.26 Charge density is the ratio of an ion's charge (regardless of sign) to its volume.

a) Both ions have a +1 charge, but the volume of **Na^+** is smaller, so it has the greater charge density.
b) **Sr^{2+}** has a greater ionic charge and a smaller size (because it Z_{eff} is greater), so it has the greater charge density.
c) **Na^+** has a smaller ion volume than Cl^-, so it has the greater charge density.
d) **O^{2-}** has a greater ionic charge and smaller ion volume, so it has the greater charge density.
e) **OH^-** has a smaller ion volume than SH^-, so it has the greater charge density.

13.28 The ion with the greater charge density will have the larger $\Delta H_{hydration}$.

a) **Na^+** would have a larger $\Delta H_{hydration}$ than Cs^+ since its charge density is greater than that of Cs^+.
b) **Sr^{2+}** would have a larger $\Delta H_{hydration}$ than Rb^+.

c) **Na^+** would have a larger $\Delta H_{hydration}$ than Cl^-.

d) **O^{2-}** would have a larger $\Delta H_{hydration}$ than F^-.

e) **OH^-** would have a larger $\Delta H_{hydration}$ than SH^-.

13.30 a) The two ions in potassium bromate are K^+ and BrO_3^-. The heat of solution for ionic compounds is $\Delta H_{soln} = -\Delta H_{lattice} + \Delta H_{hydr\ of\ the\ ions}$. Therefore, the combined heats of hydration for the ions is ($\Delta H_{soln} + \Delta H_{lattice}$) or 41.1 kJ/mol – 745 kJ/mol = **–704 kJ/mol**.

b) **K^+** ion contributes more to the heat of hydration because it has a smaller size and, therefore, a greater charge density.

13.32 Entropy increases as the possible states for a system increases.

a) Entropy **increases** as vase is shattered because the possible states for a shattered vase are greater than the one state for the vase in one piece.

b) Entropy **decreases** as the gold is separated from the ore. Pure gold has only the arrangement of gold atoms next to gold atoms while the ore mixture has a greater number of possible arrangements among the components of the mixture.

c) Entropy **increases** as a solute dissolves in the solvent.

d) Entropy **decreases** as the program has less variation in its composition.

13.35 Add a small amount of solid X to each solution. The added solid X will dissolve in the unsaturated solution, remain undissolved in the saturated solution, and cause precipitation of more solid in the supersaturated solution.

13.38 a) Increasing pressure for a gas **increases** the solubility of the gas according to Henry's law.

b) Increasing the volume of a gas causes a decrease in its pressure, which **decreases** the solubility of the gas.

13.40 a) Solubility for a gas is calculated from Henry's law: $S_{gas} = k_H \times P_{gas}$. S_{gas} is expressed as mol/L, so convert moles of O_2 to mass of O_2 using the molar mass.

$$S_{gas} = (1.28x10^{-3}\ mol/L \cdot atm)(1.00\ atm)\left(\frac{32.00\ g\ O_2}{1\ mol\ O_2}\right) = 4.10x10^{-2}\ g\ O_2/L$$

Therefore, twice that amount, or **8.19×10^{-2} g O_2**, will dissolve in 2.00 L.

b) The amount of gas that will dissolve in a given volume decreases proportionately with the partial pressure of the gas, so (0.209)(8.19×10^{-2} g) = **1.71×10^{-2} g O_2** will dissolve.

13.43 Solubility for a gas is calculated from Henry's law: $S_{gas} = k_H \times P_{gas}$.

$S_{gas} = 3.7x10^{-2}$ mol/L·atm x 5.5 atm = **0.20 mol/L**

13.46 Refer to Table 13.4 for the different methods of expressing concentration.

a) **Molarity** and **parts-by-volume** (% w/v or % v/v) include the volume of the solution.

b) **Parts-by-mass** (% w/w) and **mole fraction** include the mass of solution.

c) **Molality** includes the mass of the solvent.

13.48 Converting between molarity and molality involves conversion between volume of solution and mass of solution. Both of these quantities are given so interconversion is possible. To convert to mole fraction requires that the mass of solvent be converted to moles of solvent. Since the identity of the solvent is not given, conversion to mole fraction is not possible if the molar mass is not known. Why is the identity of the solute not necessary for conversion?

13.50 a) $$M = \frac{mol\ solute}{L\ solution} = \frac{\left(42.3\ g\ sugar\ x\ \frac{1\ mol}{342.30\ g}\right)}{\left(100.\ mL\ x\ \frac{1\ L}{1000\ mL}\right)} = \mathbf{1.24\ M}$$

b) $$M = \frac{\left(5.50\ g\ LiNO_3\ x\ \frac{1\ mol}{68.95\ g}\right)}{0.505\ L} = \mathbf{0.158\ M}$$

13.52 Dilution calculations can be done using $M_{conc}V_{conc} = M_{dil}V_{dil}$.

a) 75.0 mL of 0.250 M NaOH diluted to 0.250 L with water.

$$M_{dil} = 0.250\ M\left(\frac{0.0750\ L}{0.250\ L}\right) = \mathbf{0.0750\ M}$$

b) 35.5 mL of 1.3 M HNO_3 diluted to 0.150 L

$$M_{dil} = 1.3\ M\left(\frac{0.0355\ L}{0.150\ L}\right) = \mathbf{0.31 M}$$

13.54 a) Find the number of moles KH_2PO_4 needed to make 355 mL of this solution.

$$mol\ solute = MV_{sol'n} = (8.74x10^{-2}\ M)(0.355\ L) = 3.10_{27}x10^{-2}\ mol$$

Convert moles to mass using the molar mass of KH_2PO_4 ($\mathcal{M}$ = 136.09 g/mol)

$$mass\ KH_2PO_4 = (3.10_{27}x10^{-2}\ mol)\left(\frac{136.09\ g}{mol}\right) = 4.22\ g$$

Add **4.22 g KH_2PO_4** to enough water to make 355 mL of aqueous solution.

b) Use the relationship $M_1V_1 = M_2V_2$ to find the volume of 1.25 M NaOH needed.

$$(425\ mL)(0.315\ M) = (1.25\ M)V_2;\ V_2 = 107\ mL$$

Add **107 mL** of 1.25 M NaOH to enough water to make 425 mL of solution.

13.56 a) To find the mass of KBr needed, find the moles of KBr in 1.50 L of a 0.257 M solution and convert to grams using molar mass of KBr.

$$mass\ KBr = (0.257\ M)(1.50\ L)\left(\frac{119.0\ g\ KBr}{mol}\right) = 45.9\ g\ KBr$$

To make the solution, weigh **45.9 g KBr** and then dilute to 1.50 L with distilled water.

b) To find the volume of the concentrated solution that will be diluted to 355 mL use $M_{conc}V_{conc} = M_{dil}V_{dil}$ and solve for V_{conc}.

$$V_{conc} = (355\ mL)\left(\frac{0.0956\ M}{0.244\ M}\right) = 139\ mL$$

To make the 0.0956 M solution, measure **139 mL** of the 0.244 M solution and add distilled water to make a total of 355 mL.

13.58 a) Molality = moles solute/kg solvent.

$$m = \frac{\left(88.4\ g\ glycine\ x \dfrac{1\ mol}{75.07\ g}\right)}{1.250\ kg} = \mathbf{0.942}\ \boldsymbol{m}$$

b) $$m = \frac{\left(8.89\ g\ glycerol\ x \dfrac{1\ mol}{92.09\ g}\right)}{75.0x10^{-3}\ kg} = \mathbf{1.29}\ \boldsymbol{m}$$

13.60 Molality = moles solute/kg of solvent.

$$molality = \frac{(34.0\ mL)(0.877\ g/mL)\left(\dfrac{1\ mol\ benzene}{78.11\ g}\right)}{(187\ mL)(0.660\ g/mL)(1\ kg/1000\ g)} = \mathbf{3.09\ m}$$

13.62 a) The total weight of the solution is 3.00 x 10^2 g, so m_{solute} + $m_{solvent}$ = 3.00 x 10^2 g and $m_{solvent}$ = 3.00 x 10^2 g – m_{solute}. The moles of solute can be expressed as the product of mass of solute (m_{solute}) and reciprocal molar mass ($\mathcal{M}_{solute}$)$^{-1}$.

$$\frac{m_{solute}/\mathcal{M}_{solute}}{m_{solvent}} = 0.115\,m$$

$$\frac{m_{solute}/62.07\ g/mol}{(300.\ g - m_{solute})\left(\dfrac{1\ kg}{1000\ g}\right)} = 0.115\,m$$

$$m_{solute} = (0.115\,m)(62.07\ g/mol)(300.\ g - m_{solute})\left(\frac{1\ kg}{1000\ g}\right)$$

$$m_{solute} = 2.1414 - 0.007138 m_{solute}$$

$$m_{solute} = 2.13\ g$$

The $m_{solvent}$ = 300.g – 2.13 g = 298 g. Therefore, **add 2.13 g $C_2H_6O_2$ to 298 g of H_2O** to make a 0.115 *m* solution.

b) First determine the amount of solute in your target solution:

$$\left(\frac{2.00\ g\ solute}{100.0\ g\ solution}\right)(1.00x10^3\ g\ solution) = 20.0\ g\ solute$$

Then determine the amount of the concentrated acid solution needed to get 20.0 g solute:

$$\frac{62.00\ g\ solute}{100.0\ g\ conc\ sol'n}\ x\ (mass\ needed) = 20.0\ g\ solute$$

$$mass = 32.3\ g\ conc\ sol'n$$

Add 32.3 g of the 62.0% (w/w) HNO_3 to 968 g H_2O to make 1.00 kg of 2.00% (w/w) HNO_3.

13.64 a) Mole fraction is moles of isopropanol per total moles.

$$X_{propanol} = \frac{0.30\ mol\ isopropanol}{0.30\ mol + 0.80\ mol} = \mathbf{0.27}$$

b) Mass percent is the mass of isopropanol per 100 g of solution.

$$\%\ isopropanol = \left(\frac{0.30\ mol(60.09\ g/mol)}{0.30\ mol(60.09\ g/mol) + 0.80\ mol(18.02\ g/mol)}\right)(100) = \mathbf{56\ \%}$$

c) Molality of isopropanol is moles of isopropanol per kg of solvent.

$$molality = \frac{0.30\ mol}{(0.80\ mol)(18.02\ g/mol)(1\ kg/1000\ g)} = \mathbf{21\ \mathit{m}}$$

13.66 Since the density of water is 1.00 g/mL, the mass of water is 0.500 kg.

$$\frac{m_{CsCl}\ /\ \mathcal{M}_{CsCl}}{m_{solvent}} = 0.400\ m$$

$$\frac{m_{CsCl}\ /\ (168.4\ g/mol)}{0.500\ kg} = 0.400\ m$$

$$m_{CsCl} = 33.7\ g$$

Add **33.7 g CsCl** to 0.500 kg water to produce a 0.500 *m* solution.

Mole fraction of CsCl:

$$X_{CsCl} = \frac{(33.6_{80}\ g / 168.4\ g/mol)}{\left(\frac{33.6_{80}\ g}{168.4\ g/mol} + \frac{500.\ g}{18.02\ g/mol}\right)} = \mathbf{7.15x10^{-3}}$$

Mass % of CsCl:

$$mass\% = \frac{33.6_{80}\ g}{33.6_{80}\ g + 500.\ g} = \mathbf{6.31\%}$$

13.68 The information given is mass % as 8.00 g NH_3 per 100 g solution and density of 0.9651 g/mL. To find molality, convert grams of ammonia to moles of ammonia and subtract mass of solute from mass of solution.

$$molality = \frac{(8.00\ g\ NH_3)(1\ mol/17.03\ g)}{(100\ g - 8.00\ g)(1\ kg/1000\ g)} = \mathbf{5.11\ \mathit{m}}$$

Molarity can be calculated by converting grams of ammonia to moles and dividing the mass of solution by the density of the solution.

$$molarity = \frac{(8.00\ g)(1\ mol/17.03\ g)}{(100\ g)(1\ mL/0.9651\ g)(1\ L/1000\ mL)} = \mathbf{4.53\ M}$$

To calculate the mole fraction, convert mass of ammonia to moles and the mass of solvent to moles.

$$mole\ fraction = \frac{(8.00\ g)(1\ mol/17.03\ g)}{(8.00\ g)(1\ mol/17.03\ g) + (92.0\ g)(1\ mol/18.02\ g)} = \mathbf{0.0842}$$

13.70 The mass of 100.0 L of waste water solution is (100.0 L)(1.001 g/mL)(1000 mL/L) = 1.001×10^5 g.

$$ppm\,Ca^{2+} = \frac{0.22\ g\ Ca^{2+}}{1.001x10^5\ g\ sol'n} x10^6 = \mathbf{2.2\,ppm}$$

$$ppm\ Mg^{2+} = \frac{0.066\ g\ Mg^{2+}}{1.001x10^5\ g\ sol'n} x10^6 = \mathbf{0.66\,ppm}$$

13.74 The "strong" in "strong electrolyte" refers to the ability of an electrolyte solution to conduct a large current. This conductivity occurs because solutes that are strong electrolytes dissociate completely into ions when dissolved in water.

13.76 The boiling point temperature is higher and the freezing point temperature is lower for the solution compared to the solvent because the addition of a solute lowers the freezing point and raises the boiling point of a liquid.

13.79 The predicted value is based on ideal behavior. A dilute solution behaves more ideally than a concentrated one, so the less concentrated **NaF solution** will have a boiling point closer to the predicted value.

13.82 a) **Strong electrolyte**. When hydrogen chloride is bubbled through water, it dissolves and dissociates completely into H^+ (or H_3O^+) ions and Cl^- ions.

b) **Strong electrolyte**. Potassium nitrate is a soluble salt.

c) **Nonelectrolyte**. Glucose solid dissolves in water to form individual $C_6H_{12}O_6$ molecules, but these units are not ionic and therefore don't conduct electricity.

d) **Weak electrolyte**. Ammonia gas dissolves in water, but is a weak base that forms few NH_4^+ and OH^- ions.

13.84 To count solute particles in a solution of an ionic compound, count the number of ions per mole and multiply by the number of moles in solution. For a covalent compound the number of particles equals the number of molecules.

a) 0.2 mol KI/L x 2 mol ions/mol x 1 L = **0.4 mol** particles. KI consists of K^+ ions and I^- ions, 2 particles for each KI.

b) 0.070 mol HNO_3/L x 2 mol ions/mol x 1 L = **0.14 mol** particles. HNO_3 is a strong acid that forms hydronium ions and nitrate ions in aqueous solution.

c) 10^{-4} mol K_2SO_4 /L x 3 mol ions/mol x 1 L = **$3x10^{-4}$ mol** particles. Each K_2SO_4 forms 2 potassium ions and 1 sulfate ion in aqueous solution.

d) Ethanol is not an ionic compound so each molecule dissolves as one particle. The number of moles of particles is the same as the number of moles of molecules, **0.07 mol** in 1 L.

13.86 The magnitude of freezing point depression is directly proportional to molality.

a) The molality of methanol, CH_3OH, in water is 3.12 m whereas the molality of ethanol, CH_3CH_2OH, in water is 2.17 m. Thus, **CH_3OH/H_2O solution** has the lower freezing point.

b) The molality of H_2O in CH_3OH is 0.555 m, whereas CH_3CH_2OH in CH_3OH is 0.217 m. Therefore, **H_2O/CH_3OH solution** has the lower freezing point.

13.88 To rank the solutions in order of increasing osmotic pressure, boiling point, freezing point, and vapor pressure, convert the molality of each solute to molality of particles in the solution. The higher the molality of particles the higher the osmotic pressure, the higher the boiling point, the lower the freezing point, and the lower the vapor pressure at a given temperature.

(I) 0.100 m $NaNO_3 \times 2$ mol ions/mol $NaNO_3$ = 0.200 m ions

(II) 0.200 m glucose × 1 mol molecules/mol glucose = 0.200 m molecules

(III) 0.100 m $CaCl_2 \times 3$ mol ions/mol $CaCl_2$ = 0.300 m ions

a) Osmotic pressure: $\pi_I = \pi_{II} < \pi_{III}$

b) Boiling point: $\mathbf{bp_I = bp_{II} < bp_{III}}$

c) Freezing point: $\mathbf{fp_{III} < fp_I = fp_{II}}$

d) Vapor pressure at 50°C: $\mathbf{vp_{III} < vp_I = vp_{II}}$

13.90 The amount of solute affects the vapor pressure according to the following equation:

$$\Delta P = X_{solute} P^o_{solvent}$$

$$\Delta P = \left(\frac{\dfrac{44.0\ g}{92.09\ g/mol}}{\dfrac{44.0\ g}{92.09\ g/mol} + \dfrac{500.0\ g}{18.02\ g/mol}} \right) (23.76\ torr) = 0.402\ torr$$

$$P_{final} = P_{initial} - \Delta P = 23.76\ torr - 0.402\ torr = \mathbf{23.36\ torr}$$

13.92 The change in freezing point is calculated from $\Delta T_f = m \times K_f$, where K_f = 1.86°C/ m for aqueous solutions and m is the molality of particles in the solution. Since urea is a covalent compound the molality of urea equals the molality of particles in solution. Once ΔT_f is calculated the freezing point is determined by subtracting it from the freezing point of pure water, 0.00°C.

ΔT_f = 0.111 m × 1.86°C/ m = 0.206°C

The freezing point is 0.00°C – 0.206°C = **-0.206°C**.

13.94 The boiling point of a solution is increased relative to the pure solvent by the relationship $\Delta T = K_b m$. The molality of the vanillin/ethanol solution is

$$molality = \frac{3.4\ g / 152.14\ g/mol}{0.0500\ kg} = 0.44_{70}\ m$$

Since vanillin is a nonelectrolyte, the molality of vanillin equals the molality of particles in solution.

$\Delta T = K_b m$ = (1.22 °C/m)(0.44$_{70}$ m) = 0.545 °C

T_{final} = 78.5 °C + 0.545 °C = **79.0 °C**

13.96 The molality of the solution can be determined from the relationship $\Delta T_f = m \times K_f$ with the value 1.86°C/m inserted for K_f and the given ΔT_f of –10.0°F converted to °C. Multiply the molality by the given mass of solvent to find the mass of ethylene glycol that must be in solution. Note that ethylene glycol is a covalent compound that will form one particle per molecule when dissolved.

$$T(^\circ C) = \frac{5^\circ C}{9^\circ F}(-10.0^\circ F - 32.0^\circ F) = -23.3_{33}{}^\circ C$$

$$m = \frac{0.00^\circ C - (-23.3_{33}{}^\circ C)}{1.86^\circ C / m} = 12.5_{45}\ m$$

$$mass_{ethylene\ glycol} = 12.544\ mol/kg(14.5\ kg)(62.07\ g/mol) = 1.13x10^4\ g$$

To prevent the solution from freezing, dissolve a minimum of **11.3 kg** ethylene glycol in 14.5 kg water.

13.98 a) Convert the mass % to molality and use $\Delta T = K_b m$ to find the expected freezing point depression. The van't Hoff factor is the ratio of measured T_f to expected T_f.

$$molality = \left(\frac{1.00\ g\ NaCl}{100.00\ g - 1.00\ g}\right)\left(\frac{1\ mol\ NaCl}{58.44\ g\ NaCl}\right)\left(\frac{1000\ g}{kg}\right) = 0.172_{84}\ m$$

Molality = **0.173 *m***

Calculate ΔT as though NaCl was a nonelectrolyte.

$$\Delta T = K_f m = (1.86\ ^\circ C/m)(0.172_{84}\ m) = 0.321_{48}\ ^\circ C$$

$$T_{final} = 0^\circ C - 0.321_{48}{}^\circ C = -0.321_{48}\ ^\circ C$$

$$i = \frac{\text{measured for electrolyte}}{\text{expected for nonelectrolyte}} = \frac{-0.593\ ^\circ C}{-0.321_{48}\ ^\circ C} = \mathbf{1.84}$$

The value of *i* should be close to 2 because NaCl dissociates into 2 particles when dissolving in water.

b) For acetic acid, CH_3COOH:

$$molality = \left(\frac{0.500\ g\ acetic}{100.00\ g - 0.500\ g}\right)\left(\frac{1\ mol\ acetic}{60.05\ g\ acetic}\right)\left(\frac{1000\ g}{kg}\right) = 0.0836_{82}\ m$$

Molality = **0.0837 *m***

$$\Delta T = K_f m = (1.86\ ^\circ C/m)(0.0836_{82}\ m) = 0.155_{65}\ ^\circ C$$

$$T_{final} = 0^\circ C - 0.155_{65}{}^\circ C = -0.155_{65}\ ^\circ C$$

$$i = \frac{\text{measured}}{\text{expected}} = \frac{-0.159\ ^\circ C}{-0.155_{65}\ ^\circ C} = \mathbf{1.02}$$

Acetic acid is a weak acid and dissociates to a small extent in solution, hence a van't Hoff factor that is close to 1.

13.100 Osmotic pressure is calculated from the molarity of particles, gas constant and temperature. Convert the mass of sucrose to moles using molar mass.

$$\pi = \left(\frac{3.42\ g\ sucrose}{L}\right)\left(\frac{1\ mol}{342.3\ g}\right)(0.08206\ atm \cdot L/mol \cdot K)(273.15\ K + 20.) = 0.240\ atm$$

A pressure of **0.240 atm** must be applied to obtain pure water from a 3.42 g/L solution.

13.102 The pressure of each compound is proportional to its mole fraction according to Raoult's law: $P_A = X_A P_A^\circ$

$$X_{CH_2Cl_2} = \frac{1.50\,mol}{1.50\,mol + 1.00\,mol} = 0.600; \quad X_{CCl_4} = \frac{1.00\,mol}{1.50\,mol + 1.00\,mol} = 0.400$$

$$P_{CH_2Cl_2} = (0.600)(352\,torr) = \mathbf{211\,torr}$$

$$P_{CCl_4} = (0.400)(118\,torr) = \mathbf{47.2\,torr}$$

13.103 The fluid inside a bacterial cell is **both a solution and a colloid**. It is a solution of ions and small molecules and a colloid of large molecules, proteins and nucleic acids.

13.107 Soap micelles have nonpolar "tails" pointed inward and anionic "heads" pointed outward. The charges on the "heads" on one micelle repel the "heads" on a neighboring micelle because the charges are the same. This repulsion between soap micelles keeps them from coagulating.

Soap is more effective in **freshwater** than in seawater because the divalent cations in seawater combine with the anionic "head" to form an insoluble precipitate.

13.108 Foam comes from gas bubbles trapped in the cream. The best gas for foam would be one that is not very soluble in the cream. Since nitrous oxide is a good gas to make foam and carbon dioxide a bad one, nitrous oxide must be less soluble in the foam than carbon dioxide. (In fact, at room temperature CO_2 is almost twice as soluble in water than N_2O.) Henry's law constant is larger for a gas that is more soluble, so the constant for carbon dioxide must be larger than the constant for nitrous oxide.

13.113 To find the volume of seawater needed, substitute the given information into the equation that describes the ppb concentration, account for extraction efficiency, and convert mass to volume using the density of seawater.

$$\frac{mass\ of\ gold}{mass\ of\ seawater} x\,10^9 = 1.1x10^{-2}\ ppb$$

$$\frac{31.1\,g}{mass\ of\ seawater} x\,10^9 = 1.1x10^{-2}\ ppb$$

$$mass\ of\ seawater\,(100\%\ efficiency) = 2.8_{27}x10^{12}\ g$$

$$mass\ of\ seawater\,(79.5\%\ efficiency) = 2.827x10^{12} \div 0.795 = 3.5_{56}x10^{12}\ g$$

$$V = \left(\frac{3.556x10^{12}\ g}{1.025\,g/mL}\right)\left(\frac{L}{1000\,mL}\right) = \mathbf{3.5x10^9\ L}$$

13.115 Molality and molarity are nearly the same for dilute aqueous solutions because the density of water is 1.00 kg/L so the conversion from volume of solution to kg of solvent gives practically the same value. Most nonaqueous solvents have densities significantly different than 1.00 kg/L so the volume of a dilute solution would not equal the mass of solvent.

13.120 a) First find the molality from the freezing point depression and then use the molality and given mass of solute and volume of water to calculate the molar mass of the solute compound.

$$m = \frac{0.201^\circ C}{1.86^\circ C/m} = 0.108_{06}\ m$$

$$\left(\frac{0.243\,g}{25.0\,mL}\right)\left(\frac{1.00\,mL}{g\,H_2O}\right)\left(\frac{1000\,g}{kg}\right)\left(\frac{1\,kg\,H_2O}{0.108_{06}\,mol}\right) = \mathbf{90.0\,g/mol}$$

b) Assume that 100.00 g of the compound gives 53.31 g carbon, 11.18 g hydrogen and 100.00 – 53.31 – 11.18 = 35.51 g oxygen.

$$n_C = \frac{53.31\,g}{12.01\,g/mol} = 4.44\,mol;\; n_H = \frac{11.18\,g}{1.008\,g/mol} = 11.1\,mol;\; n_O = \frac{35.51\,g}{16.00\,g/mol} = 2.22\,mol$$

Dividing by the lowest amount of moles, 2.22, gives 2 mol C, 5 mol H and 1 mol O for an **empirical formula C_2H_5O** with molar mass 45.06 g/mol.

Since the molar mass of the compound is twice the molar mass of the empirical formula the **molecular formula is $C_4H_{10}O_2$**.

c) Possible Lewis structures:

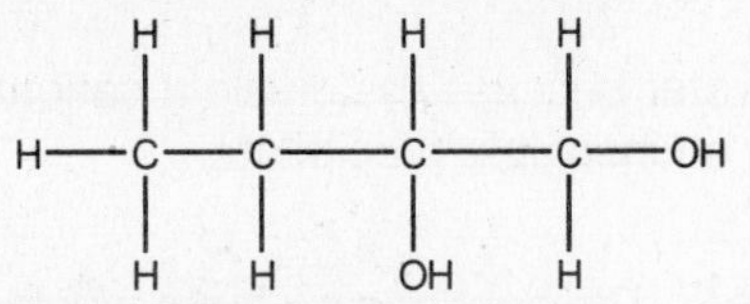

forms hydrogen bonds

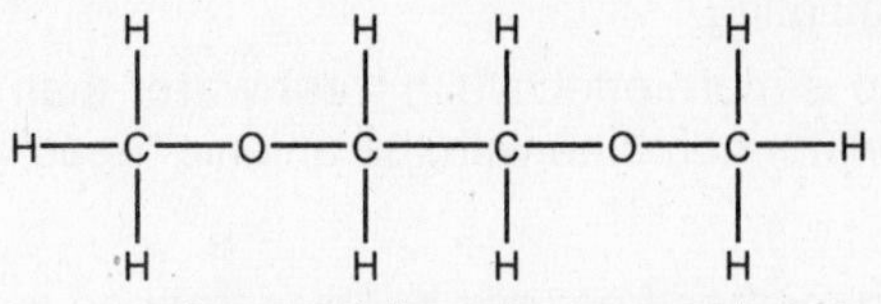

does not form hydrogen bonds

13.123 ΔH_{hydr} increases with increasing charge density.

a) **Mg^{2+}** has the higher charge density because it has a smaller ion volume.

b) **Mg^{2+}** has the higher charge density because it has both a smaller ion volume and greater charge.

c) **CO_3^{2-}** has the higher charge density for the same reasons in b).

d) **SO_4^{2-}** has the higher charge density for the same reasons in b).

e) **Fe^{3+}** has the higher charge density for the same reasons in b).

f) **Ca^{2+}** has the higher charge density for the same reasons in b).

13.125 a) Use the boiling point elevation of 0.45°C to calculate the molality of the solution. Then with molality and the mass of solute and volume of water calculate the molar mass.

$$m = \frac{0.45^\circ C}{0.512^\circ C/m} = 0.87_{89}\,m$$

$$\left(\frac{1.50\,g}{25.0\,mL}\right)\left(\frac{1\,mL}{0.997\,g}\right)\left(\frac{1000\,g}{kg}\right)\left(\frac{1\,kg}{0.87_{89}\,mol}\right) = \mathbf{68\,g/mol}$$

b) The molality calculated would be the moles of ions per kg of solvent. If the compound consists of three ions the molality of the compound would be 1/3 of $0.87_{89}\,m$ and the calculated molar mass would be three times greater: 3 x $68._{47}$ = **$2.0\text{x}10^2$ g/mol**.

c) The molar mass of CaN_2O_6 is 164.1. This molar mass is less than the $2.0\text{x}10^2$ calculated when the compound is assumed to be a strong electrolyte and is greater than the 68.5 calculated when the compound is assumed to be a nonelectrolyte. Thus, the compound is an electrolyte, but does not dissociate completely into ions in solution.

d) Use the molar mass of CaN_2O_6 to calculate the molality of the compound. Then calculate i in the boiling point elevation formula.

$$\left(\frac{1.50\,g}{25.0\,mL}\right)\left(\frac{1\,mol}{164.1\,g}\right)\left(\frac{1\,mL}{0.997\,g}\right)\left(\frac{1000g}{kg}\right) = 0.366_{73}\,m$$

$$i = \frac{0.45^\circ C}{(0.512^\circ C/m)(0.366_{73}\, m)} = \mathbf{2.4}$$

13.129 a) From the osmotic pressure the molarity of the solution can be found. The ratio of mass per volume to moles per volume gives the molar mass of the compound.

$$M = \frac{(0.340\, torr)\left(\frac{1\, atm}{760\, torr}\right)}{(0.08206\, atm \cdot L/mol \cdot K)(273.15\, K + 25)} = 1.82_{85} x10^{-5}\ M$$

$$\mathcal{M} = \frac{(0.0100\, g)/(0.0300\, L)}{1.82_{85} x10^{-5}\ mol/L} = \mathbf{1.82x10^4\ g/mol}$$

b) To find freezing point depression the molarity of the solution must be converted to molality. Then use $\Delta T_f = K_f \times m$.

$$\Delta T_f = (1.86^\circ C/m)\left(\frac{1.82_{85} x10^{-5}\ mol}{(1\, L)(0.997\, kg/L) - (1.00x10^{-5}\ kg)}\right) = \mathbf{3.41x10^{-5}\ ^\circ C}$$

13.130 Henry's law expresses the relationship between gas pressure and the gas' solubility in a given solvent. Use Henry's law to solve for pressure (assume that the constant is given at 21°C), use the ideal gas law to find moles per unit volume, and convert moles/L to ng/L.

$$S_{C_2H_2Cl_2} = k_H P_{C_2H_2Cl_2}$$

$$P_{C_2H_2Cl_2} = \frac{(0.75\, mg/L)\left(\frac{g}{1000\, mg}\right)\left(\frac{1\, mol}{96.94\, g}\right)}{0.033\, mol/L \cdot atm} = 2.3_{44} x10^{-4}\ atm$$

$$\frac{n}{V} = \frac{P}{RT} = \frac{2.3_{44} x10^{-4}\ atm}{\left(0.08206 \frac{L \cdot atm}{mol \cdot K}\right)(21 + 273)K} = 9.7_{16} x10^{-6}\ \frac{mol}{L}$$

$$9.7_{16} x10^{-6}\ \frac{mol}{L}\left(\frac{96.94\, g}{mol}\right)\left(\frac{1x10^9\ ng}{g}\right) = \mathbf{9.4x10^5\ ng/L}$$

13.134 Molality is defined as moles of solute per kg of solvent, so 0.150 m means 0.150 mol $NaHCO_3$ per kg of water. The total mass of the solution would be 1 kg + 0.150 mol x molar mass of $NaHCO_3$.

$$\left(\frac{0.150\, mol\ NaHCO_3}{1\, kg + (0.150\, mol)(0.08401\, kg/mol)}\right)(0.250\, kg) = 3.70_{33} x10^{-2}\ mol\ NaHCO_3$$

$$(3.70_{33} x10^{-2}\ mol)(84.01\, g/mol) = 3.11\, g\ NaHCO_3$$

To make 250. g of a 0.150 m solution of $NaHCO_3$, **weigh 3.11 g $NaHCO_3$ and dissolve in 247 g water.**

13.138 Molarity is moles solute/L solution and molality is moles solute/kg solvent.

Multiplying molality by concentration of solvent in kg solvent per liter of solution gives molarity:

$$\frac{mol\ solute}{L\ solution} = \left(\frac{mol\ solute}{kg\ solvent}\right)\left(\frac{kg\ solvent}{L\ solution}\right) \text{ or } M = m\left(\frac{kg\ solvent}{L\ solution}\right)$$

For a very dilute solution the assumption that mass of solvent ≅ mass of solution is valid. Equation above becomes

$$M = m\left(\frac{kg\ solution}{L\ solution}\right) = m \, x \, d_{solution}$$

Thus, for very dilute solutions **molality x density = molarity**.

In an aqueous solution, the liters of solution have approximately the same value as the kg of solvent because the density of water is close to 1 kg/L, so m = M.

13.140 a) Assume a 100 g sample of urea.

$$mol\ C = (20.1\ g\ C)\left(\frac{1\ mol\ C}{12.01\ g\ C}\right) = 1.67_{36}\ mol\ C \quad mol\ N = (46.5\ g)\left(\frac{1\ mol\ N}{14.01\ g\ N}\right) = 3.31_{90}\ mol\ N$$

$$mol\ H = (6.7\ g\ H)\left(\frac{1\ mol\ H}{1.008\ g\ H}\right) = 6.6_{47}\ mol\ H \quad mol\ O = (26.7\ g)\left(\frac{1\ mol\ O}{16.00\ g\ O}\right) = 1.66_{88}\ mol\ O$$

Dividing each mole quantity by 1.6688 yields $C_{1.00}H_{4.00}N_{1.99}O$. The empirical formula for urea is $\mathbf{CH_4N_2O}$. Its empirical weight is 60.06 g/mol.

b) Use π = MRT to solve for the molarity of the urea solution. The solution molarity is related to the concentration expressed in % w/v by using the molar mass.

$$M = \frac{\pi}{RT} = \frac{2.04\ atm}{\left(0.08206\frac{L \cdot atm}{mol \cdot K}\right)(25+273)K} = 8.34_{22} x 10^{-2}\ M$$

$$\mathcal{M} = \frac{\%w/v}{M} = \frac{5.0\frac{g}{L}}{8.34_{22} x 10^{-2}\ \frac{mol}{L}} = 60.\ g/mol$$

Because the molecular weight equals the empirical weight, the molecular formula is also $\mathbf{CH_4N_2O}$.

13.142 a) The solubility of a gas is proportional to its partial pressure. If the solubility at a pressure of 0.10 MPa is 0.147 cm^3 then the solubility at a pressure that is 0.6 of 0.10 MPa will be 0.6 of 0.147 cm^3.

0.147 cm^3/0.10 MPa x 0.060 MPa = **0.088 cm^3 per gram of H_2O**

b) Henry's law states that $S_{gas} = k_H \times P_{gas}$.

k_H = 0.147 cm^3/g H_2O ÷ 0.10 MPa = **1.4_7 cm^3/g H_2O·MPa**

c) At 5.0 MPa Henry's law would give a solubility of

1.4_7 cm^3/g H_2O·MPa x 0.50 MPa = **0.74 cm^3/g H_2O.**

%error = [(0.825 – 0.735)/0.825](100) = **11%**

An error of approximately 11% occurs when Henry's law is used to calculate solubility at high pressures around 1 MPa.

13.146 a) Yes, the phases of water can still coexist at some temperature and can therefore establish equilibrium.

b) The triple point would occur at a lower pressure and temperature because the dissolved air solute lowers the vapor-pressure of the solvent. Figure 13.16 shows a phase diagram for such a solution.

c) Yes, this is possible because the gas-solid phase boundary exists below the new triple point.

d) No, the presence of the solute lowers the vapor pressure of the liquid.

CHAPTER 14

PERIODIC PATTERNS IN MAIN-GROUP ELEMENTS: BONDING, STRUCTURE, AND REACTIVITY

END-OF-CHAPTER PROBLEMS

14.2 a) Z_{eff}, the effective nuclear charge, increases across a period and decreases down a group.

b) As you move across a period, the atomic size decreases, the IE_1 increases and the EN increases, all as a result of the increased Z_{eff}.

14.3 a) Polarity is the molecular property that is responsible for the difference in boiling points between iodine monochloride and bromine. The stronger the intermolecular forces holding molecules together, the higher the boiling point. More polar molecules have stronger intermolecular forces and higher boiling points.

Molecular polarity is a result of the difference in electronegativity, an atomic property, of the atoms in the molecule.

b) ICl is a polar molecule while Br_2 is nonpolar. The dipole-dipole forces between polar molecules are stronger than the dispersion forces between nonpolar molecules. The boiling point of the polar ICl is higher than the boiling point of Br_2.

14.7 The leftmost element in a period is an alkali metal, Group 1A(1). As the other bonding atom moved to the right, the bonding would change from metallic to polar covalent to ionic. The atomic properties (IE, EA, and EN) of leftmost element and the "other" element become increasingly different, causing changes in the character of the bond between the two elements.

14.9 The most stable ion of elements in groups 1A, 2A, and 3A is a cation with electron configuration of the noble gas from the previous period. The most stable ion of the elements in groups 5A, 6A, and 7A is an anion with electron configuration of the noble gas of the same period. For ions in the *n*th period the outer level electrons for the cations are in the *n*-1 level whereas the outer level electrons for the anions are in the *n* level. Ions with *n*-1 outer level are smaller than ions with n as the outer level since the size of the electron cloud increases as the principle quantum number, *n*, increases. Thus, the cations will be significantly smaller than the anions in a period.

As an example, look at period 3. The cations Na^+, Mg^{2+}, and Al^{3+} have electron configurations $1s^22s^22p^6$, isoelectronic with neon. The anions P^{3-}, S^{2-}, and Cl^- have electron configurations $1s^22s^22p^63s^23p^6$, isoelectronic with argon. Figure 8.29 shows the cations decrease in size from 102 pm for Na^+ to 54 pm for Al^{3+}, then a jump to 212 pm for the first anion, P^{3-}. Moving across the anions, the ionic radius decreases from the 212 pm for P^{3-} to 181 pm for Cl^-. The irregularity in the trend occurs in the change from positively charged ions to negatively charged ions.

14.11 a) Element oxides are either ionic or covalent molecules. Elements with low EN (metals) form ionic oxides whereas elements with relatively high EN (nonmetals) from covalent oxides. The higher the EN of the element, the more covalent is the bonding in its oxide.

b) Oxide acidity increases across a period and decreases down a group, so increasing acidity directly correlates with increasing EN.

14.14 Table 12.5 shows that network covalent solids differ from molecular solids in that they are harder and have higher melting points than molecular solids. A network solid is held together by covalent bonds while molecules in molecular solids are held together by intermolecular forces. Covalent bonds are much stronger than intermolecular forces and require more energy to break. To melt a network solid, more energy (higher temperature) is required to break the covalent bonds than is required to break the intermolecular forces in the molecular solid, so the network solid has a higher melting point.

The covalent bonds also hold the atoms in a much more rigid structure than the intermolecular forces hold the molecules. Thus, a network solid is much harder than a molecular solid.

14.15 a) **Mg < Sr < Ba**. Atomic size increases down a group.
b) **Na < Al < P**. IE_1 increases across a period.
c) **Se < Br < Cl**. EN increases across a period and up a group.
d) **Ga < Sn < Bi**. Ga has one p electron, whereas Sn has two p electrons and Bi has three p electrons.

14.17 a) Chlorine and bromine will not form ionic compounds since they are both nonmetals that form anions.
b) Sodium and bromine will form an ionic compound because sodium is a metal that forms a cation and bromine is a nonmetal that forms an anion.
c) Phosphorus and selenium will not form ionic compounds since they are both nonmetals that form anions.
d) Hydrogen and barium will form barium hydride, an ionic compound. Hydrogen reacts with reactive metals, such as Ba, to form ionic hydrides and reacts with nonmetals to form covalent hydrides.

14.19 Recall in the Lewis structure model, bond order is the number of electron pairs shared between 2 bonded atoms (bond order is defined differently in molecular orbital theory).

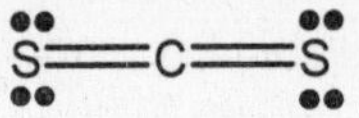

14.21 Bond length increases as the size of the atoms in the bond increases. Silicon, carbon and germanium are all in group 4A and chlorine, fluorine, and bromine are all in group 7A. Size increases down a group, so carbon is the smallest and germanium the largest in group 4A and fluorine is the smallest and bromine the largest in group 7A. The order of increasing bond length follows increasing size of atoms: $\mathbf{CF_4 < SiCl_4 < GeBr_4}$.

14.23 a) $\mathbf{Na^+ < F^- < O^{2-}}$. In an isoelectronic series, ionic size decreases due to an increase in Z_{eff}.
b) $\mathbf{Cl^- < S^{2-} < P^{3-}}$. This series is also isoelectronic. Recall that anions are larger than their respective atoms due to electron repulsion; the more electrons added the greater the repulsion and the larger space they will occupy.

14.25 The difference in the three species is in the number of electrons in the outer level. As the number of electrons increase, electron repulsion increases which increases the size of the electron cloud. Thus, O^{2-} will be the largest of the three since it contains the most electrons. In order of increasing size the rank is $O < O^- < O^{2-}$.

14.27 a) Metals form basic oxides whereas nonmetals form acidic oxides. **Calcium oxide** will form the more basic oxide.

$CaO(s) + H_2O(l) \rightarrow Ca^{2+}(aq) + 2OH^-(aq)$

$SO_3(g) + H_2O(l) \rightarrow H_2SO_4(aq) \rightarrow 2H^+(aq) + SO_4^{2-}(aq)$

b) Oxide basicity increases down a group as the element becomes less electronegative. **Barium oxide** produces the more basic solution.

c) Oxide basicity decreases across a period because the element becomes less electronegative. **Carbon dioxide** produces the more basic solution.

d) **Potassium oxide** produces the more basic solution because K is more metallic (lower EN) than P.

14.29 Covalent character increases as the electronegativity difference between the atoms decreases.

a) **Lithium chloride** will have more covalent character than potassium chloride because Li is more electronegative than K (EN increases up a group) so the EN difference with chlorine will be smaller for LiCl than for KCl.

b) **Phosphorus trichloride** will have more covalent character than aluminum chloride because P is more electronegative than Al (EN increases across a period) so the EN difference with chlorine will be smaller for PCl_3 than for $AlCl_3$.

c) **Nitrogen trichloride** will have more covalent character than arsenic trichloride since EN of N is closer to the EN of Cl. To answer this question, the trends in EN are not sufficient to answer the question since the EN needs to be compared with that of chlorine. From Figure 9.16, the EN values are 3.0 for Cl, 3.0 for N, and 2.0 for As. ΔEN for NCl_3 is 0 and ΔEN for $AsCl_3$ is 1.0.

14.31 a) **Ar < Na < Si**. Ar forms an atomic solid held together only by dispersion forces. Na is a metallic solid and Si is a network covalent solid, requiring the most energy to separate.

b) **Cs < Rb < Li**. The heat of fusion, ΔH_{fus}, is the amount of energy needed to melt the solid. As melting point increases, so does ΔH_{fus}. All three are atomic solids so the intermolecular forces are dispersion forces. Li has the highest melting point because smaller atoms attract delocalized electrons more strongly and produce stronger metallic bonds.

14.33 Ionization energy is defined as the energy to remove the outermost electron from an atom. The further the outermost electron is from the nucleus, the less energy is required to remove it from the attractive force of the nucleus. In hydrogen the outermost electron is in the n = 1 level and in lithium the outermost electron is in the n = 2 level. So, the outermost electron in lithium requires less energy to remove resulting in a lower ionization energy.

14.35 a) NH_3 will H-bond because H is attached to one of three atoms (N, O, and F) necessary for H-bonding.

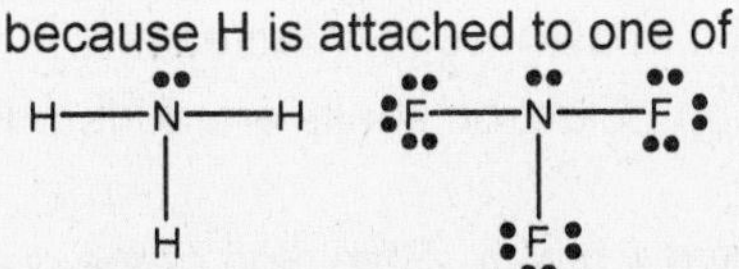

b) CH_3CH_2OH will H-bond. CH_3OCH_3 has no O–H bonds, only C–H bonds.

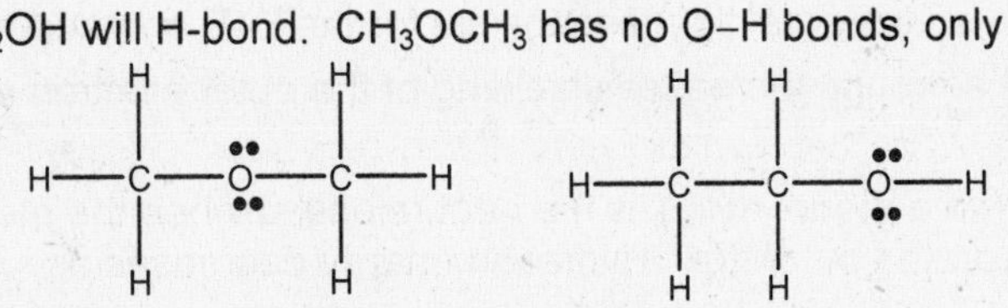

14.37 a) $2Al(s) + 6HCl(aq) \rightarrow 2AlCl_3(aq) + 3H_2(g)$

b) $LiH(s) + H_2O(l) \rightarrow LiOH(aq) + H_2(g)$

14.39 a) Oxidation numbers are determined by the set of values in Table 4.3. Hydrogen in ionic hydrides (i.e. combined with a metal or B) has an O.N. = -1.

Na = +1, B = +3, H = -1 for $NaBH_4$

Al = +3, B = +3, H = -1 for $Al(BH_4)_3$

Li = +1, Al = +3, H = -1 for $LiAlH_4$

b) The polyatomic ion in $NaBH_4$ is $[BH_4]^-$. Boron is the central atom and has four surrounding electron groups; therefore its shape is **tetrahedral**.

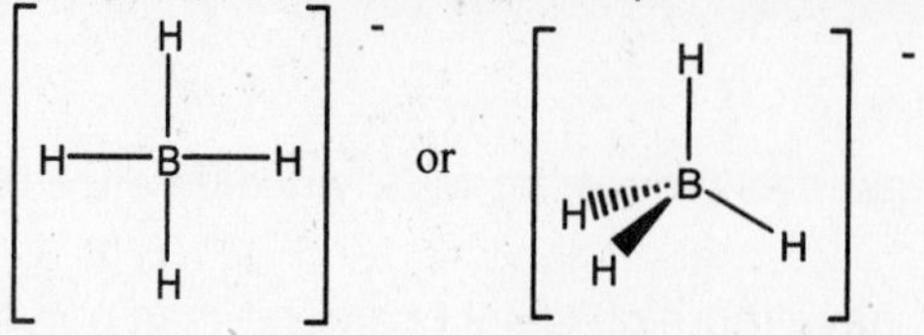

14.42 For period 2 elements in the first four groups, the number of covalent bonds equals the number of electrons in the outer level, so it increases from 1 covalent bond for lithium in group 1A(1) to 4 covalent bonds for carbon in group 4A(4). For the rest of period 2 elements, the number of covalent bonds equals the difference between 8 and the number of electrons in the outer level. So for nitrogen, 8 - 5 = 3 covalent bonds; for oxygen, 8 - 6 = 2 covalent bonds; for fluorine, 8 – 7 = 1 covalent bond; and for neon, 8 – 8 = 0, no bonds.

For elements in higher periods the same pattern exists but with exceptions for groups 3A(13) to 7A(17) when an expanded octet allows for more covalent bonds.

14.45 a) E must have an oxidation of +3 to form an oxide E_2O_3 or fluoride EF_3. E is in group **3A(13) or 3B(3)**.

b) If E were in Group 3B(3), the oxide and fluoride would have more ionic character because 3B elements have lower EN than 3A element. The Group 3B(3) oxides would be more basic.

14.48 a) Alkali metals generally lose electrons (act as reducing agents) in their reactions.

b) Alkali metals have relatively low ionization energies meaning they easily lose the outermost electron. The electron configuration of alkali metals have one more electron than a noble gas configuration so losing an electron gives a stable electron configuration.

c) $2Na(s) + 2H_2O(l) \rightarrow 2Na^+(aq) + 2OH^-(aq) + H_2(g)$

$2Na(s) + Cl_2(g) \rightarrow 2NaCl(s)$

14.50 a) Density increases down a group. The increasing atomic size (volume) is not offset by the increasing size of the nucleus (mass), so m/V increases.

b) Ionic size increases down a group. Electron shells are added down a group, so both atomic and ionic size increase.

c) E–E bond energy decreases down a group. Shielding of the outer electron increases as the atom gets larger, so the attraction responsible for the E–E bond decreases.

d) IE_1 decreases down a group. Increased shielding of the outer electron accounts for the decreasing IE_1.

e) ΔH_{hydr} decreases down a group. ΔH_{hydr} is the heat released when the metal salt dissolves in, or is hydrated by, water. Hydration energy decreases as ionic size increases.

14.52 Peroxides are oxides in which oxygen has a –1 oxidation state. Sodium peroxide has the formula Na_2O_2 and is formed from the elements Na and O_2.

$$2Na(s) + O_2(g) \rightarrow Na_2O_2(s)$$

14.54 The problem specifies that an alkali halide is the desired product. The alkali metal is K (comes from potassium carbonate, $K_2CO_3(s)$) and the halide is I (comes from hydroiodic acid, HI(aq)). Treat the reaction as a double displacement reaction.

$$K_2CO_3(s) + 2HI(aq) \rightarrow 2KI(aq) + H_2CO_3(aq)$$

However, $H_2CO_3(aq)$ is unstable and decomposes to $H_2O(l)$ and $CO_2(g)$, so the final reaction is:

$$K_2CO_3(s) + 2HI(aq) \rightarrow 2KI(aq) + H_2O(l) + CO_2(g)$$

14.58 Metal atoms are held together by metallic bonding, a sharing of valence electrons. Alkaline earth metal atoms have one more valence electron than alkali metal atoms, so the number of electrons shared is greater. Thus, metallic bonds in alkaline earth metals are stronger than in alkali metals. Melting requires weakening the metallic bonds. To weaken the stronger alkaline earth metal bonds requires more energy (higher temperature) than to weaken the alkali earth metal bonds.

First ionization energy, density, and boiling points will be larger for alkaline earth metals than for alkali metals.

14.59 a) A base forms when a basic oxide, such as CaO, is added to water.

$CaO(s) + H_2O(l) \rightarrow Ca(OH)_2(s)$

b) Alkaline earth metals reduce O_2 to form the oxide.

$2Ca(s) + O_2(g) \rightarrow 2CaO(s)$

14.62 Beryllium is generally the exception to properties exhibited by other alkaline earth elements.

a) Here Be does not behave like other alkaline earth metals: $BeO(s) + H_2O(l) \rightarrow NR$.

The oxides of alkaline earth metals are strongly basic, but BeO is amphoteric. BeO will react with both acids and bases to form salts, but an amphoteric substance does not react with water.

b) Here Be does behave like other alkaline earth metals:

$$BeCl_2(l) + 2Cl^-(solvated) \rightarrow BeCl_4^{2-}(solvated)$$

Each chloride ion donates a lone pair of electrons to form a covalent bond with the Be in $BeCl_2$. Metal ions form similar covalent bonds with ions or molecules containing a lone pair of electrons. The difference in beryllium is that the orbital involved in the bonding is a p orbital whereas in metal ions it is usually the d orbitals that are involved. See problem 10.23 for a Lewis dot depiction of the covalent bond in $BeCl_4^{2-}$.

14.65 The electron removed in Group 2A(2) atoms is from the outer level s orbital whereas in Group 3A(13) atoms the electron is from the outer level p orbital. For example, the electron configuration for Be is $1s^22s^2$ and for B is $1s^22s^22p^1$. It is easier to remove the p electron of B than the s electron of Be, because the energy of a p orbital is slightly higher than that of the s orbital from the same level. Even though the atomic size decreases from increasing Z_{eff}, the IE decreases from 2A(2) to 3A(13).

14.66 a) Compounds of Group 3A(13) elements like boron, have only six electrons in their valence shell when combined with halogens to form three bonds. Having six electrons, rather than an octet, results in an "electron deficiency."

b) As an electron deficient central atom, B is trigonal planar. Upon accepting an electron pair to form a bond, the shape changes to tetrahedral.

$BF_3(g) + NH_3(g) \rightarrow F_3B\text{–}NH_3(g)$

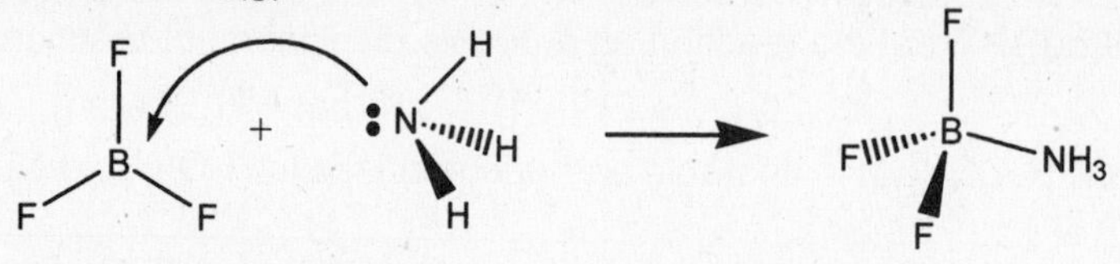

$B(OH)_3(aq) + OH^-(aq) \rightarrow B(OH)_4^-(aq)$

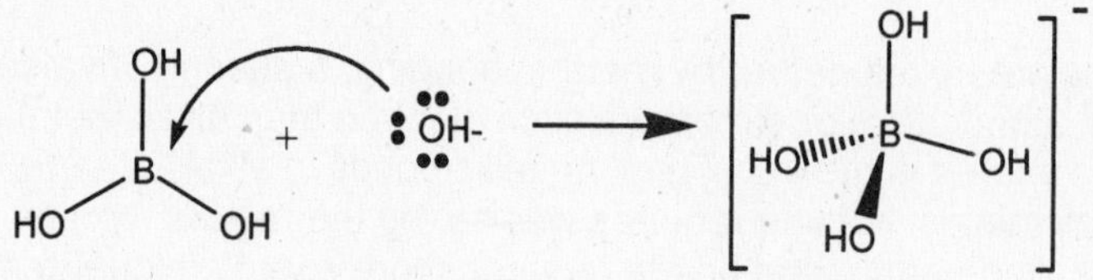

14.68 Acidity of oxides increases up a group: $In_2O_3 < Ga_2O_3 < Al_2O_3$.

14.70 Halogens typically have a –1 oxidation state in metal-halide combinations, so the apparent oxidation state of Tl = +3. However, the anion I_3^- combines with Tl in the +1 oxidation state. The anion I_3^- has 22 valence electrons, a general formula of AX_2E_3 and is a linear molecule with bond angles = 180°. $(Tl^{3+})(I^-)_3$ does not exist because of the low strength of the Tl-I bond.

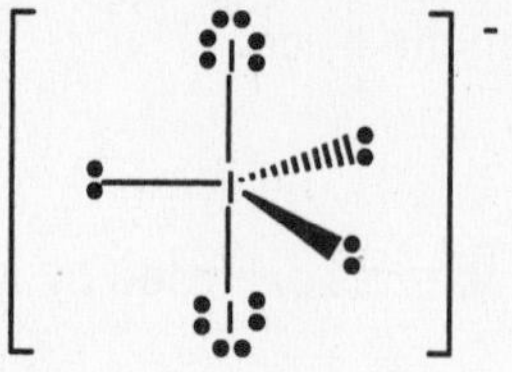

14.73 a) The indium(III) ion will combine with the As^{3-} ion in a 1:1 ratio.

$$In(s) + As(s) \rightarrow InAs(s)$$

b) Gallium antimonide would consist of one Ga^{3+} ion and one Sb^{3-} ion. When exposed to moist air, the gallium antimonide would decompose to gallium hydroxide and antimony hydride.

$$GaSb(s) + 3H_2O(g) \rightarrow Ga(OH)_3(s) + SbH_3(g)$$

14.75 Figure 14.8 shows the analogous structure of the dimer Al_2Cl_6. The $GaBr_3$ dimer, Ga_2Br_6, has two central Ga atoms and two central Br atoms. Each central atom is surrounded by 4 electron groups, so the geometry around each central atom is tetrahedral with ideal bond angles of 109.5°. The lone pairs on the central Br atoms would compress the Ga-Br-Ga angles slightly. Verify that the structure should and does contain 48 valence electrons.

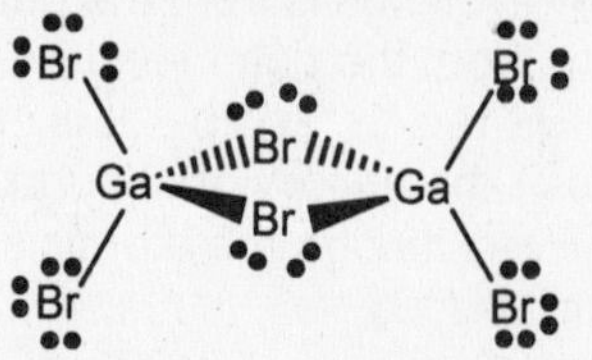

14.76 Oxide basicity is greater for the oxide of a metal atom. Tin(IV) oxide would be more basic than carbon dioxide since tin atoms have more metallic character than carbon atoms.

14.78 a) IE_1 generally decreases down a group.

b) The increase in Z_{eff} from Si to Ge is larger than the increase from C to Si because more protons have been added. Between C and Si an additional 8 protons have been added whereas between Si and Ge an additional 18 (includes the protons for the d-block) protons have been added. Same type of change occurs going from Sn to Pb when the 14 f-block protons are added.

c) Group 3A(13) would show greater deviations because the single p electron receives no shielding effect offered by other p electrons.

14.81 Atomic size increases moving down a group. As atomic size increases, ionization energy decreases so that it is easier to form a positive ion. An atom that is easier to ionize exhibits greater metallic character.

14.83 a) The silicate building unit is the tetrahedral $-SiO_4-$ unit. The structure, arranged in a ring (cyclic), has 96 valence electrons.

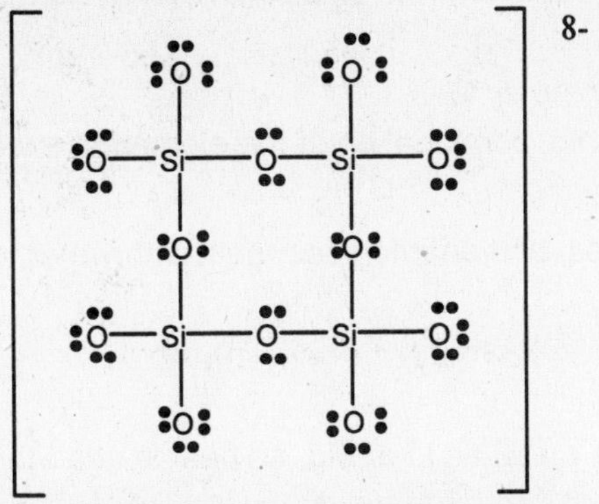

b) Carbon can exist in a 3-member or 4-member ring.

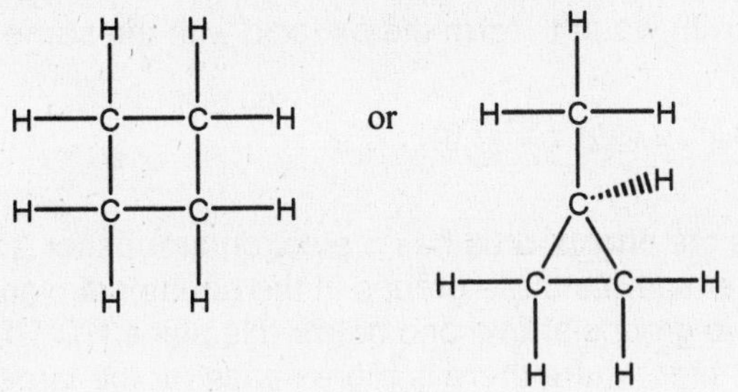

14.85 The $(CH_3)_2Si(OH)_2$ molecules will form bonds between silicon and oxygen to make the polymer. The structure could be drawn as a flat two-dimensional structure as well.

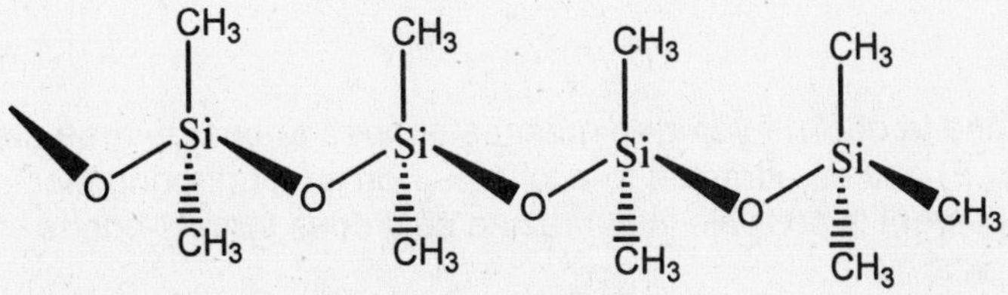

14.88 a) diamond, a network covalent solid of carbon

b) $CaCO_3$ (brands that use this compound as an antacid also advertise them as an important source of calcium)

c) CO_2 is the most widely known greenhouse gas; CH_4 is also implicated.

d) silicone, $[(CH_3)_2SiO]_n$

e) carborundum, SiC

f) CO is formed in combustion when the amount of O_2 (air) is limited.

g) Pb (in gasoline as an anti-knocking additive, in old plumbing as lead solder, and in paint as a pigment)

14.92 a) In Group 5A(15) all elements except bismuth have a range of oxidation states from –3 to +5.

b) For nonmetals the range of oxidation states is from the lowest at group(A) number – 8, which is 5 – 8 = -3 for Group 5A, to the highest equal to the group(A) number, which is +5 for Group 5A.

14.95 Arsenic is less electronegative than phosphorus, which is less electronegative than nitrogen. Therefore, arsenic acid is the acid and nitric acid is the strongest. Order of increasing strength: $\mathbf{H_3AsO_4 < H_3PO_4 < HNO_3}$.

14.97 a) With excess oxygen arsenic will form the oxide with arsenic in its highest possible oxidation state, +5.

$$4As(s) + 5O_2(g) \rightarrow 2As_2O_5(s)$$

b) Trihalides are formed by direct combination of the elements (except N).

$$2Bi(s) + 3I_2(s) \rightarrow 2BiI_3(s)$$

c) Metal phosphides, arsenides, and antimonides react with water to form Group 5A hydrides.

$$Ca_3As_2(s) + 6H_2O(l) \rightarrow 2AsH_3(g) + 3Ca(OH)_2(s)$$

14.99 a) Aluminum is not as active a metal as Li or Mg, so heat is need to drive this reaction.

$$N_2(g) + 2Al(s) \xrightarrow{\Delta} 2AlN(s)$$

b) The 5A halides react with water to form the oxoacid with the same oxidation state as the original halide.

$$PF_5(g) + 4H_2O(l) \rightarrow H_3PO_4\ (aq) + 5HF(g)$$

14.101 From the Lewis structure the phosphorus has 5 electron groups for a trigonal bipyramidal molecular shape. In this shape the three groups in the equatorial plane have greater bond angles (120°) than the two groups above and below this plane (90°). The chlorine atoms would occupy the planar sites where there is more space for the larger atoms.

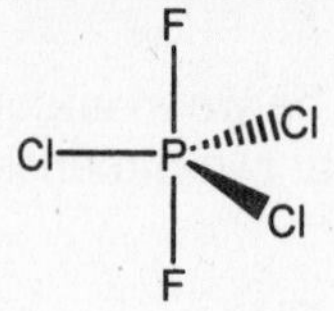

14.105 For a hydrogen bond to occur, hydrogen must be bonded to one of three electronegative elements (N, O, or F) and be attracted to a lone pair on a neighboring electronegative atom. This arrangement is possible in hydrazine so it does form H-bonds.

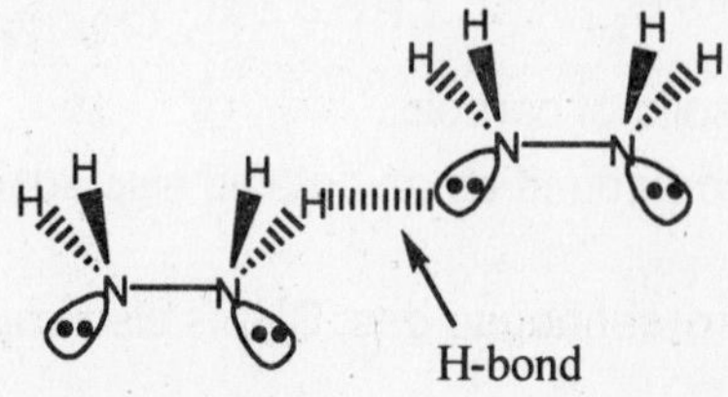

14.107 a) Thermal decomposition of KNO_3 at low temperatures

$$2KNO_3(s) \xrightarrow{\Delta} 2KNO_2(s) + O_2(g)$$

b) Thermal decomposition of KNO_3 at high temperatures

$$4KNO_3(s) \xrightarrow{\Delta} 2K_2O(s) + 2N_2(g) + 5O_2(g)$$

14.109 a) Both groups have elements that range from gas to metalloid to metal. Thus their boiling points and conductivity vary in similar ways down a group.

b) The degree of metallic character and methods of bonding vary in similar ways down a group.

c) Both P and S have allotropes and both bond covalently with almost every other nonmetal.

d) Both N and O are diatomic gases at normal temperatures and pressures. Both N and O have very low melting and boiling points.

e) Oxygen, O_2, is a reactive gas whereas nitrogen, N_2, is not. Nitrogen can exist in multiple oxidation states whereas oxygen has 2 oxidation states.

14.111 a) To decide what type of reaction will occur, examine the reactants. Notice that sodium hydroxide is a strong base. Is the other reactant an acid? If we separate the salt, sodium hydrogen sulfate, into the two ions, Na^+ and HSO_4^- then it is easier to see the hydrogen sulfate ion as the acid. The sodium ions could be left out for the net ionic reaction.

$$NaHSO_4(aq) + NaOH(aq) \rightarrow Na_2SO_4(aq) + H_2O(l)$$

b) As mentioned in the book, hexafluorides are known to exist for sulfur. These will form when excess fluorine is present.

$$S_8(s) + 24F_2(g) \rightarrow 8SF_6(g)$$

c) Group 6A(16) elements except oxygen form hydrides in the following reaction.

$$FeS(s) + 2HCl(aq) \rightarrow H_2S(g) + FeCl_2(aq)$$

d) Tetraiodides, but not hexaiodides of tellurium are known.

$$Te(s) + 2I_2(s) \rightarrow TeI_4(s)$$

14.113 a) Se is a non-metal; its oxide is **acidic**.

b) N is a non-metal; its oxide is **acidic**.

c) K is a metal; its oxide is **basic**.

d) Be is an alkaline earth metal, but all of its bonds are covalent; its oxide is **amphoteric**.

e) Ba is a metal; its oxide is **basic**.

14.115 Acid strength increases down a group: $H_2O < H_2S < H_2Te$.

14.118 a) SF_6, sulfur hexafluoride

b) O_3, ozone

c) SO_3, sulfur trioxide (+6 ox state)

d) SO_2, sulfur dioxide

e) SO_3, sulfur trioxide

f) $Na_2S_2O_3{\cdot}5H_2O$, sodium thiosulfate pentahydrate

g) H_2S, hydrogen sulfide

14.120 $S_2F_{10}(g) \rightarrow SF_4(g) + SF_6(g)$

ON of S in S_2F_{10}: -(10 x –1 for F) ÷ 2 = +5

ON of S in SF_4: -(4 x –1 for F) = +4

ON of S in SF_6: -(6 x –1 for F) = +6

14.122 a) Bonding with other halogens: +1, +3, +5, +7

Bonding with other non-halogens: -1

b) The electron configuration of Cl is [Ne]$3s^2 3p^5$. By adding on electron, Cl achieves a octet similar to the noble gas Ar. By forming covalent bonds, Cl completes or expands its octet by maintaining its electrons paired in bonds or lone pairs.

c) Fluorine only forms the –1 oxidation state because its small size and no access to d orbitals prevents it from forming multiple covalent bonds. Fluorine's high electronegativity also prevents it from sharing its electron.

14.124 a) Cl—Cl bond is stronger than Br—Br bond since the chlorine atoms are smaller than the bromine atoms so the shared electrons are held more tightly by the two nuclei.

b) Br—Br bond is stronger than I—I bond since the bromine atoms are smaller than the iodine atoms.

c) Cl—Cl bond is stronger than F—F bond. The fluorine atoms are smaller than the chlorine but they are so small that electron-electron repulsion of the lone pairs decreases the strength of the bond.

14.126 a) A substance that disproportionates serves as both an oxidizing and reducing agent. Assume that OH^- serves as the base. Write the reactants and products of the reaction, and balance like a redox reaction.

Gain 1 e^-

$$\overset{0}{Br_2} + OH^- \rightarrow \overset{-1}{Br^-} + \overset{+5}{Br}O_3$$

Lose 5 e^-

To balance the gain and loss of electrons, add the coefficient "5" to Br^-, balance bromines and balance remaining oxygen with H_2O.

$$\mathbf{3Br_2(l) + 6OH^-(aq) \rightarrow 5Br^-(aq) + BrO_3^-(aq) + 3H_2O(l)}$$

b) In the presence of base, instead of water, only the oxy*anion* (not oxo*acid*) and fluoride (not *hydro*fluoride) form. No oxidation or reduction takes place, because Cl maintains its +5 oxidation state and F maintains its –1 oxidation state.

$$\mathbf{ClF_5(l) + 6OH^-(aq) \rightarrow 5F^-(aq) + ClO_3^-(aq) + 3H_2O(l)}$$

14.127 a) $2Rb(s) + Br_2(l) \rightarrow 2RbBr(s)$

b) $I_2(s) + H_2O(l) \rightarrow HI(aq) + HIO(aq)$

c) $Br_2(l) + 2I^-(aq) \rightarrow I_2(s) + 2Br^-(aq)$

d) $CaF_2(s) + H_2SO_4(l) \rightarrow CaSO_4(s) + 2HF(g)$

14.129 Acid strength increases with increasing electronegativity of central atom and increasing number of oxygens. Therefore, $\mathbf{HIO < HBrO < HClO < HClO_2}$.

14.132 a) In the reaction between NaI and H_2SO_4 the oxidation states of iodine and of sulfur change so the reaction is an **oxidation-reduction** reaction.

b) The reducing ability of X^- increases down the group since the larger the ion the more easily it loses an electron. So I^- is more easily oxidized than Cl^-.

c) Some acids, such as HCl, are not oxidizers, so substituting an non-oxidizing acid for H_2SO_4 would produce HI.

14.134 Helium is the second most abundant element in the universe. Argon is the most abundant noble gas in Earth's atmosphere, the third most abundant constituent after N_2 and O_2.

14.136 Whether a boiling point is high or low is a result of the strength of the forces between particles. Atoms of noble gases are held together by dispersion forces, the weakest of all the intermolecular forces. Only a relatively low temperature is required for the atoms to have enough kinetic energy to break away from the attractive force of other atoms and go into the gas phase. The boiling points are so low that all the noble gases are gases at room temperature.

14.139 a) $$2\overset{+1}{Na}\overset{-1}{H}(s) + 2\overset{+4}{S}\overset{-2}{O}_2(l) \rightarrow \overset{+1}{Na}_2\overset{+3}{S}_2\overset{-2}{O}_4(s) + \overset{0}{H}_2(g)$$

b) First determine the limiting reactant. Because the $NaH:H_2$ and $SO_2:H_2$ molar relationships are the same, the reactant with the larger molar mass will produce less product and limit the reaction. Since $\mathcal{M}_{NaH}$ = 24.00 g/mol and $\mathcal{M}_{SO2}$ = 64.07 g/mol, SO_2 is the limiting reactant. Use the stoichiometric relationships and the ideal gas law to find the volume of $H_2(g)$ generated. (1 metric ton = 10^6 g)

$$mol\ H_2 = \left(10^6\ g\ SO_2\right)\left(\frac{1\ mol\ SO_2}{64.07\ g\ SO_2}\right)\left(\frac{1\ mol\ H_2}{2\ mol\ SO_2}\right) = 7.803x10^3\ mol\ H_2$$

$$V = \frac{nRT}{P} = \frac{\left(7.803x10^3\ mol\right)\left(0.08206\ L\cdot atm/mol\cdot K\right)\left((273-45)K\right)}{\left(758\ torr \big/ 760\ torr/atm\right)} = \mathbf{1.46x10^5\ L\ H_2}$$

14.141 Calculation of energy from wavelength: $\Delta E = hc/\lambda$.

$$\Delta E = \frac{\left(6.62608x10^{-34}\ J\cdot s\right)\left(2.99792x10^8\ m/s\right)}{589.2x10^{-9}\ m} = \mathbf{3.371x10^{-19}\ J}$$

14.145 a) No oxidation state changes occur in this reaction; S maintains a +4 oxidation state. The prefix "hexa-" denotes 6, in this case 6 water of hydration. Start by balancing H, then balance O. $\mathbf{MgCl_2\text{-}6H_2O(s) + 6SOCl_2(l) \rightarrow 6SO_2(g) + 12HCl(g) + MgCl_2(s)}$

b) Sulfur is the central atom because it is the least electronegative element. The structure has 26 valence e^-. Following the standard method for drawing Lewis structures (single bonds from central to surrounding atoms, remaining electrons distributed to surrounding atoms first) give structure **I**. An analysis of formal charge shows that structure **II** is more stable.

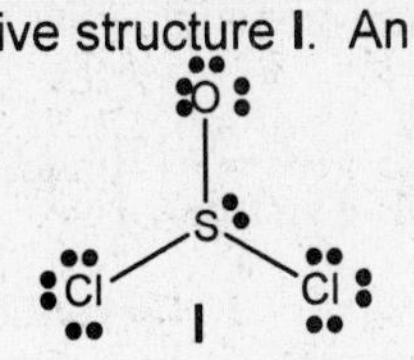

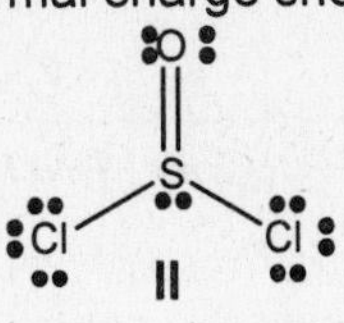

$FC_S = 6 - (2 + ½(6)) = +1$ $\qquad$ $FC_S = 6 - (2 + ½(8)) = 0$

$FC_O = 6 – (6 + ½(2)) = -1$ $\quad$ $FC_O = 6 – (4 + ½(4)) = 0$

$FC_{Cl} = 7 – (6 + ½(2)) = 0$ $\quad$ $FC_{Cl} = 7 – (6 + ½(2)) = 0$

14.146 a) Second ionization energies for alkali metals are high because the electron being removed is from the next lower energy level and electrons in a lower level are more tightly held by the nucleus.

b) Reaction is $2CsF_2(s) \rightarrow 2CsF(s) + F_2(g)$

This reaction can be obtained by combining the formation reaction of CsF with the reverse of the formation reaction of CsF_2 both multiplied by 2:

$$2Cs(s) + F_2(g) \rightarrow 2CsF(s) \qquad \Delta H = 2 \times (-530\ kJ)$$
$$2CsF_2(s) \rightarrow 2Cs(s) + 2F_2(g) \qquad \Delta H = 2 \times (+125\ kJ)$$
$$2CsF_2(s) \rightarrow 2CsF(s) + F_2(g) \qquad \Delta H = -810\ kJ$$

810 kJ of energy are released when 2 moles of CsF_2 convert to 2 moles of CsF, so heat of reaction for one mole of CsF is **–405 kJ/mol**.

14.148 a) Empirical formula HNO has molar mass 31.02, so hyponitrous acid with molar mass twice 31.02 would have molecular formula twice the empirical formula, $H_2N_2O_2$. And the molecular formula of nitroxyl would be the same as the empirical formula, HNO, since the molar mass of nitroxyl is the same.

b) Lewis structures:

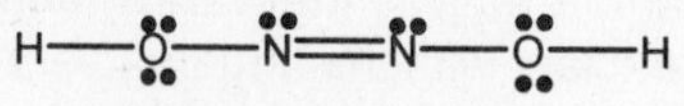

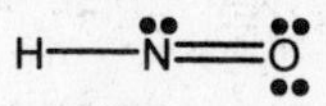

hyponitrous acid $\qquad$ nitroxyl

c) In both hyponitrous acid and nitroxyl the nitrogens are surrounded by 3 electron groups, so the *electron arrangement* is trigonal planar and the molecular shape is **bent**.

hyponitrous acid $\qquad$ nitroxyl

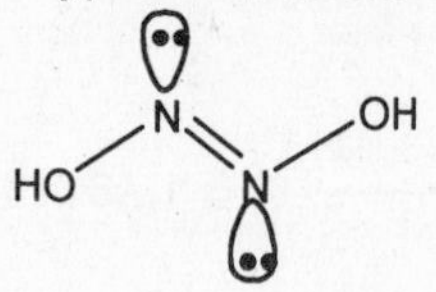

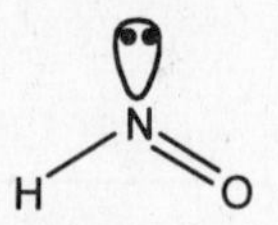

d) The hyponitrite ion would have the formula $^-ON{=}NO^-$.

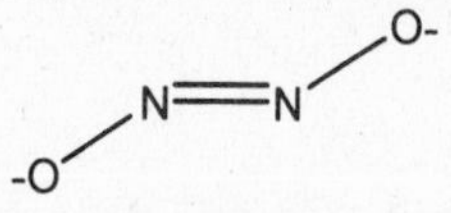

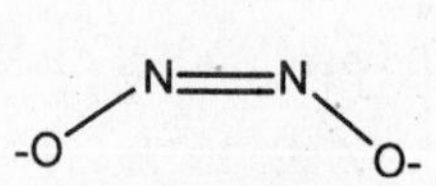

trans isomer $\qquad$ cis isomer

14.150 a) The three steps of the Ostwald process are given in section 14.7:

oxidation of NH_3 to NO: $\quad 4NH_3(g) + 5O_2(g) \rightarrow 4NO(g) + 6H_2O(g)$

oxidation of NO to NO_2: $\quad 2NO(g) + O_2(g) \rightarrow 2NO_2(g)$

disproportionation: $\quad 3NO_2(g) + H_2O(l) \rightarrow 2HNO_3(l) + NO(g)$

b) If NO is not recycled, the reaction steps proceed as written above. The molar relationships for each reaction yield NH_3 consumed per mole HNO_3 produced:

$$\frac{mol\ NH_3\ consumed}{mol\ HNO_3\ produced} = \left(\frac{3\ mol\ NO_2}{2\ mol\ HNO_3}\right)\left(\frac{2\ mol\ NO}{2\ mol\ NO_2}\right)\left(\frac{4\ mol\ NH_3}{4\ mol\ NO}\right) = 1.5\ mol\ NH_3\ /\ mol\ HNO_3$$

c) The goal is to find the mass of HNO_3 produced, which can be converted to volume of aqueous solution using the density and mass percent. To find mass of HNO_3, determine the number of moles of NH_3 present in 1 m^3 of gas mixture (ideal gas law) and convert moles of NH_3 to moles HNO_3 using the mole ratio in b). Convert moles to grams using the molar mass of HNO_3 and convert to volume.

Step 1: Moles of NH_3

V = (1 m^3)(1000 L/m^3)(0.10) = 1.0 x 10^2 L

T = 273 + 850.°C = 1123 K

P = 5.0 atm

$$mol\ NH_3 = \frac{PV}{RT} = \frac{(5.0\ atm)(1.0x10^2\ L)}{(0.08206\ L\cdot atm/mol\cdot K)(1123K)} = 5.4_{26}\ mol\ NH_3$$

Step 2: Moles of HNO_3 (reaction is 96% efficient in first step)

$$mol\ HNO_3 = (5.4_{26}\ mol\ NH_3)\left(\frac{1\ mol\ HNO_3}{1.5\ mol\ NH_3}\right)(0.96) = 3.4_{72}\ mol\ HNO_3$$

Step 3: Mass of HNO_3

$$mass\ HNO_3 = (3.4_{72}\ mol\ HNO_3)\left(\frac{63.02\ g\ HNO_3}{mol\ HNO_3}\right) = 2.1_{88}x10^2\ g\ HNO_3$$

Step 4: Volume of HNO_3 solution

$$vol\ HNO_3 = (2.1_{88}x10^2\ g\ HNO_3)\left(\frac{100\ g\ sol'n}{60.\ g\ HNO_3}\right)\left(\frac{1\ mL\ sol'n}{1.37\ g\ sol'n}\right) = 2.7x10^2\ \text{mL sol'n}$$

Therefore, **2.7x10^2 mL HNO_3 is formed per m^3** of gas mixture at the given conditions.

14.154 Carbon monoxide and carbon dioxide would be formed from the reaction of coke (carbon) with the oxygen in the air. The nitrogen in the producer gas would come from the nitrogen already in the air. So, the calculation of mass of product is based on the mass of CO and CO_2 that can be produced from 1.75 metric tons of coke.

Since 5 grams of CO_2 is produced for each 25 grams of CO we can calculate a mass ratio of carbon that produces each:

$$\frac{mass\ C\ forms\ CO}{mass\ C\ forms\ CO_2} = \frac{(25\ g\ CO)\left(\frac{12.01\ g\ C}{28.01\ g\ CO}\right)}{(5\ g\ CO_2)\left(\frac{12.01\ g\ C}{44.01\ g\ CO_2}\right)} = \frac{7.86}{1}$$

Using the ratio of carbon that reacts as 7.86:1, the total C reacting is 7.86 + 1 = 8.86. The mass fraction of the total carbon that produces CO is 7.86/8.86 and the mass fraction of the total carbon reacting that produces CO_2 is 1.00/8.86. To find the mass of CO produced from 1.75 metric tons of carbon with an 87% yield:

$$(1.75\ metric\ tons\ C)\left(\frac{7.86}{8.86}\right)\left(\frac{28.01\ metric\ tons\ CO}{12.01\ metric\ tons\ C}\right)(0.87) = 3.15\ metric\ tons\ CO$$

And to find the mass of CO_2 produced:

$$(1.75\ metric\ tons\ C)\left(\frac{1.00}{8.86}\right)\left(\frac{44.01\ metric\ tons\ CO}{12.01\ metric\ tons\ C}\right)(0.87) = 0.630\ metric\ tons\ CO_2$$

The mass of CO and CO_2 represent a total of 30% of the mass of the producer gas, so the total mass would be (3.15 + 0.63) ÷ 0.3 = **13 metric tons.**

14.156 In a disproportionation reaction, a substance acts as both a reducing agent and oxidizing agent because an atom within the substance reacts to form atoms with higher and lower oxidation states.

a) No. The *reverse* direction would be a disproportionation reaction because a single substance both oxidizes and reduces.

b) Yes. ClO_2 disproportionates.

$$2\overset{+4}{Cl}O_2 + H_2O \rightarrow H\overset{+5}{Cl}O_3 + H\overset{+3}{Cl}O_2$$

c) Yes. Cl_2 disproportionates.

$$\overset{0}{Cl_2} + 2NaOH \rightarrow Na\overset{-1}{Cl} + Na\overset{+1}{Cl}O + H_2O$$

d) Yes. NH_4NO_2 disproportionates.

$$\overset{-3}{N}H_4\overset{+3}{N}O_2 \rightarrow \overset{0}{N_2} + 2H_2O$$

e) Yes. MnO_4^{2-} disproportionates.

$$3\,\overset{+6}{Mn}O_4^{2-} + 2H_2O \rightarrow 2\,\overset{+7}{Mn}O_4^- + \overset{+4}{Mn}O_2 + 4OH^-$$

f) Yes. AuCl disproportionates.

$$3\,\overset{+1}{Au}Cl \rightarrow \overset{+3}{Au}Cl_3 + 2\,\overset{0}{Au}$$

14.158 a) Group **5A(15)** elements have 5 valence electrons and typically form three bonds with a lone pair to complete the octet.

b) Group **7A(17)** elements readily gain an electron causing the other reactant to be oxidized. They form monatomic ions of formula X^- and oxoanions, such as ClO^-.

c) Group **6A(16)** elements have six valence electrons and gain a complete octet by forming two covalent bonds.

d) Group **1A(1)** elements are the strongest reducing agents because they most easily lose an electron. As the least electronegative and most metallic of the elements, they are not likely to form covalent bonds. Group **2A(2)** have similar characteristics.

e) Group **3A(13)** elements have only three valence electrons to share in covalent bonds, but with an empty orbital they can accept an electron pair from another atom.

f) Group **8A(18)**, the noble gases, are the least reactive of all the elements.

14.160 a) The Lewis structure has 72 valence e^-. An initial structure, **I**, is obtained; however a comparison of formal charges indicates that structure **II** is more stable.

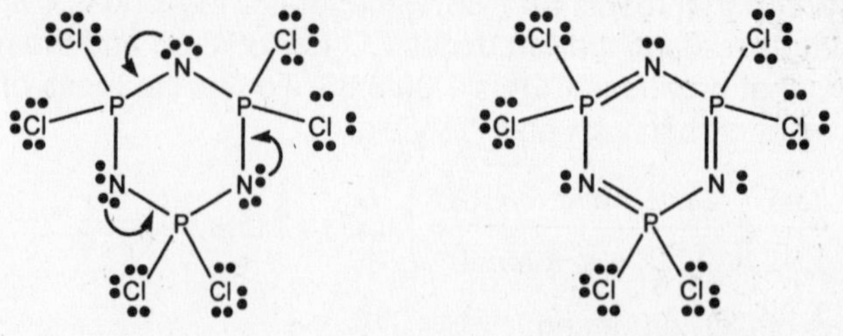

I — $FC_N = 5 - (4 + ½(4)) = -1$

II — $FC_N = 5 - (2 + ½(6)) = 0$

b) Three lone pairs appear on N ring atoms.

c) The bond order is 1.5. Structure II has a resonance structure, similar to the "alternating" bonds of benzene. The single-double bonds don't really alternate, but rather the valence electrons are delocalized on the ring, resulting in a bond order of 1.5.

14.163 a) Ions with outer levels of higher *n* values tend to be larger, but ions with greater charge tend to be smaller. These two effects cancel each other to make sodium ions approximately the same size as calcium ions, which have a higher *n* and a greater charge than Na^+.

b) Dissolution is a process that involves the breaking of the force holding ions together (lattice energy) and the formation of ion-dipole forces between the ions and water molecules. If a salt (ionic compound) is soluble, the ion-dipole forces are greater than the lattice energy and if a salt is insoluble lattice energy is greater. Lattice energy is greater with greater ionic charge. The CaF_2 salt with Ca^{2+} ions will have greater lattice energy, and lower solubility, than the NaF salt with Na^+ ions.

c) Beryllium is unusual in Group 2A(2) in that all its compounds are covalent. Covalent compounds do not conduct electricity. All other members of Group 2A(2) including calcium form ionic compounds that do conduct electricity when melted.

14.166 Hydrogen can act as a reducing agent by losing an electron and can act as an oxidizing agent by gaining an electron.

$2Na(s) + H_2(g) \rightarrow 2NaH(s)$ Na loses an electron, so H_2 is oxidizing agent.

$H_2(g) + Cl_2(g) \rightarrow 2HCl(g)$ Cl gains an electron, so H_2 is reducing agent.

14.168 Both compounds contain 2 moles of N per mole of compound. The compound with the higher percentage of N will have the higher mass of N per gram of compound.

$\mathcal{M}(NH_4NO_3) = 80.05$ g/mol $\mathcal{M}(N_2H_4) = 32.05$ g/mol

% $N/NH_4NO_3 = 2(14.01 \text{ g/mol}) \div 80.05 \text{ g/mol} = 35.00\%$

% $N/N_2H_4 = 2(14.01 \text{ g/mol}) \div 32.05 \text{ g/mol} = 87.42\%$

N_2H_4 contains the higher mass of N per gram of compound.

14.170 To answer these questions draw an initial structure for $Al_2Cl_7^-$:

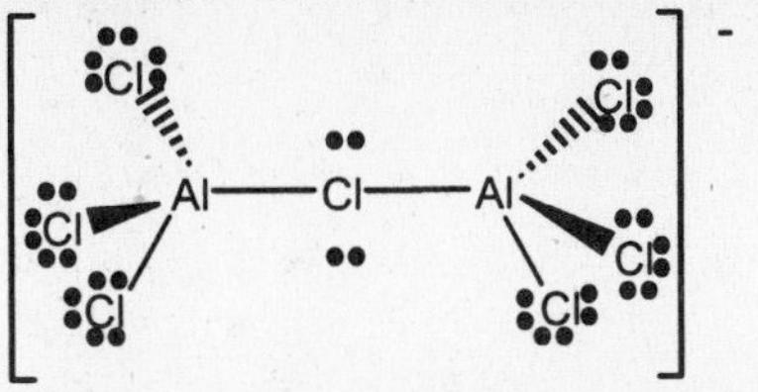

a) Aluminum uses its 3s and 3p valence orbitals to form sp^3 hybrid orbitals for bonding.

b) With formula AX_4 shape is tetrahedral around each aluminum atom.

c) Central chlorine is sp hybridized since the bond is linear.

d) The sp hybridization suggests that there are no lone pairs on the central chlorine. Instead the extra four electrons participate in bonding with the empty d-orbitals on the aluminum to form double bonds between the chlorine and each aluminum.

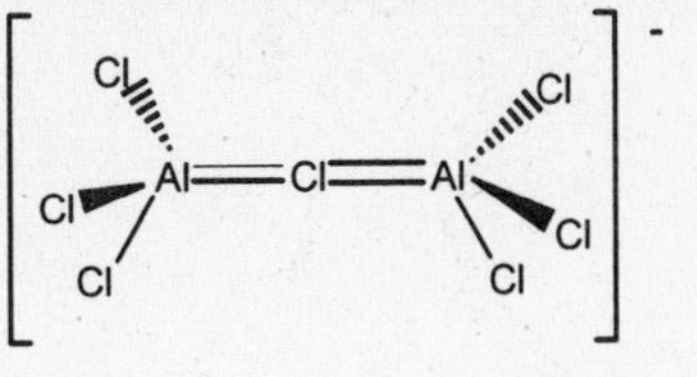

14.172 Nitrite ion, NO_2^- 18 valence e^- — nitrogen dioxide, NO_2 17 valence e^- — nitronium ion, NO_2^+ 16 valence e^-

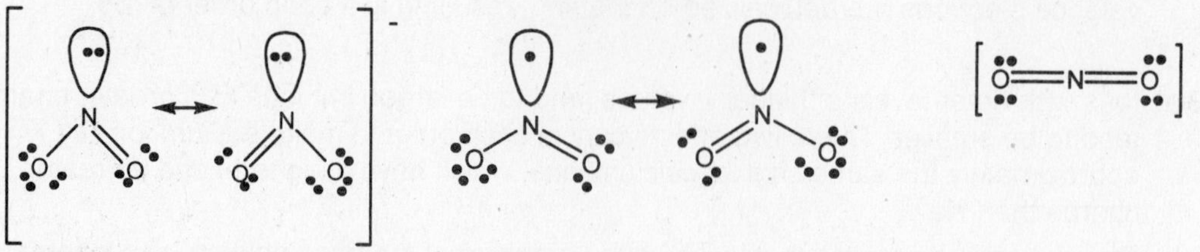

The nitronium ion has a linear shape because the central N atom has 2 surrounding electron groups, which achieve maximum repulsion at 180°. The nitrite ion's bond angle is more compressed than NO_2's bond angle because the lone pair of electrons takes up more space than the lone electron.

14.174 To find limiting reactant calculate the amount of UF_6 that would form from each reactant:

$$n_{from\,U} = (1.00\,metric\,ton\,ore)\left(\frac{10^6\,g}{1\,metric\,ton}\right)\left(\frac{1.55\,g\,U}{100\,g\,ore}\right)\left(\frac{1\,mol\,U}{238.0\,g}\right)\left(\frac{1\,mol\,UF_6}{1\,mol\,U}\right) = 65.1_{26}\,mol\,UF_6$$

$$n_{from\,ClF_3} = (12.75\,L\,ClF_3)\left(\frac{1880\,g\,ClF_3}{L}\right)\left(\frac{1\,mol\,ClF_3}{92.45\,g}\right)\left(\frac{1\,mol\,UF_6}{3\,mol\,ClF_3}\right) = 86.4_{25}\,mol\,UF_6$$

Since the amount of uranium will produce less uranium hexafluoride it is the limiting reactant and 65.1 mol is the maximum yield of UF_6. The mass of UF_6 that could form is 65.1_{26} x 352.0 g/mol = **$2.29\text{x}10^4$ g of UF_6**.

14.176 Apply Hess's Law to the two step process below. The bond energy (BE) of H_2 is exothermic because heat is given off as the two H atoms at higher energy combine to form the H_2 molecule at lower energy.

$H + H \rightarrow \cancel{H_2}$	BE = -432 kJ/mol
$\cancel{H_2} + H^+ \rightarrow H_3^+$	ΔH = -337 kJ/mol
$H + H + H^+ \rightarrow H_3^+$	**ΔH_{rxn} = -769 kJ/mol**

CHAPTER 15

ORGANIC COMPOUNDS AND THE PROPERTIES OF CARBON

FOLLOW-UP PROBLEMS

15.1 a) Plan: For compounds with seven carbons, first start with 7 C atoms in a straight chain. Then make all arrangements with 6 carbons in a straight chain and 1 carbon branched off the chain. Examine the structures to make sure they are all different. Continue in the same manner with 5 C atoms in chain and 2 branched off the chain, then 4 C atoms in chain and 3 branched off and 3 C atoms in chain and 4 branched off. Examine all structures to guarantee there are no duplicates.

Solution:

7 carbons in chain:

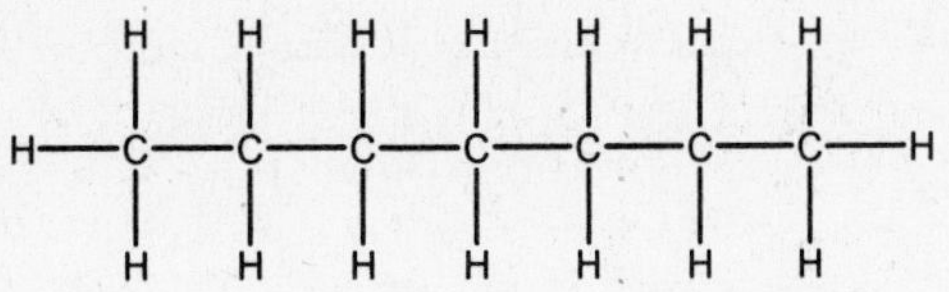

(1)

6 carbons in chain:

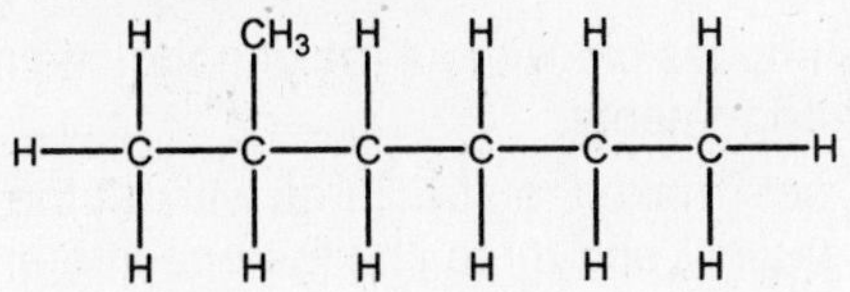

(2)

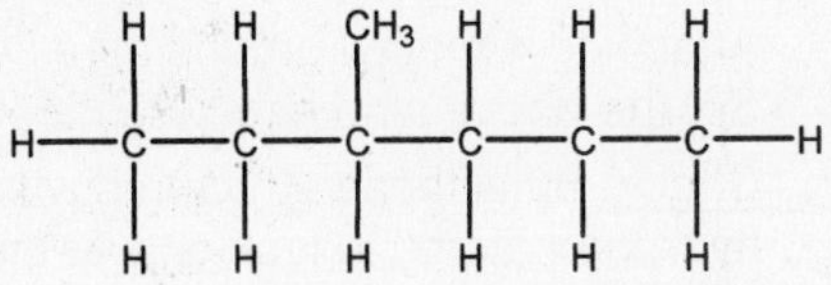

(3)

If the methyl group ($-CH_3$) is moved to the 4th carbon from the left, the structure is the same as (3). If the methyl group is moved to the 5th carbon from the left, the structure is the same as (2). And if the methyl group is moved to the 6th carbon from the left, the resulting structure has 7 carbons in a chain and is the same as (1).

5 carbons in chain:

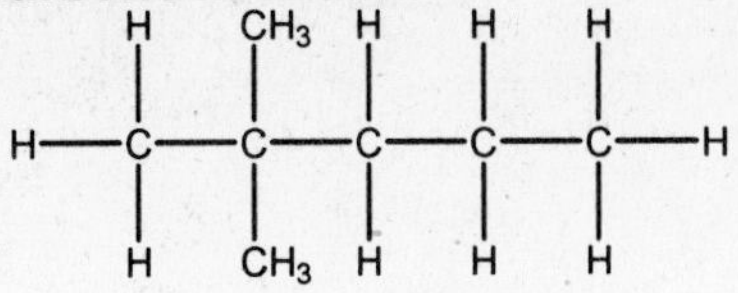

(4)

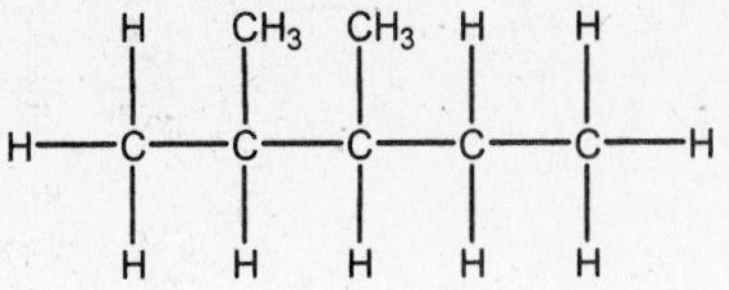

(5)

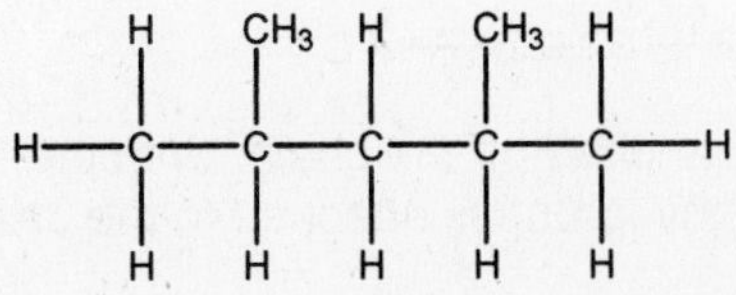

(6)

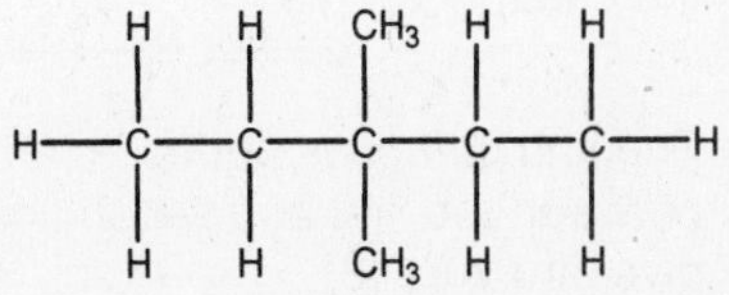

(7)

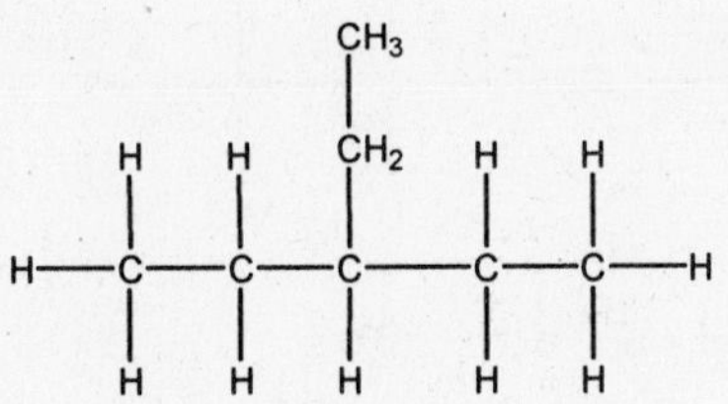

(8)

Starting with a methyl group on the 2nd carbon from the left, add another methyl group in a systematic pattern – structure (4) has the second methyl on the 2nd carbon, (5) on the 3rd carbon, (6) on the 4th carbon. These are all the unique possibilities with the first methyl group on the 2nd carbon – any other variation will produce a structure identical to a structure already drawn. Next move the first methyl group to the 3rd carbon where the only unique placement for the second methyl group is on the 3rd carbon as shown in structure (7). Which structure above would be the same as placing the first methyl group on the 3rd carbon and the second methyl group on the 4th carbon?

Are there any more arrangements with a 5-carbon chain? Think of attaching the two extra carbons as an ethyl group (-CH_2CH_3). If the ethyl group is attached to the 2nd carbon from the left the longest chain becomes 6 carbons instead of 5 and the structure is the same as (3). If the ethyl group is instead attached to the 3rd carbon the longest chain is still 5 carbons and the ethyl group is a side chain to give structure (8).

4 carbons in chain:

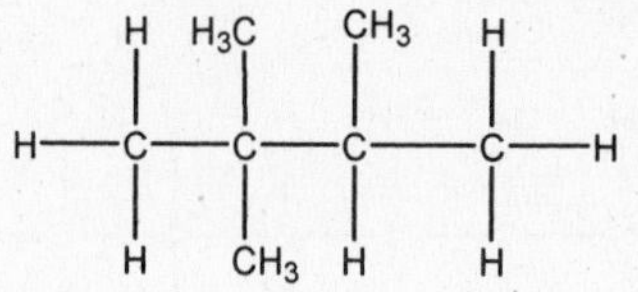

(9)

Only this arrangement of 3 methyl groups branching off the two inner carbons in a 4C chain produces a unique structure.

Attaching 4 methyl groups off a 3C chain is impossible without making the chain longer. So, the above nine structures are all the arrangements.

Check: Build these structures with a model set (make the structures without the hydrogen atoms to look only at the carbon chains) and try to change one into another by rotating the molecule, but not by breaking any bonds. If you have to break a bond to convert from one structure to another then the structures are different.

b) For a five carbon compound start with 5 C atoms in a chain and place the triple bond in as many unique places as possible:

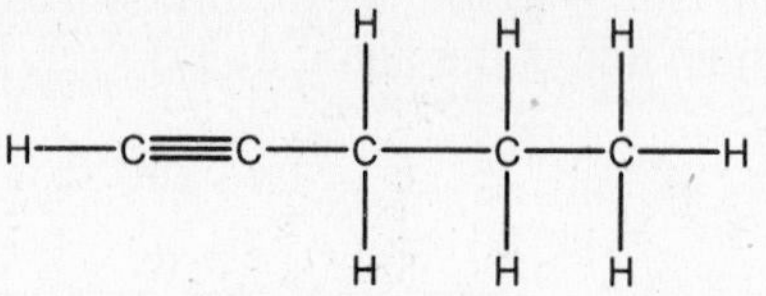

(1)

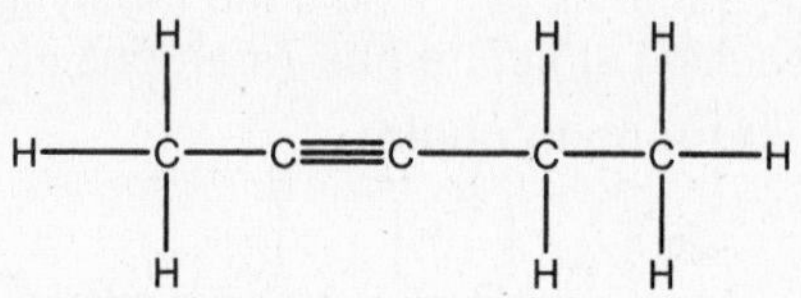

(2)

4 C atoms in chain: There are two unique placements for the triple bond – one between the first and second carbons and one between the second and third carbons in the chain.

The fifth carbon is added as a methyl group branched off the chain. With the triple bond between the first and second carbons, the methyl group is attached to the 3rd carbon to give structure (3).

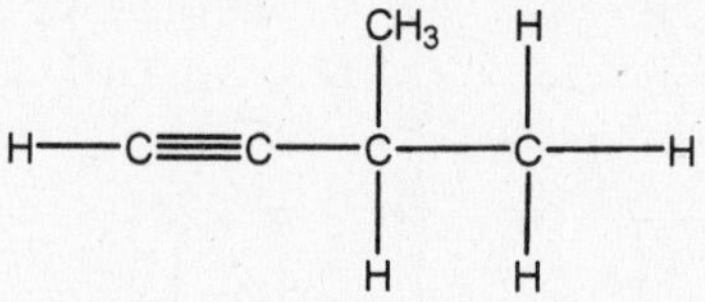

(3)

With the triple bond between the 2nd and 3rd carbons, the methyl group cannot be attached to either the 2nd or 3rd carbons because that would create 5 bonds to a carbon and carbon has only 4 bonds. Thus, a unique structure cannot be formed with a 4 C chain where the triple bond is between the 2nd and 3rd carbons.

There are only 3 structures with 5 carbon atoms, one triple bond, and no rings.

15.2 Plan: Analyze the name for chain length and side groups, then draw the structure.

Solution:

a) 3-ethyl-3-methyloctane. The end of the name, *octane*, indicates an 8 C chain (*oct-* represents 8 C) with only single bonds between the carbons (*-ane* represents alkanes, only single bonds). *3-ethyl* means an ethyl group ($-CH_2CH_3$) attached to carbon #3 and *3-methyl* means a methyl group ($-CH_3$) attached to carbon #3.

$$H_3C-CH_2-\overset{\overset{\displaystyle CH_3}{|}}{\underset{\underset{\displaystyle CH_3}{\underset{|}{CH_2}}}{C}}-CH_2-CH_2-CH_2-CH_2-CH_3$$

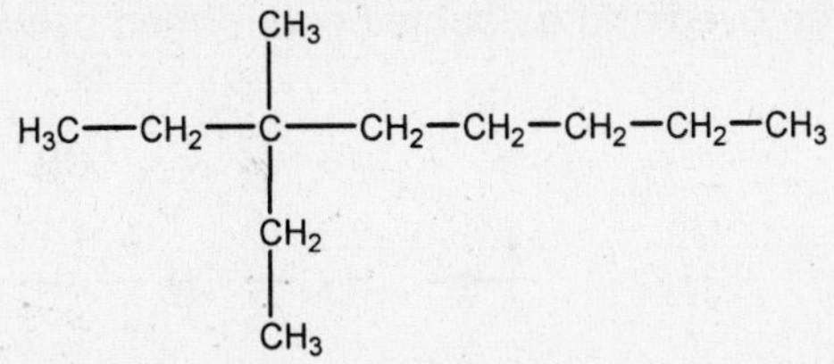

b) 1-ethyl-3-propylcyclohexane. The *hexane* indicates 6 C chain (*hex-*) and only single bonds (*-ane*). The *cyclo-* indicates that the 6 carbons are in a ring. The *1-ethyl* indicates that an ethyl group ($-CH_2CH_3$) is attached to carbon #1 – select any carbon atom in the ring as carbon #1 since all the carbon atoms in the ring are equivalent. The *3-propyl* indicates that a propyl group ($-CH_2CH_2CH_3$) is attached to carbon #3.

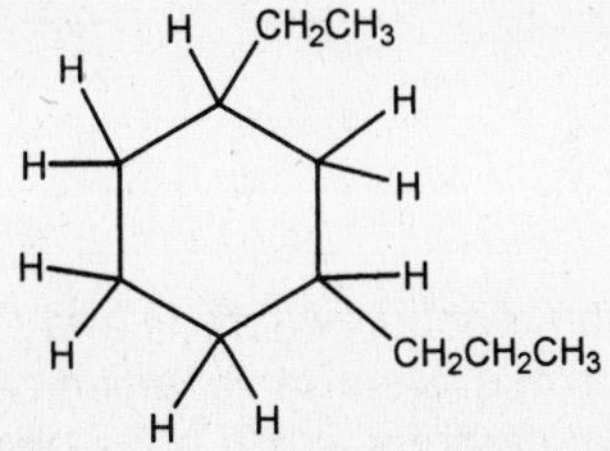

c) 3,3-diethyl-1-hexyne. The *1-hexyne* indicates a 6 C chain with a triple bond between C #1 and C #2. The *3,3-diethyl* means two (di-) ethyl groups ($-CH_2CH_3$) both attached to C #3.

$$H-C\equiv C-\overset{\overset{\displaystyle CH_3}{\overset{|}{\underset{|}{CH_2}}}}{\underset{\underset{\displaystyle CH_3}{\underset{|}{CH_2}}}{C}}-CH_2-CH_2-CH_3$$

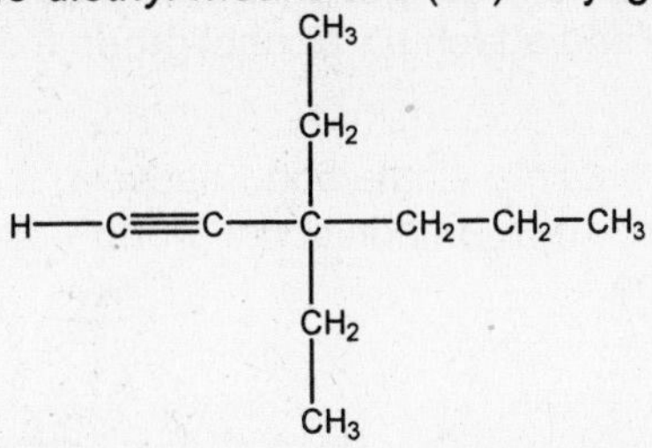

d) *trans*-3-methyl-3-heptene. The *3-heptene* indicates a 7 C chain (hept-) with one double bond (-ene) between the 3rd and 4th carbons. The *3-methyl* indicates a methyl group ($-CH_3$) attached to carbon #3. The *trans* indicates that the arrangement around the two carbons in the double bond gives the two smaller groups on opposite sides of the double bond. So, the smaller group on the 3rd carbon, which would be the methyl group, is above the double bond than the smaller group, H, on the 4th carbon is below the double bond.

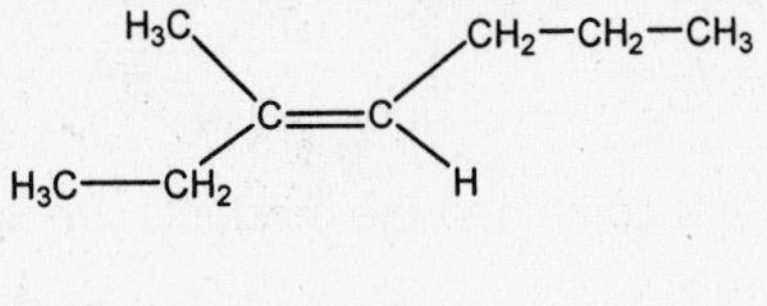

15.3 Plan:

a) An addition reaction involves breaking a multiple bond, in this case the double bond in 2-butene, and adding the other reactant to the carbons in the double bond. The reactant H_2O will add –H to one of the carbons and –OH to the other carbon. H^+ is needed for the reaction to occur, so it is listed above the arrow.

b) A substitution reaction involves removing one atom or group from a carbon chain and replacing it with another atom or group. For 1-bromopropane the bromine will be replaced by hydroxide.

c) An elimination reaction involves removing two atoms or groups, one from each of two adjacent carbon atoms, and forming a double bond between the two carbon atoms. For 2-chloropropane the chlorine from the center carbon and a hydrogen from one of the terminal carbons will be removed and a double bond formed between the two carbon atoms.

Solution:

a)

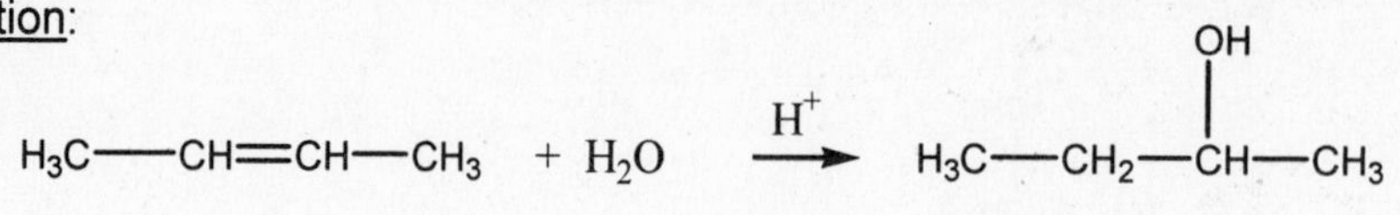

$$H_3C-CH=CH-CH_3 \ + \ H_2O \ \xrightarrow{H^+} \ H_3C-CH_2-\underset{OH}{CH}-CH_3$$

b)

$$H_3C-CH_2-\underset{Br}{CH_2} \ + \ OH^- \ \longrightarrow \ H_3C-CH_2-\underset{OH}{CH_2} \ + \ Br^-$$

c)

$$H_3C-\underset{Cl}{CH}-CH_3 \ + \ H_3C-ONa \ \longrightarrow \ H_3C-CH=CH_2 \ + \ NaCl \ + \ H_3C-OH$$

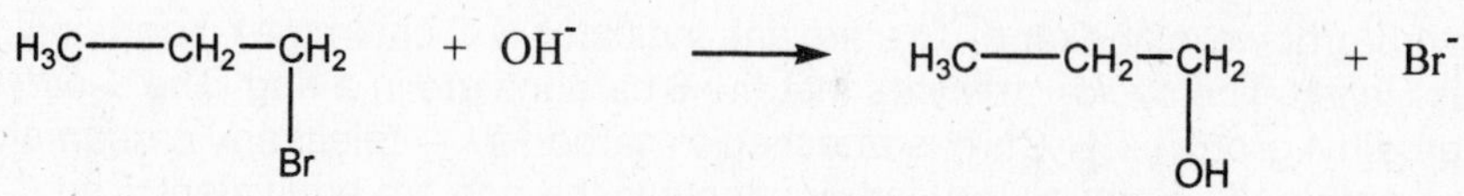

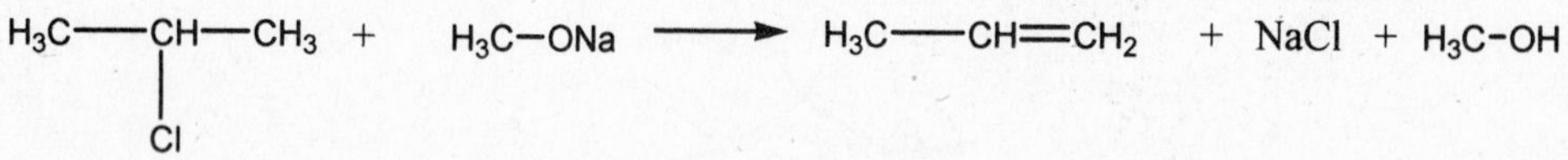

15.4 Plan: Reaction a) is an elimination reaction where H and Cl are eliminated from the organic reactant and a double bond is formed. Determine which carbons were involved in formation of the double bond and those two carbons would have been bonded to the H and Cl. Reaction b) is an oxidation reaction because $Cr_2O_7^{2-}$ and H_2SO_4 are oxidizing agents. An alcohol group (-OH) is oxidized to an acid group (-COOH), so determine which carbon in the product is in the acid group and attach an alcohol group to it for the reactant.

Solution:

a) The organic reactant must contain a single bond between the 2nd and 3rd carbons and the 2nd and 3rd carbons each have an additional group -- an H atom on one carbon and a Cl atom on the other carbon.

$$H_3C-\overset{H}{\underset{Cl}{C}}-\overset{CH_3}{\underset{H}{C}}-CH_3 \quad or \quad H_3C-CH_2-\overset{CH_3}{\underset{Cl}{C}}-CH_3$$

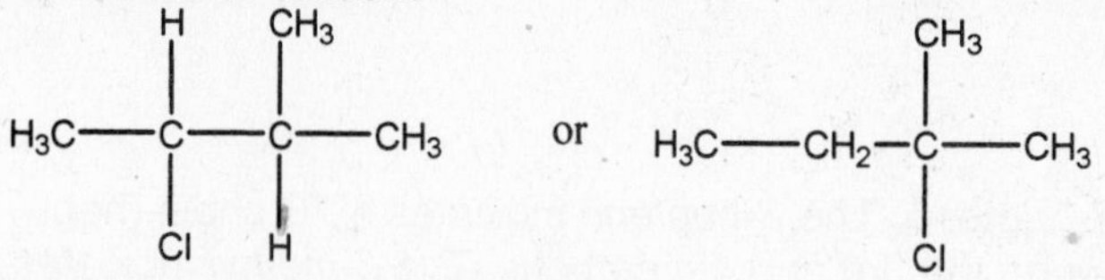

b) The 3rd carbon from the left in the product contains the acid group. So, in the reactant this carbon should have an alcohol group.

$$H_3C-CH_2-\overset{H}{\underset{H}{C}}-OH$$

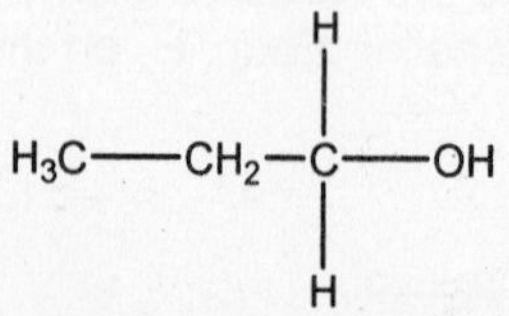

Check: When the reactants are combined the products are identical to those given.

15.5 Plan: The oxidizing agents in reaction a) indicate that the ketone group (=O) has been oxidized from an alcohol. To form the reactant, replace the C=O with an alcohol group, (C-OH). In reaction b) the reactants CH_3CH_2—Li and H_2O indicate that a ketone or aldehyde reacts to form the alcohol. To form the reactant find the carbon with the alcohol group, remove the ethyl group that came from CH_3CH_2—Li and change the –OH bond to a carbonyl group, =O.

Solution:

a) In the reactant, the double bond between oxygen and the carbon at the top of the ring will break and be replaced by a bond to H. Another H will bond to the oxygen that is now single-bonded to the carbon.

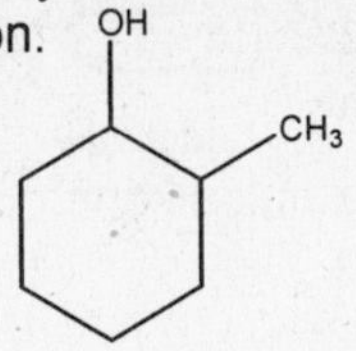

b) In the reactant, the first carbon off the ring will have a carbonyl group (-C=O) in place of the alcohol group (-OH) and the ethyl group (-CH_2CH_3).

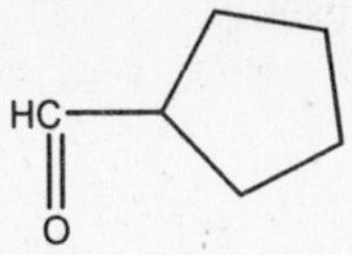

Check: When the reactants are combined, the products are identical to those given.

15.6 Plan: a) The product is an ester because it contains the 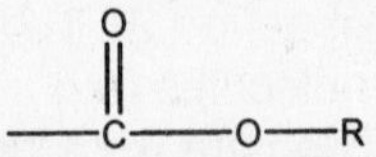 unit.

Reacting an alcohol with an acid forms an ester. Since the reactant shown is an alcohol the other reactant must be the acid. Take the –OR group off the carbon in the ester product and replace with a hydroxyl group to get the structure of the acid.

b) The product is an amide because it contains the 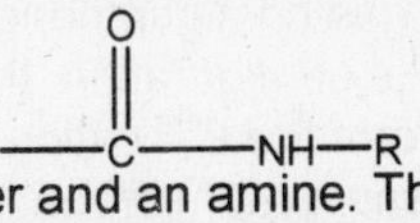 unit.

An amide forms from the reaction between an ester and an amine. This is the indicated reaction because the other product is an alcohol that results when the -OR group on the ester breaks off. To draw the reactants first take the –NH—R group off the amide and add another hydrogen to the nitrogen to form the amine. Attach the –OR group to the remaining carbonyl (-C=O) group of the amide molecule so that an ester linkage is formed.

Solution:

a) The –OR group attached to the oxygen in the product is –OCH_3. Removing this and adding the –OH gives the acid reactant:

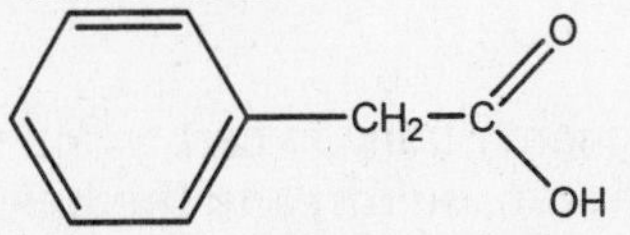

b) Take the –$NHCH_2CH_3$ group off the amide product, attach a H to the nitrogen to make the amine reactant, and attach the –OCH_3 from the alcohol product in its place to make the ester reactant.

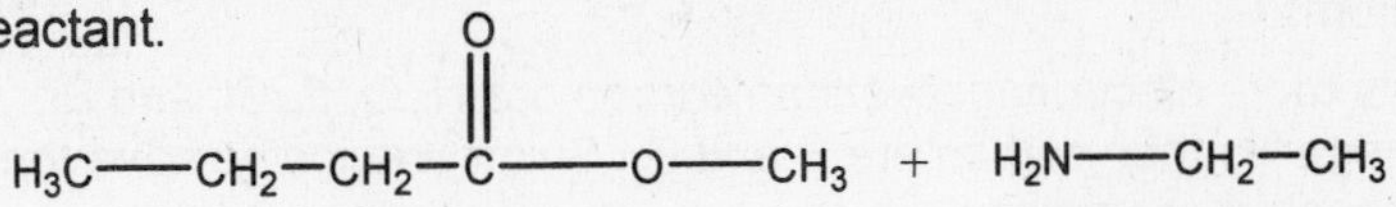

ester amine

Check: When the reactants are combined, the products are identical to those given.

15.7 Plan: Examine the structures for known functional groups: alkenes (C=C), alkynes (C≡C), haloalkanes (C-X where X is a halogen), alcohols (C-OH), esters (-COOR), ethers (C-O-C), amines (NR_3), carboxylic acids (-COOH), amides (-$CONR_2$), aldehydes (O=CHR) and ketones (O=CR_2 where R cannot be a H).

Solution:

a) The structure contains an alkene bond and an aldehyde group.

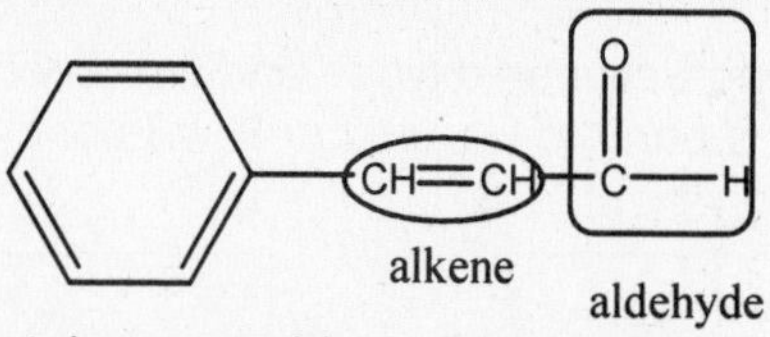

b) The structure contains an amide and a haloalkane.

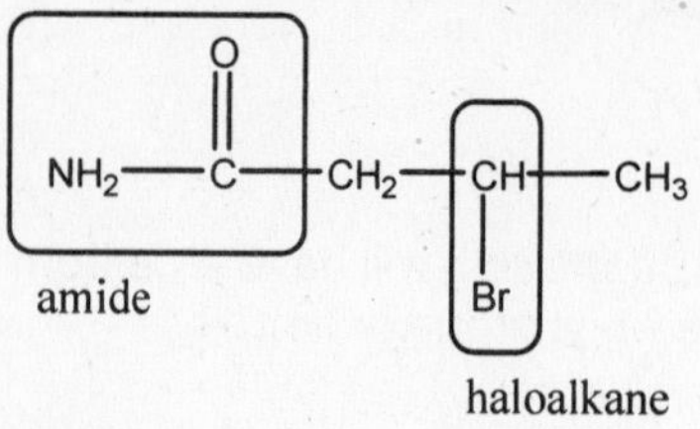

END-OF-CHAPTER PROBLEMS

15.3 a) Carbon's electronegativity is midway between the most metallic and nonmetallic elements of period 2. To attain a filled outer shell, carbon forms covalent bonds to other atoms in molecules (e.g. methane, CH_4), network covalent solids (e.g. diamond) and polyatomic ions (e.g. carbonate, CO_3^{2-}).

b) Since carbon has 4 valence shell electrons, it forms four covalent bonds to attain an octet.

c) Two noble gas configurations, He and Ne, are equally near carbon's configuration. To reach the He configuration, the carbon atom must lose 4 electrons, requiring too much energy to form the C^{4+} cation. This is confirmed by the fact that the value of the ionization energy for carbon is very high. To reach the Ne configuration, the carbon atom must gain 4 electrons, also requiring too much energy to form the C^{4-} anion. The fact that a carbon anion is unlikely to form is supported by carbon's electron affinity.

d) Carbon is able to bond to itself extensively because carbon's small size allows for closer approach and greater orbital overlap. The greater orbital overlap results in a strong, stable bond.

e) The C–C bond is short enough to allow the sideways overlap of unhybridized p orbitals on neighboring C atoms. The sideways overlap of p orbitals results in double and triple bonds.

15.4 a) The elements that most frequently bond to carbon are other **carbon atoms, hydrogen, oxygen, nitrogen, phosphorus, sulfur, and the halogens, F, Cl, Br, and I.**

b) In organic compounds heteroatoms are defined as atoms of any element other than carbon and hydrogen. The elements **O, N, P, S, F, Cl, Br, and I** listed in part a) are heteroatoms.

c) Elements more electronegative than carbon are N, O, F, Cl, and Br. Elements less electronegative than carbon are H and P. Sulfur and iodine have the same electronegativity as carbon.

d) The more types of atoms that can bond to carbon, the greater the variety of organic compounds that are possible.

15.7 Chemical reactivity occurs when unequal sharing of electrons in a covalent bond results in regions of high and low electron density. The C–H, C–C and C–I bonds are unreactive because electron density is shared equally between the two atoms. The **C=O** bond is reactive because oxygen is more electronegative than carbon and the electron rich pi bond is above and below the C–O bond axis, making it very attractive to electron-poor atoms. The **C–Li** bond is also reactive because the bond polarity results in an electron-rich region around carbon and an electron-poor region around Li.

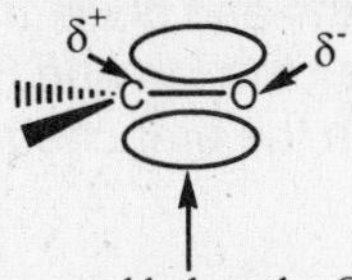

π bond above and below the C-O bond axis

δ^- δ^+
C——Li

15.8 a) An alkane is an organic compound consisting of carbon and hydrogen in which there are no multiple bonds between carbons, only single bonds. A cycloalkane is an alkane in which the carbon chain is arranged in a ring. An alkene is a hydrocarbon with at least one double bond between two carbons. An alkyne is a hydrocarbon with at least one triple bond between two carbons.

b) The general formula for an alkane is $\mathbf{C_nH_{2n+2}}$.

The general formula for a cycloalkane is $\mathbf{C_nH_{2n}}$. Elimination of two hydrogen atoms is required to form the additional bond between carbons in the ring.

For an alkene, assuming only one double bond, the general formula is $\mathbf{C_nH_{2n}}$. When a double bond is formed in an alkane, two hydrogen atoms are removed.

For an alkyne, assuming only one triple bond, the general formula is $\mathbf{C_nH_{2n-2}}$. Forming a triple bond from a double bond causes the loss of two hydrogen atoms.

c) For hydrocarbons, "saturated" is defined as a compound that cannot add more hydrogen. An unsaturated hydrocarbon contains multiple bonds that react with H_2 to form single bonds. The **alkanes and cycloalkanes** are saturated hydrocarbons since they contain only single C—C bonds.

15.11 a) A circular clock face numbered 1 to 12 o'clock is **asymmetric**. Imagine that the clock is cut in half, from 12 to 6 or from 9 to 3. The one half of the clock could never be superimposed on the other half, so the halves are not identical. Another way to visualize symmetry is to imagine cutting an object in half and holding the half up to a mirror. If the original object is "re-created" in the mirror, then the object has a plane of symmetry.

b) A football is symmetric and has two planes of symmetry – one axis along the length and one axis along the fattest part of the football.

c) A dime is **asymmetric**. Either cutting it in half or slicing it into two thin diameters results in two pieces that cannot be superimposed on one another.

d) A brick, assuming that it is perfectly shaped, is symmetric and has two planes of symmetry like the football.

e) A hammer is symmetric and has one plane of symmetry, slicing through the metal head and down through the handle.

f) A spring is **asymmetric**. Every coil of the spring is identical to the one before it, so a spring can be cut in half and the two pieces can be superimposed on one another by sliding (not flipping) the second half over the first. However, if the cut spring is held up to a mirror, the resulting image is not the same as the uncut spring. Disassemble a ball point pen and cut the spring inside to verify this explanation.

15.14 To draw the possible skeletons it is useful to have a systematic approach to make sure no structures are missed.

a) First draw the skeleton with the double bond between the 1st and 2nd carbons and place the branched carbon in all possible positions starting with C #2. Then move the double bond to between the 2nd and 3rd carbon and place the branched carbon in all possible positions. Then move the double bond to between the 3rd and 4th carbons and place the branched carbon in all possible positions. The double bond does not need to be moved further in the chain since the placement between the 2nd and 3rd carbon is equivalent to placement between the 4th and 5th carbons and placement between the 1st and 2nd carbons is equivalent to placement between the 5th and 6th carbons. The only position to consider for the double bond is between the branched carbon and the 6 C chain.

Double bond between 1st and 2nd carbons:

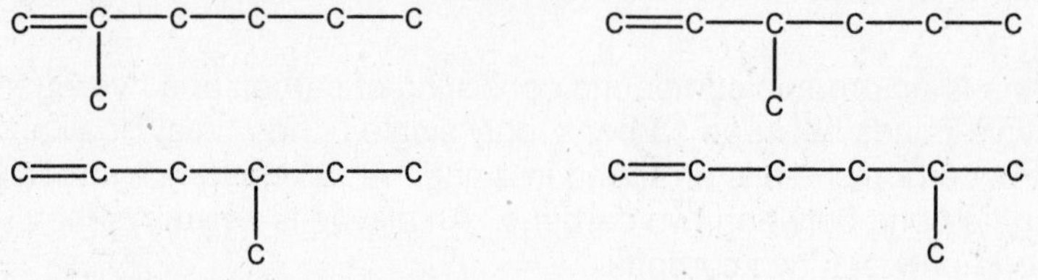

Double bond between 2nd and 3rd carbons:

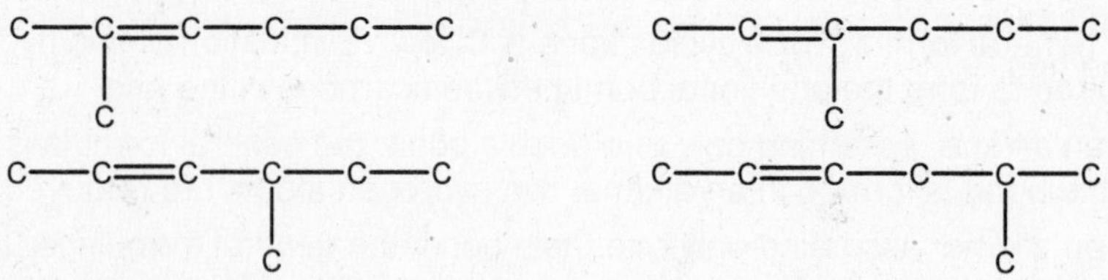

Double bond between 3rd and 4th carbons:

Double bond between branched carbon and chain:

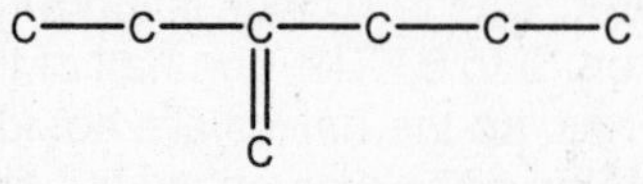

The total number of unique skeletons is 11. To determine if structures are the same, build a model of one skeleton and see if you can match the structure of the other skeleton by rotating the model and without breaking any bonds. If bonds must be broken to make the other skeleton the structures are not the same.

b) The same approach can be used here with placement of the double bond first between C #1 and C #2, then between C #2 and C #3.

Double bond between 1st and 2nd carbons:

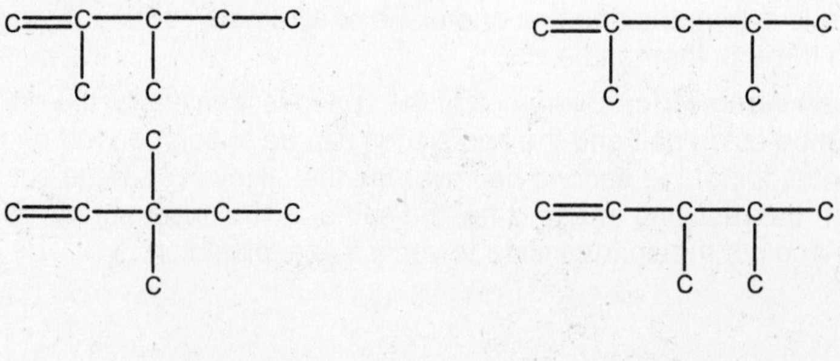

Double bond between 2nd and 3rd carbons:

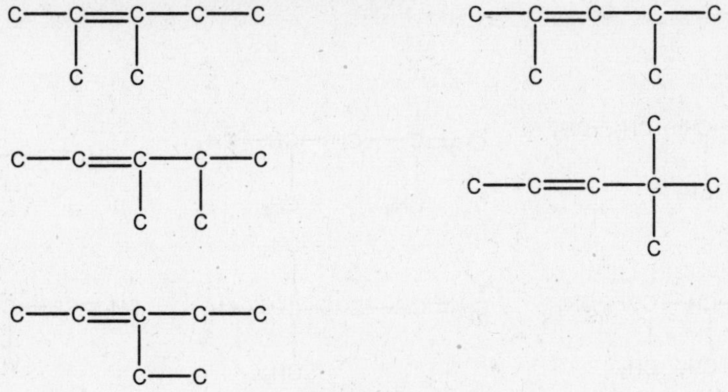

c) Five of the carbons are in the ring and two are branched off the ring. Remember that all the carbons in the ring are equivalent and there are two groups bonded to each carbon in the ring.

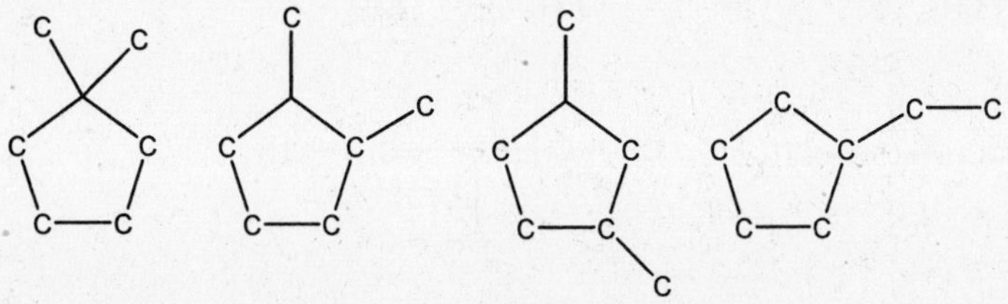

15.16 Add hydrogen atoms to make a total of 4 bonds to each carbon.

a)

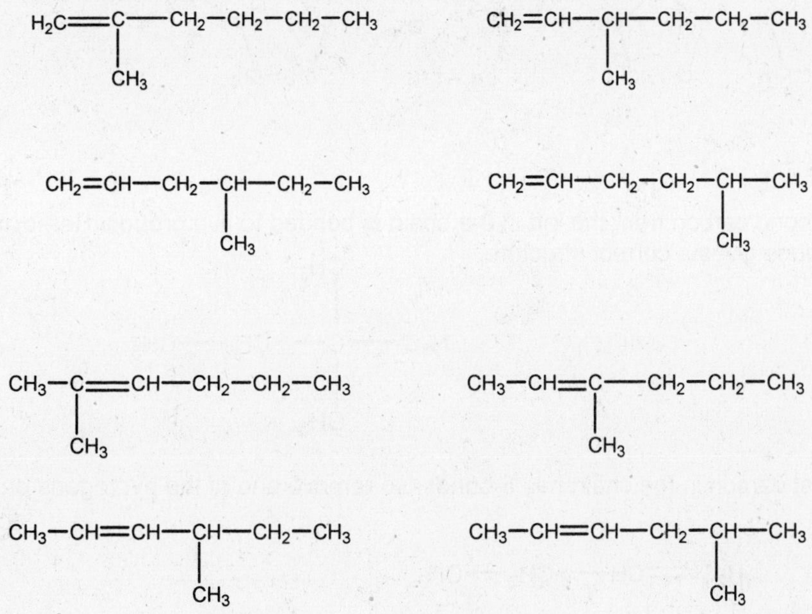

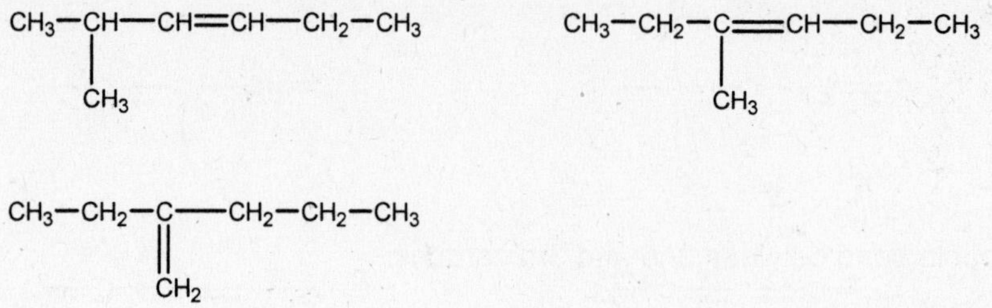

$CH_3-CH-CH=CH-CH_2-CH_3$
 $\quad\quad\quad | $
 $\quad\quad CH_3$

$CH_3-CH_2-C=CH-CH_2-CH_3$
 $\quad\quad\quad\quad\quad | $
 $\quad\quad\quad\quad CH_3$

$CH_3-CH_2-C-CH_2-CH_2-CH_3$
 $\quad\quad\quad\quad\|$
 $\quad\quad\quad\quad CH_2$

b)

$CH_2=C-CH-CH_2-CH_3$
 $\quad\quad | \quad | $
 $\quad\quad CH_3 \; CH_3$

$CH_2=C-CH_2-CH-CH_3$
 $\quad\quad | \quad\quad\quad | $
 $\quad\quad CH_3 \quad\quad CH_3$

$\quad\quad\quad\quad CH_3$
 $\quad\quad\quad\quad |$
$CH_2=CH-C-CH_2-CH_3$
 $\quad\quad\quad | $
 $\quad\quad\quad CH_3$

$CH_2=CH-CH-CH-CH_3$
 $\quad\quad\quad | \quad | $
 $\quad\quad\quad CH_3 \; CH_3$

$\quad\quad\quad\quad\quad\quad CH_3$
 $\quad\quad\quad\quad\quad\quad |$
$CH_2=CH-CH_2-C-CH_3$
 $\quad\quad\quad\quad\quad\quad | $
 $\quad\quad\quad\quad\quad\quad CH_3$

$CH_2=CH-CH-CH_2-CH_3$
 $\quad\quad\quad | $
 $\quad\quad\quad CH_2-CH_3$

$CH_3-C=C-CH_2-CH_3$
 $\quad\quad | \quad | $
 $\quad\quad CH_3 \; CH_3$

$CH_3-C=CH-CH-CH_3$
 $\quad\quad | \quad\quad\quad | $
 $\quad\quad CH_3 \quad\quad CH_3$

$CH_3-CH=C-CH-CH_3$
 $\quad\quad\quad\quad | \quad | $
 $\quad\quad\quad\quad CH_3 \; CH_3$

$\quad\quad\quad\quad\quad CH_3$
 $\quad\quad\quad\quad\quad |$
$CH_3-CH=CH-C-CH_3$
 $\quad\quad\quad\quad\quad | $
 $\quad\quad\quad\quad\quad CH_3$

$CH_3-CH=C-CH_2-CH_3$
 $\quad\quad\quad\quad | $
 $\quad\quad\quad\quad CH_2-CH_3$

c)

$\quad\quad CH_3 \quad CH_3$
 $\quad\quad\;\;\backslash \quad /$
 $\quad\quad\quad\; C$
 $CH_2 \quad\quad CH_2$
 $\;\; | \quad\quad\quad\;\; |$
 CH_2-CH_2

$\quad\quad CH_3$
 $\quad\quad |$
 $\quad\quad CH$
 $\quad\quad\quad\;\; \backslash CH_3$
 $CH_2 \quad\quad\; CH$
 $\;\; | \quad\quad\quad\;\; |$
 CH_2-CH_2

$\quad\quad CH_3$
 $\quad\quad |$
 $\quad\quad CH$
 $CH_2 \quad\quad CH_2$
 $\;\; | \quad\quad\quad\;\; |$
 CH_2-CH
 $\quad\quad\quad\quad | $
 $\quad\quad\quad\quad CH_3$

$\quad\; CH_2 \quad\;\; CH_2-CH_3$
 $\quad\;\; | \quad\quad\; /$
 $CH_2 \quad\quad\; CH$
 $\;\; | \quad\quad\quad\;\; |$
 CH_2-CH_2

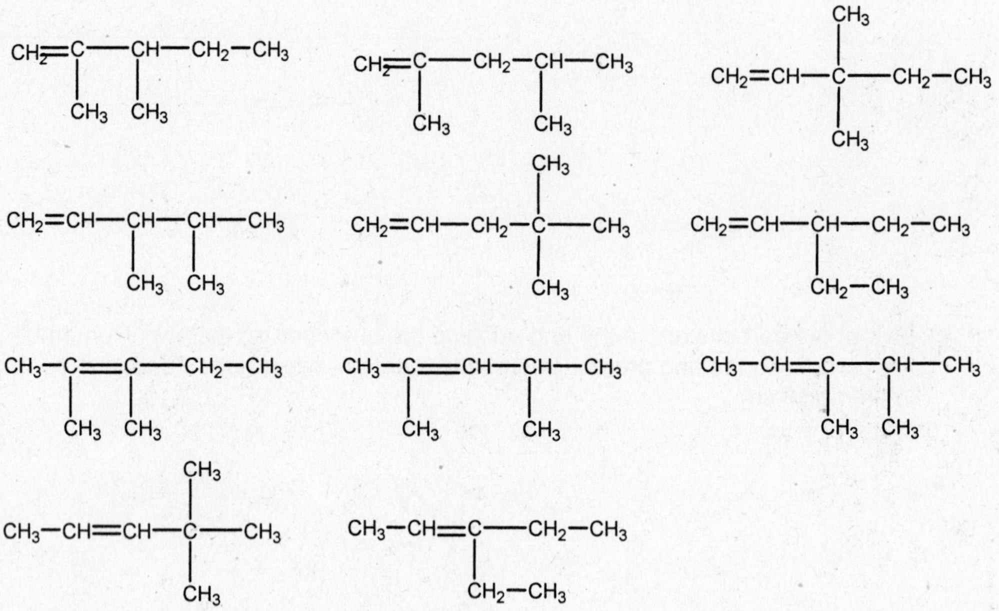

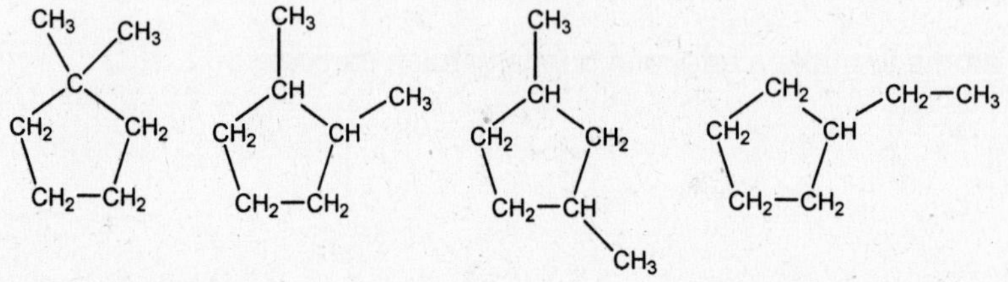

15.18 a) The second carbon from the left in the chain is bonded to five groups. Removing one of the groups gives a correct structure.

$\quad\quad\quad\quad\quad CH_3$
 $\quad\quad\quad\quad\quad |$
$H_3C-C-CH_2-CH_3$
 $\quad\quad\quad\quad | $
 $\quad\quad\quad\quad CH_3$

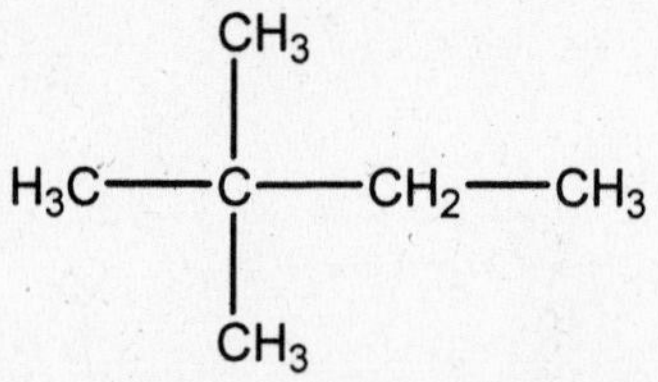

b) The first carbon in the chain has 5 bonds, so remove one of the hydrogens on this carbon.

$H_2C=CH-CH_2-CH_3$

c) The second carbon in the chain has 5 bonds, so move the ethyl group from the second carbon to the third.

$$HC\equiv C-CH-CH_3$$
$$|$$
$$CH_2$$
$$|$$
$$CH_3$$

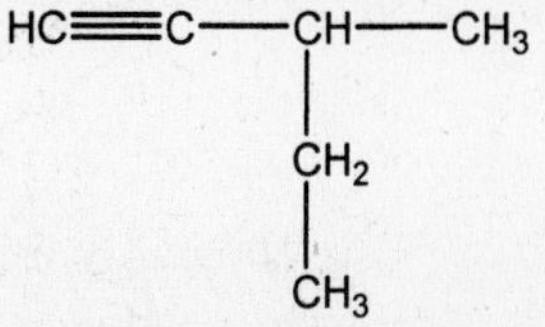

d) Structure is correct.

15.20 a) *Octane* denotes an eight carbon alkane chain. A methyl group ($-CH_3$) is located at the second and third carbon position from the left.

$$CH_3$$
$$|$$
$$CH_3-CH-CH-CH_2-CH_2-CH_2-CH_2-CH_3$$
$$|$$
$$CH_3$$

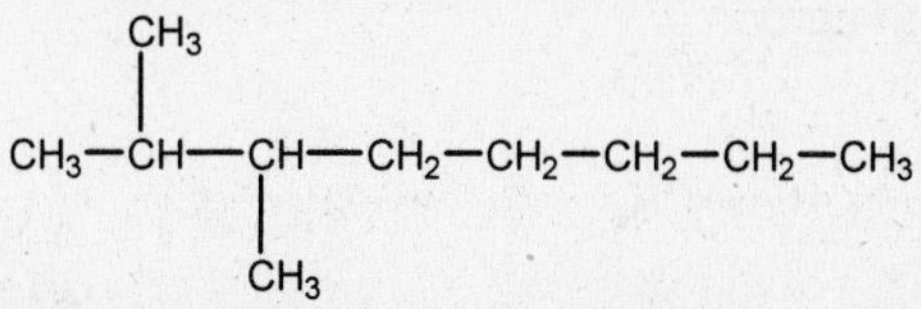

b) *Cyclohexane* denotes a six carbon ring containing only single bonds. Numbering of the carbons on the ring could start at any point, but typically numbering starts at the top carbon atom of the ring for convenience. The ethyl group ($-CH_2CH_3$) is located at position 1 and the methyl group is located at position 3.

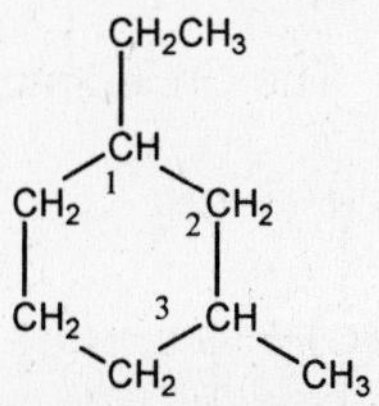

c) The longest continuous chain contains 7 carbon atom, so the root name is "hept". The molecule contains only single bonds, so the suffix is "ane". Numbering the carbon chain from the left results in side groups (methyl groups) at positions 3 and 4. Numbering the carbon chain from the other end results in side groups at positions 4 and 5. Since the goal is to obtain the lowest numbering position for a side group, the correct name is **3,4-dimethylheptane**. Note that the prefix "di" is used to denote that two methyl side groups are present in this molecule.

d) At first glance, this molecule looks like a 4 carbon *ring*, but the two $-CH_3$ groups mean that they cannot be bonded to each other. Instead, this molecule is a 4 carbon *chain*, with two methyl groups (dimethyl) located at the position 2 carbon. The correct name is **2,2-dimethylbutane**.

15.22 a) 4-methylhexane means a 6C chain with a methyl group on the 4th carbon:

$$\overset{6}{CH_3}-\overset{5}{CH_2}-\overset{4}{CH_2}-\overset{3}{CH}-\overset{2}{CH_2}-\overset{1}{CH_3}$$
$$|$$
$$CH_3$$

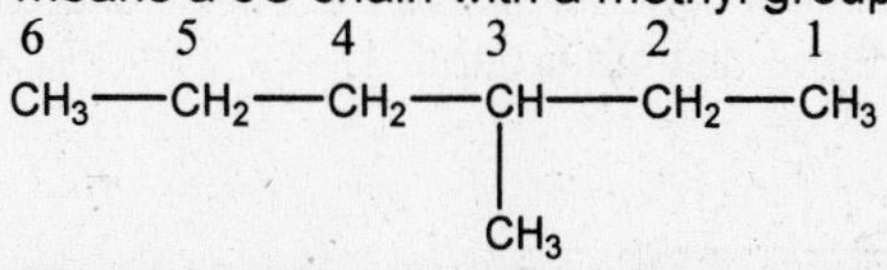

Numbering from the end carbon to give the lowest value for the methyl group gives correct name of **3-methylhexane**.

b) 2-ethylpentane means a 5C chain with an ethyl group on the 2nd carbon:

$$CH_3-\overset{3}{CH}-\overset{4}{CH_2}-\overset{5}{CH_2}-\overset{6}{CH_3}$$
$$|$$
$$\overset{}{CH_2}-CH_3$$
$$\overset{2}{}\quad\overset{1}{}$$

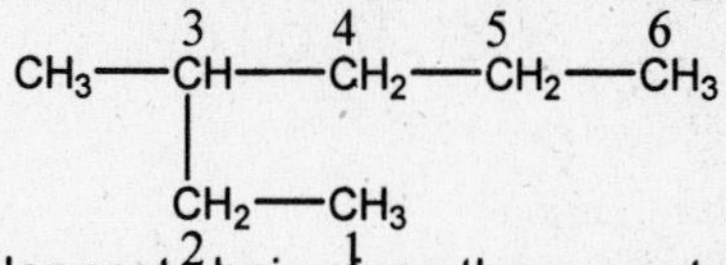

Numbering the longest chain gives the correct name, **3-methylhexane**.

c) 2-methylcyclohexane means a 6C ring with a methyl group on carbon #2:

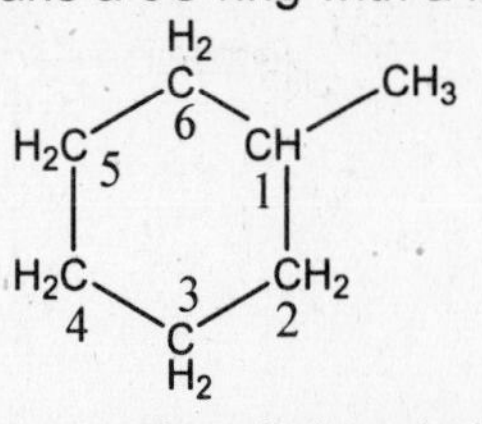

In a ring structure whichever carbon is bonded to the methyl group is automatically assigned as carbon #1. Since this is automatic it is not necessary to specify 1-methyl in the name. Correct name is **methylcyclohexane**.

d) 3,3-methyl-4-ethyloctane means an 8C chain with 2 methyl groups attached to the 3rd carbon and one ethyl group to the 4th carbon.

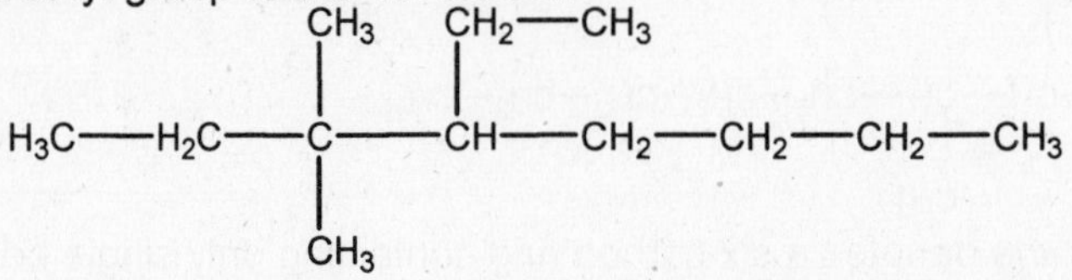

Numbering is good for this structure, but the fact that there are two methyl groups must be indicated by the prefix di- in addition to listing 3,3. Correct name is **4-ethyl-3,3-dimethyloctane**.

15.24 A carbon atom is chiral if it is attached to four different groups. The circled atoms below are chiral.

a)

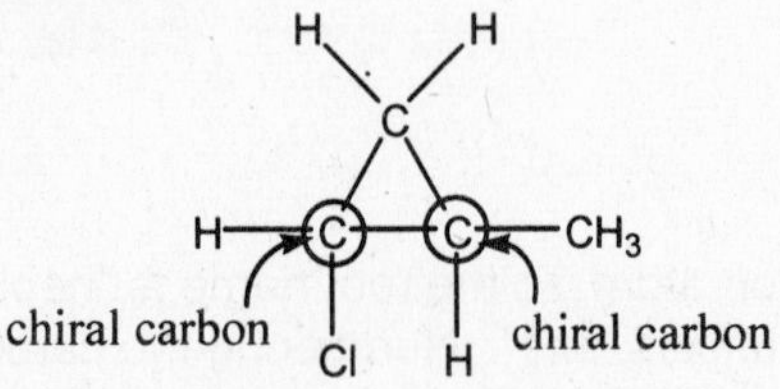

b)

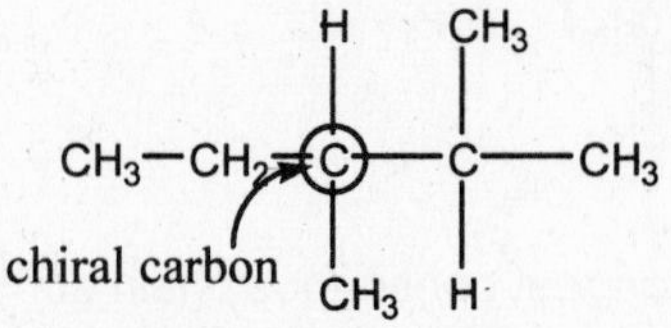

15.26 An optically active compound will contain at least one chiral center, a carbon with four distinct groups bonded to it.

a) 3-bromohexane is optically active because carbon # 3 has four distinct groups bonded to it: 1) –Br, 2) –H, 3) $-CH_2CH_3$, 4) $-CH_2CH_2CH_3$.

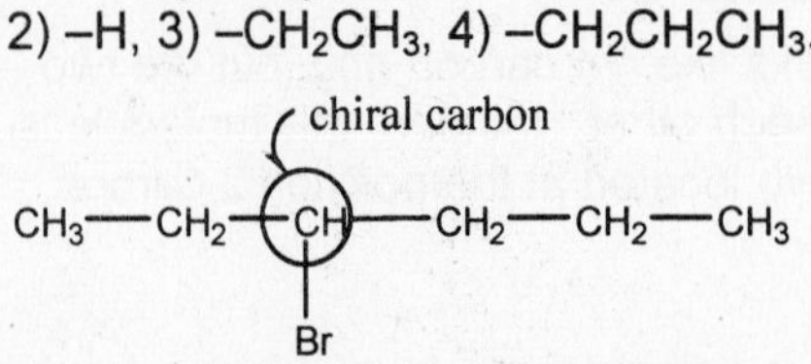

b) 3-chloro-3-methylpentane is not optically active because no carbon has four distinct groups. Third carbon has three distinct groups: 1) –Cl, 2) $-CH_3$ 3) two $-CH_2CH_3$ groups.

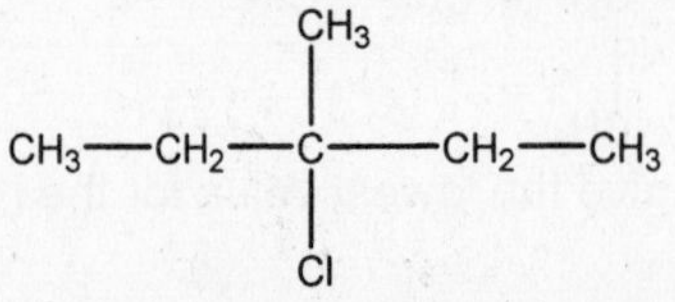

c) 1,2-dibromo-2-methylbutane is optically active because the 2nd carbon is chiral bonded to the four groups: 1) $-CH_2Br$, 2) $-CH_3$, 3) –Br, 4) $-CH_2CH_3$.

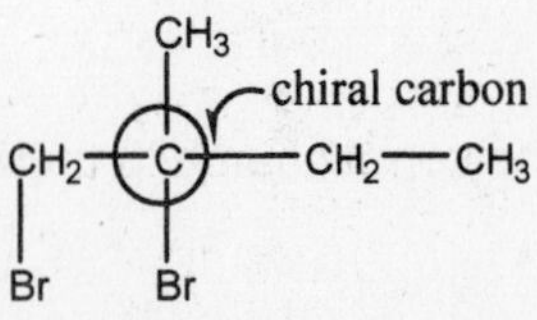

15.28 Geometric isomers are defined as compounds with the same atom sequence but different arrangements of the atoms in space. The cis-trans geometric isomers occur when rotation is restricted around a bond, as in a double bond or a ring structure, and when two different groups are bonded to each atom in the restricted bond.

a) Both carbons in the double bond are bonded to two distinct groups, so geometric isomers will occur. The double bond occurs at position 2 in a five carbon chain.

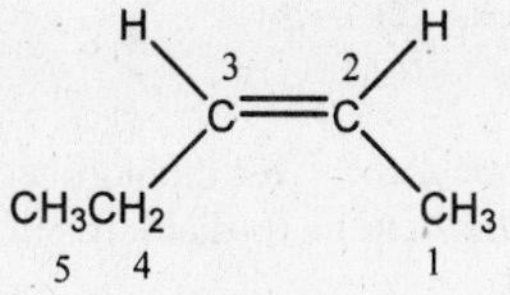

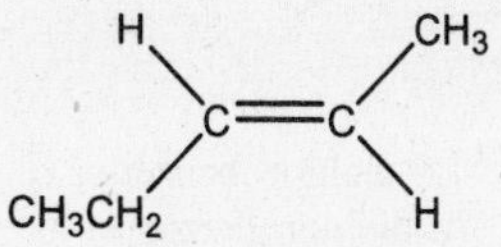

cis-2-pentene *trans*-2-pentene

b) Cis-trans geometric isomerism occurs about the double bond. The ring is named as a side group (cyclohexyl) occurring at position 1 on the propene main chain.

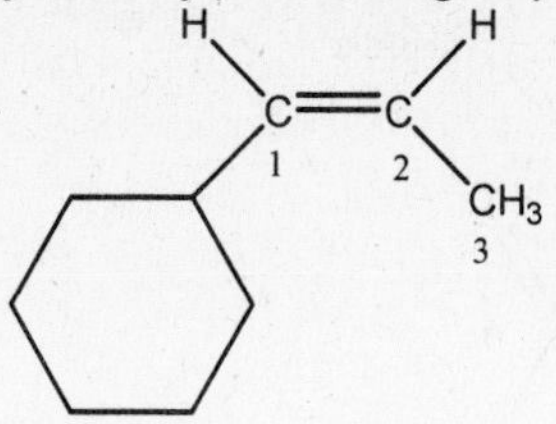

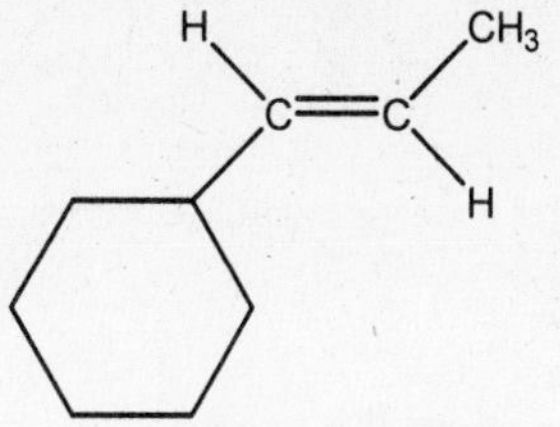

cis-1-cyclohexylpropene *trans*-1-cyclohexylpropene

c) No geometric isomers occur because the left carbon participating in the double bond is attached to two identical methyl ($-CH_3$) groups.

15.30 a) The structure of propene is $CH_2{=}CH{-}CH_3$. The first carbon that is involved in the double bond is bonded to two of the same type of group, hydrogen. Geometric isomers will not occur in this case.

b) The structure of 3-hexene is $CH_3CH_2CH{=}CHCH_2CH_3$. Both carbons in the double bond are bonded to two distinct groups, so geometric isomers will occur.

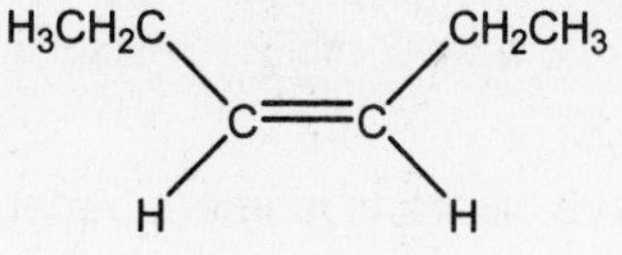

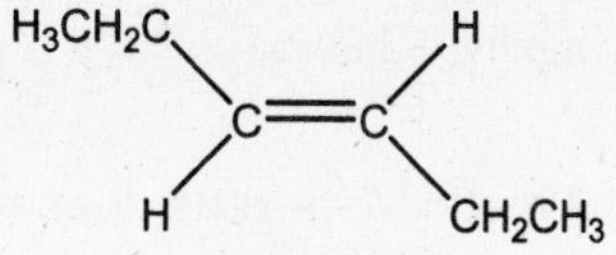

cis-3-hexene *trans*-3-hexene

c) The structure of 1,1-dichloroethene is $CCl_2{=}CH_2$. Both carbons in the double bond are bonded to two identical groups, so no geometric isomers occur.

d) The structure of 1,2-dichloroethene is $CHCl{=}CHCl$. Each carbon in the double bond is bonded to two distinct groups, so geometric isomers do exist.

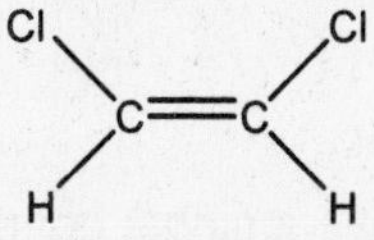

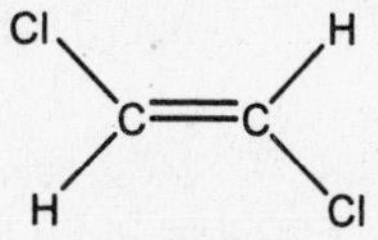

cis-1,2-dichloroethene *trans*-1,2-dichlorethene

15.32 Benzene is a planar, aromatic hydrocarbon. It is commonly depicted as a hexagon with a circle in the middle to indicate that the π bonds are delocalized around the ring and that all ring bonds are identical. The ortho, meta, para naming system is used to denote the location of attached groups in benzene compounds only, not other ring structures like the cycloalkanes.

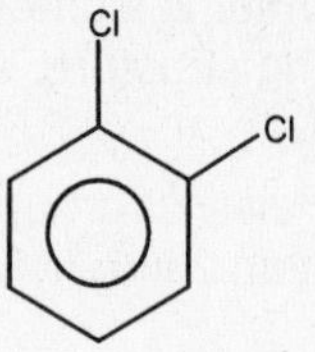

1,2 dichlorobenzene
(o-dichlorobenzene)

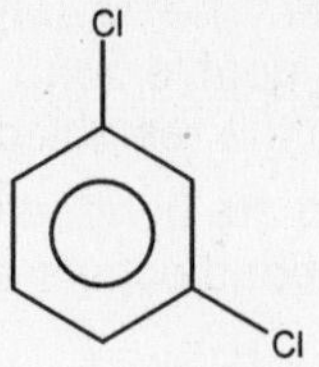

1,3 dichlorobenzene
(m-dichlorobenzene)

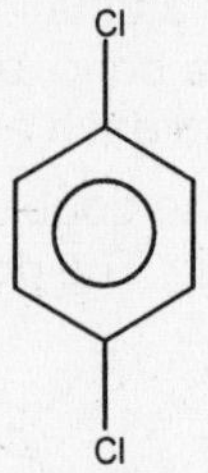

1,4 dichlorobenzene
(p-dichlorobenzene)

15.34 Analyzing the name gives benzene as the base structure with the following groups bonded to it: 1) on carbon #1 a hydroxy group, -OH; 2) on carbons #2 and #6 a *tert*-butyl group, $-C(CH_3)_3$; and 3) on carbon #4 a methyl group, $-CH_3$.

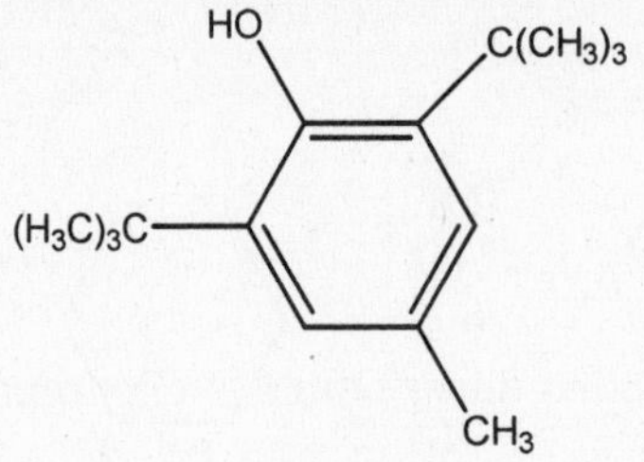

15.35 The compound 2-methyl-3-hexene has cis-trans isomers.

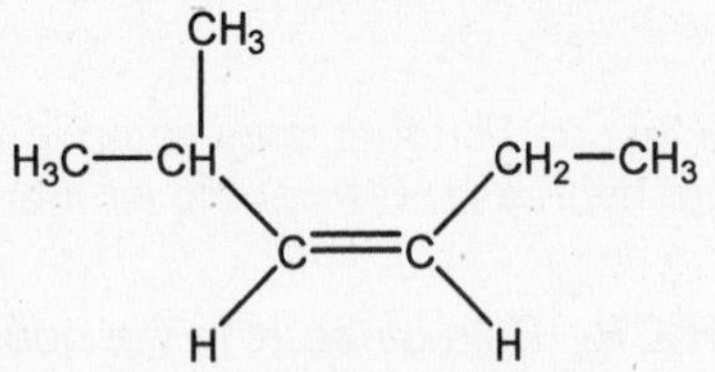

cis-2-methyl-3-hexene

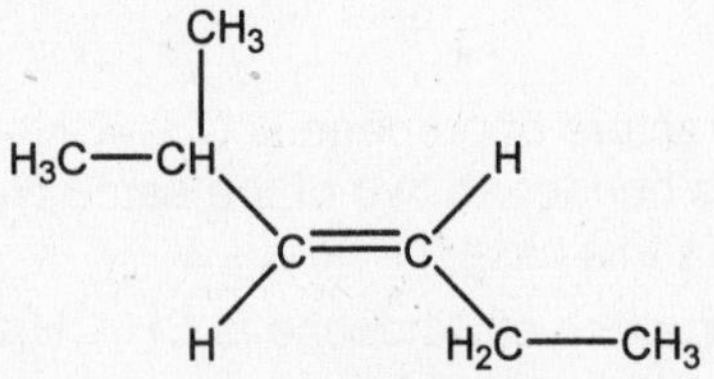

trans-2-methyl-3-hexene

The compound 2-methyl-2-hexene does not have cis-trans isomers because the #2 carbon atom is attached to two identical groups.

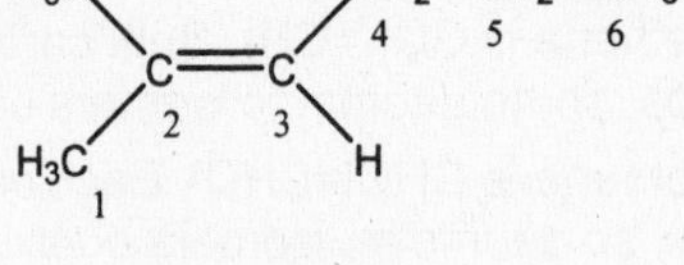

2-methyl-2-hexene

15.39 In an addition reaction a double bond is broken to leave a single bond, and in an elimination reaction a double bond is formed from a single bond. A double bond consists of a σ bond and a **π bond**. It is the π bond that breaks in the addition reaction and that forms in the elimination reaction.

15.41 a) HBr is removed in this **elimination reaction**, and an unsaturated product is formed.

b) This is an **addition reaction** in which hydrogen is added to the double bond, resulting in a saturated product.

15.43 a)

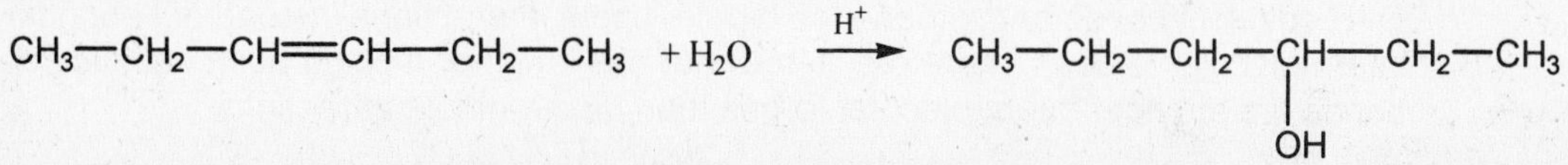

$$CH_3-CH_2-CH=CH-CH_2-CH_3 + H_2O \xrightarrow{H^+} CH_3-CH_2-CH_2-\underset{\underset{OH}{|}}{CH}-CH_2-CH_3$$

b)

$$CH_3-\underset{\underset{Br}{|}}{CH}-CH_3 + CH_3-CH_2-OK \xrightarrow{\Delta} CH_2=CH-CH_3 + CH_3-CH_2-OH + KBr$$

c)

$$CH_3-CH_3 + 2Cl_2 \xrightarrow{h\nu} \underset{\underset{Cl}{|}}{\overset{\overset{Cl}{|}}{HC}}-CH_3 + 2HCl$$

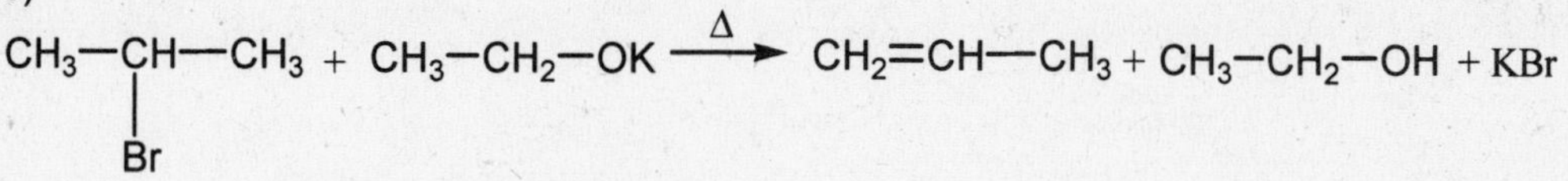

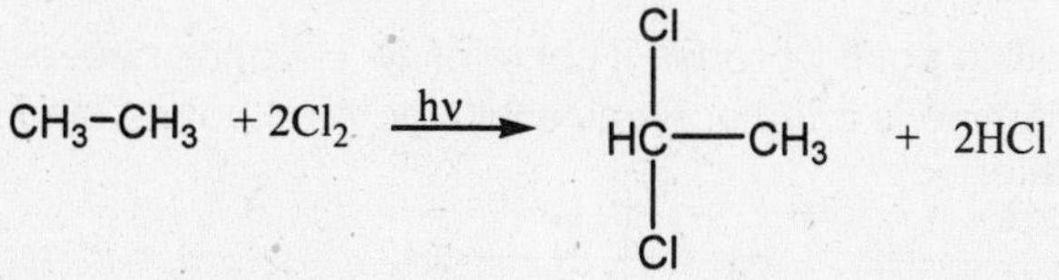

15.45 To decide whether an organic compound is oxidized or reduced in a reaction rely on the rules from section 15.3:

A C atom is oxidized when it forms more bonds to O or fewer bonds to H as a result of the reaction.

A C atom is reduced when it forms fewer bonds to O or more bonds to H as a result of the the reaction.

a) The C atom is **oxidized** because it forms more bonds to O.

b) The C atom is **reduced** because it forms more bonds to H.

c) The C atom is **reduced** because it forms more bonds to H.

15.47 a) The reaction $CH_3CH=CHCH_2CH_2CH_3 \rightarrow CH_2CH(OH)-CH(OH)CHCH_2CH_2CH_3$ shows the second and third carbon in the chain gaining a bond to oxygen: C—O—H. So, the 2-hexene compound has been **oxidized**.

b) The reaction

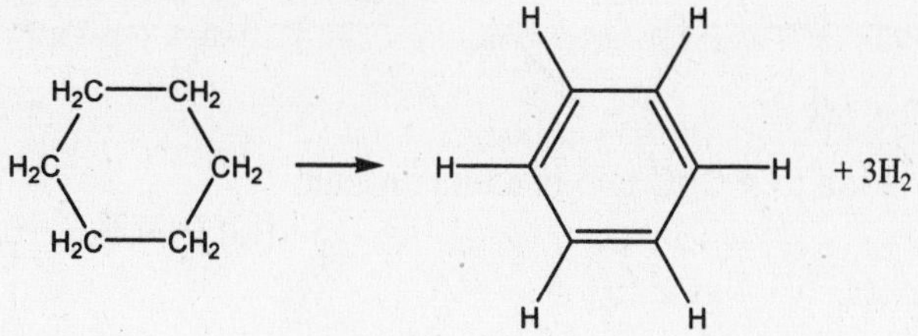

shows that each carbon atom in the cyclohexane loses a bond to hydrogen to form benzene. Fewer bonds to hydrogen in the product indicate **oxidation**.

15.50 a) The structures for chloroethane and methylethylamine are given below. The compound **methylethylamine** is more soluble due to its ability to form H-bonds with water. Recall that N–H, O–H, or F–H bonds are required for H-bonding.

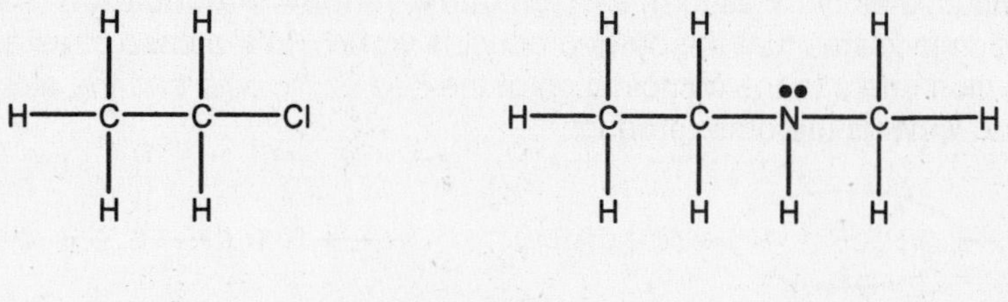

b) The compound 1-butanol is able to H-bond with itself (shown below) because it contains covalent oxygen-hydrogen bonds in the molecule. Diethylether molecules contain no O–H covalent bonds and experience dipole-dipole interactions instead of H-bonding as intermolecular forces. Therefore, **1-butanol** has a higher melting point because H-bonds are stronger intermolecular forces than dipole-dipole attractions.

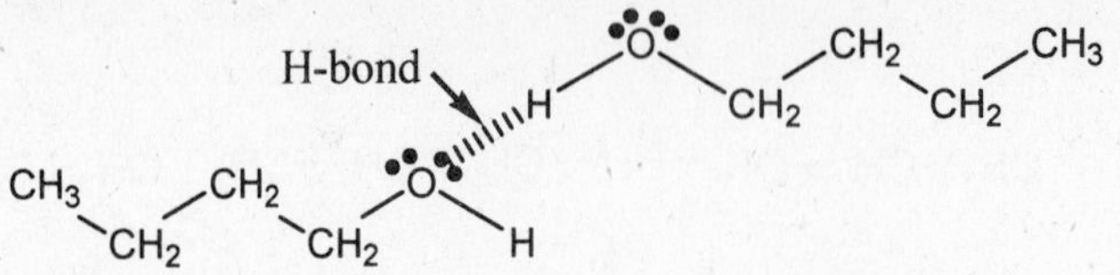

c) **Propylamine** has a higher boiling point because it contains N–H bonds necessary for hydrogen bonding. Trimethylamine is a tertiary amine with no N–H bonds, and so its intermolecular forces are weaker.

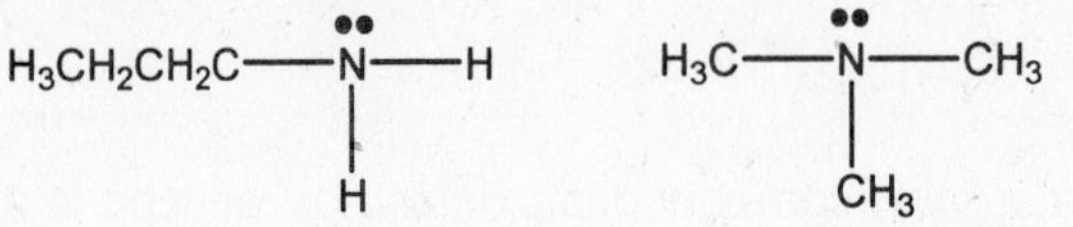

propylamine trimethylamine

15.53 The C=C bond is nonpolar while the C=O bond is polar with oxygen more electronegative than carbon. Both bonds react by addition. In the case of addition to a C=O bond, an electron-rich group will bond to the carbon and an electron-poor group will bond to the oxygen resulting in one product. In the case of addition to an alkene, the carbons are identical, or nearly so, so there will be no preference for which carbon bonds the electron-poor group and which bonds the electron-rich group. This may lead to two isomeric products, depending on the structure of the alkene.

When water is added to a double bond, –H is the electron-poor group and –OH is the electron-rich group. For a compound with a carbonyl group only one product results:

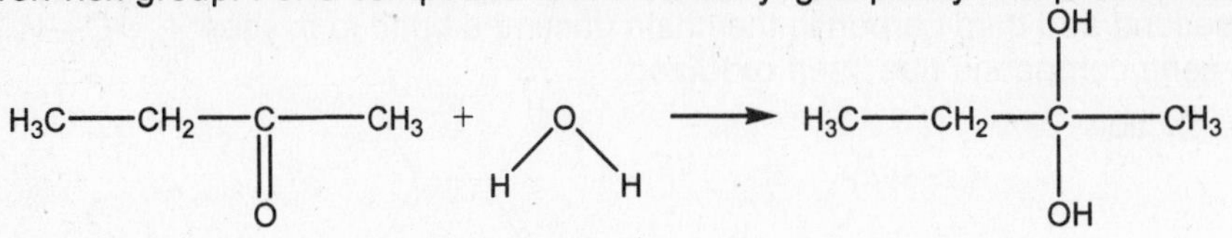

But when water adds to a C=C two products result:

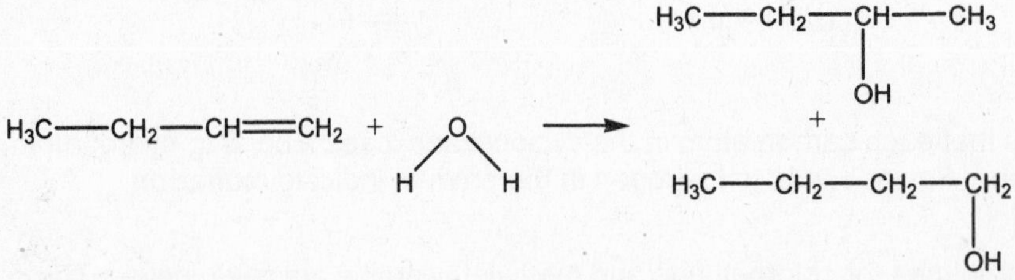

The second product, however, is practically impossible to form when adding water across a terminal double bond.

15.56 Esters and acid anhydrides form through **dehydration-condensation** reactions. Dehydration indicates that the other product is **water**. In the case of ester formation, condensation refers to the combination of the carboxylic acid and the alcohol. The ester forms, and water is the other product.

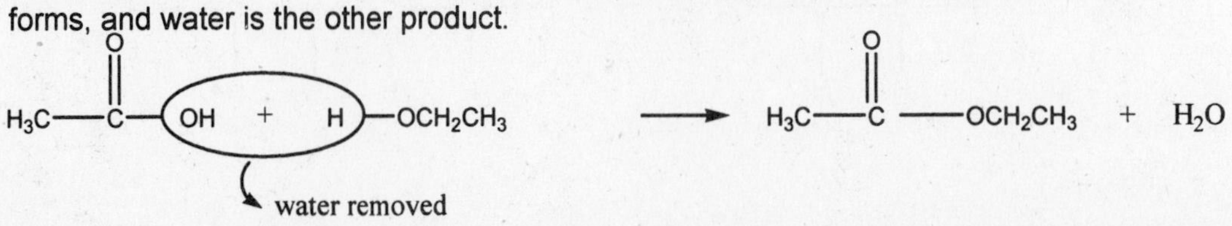

15.58 a) Halogens, except iodine, differ from carbon in electronegativity and form a single bond with carbon. The organic compound is an **alkyl halide**.

b) Carbon forms triple bonds with itself and nitrogen. For the bond to be polar, it must be between carbon and nitrogen. The compound is a **nitrile**.

c) **Carboxylic acids** contain a double bond to oxygen and a single bond to oxygen. Carboxylic acids dissolve in water to give acidic solutions.

d) Oxygen is commonly double-bonded to carbon. A carbonyl group (C=O) that is at the end of a chain is found in an **aldehyde**.

15.60 a)

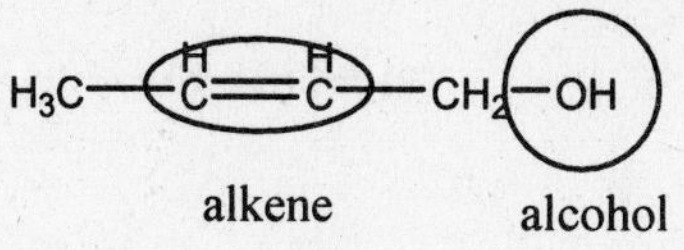

b)

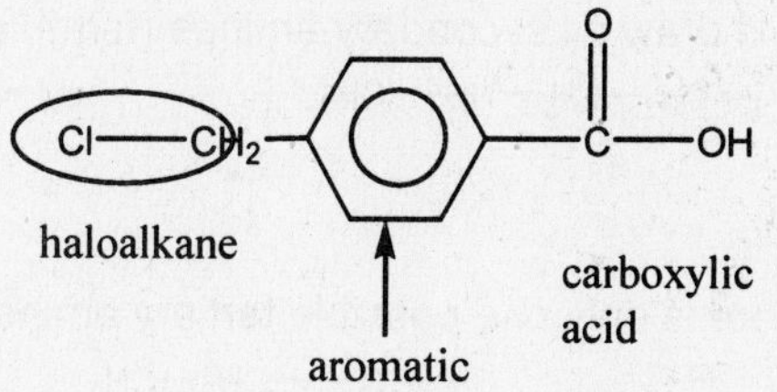

c)

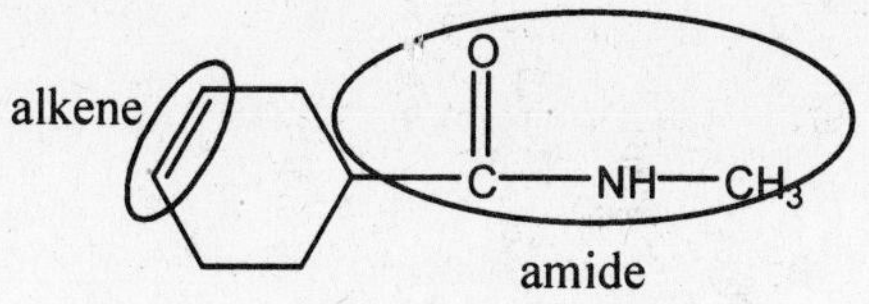

d)

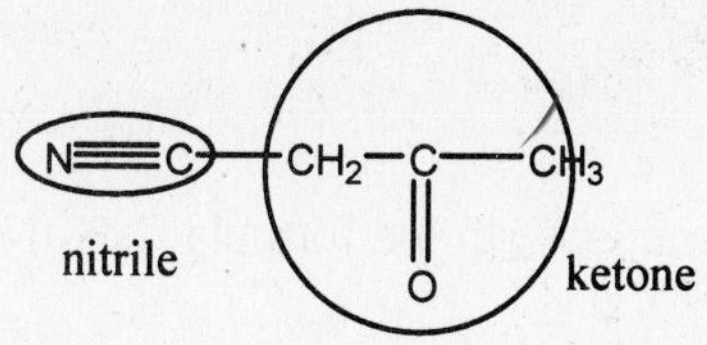

e)

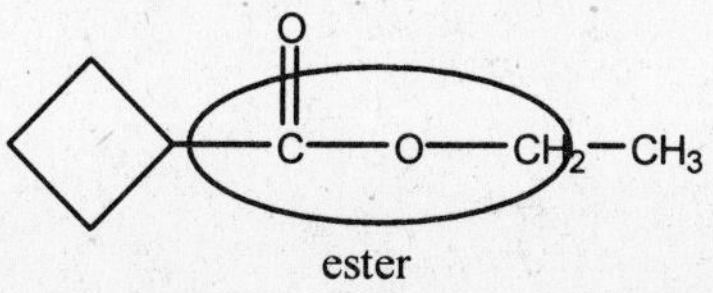

15.62 To draw all structural isomers, begin with the longest chain possible. For $C_5H_{10}O$ the longest chain is 5 carbons:

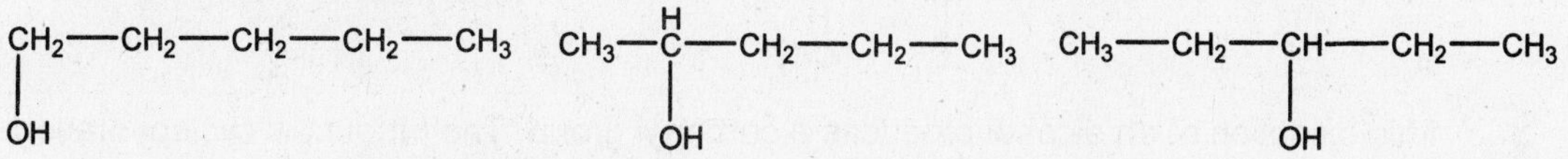

The three structures represent all the unique positions for the alcohol group on a 5C chain.

Next use a 4C chain and attach to side groups, -OH and $-CH_3$.

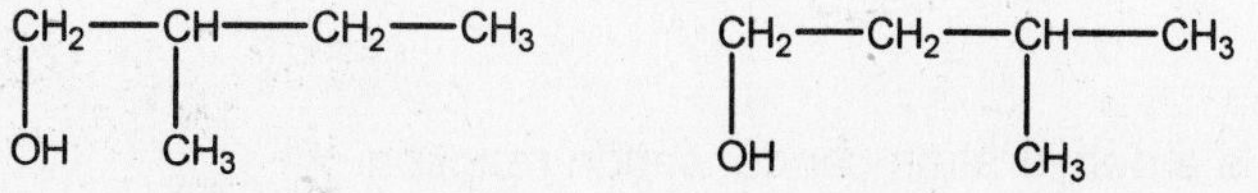

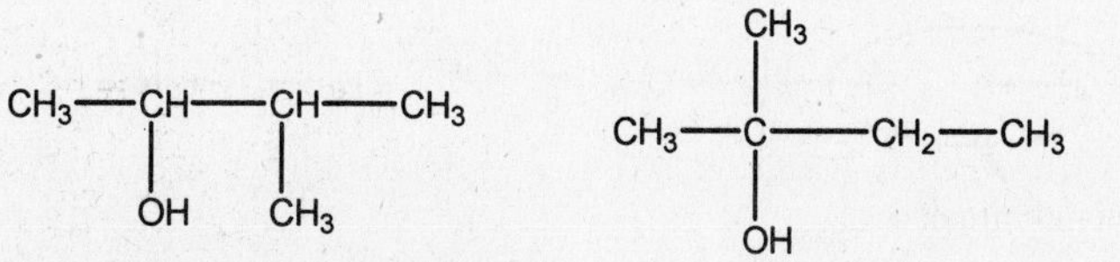

And use a 3C chain with three side groups:

(2 methyl and one –OH)

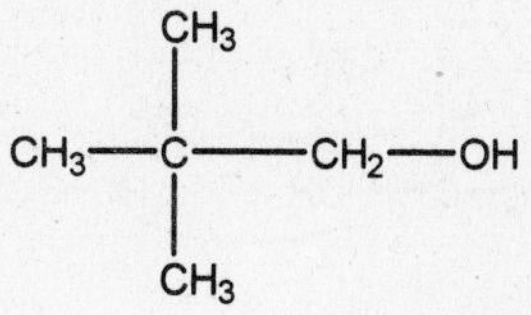

The total number of different structures is 8.

15.64 First draw all primary amines (formula R–NH₂):

$CH_3-CH_2-CH_2-CH_2-NH_2$

$CH_3-CH-CH_2-NH_2$
　　　|
　　CH_3

$CH_3-\underset{\underset{CH_3}{|}}{\overset{\overset{CH_3}{|}}{C}}-NH_2$

$CH_3-CH_2-CH-NH_2$
　　　　　　|
　　　　　CH_3

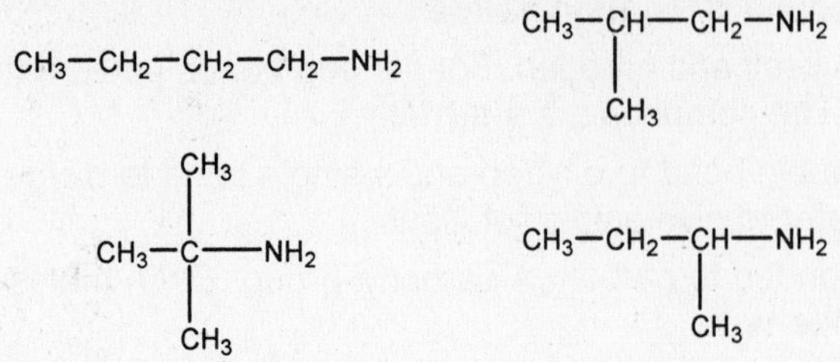

Next draw all secondary amines (formula R–NH–R'):

$CH_3-CH_2-CH_2-NH-CH_3$

$CH_3-\underset{\underset{CH_3}{|}}{CH}-NH-CH_3$

$CH_3-CH_2-NH-CH_2-CH_3$

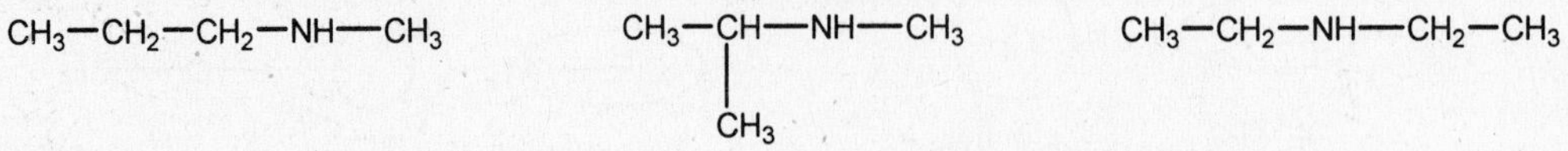

There is only one possible tertiary amine structure (formula R₃–N):

$CH_3-\underset{\underset{CH_3}{|}}{N}-CH_2-CH_3$

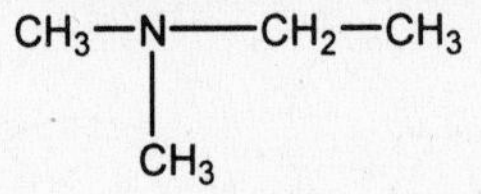

Eight amines with the formula $C_4H_{11}N$ exist.

15.66 a) With mild oxidation, an alcohol group is oxidized to a carbonyl group. The product is 2-butanone.

$CH_3-\underset{\underset{O}{\|}}{C}-CH_2-CH_3$

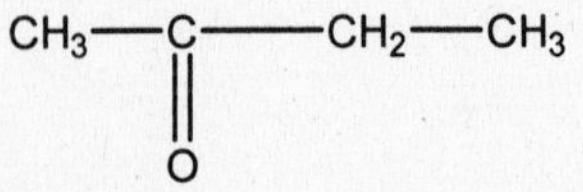

b) With mild oxidation an aldehyde is oxidized to a carboxylic acid. The product is 2-methylpropanoic acid.

$HO-\underset{\underset{O}{\|}}{C}-\underset{\underset{CH_3}{|}}{CH}-CH_3$

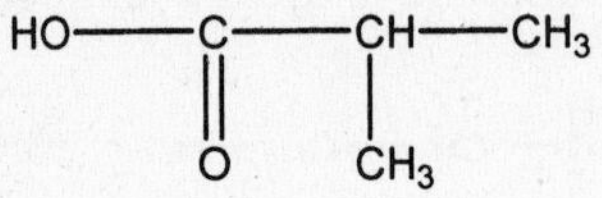

c) Mild oxidation of an alcohol produces a carbonyl group. The product is cyclopentanone.

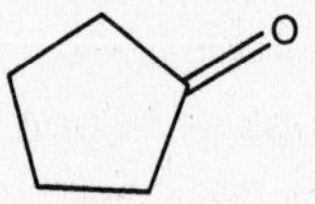

15.68 a) This reaction is a dehydration-condensation reaction.

$CH_3-\overset{\overset{O}{\|}}{C}-(O-H + H)-\underset{\underset{H}{|}}{N}-CH_3 \longrightarrow CH_3-\overset{\overset{O}{\|}}{C}-\overset{\overset{H}{|}}{N}-CH_3 + H_2O$

water eliminated

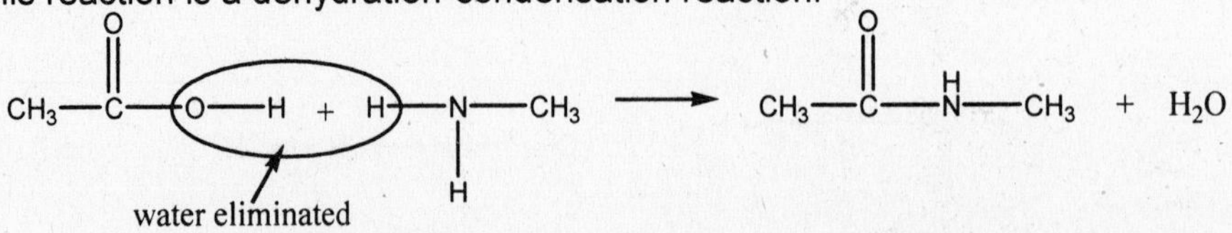

b) An alcohol and a carboxylic acid undergo dehydration-condensation to form an ester.

$CH_3-CH_2-CH_2-\overset{\overset{O}{\|}}{C}-(OH + H)O-\underset{\underset{CH_3}{|}}{\overset{\overset{CH_3}{|}}{CH}} \longrightarrow CH_3-CH_2-CH_2-\overset{\overset{O}{\|}}{C}-O-\underset{\underset{CH_3}{|}}{\overset{\overset{CH_3}{|}}{CH}}$

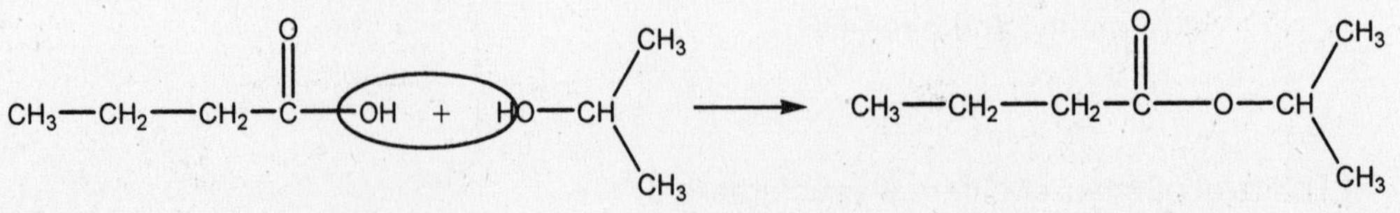

c) This reaction is also ester formation through dehydration-condensation.

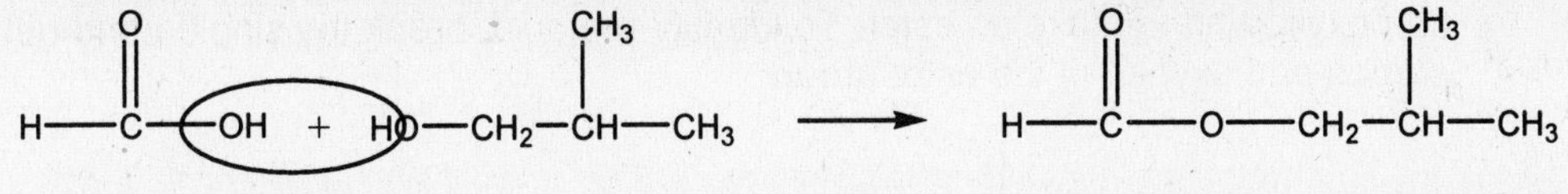

15.70 To break an ester apart, break the –C—O— single bond and add water (–H and –OH) as shown:

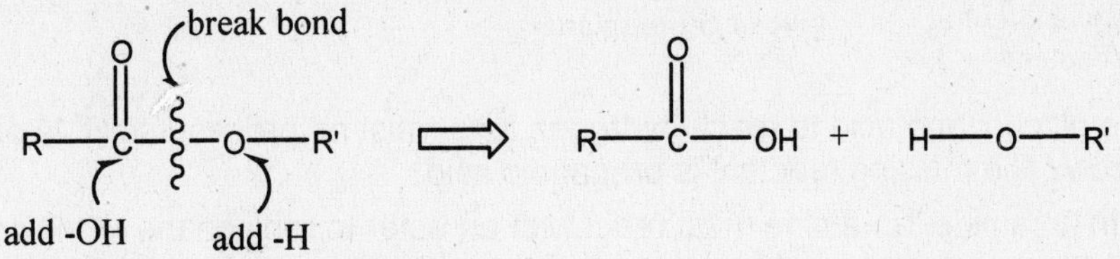

a)

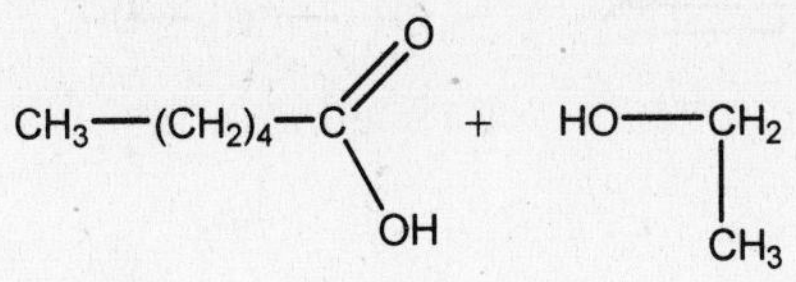

b)

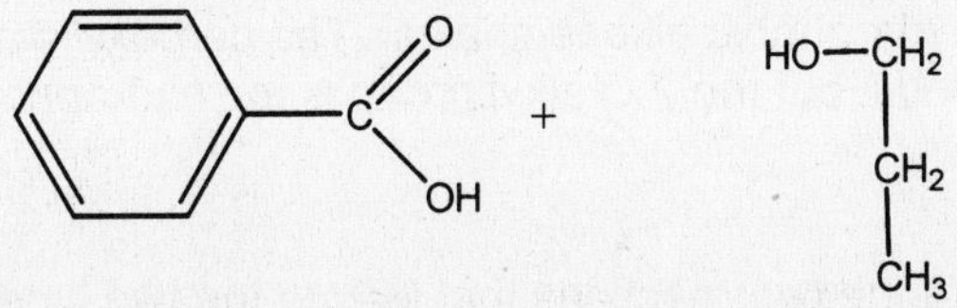

c)

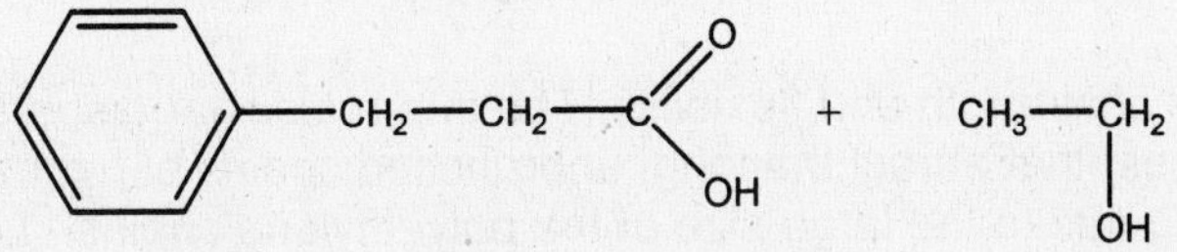

15.72 a) Substitution of Br⁻ occurs by the stronger base, OH⁻.

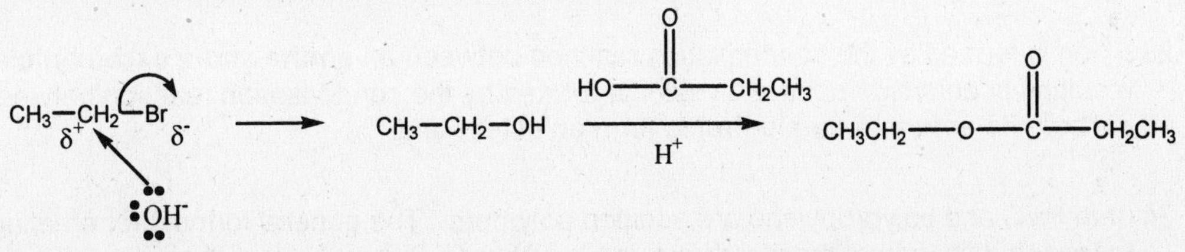

b) The strong base, CN⁻, substitutes for Br. The nitrile is then hydrolyzed to a carboxylic acid.

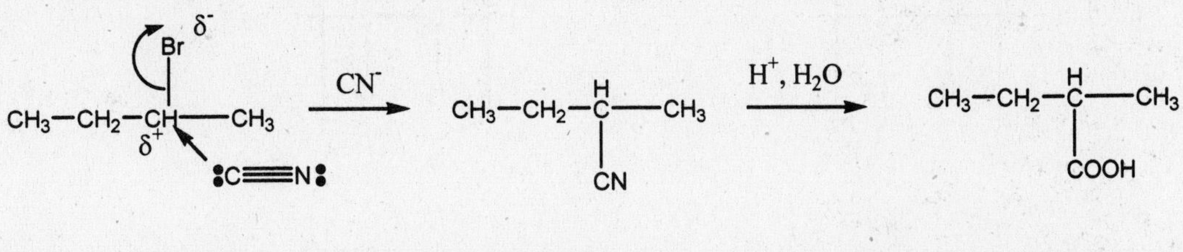

15.74 a) The product is an ester and the given reactant is an alcohol. An alcohol reacts with a carboxylic acid to make an ester. To identify the acid, break the single bond between the carbon and oxygen in the ester group.

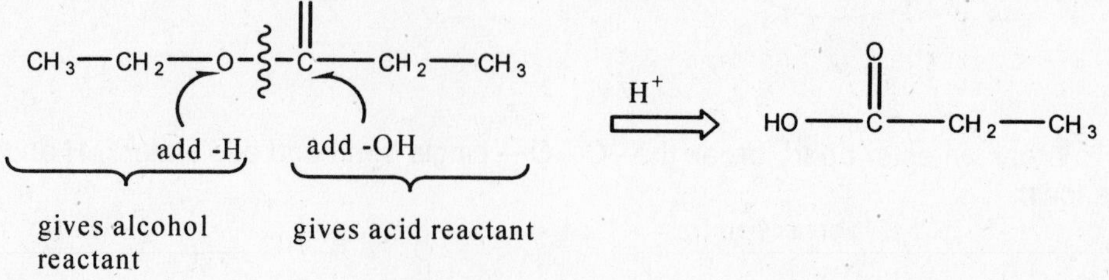

For the alcohol and acid to react, hydrogen ions must be present, so H^+ is present above the arrow. The missing reactant is propanoic acid.

b) To form an amide, an amine must react with an ester to replace the –O-R group. To identify the amine that must be added break the C—N bond in the amide.

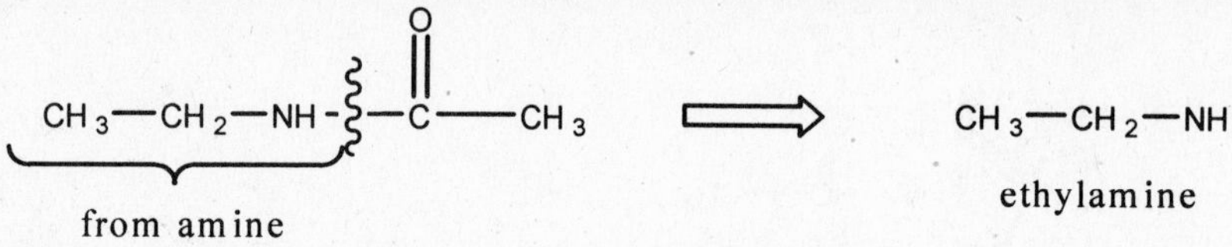

15.76 The compounds (b), (c), and (d) are highly soluble in water because they can form H-bonds with water and have sufficiently short hydrophobic alkane chains. The carboxylic acid portion in compound (e) can form H-bonds, but the 7 C alkyl group is too hydrophobic to dissolve in water.

15.78 Addition and condensation reactions are the two reactions that lead to the two types of synthetic polymers that are named for the reactions that form them.

15.81 Polyethylene comes in a range of strength and flexibility. The intermolecular dispersion forces (also called London forces) that attract the long, unbranched chains of high density polyethylene (HDPE) are strong due to the large size of the polyethylene chains. Low density polyethylene (LDPE) has increased branching that prevents packing and weakens intermolecular dispersion forces.

15.83 Nylon is formed by the condensation reaction between an **amine and a carboxylic acid** resulting in an amide bond. Polyester is formed by the condensation reaction between a **carboxylic acid and an alcohol** to form an ester bond.

15.84 Both PVC and polypropylene are addition polymers. The general formula for creating repeating units from a monomer is given in Table 15.6.

a)

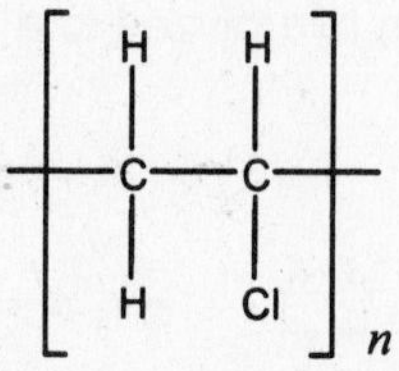

b)

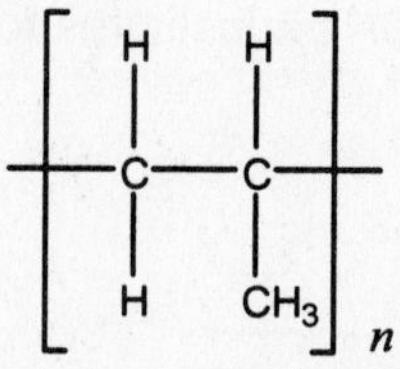

15.86 A carboxylic acid and an alcohol react to form an ester bond.

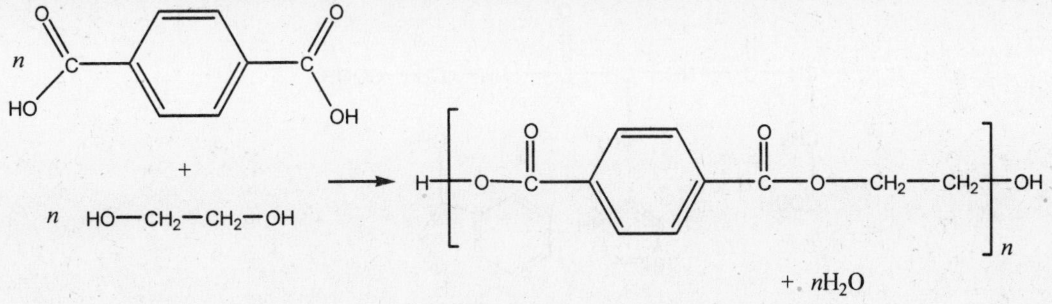

15.88 a) Amino acids form **condensation** polymers called proteins.

b) Alkenes form **addition** polymers, the simplest of which is polyethylene.

c) Simple sugars form **condensation** polymers called polysaccharides.

d) Mononucleotides form **condensation** polymers called nucleic acids.

15.90 Amino acid sequence determines the protein shape and the protein shape determines its function.

15.93 The DNA base sequence contains an information template that is carried by the RNA base sequence (messenger and transfer) to create the protein amino acid sequence. In other words, the DNA sequence determines the RNA sequence, which determines the protein amino acid sequence.

15.94 a) Alanine

R group is $-CH_3$

Complete structure is

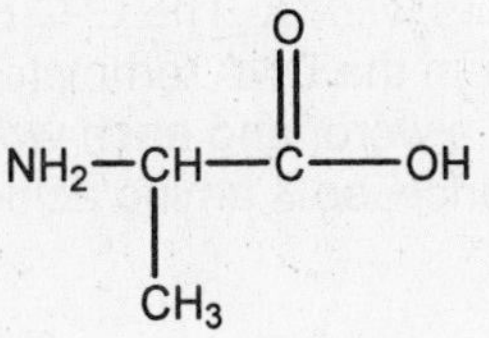

b) Histidine

R group is

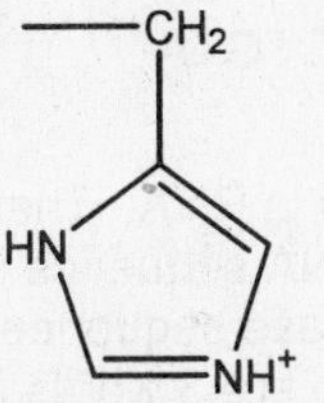

Complete structure is

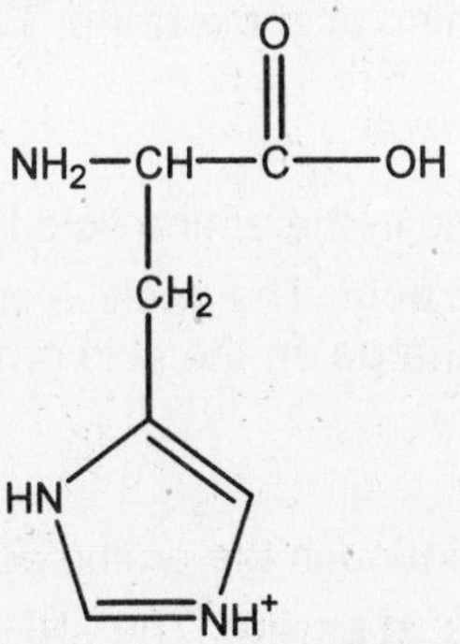

c) Methionine

R group is

$—CH_2—CH_2—S—CH_3$

Complete structure is

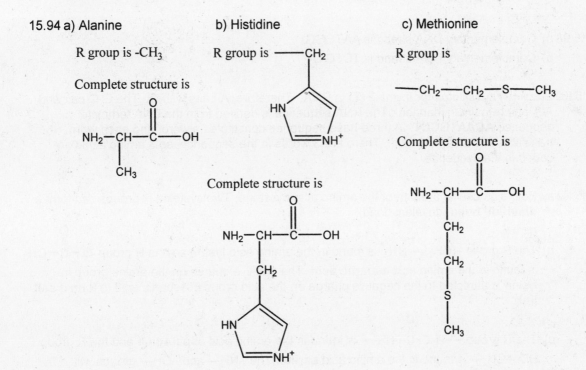

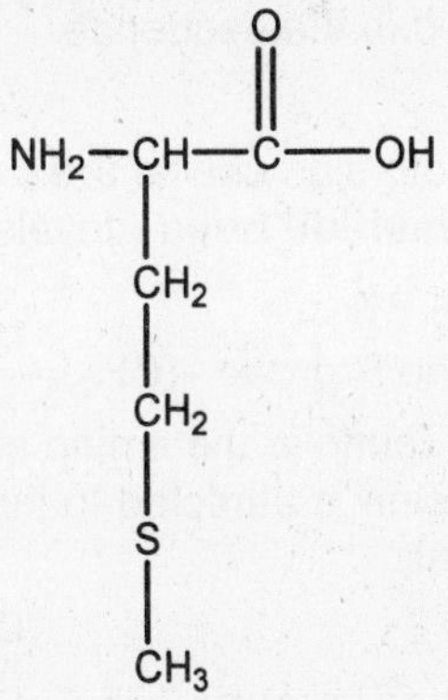

15.96 a) A tripeptide contains three amino acids and two peptide (amide) bonds.

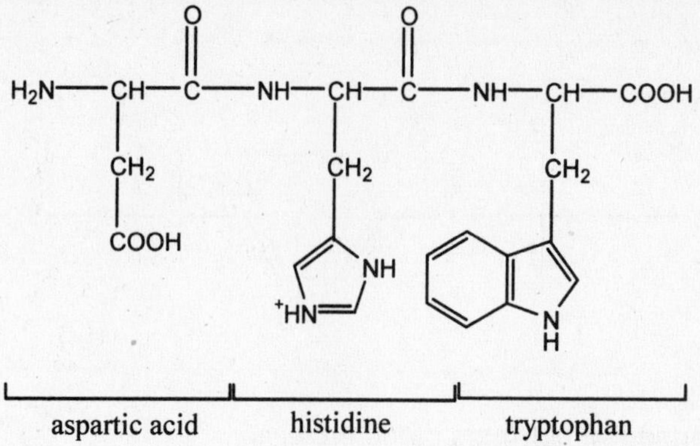

b)

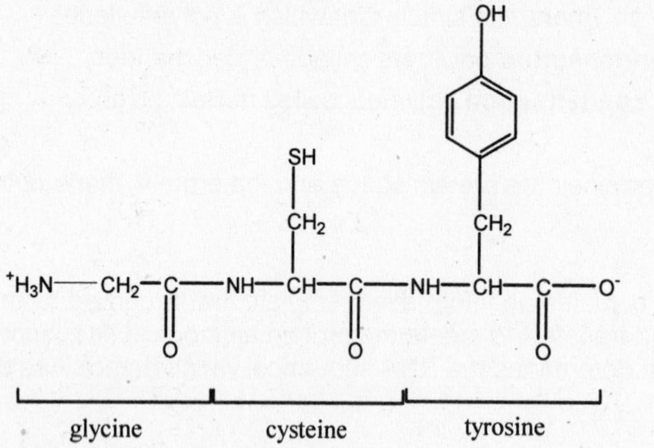

15.98 a) Complementary DNA strand is AATCGG.

b) Complementary DNA strand is TCTGTA.

15.100 Uracil (U) substitutes for thymine (T) in RNA. Therefore, A pairs with U. The G-C pair and A-T pair remain unchanged. The RNA sequence is derived from the DNA template sequence **ACAATGCCT**. A three-base sequence constitutes a word, and each word translates into an amino acid. There are 3 words in the sequence, so **3** amino acids are coded in the sequence.

15.102 a) Both side chains are part of the amino acid **cysteine**. Two cysteine R groups can form a **disulfide bond** (covalent bond).

b) The R group $-(CH_2)_4-NH_3^+$ is found in the amino acid **lysine** and the R group $O^--\overset{O}{\overset{\|}{C}}-CH_2$ is found in the amino acid **aspartic acid**. The positive charge on the amine group in lysine is attracted to the negative charge on the acid group in aspartic acid to form a **salt link**.

c) The R group $-H_2C-\overset{O}{\overset{\|}{C}}-NH-$ is found in the amino acid **asparagine** and the R group $HO-CH_2-$ is found in the amino acid **serine**. The $-NH-$ and $-OH-$ groups will **hydrogen bond**.

d) Both the R group $-CH(CH_3)-CH_3$ from **valine** and the R group $C_6H_5-CH_2-$ from **phenylalanine** are both nonpolar so their interaction is **hydrophobic forces**.

15.104 a) Perform an acid-catalyzed dehydration of the alcohol (elimination), followed by bromination of the double bond (addition of Br_2)

$$CH_3-CH_2-CH_2-OH \xrightarrow{H_2SO_4} CH_3-CH{=}CH_2 \xrightarrow{Br_2}$$

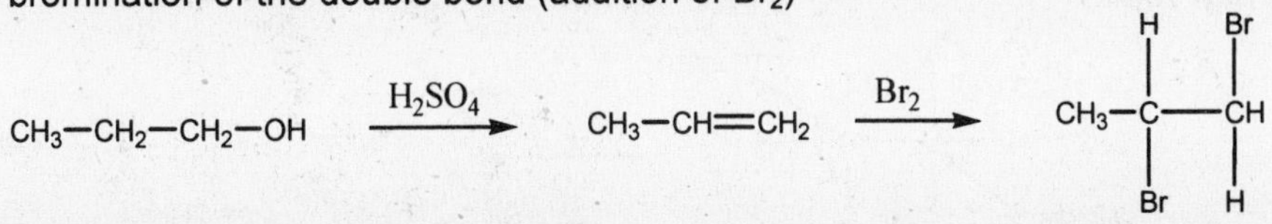

b) The product is an ester, so a carboxylic acid is needed to prepare the ester. First, oxidize one mole of ethanol to acetic acid:

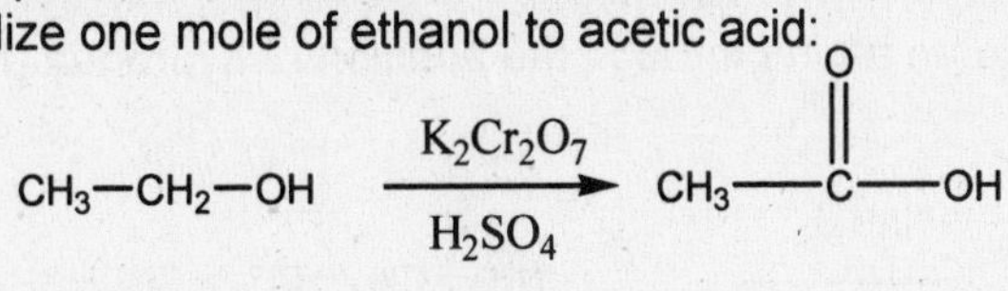

Then react one mole of acetic acid with a second mole of ethanol to form the ester:

$$+ \; H_2O$$

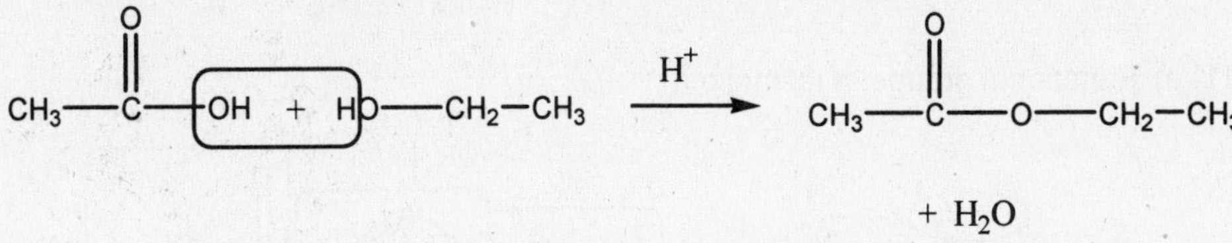

15.106 Compound X is oxidized to a carboxylic acid and is formed from an alcohol. An alcohol group –OH is oxidized to a carbonyl group, =O, either as a ketone or an aldehyde. A ketone cannot be oxidized further, but an aldehyde can be oxidized to a carboxylic acid. Compound X must be an aldehyde formed from the oxidation of 2-methyl-propanol. Structure of compound X is 2-methylpropanal:

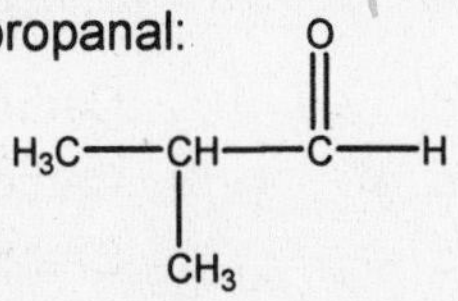

15.108 The desired reaction is the displacement of Br⁻ by OH⁻:

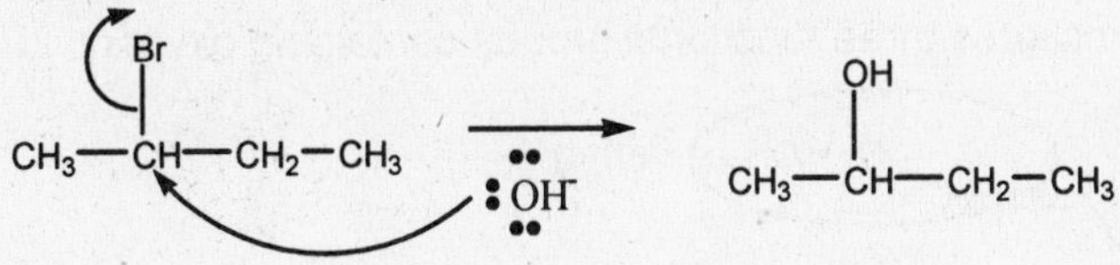

However, the elimination of HX also occurs in the presence of a strong base:

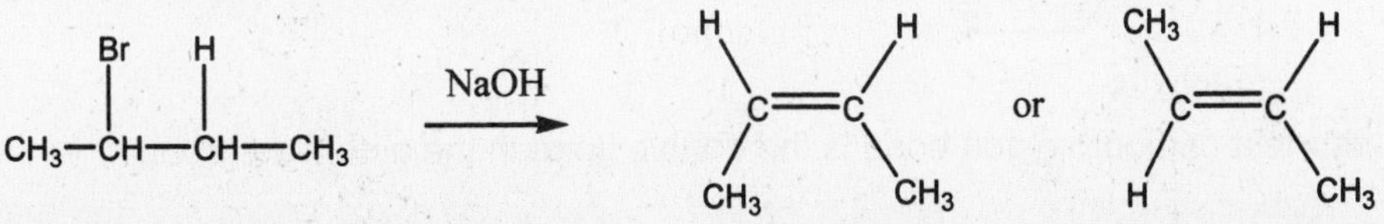

A simple test to detect the presence of a double bond is the addition of liquid Br_2. Bromine is reddish-brown in color and adds easily to the double bond to form a brominated alkane, which has no color. The reddish-brown color of liquid Br_2 disappears if added to an alkene.

15.110 The two structures are:

$$NH_2-CH_2-CH_2-CH_2-CH_2-CH_2-NH_2 \qquad NH_2-CH_2-CH_2-CH_2-CH_2-NH_2$$

cadaverine   putrescine

The addition of cyanide ($[C{\equiv}N]^-$) to form a nitrile is a convenient way to increase the length of a carbon chain:

$$+ \; 2Br^-$$

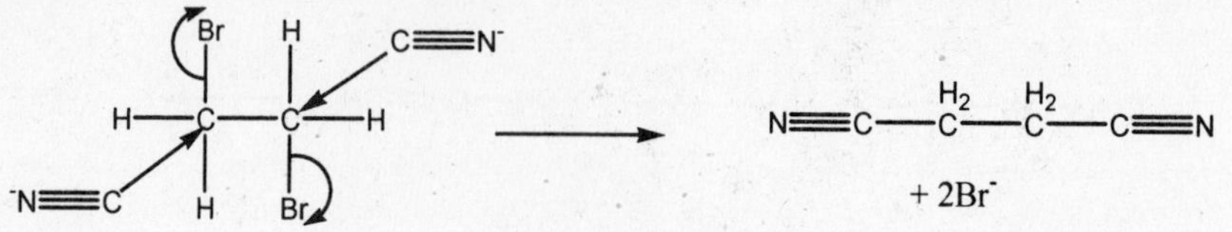

The nitrile can then be reduced to an amine through the addition of hydrogen (NaBH$_4$ is a hydrogen-rich reducing agent):

$$N{\equiv}C-CH_2-CH_2-C{\equiv}N \; \xrightarrow{NaBH_4} \; NH_2-CH_2-CH_2-CH_2-CH_2-NH_2$$

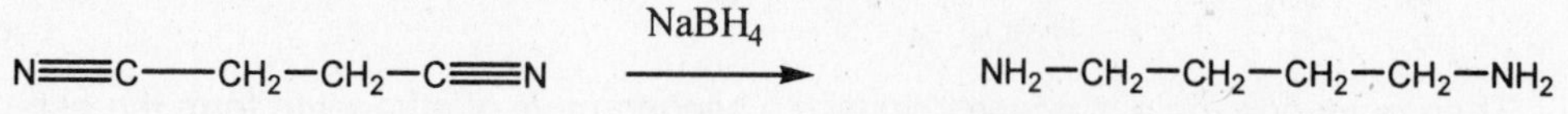

15.111 a) Functional groups in jasmolin II

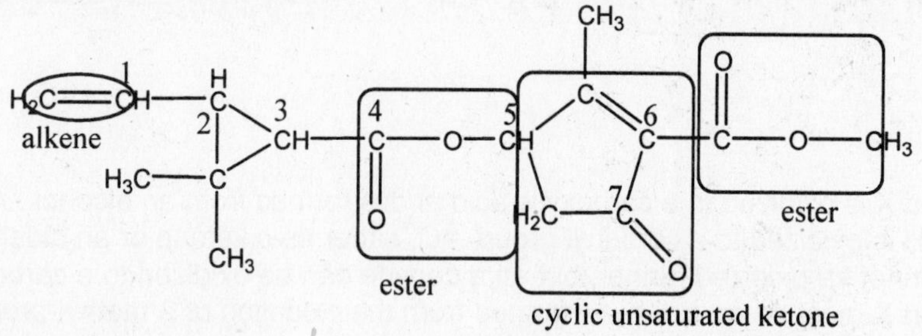

b) Carbon 1 is sp^2 hybridized.   Carbon 2 is sp^3 hybridized.
    Carbon 3 is sp^3 hybridized.   Carbon 4 is sp^2 hybridized.
    Carbon 5 is sp^3 hybridized.   Carbon 6 and 7 are sp^2 hybridized.

c) Carbons 2, 3 and 5 are chiral centers as they are each bonded to 4 different groups.

15.113 A vanillin molecule includes three functional groups containing oxygen:

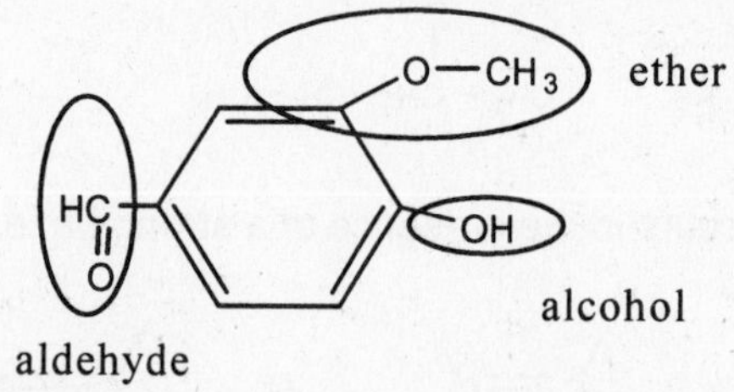

The shortest carbon-oxygen bond is the double bond in the aldehyde group.

15.116 The resonance structures show that the bond between carbon and nitrogen will have some double bond character that restricts rotation around the bond.

15.120 a) The hydrolysis process requires the addition of water to break the peptide bonds.

b) Convert each mass to moles, and divide the moles by the smallest number to determine molar ratio, and thus relative numbers of the amino acids.

(3.00 g gly) ÷ (75.07 g/mol) = 0.0399_{62} mol glycine

(0.90 g ala) ÷ (89.09 g/mol) = 0.010_{10} mol alanine

Other molar amounts are 0.0315_{83} mol valine, 0.0599_{32} mol proline, 0.0694_{64} mol serine and 0.493_{68} mol arginine. Dividing each molar amount by 0.010_{10} gives the following relative amounts: glycine = 3.95_{66} (**4**), alanine = 1.0 (**1**), valine = 3.12_{70} (**3**), proline = 5.93_{39} (**6**), serine = 6.87_{76} (**7**) and arginine = 48.8_{79} (**49**).

c) Minimum **M** = (4 x 75.07g/mol) + (1 x 89.09 g/mol) + (3 x 117.15 g/mol) + (6 x 115.13 g/mol) + (7 x 105.09 g/mol) + (49 x 174.20 g/mol) = **10700 g/mol**

15.122 When the 2-butanone is reduced, an equal amount of both isomers is produced since the reaction does not favor the production of one over the other. In a 1:1 mixture of the two stereoisomers, the light rotated to the left by one isomer is canceled by the light rotated to the right by the other isomer. The mixture is not optically active since there is no net rotation of light. The two stereoisomers of 2-butanol are shown below.

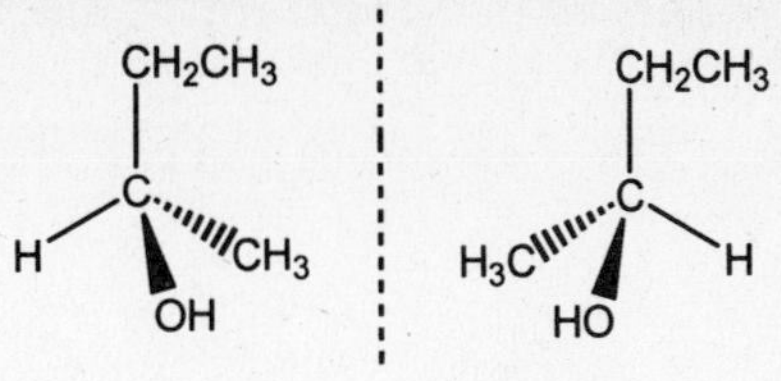

CHAPTER 16

KINETICS: RATES AND MECHANISMS OF CHEMICAL REACTIONS

FOLLOW-UP PROBLEMS

16.1 a) Plan: The balanced equation is $4NO(g) + O_2(g) \rightarrow 2N_2O_3$. Choose O_2 as the reference because its coefficient is 1. Four molecules of NO (nitrogen monoxide) are consumed for every one O_2 molecule, so the rate of O_2 disappearance is ¼ the rate of NO decrease. By similar reasoning, the rate of O_2 disappearance is ½ the rate of N_2O_3 (dinitrogen trioxide) increase.

Solution:

$$rate = -\frac{1}{4}\frac{\Delta[NO]}{\Delta t} = -\frac{\Delta[O_2]}{\Delta t} = +\frac{1}{2}\frac{\Delta[N_2O_3]}{\Delta t}$$

b) Plan: Because NO is decreasing, its rate of concentration change is negative. Substitute the negative value into the expression and solve for $\Delta[O_2]/\Delta t$.

Solution:

$$-\frac{1}{4}\frac{\Delta[NO]}{\Delta t} = -\frac{\Delta[O_2]}{\Delta t} \Rightarrow -\frac{1}{4}\left(-1.60x10^{-4}\ mol/L\cdot s\right) = -\frac{\Delta[O_2]}{\Delta t}$$

$$\frac{\Delta[O_2]}{\Delta t} = \mathbf{-4.00x10^{-5}\ mol/L\cdot s}$$

Check: a) Verify that the general formula shown in Equation 16.2 produces the same result when a = 4, b = 1 and c = 2. b) The rate should be negative because O_2 is disappearing. The rate can also be expressed as $-\Delta[O_2]/\Delta t = 4.00 \times 10^{-5}$ mol/L·sec, where the negative sign on the left side of the equation denote reactant disappearance as well. It makes sense the rate of NO disappearance is 4 times greater than the rate of O_2 disappearance, because 4 molecules of NO disappear for every one O_2 molecule.

16.2 The exponent of $[Br^-]$ is 1, so the reaction is **first order with respect to Br^-**. Similarly, the reaction is **first order with respect to BrO_3^-**, and **second order with respect to H^+**. The overall reaction order is (1+ 1 + 2) = 4, or **fourth order overall**.

16.3 Solution: Assume that the rate law takes the general form rate = $k[H_2]^m[I_2]^n$. To find how the rate varies with respect to $[H_2]$, find two experiments in which $[H_2]$ changes but $[I_2]$ remains constant. Take the ratio of rate laws for Experiments 1 and 3 to find m.

$$\frac{rate\,3}{rate\,1} = \left(\frac{[H_2]_3}{[H_2]_1}\right)^m \Rightarrow \frac{9.3x10^{-23}}{1.9x10^{-23}} = \left(\frac{0.0550}{0.0113}\right)^m$$

$4.89 = (4.87)^m$; therefore **m = 1**

If the reaction order was more complex, an alternate method of solving for m is:

log(4.89) = $m$log(4.87); m = log(4.89)/log(4.87) = 1

Take the ratio of rate laws for Experiments 2 and 4 pair to find n.

$$\frac{rate\,4}{rate\,2} = \left(\frac{[I_2]_4}{[I_2]_2}\right)^n \Rightarrow \frac{1.9x10^{-22}}{1.1x10^{-22}} = \left(\frac{0.0056}{0.0033}\right)^n \Rightarrow 1.73 = (1.70)^n;\ \text{therefore } \mathbf{n = 1}$$

The rate law is rate = k$[H_2][I_2]$ and is second order overall.

Check: The rate is first order with respect to each reactant. If one reactant doubles, the rate doubles; if one reactant triples, the rate triples. If *both* reactants double (each reactant x 2), the rate quadruples (2 x 2). Comparing reactions 1 and 4, the $[H_2]$ doubles (x 2) and the $[I_2]$ increases by 5 times (x 5). Accordingly, the rate increases by 10 (2 x 5). This exercise verifies that the determined rate law exponents are correct.

16.4 Plan: The rate expression indicates that the reaction order is two (exponent of [HI] = 2), so use the integrated second-order law. Substitute the given concentrations and the rate constant into the expression and solve for time.

Solution:

$$\frac{1}{[A]_t} - \frac{1}{[A]_0} = kt$$

$$\frac{1}{0.00900\,mol/L} - \frac{1}{0.0100\,mol/L} = \left(2.4x10^{-21}\ L/mol \cdot s\right)t$$

$$t = \frac{11.111\,L/mol}{2.4x10^{-21}\ L/mol \cdot s} = 4.6x10^{21}\ \text{sec}$$

$$\left(4.6x10^{21}\ \text{sec}\right)\left(\frac{hr}{3600\ \text{sec}}\right)\left(\frac{day}{24\,hr}\right)\left(\frac{yr}{365\,days}\right) = \mathbf{1.5x10^{14}\ yr}$$

Check: The problem indicates this decomposition reaction is very slow, so it should take a long time to decrease the HI concentration. The incredibly small rate constant also indicates that the reaction rate is very slow. A hundred thousand billion years is indeed a long time.

16.5 Plan: Rearrange the first-order half-life equation to solve for k.

Solution: $k = \frac{\ln 2}{t_{1/2}} = \frac{\ln 2}{13.1\,h} = \mathbf{5.29x10^{-2}\ h^{-1}}$

Check: Verify that the units on k reflect the data given.

16.6 Plan: Activation energy, rate constant at T_1, and a second temperature, T_2, are given. Substitute these values into Equation 16.8 and solve for k_2, the rate constant at T_2.

Solution: Rearranging Equation 16.8 to solve for k_2:

$$\ln\frac{k_2}{k_1} = -\frac{E_a}{R}\left(\frac{1}{T_2} - \frac{1}{T_1}\right)$$

$$\ln k_2 - \ln k_1 = -\frac{E_a}{R}\left(\frac{1}{T_2} - \frac{1}{T_1}\right)$$

$$\ln k_2 = \ln k_1 - \frac{E_a}{R}\left(\frac{1}{T_2} - \frac{1}{T_1}\right) = \ln\left(0.286\,s^{-1}\right) - \frac{1.00x10^2\ kJ/mol}{8.314\ J/mol \cdot K\left(\frac{kJ}{1000\ J}\right)}\left(\frac{1}{490.\ K} - \frac{1}{500.\ K}\right)$$

$$\ln k_2 = \ln\left(0.286\,s^{-1}\right) - 0.490_{34}$$

$$\ln k_2 = -1.74_{27}$$

$$k_2 = e^{-1.7427} = \mathbf{0.175\,s^{-1}}$$

Check: The temperature only changes by 10K degrees, so it is likely that the rate constant at T_2 would not change by orders of magnitude; 0.175 s^{-1} is in the same "ballpark" as 0.286 s^{-1}. Furthermore, the rate should decrease at the lower temperature, so the lower rate of 0.175 s^{-1} at 490K is expected.

16.7 Plan: Use Figure 16.19 as a guide for labeling the diagram.

Solution: The reaction energy diagram indicates that $O(g) + H_2O(g) \rightarrow 2OH(g)$ is an endothermic process, because the energy of the product is higher than the energy of the reactants. The highest point on the curve indicates the transition state. In the transition state, an oxygen atom forms a bond with one of the hydrogen atoms on the H_2O molecule (dashed line) and the O–H bond (solid line) in H_2O weakens. $E_{a(fwd)}$ is the difference between ΔH_{rxn} and $E_{a(rev)}$.

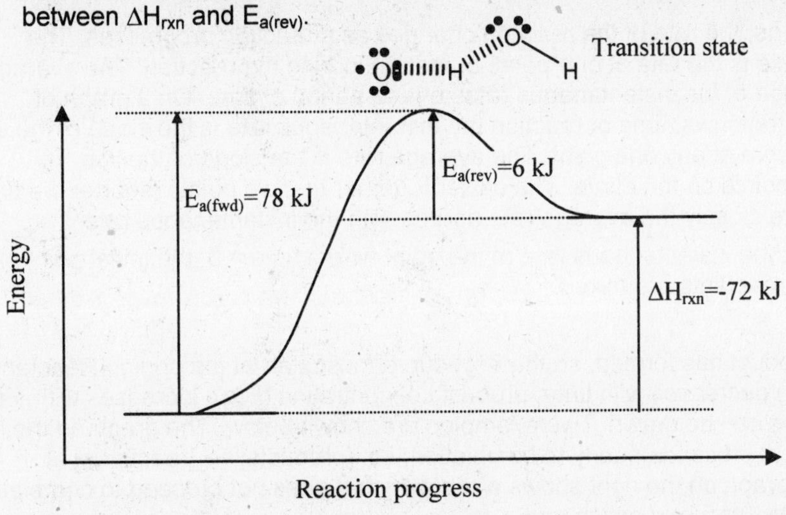

16.8 Plan: Sum the elementary steps to obtain the overall equation. Think of the individual steps as reactions along the "reaction progress" axis, i.e. sum each equation as written (this is different from the application of Hess's law in which reactions are reversed as necessary to reach a final reaction).

Solution:

	b) Molecularity	c) Rate law
(1) $2NO(g) \rightarrow$ ~~$N_2O_2(g)$~~	2	$rate_1 = k_1[NO]^2$
(2) $2[H_2(g) \rightarrow$ ~~$2H(g)$~~$]$	1	$rate_2 = k_2[H_2]$
(3) ~~$N_2O_2(g)$~~ + ~~$H(g)$~~ $\rightarrow$ ~~$N_2O(g)$~~ + ~~$HO(g)$~~	2	$rate_3 = k_3[N_2O_2][H]$
(4) 2[~~$HO(g)$~~ + ~~$H(g)$~~ $\rightarrow H_2O(g)$]	2	$rate_4 = k_4[HO][H]$
(5) ~~$H(g)$~~ + ~~$N_2O(g)$~~ $\rightarrow$ ~~$HO(g)$~~ + $N_2(g)$	2	$rate_5 = k_5[H][N_2O]$

a) $2NO(g) + 2H_2(g) \rightarrow 2H_2O(g) + N_2(g)$

Check: The overall equation is balanced.

END-OF-CHAPTER PROBLEMS

16.2 Rate is proportional to concentration. An increase in pressure would create an increase in the number of gas molecules per unit volume. In other words, the gas concentration increases due to increased pressure, so the **reaction rate increases**. Increased pressure also causes more collisions between gas molecules. Why does the increased collision rate not necessarily increase reaction rate?

16.3 The addition of water will dilute the concentrations of all solutes dissolved in the water. If any of these solutes are reactants, the **rate of the reaction will decrease**.

16.5 An increase in temperature affects the rate of a reaction by increasing the number of collisions, but more importantly the energy of collisions increases. As the energy of collisions increase, more collisions result in reaction (i.e. reactants → products) so the rate of reaction increases.

16.8 a) For most reactions, the rate of the reaction changes as a reaction progresses. The instantaneous rate is the rate at one point, or instant, during the reaction. The average rate is the average of the instantaneous rates over a period of time. On a graph of reactant concentration vs. time of reaction the instantaneous rate is the slope of the tangent to the curve at any one point. The average rate is the slope of the line connecting two points on the curve. The closer together the two points (shorter the time interval) the more closely the average rate agrees with the instantaneous rate.

b) The initial rate is the instantaneous rate at the point where time = 0, the initial point on the graph when reactants are mixed.

16.10 At time t = 0, no product has formed, so the B(g) curve must start at the origin. Reactant concentration (A(g)) decreases with time; product concentration (B(g)) increases with time. Many correct graphs can be drawn. Two examples are shown below. The graph on the left shows a reaction that proceeds nearly to completion, i.e. [products] >> [reactants] at reaction end. The graph on the right shows a reaction that does not proceed to completion, i.e. [reactants] > [products] at reaction end.

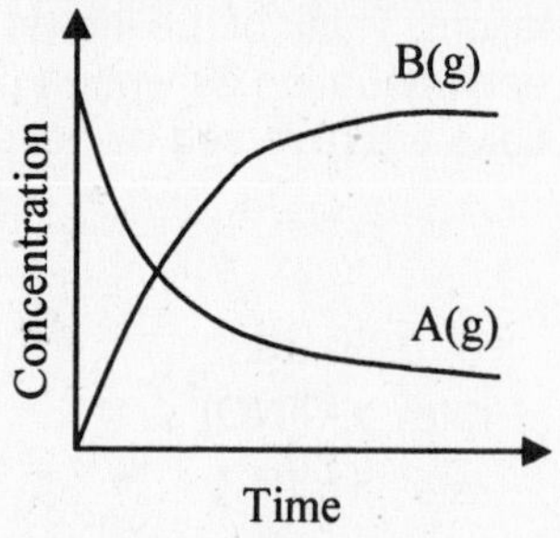

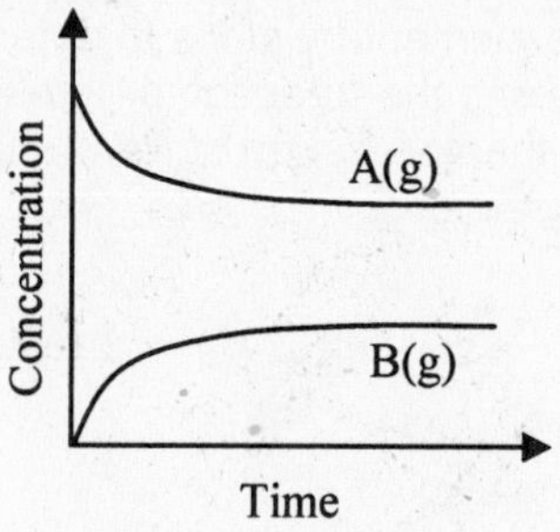

16.12 a) The average rate from t = 0 to t = 20.0 s is proportional to the slope of the line connecting these two points:

$$\text{rate} = -\tfrac{1}{2}\,\Delta[AX_2]/\Delta t = -(0.0088\text{ M} - 0.0500\text{ M})/(20.0\text{ s} - 0\text{ s}) = \mathbf{0.0010\ M/s}$$

The negative of the slope is used because rate is defined as the change in product concentration with time. If a reactant is used, the rate is the negative of the change in reactant concentration. The ½ factor is included to account for the stoichiometric coefficient of 2 for AX_2 in the reaction.

b)

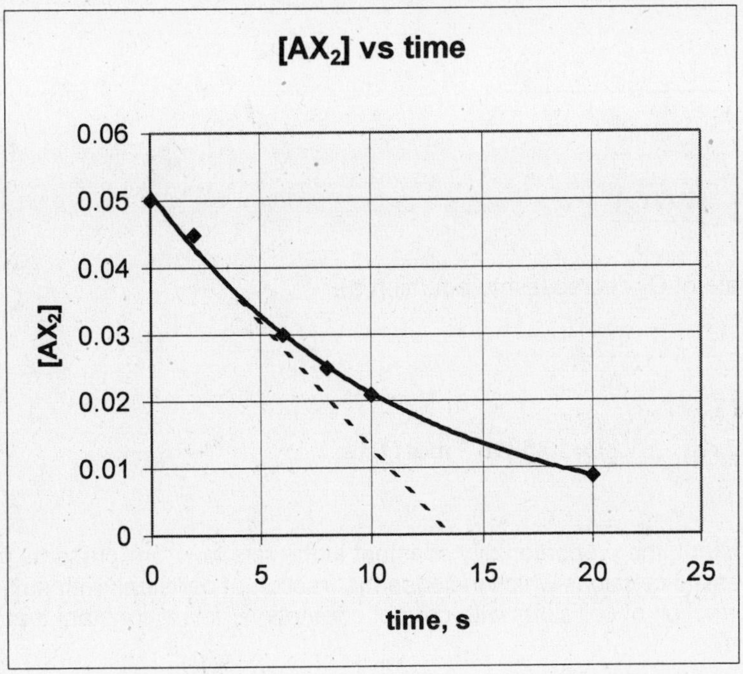

The slope of the tangent to the curve (dashed line) at t = 0 is approximately –0.004 M/s. This initial rate is greater than the average rate as calculated in part (a). The **initial rate is greater than the average rate** because rate decreases as reactant concentration decreases.

16.14 Use Equation 16.2 to describe the rate of this reaction in terms of reactant disappearance and product appearance.

$$rate = -\frac{1}{2}\frac{\Delta[A]}{\Delta t} = \frac{\Delta[B]}{\Delta t} = \frac{\Delta[C]}{\Delta t}$$

Reactant A decreases twice as fast as product C increases because 2 molecules of A disappear for every molecule of C that appears.

$$-\frac{1}{2}\frac{\Delta[A]}{\Delta t} = \frac{\Delta[C]}{\Delta t}$$

$$\frac{\Delta[A]}{\Delta t} = -2\frac{(2\ M)}{(1\ s)} = -4\ M/s \text{ or } -4\ mol/L\cdot s$$

The negative value indicates that [A] is decreasing as the reaction progresses. The rate of reaction is always expressed as a positive number, so [A] is decreasing at a rate of **4 mol/L·s**.

16.16 $rate = -\frac{\Delta[A]}{\Delta t} = -\frac{1}{2}\frac{\Delta[B]}{\Delta t} = \frac{\Delta[C]}{\Delta t}$

[B] decreases twice as fast as [A] so [A] is decreasing at a rate of ½ (0.5 mol/L·s) or **0.25 mol/L·s**.

16.18 A term with a negative sign is a reactant; a term with a positive sign is a product. The inverse of the fraction becomes the coefficient of the molecule:

$$2N_2O_5(g) \rightarrow 4NO_2(g) + O_2(g)$$

16.21 $rate = -\frac{\Delta[N_2]}{\Delta t} = -\frac{1}{3}\frac{\Delta[H_2]}{\Delta t} = \frac{1}{2}\frac{\Delta[NH_3]}{\Delta t}$

16.22 a) $rate = -\frac{1}{3}\frac{\Delta[O_2]}{\Delta t} = \frac{1}{2}\frac{\Delta[O_3]}{\Delta t}$

b) Substitute the rate of O_2 decrease into equation a).

$$-\frac{1}{3}\frac{(-2.17x10^{-5}\ M)}{1s} = \frac{1}{2}\frac{\Delta[O_3]}{\Delta t}$$

$$\frac{\Delta[O_3]}{\Delta t} = \frac{2}{3}(-2.17x10^{-5}\ M/s) = \mathbf{1.45x10^{-5}\ mol/L \cdot s}$$

16.23 a) *k* is the rate constant, the proportionality constant in the rate law. *k* represents the fraction of successful collisions which includes the fraction of collisions with sufficient energy and the fraction of collisions with correct orientation. *k* is a constant that varies with temperature.

b) *m* represents the order of the reaction with respect to [A] and *n* represents the order of the reaction with respect to [B]. The order is the exponent in the relationship between rate and reactant concentration and defines how reactant concentration impacts rate. The order of a reactant does not necessarily equal its stoichiometric coefficient in the balanced equation. If a reaction is an elementary reaction, meaning the reaction occurs in only one step, then the orders and stoichiometric coefficients are equal. However, if a reaction occurs in a series of elementary reactions, called a mechanism, then the rate law is based on the slowest elementary reaction in the mechanism. The orders of the reactants will equal the stoichiometric coefficients in the slowest elementary reaction.

c) For the rate law rate = $k[A][B]^2$ substitute in the units:

$$\frac{mol/L}{\min} = k(mol/L)(mol/L)^2;\ k = \frac{mol/L}{\min(mol/L)(mol/L)^2} = \mathbf{\frac{L^2}{mol^2 \cdot min}}$$

16.25 a) The rate doubles. If rate = $k[A]^1$ and [A] is doubled, then the rate law becomes rate = $k[2\ x\ A]^1$. The rate increases by 2^1 or **2**.

b) The rate decreases by a factor of four. If rate = $k[½\ x\ B]^2$, then rate decreases to $(½)^2$ or ¼ of its original value.

c) The rate increases by a factor of nine. If rate = $k[3\ x\ C]^2$, then rate increases to 3^2 or **9** times its original value.

16.26 The order for each reactant is the exponent on the reactant concentration in the rate law. The orders with respect to $[BrO_3^-]$ and to $[Br^-]$ are both 1. The order with respect to $[H^+]$ is 2. The overall reaction order is the sum of each reactant order: 1 + 1 + 2 = **4**.

16.28 a) The rate is first order with respect to $[BrO_3^-]$. If $[BrO_3^-]$ doubles, the rate **doubles**. (see problem 16.25a)

b) The rate is first order with respect to $[Br^-]$. If $[Br^-]$ is halved, the rate is **halved**.

c) The rate is second order with respect to $[H^+]$. If $[H^+]$ is quadrupled, the rate increases by 4^2 or **16.**

16.30 The order with respect to $[NO_2]$ is 2 and the order with respect to $[Cl_2]$ is 1. The overall order is the sum of the orders of individual reactants: 2 + 1 = **3** for the overall order.

16.32 a) The rate is second order with respect to $[NO_2]$. If $[NO_2]$ is tripled, the rate increases by a factor 3^2 or **9**.

b) If rate = $k[2 \times NO_2]^2[2 \times Cl_2]^1$, then the rate increases by a factor of $2^2 \times 2^1$ = **8**.

c) If Cl_2 is halved, the reaction rate is **halved** because the rate is first order with respect to $[Cl_2]$.

16.34 a) To find the order for reactant A, first identify the reaction experiments in which [A] changes but [B] is constant. Set up a proportionality: $rate_{exp\ a}/rate_{exp\ b} = ([A]_{exp\ a}/[A]_{exp\ b})^m$. Fill in the values given for rates and concentrations and solve for m, the order with respect to [A]. Repeat the process to find the order for reactant B.

Use experiments 1 and 2 (or 3 and 4 would work) to find the order with respect to [A].

$$\frac{rate_{exp1}}{rate_{exp2}} = \left(\frac{[A]_{exp1}}{[A]_{exp2}}\right)^m$$

$$\frac{5.00\ mol/L \cdot min}{45.00\ mol/L \cdot min} = \left(\frac{0.100\ mol/L}{0.300\ mol/L}\right)^m; \quad \frac{1}{9} = \left(\frac{1}{3}\right)^m$$

$$\log\left(\frac{1}{9}\right) = m \log\left(\frac{1}{3}\right); \quad m = 2$$

Using experiments 3 and 4 also gives **2nd order with respect to [A]**.

Use experiments 1 and 3 with [A] = 0.100 M or 2 and 4 with [A] = 0.300 M to find order with respect to [B].

$$\frac{rate_{exp1}}{rate_{exp3}} = \left(\frac{[B]_{exp1}}{[B]_{exp3}}\right)^n$$

$$n = \frac{\log\left(\dfrac{5.00\ mol/L \cdot \min}{10.0\ mol/L \cdot \min}\right)}{\log\left(\dfrac{0.100\ mol/L}{0.200\ mol/L}\right)} = 1$$

The reaction is **first order with respect to [B]**.

b) The rate law, without a value for k, is **rate = k $[A]^2[B]$.**

c) Using experiment 1 to calculate k:

$$k = \frac{rate}{[A]^2[B]} = \frac{5.00\ mol/L \cdot \min}{(0.100\ mol/L)^2(0.100\ mol/L)} = \mathbf{5.00x10^3\ L^2/mol^2 \cdot min}$$

16.36 a) A first order rate law follows the general expression, rate = k[A]. The reaction rate is expressed as a change in concentration per unit time with units of M/time or mol/L·time. Since [A] has units of M (the brackets stand for concentration), then k **has units of time^{-1}**:

rate = k[A]

(mol/L)/time = time^{-1}(mol/L)

b) Second order: rate = k[A]^2

(mol/L)/time = k(mol/L)^2

k = (mol/L)/time ÷ (mol/L)^2 = **L/mol·time** or M^{-1} time^{-1}

c) Third order: rate = k[A]^3

(mol/L)/time = k((mol/L))^3

k = (mol/L)/time ÷ (mol/L)^3 = **L^2/mol^2·time** or M^{-2} time^{-1}

d) 5/2 order: rate = k[A]$^{5/2}$

(mol/L)/time = k(mol/L)$^{5/2}$

k = (mol/L)/time ÷ (mol/L)$^{5/2}$ = **L$^{3/2}$/mol$^{3/2}$·time** or M$^{-3/2}$ time^{-1}

16.39 The integrated rate law can be used to plot a linear graph. If the plot of [reactant] vs. time is linear, the order is zero. If the plot of ln[reactant] vs. time is linear the order is first. If the plot of inverse concentration vs. time is linear the order is second.

a) Reaction is first order since ln[reactant] vs. time is linear.

b) Reaction is second order since 1/[reactant] vs. time is linear.

c) Reaction is zero order since [reactant] vs. time is linear.

16.41 The rate expression indicates that this reaction is second order overall, so use the second order integrated rate law (Equation 16.5) to find time.

$$\frac{1}{[A]_t} - \frac{1}{[A]_0} = kt$$

$$\frac{1}{\frac{1}{3}(1.50\ M)} - \frac{1}{1.50\ M} = (0.2\ L/mol \cdot s)t$$

t = **7 s**

16.43 a) The given information is the amount that has reacted in a specified amount of time. With this information the integrated rate law must be used to find a value for the rate constant. For a first order reaction the rate law is ln([A]$_0$/[A]$_t$) = kt. Using the fact that 50.0% has decomposed the concentration of A has decreased by half, giving a value for the ratio of ([A]$_0$/[A]$_t$) = (1/½) = 2.

$$\ln(2.00) = k(10.5\ \text{min});\ k = \mathbf{6.60x10^{-2}\ min^{-1}}$$

b) Use the value for k calculated in part a, use ([A]$_0$/[A]$_t$) = (1/¼) = 4.00 and solve for time.

$$\ln(4.00) = (6.60x10^{-2}\ \text{min}^{-1})t;\ t = \mathbf{21.0\,min}$$

16.45 a) In a first order reaction, $\ln[NH_3]$ vs. time is a straight line with slope equal to k. A new data table is constructed below.

Note that additional significant figures retained in the calculation are shown as subscripts.

x-axis (time,s)	$[NH_3]$	y-axis ($\ln[NH_3]$)
0	4.000 M	1.386_{29}
1.000	3.986 M	1.382_{79}
2.000	3.974 M	1.379_{77}

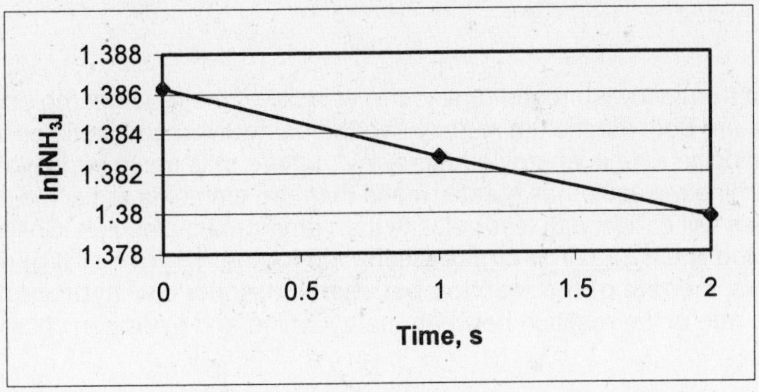

k = slope = rise/run = $(y_2 - y_1)/(x_2 - x_1) = (1.379_{77} - 1.386_{29})/(2.000 - 0) = (0.006_{52})/(2.000) = 3._{260} \times 10^{-3}\ s^{-1}$ = $\mathbf{3 \times 10^{-3}\ s^{-1}}$

b) $t_{1/2} = \ln 2/k$ = $\mathbf{2 \times 10^2\ s}$

16.47 The Arrhenius equation, $k = Ae^{-Ea/RT}$, can be used directly to solve for activation energy at a specified temperature if the rate constant, k, and the frequency factor, A, are known. However, the frequency factor is usually not known. To find E_a without knowing A, rearrange the Arrhenius equation to put it in the form of a linear plot: $\ln k = \ln A - E_a/RT$ where the y value is ln k and the x value is 1/T. Measure the rate constant at a series of temperatures and plot lnk vs 1/T. The slope equals $-E_a/R$.

16.49 Substitute the give values into Equation 16.8 and solve for k_2.

$k_1 = 4.7 \times 10^{-3}\ s^{-1}$ $T_1 = 25\ ^\circ C = 298\ K$

k_2 = ? $T_2 = 75\ ^\circ C = 348\ K$

$$\ln\frac{k_2}{k_1} = -\frac{E_a}{R}\left(\frac{1}{T_2} - \frac{1}{T_1}\right)$$

$$\ln k_2 - \ln(4.7x10^{-3}) = -\frac{33.6\ kJ/mol\left(\frac{1000\ J}{kJ}\right)}{8.314\ J/mol \cdot K}\left(\frac{1}{348} - \frac{1}{298}\right)$$

$$\ln k_2 - (-5.3_{60}) = 1.94_{85}$$

$$\ln k_2 = -3.4_{12}$$

$$k_2 = e^{-3.412} = \mathbf{3.3x10^{-2}\ s^{-1}}$$

16.53 No, collision frequency is not the only factor affecting reaction rate. The collision frequency is a count of the total number of collisions between reactant molecules. Only a small number of these collisions lead to a reaction. Other factors that impact the fraction of collisions that lead to reaction are the energy and orientation of the collision. A collision must occur with a minimum energy (activation energy) to be successful. In a collision the orientation - which ends of the reactant molecules collide – must bring the reacting atoms in the molecules together in order for the collision to lead to a reaction.

16.56 For 4×10^{-5} moles of EF to form, every collision must result in a reaction and no EF molecule can decompose back to AB and CD. Neither condition is likely. All reactions are reversible, so some EF molecules decompose. Even if all AB and CD molecules did combine, the reverse decomposition rate would result in an amount of EF that is less than 4×10^{-5} moles.

16.57 Collision frequency is proportional to the velocity of the reactant molecules. At the same temperature both reaction mixtures have the same average kinetic energy, but not the same velocity. Kinetic energy equals $\frac{1}{2}mv^2$, where m is mass and v velocity. The methylamine molecule has greater mass than the ammonia molecule, so methylamine molecules will collide with less velocity than ammonia molecules. Collision energy thus is less for the $N(CH_3)_3(g) + HCl(g)$ reaction than for the $NH_3(g) + HCl(g)$ reaction. Therefore, the **rate of the reaction between ammonia and hydrogen chloride is greater** than the rate of the reaction between methylamine and hydrogen chloride.

The fraction of successful collisions also differs between the two reactions. In both reactions the hydrogen from HCl is bonding to the nitrogen in NH_3 or $N(CH_3)_3$. The difference between the reactions is in how easily the H— can collide with the N—, the correct orientation for a successful reaction. The groups bonded to nitrogen in ammonia are less bulky than the groups bonded to nitrogen in trimethylamine. So, collisions with correct orientation between HCl and NH_3 occur more frequently than between HCl and $N(CH_3)_3$ and $NH_3(g) + HCl(g) \rightarrow NH_4Cl(s)$ occurs at a higher rate than $N(CH_3)_3(g) + HCl(g) \rightarrow (CH_3)_3NHCl(s)$.

16.58 Each A particle can collide with three B particles, so there are (4 x 3) = 12 possible unique collisions.

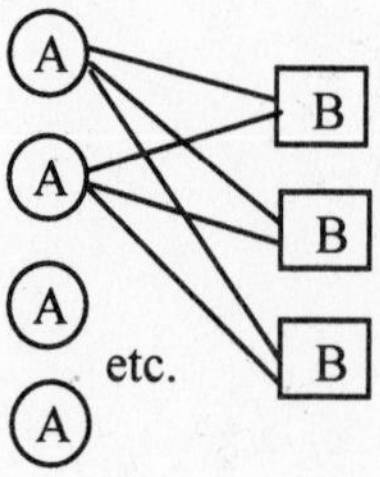

16.60The fraction of collisions with a specified energy is equal to the $e^{-Ea/RT}$ term in the Arrhenius equation.

$$e^{-E_a/RT} = e^{-\left(100.x10^3\,J/mol \big/ (8.314\,J/mol\cdot K)(298\,K)\right)} = \mathbf{2.96x10^{-18}}$$

16.62 a) The reaction is exothermic, so the energy of the reactants is higher than the energy of the products.

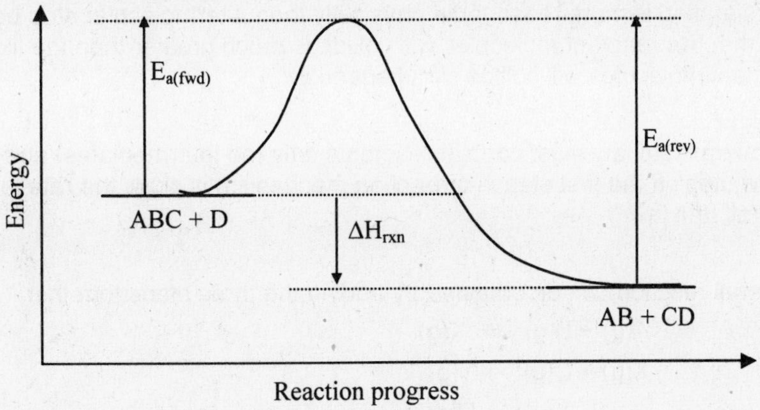

b) $E_a(rev) = E_a(fwd) + \Delta H_{rxn} = 215 + 55 =$ **270. kJ/mol.**

c)

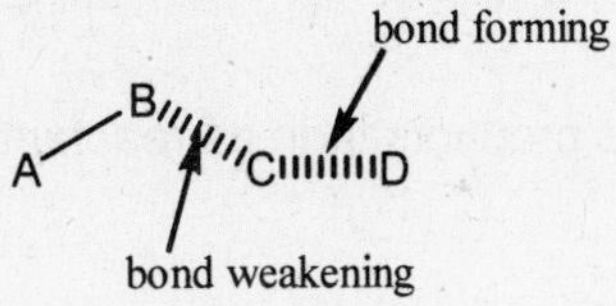

16.64 a) The reaction is endothermic since the enthalpy change is positive.

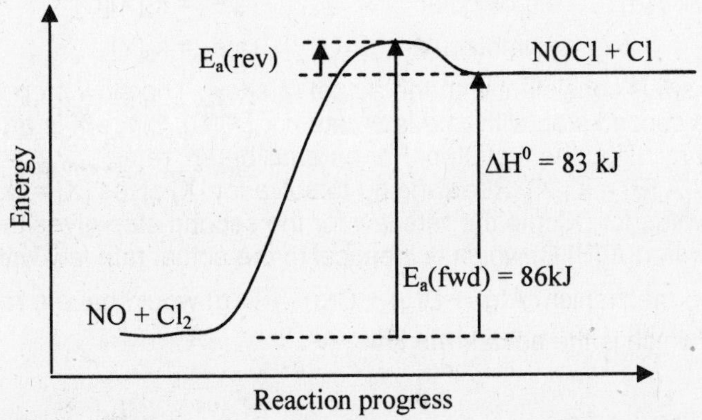

b) Activation energy for the reverse reaction: $E_a(rev) = E_a(fwd) - \Delta H^0 = 86\ kJ - 83\ kJ = 3\ kJ$.

c) To draw the transition state, look at structures of reactants and products:

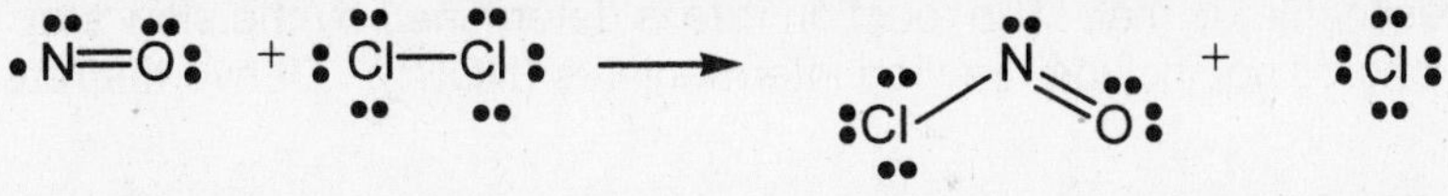

The collision must occur between one of the chlorines and the nitrogen. The transition state would have weak bonds between the nitrogen and chlorine and between the two chlorines.

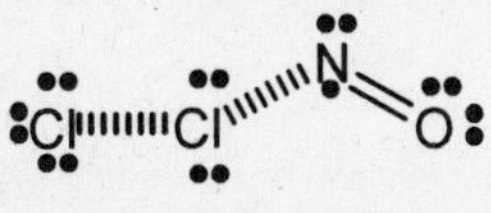

16.65 The rate of an overall reaction depends on the slowest step. Each individual step's reaction rate can vary widely, so the rate of the slowest step, and hence the overall reaction, will be **slower than the average of the individual rates** because the average contains faster rates as well.

16.69 A bimolecular step is more reasonable physically than a termolecular step because the likelihood that two reactant molecules will collide is much greater than the likelihood that three reactant molecules will collide simultaneously.

16.70 No. The overall rate law must contain reactants only (no intermediates) and is determined by the slow step. If the first step in a reaction mechanism is slow, the rate law for that step is the overall rate law.

16.72 a) The overall reaction can be obtained by adding the three steps together:

$$A(g) + B(g) \rightleftharpoons X(g)$$
$$X(g) + C(g) \rightarrow Y(g)$$
$$\underline{Y(g) \rightarrow D(g)}$$
$$A(g) + B(g) + \cancel{X(g)} + C(g) + \cancel{Y(g)} \rightarrow \cancel{X(g)} + \cancel{Y(g)} + D(g)$$
$$A(g) + B(g) + C(g) \rightarrow D(g)$$

b) Intermediates appear in the mechanism first as products then as reactants. Both X and Y are intermediates in the given mechanism.

c)

Step	Molecularity	Rate law
$A(g) + B(g) \rightleftharpoons X(g)$	bimolecular	$rate_1 = k_1[A][B]$
$X(g) + C(g) \rightarrow Y(g)$	bimolecular	$rate_2 = k_2[X][C]$
$Y(g) \rightarrow D(g)$	unimolecular	$rate_3 = k_3[Y]$

d) Yes, the mechanism is consistent with the actual rate law. The slow step in the mechanism is the second step with rate law: rate = $k_2[X][C]$. Since X is an intermediate it must be replaced by using the first step. For an equilibrium, $rate_{forward\ rxn} = rate_{reverse\ rxn}$. For step 1 then $k_1[A][B] = k_{-1}[X]$. Rearranging to solve for [X] gives $[X] = (k_1/k_{-1})[A][B]$. Substituting this value for [X] into the rate law for the second step gives the overall rate law as rate = $(k_2 k_1/k_{-1})[A][B][C]$ which is identical to the actual rate law with $k = k_2 k_1/k_{-1}$.

e) Yes. The one step mechanism $A(g) + B(g) + C(g) \rightarrow D(g)$ would have a rate law of rate = k[A][B][C], which is the actual rate law.

16.74 Nitrosyl bromide is NOBr(g). The reactions sum to the equation $2NO(g) + Br_2(g) \rightarrow 2NOBr(g)$, so criterion 1 (elementary steps must add to overall equation) is satisfied. Both elementary steps are bimolecular and physically reasonable, so criterion 2 (steps are physically reasonable) is met. The reaction rate is determined by the slow step, however rate expressions do not include reaction intermediates ($NOBr_2$). Derive the rate law as follows:

$$rate_1(fwd) = k_1[NO][Br_2] \qquad (1)$$
$$rate_1(rev) = k_{-1}[NOBr_2] \qquad (2)$$
$$rate_2(fwd) = k_2[NOBr_2][NO] \qquad (3)$$

Solve for $[NOBr_2]$:

$$rate_1(fwd) = rate_1(rev)$$
$$k_1[NO][Br_2] = k_{-1}[NOBr_2]$$
$$[NOBr_2] = \frac{k_1}{k_{-1}}[NO][Br_2]$$

Substitute the expression for $[NOBr_2]$ into equation (3):

$$rate_2 = k_2 \frac{k_1}{k_{-1}}[NO][Br_2][NO]$$

$$rate_2 = k[NO]^2[Br_2]$$

The derived rate law equals the known rate law, so criterion 3 is satisfied. The proposed mechanism is valid.

16.77 No, a catalyst changes the mechanism of a reaction to one with lower activation energy. Lower activation energy means a faster reaction. An increase in temperature does not impact the activation energy, but instead increases the fraction of collisions with sufficient energy to react.

16.78 a) No. By definition, a catalyst is a substance that increases reaction rate without being consumed. The spark provides energy that is absorbed by the H_2 and O_2 molecules to achieve the threshold energy needed for reaction.

b) Yes. The powdered metal acts like a heterogeneous catalyst, providing a surface upon which the O_2-H_2 reaction becomes more favorable because the activation energy is lowered.

16.82 a) Water does not appear as a reactant in the rate-determining step.

Note that as a solvent in the reaction the concentration of the water is assumed not to change even though some water is used up as a reactant. This assumption is valid as long as the solute concentrations are low (~1M or less). So, even if water did appear as a reactant in the rate-determining step it would not appear in the rate law. See rate law for step II below.

b) Rate law for step I: $rate_1 = k_1[(CH_3)_3CBr]$

Rate law for step II: $rate_2 = k_2[(CH_3)_3C^+]$

Rate law for step III: $rate_3 = k_3[(CH_3)_3COH_2^+]$

c) The intermediates are $(CH_3)_3C^+$ and $(CH_3)_3COH_2^+$.

d) The rate-determining step is the slow step, I. The rate law for this step agrees with the actual rate law with $k = k_1$.

16.83 The structure of nitrous oxide is •N=O with an unpaired electron on the nitrogen similar to the unpaired electron on the chlorine atom. The unpaired electron is shown in the formula to emphasize its importance in the reactivity of NO. Mechanism for catalysis by •N=O:

$$O_3(g) + \bullet NO(g) \rightarrow \bullet NO_2(g) + O_2(g)$$
$$O_3(g) \rightarrow O_2(g) + O(g)$$
$$\bullet NO_2(g) + O(g) \rightarrow \bullet NO(g) + O_2(g)$$

The overall reaction is $2O_3(g) \rightarrow 3O_2(g)$. Nitrous oxide provides an alternative mechanism to convert ozone to oxygen.

Nitrous oxide appears first as a reactant in the mechanism and then as a product, so it is a catalyst because it is not used up in the reaction.

16.85 The activation energy can be calculated using Equation 16.8. Although the rate constants, k_1 and k_2, are not expressly stated, the rot times give an idea of the rate. The reaction rate is proportional to the rate constant (rate $\propto k$). At $T_2 = 0\ ^\circ C = 273$ K, the rate is slow relative to the rate at T_1 = 20 °C = 293 K. Therefore, $rate_2$ = 1/16 and $rate_1$ = ¼ are substituted for k_2 and k_1, respectively.

$$\ln\frac{k_2}{k_1} = -\frac{E_a}{R}\left(\frac{1}{T_2} - \frac{1}{T_1}\right)$$

$$\ln\left(\frac{1/16\,day}{1/4\,day}\right) = -\frac{E_a}{8.314\,J/mol\cdot K}\left(\frac{1}{273} - \frac{1}{293}\right)$$

$$-1._{386} = -\frac{E_a}{8.314\,J/mol\cdot K}(0.00250_{03})$$

$$E_a = 4.6x10^4\,J/mol = \mathbf{46\,kJ/mol}$$

16.87 a)

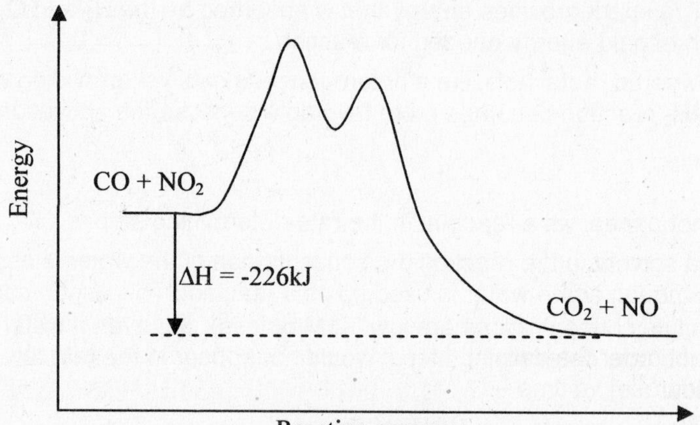

b) Yes, the alternative mechanism is consistent with the rate law since the rate of the mechanism is based on the slowest step, $2NO_2(g) \rightarrow N_2(g) + 2O_2(g)$. The rate law for this step is rate = $k_1[NO_2]^2$, the same as the actual rate law.

The alternative mechanism includes an elementary reaction (step 2) that is a termolecular reaction. Thus, the original mechanism given in the text is more reasonable physically since it involves only bimolecular reactions.

16.89 a) Decreasing the pressure for a reaction that includes a gas reactant decreases the concentration of the gas by increasing the volume. The decreased gas concentration **decreases** the rate of the reaction.

b) Concentration of reactants is steps preceding the rate-limiting step do influence the rate of the reaction. Increasing the concentration of the reactant **increases** the rate of the reaction.

c) Concentrations of reactants in steps following the rate-limiting step do not influence the rate of the reaction, so increasing the concentration of such a reactant has **no effect** on the rate of the reaction.

d) Increasing the surface area of a solid reactant **increases** the rate of the reaction.

e) Rapid stirring will increase the likelihood that two reactants collide, so reaction rate will **increase** with stirring.

16.92 The overall order is equal to the sum of the individual orders, i.e. the sum of the exponents equals 11: 11 = 1 + 4 + 2 + m + 2, so m = 2. The reaction is **second order with respect to NAD.**

16.94 Rate is proportional to the rate constant, so if the rate constant increases by a certain factor the rate increases by the same factor. Thus, to calculate the change in rate the Arrhenius equation can be used and substitute $rate_{cat}/rate_{uncat} = k_{cat}/k_{uncat}$.

$$\frac{rate_{cat}}{rate_{uncat}} = \frac{Ae^{-E_a(cat)/RT}}{Ae^{-E_a(uncat)/RT}}; \quad \frac{rate_{cat}}{rate_{uncat}} = e^{-\Delta E_a/RT}$$

$$\frac{rate_{cat}}{rate_{uncat}} = e^{-(-5000\,J/mol)/(8.314\,J/mol\cdot K)(310\,K)} = 7$$

The rate of the enzyme-catalyzed reaction occurs at a rate **7 times faster** than the rate of the uncatalyzed reaction at 37°C.

16.96 First find the rate constant, *k*, for the reaction by solving the first order half-life equation for *k*. Then use the first order integrated rate law expression to find t, the time for decay. Note that additional siginificant figures are retained in the calculation and shown as subscripted numbers.

$$k = \frac{\ln 2}{12\,yr} = 0.057_{75}\,yr^{-1}$$

$$\ln\frac{[A]_o}{[A]_t} = kt \quad \Rightarrow \quad \ln\left(\frac{275}{10.}\right) = 0.05775\,yr^{-1}(t) \quad \Rightarrow \quad t = \mathbf{57\,years}$$

16.99 a) The rate constant can be determined from the slope of the integrated rate law plot. To find the correct order the data should be plotted as 1) [sucrose] vs time – linear for zero order, 2) ln[sucrose] vs time – linear for first order, and 3) 1/[sucrose] vs time – linear for second order.

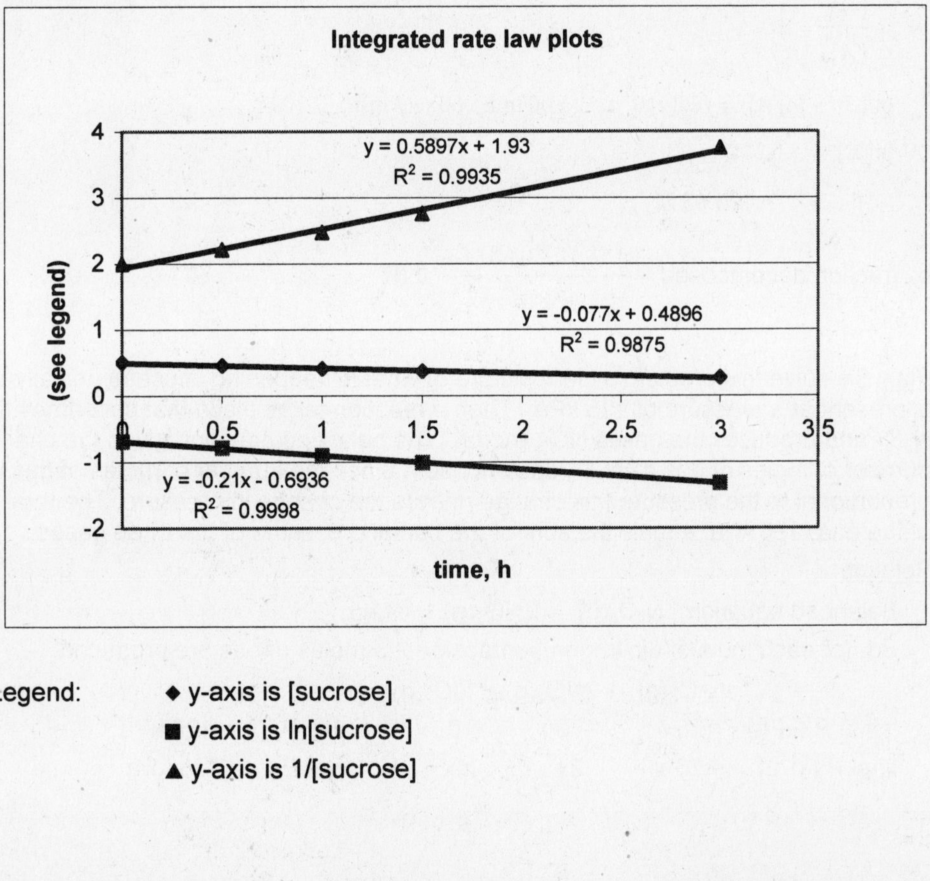

All three graphs are linear, so picking the correct order is difficult. One way to select the order is to compare correlation coefficients (R^2) – you may or may not have experience with this. The best correlation coefficient is the one closest to a value of 1.00. Based on this selection criterion, the plot of ln[sucrose] vs time for the first order reaction is the best.

Another method when linearity is not obvious from the graphs is to examine the reaction and decide which order fits the reaction. For the reaction of one molecule of sucrose with one molecule of liquid water, the rate law would most likely include sucrose with an order of one and would not include water.

The plot for a first order reaction is described by the equation $\ln[A]_t = -kt + \ln[A]_0$. The slope of the plot of ln[sucrose] vs t equals –k. So, $k = -(-0.21\ h^{-1}) = \mathbf{0.21\ h^{-1}}$.

Half-life for a first order reaction equals ln2/k, so $t_{½} = \ln 2 \div 0.21\ h^{-1} = \mathbf{3.3\ h}$.

b) The time to hydrolyze 75% of the sucrose can be calculated by substituting 1.0 for $[A]_0$ and 0.25 for $[A]_t$ in the integrated rate law equation: $\ln[A]_t = (-0.21\ h^{-1})t + \ln[A]_0$.

$$t = \frac{\ln(1.0) - \ln(0.25)}{0.21\,h^{-1}} = \mathbf{6.6\,h}$$

c) The reaction might be second order overall with first order in sucrose and first order in water. If the concentration of sucrose is relatively low, the concentration of water remains constant even with small changes in the amount of water. This yields an apparent zero-order behavior with respect to water. Thus, the reaction appears to be first order because the rate does not change with changes in the amount of water.

16.101 a) To find concentration at a later time, given starting concentration and the rate constant, use an integrated rate law expression. Since the units on *k* are s^{-1}, this is a first order reaction. Subscripted numbers are additional significant figures retained in preliminary steps of calculation.

$$\ln\frac{[A]_0}{[A]_t} = kt$$

$$\ln 1.58 - \ln[A]_t = (2.8x10^{-3}s^{-1})(5.00\,\text{min})(60\,\text{sec/ min})$$

$$\ln[A]_t = -0.38_{26}$$

$$[A]_t = e^{-0.38_{26}} = \mathbf{0.68\,M}$$

b) fraction decomposed = $\dfrac{1.58\,M - 0.68\,M}{1.58\,M} = \mathbf{0.57}$

16.104 Plan: To solve this problem a clear picture of what is happening is useful. Initially only N_2O_5 is present at a pressure of 125 kPa. Then a reaction takes place that consumes the gas N_2O_5 and produces the gases NO_2 and O_2. The balanced equation gives the change in the number of moles of gas as N_2O_5 decomposes. Since the number of moles of gas is proportional to the pressure this change mirrors the change in pressure. The total pressure at the end, 178 kPa, equals the sum of the partial pressures of the three gases.

Solution:

Balanced equation: $N_2O_5(g) \rightarrow 2NO_2(g) + ½O_2(g)$

So, for each mole of dinitrogen pentaoxide 2.5 moles of gas are produced.

	$N_2O_5(g)$ →	$2NO_2(g)$ +	$½O_2(g)$	
initial P (kPa)	125	0	0	total $P_{initial}$ = 125 kPa
final P (kPa)	125-x	2x	½x	total P_{final} = 178 kPa

Solve for x:

$P_{N2O5} + P_{NO2} + P_{O2} = (125 - x) + 2x + ½x = 178$

$x = 35._{33}$ kPa

Partial pressure of NO_2 equals 2x or **71 kPa**.

Check: Substitute values for all partial pressures to find total final pressure:

$(125 - 35._{33}) + (2 \cdot 35._{33}) + (½ \cdot 35._{33}) = 178$ kPa

The result agrees with the given total final pressure.

16.108 The reaction can be described by both the disappearance of reactant and appearance of product using the general form of Equation 16.2.

$$rate = -\frac{\Delta[C_2H_6]}{\Delta t} = -\frac{1}{4}\frac{\Delta[Cl_2]}{\Delta t} = \frac{\Delta[C_2H_2Cl_4]}{\Delta t} = \frac{1}{4}\frac{\Delta[HCl]}{\Delta t}$$

16.110 a) Starting with the fact that rate of formation of O (rate of step 1) equals the rate of consumption of O (rate of step 2), set up an equation to solve for [O] using the given values of k_1, k_2, $[NO_2]$, and $[O_2]$.

$$rate_{step 1} = rate_{step 2}$$

$$k_1[NO_2] = k_2[O][O_2]$$

$$[O] = \frac{k_1[NO_2]}{k_2[O_2]} = \frac{(6.0x10^{-3}\ s^{-1})(4.0x10^{-9}\ M)}{(1.0x10^{6}\ M^{-1}s^{-1})(1.0x10^{-2}\ M)} = \mathbf{2.4x10^{-15}\ M}$$

b) Since the rate of the two steps is equal either can be used to determine rate of formation of ozone.

$rate_{step 2} = k_2[O][O_2] = (1.0x10^6\ M^{-1}s^{-1})(2.4x10^{-15}\ M)(1.0x10^{-2}\ M) = \mathbf{2.4x10^{-11}\ M/s}$

16.114 The rate of the reaction between I^- and $S_2O_8^{2-}$ is determined in this analysis. The rate of the reaction between I_2 and $S_2O_3^{2-}$ is so fast that it is not influenced by how long it takes to see the blue-black color. The blue-black color of the starch·I_2 complex appears when all of the $Na_2S_2O_3$ (sodium thiosulfate) is consumed. In each case, 10.0 mL of 0.0050 M $Na_2S_2O_3$, or (0.0100 L)(0.0050 M) = 5.0×10^{-5} moles are consumed. This translates to a concentration of 0.0010 M, because the total volume is 50.0 mL in each experiment:

$$\left(\frac{5.0x10^{-5}\ mol}{0.0500\ L}\right) = 0.0010\ M$$

a) Since the measured quantity in each experiment is the "time to color" for the starch-iodine reaction, the reaction rate is based on the appearance of iodine:

$$rate = \frac{\Delta[I_2]}{\Delta t}$$

But since this reaction would occur as soon as a small amount of iodine is produced the time elapsed to see the color would be so short that it could not be measured. The thiosulfate ion ($S_2O_3^{2-}$) is added because it will react with the iodine produced before the iodine reacts with the starch. Only after the iodine has reacted with all the thiosulfate ions present will the iodine-starch reaction take place and the solution change color. So the change in concentration of iodine is proportional to the concentration of thiosulfate ion added to the solution. The stoichiometry of the iodine – thiosulfate reaction gives a ratio of 1 I_2 : 2 $S_2O_3^{2-}$, so $\Delta[I_2] = ½\Delta[S_2O_3^{2-}]$. The rate is calculated as $-½\Delta[S_2O_3^{2-}]/\Delta t$ which equals $5.0x10^{-4}$ M/Δt .

$rate_{exp\ 1}$ = 5.0x10^{-4} M/29.0 s = **1.7$_{24}$x10^{-5} M/s**
$rate_{exp\ 2}$ = 5.0x10^{-4} M/14.5 s = **3.4$_{48}$x10^{-5} M/s**
$rate_{exp\ 3}$ = 5.0x10^{-4} M/14.5 s = **3.4$_{48}$x10^{-5} M/s**

Subscripted numbers are additional significant figures retained for following calculations.

b) The rate law is rate = $k[I^-]^m[S_2O_8^{2-}]^n$. To find *m*, the order for I^-, use experiments 1 and 2 since the $[I^-]$ doubles between the two experiments while $[S_2O_8^{2-}]$ is constant.

$$\frac{rate_{exp1}}{rate_{exp2}} = \left(\frac{[I^-]_{exp1}}{[I^-]_{exp2}}\right)^m \Rightarrow \frac{1.7_{24}x10^{-5}\ M^{-1}s^{-1}}{3.4_{48}x10^{-5}\ M^{-1}s^{-1}} = \left(\frac{0.0400\ M}{0.0800\ M}\right)^m$$

$$0.5 = (0.5)^m;\ so\ \mathbf{m = 1}$$

For *n*, order for $S_2O_8^{2-}$, use trials 2 and 3 for which $[S_2O_8^{2-}]$ varies while $[I^-]$ is constant.

$$\frac{rate_{exp2}}{rate_{exp3}} = \left(\frac{[S_2O_8^{2-}]_{exp2}}{[S_2O_8^{2-}]_{exp3}}\right)^n \Rightarrow \frac{3.4_{48}x10^{-5}\ M^{-1}s^{-1}}{3.4_{48}x10^{-5}\ M^{-1}s^{-1}} = \left(\frac{0.0400\ M}{0.0200\ M}\right)^n$$

$$1.0 = (0.5)^n;\ so\ \mathbf{n = 0}$$

The reaction is first order with respect to I^- and zero order with respect to $S_2O_8^{2-}$.

c) With the orders the rate law can be used to find the rate constant.
k = rate/$[I^-]$

Experiment #	rate ($M^{-1}s^{-1}$)	$[I^-]$	k (s^{-1})
1	1.7$_{24}$x10^{-5}	0.0400	**4.3x10^{-4}**
2	3.4$_{48}$x10^{-5}	0.0800	**4.3x10^{-4}**
3	3.4$_{48}$x10^{-5}	0.0800	**4.3x10^{-4}**

d) The complete rate law at 23°C is **rate = (4.3x10^{-4} s^{-1})[KI].**

CHAPTER 17

EQUILIBRIUM: THE EXTENT OF CHEMICAL REACTIONS

FOLLOW-UP PROBLEMS

17.1 Plan: First balance the equations and then use the coefficients to write the reaction quotient as outlined in Equation 17.4.

Solution:

a) Balanced equation: $4NH_3(g) + 5O_2(g) \rightleftharpoons 4NO(g) + 6H_2O(g)$

Reaction quotient: $$Q_c = \frac{[NO]^4[H_2O]^6}{[NH_3]^4[O_2]^5}$$

b) Balanced equation: $3NO(g) \rightleftharpoons N_2O(g) + NO_2(g)$

Reaction quotient: $$Q_c = \frac{[N_2O][NO_2]}{[NO]^3}$$

Check: Are products in numerator and reactants in denominator? Are all reactants and products expressed as concentrations by using brackets? Are all exponents equal to the coefficient of the reactant or product in the balanced equation?

17.2 Plan: Add the individual steps to find the overall equation. Write the reaction quotient for each step and the overall equation. Multiply the reaction quotients for each step and cancel terms to obtain the overall reaction quotient.

Solution:

(1) $Br_2(g) \rightleftharpoons 2Br(g)$

(2) $Br(g) + H_2(g) \rightleftharpoons HBr(g) + H(g)$

(3) $H(g) + Br(g) \rightleftharpoons HBr(g)$

$$Br_2(g) + \cancel{2Br(g)} + H_2(g) + \cancel{H(g)} \rightleftharpoons \cancel{2Br(g)} + 2HBr(g) + \cancel{H(g)}$$

Canceling the reactants leaves the overall equation as $Br_2(g) + H_2(g) \rightleftharpoons 2HBr(g)$.

Write the reaction quotients for each step:

$$Q_{c(1)} = \frac{[Br]^2}{[Br_2]};\quad Q_{c(2)} = \frac{[HBr][H]}{[Br][H_2]};\quad Q_{c(3)} = \frac{[HBr]}{[Br][H]}$$

and for the overall equation:

$$Q_c = \frac{[HBr]^2}{[H_2][Br_2]}$$

Multiplying the individual Q_c's and canceling terms gives

$$\left(Q_{c(1)}\right)\left(Q_{c(2)}\right)\left(Q_{c(3)}\right) = \left(\frac{[Br]^2}{[Br_2]}\right)\left(\frac{[HBr][H]}{[Br][H_2]}\right)\left(\frac{[HBr]}{[Br][H]}\right) = \frac{[HBr]^2}{[H_2][Br_2]}$$

The product of the multiplication of the three individual reaction quotients equals the overall reaction quotient.

17.3 Plan: When a reaction is multiplied by a factor, the equilibrium constant is raised to a power equal to the factor.

Solution:

a) All coefficients have been multiplied by the factor ½ so the equilibrium constant should be raised to the ½ power (which is the square root).

for reaction a) $K_c = (7.6x10^8)^{1/2}$ = **$2.8x10^4$**

b) The reaction has been reversed so the new K_c is the inverse of the original equilibrium constant. The reaction has also been multiplied by a factor of 2/3, so the inverse of the original K_c must be raised to the 2/3 power.

for reaction b) $K_c = \left(\frac{1}{7.6x10^8}\right)^{2/3}$ = **$1.2x10^{-6}$**

Check: The order of magnitude of each answer can be estimated by examining the exponents: for part a) $(10^8)^{1/2} = 10^4$ and for part b) rounding $7.6x10^8$ to 10^9 gives $1/10^9 = 10^{-9}$ and $(10^{-9})^{2/3} = 10^{-6}$. These estimates are within the same order of magnitude as the answers under solution.

17.4 Plan: K_p and K_c for a reaction are related through the ideal gas equation as shown in Equation 17.8. Find Δn_{gas}, the change in the number of moles of gas between reactants and products. Then use the given K_c to solve for K_p.

Solution: The total number of product moles of gas is 1 and the total number of reactant moles of gas is 2. $\Delta n = 1 - 2 = -1$.

$$K_p = K_c(RT)^{\Delta n} = (1.67)((0.08206\, atm \cdot L/mol \cdot K)(500\, K))^{-1} = \mathbf{4.07x10^{-2}}$$

Check: Although equilibrium constants do not have units, it is useful to check this type of calculation by doing unit analysis. For this purpose add units for the equilibrium constants and substitute M for mol/L in gas constant.

K_c = (M/M·M) = M and K_p = (atm/atm·atm) = atm

K_p = (M)(atm/M·K)$(K)^{-1}$ = atm

The units check out to convert from M to atm.

17.5 Plan: To decide whether CH_3Cl or CH_4 are forming while the reaction system moves toward equilibrium, calculate Q_p and compare it to K_p. If $Q_p > K_p$, reactants are forming. If $Q_p < K_p$, products are forming.

Solution:

$$Q_p = \frac{(P_{CH_3Cl})(P_{HCl})}{(P_{CH_4})(P_{Cl_2})} = \frac{(0.24\, atm)(0.47\, atm)}{(0.13\, atm)(0.035\, atm)} = 25$$

K_p for this reaction is given as $1.6x10^4$. Q_p is smaller than K_p ($Q_p < K_p$) so more products will form. **CH_3Cl** is one of the products forming.

Check: Since number of moles of gas is proportional to pressure, a new calculation of Q_p can be done with increased pressures for products and decreased pressures for reactants. If the conclusion that formation of products moves the reaction system closer to equilibrium then the new Q_p should be closer to K_p than the 25 found for the original Q_p.

Increasing and decreasing by 0.01 atm:

$$Q_p = \frac{(0.25\, atm)(0.48\, atm)}{(0.12\, atm)(0.025\, atm)} = 40$$

40 is closer to 16000 than 25, so the answer that CH_3Cl is forming is correct.

17.6 Plan: The information given includes the equilibrium constant, initial pressures of both reactants and equilibrium pressure for one reactant. First set up a reaction table showing initial partial pressures for reactants and 0 for product. The change to get to equilibrium is to react some of reactants to form some product. Use the equilibrium quantity for O_2 and the expression for O_2 at equilibrium to solve for the change. From the change find the equilibrium partial pressure for NO and NO_2. Calculate K_p using the equilibrium values.

Solution:

Pressures (atm)	2NO(g)	+	O_2(g)	$\rightleftharpoons$	$2NO_2$(g)
Initial	1.000		1.000		0
Change	-2x		-x		+2x
Equilibrium	1.000 – 2x		1.000 – x		2x

At equilibrium P_{O2} = 0.506 atm = 1.000 – x; so x = 1.000 – 0.506 = 0.494 atm

P_{NO} = 1.000 – 2x = 1.000 – 2(0.494) = 0.012 atm

P_{NO2} = 2x = 2(0.494) = 0.988 atm

Use the equilibrium pressures to calculate K_p.

$$K_p = \frac{(P_{NO_2})^2}{(P_{NO})^2(P_{O_2})} = \frac{(0.988)^2}{(0.012)^2(0.506)} = \mathbf{1.3x10^4}$$

Check: One way to check is to plug the equilibrium expressions for NO and NO_2 into the equilibrium expression along with 0.506 atm for the equilibrium pressure of O_2. Then solve for x and make sure value of x by this calculation is the same as the 0494 atm calculated above.

$$1.3x10^4 = \frac{(2x)^2(0.506)}{(1.000-2x)^2}; \sqrt{2.5_{69}x10^4} = \frac{(2x)}{(1.000-2x)}; \quad x = 0.497$$

The two values for x agree to give K_p = 1.3x10^4.

17.7 Plan: Convert K_c to K_p for the reaction. Write the equilibrium expression for K_p and insert the atmospheric pressures for P_{N2} and P_{O2} as their equilibrium values. Solve for P_{NO}.

Solution: The conversion of K_c to K_p: $K_p = K_c(RT)^{\Delta n}$. For this reaction Δn = 0, so $K_p = K_c$.

$$K_p = \frac{(P_{N_2})(P_{O_2})}{(P_{NO})^2}$$

$$2.3x10^{30} = \frac{(0.781)(0.209)}{x^2}; \quad x = 2.7x10^{-16}\ atm$$

The equilibrium partial pressure of NO in the atmosphere is **2.7x10^{-16} atm**.

Check: Insert the pressure of NO and calculate the equilibrium constant to check the result.

$$\frac{(0.781)(0.209)}{(2.7x10^{-16})^2} = 2.2x10^{30}$$

17.8 Plan: Set up a reaction table and use the variables to find equilibrium concentrations in the equilibrium expression.

Solution:

Concentration (M)	$2HI(g) \rightleftharpoons$	$H_2(g)$	+ $I_2(g)$
Initial	2.50/10.32	0	0
Change	-2x	+x	+x
Equilibrium	$0.242_{25} - 2x$	x	x

Set up equilibrium expression:

$$K_c = 1.26x10^{-3} = \frac{[H_2][I_2]}{[HI]^2} = \frac{(x)(x)}{(0.242_{25} - 2x)^2}$$

Solving for x gives $8.02x10^{-3}$

$[H_2] = [I_2] =$ **$8.02x10^{-3}$ M**

Check: Plug equilibrium concentrations into the equilibrium expression and calculate K_c.

$[HI] = 0.242 - 2(8.02x10^{-3}) = 0.226$ M

$$\frac{(8.02x10^{-3})^2}{(0.226)^2} = 1.26x10^{-3}$$

17.9 Plan: First set up the reaction table, then set up the equilibrium expression. To solve for variable, x, first assume that x is negligible with respect to initial concentration of I_2. Check the assumption by calculating the % error. If the error is greater than 5%, calculate x using quadratic equation. Then use x to determine equilibrium concentrations of I_2 and I.

Solution: $[I_2]_{init} = 0.50$ mol I_2/2.5 L = 0.20 M

a) Equilibrium at 600 K

Concentration (M)	$I_2(g) \rightleftharpoons$	$2I(g)$
Initial	0.20	0
Change	-x	+2x
Equilibrium	0.20 – x	2x

Equilibrium expression: $$2.94x10^{-10} = \frac{(2x)^2}{0.20 - x}$$

Assume x is negligible so $0.20 - x \approx 0.20$; $4x^2 = (2.94x10^{-10})(0.20)$; $x = 3.8x10^{-6}$.

Check assumption by calculating % error: $3.8x10^{-6} \div 0.20$ x 100% = $1.9x10^{-3}$% which is smaller than 5%, so assumption is valid.

At equilibrium $[I]_{eq} = 2x =$ **$7.6x10^{-6}$ M** and $[I_2]_{eq} = 0.20 - 3.8x10^{-6} =$ **0.20 M**

b) Equilibrium at 2000 K

Equilibrium expression: $$0.209 = \frac{(2x)^2}{0.20 - x}$$

Assume x is negligible so $0.20 - x \approx 0.20$, so $4x^2 = (0.209)(0.20)$; x = 0.102

Check assumption by calculating % error: $0.102 \div 0.20 = 51\%$ which is larger than 5% so assumption is not valid. Solve using quadratic equation.

$$4x^2 + 0.209x - 0.041_8 = 0$$

$$x = \frac{-0.209 \pm \sqrt{(0.209)^2 - 4(4)(-0.042)}}{2(4)} = -0.13 \text{ or } 0.079$$

Choose the positive value, x = 0.079

At equilibrium $[I]_{eq}$ = 2x = **0.16 M** and $[I_2]_{eq}$ = 0.20 – 0.079 = **0.12 M**

Check: Calculate equilibrium constants using calculated equilibrium concentrations:

a) K = $(7.7\times10^{-6})^2/(0.20) = 2.96\times10^{-10}$

b) K = $(0.16)^2/(0.12) = 0.213$

17.10 Plan: Calculate initial concentrations of each substance. For part (a), calculate Q_c and compare to given K_c. If $Q_c > K_c$ then reaction proceeds to the left to make reactants from products. If $Q_c < K_c$ then reaction proceeds to right to make products from reactants. For part (b), use the result of part (a) and the given equilibrium concentration of PCl_5 to find equilibrium concentrations of PCl_3 and Cl_2.

Solution:

Initial concentrations: $[PCl_5]$ = 0.1050 mol/0.5000 L = 0.2100 M

$[PCl_3] = [Cl_2]$ = 0.0450 mol/0.5000 L = 0.0900 M

a) $$Q_c = \frac{[PCl_3][Cl_2]}{[PCl_5]} = \frac{(0.0900)(0.0900)}{(0.2100)} = 0.0386$$

Q_c, 0.0386, is less than K_c, 0.042, so **the reaction will proceed to the right to make more products**.

b) To reach equilibrium, concentrations will increase for the products, PCl_3 and Cl_2, and decrease for the reactant, PCl_5.

$[PCl_5]$ = 0.2065 = 0.2100 – x; x = 0.0035

$[PCl_3] = [Cl_2]$ = 0.0900 + x = 0.0900 + 0.0035 = **0.0935 M**

Check: The equilibrium concentrations could also be calculated from the initial concentrations using the quadratic equation:

$$K_c = 4.2x10^{-2} = \frac{(0.0900)^2}{0.2100 - x};\ 0.00882 - 0.042x = 0.0081 + 0.18x + x^2;\ x = 0.003197$$

So equilibrium concentrations are $[PCl_5]$ = 0.2100 – 0.0032 = 0.2068 M

$[PCl_3] = [Cl_2]$ = 0.0900 + 0.0032 = 0.0932 M

These concentrations are essentially equal to those determined above.

17.11 Plan: Examine each change for its impact on Q_c. Then decide how the system would respond to re-establish equilibrium.

Solution: $$Q_c = \frac{[SiF_4][H_2O]^2}{[HF]^4}$$

a) Decreasing $[H_2O]$ leads to $Q_c < K_c$, so the reaction would shift to make more products from reactants. So, $[SiF_4]$ as a product would **increase**.

b) Adding liquid water to this system at a temperature above the boiling point of water would result in an increase in the concentration of water vapor. The increase in $[H_2O]$ increases Q_c to make it greater than K_c To re-establish equilibrium products will be converted to reactants and the $[SiF_4]$ will **decrease**.

c) Removing the reactant HF increases Q_c, which causes the products to react to form more reactants. Thus, $[SiF_4]$ **decreases**.

d) Removal of a solid product has no impact on the equilibrium; $[SiF_4]$ **does not change**.

Check: Look at each change and decide which direction the equilibrium would shift using Le Châtelier's principle to check the changes predicted above.

a) Remove product, equilibrium shifts to right.

b) Add product, equilibrium shifts to left.

c) Remove reactant, equilibrium shifts to left.

d) Remove solid reactant, equilibrium does not shift.

17.12 Plan: Changes in pressure (and volume) impact the concentration of gaseous reactants and products. A decrease in pressure, i.e. increase in volume, favors the production of more gas molecules whereas an increase in pressure favors the production of fewer gas molecules. Examine each reaction to decide whether more or fewer gas molecules will result from producing more product. If more gas molecules result, then the pressure should be increased (volume decreased) to reduce product formation. If fewer gas molecules result, then pressure should be decreased to produce more reactants.

Solution:

a) In $2SO_2(g) + O_2(g) \rightleftharpoons 2SO_3(g)$ three molecules of gas form two molecules of gas, so there are fewer gas molecules in the product. **Decreasing pressure** (increasing volume) will decrease the product yield.

b) In $4NH_3(g) + 5O_2(g) \rightleftharpoons 4NO(g) + 6H_2O(g)$ 9 molecules or reactant gas convert to 10 molecules of product gas. **Increasing pressure** (decreasing volume) will favor the reaction direction that produces fewer moles of gas, towards the reactants and away from products.

c) In $CaC_2O_4(s) \rightleftharpoons CaCO_3(s) + CO(g)$ there are no reactant gas molecules and one product gas molecules. Yield of the products will decrease when volume decreases, which corresponds to a **pressure increase**.

17.13 Plan: A decrease in temperature favors the exothermic direction of an equilibrium reaction. First, identify whether the forward or reverse reaction is exothermic from the given enthalpy change. $\Delta H < 0$ means the forward reaction is endothermic, and $\Delta H > 0$ means the reverse reaction is exothermic. If the forward reaction is exothermic then a decrease in temperature will shift the equilibrium to make more products from reactants and increase K_p. If the reverse reaction is exothermic then a decrease in temperature will shift the equilibrium to make more reactants from products and decrease K_p

Solution:

a) $\Delta H < 0$ so forward reaction is exothermic. Decrease in temperature increases partial pressure of products and decreases partial pressures of reactants, so P_{H2} **decreases**. With increases in product pressures and decreases in reactant pressures, K_p **increases**.

b) $\Delta H > 0$ so reverse reaction is exothermic. Decrease in temperature decreases partial pressure of products and increases partial pressures of reactants, so P_{N2} **increases**. K_p **decreases** with decrease in product pressures and increase in reactant pressures.

c) $\Delta H < 0$ so forward reaction is exothermic. Decreasing temperature **increases** P_{PCl5} and **increases** K_p.

Check: Solve the problem for increasing instead of decreasing temperature and make sure the answers are opposite.

a) Increased temperature favors the reverse, endothermic reaction leading to increase in reactant pressures, so P_{H2} increases and K_p decreases.

b) Increased temperature favors the forward, endothermic reaction leading to decrease in reactant pressures. P_{N2} decreases and K_p increases.

c) Both P_{PCl5} and K_p decrease with increased temperature.

End-of-Chapter Problems

17.1 If the rate of the forward reaction exceeds the rate of reverse reaction, products are formed faster than they are consumed. The change in reaction conditions results in more products and less reactants. A change in reaction conditions can result from a change in concentration or a change in temperature. If concentration changes, product concentration increases while reactant concentration decreases, but the K_c remains unchanged because the *ratio* of products and reactants remains the same. If the increase in the forward rate is due to a change in temperature, the rate of the reverse reaction also increases. The equilibrium ratio of product concentration to reactant concentration is no longer the same. Since the rate of the forward reaction increases more than the rate of the reverse reaction K_c increases (numerator, [products], is larger and denominator, [reactants], is smaller).

$$K_c = \frac{[products]}{[reactants]}$$

17.5 The reaction is exothermic because the energy of the reactants is greater than the energy of products. The products would be favored and be present at equilibrium at higher concentrations than the reactants because a lower energy state is favored. The change in energy looks large on the graph so the product concentrations are much larger than the reactant concentrations and K is large.

17.7 The equilibrium expression for this reaction is $K = [O_2]$. If the temperature remains constant, K remains constant as well. Therefore, as long as sufficient Li_2O_2 is present, the amount of $O_2(g)$ that forms will also remain constant.

17.8 a)

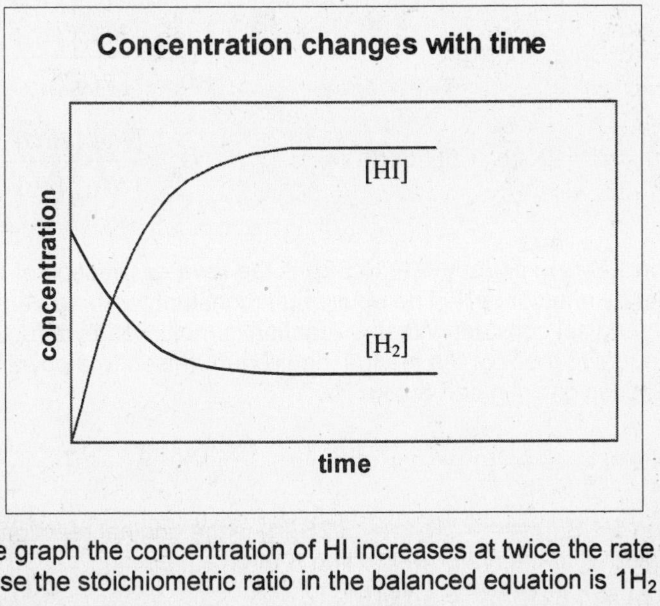

On the graph the concentration of HI increases at twice the rate that H_2 decreases because the stoichiometric ratio in the balanced equation is $1H_2 : 2HI$.

Q for a reaction is the ratio of concentrations of products to concentrations of reactants. As the reaction progresses the concentration of reactants decrease and the concentration of products increase, which means that Q increases as a function of time.

b) No, Q would still increase with time because the $[I_2]$ would decrease in exactly the same way as $[H_2]$ decreases.

17.11 Yes, the Q's for the two reactions do differ. The balanced equation for the first reaction is

$^{3}/_{2}H_2(g) + \frac{1}{2}N_2(g) \rightleftharpoons NH_3(g)$ (1)

The coefficient in front of NH_3 is fixed at 1 mole according to the description. In the second reaction, the coefficient in front of N_2 is fixed at one mole.

$3H_2(g) + N_2(g) \rightleftharpoons 2NH_3(g)$ (2)

The reaction quotients for the two equations and their relationship are:

$$Q_1 = \frac{[NH_3]}{[H_2]^{3/2}[N_2]^{1/2}} \quad and \quad Q_2 = \frac{[NH_3]^2}{[H_2]^3[N_2]}$$

$$(Q_1)^2 = Q_2$$

17.12 Check that correct coefficients from balanced equation are included as exponents in mass action expression.

a) $4NO(g) + O_2(g) \rightleftharpoons 2N_2O_3(g)$ $\qquad Q_c = \frac{[N_2O_3]^2}{[NO]^4[O_2]}$

b) $SF_6(g) + 2SO_3(g) \rightleftharpoons 3SO_2F_2(g)$ $\qquad Q_c = \frac{[SO_2F_2]^3}{[SF_6][SO_3]^2}$

c) $2SClF_5(g) + H_2(g) \rightleftharpoons S_2F_{10}(g) + 2HCl(g)$ $\qquad Q_c = \frac{[S_2F_{10}][HCl]^2}{[SClF_5]^2[H_2]}$

17.14 a) $2NO_2Cl(g) \rightleftharpoons 2NO_2(g) + Cl_2(g)$ $\qquad Q_c = \frac{[NO_2]^2[Cl_2]}{[NO_2Cl]^2}$

b) $2POCl_3(g) \rightleftharpoons 2PCl_3(g) + O_2(g)$ $\qquad Q_c = \frac{[PCl_3]^2[O_2]}{[POCl_3]^2}$

c) $4NH_3(g) + 3O_2(g) \rightleftharpoons 2N_2(g) + 6H_2O(g)$ $\qquad Q_c = \frac{[N_2]^2[H_2O]^6}{[NH_3]^4[O_2]^3}$

17.16 a) The given reaction $\frac{1}{2}S_2(g) + H_2(g) \rightleftharpoons H_2S(g)$ is the reverse reaction of the original reaction multiplied by a factor of ½. The equilibrium constant for the reverse reaction is the inverse of the original constant. When a reaction is multiplied by a factor K of the new equation is equal to the K of the original equilibrium raised to a power equal to the factor. For the reaction given in part a take $(1/K)^{1/2}$.

$$K_c \; for \; rxn \; a = \left(\frac{1}{1.6x10^{-2}}\right)^{1/2} = \mathbf{7.9}$$

b) The given reaction $5H_2S(g) \rightleftharpoons 5H_2(g) + 5/2S_2(g)$ is the original reaction multiplied by 5/2. Take the original K to the 5/2 power to find K of given reaction.

K_c for rxn b) = $(1.6x10^{-2})^{5/2}$ = $\mathbf{3.2x10^{-5}}$

17.18 The concentration of solids and pure liquids do not change, so their concentration terms are not written in the reaction quotient expression.

a) $2Na_2O_2(s) + 2CO_2(g) \rightleftharpoons 2Na_2CO_3(s) + O_2(g)$ $\qquad Q_c = \frac{[O_2]}{[CO_2]^2}$

b) $H_2O(l) \rightleftharpoons H_2O(g)$ $\qquad Q_c = [H_2O(g)]$

c) $NH_4Cl(s) \rightleftharpoons NH_3(g) + HCl(g)$ $\qquad Q_c = [NH_3][HCl]$

17.20 Make sure not to include solids and liquids in reaction quotient.

a) $2NaHCO_3(s) \rightleftharpoons Na_2CO_3(s) + CO_2(g) + H_2O(g)$ $\qquad Q_c = [CO_2][H_2O]$

b) $SnO_2(s) + 2H_2(g) \rightleftharpoons Sn(s) + 2H_2O(g)$ $\qquad Q_c = \frac{[H_2O]^2}{[H_2]^2}$

c) $H_2SO_4(l) + SO_3(g) \rightleftharpoons H_2S_2O_7(l)$ $\qquad Q_c = \frac{1}{[SO_3]}$

17.23 a) The balanced equations and corresponding reaction quotients are:

$Cl_2(g) + F_2(g) \rightleftharpoons 2ClF(g)$ $\qquad Q_{c1} = \frac{[ClF]^2}{[Cl_2][F_2]}$

$\underline{2\{ClF(g) + F_2(g) \rightleftharpoons ClF_3(g)\}}$ $\qquad Q_{c2} = \left(\frac{[ClF_3]}{[ClF][F_2]}\right)^2$

$Cl_2(g) + 3F_2(g) \rightleftharpoons 2ClF_3(g)$

b) The second equation is multiplied by a factor of 2 so that the two equations can be added to yield the overall equation. The reaction quotient, Q_{c2}, is raised to the power of 2, reflecting the change in coefficients. The reaction quotient for the overall reaction, $Q_{overall}$, determined from the reaction is:

$$Q_{overall} = \frac{[ClF_3]^2}{[Cl_2][F_2]^3}$$

Alternatively, $Q_{overall}$ is calculated as the product of Q_{c1} and Q_{c2}.

$$Q_{overall} = Q_{c1} x Q_{c2} = \left(\frac{[ClF]^2}{[Cl_2][F_2]}\right)\left(\frac{[ClF_3]}{[ClF][F_2]}\right)^2 = \frac{[ClF]^2[ClF_3]^2}{[Cl_2][F_2][ClF]^2[F_2]^2} = \frac{[ClF_3]^2}{[Cl_2][F_2]^3}$$

The two methods result in the same reaction quotient.

17.25 K_c and K_p are related by the equation $K_c = K_p(RT)^{-\Delta n}$ where Δn represents the change in the number of moles of gas in the reaction (moles gaseous products – moles gaseous reactants). When Δn is zero (no change in number of moles of gas) the term $(RT)^{\Delta n}$ equals 1 and $K_c = K_p$. When Δn is not zero, meaning that there is a change in the number of moles of gas in the reaction, then $K_c \neq K_p$.

17.26 a) Since Δn = # of moles gaseous products – # of moles gaseous reactants, Δn is a positive integer. According to the relationship $K_c = K_p(RT)^{-\Delta n}$, K_p must be larger since it is multiplied by a fraction that is less than 1. **K_c is smaller than K_p.**

b) Assuming that RT > 1 (which occurs when T > 12.2 K, because 0.0821 x 12.2 = 1), $K_p > K_c$ if the number of moles of gaseous products exceeds the number of moles of gaseous reactants. $K_p < K_c$ when the number of moles of gaseous reactants exceeds the number of moles of gaseous product.

17.27 a) # moles of gaseous reactants = 0; # moles of gaseous products = 3; $\Delta n = 3 - 0 =$ **3**

b) # moles of gaseous reactants = 1; # moles of gaseous products = 0; $\Delta n = 0 - 1 =$ **-1**

c) # moles of gaseous reactants = 0; # moles of gaseous products = 3; $\Delta n = 3 - 0 =$ **3**

17.29 First determine Δn for the reaction and then calculate K_c using $K_c = K_p(RT)^{-\Delta n}$.

a) Δn = # of product gas moles – # of reactant gas moles = 1 – 2 = -1

$K_c = K_p(RT)^{-\Delta n}$

$$K_c = \left(3.9x10^{-2}\right)\left\{\left(0.08206\frac{atm\cdot L}{mol\cdot K}\right)(1000.\,K)\right\}^{-(-1)} = \mathbf{3.2}$$

b) $\Delta n = 1 - 1 = 0$. When $\Delta n = 0$, $K_c = K_p =$ **28.5**

17.31 First determine Δn for the reaction and then calculate K_p using $K_p = K_c(RT)^{\Delta n}$ (note change in equation from 17.29).

a) $\Delta n = 2 - 1 = 1$. So $K_p = K_c(RT)^1$.

$$K_p = \left(6.1x10^{-3}\right)\left\{\left(0.08206\frac{atm\cdot L}{mol\cdot K}\right)(298\,K)\right\}^{1} = \mathbf{0.15}$$

b) $\Delta n = 2 - 4 = -2$. So $K_p = K_c(RT)^{-2}$.

$$K_p = \left(2.4x10^{-3}\right)\left\{\left(0.08206\frac{atm\cdot L}{mol\cdot K}\right)(1000\,K)\right\}^{-2} = \mathbf{3.6x10^{-7}}$$

17.33 When $Q < K$, the reaction proceeds to the right to form more products. The reaction quotient and equilibrium constant are determined by [products]/[reactants]. For Q to increase and reach the value of K, the concentration of products (numerator) must increase in relation to the concentration of reactants (denominator).

17.35 To decide if the reaction is at equilibrium, calculate Q_p and compare it to K_p. If $Q_p = K_p$ then reaction is at equilibrium. If $Q_p > K_p$ reaction proceeds to left to produce more reactants. If $Q_p < K_p$ reaction proceeds to right to produce more products.

$$Q_p = \frac{\left(P_{H_2}\right)\left(P_{Br_2}\right)}{\left(P_{HBr}\right)^2} = \frac{(0.010)(0.010)}{(0.20)^2} = 2.5x10^{-3}$$

$2.5x10^{-3} > 4.18x10^{-9}$ so reaction is not at equilibrium. Reaction will proceed to **left** to reach equilibrium.

17.38 At equilibrium, equal concentrations of $CFCl_3$ and HCl exist, regardless of starting reactant concentrations. The equilibrium concentrations of $CFCl_3$ and HCl would still be equal if unequal concentrations of CCl_4 and HF were used. This occurs only when the two products have the same coefficients in the balanced equation. Otherwise more of the product with the larger coefficient will be produced.

17.40 a) The approximation applies when the change in concentration from initial to equilibrium is so small that it is insignificant, i.e. K is small and initial concentration is large.

b) This approximation will not work when the change in concentration, $\frac{[\text{reactant}]_{change}}{[\text{reactant}]_{initial}}$, is greater than 5%. This can occur when $[\text{reactant}]_{\text{initial}}$ is very small, or when $[\text{reactant}]_{\text{change}}$ is relatively large due to a large K_{eq}.

17.41 Since all equilibrium concentrations are given in molarities and the reaction is balanced, construct an equilibrium expression and substitute the equilibrium concentrations to find K_c.

$$K_c = \frac{[HI]^2}{[H_2][I_2]} = \frac{(1.87x10^{-3})^2}{(6.50x10^{-5})(1.06x10^{-3})} = \mathbf{50.8}$$

17.43 Reaction table requires that the initial $[PCl_5]$ be calculated: $[PCl_5]$ = 0.15mol/2.0 L = 0.075 M

Concentration (M)	$PCl_5(g)$ $\rightleftharpoons$	$PCl_3(g)$ +	$Cl_2(g)$
Initial	0.075	0	0
Change	-x	+x	+x
Equilibrium	0.075 - x	x	x

17.45 Construct an equilibrium expression and solve for P_{NOCl}.

$$K_p = \frac{P^2_{NOCl}}{P^2_{NO}P_{Cl_2}} \Rightarrow P_{NOCl} = \sqrt{K_p P^2_{NO} P_{Cl_2}}$$

$$P_{NOCl} = \sqrt{(6.5x10^4)(0.35)^2(0.10)} = \mathbf{28\ atm}$$

A high pressure for NOCl is expected because the K_p indicates that the reaction proceeds largely to the right, i.e. to the formation of products.

17.47 The solid ammonium hydrogen sulfide will decompose to produce hydrogen sulfide and ammonia gas until $K_p = (P_{H2S})(P_{NH3}) = 0.11$.

Pressures (atm)	$NH_4HS(s)$ $\rightleftharpoons$	$H_2S(g)$ +	$NH_3(g)$
Initial	55.0 g	0	0
Change	---	+x	+x
Equilibrium	---	x	x

$x^2 = 0.11$; x = 0.33

The equilibrium partial pressures of H_2S and NH_3 are both **0.33 atm**.

17.49 The initial concentrations of N_2 and O_2 are (0.20 mol/1.0 L) = 0.20 M and (0.15 mol/1.0 L) = 0.15 M, respectively.

Conc (M)	$N_2(g)$ +	$O_2(g)$ $\rightleftharpoons$	$2NO(g)$
Initial	0.20	0.15	0
Change	-x	-x	+2x
Equilibrium	0.20-x	0.15-x	2x

$$K_c = \frac{[NO]^2}{[N_2][O_2]}$$

$$4.10x10^{-4} = \frac{(2x)^2}{(0.20 - x)(0.15 - x)}$$

Assume that 0.20 – x ≈ 0.20

and 0.15 – x ≈ 0.15

$$4.10x10^{-4} = \frac{(2x)^2}{(0.20)(0.15)} \Rightarrow x = 1.8x10^{-3}$$

Because x represents the change in concentration, check that $\frac{[conc]_{change}}{[conc]_{init}}$ x 100% is less than 5%: (1.8 x 10^{-3}/0.15) x 100% = 1.2% so the assumption is valid.

Therefore, $[NO]_{eq}$ = 2x = 2(1.8 x 10^{-3}) = **3.6 x 10^{-3} M**.

Verify that K_c = 4.10 x 10^{-4} by substituting the equilibrium concentrations into the equilibrium expression:

$$K_c = \frac{[NO]^2}{[N_2][O_2]} = \frac{(3.6x10^{-3})^2}{(0.20)(0.15)} = 4.3x10^{-4}$$

17.51 First determine the initial concentration of H_2 and the change, x, to reach equilibrium. Note that additional significant figures are retained in calculation until the last step. These additional significant figures are shown as subscripted numbers in the calculations.

$[H_2]_{initial}$ = 0.00623 mol/1.50 L = 0.00415_{33} M

x = 0.00467 M – 0.00415_{33} = $+0.00051_{67}$

Use the value of x to find equilibrium concentrations of I_2 and HI.

$[I_2]_{initial}$ = 0.00414 mol/1.50 L = 0.00276_{00} M

$[I_2]_{eq}$ = 0.00276_{00} + 0.00051_{67} = **0.00328 M**

$[HI]_{intial}$ = 0.0244 mol/1.50 L = 0.0162_{67} M

$[HI]_{eq}$ = 0.0162_{67} - 2(0.00051_{67}) = **0.0152 M**

17.53 Construct a reaction table, using $[ICl]_{init}$ = (0.500 mol/5.00 L) = 0.100 M, and substitute the equilibrium concentrations into the equilibrium expression.

Conc (M)	2ICl(g) ⇌	I_2(g)	+	Cl_2(g)
Initial	0.100	0		0
Change	-2x	+x		+x
Equilibrium	0.100-2x	x		x

$$K_c = \frac{[I_2][Cl_2]}{[ICl]^2} \Rightarrow 0.110 = \frac{(x)(x)}{(0.100 - 2x)^2}$$

$$\sqrt{0.110} = \frac{x}{0.100 - 2x}$$

$$0.0331_{66} - 0.663_{32}x = x$$

$$x = 0.0199_{40}$$

Therefore, $[I_2]_{eq}$ = $[Cl_2]_{eq}$ = **0.0199 M** and $[ICl]_{eq}$ = 0.100 – 2(0.0199_{40}) = **0.060 M**.

17.55 To find the equilibrium constant, determine the equilibrium concentrations of each reactant and product and insert in the equilibrium expression. Since $[N_2]$ increases from 0 to 1.96×10^{-3} M, the concentration of H_2O gas will increase by 3 times as much (stoichiometric ratio is 6 H_2O : 2 N_2) and concentration of products will decrease by a factor equivalent to the stoichiometric ratio (2 for NH_3 and 3/2 for O_2).

$[H_2O]_{eq}$ = (6 mol H_2O/2 mol N_2)(1.96×10^{-3}) = $5.88_{00} \times 10^{-3}$ M

$[NH_3]_{eq}$ = (0.0150 mol NH_3/1.00 L) - (4 mol NH_3/2 mol N_2)(1.96×10^{-3}) = $1.10_{80} \times 10^{-2}$ M

$[O_2]_{eq}$ = (0.0150 mol O_2/1.00 L) – (3 mol O_2/2 mol N_2)(1.96×10^{-3}) = $1.20_{60} \times 10^{-2}$ M

$$K_c = \frac{[N_2]^2[H_2O]^6}{[NH_3]^4[O_2]^3} = \frac{(1.96x10^{-3})^2(5.88_{00}x10^{-3})^6}{(1.10_{80}x10^{-2})^4(1.20_{60}x10^{-2})^3} = \mathbf{6.01x10^{-6}}$$

If values for concentrations were rounded to calculate K_c, the answer is 5.90×10^{-6}.

17.58 Equilibrium position refers to the specific concentrations or pressures of reactants and products that exist at equilibrium whereas equilibrium constant refers to the overall ratio of equilibrium concentrations and not to specific concentrations. Reactant concentration changes cause changes in the specific equilibrium concentrations of reactants and products (equilibrium position), but not in the equilibrium constant.

17.59 A positive ΔH_{rxn} indicates that the reaction is endothermic, and that heat is consumed in the reaction: $NH_4Cl(s)$ + heat $\rightleftharpoons$ $NH_3(g)$ + $HCl(g)$

a) The addition of heat causes the reaction to proceed to the right to counterbalance the effect of the added heat. Therefore, more products form at a higher temperature and container **(B)** best represents the mixture.

b) When heat is removed, the reaction shifts to the left to offset that disturbance. Therefore, NH_3 and HCl molecules combine to form more reactant and container **(A)** best represents the mixture.

17.62 An endothermic reaction can be written as: reactants + heat $\rightleftharpoons$ products. A rise in temperature favors the forward direction of the reaction, i.e. the formation of products and consumption of reactants. Since *K* = [products]/[reactants], the addition of heat increases the numerator and decreases the denominator, making K_2 larger than K_1.

17.63 a) Equilibrium position shifts **toward products**. Adding a reactant (CO) causes production of more products.

b) Equilibrium position shifts **toward products**. Removing a product (CO_2) causes production of more products.

c) Equilibrium position **does not shift**. The amount of a solid reactant or product does not impact the equilibrium as long as there is some solid present.

d) Equilibrium position shifts **toward reactants**. Product is added - dry ice is solid carbon dioxide that sublimes to carbon dioxide gas. At very low temperatures CO_2 solid will not sublime, but since the reaction lists carbon dioxide as a gas, the assumption that sublimation takes place is reasonable.

17.65 An increase in container volume results in a decrease in pressure. Le Châtelier's principle states that the equilibrium will shift in the direction that forms more moles of gas to offset the decrease in pressure.

a) **More F** forms (2 moles of gas) and **less F_2** (1 mole of gas) is present.

b) **More C_2H_2 and H_2** form (4 moles of gas) and **less CH_4** (2 moles of gas) is present.

17.67 Decreasing container volume increases the concentration of all gaseous reactants and products.

a) With a decrease in container volume, the $[H_2]$, $[Cl_2]$, and [HCl] will increase by the same factor, x, so calculating the new value for Q_c in terms of the original equilibrium concentrations designated as "eq1" below gives

$$Q_c = \frac{(x[H_2]_{eq1})(x[Cl_2]_{eq1})}{(x[HCl]_{eq1})^2} = \frac{x^2([H_2]_{eq1})([Cl_2]_{eq1})}{x^2[HCl]^2_{eq1}} = \frac{([H_2]_{eq1})([Cl_2]_{eq1})}{[HCl]^2_{eq1}} = K_c$$

The increase in concentrations of reactants is offset by the increase in concentration of the product, so there is **no effect** on the amounts of reactants or products in this reaction.

b) In this equilibrium, a decrease in container volume will increase the concentrations of the reactant gases, H_2 and O_2. In this case $Q_c = 1/([H_2]^2[O_2])$ so Q_c decreases with the increased reactant concentrations. To return to equilibrium, the concentrations and amounts of **H_2 and O_2 will decrease** from their initial values before the volume was changed. **More H_2O** will form as a result of the shift in equilibrium position.

17.69 The purpose of adjusting the volume is to cause a shift in equilibrium.

a) Because the number of reactant gaseous moles equals the product gaseous moles, a change in volume will have **no effect** on the yield.

b) The moles of gaseous product (2CO) exceed the moles of gaseous reactant ($1O_2$). A decrease in pressure favors the reaction direction that forms more moles of gas, so **increase** the reaction vessel volume.

17.71 Increases in temperature causes a shift in the endothermic direction of the equilibrium.

a) Reverse reaction is endothermic, so amount of product **decreases**.

b) Forward reaction is endothermic, so amounts of products **increase**.

c) Forward reaction is endothermic, so amounts of products **increase**.

d) Reverse reaction is endothermic, so amount of product **decreases**.

17.73 The van't Hoff equation shows how the equilibrium constant is affected by a change in temperature. Substitute the given variables into the equation and solve for K_2.

$K_1 = 1.80$ $\quad T_1 = 298$ K $\quad \Delta H^0_{rxn} = 0.32$ kJ/mol × 2 mol DH = 640 J

$K_2 = ?$ $\quad T_2 = 500.$ K $\quad R = 8.314$ J/mol·K

$$\ln\frac{K_2}{K_1} = -\frac{\Delta H^0_{rxn}}{R}\left(\frac{1}{T_2} - \frac{1}{T_1}\right)$$

$$\ln\frac{K_2}{1.80} = -\frac{640\ J}{8.314\ J/K}\left(\frac{1}{500.} - \frac{1}{298}\right)$$

$$\ln\frac{K_2}{1.80} = -(-0.104_{36})$$

$$\frac{K_2}{1.80} = e^{0.10436} \Rightarrow K_2 = \mathbf{2.0}$$

17.76 a) The forward reaction is exothermic so it is favored by **lower temperatures**.

There are fewer moles of gas as product than as reactants, so products are favored by **higher pressure**.

b) Addition of O_2 would **decrease Q** and have **no impact on K**.

c) To enhance yield of SO_3, a low temperature is used. Reaction rates are slower at lower temperatures so a catalyst is used to speed up the reaction.

17.79 a) Initially, a total of 6.00 volumes of gas (5.00 N_2 + 1.00 O_2) exist at 5.00 atm and 900. K. The volumes are proportional to moles of gas, so volume fraction equals mole fraction. The initial partial pressures of the two gases are:

$P_{N2} + P_{H2} = 5.00$ atm

$P_{N2} = X_{N2}P_{tot} = {}^5/_6(5.00 \text{ atm}) = 4.16_{67}$ atm

$P_{H2} = X_{H2}P_{tot} = {}^1/_6(5.00 \text{ atm}) = 0.833_{33}$ atm

The reaction table describing equilibrium is written as follows:

P (atm)	$N_2(g)$	+	$O_2(g)$	$\rightleftharpoons$	$2NO(g)$
Initial	4.16_{67}		0.833_{33}		0
Change	-x		-x		+2x
Equilibrium	$(4.16_{67} - x)$		$(0.833_{33} - x)$		2x

Since K_p is very small, assume that $[N_2]_{eq} = 4.17 - x \approx 4.17$ and $[H_2]_{eq} = 0.833 - x \approx 0.833$.

$$K_p = \frac{P_{NO}^2}{P_{N_2}P_{O_2}} = \frac{(2x)^2}{(4.16_{67})(0.833_{33})} = 6.70x10^{-10}$$

$x = 2.41x10^{-5}$

Check assumption: % error = $(4.82 \times 10^{-5}/4.16)100 < 5\%$. Assumption is good.

Therefore, $[NO]_{eq} = 2x = 2(2.41 \times 10^{-5}) =$ **4.82×10^{-5} atm**

b) Convert atm (P) to concentration (mol/L) using the ideal gas law, assuming that it is applicable at 900. K.

$$\frac{n}{V} = \frac{P}{RT} = \frac{4.82x10^{-5}\ atm}{\left(0.08206\frac{L \cdot atm}{mol \cdot K}\right)(900.\ K)} = 6.53_{08}x10^{-7}\ mol / L$$

$$Conc\,(\mu g / L) = \left(\frac{6.53_{08}x10^{-7}\ mol\ NO}{L}\right)\left(\frac{30.01\ g\ NO}{mol\ NO}\right)\left(\frac{10^6\ \mu g}{1\ g}\right) = \mathbf{19.6\ mg / L}$$

17.81 Set up reaction table:

Concentration (M)	$NH_2COONH_4(s)$	$\rightleftharpoons$	$2NH_3$	+	CO_2
Initial	7.80 g		0		0
Change	---		+2x		+x
Equilibrium	---		2x		x

$K_c = 1.58x10^{-8} = (2x)^2(x)$; $x = 1.58_{08}x10^{-3}$ M

Total concentration = $2(1.58_{08}x10^{-3}\ M) + 1.58_{08}x10^{-3}\ M = 4.74_{23}x10^{-3}$ M

To find total pressure use the ideal gas equation:

$$P = \left(\frac{n}{V}\right)RT = (4.74_{23}x10^{-3}\ M)(0.08206\ atm \cdot L / mol \cdot K)(523\ K) = \mathbf{0.204\ atm}$$

Alternate solution: Convert K_c to K_p then solve for equilibrium pressures.

$K_p = K_c(RT)^3 = (1.58x10^{-8})[(0.08206\ atm \cdot L/mol \cdot K)(523\ K)]^3 = 1.24_{90}x10^{-3}$

$(2x)^2(x) = 1.24_{90}x10^{-3}$; $x = 6.78_{42}x10^{-2}$ atm

Total pressure = $2(6.78_{42}x10^{-2}$ atm) + $(6.78_{42}x10^{-2}$ atm) = **0.204 atm**

17.84 a) Convert K_c to K_p, substitute given values into the equilibrium expression and solve for P_{C2H4}.

$K_p = K_c(RT)^{-1} = (9 \times 10^3)\{(0.08206\ L{\cdot}atm/mol{\cdot}K)(600.\ K)\}^{-1} = 1._{83} \times 10^2$

$$K_p = \frac{P_{C_2H_5OH}}{P_{C_2H_4}P_{H_2O}} = \frac{200.\,atm}{P_{C_2H_4}(400.\,atm)} = 1._{83}\,x10^2$$

$P_{C_2H_4}$ = **3x10^{-3} atm**

b) The forward direction, towards the production of ethanol, produces the least number of moles of gas and is favored by **high pressure**. A **low temperature** favors an exothermic reaction.

c) Use van't Hoff's equation to find K_c at a different temperature.

$K_1 = 9 \times 10^3$ T_1 = 600. K ΔH^0_{rxn} = -47.8 kJ/mol = -47800 J/mol

K_2 = ? T_2 = 450. K   R = 8.314 J/mol·K

$$\ln\frac{K_2}{9x10^3} = -\frac{-47800\,J/mol}{8.314\,J/mol\cdot K}\left(\frac{1}{450.} - \frac{1}{600.}\right)$$

$$\ln\frac{K_2}{9x10^3} = 3.19_{41}$$

$$\frac{K_2}{9x10^3} = e^{3.19} \Rightarrow K_2 = \mathbf{2x10^5}$$

d) No, condensing the C_2H_5OH would not increase the yield. Ethanol has a lower boiling point (78.5°C) than water (100°C). Decreasing the temperature to condense the ethanol would also condense the water, so moles of gas from each side of the reaction are removed. The direction of equilibrium (yield) is unaffected when there is no net change in the number of moles of gas.

17.89 a) Convert K_c to K_p, substitute given values into the equilibrium expression and solve for P_{SO2} ($\Delta n = 2 - 3 = -1$).

$K_p = K_c(RT)^{-1} = (1.7 \times 10^8)\ \{(0.08206\ L{\cdot}atm/mol{\cdot}K)(600.\ K)\}^{-1} = 3.4_{53} \times 10^6$.

$$K_p = \frac{P^2_{SO_3}}{P^2_{SO_2}P_{O_2}} = \frac{(300.\,atm)^2}{P^2_{SO_2}(100.\,atm)} = 3.4_{53}x10^6$$

P_{SO_2} = **1.6x10^{-2} atm**

b) Create a reaction table that describes the reaction conditions.

Conc (M)	$2SO_2(g)$	+	$O_2(g)$	⇌	$2SO_3(g)$
Initial	0.0040		0.0028		0
Change	-0.0020		-0.0010		+0.0020
Equilibrium	0.0020		0.0018		0.0020

The change in concentration for O_2 is half the amount of SO_3 because 1 mol of O_2 reacts to form 2 moles of SO_3. Substitute equilibrium concentrations into the equilibrium expression and solve for K_c.

$$K_c = \frac{[SO_3]^2}{[SO_2]^2[O_2]} = \frac{(0.0020)^2}{(0.0020)^2(0.0018)} = \mathbf{5.6x10^2}$$

The pressure of SO_2 is estimated using the concentration of SO_2 and the ideal gas law (although the ideal gas law is not well behaved at high pressures and temperatures).

$$P_{SO_2} = \frac{n_{SO_2}RT}{V} = \frac{(0.0020\,mol)\left(0.08206\frac{atm\cdot L}{mol\cdot K}\right)(1000.\,K)}{1.0\,L} = \mathbf{0.16\ atm}$$

17.91 Equilibrium constant for the reaction is $K_p = P_{CO2} = 0.220$ atm.

The amount of calcium carbonate solid in the container at the first equilibrium equals the original amount, 0.100 mol, minus the amount reacted to form 0.220 atm of carbon dioxide. Moles of $CaCO_3$ reacted is equal to the number of moles of carbon dioxide produced.

$$n_{CO_2} = n_{CaCO_3} = \frac{PV}{RT} = \frac{(0.220\,atm)(10.0\,L)}{(0.08206\,atm\cdot L/mol\cdot K)(385\,K)} = 6.96_{35}x10^{-2}\ mol\ CaCO_3$$

0.100 mol $CaCO_3$ - 6.96_{35} x 10^{-2} mol $CaCO_3$ = 3.03_{64} x 10^{-2} mol $CaCO_3$

As more carbon dioxide gas is added the system returns to equilibrium by converting the added carbon dioxide to calcium carbonate to maintain the partial pressure of carbon dioxide at 0.220 atm. Convert the added 0.300 atm of CO_2 to moles. The moles of CO_2 reacted equals the moles of $CaCO_3$ formed. Add to moles of $CaCO_3$ at first equilibrium position and convert to grams.

$$n_{CO_2} = \frac{PV}{RT} = \frac{(0.300\,atm)(10.0\,L)}{(0.08206\,atm\cdot L/mol\cdot K)(385\,K)} = 9.49_{57}x10^{-2}\ mol\ CO_2$$

$$\left(3.03_{64}x10^{-2}\ mol\ CaCO_3 + 9.49_{57}x10^{-2}\ mol\ CaCO_3\right)\left(\frac{100.09\ g\ CaCO_3}{mol}\right) = \mathbf{12.5\ g}$$

17.95 The reaction is described by the following equilibrium expression: $K_{eq} = \frac{[SF_4][SF_6]}{[S_2F_{10}]}$.

At the first equilibrium, $[S_2F_{10}]$ = 0.50M and $[SF_4] = [SF_6] = x$. Therefore, the concentrations of SF_4 and SF_6 can be expressed mathematically as $[SF_4] = [SF_6] = \sqrt{0.50\,K_{eq}}$. The concentrations of SF_4 and SF_6 at the second equilibrium are $[SF_4] = [SF_6] = \sqrt{2.5\,K_{eq}}$. The concentrations of SF_4 and SF_6 increase by a factor of $\frac{\sqrt{2.5\,K_{eq}}}{\sqrt{0.50\,K_{eq}}} = \mathbf{2.2}$.

17.97 Calculate the partial pressures of oxygen and carbon dioxide based on the fact that volumes are proportional to moles of gas so volume fraction equals mole fraction. Assume that the amount of carbon monoxide gas is small relative to the other gases so total volume of gases equals $V_{CO2} + V_{O2} + V_{N2}$ = 10.0 + 1.00 + 50.0 = 61.0.

P_{CO2} = (10.0 mol CO_2/61.0 mol gas)(4.0 atm) = 0.65_{57} atm

P_{O2} = (1.00 mol O_2/61.0 mol gas)(4.0 atm) = 0.065_{57} atm

a) Use the partial pressures and given K_p to find P_{CO}.

$$1.4x10^{-28} = \frac{(P_{CO})^2(P_{O_2})}{(P_{CO_2})^2}; \; P_{CO} = \sqrt{\frac{(1.4x10^{-28})(0.65_{57})^2}{(0.065_{57})}} = \mathbf{3.0x10^{-14}\ atm}$$

b) Convert partial pressure to moles per liter using the ideal gas equation and then convert moles of CO to grams.

$$\frac{n_{CO}}{V} = \frac{P}{RT} = \frac{3.0x10^{-14}\ atm}{(0.08206\ atm \cdot L / mol \cdot K)(800\ K)} = 4.5_{70}x10^{-16}\ mol / L$$

($4.5_{70}x10^{-16}$ mol CO/L)(28.01 g CO/mol CO)($1x10^{12}$ pg/g) = **$1.3x10^{-2}$ pg CO/L**

17.100 a) Write a reaction table given that P_{CH4}(init) = P_{CO2}(init) = 10.0 atm, substitute equilibrium values into the equilibrium expression, and solve for P_{H2}.

Pressure (atm)	$CH_4(g)$ +	$CO_2(g)$ ⇌	$2CO(g)$ +	$2H_2(g)$
Initial	10.0	10.0	0	0
Change	-x	-x	+2x	+2x
Equilibrium	10.0-x	10.0-x	2x	2x

$$K_p = \frac{P_{CO}^2 P_{H_2}^2}{P_{CH_4} P_{CO_2}} = \frac{(2x)^2(2x)^2}{(10.0-x)(10.0-x)} = 3.548x10^6$$

$$\sqrt{\frac{(2x)^4}{(10.0-x)^2}} = \sqrt{3.548x10^6}$$

$$\frac{(2x)^2}{10.0-x} = 1.883_{61}x10^3$$

The assumption that 10.0 – x ≈ 10.0 is not valid because K_p is so large.

$$4x^2 + 1.883_{61}x10^3 x - 1.88_{36}x10^4 = 0$$

$$x = \frac{-1.883_{61}x10^3 \pm \sqrt{(1.883_{61}x10^3)^2 - 4(4)(-1.88_{36}x10^4)}}{2(4)} = 9.79_{62}$$

Therefore, P_{H2} = 2x = 2(9.79_{62}) = 19.6 atm is the actual yield.

If the reaction proceeded entirely to completion, the partial pressure of H_2 would be 20.0 atm (pressure is proportional to moles, and twice as many moles of H_2 form for each mole of CH_4 or CO_2 that reacts).

The percent yield is (19.6 atm/20.0 atm) x 100% = **97.9%**.

b) $$\sqrt{\frac{(2x)^4}{(10.0-x)^2}} = \sqrt{2.626x10^7}$$

$$\frac{(2x)^2}{10.0-x} = 5.124_{45}x10^3$$

$$4x^2 + 5.124_{45}x10^3 x - 5.12_{45}x10^4 = 0$$

$$x = \frac{-5.124_{45}x10^3 \pm \sqrt{(5.124_{45}x10^3)^2 - 4(4)(-5.12_{45}x10^4)}}{2(4)} = 9.92_{34}$$

Therefore, $P_{H2} = 2x = 2(9.92_{34}) = 19.8_{47}$ atm. The % yield = (19.8 atm/20.0 atm) x 100% = **99.2%**.

c) van't Hoff equation:

$$\ln\frac{K_2}{K_1} = -\frac{\Delta H^o_{rxn}}{R}\left(\frac{1}{T_2} - \frac{1}{T_1}\right)$$

$$\ln\left(\frac{3.548x10^6}{2.626x10^7}\right) = -\frac{\Delta H^o_{rxn}}{8.314\ J/mol \cdot K}\left(\frac{1}{1200.\ K} - \frac{1}{1300.\ K}\right)$$

$\Delta H^o_{rxn} = 2.596x10^5\ J$ *or* **259.6 kJ**

17.102 a) Multiply the second equation by 2 to cancel the moles of CO produced in the first reaction.

$$2CH_4(g) + O_2(g) \rightleftharpoons \cancel{2CO}(g) + 4H_2(g)$$
$$\cancel{2CO}(g) + 2H_2O(g) \rightleftharpoons 2CO_2(g) + 2H_2(g)$$
$$2CH_4(g) + O_2(g) + 2H_2O(g) \rightleftharpoons 2CO_2(g) + 6H_2(g)$$

b) $K_p = (9.34x10^{28})(1.374)^2 = 1.76x10^{29}$

c) $\Delta n = 8 - 5 = 3$

$K_c = (1.76x10^{29})[(0.08206\ atm \cdot L/mol \cdot K)(1000\ K)]^{-3} = 3.19x10^{23}$

d) The initial total pressure is given as 30. atm. To find the final pressure use relationship between pressure and number of moles of gas: $n_{initial}/P_{initial} = n_{final}/P_{final}$

Total mol of gas initial = 2.0 mol CH_4 + 1.0 mol O_2 + 2.0 mol H_2O = 5.0 mol

Total mol of gas final = 2.0 mol CO_2 + 6.0 mol H_2 = 8.0 mol

$$P_{final} = (30.\ atm)\left(\frac{8.0\ mol}{5.0\ mol}\right) = \mathbf{48\ atm}$$

17.105 a) Careful reading of the problem indicates that the given K_p occurs at 1000. K and that the initial pressure of N_2 is 200. atm. Solve the equilibrium expression for P_N, assuming that $P_{N2}(eq) = 200. - x \approx 200.$

$$K_p = \frac{P_N^2}{P_{N_2}} = 10^{-43.10}$$

$$P_N = \sqrt{(200.)(10^{-43.10})} = \mathbf{4.0x10^{-21}\ atm}$$

(*sig figs in log values limited to digits after decimal point*)

b) $$K_p = \frac{P_H^2}{P_{H_2}} = 10^{-17.30}$$

$$P_H = \sqrt{(600.)(10^{-17.30})} = \mathbf{5.5x10^{-8}\ atm}$$

c) Convert pressures to moles using the ideal gas law. Convert moles to atoms using Avogadro's number.

$$n_N = \frac{PV}{RT} = \frac{(3.9_{86} x10^{-21}\ atm)(1.0\ L)}{(0.08206\ atm \cdot L/mol \cdot K)(1000.\ K)} = 4.8_{57} x10^{-23}\ mol \left(\frac{6.022x10^{23}\ H\ atoms}{mol\ H}\right) = \mathbf{29\ N\ atoms}$$

$$n_H = \frac{PV}{RT} = \frac{(5.4_{84} x10^{-8}\ atm)(1.0\ L)}{(0.08206\ atm \cdot L/mol \cdot K)(1000.\ K)} = 6.6_{83} x10^{-10}\ mol \left(\frac{6.022x10^{23}\ H\ atoms}{mol\ H}\right) = \mathbf{4.0x10^{14}\ H\ atoms}$$

d) The more reasonable step is **N_2(g) + H(g) → NH(g) + N(g).** With only 29 N atoms in 1.0 L, the first reaction would produce virtually no NH(g) molecules. There are orders of magnitude more N_2 molecules than N atoms, so the second reaction is the more reasonable step.

17.108 a) Equilibrium partial pressures for the reactants, nitrogen and oxygen, can be assumed to equal their initial partial pressures because the equilibrium constant is so small that very little nitrogen and oxygen will react to form nitrogen monoxide. After calculating the equilibrium partial pressure of nitrogen monoxide, test this assumption by comparing the partial pressure of nitrogen monoxide with that of nitrogen and oxygen.

Use the equilibrium expression to find P_{NO}:

$$P_{NO} = \sqrt{(K_p)(P_{N_2})(P_{O_2})} = \sqrt{(4.35x10^{-31})(0.780)(0.210)} = 2.67x10^{-16}\ atm$$

Based on the small amount of nitrogen monoxide formed, the assumption that the partial pressures of nitrogen and oxygen change to an insignificant degree holds.

Equilibrium partial pressures: P_{N2} = **0.780 atm**, P_{O2} = **0.210 atm**, P_{NO} = **2.67 x 10^{-16} atm**

b) Total pressure is the sum of the three partial pressures:

0.780 atm + 0.210 atm + 2.67 x 10^{-16} atm = **0.990 atm**

c) $K_c = K_p$ = **4.35 x 10^{-31}** because there is no net increase or decrease in the number of moles of gas in the course of the reaction. (See Problem 17.25)

17.111 a) The enzyme that inhibits F is the one that catalyzes the reaction that produces F. The enzyme is number **5** that catalyzed the reaction from E to F.

b) Enzyme **8** is inhibited by I.

c) If F inhibited enzyme 1 then neither branch of the reaction would take place once enough F was produced.

d) If F inhibited enzyme 6 then the second branch would not take place when enough F was made.

CHAPTER 18

ACID-BASE EQUILIBRIA

FOLLOW-UP PROBLEMS

Format reminder: Additional significant figures are retained until the final calculation. The additional significant figures are shown as subscripts throughout the calculation.

18.1 a) Chloric acid, **$HClO_3$**, is the stronger acid because the strength increases as the number of O atoms increasingly exceeds the ionizable H atom.

b) Hydrochloric acid, **HCl**, is one of the strong hydrohalic acids whereas acetic acid, CH_3COOH, is a weak carboxylic acid.

c) Sodium hydroxide, **NaOH**, is a strong base because Na is a Group 1A(1) metal. Methyl amine, CH_3NH_2, is an organic amine and therefore a weak base.

18.2 Plan: The product of $[H_3O^+]$ and $[OH^-]$ remains constant at 25°C because the value of K_w is constant at a given temperature. Use Equation 18.2 to solve for $[H_3O^+]$.

Solution: Calculating $[H_3O^+]$:

$$[H_3O^+] = \frac{K_w}{[OH^-]} = \frac{1.0x10^{-14}\ M^2}{6.7x10^{-2}\ M} = \mathbf{1.5x10^{-13}\ M}$$

Since $[OH^-] > [H_3O^+]$, the solution is basic.

Check: The solution is neutral when either species concentration is 1 x 10^{-7} M. Since the problem specifies an $[OH^-]$ much greater than this, the solution is basic.

18.3 Plan: As determined in 18.1c), NaOH is a strong base that dissociates completely in water. Subtract pH from 14.00 to find the pOH, and calculate inverse logs of pH and pOH to find $[H_3O^+]$ and $[OH^-]$, respectively.

Solution: pH + pOH = 14.00

pOH = 14.00 – 9.52 = **4.48**

$$pH = -\log[H_3O^+]$$

$$[H_3O^+] = 10^{-pH} = 10^{-9.52} = \mathbf{3.0x10^{-10}\ M}$$

$$pOH = -\log[OH^-]$$

$$[OH^-] = 10^{-pOH} = 10^{-4.48} = \mathbf{3.3x10^{-5}\ M}$$

Check: At 25°C, $[H_3O^+][OH^-]$ should equal 1.0 x 10^{-14} M^2. (3.0 x 10^{-10} M)(3.3 x 10^{-5} M) = 9.9 x 10^{-15} M ≈ 1.0 x 10^{-14}. The significant figures in the concentrations reflect the number of significant figures after the decimal point in the pH and pOH values.

18.4 Plan: Identify the conjugate pairs by first identifying the species that donates H^+ (the acid) in the either reaction direction. The other reactant accepts the H^+ and is the base. The acid has one more H and +1 greater charge than its conjugate base.

Solution:

a) CH_3COOH has one more H^+ than CH_3COO^-. H_3O^+ has one more H^+ than H_2O. Therefore, CH_3COOH and H_3O^+ are the acids, and CH_3COO^- and H_2O are the bases. The conjugate acid/base pairs are **CH_3COOH/CH_3COO^-** and **H_3O^+/H_2O**.

b) H_2O donates a H^+ and acts as the acid. F^- accepts the H^+ and acts as the base. In the reverse direction, HF acts as the acid and OH^- acts as the base. The conjugate acid/base pairs are **H_2O/OH^-** and **HF/F^-**.

18.5 a) The following equation describes the dissolution of ammonia in water:

$$NH_3(g) + H_2O(l) \rightleftharpoons NH_4^+(aq) + OH^-(aq)$$

weak base weak acid ⟵ *stronger acid strong base*

Ammonia is a known weak base, so it makes sense that it accepts a H^+ from H_2O. The reaction arrow indicates that the equilibrium lies to the left because the question states that ammonia is smelled (NH_4^+ and OH^- are odorless). According to Figure 18.10, NH_4^+ and OH^- are the stronger acid and base, so the reaction proceeds to the formation of the weaker acid and base.

b) The addition of excess HCl results in the following equation:

$$NH_3(g) + H_3O^+(aq;\ \text{from HCl}) \rightleftharpoons NH_4^+(aq) + H_2O(l)$$

stronger base strong acid ⟶ *weak acid weak base*

HCl is a strong acid and is much stronger than NH_4^+ according to Figure 18.9. Similarly, NH_3 is a stronger base than H_2O. The reaction proceeds to produce the weak acid and base, and thus the odor from NH_3 disappears.

c) The solution in (b) is mostly NH_4^+ and H_2O. The addition of excess NaOH results in the following equation:

$$NH_4^+(aq) + OH^-(aq;\ \text{from NaOH}) \rightleftharpoons NH_3(g) + H_2O(l)$$

stronger acid strong base ⟶ *weak base weak acid*

NH_4^+ and OH^- are the strong acid and base, respectively, and drive the reaction towards the formation of the weak base and acid, $NH_3(g)$ and H_2O, respectively. The reaction direction explains the return of the ammonia odor.

18.6 Plan: Write a balanced equation for the dissociation of NH_4^+ in water. Using the given information, construct a table that describes the initial and equilibrium concentrations. Construct an equilibrium expression and make assumptions where possible to simplify the calculations.

Solution:

	$NH_4^+(aq)$ +	$H_2O(l)$	$\rightleftharpoons$	$H_3O^+(aq)$	+	$NH_3(g)$
Initial	0.2 M	-------		0		0
Change	-x	-------		+x		+x
Eqlb	0.2 – x	-------		x		x

The initial concentration of NH_4^+ = 0.2 M because each mole of NH_4Cl completely dissociates to form one mole of NH_4^+.

$x = [H_3O^+] = [NH_3] = 10^{-pH} = 10^{-5.0} = 1.00 \times 10^{-5}$ M

$$K_a = \frac{[H_3O^+][NH_3]}{[NH_4^+]} = \frac{x^2}{(0.2 - x)}$$

Assume that the amount of NH_4^+ that dissociates is very small, so that x is negligible in comparison to 0.2. In other words, $0.2 - x \approx 0.2$ and $[NH_4^+]_{eq} \approx [NH_4^+]_{init}$. Check this assumption: $(1.00 \times 10^{-5}/0.2) \times 100\% = 0.005\% < 5\%$; assumption is justified.

The K_a expression simplifies to:

$$K_a = \frac{x^2}{0.2} = \frac{(1.00x10^{-5})^2}{0.2} = \mathbf{5x10^{-10}}$$

Check: The small value of K_a is consistent with the value expected for a weak acid.

18.7 Plan: Write a balanced equation for the dissociation of HOCN in water. Using the given information, construct a table that describes the initial and equilibrium concentrations. Construct an equilibrium expression and solve the quadratic expression for x, the concentration of H_3O^+.

Solution:

	HOCN(aq) +	H_2O(l)	⇌	H_3O^+(aq)	+	OCN^-(aq)
Initial	0.10 M	-------		0		0
Change	-x	-------		+x		+x
Eqlb	0.10 – x	-------		x		x

$$K_a = \frac{[H_3O^+][OCN^-]}{[HOCN]} = 3.5x10^{-4}$$

In this example, the dissociation of HOCN is not negligible in comparison to the initial concentration: (0.10)/(3.5 x 10^{-4}) = 290 < 400. Therefore, $[HOCN]_{eq}$ = 0.10 – x and the equilibrium expression is solved using the quadratic formula.

$$K_a = \frac{(x)(x)}{(0.10 - x)} = 3.5x10^{-4}$$

$$x^2 = 3.5x10^{-5} - 3.5x10^{-4}x$$

$$x^2 + 3.5x10^{-4}x - 3.5x10^{-5} = 0 \quad (general\ formula:\ ax^2 + bx + c = 0)$$

$$x = \frac{-b \pm \sqrt{b^2 - 4ac}}{2a} = \frac{-3.5x10^{-4} \pm \sqrt{(3.5x10^{-4})^2 - 4(1)(-3.5x10^{-5})}}{2(1)}$$

$$= \frac{-3.5x10^{-4} \pm \sqrt{1.4_{01}x10^{-4}}}{2} = 5.7x10^{-3}$$

Therefore, $[H_3O^+]$ = x = **5.7 x 10^{-3} M** and pH = -log(5.7 x 10^{-3}) = **2.24**.

18.8 Plan: Write the balanced equation and corresponding equilibrium expression for each dissociation reaction. Calculate the equilibrium concentrations of all species and convert $[H_3O^+]$ to pH. From Table 18.5, K_{a1} = 5.6 x 10^{-2} and K_{a2} = 5.4 x 10^{-5}.

Solution:

HOOC–COOH(aq) + H_2O(l) ⇌ HOOC–COO^-(aq) + H_3O^+(aq)

$$K_{a1} = \frac{[HC_2O_4^-][H_3O^+]}{[H_2C_2O_4]} = 5.6x10^{-2}$$

HOOC–COO^-(aq) + H_2O(l) ⇌ $^-$OOC–COO^-(aq) + H_3O^+(aq)

$$K_{a2} = \frac{[C_2O_4^{2-}][H_3O^+]}{[HC_2O_4^-]} = 5.4x10^{-5}$$

Assumptions:

1) Since $K_{a1} >> K_{a2}$, the first dissociation produces almost all of the H_3O^+, so $[H_3O^+]_{eq}$ = $[H_3O^+]_{from\ C2H2O4}$.

2) Since $[HA]_{init}/K_{a1} = 0.150/5.6x10^{-2} < 400$, solve the first equilibrium expression using the quadratic equation.

$$\frac{[HC_2O_4^-][H_3O^+]}{[H_2C_2O_4]} = 5.6x10^{-2}$$

$$\frac{x^2}{0.150 - x} = 5.6x10^{-2}$$

$$x^2 + 5.6x10^{-2}x - 8.4x10^{-3} = 0$$

$$x = \frac{-5.6x10^{-2} \pm \sqrt{(5.6x10^{-2})^2 - 4(-8.4x10^{-3})}}{2} = 0.068$$

Therefore, $\mathbf{[H_3O^+] = [HC_2O_4^-] = 0.068\ M}$ and **pH** = -log(0.068) = **1.17**. The oxalic acid concentration at equilibrium is $[H_2C_2O_4]_{init} - [H_2C_2O_4]_{dissoc} = 0.150 - 0.068 =$ **0.082 M**. Solve for $[C_2O_4^{2-}]$ by rearranging the K_{a2} expression:

$$[C_2O_4^{2-}] = \frac{K_{a2} \times [HC_2O_4^-]}{[H_3O^+]} = \frac{(5.4x10^{-5})(0.068)}{(0.068)} = \mathbf{5.4x10^{-5}\ M}$$

Check: The K_{a1} indicates that this weak acid undergoes significant dissociation, so it makes sense that the pH is strongly acidic. The $C_2O_4^{2-}$ concentration should be very small in comparison to the $HC_2O_4^-$ concentration, because the second dissociation occurs to a much smaller extent.

18.9 Plan: Pyridine contains a nitrogen atom that accepts H^+ from water to form OH^- ions in aqueous solution. Write a balanced equation and equilibrium expression for the reaction, convert pK_b to K_b, make simplifying assumptions (if valid) and solve for $[OH^-]$. Calculate $[H_3O^+]$ using Equation 18.2 and convert to pH.

Solution: $C_5H_5N(aq) + H_2O(l) \rightleftharpoons C_5H_5NH^+(aq) + OH^-(aq)$

$$K_b = \frac{[C_5H_5NH^+][OH^-]}{[C_5H_5N]} = 10^{-8.77} = 1.6_{98}x10^{-9}$$

Assumptions:

1) Since $K_b >> K_w$, disregard the $[OH^-]$ that results from the autoionization of water.

2) Since K_b is small, $[C_5H_5N] = 0.10 - x \approx 0.10$. Check: $0.10/(1.7x10^{-9}) > 400$, so assumption is valid.

$$K_b = \frac{[C_5H_5NH^+][OH^-]}{[C_5H_5N]} = \frac{x^2}{0.10} = 1.6_{98}x10^{-9}$$

$$x = 1.3_{03}x10^{-5}$$

Therefore, $[C_5H_5NH^+] = [OH^-] = 1.3 \times 10^{-5}$ M.

$$[H_3O^+] = \frac{K_w}{[OH^-]} = \frac{1.0x10^{-14}}{1.3_{03}x10^{-5}} = 7.6_{74}x10^{-10}\ M$$

pH = $-\log(7.6_{74} \times 10^{-10}) = 9.115 =$ **9.12**

Check: Since pyridine is a weak base, a pH > 7 is expected.

18.10 Plan: The hypochlorite ion, ClO^-, acts as a weak base in water. Write a balanced equation and equilibrium expression for this reaction. The K_b of ClO^- is calculated from the K_a of its conjugate acid, hypochlorous acid, HClO (from Table 18.2, $K_a = 2.9 \times 10^{-8}$). Make simplifying assumptions (if valid), solve for $[OH^-]$, convert to $[H_3O^+]$ and calculate pH.

Solution:

	$ClO^-(aq)$ +	$H_2O(l)$	$\rightleftharpoons$	$HClO(aq)$	+	$OH^-(aq)$
Initial	0.20 M	-------		0		0
Change	-x	-------		+x		+x
Eqlb	0.20 – x	-------		x		x

$$K_b = \frac{[HClO][OH^-]}{[ClO^-]}$$

$$K_b = \frac{K_w}{K_a} = \frac{1.0x10^{-14}}{2.9x10^{-8}} = 3.4_{48}x10^{-7}$$

Assumptions:

1) Since $K_b >> K_w$, the dissociation of water provides a negligible amount of $[OH^-]$.
2) Since K_b is very small, $[ClO^-]_{eq} = 0.20 - x \approx 0.2$. Check: $0.20/(3.5 \times 10^{-7}) > 400$, so assumption is valid.

$$K_b = \frac{[HClO][OH^-]}{[ClO^-]} = \frac{x^2}{0.20} = 3.4_{48}x10^{-7}$$

$$x = 2.6_{26}x10^{-4}$$

Therefore, $[HClO] = [OH^-] = 2.6 \times 10^{-4}$ M.

$[H_3O^+] = K_w/(2.6 \times 10^{-4}) = 3.8 \times 10^{-11}$ M and pH = $-\log(3.8 \times 10^{-11})$ = **10.42**.

Check: Since hypochlorite ion is a weak base, a pH > 7 is expected.

18.11 a) The ions are K^+ and ClO_2^-. Add "OH^-" to the metal cation and "H^+" to the anion to determine from which base and acid the ions are derived: KOH and $HClO_2$. Since the base is strong and the acid is weak, the salt derived from this combination will produce a **basic** solution.

K^+ does not react with water

$ClO_2^-(aq) + H_2O(l) \rightleftharpoons HClO_2(aq) + OH^-(aq)$

b) The ions are $CH_3NH_3^+$ and NO_3^-. $CH_3NH_3^+$ is derived from the weak base methyl amine, CH_3NH_2. Nitrate ion, NO_3^-, is derived from the strong acid HNO_3 (nitric acid). A salt derived from a weak base and strong acid produces an **acidic** solution.

NO_3^- does not react with water

$CH_3NH_3^+(aq) + H_2O(l) \rightleftharpoons CH_3NH_2(aq) + H_3O^+(aq)$

c) The ions are Cs^+ and I^-. Cesium ion is derived from cesium hydroxide, CsOH, which is a strong base because Cs is a Group 1A(1) metal. Iodide ion is derived from hydroiodic acid, HI, a strong hydrohalic acid. Since the both the base and acid are strong, the salt derived from this combination will produce a **neutral** solution.

Neither Cs^+ nor I^- react with water.

18.12 a) The two ions that comprise this salt are cupric ion, Cu^{2+}, and acetate ion, CH_3COO^-. Metal ions are acidic in water. Assume that the hydrated cation is $Cu(H_2O)_6^{2+}$. The K_a is found in Table 18.7.

$Cu(H_2O)_6^{2+}(aq) + H_2O(l) \rightleftharpoons Cu(H_2O)_5OH^+(aq) + H_3O^+(aq) \qquad K_a = 3 \times 10^{-8}$

Acetate ion acts likes a base in water. It's K_b is calculated from the K_a of acetic acid (1.8×10^{-5}): $K_b = K_w/K_a = (1.0 \times 10^{-14})/(1.8 \times 10^{-5}) = 5.6 \times 10^{-10}$.

$CH_3COO^-(aq) + H_2O(l) \rightleftharpoons CH_3COOH(aq) + OH^-(aq) \qquad K_b = 5.6 \times 10^{-10}$

$Cu(H_2O)_6^{2+}$ is a better proton donor than CH_3COO^- is a proton acceptor (i.e. $K_a > K_b$), so a solution of $Cu(CH_3COO)_2$ is **acidic**.

b) The two ions that comprise this salt are ammonium ion, NH_4^+, and fluoride ion, F^-. Ammonium ion is the acid with $K_a = K_w/K_b(NH_3) = (1.0 \times 10^{-14})/(1.76 \times 10^{-5}) = 5.7 \times 10^{-10}$.

$NH_4^+(aq) + H_2O(l) \rightleftharpoons NH_3(aq) + H_3O^+(aq) \qquad K_a = 5.7 \times 10^{-10}$

Fluoride ion is the base with $K_b = K_w/K_a(HF) = (1.0 \times 10^{-14})/(6.8 \times 10^{-4}) = 1.5 \times 10^{-11}$.

$F^-(aq) + H_2O(l) \rightleftharpoons HF(aq) + OH^-(aq) \qquad K_b = 1.5 \times 10^{-11}$

Since $K_a > K_b$, a solution of (NH_4F) is **acidic**.

18.13 Plan: Draw Lewis dot structures for the reactants to see which molecule donates the electron pair (Lewis base) and which molecule accepts the electron pair (Lewis acid).

Solution:

a)

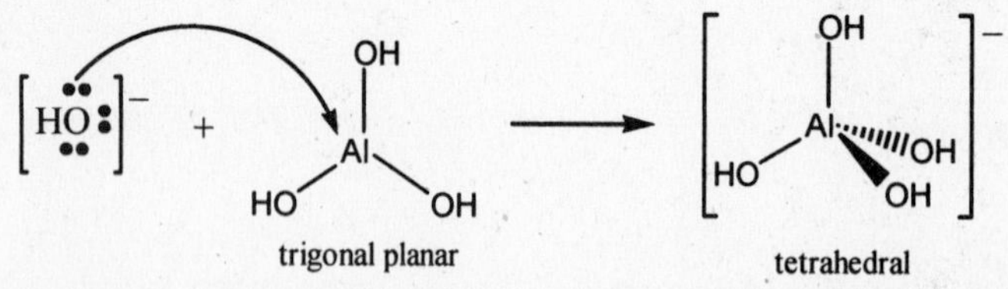

Hydroxide ion, OH^-, donates an electron pair and is the Lewis base; $Al(OH)_3$ accepts the electron pair and is the Lewis acid. Note the change in geometry in forming the adduct.

b)

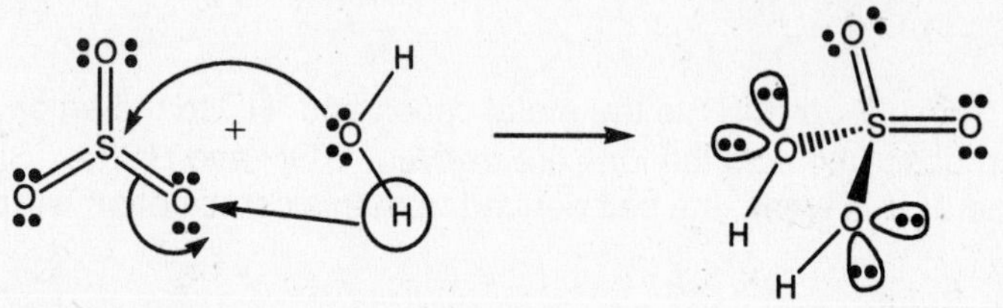

Sulfur trioxide accepts the electron pair and is the Lewis acid. Water donates an electron pair and is the Lewis base.

c)

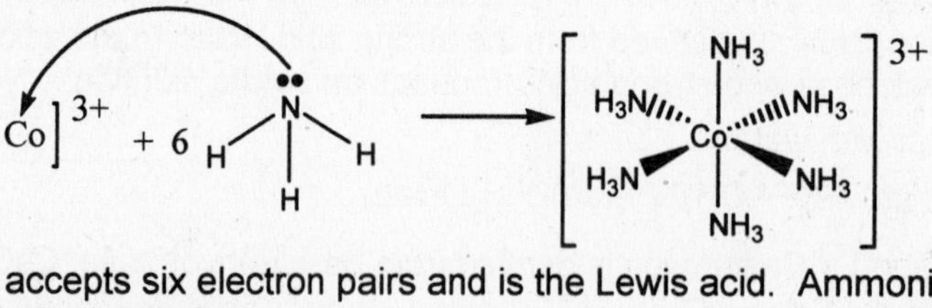

Co^{3+} accepts six electron pairs and is the Lewis acid. Ammonia donates an electron pair and is the Lewis base.

END-OF-CHAPTER PROBLEMS

18.2 All Arrhenius acids contain hydrogen and produce hydronium ion (H_3O^+) in aqueous solution. All Arrhenius bases contain an OH group and produce hydroxide ion (OH^-) in aqueous solution. Neutralization occurs when each H_3O^+ molecule combines with an OH^-

molecule to form 2 molecules of H_2O. Chemists found that the ΔH_{rxn} was independent of the combination of strong acid with strong base. In other words, the reaction of any strong base with any strong acid always produced 56 kJ/mol (ΔH = -56 kJ/mol). This was consistent with Arrhenius' hypothesis describing neutralization, because all other counter ions (those present from the dissociation of the strong acid and base) were spectators and did not participate in the overall reaction.

18.4 Strong acids and bases dissociate completely into their ions when dissolved in water. Weak acids only partially dissociate. The characteristic common to all weak acids is that a significant number of the acid molecules are dissolved and not dissociated. For a strong acid, the concentration of hydronium ion produced by dissolving the acid *is equal to* the initial concentration of the undissociated acid. For a weak acid, the concentration of hydronium ions produced when the acid dissolves *is less than* the initial concentration of the acid.

18.5 a) Water, H_2O, is an Arrhenius acid because it produces H_3O^+ ion in aqueous solution. Water is also an Arrhenius base because it produces OH^- ion as well.

b) Calcium hydroxide, $Ca(OH)_2$ is a base, not an acid.

c) Phosphoric acid, H_3PO_4, is a weak Arrhenius acid. It is weak because the number of O atoms exceeds the ionizable H atoms by 1.

d) Hydroiodic acid, HI, is a strong Arrhenius acid.

18.7 Barium hydroxide, $Ba(OH)_2$, and potassium hydroxide, KOH, (b and d) are Arrhenius bases because they contain hydroxide ions and form OH^- when dissolved in water. H_3AsO_4 and HOCl (a and c) are Arrhenius acids, not bases.

18.9 a) $HCN(aq) + H_2O(l) \rightleftharpoons H_3O^+(aq) + CN^-(aq)$

$$K_a = \frac{[H_3O^+][CN^-]}{[HCN]}$$

b) $HCO_3^-(aq) + H_2O(l) \rightleftharpoons H_3O^+(aq) + CO_3^{2-}(aq)$

$$K_a = \frac{[H_3O^+][CO_3^{2-}]}{[HCO_3^-]}$$

c) $HCOOH(l) + H_2O(l) \rightleftharpoons H_3O^+(aq) + HCOO^-(aq)$

$$K_a = \frac{[H_3O^+][HCOO^-]}{[HCOOH]}$$

Note: The acidic proton on HCOO**H** is shown in bold. A H atom bonded to a carbon is not acidic because the electronegativity difference between C and H is so small. So the dissociated molecule is not shown as $COOH^-(aq)$.

18.11 a) $$K_a = \frac{[H_3O^+][NO_2^-]}{[HNO_2]}$$

b) $$K_a = \frac{[H_3O^+][CH_3COO^-]}{[CH_3COOH]}$$

c) $K_a = \frac{[H_3O^+][BrO_2^-]}{[HBrO_2]}$

18.13 Table 18.2 ranks acids by K_a: the larger the K_a value, the stronger the acid. Hydroiodic acid, HI, is not shown in Table 18.2 because K_a approaches infinity for strong acids and is not meaningful. Therefore, HI is the strongest acid and acetic acid, CH_3COOH, is the weakest: **CH_3COOH < HF < HIO_3 < HI.**

18.15 a) Arsenic acid, H_3AsO_4, is a **weak acid**. The number of O atoms is 4, which exceeds the number of ionizable H atoms, 3, by one. This identifies H_3AsO_4 as a weak acid.

b) Strontium hydroxide, $Sr(OH)_2$, is a **strong base**. Soluble compounds containing OH^- ions are strong bases.

c) HIO is a **weak acid**. The number of O atoms is 1, which is equal to the number of ionizable H atoms identifying HIO as a weak acid.

d) Perchloric acid, $HClO_4$, is a **strong acid**. $HClO_4$ is one example of the type of strong acid in which the number of O atoms exceeds the number of ionizable H atoms by more than 2.

18.17 a) Rubidium hydroxide, RbOH, is a **strong base** because Rb is a Group 1A(1) metal.

b) Hydrobromic acid, HBr, is a **strong acid**, because it is one of the listed hydrohalic acids.

c) Hydrogen telluride, H_2Te, is a **weak acid**, because H is not bonded to a O or halide.

d) Hypochlorous acid, HClO, is a **weak acid** for the same reason identified in 18.15 c).

18.22 a) At equal concentrations the acid with the larger K_a will ionize to produce more hydronium ions than the acid with the smaller K_a. The solution of an acid with the lower **$K_a = 4x10^{-5}$** has a lower $[H_3O^+]$ and higher pH.

b) pK_a is equal to $-logK_a$. The smaller the K_a the larger the pK_a. So the acid with the larger **pK_a, 3.5**, has a lower $[H_3O^+]$ and higher pH.

c) Lower concentration of the same acid means lower concentration of hydronium ions produced. The **0.01 M** solution has a lower $[H_3O^+]$ and higher pH.

d) At the same concentration strong acids dissociate to produce more hydronium ions than weak acids. The 0.1 M solution of a **weak acid** has a lower $[H_3O^+]$ and higher pH.

e) Bases produce OH^- ions in solution, so the concentration of hydronium ion for a solution of a base solution is lower than that for a solution of an acid. The 0.1 M **base solution** has the higher pH.

f) pOH equals $-log[OH^-]$. At 25°C the equilibrium constant for water ionization, K_w, equals $1x10^{-14}$ so 14 = pH + pOH. As pOH decreases pH increases. The solution of **pOH = 6.0** has the higher pH.

18.23 a) Plan: This problem can be approached two ways. Because NaOH is a strong base, the $[OH^-]_{eq} = [NaOH]_{init}$. Alternative 1 involves calculating $[H_3O^+]$ using Equation 18.2, then calculating pH using Equation 18.3. Alternative 2 involves calculating pOH and then using Equation 18.4 to calculate pH.

Solution: Alternative 1:

$$[H_3O^+] = \frac{K_w}{[OH^-]} = \frac{1.00x10^{-14}\ M^2}{0.0111\ M} = 9.00_{90}x10^{-13}\ M$$

$pH = -\log[H_3O^+] = -\log(9.00_{90}x10^{-13}) = \mathbf{12.045}$

Alternative 2:

$pOH = -\log[OH^-] = -\log(0.0111) = 1.955$

$pH = 14 - pOH = \mathbf{12.045}$

With a pH > 7, the solution is **basic**.

<u>Check</u>: Both solution methods match each other, so the answer is correct.

b) For strong acid $[H_3O^+] = [HCl] = 1.23x10^{-3}$ M. pH = $-\log(1.23x10^{-3}) = 2.910$
pOH = 14 – 2.910 = **11.090**. Solution is **acidic**.

18.25 a) HI is a strong acid, so $[H_3O^+] = [HI] = 5.04x10^{-3}$ M.
pH = $-\log(5.04x10^{-3})$ = **2.298.** Solution is **acidic**.

b) $Ba(OH)_2$ is a strong base, so $[OH^-] = 2 \times [Ba(OH)_2] = 5.10$ M
pOH = -log(5.10) = **-0.708**. Solution is **basic**.

18.27 a) $[H_3O^+] = 10^{-pH} = 10^{-9.78} = \mathbf{1.7x10^{-10}\ M}$

$$[OH^-] = \frac{K_w}{[H_3O^+]} = \frac{1.0x10^{-14}\ M^2}{1.6_{60}x10^{-10}\ M} = \mathbf{6.0x10^{-5}\ M}$$

$pOH = -\log[OH^-] = -\log(6.0_{26}x10^{-5}) = \mathbf{4.22}$ (*or* $pOH = 14.00 - 9.78 = 4.22$)

b) $pH = 14.00 - pOH = 14.00 - 10.43 = \mathbf{3.57}$

$[H_3O^+] = 10^{-pH} = 10^{-3.57} = \mathbf{2.7x10^{-4}\ M}$

$[OH^-] = 10^{-pOH} = 10^{-10.43} = \mathbf{3.7x10^{-11}\ M}$

18.29 a) $[H_3O^+] = 10^{-pH} = 10^{-2.77} = \mathbf{1.7x10^{-3}\ M}$

$$[OH^-] = \frac{K_w}{[H_3O^+]} = \frac{1.0x10^{-14}\ M^2}{1.6_{98}x10^{-3}\ M} = \mathbf{5.9x10^{-12}\ M}$$

$pOH = -\log[OH^-] = -\log(5.8_{88}x10^{-12}) = \mathbf{11.23}$ (*or* $pOH = 14.00 - 2.77 = 11.23$)

b) $pH = 14.00 - 5.18 = \mathbf{8.82}$

$[OH^-] = 10^{-pOH} = 10^{-5.18} = \mathbf{6.6x10^{-6}\ M}$

$[H_3O^+] = 10^{-pH} = 10^{-8.82} = \mathbf{1.5x10^{-9}\ M}$

18.31 The pH is increasing so the solution is becoming more basic. Therefore, OH^- ion is added to increase the pH. Since 1 mole of H_3O^+ reacts with 1 mole of OH^-, the difference in $[H_3O^+]$ would be equal to the $[OH^-]$ added.

$[H_3O^+]_{init} = 10^{-pH} = 10^{-3.25} = 5.6_{23}x10^{-4}\ M$

$[H_3O^+]_{final} = 10^{-pH} = 10^{-3.65} = 2.2_{39}x10^{-4}\ M$

Add ($5.6_{23} \times 10^{-4}$ M – $2.2_{39} \times 10^{-4}$ M) = **3.4×10^{-4} mole of OH^- per liter.**

18.33 The pH is increasing so OH^- must be added.

$$[H_3O^+]_{init} = 10^{-pH} = 10^{-4.82} = 1.5_{14} x 10^{-5}\ M$$

$$[H_3O^+]_{final} = 10^{-pH} = 10^{-5.22} = 6.0_{26} x 10^{-6}\ M$$

Add (6.0_{26} x 10^{-6} M – 1.5_{14} x 10^{-5} M) = 9.1_{10} x 10^{-6} mole of OH^- per liter.

9.1_{10} x 10^{-6} mole of OH^-/L x 6.5 L = **5.9 x 10^{-5} mole OH^- added to 6.5 L.**

18.36 a) Heat is absorbed in an endothermic process: $2H_2O(l)$ + heat → $H_3O^+(aq)$ + $OH^-(aq)$. As the temperature increases, the reaction shifts to the formation of products. Since the products are in the numerator of the K_w expression, rising temperature **increases** the value of K_w.

b) Given that the pH is 6.80, the $[H^+]$ can be calculated. The problem specifies that the solution is neutral, meaning $[H^+] = [OH^-]$. A new K_w can then be calculated.

$[H^+] = 10^{-pH} = 10^{-6.80} = 1.5_{85}$ x 10^{-7} =

Therefore, **$[H^+] = [OH^-] = 1.6$ x 10^{-7} M.**

$K_w(37°C) = [H^+][OH^-] = (1.5_{85}$ x $10^{-7})(1.5_{85}$ x $10^{-7})$ = **2.5 x 10^{-14}**.

pOH = -log$[OH^-]$ = -log(1.5_{85} x 10^{-7}) = **6.80**

Alternate method: The van't Hoff equation describes how an increasing temperature affects the equilibrium constant. In this case, the ΔH_{rxn} = +55.9 kJ/mol, same magnitude but opposite sign of the heat of neutralization. T_2 = 273 + 37°C = 310 K and T_1 = 273 + 25°C = 298 K.

$$\ln\frac{K_2}{K_1} = -\frac{\Delta H^0_{rxn}}{R}\left(\frac{1}{T_2} - \frac{1}{T_1}\right)$$

$$\ln\frac{K_2}{1.0x10^{-14}} = -\frac{(55.9\ kJ/mol)(1000\ J/kJ)}{8.314\ J/mol \cdot K}\left(\frac{1}{310\ K} - \frac{1}{298\ K}\right)$$

$$\ln\frac{K_2}{1.0x10^{-14}} = 0.873_{38}$$

$$\frac{K_2}{1.0x10^{-14}} = e^{0.87338};\ K_2 = 2.4x10^{-14}$$

Calculate pOH and $[OH^-]$ as follows:

pOH = pK_w – pH where pK_w = -log(2.4 x 10^{-14}) = 13.62

pOH = 13.62 – 6.80 = 6.82

$[OH^-] = 10^{-pOH} = 10^{-6.82}$ = 1.5 x 10^{-7} M

18.37 The Brønsted-Lowry theory defines acids as proton donors and bases as proton acceptors while the Arrhenius definition looks at acids as containing ionizable H atoms and at bases as containing hydroxide ions. In both definitions an acid produces hydronium ions and a base produces hydroxide ions when added to water.

Ammonia and carbonate ion are two Brønsted-Lowry bases that are not Arrhenius bases because they do not contain hydroxide ions. Brønsted-Lowry acids must contain an ionizable H atom in order to be proton donors, so a Brønsted-Lowry acid that is not an Arrhenius acid can not be identified. (Other examples also acceptable.)

18.40 An amphoteric substance can act as either an acid or a base. In the presence of a strong base (OH^-), the dihydrogen phosphate ion acts like an acid by donating hydrogen:

$H_2PO_4^-(aq) + OH^-(aq) \rightarrow H_2O(aq) + HPO_4^{2-}(aq)$

In the presence of a strong acid (HCl), the dihydrogen phosphate ion acts like a base by accepting hydrogen:

$H_2PO_4^-(aq) + HCl(aq) \rightarrow H_3PO_4(aq) + Cl^-(aq)$

18.41 a) When phosphoric acid is dissolved in water, a proton is donated to the water and dihydrogen phosphate ions are generated (see pp. 776-7).

$H_3PO_4(aq) + H_2O(l) \rightleftharpoons H_2PO_4^-(aq) + H_3O^+(aq)$

$$K_{a1} = \frac{[H_3O^+][H_2PO_4^-]}{[H_3PO_4]}$$

b) Benzoic acid is an organic acid has only one proton to donate – that from the carboxylic acid group. The H atoms bonded to the benzene ring are not acidic hydrogens.

$C_6H_5COOH(aq) + H_2O(l) \rightleftharpoons C_6H_5COO^-(aq) + H_3O^+(aq)$

$$K_a = \frac{[H_3O^+][C_6H_5COO^-]}{[C_6H_5COOH]}$$

c) Hydrogen sulfate ion donates a proton to water and forms the sulfate ion.

$HSO_4^-(aq) + H_2O(l) \rightleftharpoons SO_4^{2-}(aq) + H_3O^+(aq)$

$$K_a = \frac{[H_3O^+][SO_4^{2-}]}{[HSO_4^-]}$$

18.43 To derive the conjugate base, remove one H and decrease the charge by 1. Since each formula is neutral, the conjugate base will have a charge of –1.

a) $HCl \Rightarrow$ **Cl^-**

b) $H_2CO_3 \Rightarrow$ **HCO_3^-**

c) $H_2O \Rightarrow$ **OH^-**

18.45 To derive the conjugate acid, add an H and increase the charge by 1.

a) $NH_3 \Rightarrow NH_4^+$

b) $NH_2^- \Rightarrow NH_3$

c) $C_{10}H_{14}N_2 \Rightarrow C_{10}H_{15}N_2^+$

18.47 a) $HCl + H_2O \rightleftharpoons Cl^- + H_3O^+$

acid   *base*   *conj base*   *conj acid*

Conjugate acid/base pairs: HCl/Cl^- and H_3O^+/H_2O

b) $HClO_4 + H_2SO_4 \rightleftharpoons ClO_4^- + H_3SO_4^+$

acid   *base*   *conj base*   *conj acid*

Conjugate acid/base pairs: $HClO_4/ClO_4^-$ and $H_3SO_4^+/H_2SO_4$

Note: Perchloric acid is able to protonate another strong acid, H_2SO_4, because it is a stronger acid ($HClO_4$'s oxygen atoms exceed its hydrogen atoms by one more than H_2SO_4).

c) HPO_4^{2-} + H_2SO_4 $\rightleftharpoons$ $H_2PO_4^-$ + HSO_4^-

base *acid* *conj acid* *conj base*

Conjugate acid/base pairs: H_2SO_4/ HSO_4^- and $H_2PO_4^-$/ HPO_4^{2-}

18.49 a) NH_3 + H_3PO_4 $\rightleftharpoons$ NH_4^+ + $H_2PO_4^-$

base *acid* *conj acid* *conj base*

Conjugate acid-base pairs: H_3PO_4/$H_2PO_4^-$ and NH_4^+/NH_3

b) CH_3O^- + NH_3 $\rightleftharpoons$ CH_3OH + NH_2^-

base *acid* *conj acid* *conj base*

Conjugate acid-base pairs: NH_3/NH_2^- and CH_3OH/CH_3O^-

c) HPO_4^{2-} + HSO_4^- $\rightleftharpoons$ $H_2PO_4^-$ + SO_4^{2-}

base *acid* *conj acid* *conj base*

Conjugate acid-base pairs: HSO_4^-/SO_4^{2-} and $H_2PO_4^-$/ HPO_4^{2-}

18.51 Write total ionic equations and then remove spectator ions. The (aq) subscript denotes that each species is soluble and dissociates in water.

a) ~~$Na^+(aq)$~~ + $OH^-(aq)$ + ~~$Na^+(aq)$~~ + $H_2PO_4^-(aq)$ $\rightleftharpoons$ $H_2O(l)$ + ~~$2Na^+(aq)$~~ + $HPO_4^{2-}(aq)$

Net: $OH^-(aq)$ + $H_2PO_4^-(aq)$ $\rightleftharpoons$ $H_2O(l)$ + $HPO_4^{2-}(aq)$

base *acid* *conj acid* *conj base*

Conjugate acid/base pairs: $H_2PO_4^-$/ HPO_4^{2-} and H_2O/OH^-

b) ~~$K^+(aq)$~~ + $HSO_4^-(aq)$ + ~~$2K^+(aq)$~~ + $CO_3^{2-}(aq)$ $\rightleftharpoons$ ~~$2K^+(aq)$~~ + $SO_4^{2-}(aq)$ + ~~$K^+(aq)$~~ + $HCO_3^-(aq)$

Net: $HSO_4^-(aq)$ + $CO_3^{2-}(aq)$ $\rightleftharpoons$ $SO_4^{2-}(aq)$ + $HCO_3^-(aq)$

acid *base* *conj base* *conj acid*

Conjugate acid/base pairs: HSO_4^-/ SO_4^{2-} and HCO_3^-/ CO_3^{2-}

18.53 The conjugate pairs are H_2S (acid) / HS^- (base) and HCl (acid)/Cl^- (base). The reactions involve reacting one acid from one conjugate pair with the base from the other conjugate pair. Two reactions are possible:

(1) HS^- + HCl $\rightleftharpoons$ H_2S + Cl^- and (2) H_2S + Cl^- $\rightleftharpoons$ HS^- + HCl

The first reaction is the reverse of the second. To decide which will have an equilibrium constant greater than 1 look for the stronger acid producing a weaker acid. HCl is a strong acid and H_2S a weak acid. The reaction that favors the products ($K_c > 1$) is the first one where the strong acid produces the weak acid. Reaction (2) with a weaker acid forming a stronger acid favors the reactants and $K_c < 1$.

18.55 a) HCl + NH_3 $\rightleftharpoons$ NH_4^+ + Cl^-

strong acid *stronger base* *weak acid* *weaker base*

HCl is ranked above NH_4^+ in Figure 18.10 and is the stronger acid. NH_3 is ranked above Cl^- and is the stronger base. NH_3 is shown as a "stronger" base because it is stronger than Cl^-, but is not considered a "strong" base. The reaction proceeds towards the production of the weaker acid and base, i.e. the reaction as written proceeds to the right and **K_c > 1**.

b) H_2SO_3 + NH_3 $\rightleftharpoons$ HSO_3^- + NH_4^+

stronger acid *stronger base* *weaker base* *weaker acid*

H_2SO_3 is ranked above NH_4^+ and is the stronger acid. NH_3 is a stronger base than HSO_3^-. The reaction proceeds towards the production of the weaker acid and base, i.e. the reaction as written proceeds to the right and **$K_c > 1$**.

18.57 a) NH_4^+ + HPO_4^{2-} $\rightleftharpoons$ NH_3 + $H_2PO_4^-$

weaker acid *weaker base* *stronger base* *stronger acid*

$K_c < 1$ because acid-base reaction favors stronger acid and base.

b) HSO_3^- + HS^- $\rightleftharpoons$ H_2SO_3 + S^{2-} **$K_c < 1$**

weaker base *weaker acid* *stronger acid* *stronger base*

18.59 a) The concentration of a strong acid is very different before and after dissociation. After dissociation, the concentration of the strong acid approaches 0, or [HA] ≈ 0.

b) A weak acid dissociates to a very small extent, so the acid concentration after dissociation is nearly the same as before dissociation.

c) Same as (b), but the percent, or extent, of dissociation is greater than in (b).

d) Same as (a).

18.60 No, HCl and CH_3COOH are never of equal strength because HCl is a strong acid with $K_a > 1$ and CH_3COOH is a weak acid with $K_a < 1$. The K_a of the acid, not the concentration of H_3O^+ in a solution of the acid, determines the strength of the acid.

18.63 Butanoic acid dissociates according to the following equation:

$$CH_3CH_2CH_2COOH(aq) + H_2O(l) \rightleftharpoons CH_3CH_2CH_2COO^-(aq) + H_3O^+(aq)$$

$$K_a = \frac{[CH_3CH_2CH_2COO^-][H_3O^+]}{[CH_3CH_2CH_2COOH]} = \frac{(1.51x10^{-3})^2}{(0.15-x)}$$

Can we assume that $[CH_3CH_2CH_2COOH]_{eq} = 0.15 - x \approx 0.15$? If the extent of dissociation is less than 5%, then the assumption is correct. The $[H_3O^+]$ represents the extent of dissociation, so $(1.51x10^{-3}/0.15) \times 100\% = 1.0\%$ and the assumption is correct.

$$K_a = \frac{(1.51x10^{-3})^2}{(0.15)} = \mathbf{1.5x10^{-5}}$$

18.65 For a solution of a weak acid, the acid dissociation equilibrium determines the concentrations of the weak acid, its conjugate base and H_3O^+. The acid dissociation reaction for HNO_2 is

Conc(M)	$HNO_2(aq)$ +	$H_2O(l)$ $\rightleftharpoons$	$NO_2^-(aq)$ +	$H_3O^+(aq)$
initial	0.50	--	0	0
change	-x	--	+x	+x
equilibrium	0.50 – x	--	x	x

$$K_a = 7.1x10^{-4} = \frac{x^2}{0.50-x}$$

Assume that $0.50 - x \approx 0.50$.

Solving the K_a expression yields $x = 1.8_{84} \times 10^{-2}$

Check assumption: $1.8_{84} \times 10^{-2} \div 0.50 = 0.04$ or 4% error, so assumption is valid. If the quadratic equation is used: $x = 1.8_{48} \times 10^{-2}$.

$[H_3O^+] = [NO_2^-] = x =$ **1.9×10^{-2} M** (1.8×10^{-2} M solved via quadratic equation)

The concentration of hydroxide ion is related to concentration of hydronium ion through the equilibrium for water : $2H_2O \rightleftharpoons H_3O^+ + OH^-$ with $K_w = 1.0 \times 10^{-14}$

$[OH^-] = 1.0 \times 10^{-14} \div 1.9 \times 10^{-2} =$ **5.3×10^{-13} M** (5.4×10^{-13} M solved via quadratic equation)

18.67 Write a balanced chemical equation and equilibrium expression for the dissociation of chloroacetic acid and convert pK_a to K_a.

$ClCH_2COOH(aq) + H_2O(l) \rightleftharpoons ClCH_2COO^-(aq) + H_3O^+(aq)$

$$K_a = \frac{[ClCH_2COO^-][H_3O^+]}{[ClCH_2COOH]}$$

$K_a = 10^{-pKa} = 10^{-2.87} = 1.3_{49} \times 10^{-3}$

Since $1.05/(1.349 \times 10^{-3}) = 778 > 400$, assume that $[ClCH_2COOH] = 1.05 - x \approx 1.05$.

$$1.3_{49} x10^{-3} = \frac{x^2}{1.05}$$

$$x = 3.8x10^{-2}$$

$[H_3O^+] = [ClCH_2COO^-] =$ **3.8×10^{-2} M**

Check assumption: $3.8 \times 10^{-2} \div 1.05 = .036$ or 3.6%. Assumption is good.

$pH = -\log(3.7_{63} \times 10^{-2}) =$ **1.42**

$[ClCH_2COOH] = 1.05 - 0.038 =$ **1.01 M**

18.69 Percent dissociation refers to the amount of the initial concentration of the acid that dissociates into ions. Use the percent dissociation to find the concentration of acid dissociated.

$$\% \ dissociated = \frac{[HA]_{dissociated}}{[HA]_{initial}} x\,100$$

$$[HA]_{dissociated} = \frac{(3.0)(0.25)}{(100)} = \mathbf{0.0075}$$

a) The concentration of acid dissociated is equal to the equilibrium concentrations of A^- and H_3O^+. And pH and $[OH^-]$ is determined from $[H_3O^+]$.

$[A^-] = [H_3O^+] =$ **7.5×10^{-3} M**; $pH = -\log(7.5 \times 10^{-3}) =$ **2.12**

$[OH^-] = 1.0 \times 10^{-14} \div 7.5 \times 10^{-3} =$ **1.3×10^{-12} M**; $pOH = -\log(1.3 \times 10^{-12}) =$ **11.89**

b) In the equilibrium expression, substitute the concentrations above and calculate K_a.

$$K_a = \frac{[H_3O^+][A^-]}{[HA]} = \frac{(7.5x10^{-3})(7.5x10^{-3})}{(0.25)-(7.5x10^{-3})} = \mathbf{2.3x10^{-4}}$$

18.71 Calculate the molarity of HX by dividing moles by volume. Convert pH to $[H_3O^+]$ and substitute into the equilibrium expression.

Concentration (M) = 0.250 mol/0.655 L = 0.381_{68} M

$[H_3O^+] = 10^{-pH} = 10^{-3.44} = 3.6_{31} \times 10^{-4}$ M

$$K_a = \frac{[X^-][H_3O^+]}{[HX]} = \frac{(3.6_{31} x 10^{-4})^2}{0.381_{68} - x}$$

Can $[HX]_{eq}$ be simplified? The extent of dissociation is $(3.6 \times 10^{-4})/(0.382) \times 100\% = 0.09\%$ < 5%. Therefore, $[HX]_{eq} = 0.382$.

$$K_a = \frac{(3.6_{31} x 10^{-4})^2}{0.381_{68}} = 3.4_{54} x 10^{-7} = \mathbf{3.5x10^{-7}}$$

18.73 a) Reaction table:

Conc (M)	HZ(aq)	+ $H_2O(l)$ ⇌	$Z^-(aq)$	+ $H_3O^+(aq)$
initial	0.075	---	0	0
change	-x	---	+x	+x
equilibrium	0.075-x	---	x	x

$$K_a = \frac{x^2}{0.075 - x} = 1.55 x 10^{-4}$$

Assume $0.075 - x \approx 0.075$

$$x = \sqrt{(0.075)(1.55x10^{-4})} = 3.4_{10} x 10^{-3}$$

Check assumption: $3.4 \times 10^{-3}/0.075 = 0.045$ or 4.5% so assumption is valid.

pH = $-\log(1.0_{78} \times 10^{-3})$ = **2.47**

b) Set up the reaction table as above but substitute 0.045 M for the initial concentration of HZ.

$$x = \sqrt{(0.045)(1.55x10^{-4})} = 2.6_{41} x 10^{-3}$$

$$pOH = -\log\left(\frac{1.0x10^{-14}}{2.6_{41} x10^{-3}}\right) = \mathbf{11.42}$$

18.75 a) Assume $[HY]_{eq} = 0.175 - x \approx 0.175$ because $0.175/1.00x10^{-4} > 400$.

$$K_a = \frac{[Y^-][H_3O^+]}{[HY]} = 1.00x10^{-4}$$

$$\frac{x^2}{0.175} = 1.00x10^{-4} \Rightarrow x = 4.18x10^{-3}$$

pH = $-\log(4.18 \times 10^{-3})$ = **2.378**

b) Since $0.175/1.00x10^{-2} < 400$, $[HX]_{eq} = 0.175 - x$ and quadratic equation must be solved.

$$K_a = \frac{[X^-][H_3O^+]}{[HX]} = 1.00x10^{-2}$$

$$\frac{x^2}{(0.175 - x)} = 1.00x10^{-2}$$

$$x^2 + 1.00x10^{-2}x - 1.75x10^{-3} = 0$$

$$x = \frac{-(1.00x10^{-2}) \pm \sqrt{(1.00x10^{-2})^2 - 4(-1.75x10^{-3})}}{2} = 3.71_{30}x10^{-2}$$

pH = -log[H_3O^+] = -log(3.71_{30} x 10^{-2}) = 1.430

pOH = 14.00 – 1.430 = **12.57**

18.77 First find the concentration of benzoate ion at equilibrium. Then use the initial concentration of benzoic acid and equilibrium concentration of benzoate to find % dissociation.

$$x = [C_6H_5COOH]_{dissociated} = \sqrt{(6.3x10^{-5})(0.25)} = 3.9_{69}\ x10^{-3}$$

$$\%\ dissociation = \frac{3.9_{69}\ x10^{-3}\ M}{0.25\ M} x\,100 = \mathbf{1.6\%}$$

18.79 Write balanced chemical equations and corresponding equilibrium expressions for dissociation of hydrosulfuric (H_2S) acid.

$H_2S(aq) + H_2O(l) \rightarrow HS^-(aq) + H_3O^+(aq)$ $\quad$ $HS^-(aq) + H_2O(l) \rightarrow S^{2-}(aq) + H_3O^+(aq)$

$$K_{a1} = \frac{[HS^-][H_3O^+]}{[H_2S]} = 9x10^{-8} \qquad K_{a2} = \frac{[S^{2-}][H_3O^+]}{[HS^-]} = 1x10^{-17}$$

Assumptions:

1) Since $K_{a1} >> K_{a2}$, assume that almost all of the H_3O^+ comes from the first dissociation.
2) Since K_{a1} is so small, assume that the dissociation of H_2S is negligible and $[H_2S]_{eq}$ = 0.10 – x ≈ 0.10.

$$K_{a1} = \frac{[HS^-][H_3O^+]}{[H_2S]} = 9x10^{-8}$$

$$\frac{x^2}{0.10} = 9x10^{-8} \Rightarrow x = 9._{49}\ x10^{-5} = 9x10^{-5}\ M$$

Therefore, **[H_3O^+] = [HS^-] = 9 x 10^{-5} M** and **pH = 4.0**.

Concentration is limited to one significant figure because K_a is given to only one significant figure. The pH is given to what appears to be 2 significant figures because the number before the decimal point (4) represents the exponent and the number after the decimal point represents the significant figures in the concentration.

[H_2S] = 0.10 – (9 x 10^{-5}) = **0.10 M**.

Calculate [OH^-] by using Equation 18.2:

[OH^-] = K_w/[H_3O^+] = (1.0 x10^{-14})/(9.$_{49}$ x 10^{-5}) = **1 x 10^{-10} M** and **pOH = 10.0**.

Calculate [S^{2-}] by rearranging the K_{a2} expression and assuming that [HS^-] and [H_3O^+] come mostly from the first dissociation:

$$[S^{2-}] = \frac{K_{a2}[HS^-]}{[H_3O^+]} = \frac{(1x10^{-17})(9._{49}\ x10^{-5})}{9._{49}\ x10^{-5}} = \mathbf{1x10^{-17}\ M}$$

18.82 Concentration of dissociated HCOOH:

$$x = [HCOOH]_{dissociated} = \sqrt{(1.8x10^{-4})(0.50)} = 9.4_{87}\ x10^{-3}$$

$$\% \, dissociation = \frac{9.4_{87} \, x 10^{-3} \, M}{0.50 \, M} x \, 100 = \mathbf{1.9\%}$$

18.83 All Brønsted-Lowry bases contain at least one lone pair of electrons. This lone pair binds with an H^+ and allows the base to act as a proton-acceptor.

18.86 A base accepts a proton from water in the base dissociation reaction:

a) $C_5H_5N(aq) + H_2O(l) \rightleftharpoons C_5H_5NH^+(aq) + OH^-(aq)$

$$K_b = \frac{[C_5H_5NH^+][OH^-]}{[C_5H_5N]}$$

b) The primary reaction are involved in base dissociation of carbonate ion is

$CO_3^{2-}(aq) + H_2O(l) \rightleftharpoons HCO_3^-(aq) + OH^-(aq)$

$$K_b = \frac{[HCO_3^-][OH^-]}{[CO_3^{2-}]}$$

The bicarbonate can then also dissociate as a base, but this occurs to an insignificant amount in a solution of carbonate ions.

18.88 a) Hydroxylamine has a lone pair of electrons on the nitrogen atom that acts like the Lewis base:

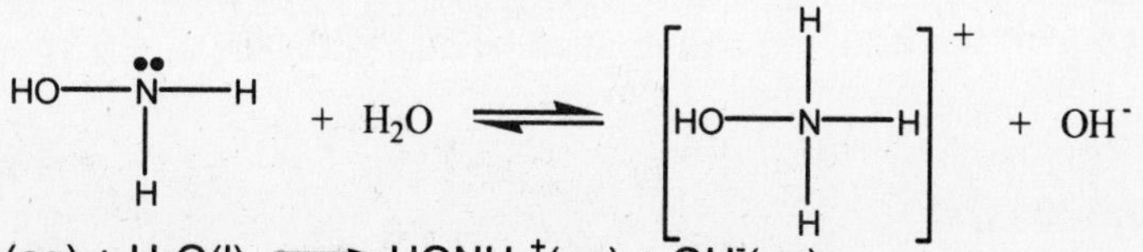

$HONH_2(aq) + H_2O(l) \rightleftharpoons HONH_3^+(aq) + OH^-(aq)$

$$K_b = \frac{[HONH_3^+][OH^-]}{[HONH_2]}$$

b) The diphosphate ion contains oxygen atoms with lone pairs of electron that act as proton acceptors.

$HPO_4^{2-}(aq) + H_2O(l) \rightleftharpoons H_2PO_4^-(aq) + OH^-(aq)$

$$K_b = \frac{[H_2PO_4^-][OH^-]}{[HPO_4^{2-}]}$$

18.90 The formula of dimethylamine has two methyl groups attached to a nitrogen:

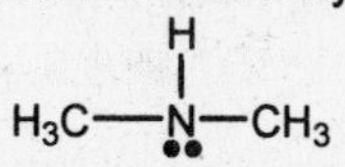

The nitrogen has a lone pair of electrons that will accept the proton from water in the base dissociation reaction:

$(CH_3)_2NH(aq) + H_2O(l) \rightleftharpoons (CH_3)_2NH_2^+(aq) + OH^-(aq)$

$$K_b = \frac{[(CH_3)_3 NH_2^+][OH^-]}{[(CH_3)_3 NH]}$$

Using the value for K_b from Table 18.6 calculate the concentration of OH^- assuming that $0.050 - x \approx 0.050$.

$$x = \sqrt{(5.9 \times 10^{-4})(0.050)} = 5.4_{31} \times 10^{-3}$$

Check assumption: $5.4 \times 10^{-3} \div 0.050 = 0.108$ or 11%, so assumption is not valid.
Solve using quadratic equation solution:

$$\frac{x^2}{0.050 - x} = 5.9 \times 10^{-4};\ x^2 + 5.9x10^{-4}x - 2.95x10^{-5} = 0$$

$$x = [OH^-] = \frac{-5.9x10^{-4} \pm \sqrt{(5.9x10^{-4})^2 - 4(1)(-2.95x10^{-5})}}{2(1)} = 5.1_{44} \times 10^{-3}$$

pOH = $-\log(5.1_{44} \times 10^{-3}) = 2.28_{87}$
pH = $14.00 - 2.28_{87}$ = **11.71**

18.92 The structure and K_b of ethanolamine are given in Table 18.6. Write a balanced equation and equilibrium expression for the reaction. Make simplifying assumptions (if valid), solve for $[OH^-]$, convert to $[H_3O^+]$ and calculate pH.

$HOCH_2CH_2NH_2(aq) + H_2O(l) \rightleftharpoons HOCH_2CH_2NH_3^+(aq) + OH^-(aq)$

$$K_b = \frac{[HOCH_2CH_2NH_3^+][OH^-]}{[HOCH_2CH_2NH_2]} = \frac{x^2}{(0.15 - x)} = 3.2x10^{-5}$$

Because $0.15/(3.2 \times 10^{-5}) > 400$, the assumption $[HOCH_2CH_2NH_2]_{eq} = 0.15 - x \approx 0.15$ is valid.

$$\frac{x^2}{0.15} = 3.2x10^{-5};\ x = 2.1_{91}x10^{-3}$$

Therefore, $[HOCH_2CH_2NH_3^+] = [OH^-] = 2.1_{91} \times 10^{-3}$.
$[H_3O^+] = K_w/(2.1_{91} \times 10^{-3}) = 4.5_{64} \times 10^{-12}$ M and pH = $-\log(4.5_{64} \times 10^{-12})$ = **11.34**.

18.94 a) Acetate ion, CH_3COO^-, is the conjugate base of acetic acid, CH_3COOH. The K_b for acetate ion is related to the K_a for acetic acid through the equation $K_w = K_a \times K_b$.

$K_b = 1.0 \times 10^{-14} \div 1.8 \times 10^{-5} =$ **5.6×10^{-10}**

b) Anilinium ion is the conjugate acid of aniline so the K_a for anilinium ion is related to the K_b of aniline (4.0×10^{-10} from Table 18.6) by the relationship $K_w = K_a \times K_b$.

$K_a = 1.0 \times 10^{-14} \div 4.0 \times 10^{-10} =$ **2.5×10^{-5}**

18.96 a) The K_a of chlorous acid, $HClO_2$, is reported in Table 18.2. $HClO_2$ is the conjugate acid of chlorite ion, ClO_2^-. Since $K_a \times K_b = K_w$, $K_b = (1.0 \times 10^{-14})/(1.12 \times 10^{-2}) = 8.9_{29} \times 10^{-13}$ and $pK_b = -\log K_b$ = **12.05**.

b) The K_b of dimethylamine, $(CH_3)_2NH$, is reported in Table 18.6. $(CH_3)_2NH$ is the conjugate base of $(CH_3)_2NH_2^+$. Since $K_a \times K_b = K_w$, $K_a = (1.0 \times 10^{-14})/(5.9 \times 10^{-4}) = 1.6_{95} \times 10^{-11}$ and $pK_a = -\log K_a$ = **10.77**.

18.98 a) Potassium cyanide, when placed in water, dissociates into potassium ions, K^+, and cyanide ions, CN^-. Potassium ion is the conjugate acid of a strong base, KOH, so K^+ does not react with water. Cyanide ion is the conjugate base of a weak acid, HCN, so it does react with the base dissociation reaction:

$$CN^-(aq) + H_2O(l) \rightleftharpoons HCN(aq) + OH^-(aq)$$

To find the pH first set up a reaction table and use K_b for CN^- to calculate $[OH^-]$.

Conc (M)	$CN^-(aq)$	+ $H_2O(l)$ ⇌	$HCN(aq)$	+ $OH^-(aq)$
initial	0.050	---	0	0
change	-x	---	+x	+x
equilibrium	0.050-x	---	x	x

$$K_b = \frac{K_w}{K_a} = \frac{1.0x10^{-14}}{6.2x10^{-10}} = 1.6_{13}\ x\,10^{-5} = \frac{[HCN][OH^-]}{[CN^-]} = \frac{x^2}{0.050 - x}$$

Assuming that $0.050 - x \approx 0.050$, solve for x.

$$x = [OH^-] = \sqrt{(1.6_{13}x10^{-10})(0.050)} = 8.9_{81}\ x\,10^{-4}$$

Check assumption: $9 \times 10^{-4}/0.050 = 0.018$ or 1.8%. Assumption ok.

pOH = $-\log(8.9_{81} \times 10^{-4}) = 3.04_{67}$

pH = $14.00 - 3.04_{67}$ = **10.95**

b) The salt triethylammonium chloride in water dissociates into two ions: $(CH_3CH_2)_3NH^+$ and Cl^-. Chloride ion is the conjugate base of a strong acid so it will not influence the pH of the solution. Triethylammonium ion is the conjugate acid of a weak base, so the acid dissociation reaction below determines the pH of the solution.

Conc(M)	$(CH_3CH_2)_3NH^+(aq)$	+ $H_2O(l)$ ⇌	$(CH_3CH_2)_3N(aq)$	+ $H_3O^+(aq)$
initial	0.30	---	0	0
change	-x	---	+x	+x
equilibrium	0.30-x	---	x	x

$$K_a = \frac{1.0x10^{-14}}{5.2x10^{-4}} = 1.9_{23}\ x\,10^{-11} = \frac{x^2}{0.30 - x}$$

Assume that x is negligible.

$$x = [H_3O^+] = \sqrt{(1.9_{23}x10^{-11})(0.30)} = 2.4_{02}\ x\,10^{-6}$$

Check assumption: $2.4x10^{-6}/0.3 = 8 \times 10^{-6}$ or 0.0008%, so assumption ok.

pH = $-\log(2.4_{02} \times 10^{-6})$ = **5.62**

18.100 a) The formate ion, $HCOO^-$, acts as the base shown by the following equation:

$$HCOO^-(aq) + H_2O(l) \rightleftharpoons HCOOH(aq) + OH^-(aq)$$

$$K_b = \frac{[HCOOH][OH^-]}{[HCOO^-]}$$

Because HCOOK is a soluble salt, $[HCOO^-]$ = [HCOOK].

The K_a of formic acid, HCOOH, is reported in Table 18.2.

Therefore, $K_b = (1.0 \times 10^{-14})/(1.8 \times 10^{-4}) = 5.5_{56} \times 10^{-11}$.

Because the K_b is very small, the assumption $[HCOO^-]_{eq} = 0.53 - x \approx 0.53$ is valid.

$$K_b = \frac{[HCOOH][OH^-]}{[HCOO^-]} = \frac{x^2}{0.53} = 5.5_{56}x10^{-11}$$

$$x = 5.4_{26}x10^{-6}$$

Therefore, $[HCOOH] = [OH^-] = 5.4_{26} \times 10^{-6}$ M.

$[H_3O^+] = K_w/(5.4_{26} \times 10^{-6}) = 1.8_{43} \times 10^{-9}$ M and pH = $-\log(1.8_{43} \times 10^{-9})$ = **8.73**.

b) The ammonium ion, NH_4^+, acts as an acid shown by the following equation:

$NH_4^+(aq) + H_2O(l) \rightleftharpoons NH_3(aq) + H_3O^+(aq)$

$$K_a = \frac{[NH_3][H_3O^+]}{[NH_4^+]}$$

Because NH_4Br is a soluble salt, $[NH_4^+] = [NH_4Cl]$.

The K_b of ammonia, NH_3, is reported in Table 18.6.

Therefore, $K_a = (1.0 \times 10^{-14}) / (1.76 \times 10^{-5}) = 5.6_{82} \times 10^{-10}$.

Because the K_a is very small, the assumption $[NH_4^+]_{eq} = 1.22 - x \approx 1.22$ is valid.

$$K_a = \frac{[NH_3][H_3O^+]}{[NH_4^+]} = \frac{x^2}{1.22} = 5.6_{82} x10^{-10}$$

$$x = 2.6_{33} x10^{-5}$$

Therefore, $[NH_3] = [H_3O^+] = 2.6_{33} \times 10^{-5}$ M and pH = $-\log(2.6_{33} \times 10^{-5})$ = **4.58**.

18.102 See Follow-Up problem 18.10 for another problem on bleach.

First calculate the initial molarity of ClO^-.

$$\left(\frac{5.0\ g\ NaOCl}{100\ g\ sol'n}\right)\left(\frac{1.0\ g\ sol'n}{mL\ sol'n}\right)\left(\frac{1\ mol\ NaOCl}{74.44\ g\ NaOCl}\right)\left(\frac{1\ mol\ OCl^-}{1\ mol\ NaOCl}\right)\left(\frac{1000\ mL}{1\ L}\right) = 0.67_{17}\ M$$

Then set up reaction table with base dissociation of OCl^-

Conc (M)	$OCl^-(aq)$	+ $H_2O(l)$	$\rightleftharpoons$ $HOCl(aq)$	+ $OH^-(aq)$
initial	0.67_{17}	---	0	0
change	-x	---	+x	+x
equilibrium	0.067_{17}-x	---	x	x

$K_b = 1.0 \times 10^{-14}/2.9 \times 10^{-8} = 3.4_{48} \times 10^{-7}$

Assuming that x is negligible with respect to $[OCl^-]$,

$$x = [OH^-] = \sqrt{(3.4_{48} x10^{-7})(0.67_{17})} = 4.8_{12}\ x10^{-4}\ M$$

The $[OH^-]$ in an aqueous solution of 5.0% NaClO is **4.8 x 10^{-4} M**

Check assumption: $4.8 \times 10^{-4}/0.67 = 0.0007$ or 0.07% $\Rightarrow$ assumption ok.

pOH = $-\log(4.8_{12} \times 10^{-4}) = 3.31_{77}$; pH = $14.00 - 3.31_{77}$ = **10.68**

18.104 As the nonmetal becomes more electronegative, the acidity of the binary hydride increases. The electronegative nonmetal attracts the electrons more strongly in the polar bond, shifting the electron density away from H^+ and making the H^+ more easily transferred to a surrounding water molecule to make H_3O^+.

18.107 The two factors that explain the greater acid strength of $HClO_4$ are:

1) Chlorine is more electronegative than iodine, so chlorine more strongly attracts the electrons in the bond with oxygen. This makes the H in $HClO_4$ less tightly held by the oxygen than the H in HIO.
2) $HClO_4$ has more oxygen atoms than HIO which leads to a greater shift in electron density making the H in $HClO_4$ more susceptible to transfer to a base.

18.108 a) Selenic acid, **H_2SeO_4**, is the stronger acid because it contains more oxygen atoms (apply the second rule of oxoacids).

b) Phosphoric acid, **H_3PO_4**, is the stronger acid because P is more electronegative than As (apply the first rule of oxoacids).

c) Hydrotelluric acid, **H_2Te**, is the stronger acid because Te is larger than S and so the E-H bond is weaker (apply the second rule of nonmetal hydrides).

18.110 a) **H_2Se**, selenium hydride, is a stronger acid than H_3As, arsenic hydride, because Se is more electronegative than As.

b) **$B(OH)_3$**, boric acid also written as H_3BO_3, is a stronger acid than $Al(OH)_3$, aluminum hydroxide, because boron is more electronegative than aluminum.

c) **$HBrO_2$**, bromous acid, is a stronger acid than HBrO, hypobromous acid, because there are more oxygen atoms in $HBrO_2$ than in HBrO.

18.112 Acidity increases as the value of K_a increases. Determine the ion formed from each salt and compare the corresponding K_a values.

a) Copper(II) sulfate, $CuSO_4$, contains Cu^{2+} ion with $K_a = 3 \times 10^{-8}$. Aluminum sulfate, $Al_2(SO_4)_3$, contains Al^{3+} ion with $K_a = 1 \times 10^{-5}$. The concentrations of Cu^{2+} and Al^{3+} are equal, but the K_a of $Al_2(SO_4)_3$ is almost three orders of magnitude greater. Therefore, **0.05 M $Al_2(SO_4)_3$** is the stronger acid.

b) Zinc chloride, $ZnCl_2$, contains the Zn^{2+} ion with $K_a = 1 \times 10^{-9}$. Lead chloride, $PbCl_2$, contains the Pb^{2+} ion with $K_a = 3 \times 10^{-8}$. Since both solutions have the same concentration, and $K_a(Pb^{2+}) > K_a(Zn^{2+})$, **0.1 M $PbCl_2$** is the stronger acid.

18.114 A higher pH (more basic solution) results when an acid has a lower K_a.

a) The **$Ni(NO_3)_2$** solution has a higher pH than the $Co(NO_3)_2$ solution because K_a of Ni^{2+} is smaller than the K_a of Co^{2+}. Note that nitrate ions are the conjugate bases of a strong acid and therefore do not impact the pH of the solution.

b) The **$Al(NO_3)_3$** solution has a higher pH than the $Cr(NO_3)_2$ solution because K_a of Al^{3+} is smaller than the K_a of Cr^{2+}.

18.117 Sodium fluoride, NaF, contains the cation of a strong base, NaOH, and anion of a weak acid, HF. This combination yields a salt that is basic in aqueous solution. Sodium chloride, NaCl, is the salt of a strong base, NaOH, and strong acid, HCl. This combination yields a salt that is neutral in aqueous solution.

18.119 For each salt, first break into the ions present in solution and then determine if either ion acts as a weak acid or weak base to change the pH of the solution.

a) $KBr(s) + H_2O(l) \rightarrow K^+(aq) + Br^-(aq)$

K^+ is the conjugate acid of a strong base so it does not impact pH.

Br^- is the conjugate base of a strong acid so it does not impact pH.

Since neither ion impacts the pH of the solution it will remain at the pH of pure water with a **neutral** pH.

b) $NH_4I(s) + H_2O(l) \rightarrow NH_4^+(aq) + I^-(aq)$

NH_4^+ is the conjugate acid of a weak base, so it will act as a weak acid in solution and produce H_3O^+ as represented by the acid dissociation reaction:

$$NH_4^+(aq) + H_2O(l) \rightleftharpoons NH_3(aq) + H_3O^+(aq)$$

I^- is the conjugate base of a strong acid so it will not impact the pH.

The production of H_3O^+ from the ammonium ion makes the solution of NH_4I **acidic**.

c) $KCN(s) + H_2O(l) \rightarrow K^+(aq) + CN^-(aq)$

K^+ is the conjugate acid of a strong base so it does not impact pH.

CN^- is the conjugate base of a weak acid so it will act as a weak base in solution and impact pH by the base dissociation reaction:

$$CN^-(aq) + H_2O(l) \rightleftharpoons HCN(aq) + OH^-(aq)$$

Hydroxide ions are produced in this equilibrium so solution will be **basic**.

18.121 a) The two ions that comprise sodium carbonate, Na_2CO_3, are sodium ion, Na^+, and carbonate ion, CO_3^{2-}.

$Na_2CO_3(s) + H_2O(l) \rightarrow 2Na^+(aq) + CO_3^{2-}(aq)$

Sodium ion is derived from the strong base NaOH. Carbonate ion is derived from the weak acid HCO_3^-. A salt derived from a strong base and a weak acid produces a solution that is **basic**.

Na^+ does not react with water

$CO_3^{2-}(aq) + H_2O(l) \rightleftharpoons HCO_3^-(aq) + OH^-(aq)$

b) The two ions that comprise calcium chloride, $CaCl_2$, are calcium ion, Ca^{2+}, and chloride ion, Cl^-.

$CaCl_2(s) + H_2O(l) \rightarrow Ca^{2+}(aq) + 2Cl^-(aq)$

Calcium ion is derived from the strong base $Ca(OH)_2$. Chloride ion is derived from the strong acid HCl. A salt derived from a strong base and strong acid produces a solution that is **neutral**.

Neither Ca^{2+} nor Cl^- react with water.

c) The two ions that comprise cupric nitrate, $Cu(NO_3)_2$, are the cupric ion, Cu^{2+}, and the nitrate ion, NO_3^-.

$Cu(NO_3)_2(s) + H_2O(l) \rightarrow Cu^{2+}(aq) + 2NO_3^-(aq)$

Small metal ions are acidic in water (assume the hydration of Cu^{2+} is 6):

$Cu(H_2O)_6^{2+}(aq) + H_2O(l) \rightleftharpoons Cu(H_2O)_5OH^+(aq) + H_3O^+(aq)$

Nitrate ion is derived from the strong acid HNO_3. Therefore, NO_3^- does not react with water. A solution of cupric nitrate is **acidic**.

18.123 a) A solution of strontium bromide is **neutral** because Sr^{2+} is the conjugate acid of a strong base, $Sr(OH)_2$ and Br^- is the conjugate base of a strong acid, HBr, so neither change the pH of the solution.

b) A solution of barium acetate is **basic** because CH_3COO^- is the conjugate base of a weak acid and therefore forms OH^- in solution whereas Ba^{2+} is the conjugate acid of a strong base, $Ba(OH)_2$, and does not impact solution pH. The base dissociation reaction of acetate ion is

$$CH_3COO^-(aq) + H_2O(l) \rightleftharpoons CH_3COOH(aq) + OH^-(aq)$$

c) A solution of dimethylammonium bromide is **acidic** because $(CH_3)_2NH_2^+$ is the conjugate acid of a weak base and therefore forms H_3O^+ in solution whereas Br^- is the conjugate base of a strong acid and does not influence the pH of the solution. The acid dissociation reaction for methylammonium ion is

$$(CH_3)_2NH_2^+(aq) + H_2O(l) \rightleftharpoons (CH_3)_2NH(aq) + H_3O^+(aq)$$

18.125 a) The two ions that comprise ammonium phosphate, $(NH_4)_3PO_4$, are ammonium ion, NH_4^+, and phosphate ion, PO_4^{3-}.

$NH_4^+(aq) + H_2O(l) \rightleftharpoons NH_3(aq) + H_3O^+(aq)$ $\quad K_a = K_w/K_b(NH_3) = 5.7 \times 10^{-10}$

$PO_4^{3-}(aq) + H_2O(l) \rightleftharpoons HPO_4^{2-}(aq) + OH^-(aq)$ $\quad K_b = K_w/K_{a3}(H_3PO_4) = 2.4 \times 10^{-2}$

A comparison of K_a and K_b is necessary since both ions are derived from a weak base and weak acid. The K_a of NH_4^+ is determined by using the K_b of its conjugate base, NH_3 (Table 18.6). The K_b of PO_4^{3-} is determined by using the K_a of its conjugate acid, HPO_4^{2-}. The K_a of HPO_4^{2-} comes from K_{a3} of H_3PO_4 (Table 18.5). Since $K_b > K_a$, a solution of $(NH_4)_3PO_4$ is **basic**.

b) The two ions that comprise sodium sulfate, Na_2SO_4, are sodium ion, Na^+, and sulfate ion, SO_4^{2-}. The sodium ion is derived from the strong base NaOH. The sulfate ion is derived from the weak acid, HSO_4^-.

$SO_4^{2-}(aq) + H_2O(l) \rightleftharpoons HSO_4^-(aq) + OH^-(aq)$

A solution of sodium sulfate is **basic**.

c) The two ions that comprise lithium hypochlorite, LiClO, are lithium ion, Li^+, and hypochlorite ion, ClO^-. Lithium ion is derived from the strong base LiOH. Hypochlorite ion is derived from the weak acid, HClO (hypochlorous acid).

$ClO^-(aq) + H_2O(l) \rightleftharpoons HClO(aq) + OH^-(aq)$

A solution of lithium hypochlorite is **basic**.

18.127 a) Order of increasing pH: **$Fe(NO_3)_2 < KNO_3 < K_2SO_3 < K_2S$** (assuming concentrations equivalent)

Iron(II) nitrate, $Fe(NO_3)_2$, is an acidic solution because the iron ion is a small highly charged metal ion that acts as a weak acid and nitrate ion is the conjugate base of a strong acid so it does not influence pH.

Potassium nitrate, KNO_3, is a neutral solution because potassium ion is the conjugate acid of a strong base and nitrate ion is the conjugate base of a strong acid so neither impact solution pH.

Potassium sulfite, K_2SO_3, and potassium sulfide, K_2S, are similar in that the potassium ion does not impact solution pH but the anions do because they are conjugate bases of weak acids. K_a for HSO_3^- is 6.5×10^{-8} so K_b for SO_3^- is 1.5×10^{-7} which indicates that sulfite ion is a weak base. K_a for HS^- is 1×10^{-17} from Table 18.5 so sulfide ion has a K_b equal to 1×10^3. Sulfide ion is thus a strong base. The solution of a strong base will have a greater concentration of hydroxide ions (and higher pH) than a solution of a weak base of equivalent concentrations.

b) In order of increasing pH: **$NaHSO_4 < NH_4NO_3 < NaHCO_3 < Na_2CO_3$**

In solutions of ammonium nitrate, only the ammonium will influence pH by dissociating as a weak acid:

$NH_4^+(aq) + H_2O(l) \rightleftharpoons NH_3(aq) + H_3O^+(aq)$

with $K_a = 1.0 \times 10^{-14}/1.8 \times 10^{-5} = 5.6 \times 10^{-10}$

So solution of ammonium nitrate is acidic.

In solutions of sodium hydrogen sulfate, only HSO_4^- will influence pH. The hydrogen sulfate ion is amphoteric so both the acid and base dissociations must be evaluated for influence on pH. As a base, HSO_4^- is the conjugate base of a strong acid so it will not influence pH. As an acid, HSO_4^- is the conjugate acid of a weak base so the acid dissociation applies

$HSO_4^-(aq) + H_2O(l) \rightleftharpoons SO_4^{2-}(aq) + H_3O^+(aq)$ $\quad K_{a2} = 1.2 \times 10^{-2}$

In solutions of sodium hydrogen carbonate only the HCO_3^- will influence pH and it, like HSO_4^-, is amphoteric:

As an acid: $HCO_3^-(aq) + H_2O(l) \rightleftharpoons CO_3^{2-}(aq) + H_3O^+(aq)$

$K_a = 4.7 \times 10^{-11}$, the second K_a for carbonic acid from Table 18.5

As a base: $HCO_3^-(aq) + H_2O(l) \rightleftharpoons H_2CO_3(aq) + OH^-(aq)$

$K_b = 1.0 \times 10^{-14}/4.5 \times 10^{-7} = 2.2 \times 10^{-8}$

Since $K_b > K_a$ a solution of sodium hydrogen carbonate is basic.

In a solution of sodium carbonate only CO_3^{2-} will influence pH by acting as a weak base:

$CO_3^{2-}(aq) + H_2O(l) \rightleftharpoons HCO_3^-(aq) + OH^-(aq)$

$K_b = 1.0 \times 10^{-14}/4.7 \times 10^{-11} = 2.1 \times 10^{-4}$

Therefore, the solution of sodium carbonate is basic.

Two of the solutions are acidic. Since the K_a of HSO_4^- is greater than that of NH_4^+ the solution of sodium hydrogen sulfate has a lower pH than solution of ammonium nitrate, assuming concentrations are relatively close.

Two of the solutions are basic. Since the K_b of CO_3^{2-} is greater than that of HCO_3^- so the solution of sodium carbonate has a higher pH than solution of sodium hydrogen carbonate, assuming concentrations are not extremely different.

18.129 Both methoxide ion and amide ion produce OH^- in aqueous solution. In water, the strongest base possible is OH^-. Since both bases produce OH^- in water, all bases appear equally strong.

$CH_3O^-(aq) + H_2O(l) \rightarrow CH_3OH(aq) + OH^-(aq)$

$NH_2^-(aq) + H_2O(l) \rightarrow NH_3(aq) + OH^-(aq)$

18.131 Ammonia, NH_3, is a more basic solvent than H_2O. In a more basic solvent, weak acids like HF act like strong acids and are 100% dissociated.

18.133 A Lewis acid is defined as an electron pair acceptor while a Brønsted-Lowry acid is a proton donor. If only the proton in a Brønsted-Lowry acid is considered then every Brønsted-Lowry acid fits the definition of a Lewis acid since the proton is accepting an electron pair when it bonds with a base. There are Lewis acids that do not include a proton so all Lewis acids are not Brønsted-Lowry acids.

A Lewis base is defined as an electron pair donor and Brønsted-Lowry base is a proton acceptor. In this case the two definitions are essentially the same except that for a Brønsted-Lowry base the acceptor is a proton.

18.134 a) No, a weak Brønsted-Lowry base is not necessarily a weak Lewis base. For example, water molecules solvate metal ions very well:

$Al^{3+}(aq) + 6H_2O(l) \rightarrow Al(H_2O)_6^{3+}(aq)$

Water is a very weak Brønsted-Lowry base, but forms the Al-adduct fairly well and is a reasonably strong Brønsted-Lowry base.

b) The **[:C≡N:]**$^-$ ion donates an electron pair to the $Cu(H_2O)_6^{2+}$ complex and is the Lewis base for the forward direction of this reaction. In the reverse direction, $\mathbf{H_2O}$ donates an electron pair to the $Cu(CN)_4^{2-}$ and is the Lewis base.

c) Because $K_c > 1$, the reaction proceeds in the direction written (left to right) and is driven by the stronger Lewis base, **[:C≡N:]**$^-$ ion.

18.137 a) Cu^{2+} is a Lewis **acid** because it accepts electron pairs from molecules such as water.

b) Cl^- is a Lewis **base** because it has lone pairs of electrons it can donate to a Lewis acid.

c) Tin(II) chloride, $SnCl_2$, is a compound with a structure similar to carbon dioxide so it will act as a Lewis **acid** to form an additional bond to the tin.

d) Oxygen difluoride, OF_2, is a Lewis **base** with a structure similar to water where the oxygen has lone pairs of electrons that it can donate to a Lewis acid.

18.139 a) The boron atom in boron trifluoride, BF_3, is electron deficient (has 6 electrons instead of 8) and can accept an electron pair; it is a **Lewis acid**.

b) The sulfide ion, 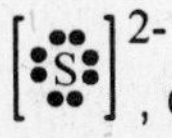, can donate an electron pair and is a **Lewis base**.

c) The Lewis dot structure for one resonance form of the sulfite ion, SO_3^{2-} is shown below.

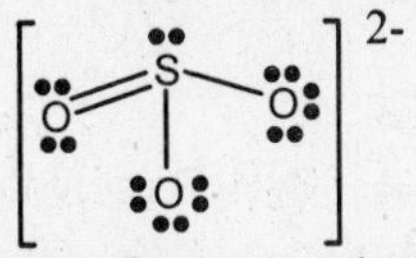

The sulfur atom has a lone electron pair that it can donate in the formation of an adduct. The sulfite ion is a **Lewis base**.

d) Sulfur trioxide, SO_3, acts as a **Lewis acid**. See Follow-up problem 18.13b) for the Lewis dot structure and an example reaction of SO_3.

18.141 a) Sodium ion is the Lewis acid because it is accepting electron pairs from water, the Lewis base.

Na^+	+	$6H_2O$	$\rightleftharpoons$	$Na(H_2O)_6^+$
Lewis acid		Lewis base		adduct

b) The oxygen from water donates a lone pair to the carbon in carbon dioxide. Water is the Lewis base and carbon dioxide the Lewis acid.

CO_2	+	H_2O	$\rightleftharpoons$	H_2CO_3
Lewis acid		Lewis base		adduct

c) Fluoride ion donates an electron pair to form a bond with boron in BF_4^-. The fluoride ion is the Lewis base and the boron trifluoride is the Lewis acid.

F^-	+	BF_3	$\rightleftharpoons$	BF_4^-
Lewis base		Lewis acid		adduct

18.143 a) Since neither H^+ nor OH^- is involved, this is not an Arrhenius acid-base reaction. Since there is no exchange of protons, this is not a Brønsted-Lowry reaction. This reaction is only classified as **Lewis acid-base reaction**, where Ag^+ is the acid and NH_3 is the base.

b) Again, no OH^- is involved so this is not an Arrhenius acid-base reaction. This is an exchange of a proton, from H_2SO_4 to NH_3, so it is a **Brønsted-Lowry acid-base reaction**. Since the Lewis definition is most inclusive, anything that is classified as a Brønsted-Lowry (or Arrhenius) reaction is automatically classified as a **Lewis acid-base reaction**.

c) This is not an acid-base reaction.

d) For the same reasons listed in (a), this reaction is only classified as **Lewis acid-base reaction**, where $AlCl_3$ is the acid and Cl^- is the base.

18.146 Acetic acid is stronger in sea water since its K_a in sea water is greater than its K_a in pure water.

18.148 a) Acids will vary in the amount they dissociate (acid strength) depending on the acid-base character of the solvent. Water and methanol have different acid-base characters.

b) The K_a is the measure of an acid's strength. Table 18.3 demonstrates that a stronger acid has a smaller pK_a. Therefore, phenol is a stronger acid in water than it is in methanol. In other words, water more readily accepts a proton from phenol than does methanol, i.e. water is a stronger base than methanol.

c)

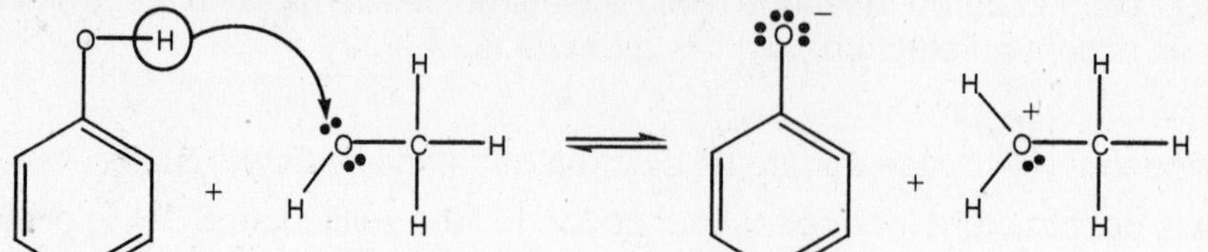

C_6H_5OH(solvated) + CH_3OH(l) $\rightleftharpoons$ $C_6H_5O^-$(solvated) + $CH_3OH_2^+$(solvated)

The term "solvated" is analogous to "aqueous". "Aqueous" would be incorrect in this case because the reaction does not take place in water.

d) In the autoionization process, one methanol molecule is the proton donor while another methanol molecule is the proton acceptor.

CH_3OH(l) + CH_3OH(l) $\rightleftharpoons$ CH_3O^-(solvated) + $CH_3OH_2^+$(solvated)

where (solvated) indicates the molecules are solvated by methanol.

The equilibrium constant for this reaction is the autoionization constant of methanol:

$$K = [CH_3O^-][CH_3OH_2^+]$$

18.151 a) $SnCl_4$ is the Lewis acid accepting an electron pair from $(CH_3)_3N$ as the Lewis base.

b) Tin is the element in the Lewis acid accepting the electron pair. The electron configuration of tin is $[Kr]5s^24d^{10}5p^2$. The four bonds to tin are formed by sp^3 hybrid orbitals which completely fill the 5s and 5p orbitals. The **5d** orbitals are empty and available for the bond with trimethylamine.

18.152 Hydrochloric acid is a strong acid that almost completely dissociates in water. Therefore, the concentration of H_3O^+ is the same as the starting acid concentration: $[H_3O^+] = [HCl]$. The original solution's **pH** = $-\log(1.0 \times 10^{-5})$ = **5.00**.

A 1:10 dilution means that the chemist takes 1 mL of the 1×10^{-5} M solution and dilutes it to 10 mL (or dilute 10 mL to 100 mL). The chemist then dilutes the diluted solution in a 1:10 ratio, and repeats this process for the next two successive dilutions.

Dilution 1:

$$[H_3O^+] = (1.0x10^{-5}\ M)\left(\frac{1.0\ mL}{10.\ mL}\right) = 1.0x10^{-6}\ M$$

pH = $-\log(1.0 \times 10^{-6})$ = **6.00**.

Dilution 2:

$$[HCl] = (1.0x10^{-6}\ M)\left(\frac{1.0\ mL}{10.\ mL}\right) = 1.0x10^{-7}\ M$$

Once the concentration of strong acid is close to the concentration of H_3O^+ from water autoionization the $[H_3O^+]$ in the solution does not equal the initial concentration of the strong acid. The calculation of $[H_3O^+]$ must be based on the water ionization equilibrium:

$2H_2O$(l) $\rightleftharpoons$ H_3O^+(aq) + OH^-(aq) with $K_w = 1.0 \times 10^{-14}$ at 25°C.

The dilution gives an initial $[H_3O^+]$ of 1.0×10^{-7} M. Assuming that the initial concentration of hydroxide ions is zero a reaction table is set up.

Conc(M)	$2H_2O(l) \rightleftharpoons$	$H_3O^+(aq)$ +	$OH^-(aq)$
Initial	---	$1x10^{-7}$	0
Change	---	+ x	+ x
Equilibrium	---	$1x10^{-7}$ + x	x

$K_w = [H_3O^+][OH^-] = (1x10^{-7} + x)(x) = 1.0x10^{-14}$

Set up as a quadratic equation: $x^2 + 1.0x10^{-7}x - 1.0x10^{-14} = 0$

$$x = \frac{-(1.0x10^{-7}) \pm \sqrt{(1.0x10^{-7})^2 - 4(1)(-1.0x10^{-14})}}{2(1)} = 6.18x10^{-8}$$

$[H_3O^+] = 1.0x10^{-7}$ M + $6.2x10^{-8}$ M = $1.62x10^{-7}$ M

pH = $-\log(1.62x10^{-7})$ = **6.79**

Dilution 3:

$$[HCl] = [H_3O^+]_0 = (1.0x10^{-7}\ M)\left(\frac{1.0\ mL}{10.\ mL}\right) = 1.0x10^{-8}\ M$$

Conc(M)	$2H_2O(l) \rightleftharpoons$	$H_3O^+(aq)$ +	$OH^-(aq)$
Initial	---	$1x10^{-8}$	0
Change	---	+ x	+ x
Equilibrium	---	$1x10^{-8}$ + x	x

$K_w = [H_3O^+][OH^-] = (1x10^{-8} + x)(x) = 1.0x10^{-14}$

Set up as a quadratic equation: $x^2 + 1.0x10^{-8}x - 1.0x10^{-14} = 0$

$$x = \frac{-(1.0x10^{-8}) \pm \sqrt{(1.0x10^{-8})^2 - 4(1)(-1.0x10^{-14})}}{2(1)} = 9.5x10^{-8}$$

$[H_3O^+] = 1.0x10^{-8}$ M + $9.5x10^{-8}$ M = $1.05x10^{-7}$ M

pH = $-\log(1.05x10^{-7})$ = **6.98**

Dilution 4:

$$[HCl] = [H_3O^+]_0 = (1.0x10^{-8}\ M)\left(\frac{1.0\ mL}{10.\ mL}\right) = 1.0x10^{-9}\ M$$

Conc(M)	$2H_2O(l) \rightleftharpoons$	$H_3O^+(aq)$ +	$OH^-(aq)$
Initial	---	$1x10^{-9}$	0
Change	---	+ x	+ x
Equilibrium	---	$1x10^{-9}$ + x	x

$K_w = [H_3O^+][OH^-] = (1x10^{-9} + x)(x) = 1.0x10^{-14}$

Set up as a quadratic equation: $x^2 + 1.0x10^{-9}x - 1.0x10^{-14} = 0$

$$x = \frac{-(1.0x10^{-9}) \pm \sqrt{(1.0x10^{-9})^2 - 4(1)(-1.0x10^{-14})}}{2(1)} = 9.95x10^{-8}$$

$[H_3O^+] = 1.0x10^{-9}$ M + $9.95x10^{-8}$ M = $1.0x10^{-7}$ M
pH = -log($1.0x10^{-7}$) = **7.00**

As the HCl solution is diluted the pH of the solution becomes closer to 7.0. Continued dilutions will not change the pH from 7.0. Thus, a solution with a basic pH cannot be made by adding acid to water.

18.155 Compare the contribution of each acid by calculating the concentration of H_3O^+ produced by each.
For 3% hydrogen peroxide, first find initial molarity of H_2O_2 assuming the density is 1g/mL.

$$\left(\frac{3\,g\,H_2O_2}{100\,g\,sol'n}\right)\left(\frac{1\,g\,sol'n}{mL\,sol'n}\right)\left(\frac{1\,mol\,H_2O_2}{34.02\,g\,H_2O_2}\right)\left(\frac{1000\,mL}{1\,L}\right) = 0.8_{82}\ M$$

Find K_a from pK_a: $K_a = 10^{-11.75} = 1.7_{78} \times 10^{-12}$
Calculate $[H_3O^+]$ assuming that x is negligible with respect to 0.9 M

$$[H_3O^+] = \sqrt{(1.7_{78}\,x10^{-12})(0.8_{82})} = 1._{25}\,x\,10^{-6}\ M$$

For 0.001% find the initial molarity again assuming density is 1g/mL

$$\left(\frac{0.001\,g\,H_3PO_4}{100\,g\,sol'n}\right)\left(\frac{1\,g\,sol'n}{mL\,sol'n}\right)\left(\frac{1\,mol\,H_2O_2}{97.99\,g\,H_3PO_4}\right)\left(\frac{1000\,mL}{1\,L}\right) = 1._{02}\ x10^{-4}\ M$$

From Table 18.5 K_a for phosphoric acid is 7.2×10^{-3}. In this calculation x is not negligible since the initial concentration of acid is less than the K_a.

$$7.2\,x\,10^{-3} = \frac{x^2}{1x10^{-4} - x};\ x^2 + 7.2x10^{-3}x - 7.2x10^{-7} = 0$$

$$x = [H_3O^+] = \frac{-7.2x10^{-3} \pm \sqrt{(7.2x10^{-3})^2 - ((4)(1)(-7.2x10^{-7}))}}{2(1)} = 9._{86}\,x\,10^{-5}\ M$$

The concentration of hydronium ion produced by the phosphoric acid, 1×10^{-4} M, is greater than the concentration produced by the hydrogen peroxide, 1×10^{-6} M. So, the **phosphoric acid** contributes more H_3O^+ to the solution.

18.156 a) Electrical conductivity of 0.1 M HCl is higher than that of 0.1 M CH_3COOH. Conductivity is proportional to the concentration of charge in the solution. Since HCl dissociates to a greater extent than CH_3COOH the concentration of ions, and thus charge, is greater in 0.1 M HCl than in 0.1 M CH_3COOH.

b) The electrical conductivity of the two solutions will be approximately the same because at low concentrations the autoionization of water is significant causing the concentration of ions, and thus charge, to be about the same in the two solutions. Also, the % dissociation of a weak electrolyte such as acetic acid increases with decreasing concentration.

18.159 The product of $[H_3O^+]$ and $[OH^-]$ in solution remains constant as defined by the ion-product constant for water, K_w. This constant has a value of 1.0×10^{-14} at 25°C and 1 atm pressure. However, the constant changes at a different temperature. At 200°C, $K_w = 10^{-9.25}$ = 5.6×10^{-10}. HCl completely dissociates in water, so $[H_3O^+]$ = [HCl] = 0.010 M.

$$[OH^-] = \frac{K_w}{[H_3O^+]} = \frac{5.6x10^{-10}}{0.010} = \mathbf{5.6x10^{-8}\ M}$$

18.163 $K_b = 10^{-5.91} = 1.23_{03} \times 10^{-6}$

Since [TRIS] >> K_b assume that $0.060 - x \approx 0.060$

$$[OH^-] = \sqrt{(1.23_{03}x10^{-6})(0.060)} = 2.7_{17}\ x10^{-4}\ M$$

Check assumption: $2.7 \times 10^{-4}/0.060 = 0.0045$ or 0.45% ⇒ assumption good.

pH = $-\log(1.0 \times 10^{-14}/2.7_{17} \times 10^{-4})$ = **10.43**

18.165 The pH is dependent on the *molar* concentration of H_3O^+. Convert %w/v to molarity and use the K_a of acetic acid to determine $[H_3O^+]$ from the equilibrium expression.

Convert %w/v to molarity using the molecular weight of acetic acid (CH_3COOH):

$$conc(M) = \left(\frac{5.0\,g\,CH_3COOH}{100\,mL\text{ soln}}\right)\left(\frac{1\,mol\,CH_3COOH}{60.05\,g\,CH_3COOH}\right)\left(\frac{1000\,mL}{L}\right) = 0.83_{26}\ M\,CH_3COOH$$

Acetic acid dissociates in water according to following equation and equilibrium expression:

$CH_3COOH(aq) + H_2O(l) \rightleftharpoons CH_3COO^-(aq) + H_3O^+(aq)$

$$K_a = \frac{[CH_3COO^-][H_3O^+]}{[CH_3COOH]} = 1.8x10^{-5}$$

Since $0.8326/(1.8 \times 10^{-5}) > 400$, the assumption that $[CH_3COOH]_{eq} = 0.83 - x \approx 0.83$ is valid.

$$\frac{x^2}{0.83_{26}} = 1.8x10^{-5} \Rightarrow x = 3.8_{71}x10^{-3}\ M$$

Therefore, $[CH_3COO^-] = [H_3O^+] = 3.8_{71} \times 10^{-3}$ M and pH = $-\log(3.8_{71} \times 10^{-3})$ = **2.41**.

18.168 Assuming that the pH in the specific cellular environment is equal to the optimum pH for the enzyme, the hydronium ion concentrations are

salivary amylase, mouth: $[H_3O^+] = 10^{-6.8}$ = $\mathbf{2 \times 10^{-7}\ M}$

pepsin, stomach: $[H_3O^+] = 10^{-2.0}$ = $\mathbf{1 \times 10^{-2}\ M}$

trypsin, pancreas: $[H_3O^+] = 10^{-9.5}$ = $\mathbf{3 \times 10^{-10}\ M}$

18.173 a) The two ions that comprise this salt are Ca^{2+} (derived from the strong base $Ca(OH)_2$) and $CH_3CH_2COO^-$ (derived from the weak acid, proprionic acid, CH_3CH_2COOH). A salt derived from a strong base and weak acid produces a **basic** solution.

Ca^{2+} does not react with water

$CH_3CH_2COO^-(aq) + H_2O(l) \rightleftharpoons CH_3CH_2COOH(aq) + OH^-(aq)$

b) The molarity of the solution is:

$$conc(M) = \left(\frac{7.05\,g\,Ca(CH_3CH_2COO)_2}{0.500\,L}\right)\left(\frac{1\,mol\,Ca(CH_3CH_2COO)_2}{186.22\,g\,Ca(CH_3CH_2COO)_2}\right) = 0.0757_{17}\ M$$

Calcium proprionate is a soluble salt and dissolves in water to yield two proprionate ions:

$Ca(CH_3CH_2COO)_2(s) + H_2O(l) \rightarrow Ca^{2+}(aq) + 2CH_3CH_2COO^-(aq)$

Therefore, $[CH_3CH_2COO^-] = 2(0.0757_{17}) = 0.151_{43}$ M.

Construct an equilibrium expression based on the equation in (a).

$$K_b = \frac{[CH_3CH_2COOH][OH^-]}{[CH_3CH_2COO^-]}$$

According to Table 18.2, $K_a(CH_3CH_2COOH) = 1.3 \times 10^{-5}$. Thus, $K_b = K_w/K_a = 7.6_{92} \times 10^{-10}$.

It is reasonable to assume that $[CH_3CH_2COO^-]_{eq} = 0.151 - x \approx 0.151$ because $0.151/K_b$ is greater than 400.

$$\frac{x^2}{0.151_{43}} = 7.6_{92} x10^{-10} \Rightarrow x = 1.0_{79} x10^{-5}$$

Therefore, $[OH^-] = 1.0_{79} \times 10^{-5}$ M.

$pOH = -\log(1.0_{79} \times 10^{-5}) = 4.97$.

pH = 14.00 – 4.97 = **9.03**.

18.175 Beryllium exhibits behavior that is different than the typical behavior of group 2A(2) elements. The difference is due to the high charge density that leads it to form covalent compounds. These beryllium compounds generally are Lewis acids since beryllium has empty orbitals for accepting lone pairs of electrons. Also, the hydrated beryllium ion is able to transfer a proton to water and form an acidic solution. The other Group 2A(2) elements tend to form ionic compounds and the ions, such as Mg^{2+}, are extremely weak acids as the conjugate acids of strong bases.

18.178 Putrescine can be abbreviated to $NH_2(CH_2)_4NH_2$. Its reaction in water is written as follows:

$$NH_2(CH_2)_4NH_2(aq) + H_2O(l) \rightleftharpoons NH_2(CH_2)_4NH_3^+(aq) + OH^-(aq)$$

$$K_b = \frac{[NH_2(CH_2)_4NH_3^+][OH^-]}{[NH_2(CH_2)_4NH_2]} = \frac{(2.1x10^{-3})(2.1x10^{-3})}{(0.10 - 2.1x10^{-3})} = \mathbf{4.5x10^{-5}}$$

18.180 a) Use the pK_b of the tertiary amine nitrogen to calculate the pH.

$K_b = 10^{-5.1} = 7._{94} \times 10^{-6}$

Use assumption that x is negligible since 1.6×10^{-3} M >> 8×10^{-6}

$$[OH^-] = \sqrt{(7._{94}\, x10^{-6})(1.6x10^{-3})} = 1._{13}\, x10^{-4}\ M$$

Assumption is good since $1 \times 10^{-4}/1.6 \times 10^{-3} = 0.06$ or 6%, close enough for a calculation with one significant figure.

$pH = -\log(1.0x10^{-14}/1 \times 10^{-4}) =$ **10.0**

b) Calculate the concentration of H_3O^+ from the aromatic nitrogen.

$K_b = 10^{-9.7} = 2._{00} \times 10^{-10}$

$$[OH^-] = \sqrt{(2._{00}\, x10^{-10})(1.6x10^{-3})} = 5._{65}\, x10^{-7}\ M$$

The contribution of hydroxide ions from the aromatic nitrogen is significantly less than that from the tertiary amine nitrogen, 6×10^{-7} M << 1×10^{-4} M.

c) Quinine hydrochloride contains the conjugate acid of quinine with $pK_a = 14 - 5.1 = 8.9$.

$K_a = 10^{-8.9} = 1._{26} \times 10^{-9}$

Assume that $0.53 - x \approx 0.53$.

$$[H_3O^+] = \sqrt{(1._{26}\ x10^{-5})(0.53)} = 2._{58}\ x\,10^{-5}\ M$$

% error with assumption is 3 x 10^{-5}/0.53 = 0.00005 or 0.005%, so assumption is valid.

pH = -log(2.$_{58}$ x 10^{-5}) = **4.6**

d) First find the initial concentration of quinine hydrochloride.

$$[quinine\ HCl] = \left(\frac{1.5\ g\ quinine\ HCl}{100\ g\ sol'n}\right)\left(\frac{1.0\ g\ sol'n}{mL\ sol'n}\right)\left(\frac{1\ mol\ quinine\ HCl}{324.41\ g\ quinine\ HCl}\right)\left(\frac{1000\ mL}{1\ L}\right) = 4.6_{24}\ x\,10^{-2}\ M$$

Then use the concentration and K_a to find the pH. Assume that x is negligible.

$$[H_3O^+] = \sqrt{(1._{26}\ x10^{-9})(4.6_{24}\ x10^{-2})} = 7._{63}\ x\,10^{-6}\ M$$

Assumption is valid since 7.$_{63}$ x 10^{-6}/4.6$_{24}$ x 10^{-2} = 0.02 or 2%.

pH = -log(7.$_{63}$ x 10^{-6}) = **5.1**.

CHAPTER 19

IONIC EQUILIBRIA IN AQUEOUS SYSTEMS

FOLLOW-UP PROBLEMS

19.1 Plan: The problems are both equilibria with the initial concentration of reactant and product given. For part (a), set up a reaction table for the dissociation of HF. Set up equilibrium expression and solve for $[H_3O^+]$ assuming that the change in [HF] and $[F^-]$ is negligible. Check this assumption after finding $[H_3O^+]$. Convert $[H_3O^+]$ to pH. For part (b), first find the concentration of OH^- added. Then use the neutralization reaction to find the change in initial [HF] and $[F^-]$. Repeat the solution in part a to find pH.

a) Solution:

Conc (M)	HF(aq) +	$H_2O(l)$ ⇌	F^-(aq) +	H_3O^+(aq)
initial	0.50	---	0.45	0
change	-x	---	+x	+x
equilibrium	0.50-x	---	0.45+x	x

Assumptions: 1) initial $[H_3O^+]$ at 1.0×10^{-7} M can be assumed to be zero and 2) x is negligible with respect to 0.50 M and 0.45 M.

$$[H_3O^+] = K_a\left(\frac{[HF]}{[F^-]}\right) = \left(6.8 \times 10^{-4}\right)\left(\frac{0.50}{0.45}\right) = 7.5_{55} \times 10^{-4}$$

Check assumptions:

1) $1.0 \times 10^{-7} << 7.6 \times 10^{-4}$ so assumption to set initial concentration of H_3O^+ to zero is valid.

2) % error in assuming x is negligible: $7.6 \times 10^{-4}/0.45 \times 100 = 0.17\%$. Error is less than 5% so assumption is valid.

Solve for pH:

pH = $-\log(7.5_{55} \times 10^{-4})$ = **3.12**

Check: Since [HF] and $[F^-]$ are similar, the pH should be close to pK_a which equals $\log(6.8 \times 10^{-4}) = 3.17$. The pH should be slightly less (more acidic) than pK_a because [HF] > $[F^-]$. The calculated pH of 3.12 is slightly less than pK_a of 3.17.

b) Solution: How many moles of NaOH are added per liter?

$$\left(\frac{0.40\ g\ NaOH}{L}\right)\left(\frac{1\ mol\ NaOH}{40.00\ g\ NaOH}\right)\left(\frac{1\ mol\ OH^-}{1\ mol\ NaOH}\right) = 0.010\ M$$

Set up reaction table for neutralization of 0.010 M OH^-

Conc (M)	HF(aq) +	OH^-(aq) →	F^-(aq) +	$H_2O(l)$
before addition	0.50	---	0.45	---
addition	---	0.010	---	---
after addition	0.49	0	0.46	---

Following the same solution path with the same assumptions as part (a):

$$[H_3O^+] = \left(6.8x10^{-4}\right)\left(\frac{0.49}{0.46}\right) = 7.2_{43} \times 10^{-4}\ M$$

Check assumptions:

1) $7.2 \times 10^{-4} >> 1.0 \times 10^{-7}$ ⇒ assumption ok.

2) $(7.2 \times 10^{-4}/0.46)100 = 0.16\%$ which is less than the 5% maximum ⇒ assumption ok.

Solve for pH :

pH = $-\log(7.2_{43} \times 10^{-4})$ = **3.14**

Check: With addition of base the pH should increase and it does from 3.12 to 3.14. However, the pH should still be slightly less than pK_a. 3.14 is still less than 3.17.

19.2 Plan: Sodium benzoate is a salt so it dissolves in water to form Na^+ ions and $C_6H_5COO^-$ ions. Only the benzoate ion is involved in the buffer system represented by the equilibrium

$$C_6H_5COOH(aq) + H_2O(l) \rightleftharpoons C_6H_5COO^-(aq) + H_3O^+(aq)$$

Given in the problem are the volume and pH of the buffer and the concentrations of the base, benzoate ion. The question asks for mass of benzoic acid to add to the sodium benzoate solution. First find the concentration of C_6H_5COOH needed to make a buffer with a pH of 4.25. Multiply the volume by the concentration to find moles of C_6H_5COOH and use molar mass to find grams of benzoic acid.

Solution: From given pH calculate $[H_3O^+]$.

$[H_3O^+] = 10^{-4.25} = 5.6_{23} \times 10^{-5}$

The concentration of benzoic acid is calculated from equilibrium expression.

$$[C_6H_5COOH] = \frac{[H_3O^+][C_6H_5COO^-]}{K_a} = \frac{(5.6_{23} x10^{-5})(0.050)}{6.3x10^{-5}} = 0.044_{63}\ M$$

The number of moles of benzoic acid is determined from the concentration and volume.

$(0.044_{63}$ M $C_6H_5COOH) \times (5.0$ L$) = 0.22_{32}$ mol C_6H_5COOH

Mass of C_6H_5COOH:

$(0.22_{32}$ mol $C_6H_5COOH)(122.12$ g $C_6H_5COOH/$mol $C_6H_5COOH)$ = **27 g** C_6H_5COOH

Prepare a benzoic acid/benzoate buffer by dissolving 27 g of C_6H_5COOH into 5.0 L of 0.050 M C_6H_5COONa. Using a pH meter, adjust the pH to 4.25 with strong acid or base.

Check: pK_a for benzoic acid is 4.20. The buffer pH is slightly greater than the pK_a so the concentration of the acid should be slightly less than the concentration of the base. The calculation confirms this with $[C_6H_5COOH]$ at 0.045 M being slightly less than $[C_6H_5COO^-]$ at 0.050 M.

19.3 Plan: The titration is of a weak acid, HBrO, with a strong base, NaOH. The reactions involved are

1) Neutralization of weak acid with strong base:

$$HBrO(aq) + OH^-(aq) \rightarrow H_2O(l) + BrO^-(aq)$$

Note that the reaction goes to completion and produces the conjugate base, BrO^-.

2) The weak acid and its conjugate base are in equilibrium based on the acid dissociation reaction:

a) $HBrO(aq) + H_2O(aq) \rightleftharpoons BrO^-(aq) + H_3O^+(aq)$

or one could also use the base hydrolysis reaction

b) $BrO^-(aq) + H_2O(l) \rightleftharpoons HBrO(aq) + OH^-(aq)$

The pH of the solution is controlled by the relative concentrations of HBrO and BrO^-.

For each step in the titration, first think about what is present initially in the solution. Then use the two reactions to determine solution pH.

It is useful in a titration problem to first determine at what volume of titrant the equivalence point occurs.

a) Before any base is added the solution initially contains only HBrO and water. Equilibrium reaction 2(a) applies and pH can be found in the same way as for a weak acid solution.

b) When [HBrO] = [BrO^-] the solution contains significant concentrations of both the weak acid and its conjugate base. For the concentrations to be equal, half of the equivalence point volume has been added. Calculate the concentration of the acid and base and use equilibrium expression for reaction 2(a) to find pH.

c) At the equivalence point the total number of moles of HBrO present initially in solution equals the number of moles of base added. So reaction 1 goes to completion to produce that number of moles of BrO^-. The solution consists of BrO^- and water. Calculate the concentration of BrO^- then find pH using the base dissociation equilibrium, reaction 2b.

d) After the equivalence point the concentration of excess strong base determines the pH. Find the concentration of excess base and use to calculate pH.

e) Use pH values from a – d to plot S-shaped titration curve.

Solution:

First find equivalence point volume of NaOH.

(0.2000 M HBrO)(20.00 mL HBrO) = V_{NaOH}(0.1000 M NaOH) $\Rightarrow$ V_{NaOH} = 40.00 mL

a) Set up a reaction table for the weak acid solution.

Conc (M)	HBrO(aq)	+ H_2O(aq) $\rightleftharpoons$	BrO^-(aq)	+ H_3O^+(aq)
initial	0.2000	---	0	0
change	-x	---	+x	+x
equilibrium	0.2000-x	---	x	x

$K_a = \frac{x^2}{0.2000 - x}$ Since 0.2000 >> K_a use assumption that 0.2000-x ≈ 0.2000.

$[H_3O^+] = x = \sqrt{(2.3x10^{-9})(0.2000)} = 2.1_{45}\ x\ 10^{-5}\ M;\ pH = -\log(2.1_{45}x10^{-5}) = \mathbf{4.67}$

This calculation can be represented by the formula [H_3O^+] = $\sqrt{K_a[HA]}$.

Check the assumption by calculating %error in $[HBrO]_{eq}$.

%error = (2.1 x 10^{-5}/0.2000)100 = 0.01%, well below the 5% maximum.

b) Since [HBrO] = [BrO^-] their ratio equals 1.

$[H_3O^+] = K_a \frac{[HBrO]}{[BrO^-]} = (2.3\,x\,10^{-9})(1) = 2.3\,x\,10^{-9}\ M$ $\Rightarrow$ pH = -log(2.3 x 10^{-9}) = **8.64**

Note that when [HBrO] = [BrO^-] the titration is at the midpoint (half the volume to equivalence point) and pH = pK_a.

c) At the equivalence point, 40.00 mL of NaOH solution has been added (see calculation above) to make the total volume of the solution 60.00 mL. All of the HBrO present at the beginning of the titration is neutralized and converted to BrO^- at the equivalence point. Calculate the concentration of BrO^-.

$$[BrO^-] = (0.2000\ M\ HBrO)\left(\frac{1\,mol\ BrO^-}{1\,mol\ HBrO}\right)\left(\frac{0.02000\ L}{0.06000\ L}\right) = 0.06666_{67}\ M$$

Set up reaction table with reaction 2b since only BrO^- and water present initially

Conc (M)	BrO^- (aq)	+ H_2O(l) $\rightleftharpoons$	HBrO(aq)	+ OH^- (aq)
initial	0.06666_{67}	---	0	0
change	-x	---	+x	+x
equilibrium	0.06666_{67}-x	---	x	x

$$K_b = \frac{1.0x10^{-14}}{2.3x10^{-9}} = 4.3_{48}\ x\ 10^{-6} = \frac{x^2}{0.06666_{67} - x}$$

Assume that x is negligible since [BrO^-] >> K_b.

$$[OH^-] = \sqrt{(4.3_{48}x10^{-6})(0.06666_{67})} = 5.3_{84}\ x\ 10^{-4}\ M$$

Check the assumption by calculating %error in $[BrO^-]_{eq}$.

%error = (5.3_{84} x 10^{-4}/0.06667)100 = 0.3%, well below the 5% maximum.

pH = -log(1.0 x 10^{-14}/5.3_{84} x 10^{-4}) = **10.73**

d) Calculate the number of moles of OH^- left after the neutralization is complete.

mol OH^- needed to neutralize HBrO:

$$(0.2000\ M\ HBrO)(0.02000\ L)\left(\frac{1\,mol\ OH^-}{1\,mol\ HBrO}\right) = 4.000\,x\,10^{-3}\ mol\ OH^-$$

mol OH^- added is twice the moles of HBrO so 2(4.000 x 10^{-3} mol) = 8.000 x 10^{-3} mol.

The amount of excess OH^- is the difference between the mol OH^- added and the mol OH^- used to neutralize HBrO: 8.000 x 10^{-3} – 4.000 x 10^{-3} = 4.000 x 10^{-3} mol OH^- left.

Volume of NaOH solution added = 8.000 x 10^{-3} mol/0.1000 M = 0.08000 L

Total volume of solution is 0.08000 L NaOH solution + 0.02000 L HBrO solution = 0.10000 L.

Concentration of OH^- is calculated by dividing mol of OH^- left by total volume.

$$[OH^-] = \frac{4.000x10^{-3}\ mol\ OH^-}{0.10000\ L} = 0.04000\ M$$

pH = -log(1.0 x 10^{-14}/4.000 x 10^{-2}) = **12.60**.

e) Plot the pH values calculated above.

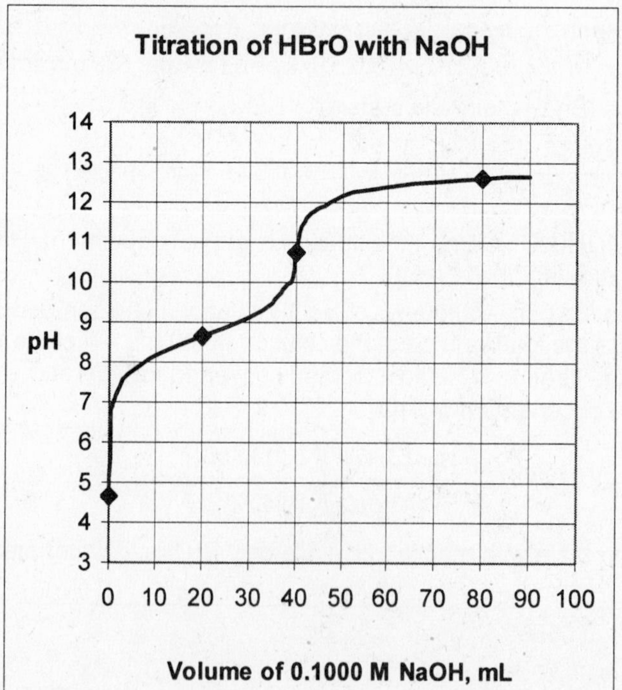

Check: The plot and pH values follow the pattern for a weak acid vs strong base titration. The pH at the midpoint of the titration does equal pK_a. The equivalence point should be and is greater than 7.

19.4 Plan: Write the formula of the salt and the reaction showing the equilibrium of a saturated solution. The ion-product expression can be written from the stoichiometry of the solution reaction.

Solution:

a) Calcium sulfate formula is $CaSO_4$. Equilibrium reaction is

$CaSO_4(s) \rightleftharpoons Ca^{2+}(aq) + SO_4^{2-}(aq)$

Ion-product expression: $K_{sp} = [Ca^{2+}][SO_4^{2-}]$

b) Chromium(III) carbonate is $Cr_2(CO_3)_3$

$Cr_2(CO_3)_3(s) \rightleftharpoons 2Cr^{3+}(aq) + 3CO_3^{2-}(aq)$

Ion-product expression: $K_{sp} = [Cr^{3+}]^2[CO_3^{2-}]^3$

c) Magnesium hydroxide is $Mg(OH)_2$

$Mg(OH)_2(s) \rightleftharpoons Mg^{2+}(aq) + 2OH^-(aq)$

Ion-product expression: $K_{sp} = [Mg^{2+}][OH^-]^2$

d) Arsenic(III) sulfide is As_2S_3.

$$As_2S_3(s) \rightleftharpoons 2As^{3+}(aq) + 3S^{2-}(aq)$$

$$3\{S^{2-}(aq) + H_2O(l) \rightleftharpoons HS^-(aq) + OH^-(aq)\}$$

$$As_2S_3(s) + 3H_2O(l) \rightleftharpoons 2As^{3+}(aq) + 3HS^-(aq) + 3OH^-(aq)$$

The second equilibrium must be considered in this case because its equilibrium constant is large, so essentially all of the sulfide ion is converted to HS^- and OH^-.

Ion-product expression: $K_{sp} = [As^{3+}]^2[HS^-]^3[OH^-]^3$

Check: All equilibria agree with formulas of salts.

19.5 Plan: Calculate solubility of CaF_2 as molarity and use molar ratios to find molarity of Ca^{2+} and F^- dissolved in solution. Calculate K_{sp} from $[Ca^{2+}]$ and $[F^-]$ using the ion-product expression.

Solution: Convert the solubility to molar solubility.

$$\left(\frac{1.5x10^{-4}\ g\,CaF_2}{10.0\,mL\ sol'n}\right)\left(\frac{1000\,mL}{1\,L}\right)\left(\frac{1\,mol\,CaF_2}{78.08\,g\,CaF_2}\right) = 1.9_{21}\ x\,10^{-4}\ M\,CaF_2$$

$[Ca^{2+}] = [CaF_2] = 1.9_{21} \times 10^{-4}$ M because there is 1 mol calcium ions in each mol of CaF_2.

$[F^-] = 2[CaF_2] = 3.8_{42} \times 10^{-4}$ M because there are 2 mol fluoride ions in each mol of CaF_2.

The solubility equilibrium:

$$CaF_2(s) \rightleftharpoons Ca^{2+}(aq) + 2F^-(aq) \qquad K_{sp} = [Ca^{2+}][F^-]^2$$

Calculate K_{sp} using the solubility product expression from above and the saturated concentrations of calcium and fluoride ions.

$$K_{sp} = [Ca^{2+}][F^-]^2 = (1.9_{21}x10^{-4})(3.8_{42}x10^{-4})^2 = 2.8\,x\,10^{-11}$$

The K_{sp} for CaF_2 is **2.8 x 10^{-11}** at 18°C.

Check: Table 19.2 gives a K_{sp} value at 25°C for CaF_2 to be 3.2 x 10^{-11}. The calculated value at 18°C is very close to this literature value.

19.6 Plan: Write the solubility reaction for $Mg(OH)_2$ and set up a reaction table, where S is the unknown molar solubility of the Mg^{2+} ion. Use the ion-product expression to solve for the concentration of $Mg(OH)_2$ in a saturated solution – also called the solubility of $Mg(OH)_2$.

Solution:

Conc (M)	$Mg(OH)_2(s)$	$\rightleftharpoons$	$Mg^{2+}(aq)$	+	$2OH^-(aq)$
initial	-----		0		0
change	-----		+S		+2S
equilibrium	-----		S		2S

$$K_{sp} = [Mg^{2+}][OH^-]^2 = S(2S)^2$$

$$6.3\,x\,10^{-10} = 4S^3;\ S = 5.4\,x\,10^{-4}$$

The solubility of $Mg(OH)_2$ is equal to S, the concentration of magnesium ions at equilibrium, so molar solubility of magnesium hydroxide in pure water is **5.4 x 10^{-4} M**.

Check: To check calculation work it backwards (also a good technique to use problems as study tools). From the calculated solubility find K_{sp}.

$[Mg^{2+}]$ = 5.4 x 10^{-4} M and $[OH^-]$ = 2(5.4 x 10^{-4} M) = 1.0_8 x 10^{-4} M

K_{sp} = (5.4 x 10^{-4})(1.0_8 x 10^{-4})^2 = 6.5 x 10^{-10}

Backwards calculation gives back given K_{sp}.

19.7 Plan: Write the solubility reaction of $BaSO_4$. For part (a) set up a reaction table in which $[Ba^{2+}] = [SO_4^{2-}]$ = S which also equals the solubility of $BaSO_4$. Then solve for S using the ion-product expression. For part (b) there is an initial concentration of sulfate, so set up the reaction table including this initial $[SO_4^{2-}]$. Solve for solubility, S, which equals $[Ba^{2+}]$ at equilibrium.

Solution:

a) Set up reaction table.

Conc (M)	$BaSO_4(s)$	$\rightleftharpoons$	$Ba^{2+}(aq)$	+	$SO_4^{2-}(aq)$
initial	------		0		0
change	------		+S		+S
equilibrium	------		S		S

$$K_{sp} = 1.1\,x\,10^{-10} = [Ba^{2+}][SO_4^{2-}] = S^2;\ S = 1.0_{49}\,x\,10^{-5}\ M$$

Molar solubility of $BaSO_4$ in pure water is **1.0 x 10^{-5} M**.

b) Set up another reaction table with initial $[SO_4^{2-}]$ = 0.10 M.

Conc (M)	$BaSO_4(s)$	$\rightleftharpoons$	$Ba^{2+}(aq)$	+	$SO_4^{2-}(aq)$
initial	-----		0		0.10
change	-----		+S		+S
equilibrium	-----		S		0.10+S

$$K_{sp} = 1.1\,x\,10^{-10} = [Ba^{2+}][SO_4^{2-}] = S(0.10 + S)$$

Assuming that 0.10 + S $\approx$ 0.10 which appears to be a good assumption based on the fact that 0.10 >> 1 x 10^{-10}, K_{sp}.

Solving 1.1 x 10^{-10} = 0.10S gives S = 1.1 x 10^{-9} M

Molar solubility of $BaSO_4$ in 0.10 M Na_2SO_4 is **1.1 x 10^{-9} M**.

Check: The solubility of $BaSO_4$ decreases when sulfate ions are already present in the solution. The calculated decrease is from 10^{-5} M to 10^{-9} M for a 10000-fold decrease. This decrease is expected to be large because of the high concentration of sulfate ions.

19.8 Plan: First write the solubility reaction for the salt. Then check the ions produced by dissolving the salt for reaction with acid. Three cases are possible:

1) If OH^- is produced, then addition of acid will neutralize the hydroxide ions and shift the solubility equilibrium towards the products. This causes more salt to dissolve. Write the solubility and neutralization reactions.
2) If the anion from the salt is the conjugate base of a weak acid, it will react with the added acid in a neutralization reaction. Solubility of the salt increases as the anion is neutralized. Write the solubility and neutralization reaction.
3) If the anion from the salt is the conjugate base of a strong acid, it does not react with a strong acid. The solubility of the salt is unchanged by the addition of acid. Write the solubility reaction.

Solution:

a) Calcium fluoride, CaF_2.

Solubility reaction: $CaF_2(s) \rightleftharpoons Ca^{2+}(aq) + 2F^-(aq)$

Fluoride ion is the conjugate base of HF, a weak acid. Thus it will react with H_3O^+ from strong acid, HNO_3.

Neutralization reaction: $F^-(aq) + H_3O^+(aq) \rightarrow HF(aq) + H_2O(l)$

The neutralization reaction decreases the concentration of fluoride ions, which causes the solubility equilibrium to shift to the right and more CaF_2 dissolves. Solubility of CaF_2 increases with the addition of HNO_3.

b) Zinc sulfide, ZnS.

Solubility reaction: $ZnS(s) + H_2O(l) \rightleftharpoons Zn^{2+}(aq) + HS^-(aq) + OH^-(aq)$

Two anions are formed because the sulfide ion from ZnS reacts almost completely with water to form HS^- and OH^-.

The hydroxide ion reacts with the added acid:

Neutralization reaction: $OH^-(aq) + H_3O^+(aq) \rightarrow 2H_2O(l)$

And the hydrogen sulfide ion, conjugate base of a weak acid H_2S, reacts with the added acid:

Neutralization reaction: $HS^-(aq) + H_3O^+(aq) \rightarrow H_2S(aq) + H_2O(l)$

Both neutralization reactions decrease the concentration of products in the solubility equilibrium which causes a shift to the right and more ZnS dissolves. The addition of HNO_3 increases the solubility of ZnS.

c) Silver iodide, AgI.

Solubility reaction: $AgI(s) \rightleftharpoons Ag^+(aq) + I^-(aq)$

The iodide ion is the conjugate base of a strong acid, HI. So, I^- will not react with added acid. Solubility of AgI will not change with added HNO_3.

19.9 Plan: First, write the solubility equilibrium equation and ion-product expression. Use given concentrations of calcium and phosphate ions to calculate Q_{sp}. Compare Q_{sp} to K_{sp}.

If $K_{sp} < Q_{sp}$ precipitation occurs. If $K_{sp} \geq Q_{sp}$ precipitation does not occur.

Solution:

Write the solubility equation:

$$Ca_3(PO_4)_2(s) \rightleftharpoons 3Ca^{2+}(aq) + 2PO_4^{3-}(aq)$$

and ion-product expression: $Q_{sp} = [Ca^{2+}]^3[PO_4^{3-}]^2$

$$Q_{sp} = \left(1.0x10^{-9}\right)^3\left(1.0x10^{-9}\right)^2 = 1.0 \times 10^{-45}$$

Compare K_{sp} and Q_{sp}. $K_{sp} = 1.2 \times 10^{-29} > 1.0 \times 10^{-45} = Q_{sp}$

Precipitation will not occur because concentrations are below the level of a saturated solution as shown by the value of Q_{sp}.

Check: The concentrations of calcium and phosphate ions are quite low, but K_{sp} is also small so it is difficult to predict whether precipitation will occur by just looking at the numbers. One way to check is with an estimate of the concentrations of ions in a saturated solution.

$K_{sp} = [Ca^{2+}]^3[PO_4^{3-}]^2 = (3S)^3(2S)^2 = 18S^5 = 1 \times 10^{-29}$; $S = 9 \times 10^{-7}$ M

In a saturated solution

$[Ca^{2+}] = 3(9 \times 10^{-7}$ M$) = 3 \times 10^{-6}$ M and $[PO_4^{3-}] = 2(9 \times 10^{-7}) = 2 \times 10^{-6}$ M

Both of these concentrations are much greater than the actual concentrations of 1×10^{-9} M for both ions, so the solution is unsaturated and no $Ca_3(PO_4)_2$ will precipitate.

19.10 Plan: Write the complex-ion formation equilibrium reaction. Calculate the initial concentrations of $Fe(H_2O)_6^{3+}$ and CN^-. The approach to complex ion equilibria problems is slightly different than for solubility equilibria because formation constants are generally large while solubility product constants are generally very small. The best mathematical approach is to assume that the equilibrium reaction goes to completion and then calculate back to find the equilibrium concentrations of reactants. So, assume that all of the $Fe(H_2O)_6^{3+}$ reacts to form $Fe(CN)_6^{3-}$ and calculate the concentration of $Fe(CN)_6^{3-}$ formed from the given concentrations of $Fe(H_2O)_6^{3+}$ and CN^- and the concentration of the excess reactant. Then use the complex ion formation equilibrium to find the equilibrium concentration of $Fe(H_2O)_6^{3+}$.

Solution:

Equilibrium reaction: $Fe(H_2O)_6^{3+}(aq) + 6CN^-(aq) \rightleftharpoons Fe(CN)_6^{3-}(aq) + 6H_2O(l)$

Initial concentrations:

$$[Fe(H_2O)_6^{3+}] = (3.1 \times 10^{-2}\ M)\left(\frac{25.5\ mL}{60.5\ mL}\right) = 0.013_{07}\ M;\ [CN^-] = (1.5\ M)\left(\frac{35.0\ mL}{60.5\ mL}\right) = 0.86_{78}\ M$$

Set up reaction table:

Conc (M)	$Fe(H_2O)_6^{3+}(aq)$ +	$6CN^-(aq)$ $\rightleftharpoons$	$Fe(CN)_6^{3-}(aq)$ +	$6H_2O(l)$
initial	0.013_{07}	0.86_{78}	0	---
change	-0.013_{07}+x	$-6(0.013_{07})$	$+0.013_{07}$	---
equilibrium	x	0.78_{93}	0.013_{07}	---

$$K_f = 4.0 \times 10^{43} = \frac{[Fe(CN)_6^{3-}]}{[Fe(H_2O)_6^{3+}][CN^-]^6} = \frac{0.013_{07}}{x(0.78_{93})^6};\ x = 1.3_{51} \times 10^{-45}$$

The concentration of $Fe(H_2O)_6^{3+}$ at equilibrium is **1.4×10^{-45} M**. This concentration is so low that it is impossible to calculate it using the initial concentrations minus a variable x. The variable would have to be so close to the initial concentration that the initial concentration of x cannot be calculated to enough significant figures (43 in this case) to get a difference in concentrations of 1×10^{-45} M. Thus, the approach above is the best to calculate the very low equilibrium concentration of $Fe(H_2O)_6^{3+}$.

Check: The equilibrium concentration of $Fe(H_2O)_6^{3+}$ should be low because K_f is very large.

19.11 Plan: Write equations for solubility equilibrium and formation of silver-ammonia complex ion. Add the two equations to get the overall reaction. Set up a reaction table for the overall reaction with the given value for initial $[NH_3]$. Write equilibrium expressions from the overall balanced reaction and calculate $K_{overall}$ from K_f and K_{sp} values. Insert equilibrium concentration values from the reaction table into equilibrium expression and calculate solubility.

Solution: Equilibria:

$$AgBr(s) \rightleftharpoons Ag^+(aq) + Br^-(aq)$$
$$Ag^+(aq) + 2NH_3(aq) \rightleftharpoons Ag(NH_3)_2^+(aq)$$
$$AgBr(s) + 2NH_3(aq) \rightleftharpoons Ag(NH_3)_2^+(aq) + Br^-(aq)$$

Set up reaction table:

Conc (M)	AgBr(s) +	$2NH_3(aq)$	$\rightleftharpoons$	$Ag(NH_3)_2^+(aq)$ +	$Br^-(aq)$
initial	---	1.0		0	0
change	---	-2S		+S	+S
equilibrium	---	1.0-2S		S	S

Equilibrium expression:

$$K_{overall} = \frac{[Ag(NH_3)_2^+][Br^-]}{[NH_3]^2} = K_{sp} \cdot K_f = (5.0 x 10^{-13})(1.7 x 10^7) = 8.5 x 10^{-6}$$

Calculate solubility of AgBr.

$$\frac{S^2}{(1.0 - 2S)^2} = 8.5 x 10^{-6}; \; S = 2.9 x 10^{-3} \; M$$

The solubility of AgBr in ammonia is less than its solubility in hypo (sodium thiosulfate).
Check: Since the formation constant for $Ag(NH_3)_2^+$ is less than the formation constant of $Ag(S_2O_3)_2^{3-}$, the addition of ammonia will increase the solubility of AgBr less than the addition of thiosulfate ion increases its solubility.

19.12 Plan: Compare the K_{sp}'s for the two salts. Since $CaSO_4$ is more soluble, calculate what concentration of sulfate ions that would be added to decrease the concentration of calcium ions to 99.99% of 0.025 M.

Solution:

The solubility equilibrium for $CaSO_4$ is

$$CaSO_4(s) \rightleftharpoons Ca^{2+}(aq) + SO_4^{2-}(aq)$$

so $K_{sp} = [Ca^{2+}][SO_4^{2-}] = 2.4 \times 10^{-5}$

The concentration of calcium ions will be 99.99% of 0.025 M. Calculate $[SO_4^{2-}]$:

$[SO_4^{2-}] = (2.4 \times 10^{-5}) \div (0.9999 \times 0.025) =$ **9.6×10^{-4} M**

Check: An estimate of $[SO_4^{2-}]$ can be calculated by dividing 2.4×10^{-5} by 0.025 to give 1×10^{-4} M. This is approximately equal to the calculated result of 9.6×10^{-4} M.

END-OF-CHAPTER PROBLEMS

19.2 The weak acid component neutralizes added base and the weak base component neutralizes added acid so that the pH of the buffer solution remains relatively constant. The components of a buffer do not neutralize one another when they are a conjugate acid/base pair.

19.4 A buffer is a mixture of a weak acid and its conjugate base (or weak base and its conjugate acid). The pH of a buffer changes only slightly with added H_3O^+ because the added H_3O^+

reacts with the base of the buffer. The net result is that the concentration of H_3O^+ does not change much from the original concentration, keeping the pH fairly constant.

19.7 The buffer component ratio refers to the ratio of concentrations of the acid and base that make up the buffer. When this ratio is equal to 1 the buffer resists changes in pH with added acid to the same extent that it resists changes in pH with added base. The buffer range extends equally in both the acidic and basic direction. When the ratio shifts with higher [base] than [acid], the buffer is more effective at neutralizing added acid than base so the range extends further in the acidic than basic direction. The opposite is true for a buffer where [acid] > [base]. Buffers with a ratio equal to 1 have the greatest buffer range. The more the buffer component ratio deviates from 1, the smaller the buffer range.

19.9 a) The buffer component ratio and pH **increase** with added base. The OH^- reacts with HA to decrease its concentration and increase [NaA]. The ratio [NaA]/[HA] thus increases. The pH of the buffer will be more basic because the concentration of base, A^-, has increased and the concentration of acid, HA, decreased.

b) Buffer component ratio and pH **decrease** with added acid. The H_3O^+ reacts with A^- to decrease its concentration and increase [HA]. The ratio [NaA]/[HA] thus decreases. The pH of the buffer will be more acidic because the concentration of base, A^-, has decreased and the concentration of acid, HA, increased.

c) Buffer component ratio and pH **increase** with the added sodium salt. The additional NaA increases the concentration of both NaA and HA, but the relative increase in [NaA] is greater. Thus the ratio increases and the solution becomes more basic. Whenever base is added to a buffer the pH always increases, but only slightly if the amount of base is not too large.

d) Buffer component ratio and pH **decrease**. The concentration of HA increases more than the concentration of NaA so the ratio is less and the solution more acidic.

19.11 The buffer components are propanoic acid and propanoate ion. The sodium ions are ignored because they are not involved in the buffer. The reaction table that describes this buffer is:

Conc (M)	$C_2H_6O_2(aq)$ +	$H_2O(l)$ $\rightleftharpoons$	$C_2H_5O_2^-(aq)$ +	$H_3O^+(aq)$
initial	0.15	---	0.25	0
change	-x	---	+x	+x
equilibrium	0.15-x	---	0.25+x	x

Assume that x is negligible with respect to both 0.15 and 0.25 because both concentrations are much larger than K_a.

$$[H_3O^+] = (1.3 x 10^{-5})\left(\frac{0.15}{0.25}\right) = \mathbf{7.8 x 10^{-6} M}$$

Check assumption: %error = $(7.8 \times 10^{-6}/0.15)100$ = 0.005%. Assumption is valid.

pH = $-\log[H_3O^+] = -\log(7.8 \times 10^{-6})$ = **5.11**.

Another solution path to find pH is using the Henderson-Hasselbalch equation:

$$pH = pK_a + \log\frac{[base]}{[acid]} = 4.89 + \log\frac{0.25\ M}{0.15\ M} = 5.11$$

19.13 The buffer components are nitrous acid, HNO_2, and nitrite ion, NO_2^-. The potassium ions are ignored because they are not involved in the buffer. Set up the problem with a reaction table.

Conc (M)	$HNO_2(aq)$ +	$H_2O(l)$	⇌	$NO_2^-(aq)$ +	$H_3O^+(aq)$
initial	0.50	---		0.65	0
change	-x	---		+x	+x
equilibrium	0.50-x	---		0.65+x	x

Assume that x is negligible with respect to both 0.50 and 0.65 because both concentrations are much larger than K_a.

$$[H_3O^+] = (7.1 x 10^{-4})\left(\frac{0.50}{0.65}\right) = 5.4_{62} x 10^{-4}\ M = \mathbf{5.5x10^{-4}\ M}$$

Check assumption: %error = (5.5 x 10^{-4}/0.50)100 = 0.1%. Assumption is valid.

pH = -log(5.4$_{62}$ x 10^{-4}) = **3.26**. Verify the pH using the Henderson-Hasselbalch equation.

$$pH = pK_a + \log\frac{[base]}{[acid]} = 3.15 + \log\frac{0.65\ M}{0.50\ M} = 3.26$$

19.15 The buffer components are formic acid, HCOOH, and formate ion, $HCOO^-$. The sodium ions are ignored because they are not involved in the buffer. Calculate K_a from pK_a and write a reaction table for the dissociation of formic acid.

$K_a = 10^{-3.74} = 1.8_{20} \text{ x } 10^{-10}$

Conc (M)	HCOOH (aq) +	$H_2O(l)$	⇌	$HCOO^-(aq)$ +	$H_3O^+(aq)$
initial	0.55	---		0.63	0
change	-x	---		+x	+x
equilibrium	0.55-x	---		0.63+x	x

Assume that x is negligible because both concentrations are much larger than K_a.

$$[H_3O^+] = (1.8_{20}\ x\ 10^{-4})\left(\frac{0.55}{0.63}\right) = 1.5_{89}\ x\ 10^{-4}\ M = \mathbf{1.6x10^{-4}\ M}$$

Check assumption: %error = (1.6 x 10^{-4}/0.55)100 = 0.03%. Assumption is valid.

pH = -log(1.5$_{89}$ x 10^{-4}) = **3.80**. Verify the pH using the Henderson-Hasselbalch equation:

$$pH = pK_a + \log\frac{[base]}{[acid]} = 3.74 + \log\frac{0.63\ M}{0.55\ M} = 3.80$$

19.17 The buffer components are phenol, C_6H_5OH, and phenolate ion, $C_6H_5O^-$. The sodium ions are ignored because they are not involved in the buffer. Calculate K_a from pK_a and set up the problem with a reaction table.

$K_a = 10^{-10.00} = 1.0 \text{ x } 10^{-10}$

Conc (M)	$C_6H_5OH(aq)$ +	$H_2O(l)$	⇌	$C_6H_5O^-(aq)$ +	$H_3O^+(aq)$
initial	1.2	---		1.0	0
change	-x	---		+x	+x
equilibrium	1.2-x	---		1.0+x	x

Assume that x is negligible with respect to both 1.0 and 1.2 because both concentrations are much larger than K_a.

$$[H_3O^+]=(1.0\,x\,10^{-10})\left(\frac{1.2}{1.0}\right)=1.2_{00}\,x\,10^{-10}\ M$$

Assumption is good since concentration of hydronium ions is so low.

pH = -log(1.2_{00} x 10^{-10}) = **9.92**. Verify the pH using the Henderson-Hasselbalch equation:

$$pH = pK_a + \log\frac{[base]}{[acid]} = 10.00 + \log\frac{1.0\ M}{1.2\ M} = 9.92$$

19.19 Use Equation 19.1, the Henderson-Hasselbalch equation, to solve for pH. Convert pK_b of NH_3 to pK_a of NH_4^+, given that $pK_a + pK_b = 14$: $pK_a = 14 – 4.75 = 9.25$.

$$pH = pK_a + \log\left(\frac{[base]}{[acid]}\right) = pK_a + \log\left(\frac{[NH_3]}{[NH_4^+]}\right)$$

$$pH = 9.25 + \log\left(\frac{0.20}{0.10}\right) = \mathbf{9.55}$$

or convert Henderson Hasselbach equation to solve for [OH^-]:

$$pOH = pK_b + \log\frac{[acid]}{[base]} = 4.75 + \log\frac{0.10\ M}{0.20\ M} = 4.45;\ pH = 14.00 - 4.45 = 9.55$$

19.21 a) The buffer components are HCO_3^- from the salt $KHCO_3$ and CO_3^{2-} from the salt K_2CO_3. Choose the K_a value that corresponds to the equilibrium with these two components. K_{a1} refers to carbonic acid, H_2CO_3 losing one proton to produce HCO_3^-. This is not the correct K_a because H_2CO_3 is not involved in the buffer. K_{a2} is the correct K_a to choose because it is the equilibrium constant for the loss of the second proton to produce CO_3^{2-} from HCO_3^-.

b) Set up the reaction table and use K_{a2} to calculate pH.

Conc (M)	HCO_3^-(aq)	+ H_2O(l)	⇌ CO_3^{2-}(aq)	+ H_3O^+(aq)
initial	0.25	---	0.32	0
change	-x	---	+x	+x
equilibrium	0.25-x	---	0.32+x	x

Assume that x is negligible with respect to both 0.25 and 0.32 because both concentrations are much larger than K_a.

$$[H_3O^+]=(4.7\,x\,10^{-11})\left(\frac{0.25}{0.32}\right)=3.6_{72}\,x\,10^{-11}\ M$$

Assumption is valid because x << 0.25.

pH = -log(3.6_{72} x 10^{-11}) = **10.44**

19.23 Given the pH and pK_a of an acid, the buffer–component ratio can be calculated from Equation 19.1.

$pK_a = -\log K_a = -\log(1.3 \times 10^{-5}) = 4.88_{61}$

pH = pK_a + log([base]/[acid])

5.11 = 4.88_{61} + log([Pr^-]/[HPr])

log([Pr^-]/[HPr]) = 0.22_{39}

[Pr^-]/[HPr] = **1.7**.

19.25 The buffer–component ratio can also be calculated from the equilibrium expression.

$[H_3O^+] = 10^{-7.88} = 1.3_{18} \times 10^{-8}$ M

$$\frac{[OBr^-]}{[HOBr]} = \frac{K_a}{[H_3O^+]} = \frac{2.3 x 10^{-9}}{1.3_{18} x 10^{-8}} = \mathbf{0.17}$$

19.27 First find the concentration of hydroxide added, $[OH^-]_{added}$, and adjust the starting concentrations of acid, HA, and base, A^-, based on their reaction with the added OH^-. Then set up a reaction table for the acid dissociation using the new initial concentrations. Solve for $[H_3O^+]$ and pH.

$$[OH^-]_{added} = \frac{0.0015\, mol\ NaOH}{0.5000\, L\ sol'n} = 0.0030\ M$$

The added $[OH^-]$ reacts with the HA with a 1:1 mole ratio according to the reaction $HA + OH^- \rightarrow A^- + H_2O$. Therefore [HA] decreases to (0.2000 – 0.0030) = 0.1970 M and $[A^-]$ increases to (0.1500 + 0.0030) = 0.1530 M.

The reaction table reflecting these new concentrations is:

Conc (M)	HA(aq) +	$H_2O(l)$ ⇌	A^-(aq) +	H_3O^+(aq)
initial	0.1970	---	0.1530	0
change	-x	---	+x	+x
equilibrium	0.1970-x	---	0.1530+x	x

Assume that x is negligible with respect to both 0.1970 and 0.1530.

$$K_a = \frac{[A^-][H_3O^+]}{[HA]} \Rightarrow [H_3O^+] = \frac{K_a[HA]}{[A^-]}$$

What is the K_a for this acid? It is not explicitly given and must be solved for given the *initial* conditions of the problem, where [HA] = 0.2000 M, $[A^-]$ = 0.1500 M and $[H_3O^+] = 10^{-3.35}$.

$$K_a = \frac{(0.1500)(10^{-3.35})}{0.2000} = 3.3_{50} x 10^{-4}$$

Substitute this value into the equation above to solve for $[H_3O^+]$.

$$[H_3O^+] = \frac{K_a[HA]}{[A^-]} = \frac{(3.3_{50} x 10^{-4})(0.1970)}{(0.1530)} = 4.3_{14} x 10^{-4}\ M$$

pH = -log($4.3_{14} \times 10^{-4}$) = **3.37**.

Does this answer seem reasonable? The initial pH was 3.35, so one would expect the pH to rise slightly upon addition of a base. The answer can also be verified using the Henderson-Hasselbalch equation. An alternative approach is shown in Problem 19.29.

19.29 Set up a relationship between the concentration of H_3O^+ and the buffer-component ratio before and after the addition of base to find the new pH.

$$[H_3O^+]_{before} \frac{[Y^-]_{before}}{[HY]_{before}} = [H_3O^+]_{after} \frac{[Y^-]_{after}}{[HY]_{after}}$$

$[H_3O^+]_{before} = 10^{-8.77} = 1.6_{98} \times 10^{-9}$ M

The OH^- from the $Ba(OH)_2$ added reacts with HY to decrease its concentration and to increase $[Y^-]$. The moles of OH^- added are twice the number of moles of $Ba(OH)_2$.

$[HY]_{after}$ = (mol HY before – mol OH^- added)/0.750 L

mol HY before = (0.110 M)(0.750 L) = 0.0825_{00} mol

$[HY]_{after}$ = (0.0825_{00} mol HY before – 0.0020 mol OH^- added)/0.750 L = 0.107_{33} M

$[Y^-]_{after}$ = (mol Y^- before + mol OH^- added)/0.750 L

mol Y^- before = (0.220 M)(0.750 L) = 0.165_{00} mol

$[Y^-]_{after}$ = (0.165_{00} mol Y^- before + 0.0020 mol OH^- added)/0.750 L = 0.222_{67} M

$$[H_3O^+]_{after} = (1.6_{98} \times 10^{-9})\left(\frac{0.220}{0.110}\right)\left(\frac{0.107_{33}}{0.222_{67}}\right) = 1.6_{37} \times 10^{-9}\ M$$

pH = -log(1.6_{37} x 10^{-9}) = **8.79**

19.31 a) The hydrochloric acid will react with the sodium acetate, abbreviated NaAc, to form acetic acid, abbreviated HAc: HCl + NaAc → HAc + NaCl

Calculate the number of moles of HCl and NaAc. All of the HCl will be consumed to form HAc, and the number of moles of Ac^- will decrease.

Initial moles HCl = (0.184 L)(0.442 M) = 0.0813_{28} mol

Initial moles NaAc = (0.500 L)(0.400 M) = 0.200 mol = mol Ac^-

Moles HAc formed = 0.0813_{28} mol

Moles Ac^- remaining = 0.200 mol – 0.0813_{28} mol = 0.118_{67} mol

[HAc] = (0.0813_{28} mol)/ (0.184 L + 0.500 L) = 0.118_{90} M

$[Ac^-]$ = (0.118_{67} mol)/(0.184 L + 0.500 L) = 0.173_{50} M

pH = pK_a + log($[Ac^-]$/[HAc]) = -log(1.8 x 10^{-5}) + log(0.173_{50}/0.118_{90}) = **4.91**.

b) The addition of base would increase the pH, so the new pH is (4.91 + 0.15) = 5.06. The new $[Ac^-]$/[HAc] ratio is calculated using the Henderson-Hasselbalch equation.

pH = pK_a + log($[Ac^-]$/[HAc])

5.06 = 4.74_{47} + log($[Ac^-]$/[HAc])

log($[Ac^-]$/[HAc]) = 0.31_{53}

$[Ac^-]$/[HAc] = 2.0_{67}

From part (a), we know that [HAc] + $[Ac^-]$ = (0.118_{90} + 0.173_{50}) = 0.292_{40} M. Although the *ratio* of $[Ac^-]$ to [HAc] can change when acid or base is added, the *absolute amount* does not change unless acetic acid or an acetate salt is added.

Given that $[Ac^-]$/[HAc] = 2.0_{67} and [HAc] + $[Ac^-]$ = 0.292_{40} M, solve for $[Ac^-]$ and substitute into the second equation.

$[Ac^-]$ = 2.0_{67}[HAc] ⇒ [HAc] + 2.0_{67}[HAc] = 0.292_{40} M

[HAc] = 0.095_{34} M and $[Ac^-]$ = 0.197_{1} M.

Moles of Ac^- needed = (0.197_{1} M)(0.500 L) = 0.098_{53} mol

Moles of Ac^- initially = (0.173_{50} M)(0.500 L) = 0.0867_{50} mol

This would require the addition of (0.098_{53} – 0.0867_{50}) = 0.011_{78} mol KOH.

$$mass\ KOH = (0.011_{78}\ mol\ KOH)\left(\frac{56.11\,g\ KOH}{mol\ KOH}\right) = \mathbf{0.66\,g\,KOH}$$

19.33 Select conjugate pairs with K_a values close to the desired $[H_3O^+]$.

a) For pH ≈ 4.0, the best selection is the HCOOH/$HCOO^-$ conjugate pair with K_a equal to 1.8×10^{-4}. From the base list, only the $C_6H_5NH_2/C_6H_5NH_3^+$ conjugate pair comes close with $K_a = 1 \times 10^{-14}/4.0 \times 10^{-10} = 2.5 \times 10^{-5}$.

b) For pH ≈ 7.0, two choices are the $H_2PO_4^-/HPO_4^{2-}$ conjugate pair with K_a of 1.7×10^{-7} and the $H_2AsO_4^-/HAsO_4^{2-}$ conjugate pair with K_a of 1.1×10^{-7}.

19.35 Select conjugate pairs with pK_a values close to the desired pH. Convert pH to $[H_3O^+]$ for easy comparison to K_a values.

a) For pH ≈ 2.5 ($[H_3O^+] = 3 \times 10^{-3}$), the best selection is the $H_3AsO_4/H_2AsO_4^-$ conjugate pair with a pK_a = 2.22. The $H_3PO_4/H_2PO_4^-$ pair, with pK_a = 2.14, is also a good choice.

b) For pH ≈ 5.5 ($[H_3O^+] = 3 \times 10^{-6}$), the best selection is the $C_5H_5N/C_5H_5NH^+$ pair with a $K_a = 1 \times 10^{-14}/1.7 \times 10^{-9} = 6 \times 10^{-6}$ and pK_a = 5.2. None of the other acids in Tables 18.2 and 18.5 have sufficiently close pK_a's.

19.38 Convert pH to $[H_3O^+]$ and use equilibrium expression to calculate ratio.

$[H_3O^+] = 10^{-7.40} = 3.9_{81} \times 10^{-8}$ M

K_{a2} for $H_2PO_4^-/HPO_4^{2-}$ conjugate pair is 6.3×10^{-8}: $\frac{[HPO_4^{2-}]}{[H_2PO_4^-]} = \frac{K_a}{[H_3O^+]} = \frac{6.3 \times 10^{-8}}{3.9_{81} \times 10^{-8}} = \mathbf{1.6}$

Alternatively, the ratio can be calculated using the Henderson-Hasselbalch equation.

19.40 A person's ability to perceive a color change of a mixture of two colored species occurs over a 100-fold range in the [HIn]/$[In^-]$ ratio. This translates on a logarithmic scale to 2 units (log100).

19.42 The equivalence point in a titration is the point at which the number of moles of base equals the number of moles of acid (be sure to account for stoichiometric ratios, e.g. 1 mol of $Ca(OH)_2$ produces 2 moles of OH^-). The endpoint is the point at which the added indicator changes color. If an appropriate indicator is selected the endpoint is close to the equivalence point, but not usually exactly the same. Using an indicator that changes color at a pH after the equivalence point means the equivalence point is reached first. But, if an indicator is selected that changes color at a pH before the equivalence point then the endpoint is reached first.

19.44 a) The initial pH is lowest for flask solution of the strong acid, followed by the weak acid and then the weak base. In other words, *strong acid* – strong base < *weak acid* – strong base < strong acid – *weak base*.

b) At the equivalence point, the moles of acid equal the moles of base, regardless of the type of titration. However, the strong acid – strong base equivalence point occurs at pH = 7.00 because the resulting cation-anion combination does not react with water. For example, NaOH + HCl → H_2O + NaCl. Neither Na^+ and Cl^- ions dissociate in water.

The weak acid – strong base equivalence point occurs at pH > 7, because the anion of the weak acid is weakly basic, whereas the cation of the strong base does not react with water. For example, HCOOH + NaOH → $HCOO^-$ + H_2O + Na^+. The conjugate base, $HCOO^-$, reacts with water according to this reaction: $HCOO^-$ + H_2O → HCOOH + OH^-.

The strong acid – weak base equivalence point occurs at pH < 7, because the anion of the strong acid does not react with water, whereas the cation of the weak base is weakly acidic. For example, HCl + NH_3 → NH_4^+ + Cl^-. The conjugate acid, NH_4^+, dissociates slightly in water: NH_4^+ + H_2O → NH_3 + H_3O^+.

In rank order of pH of equivalence point, strong acid – *weak base* < *strong acid* – strong base < *weak acid* – strong base.

19.46 At the very center of the buffer region of a weak acid – strong base titration the concentration of the weak acid and its conjugate base are equal, which means that at this point the pH of the solution equals the pK_a of the weak acid.

19.48 Indicators have pH range that is approximated by pK_a ± 1. The pK_a of cresol red is $-\log(5.0 \times 10^{-9}) = 8.3$, so the indicator changes color over an approximate range of 7.3 to 9.3.

19.50 Choose an indicator that changes color at a pH close to the pH of the equivalence point.

a) The equivalence point for a strong acid – strong base titration occurs at pH = 7.0. **Bromthymol blue** is an indicator that changes color around pH 7.

b) The equivalence point for a weak acid – strong base is above pH 7. Estimate the pH at equivalence point from equilibrium calculation.

At the equivalence point, the solution is 0.050 M $HCOO^-$. (The volume doubles because equal volumes of base and acid are required to reach the equivalence point. When the volume doubles, the concentration is halved.)

K_b for $HCOO^-$ = $1 \times 10^{-14}/1.8 \times 10^{-4} = 5.6 \times 10^{-11}$

$$[OH^-] = \sqrt{K_b C_{HCOO^-}} = \sqrt{(5.6 x 10^{-11})(0.050)} = 1.7 x 10^{-6}\ M$$

pH = $-\log(1 \times 10^{-14}/1.7 \times 10^{-6}) = 8.2$

Thymol blue, phenol red and phenolphthalein are appropriate choices for indicator.

19.52 a) The equivalence point for a weak base – strong acid is below pH 7. Estimate the pH at equivalence point from equilibrium calculation.

At the equivalence point, the solution is 0.25 M $(CH_3)_2NH_2^+$. (The volume doubles because equal volumes of base and acid are required to reach the equivalence point. When the volume doubles, the concentration is halved.)

K_a for $(CH_3)_2NH_2^+ = K_w/K_b((CH_3)_2NH) = 1.0 \times 10^{-14}/5.9 \times 10^{-4} = 1.7 \times 10^{-11}$

$$K_a = \frac{[(CH_3)_2NH][H_3O^+]}{[(CH_3)_2NH_2^+]} = \frac{x^2}{0.25}$$

$$x = [H_3O^+] = \sqrt{K_a(0.25)} = 2.1x10^{-6}$$

$$pH = -\log(2.1x10^{-6}) = 5.7$$

Methyl red is an indicator that changes color around pH 5.7.

b) Equivalence point for a strong acid – strong base titration occurs at pH = 7.0. **Bromthymol blue** is an indicator that changes color around pH 7.

19.54 The reaction occurring in the titration is the neutralization of H_3O^+ (from HCl) by OH^- (from NaOH): $H_3O^+(aq) + OH^-(aq) \rightarrow 2H_2O(l)$

For the titration of a strong acid with a strong base, the pH before the equivalence point depends on the excess concentration of acid and the pH after the equivalence point depends on the excess concentration of base. At the equivalence point there is not an excess of either acid or base so pH is 7.0. The equivalence point occurs when 50.00 mL of base has been added.

a) At 0 mL of base added, the concentration of hydronium ion equals the original concentration of HCl. pH = -log(0.1000 M) = **1.00** (recall that pH's are usually reported to the hundredths place, even when more significant figures are warranted).

b) At 25.00 mL of base added, half of the HCl has been neutralized.

$[HCl]_{left}$ = ½ (0.1000 M x 50.00 mL)/75.00 mL = 0.03333 M

pH = -log(0.03333) = **1.48**.

c) At 49.00 mL of base added, the moles of H_3O^+ reacted is 0.1000 M x 0.04900 L = 0.004900 mol. Subtracting this from the original 0.005000 mol of H_3O^+ gives 0.000100 mol H_3O^+. $[H_3O^+]$ = 0.000100 mol/0.09900 = 1.01 x 10^{-3} M

pH = -log(1.01 x 10^{-3}) = **3.00**.

d) At 49.90 mL of base added,

$[H_3O^+]$ = [0.005000 mol - (0.1000 M x 0.04990 L)]/0.09990 L = 1.0 x 10^{-4} M

pH = -log(1.0 x 10^{-4}) = **4.00**.

e) At 50.00 mL base added the titration is at the equivalence point so pH = **7.00**.

f) At 50.10 mL base added, all of the H_3O^+ has been neutralized so pH is determined by the excess OH^- present after neutralization.

$[OH^-]$ = [(0.05010 L x 0.1000 M) – 0.005000 mol)]/0.1010 L = 9.90 x 10^{-5} M

pH = -log(1.0 x 10^{-14}/9.90 x 10^{-5}) = **10.00**.

g) At 60.00 mL added base the pH is determined by excess OH^-.

$[OH^-]$ = [(0.06000 L x 0.1000 M) – 0.005000 mol)]/0.1100 L = 9.091 x 10^{-3} M

pH = -log(1.0 x 10^{-14}/9.091 x 10^{-3}) = **11.96**.

19.56 This is a titration between a weak acid and a strong base. The pH before addition of the base is dependent on the K_a of the acid. Prior to reaching the equivalence point, the added base reacts with the acid to form butanoate ion. The equivalence point occurs when 20.00 mL of base is added to the acid because at this point, moles acid = moles base. Addition of base beyond the equivalence point is simply the addition of excess OH^-.

a) At 0 mL of base added, the concentration of $[H_3O^+]$ is dependent on the dissociation of butanoic acid: $C_3H_7COOH + H_2O \rightleftharpoons C_3H_7COO^- + H_3O^+$.

$$K_a = \frac{[C_3H_7COO^-][H_3O^+]}{[C_3H_7COOH]} = \frac{x^2}{0.1000 - x} \approx \frac{x^2}{0.1000}$$

$$x = [H_3O^+] = \sqrt{K_a(C_3H_7COOH)} = \sqrt{(1.54x10^{-5})(0.1000)} = 0.00124_{10}\ M$$

pH = -log(0.00124_{10}) = **2.91**.

b) The added NaOH reacts with C_3H_7COOH to form $C_3H_7COO^-$. The K_a describes the equilibrium between C_3H_7COOH, $C_3H_7COO^-$ and H_3O^+.

moles added NaOH = moles $C_3H_7COO^-$ formed = (0.01000 L)(0.1000 M) = 1.000 x 10^{-3}

moles of C_3H_7COOH remaining = (0.02000 L)(0.1000 M) – 1.000 x 10^{-3} moles = 1.000 x 10^{-3}

Again, $[H_3O^+]$ is dictated by the K_a expression, which can be mathematically manipulated to give the Henderson-Hasselbalch equation.

$$pH = pK_a + \log\left(\frac{mol\ base}{mol\ acid}\right)$$

$$pH = -\log(1.54x10^{-5}) + \log\left(\frac{1.000x10^{-3}}{1.000x10^{-3}}\right)$$

pH = **4.81**

Note that this is the midpoint of the buffer region, where pH = pK_a.

c) At 15 mL of base added, more $C_3H_7COO^-$ is formed, but not all of the C_3H_7COOH is consumed.

moles added NaOH = moles $C_3H_7COO^-$ formed = (0.01500 L)(0.1000 M) = 1.500 x 10^{-3}

moles of C_3H_7COOH remaining = (0.02000 L)(0.1000 M) – 1.500 x 10^{-3} moles = 5.000 x 10^{-4}

$$pH = -\log(1.54x10^{-5}) + \log\left(\frac{1.500x10^{-3}}{5.000x10^{-4}}\right)$$

$pH =$ **5.29**

d) At 19 mL of base added, still more $C_3H_7COO^-$ is formed, but not all of the C_3H_7COOH is consumed.

moles added NaOH = moles $C_3H_7COO^-$ formed = (0.01900 L)(0.1000 M) = 1.900 x 10^{-3}

moles of C_3H_7COOH remaining = (0.02000 L)(0.1000 M) – 1.900 x 10^{-3} moles = 1.000 x 10^{-4}

$$pH = -\log(1.54x10^{-5}) + \log\left(\frac{1.900x10^{-3}}{1.000x10^{-4}}\right)$$

$pH =$ **6.09**

e) At 19.95 mL of base added, still more $C_3H_7COO^-$ is formed, but not all of the C_3H_7COOH is consumed.

moles added NaOH = moles $C_3H_7COO^-$ formed = (0.01995 L)(0.1000 M) = 1.995 x 10^{-3}

moles of C_3H_7COOH remaining = (0.02000 L)(0.1000 M) – 1.995 x 10^{-3} moles = 5.000 x 10^{-6}

$$pH = -\log(1.54x10^{-5}) + \log\left(\frac{1.995x10^{-3}}{5.000x10^{-6}}\right)$$

$pH =$ **7.41**

f) At 20.00 mL of base added, the equivalence point is reached because the concentration of the acid and base are the same (0.1000 M). All of the C_3H_7COOH has been converted to $C_3H_7COO^-$, which reacts with water to form OH^-:

$$C_3H_7COO^- + H_2O \rightleftharpoons C_3H_7COOH + OH^-$$

$$K_b = \frac{[C_3H_7COOH][OH^-]}{[C_3H_7COO^-]} \quad where\ K_b = \frac{1.0x10^{-14}}{1.54x10^{-5}} \quad and\ [C_3H_7COO^-] = \frac{mol\ C_3H_7COOH_{init}}{total\ volume}$$

moles of $C_3H_7COOH_{init}$ = moles of $C_3H_7COO^-$ formed = (0.02000 L)(0.1000 M) = 2.000 x 10^{-3}

total volume = 20.00 mL acid + 20.00 mL base = 40.00 mL = 0.04000 L

$[C_3H_7COO^-]$ = 2.000 x 10^{-3} mol/0.04000 L = 0.05000 M

$$[OH^-] = \sqrt{K_b[C_3H_7COO^-]} = \sqrt{\left(\frac{1.0x10^{-14}}{1.54x10^{-5}}\right)(0.05000)} = 5.69_{80}x10^{-6}$$

pH = 14 – pOH = 14 – log(5.69$_{80}$ x 10^{-6}) = **8.76**.

g) At 20.05 mL, 0.05 mL (5.0 x 10^{-5} L) of excess OH^- have been added.

$[OH^-]$ = mol excess OH^-/total volume = (5.0 x 10^{-5} L)(0.1000 M)/(0.04005 L) = 1.2$_{48}$x10^{-4} M

pH = 14 – pOH = 14 – log(1.2$_{48}$x10^{-4}) = **10.10**.

h) At 25.00 mL, 5.00 mL (5.00 x 10^{-3} L) of excess OH^- have been added.

$[OH^-]$ = mol excess OH^-/total volume = (5.0 x 10^{-3} L)(0.1000 M)/(0.04500 L) = 1.1$_{11}$x10^{-2} M

pH = 14 – pOH = 14 – log(1.1$_{11}$x10^{-2}) = **12.05**.

19.58 At the equivalence point the volume of titrant can be calculated by $M_{acid}V_{acid} = M_{base}V_{base}$. At the equivalence point the solution contains only the conjugate base of the titrated acid. Calculate the pH from the base equilibrium.

a) Volume of NaOH to reach equivalence point:

V_{NaOH} = (42.2 mL CH_3COOH)(0.0520 M CH_3COOH)/(0.0372 M NaOH) = 58.9$_{89}$ mL

K_b for CH_3COO^- = 1 x 10^{-14}/1.8 x 10^{-5} = 5.6 x 10^{-10}

[CH_3COOH] = (42.2 mL x 0.0520 M)/(42.2 mL + 58.9$_{89}$ mL) = 0.0216$_{86}$ M

$$[OH^-] = \sqrt{(5.6 x 10^{-10})(0.0216_{86})} = 3.4_{85} x 10^{-6}$$

pH = -log(1 x 10^{-14}/3.4$_{85}$ x 10^{-6}) = 8.54

To reach the equivalence point requires **59.0 mL** of 0.0372 M NaOH. The pH at the equivalence point is **8.54**.

b) Volume of NaOH to reach first equivalence point:

V_{NaOH} = (18.9 mL H_2SO_3)(0.0890 M H_2SO_3)/(0.0372 M NaOH) = 45.2$_{18}$ mL

The equivalence point pH for the first equivalence point can be calculated as the average of the two pK_a values no matter what the concentrations of acid and base.

pH = (1.85 + 7.19)/2 = 4.52

Volume of NaOH to reach second equivalence point is twice the volume to reach the first. V_{NaOH} = 90.4$_{36}$ mL

K_b for SO_3^{2-} = 1 x 10^{-14}/6.5 x 10^{-8} = 1.5$_{38}$ x 10^{-7}

C_{SO32^-} = (18.9 mL x 0.0890 M)/(18.9 mL + 90.4$_{36}$ mL) = 0.0153$_{85}$ M

$$[OH^-] = \sqrt{(1.5_{38} x 10^{-7})(0.0153_{85})} = 4.8_{64} x 10^{-5} M$$

pH = -log(1 x 10^{-14}/4.8$_{64}$ x 10^{-5}) = 9.69

To reach the first equivalence point requires **45.2 mL** of 0.0372 M NaOH. The pH at the first equivalence point is **4.52**. To reach the second equivalence point requires **90.4 mL** of 0.0372 M NaOH. The pH at the second equivalence point is **9.69**.

19.60 At the equivalence point, the moles of base = moles of acid, so the volume of titrant can be calculated by $M_{acid}V_{acid} = M_{base}V_{base}$.

a) $$V_{acid} = \frac{M_{base}V_{base}}{M_{acid}} = \frac{(0.234\ M)(55.5\ mL)}{0.135\ M} = 96.2\ mL$$

To reach the equivalence point, **96.2 mL** of 0.135 M HCl is needed. At the equivalence point, all of the NH_3 has been converted to NH_4^+, which reacts with water as follows:

$NH_4^+ + H_2O \rightleftharpoons NH_3 + H_3O^+$ where $K_a = K_w/K_b$ = 1.0x10^{-14}/1.76x10^{-5} = 5.68x10^{-10}

$$[H_3O^+] = \sqrt{K_a[NH_4^+]} = \sqrt{(5.68_{18} x 10^{-10})\left(\frac{(55.5\ mL)(0.234\ M)}{(55.5 + 96.2\ mL)}\right)} = 6.97_{44} x 10^{-6}$$

pH = -log(6.97$_{44}$ x 10^{-6}) = **5.16**.

The equivalence point pH for a weak base – strong acid combination should be less than 7.

b) $$V_{acid} = \frac{M_{base}V_{base}}{M_{acid}} = \frac{(1.11\ M)(17.8\ mL)}{0.135\ M} = 146._{36}\ mL$$

To reach the equivalence point, **146 mL** of 0.135 M HCl is needed.

$CH_3NH_3^+ + H_2O \rightleftharpoons CH_3NH_2 + H_3O^+$ where $K_a = K_w/K_b$ = 1.0x10^{-14}/4.4x10^{-4} = 2.27x10^{-11}

$$[H_3O^+] = \sqrt{K_a[CH_3NH_3^+]} = \sqrt{(2.27_{27} x10^{-11})\left(\frac{(17.8\,mL)(1.11\,M)}{(17.8+146\,mL)}\right)} = 1.65_{57} x10^{-6}$$

pH = -log(1.65_{57} x 10^{-6}) = **5.78**.

19.63 Fluoride ion is the conjugate base of a weak acid. The base hydrolysis reaction of fluoride ion: F^- (aq) + H_2O(l) $\rightleftharpoons$ HF(aq) + OH^- (aq) is influenced by the pH of the solution. As the pH increases, the equilibrium shifts to the left to increase the [F^-]. As the pH decreases, the equilibrium shifts to the right to decrease [F^-]. The changes in [F^-] influence the solubility of CaF_2.

Chloride ion is the conjugate base of a strong acid so it does not react with water. Thus, its concentration is not influenced by pH, and solubility of $CaCl_2$ does not change with pH.

19.65 Consider the reaction AB(s) $\rightleftharpoons$ A^+(aq) + B^- (aq), where Q_{sp} = [A^+][B^-]. If $Q_{sp} > K_{sp}$, then there are more ions dissolved than expected at equilibrium, and the equilibrium shifts to the left and the compound AB precipitates.

19.66 a) Ag_2CO_3(s) $\rightleftharpoons$ $2Ag^+$(aq) + CO_3^{2-} (aq)
Ion-product expression: $\boldsymbol{Q_{sp} = [Ag^+]^2[CO_3^{2-}]}$
b) BaF_2(s) $\rightleftharpoons$ Ba^{2+}(aq) + $2F^-$(aq)
Ion-product expression: $\boldsymbol{Q_{sp} = [Ba^{2+}][F^-]^2}$
c) CuS(s) $\rightleftharpoons$ Cu^{2+}(aq) + HS^-(aq) + OH^-(aq)
Ion-product expression: $\boldsymbol{Q_{sp} = [Cu^{2+}][HS^-][OH^-]}$

19.68 a) $CaCrO_4$(s) $\rightleftharpoons$ Ca^{2+}(aq) + CrO_4^{2-}(aq) $\boldsymbol{Q_{sp} = [Ca^{2+}][CrO_4^{2-}]}$
b) AgCN(s) $\rightleftharpoons$ Ag^+(aq) + CN^- (aq) $\boldsymbol{Q_{sp} = [Ag^+][CN^-]}$
c) NiS(s) + H_2O(l) $\rightleftharpoons$ Ni^{2+}(aq) + HS^-(aq) + OH^-(aq) $\boldsymbol{Q_{sp} = [Ni^{2+}][HS^-][OH^-]}$

19.70 Write a reaction table, where S is the molar solubility of Ag_2CO_3:

Conc (M)	Ag_2CO_3(s) $\rightleftharpoons$	$2Ag^+$(aq) +	CO_3^{2-} (aq)
initial	---	0	0
change	---	+2S	+S
equilibrium	---	2S	S

S = [Ag_2CO_3] = 0.032 M so [Ag^+] = 2S = 0.064 M and [CO_3^{2-}] = 0.032 M
K_{sp} = [Ag^+]^2[CO_3^{2-}] = $(0.064)^2(0.032)$ = **1.3 x 10^{-4}**

19.72 The equation and ion-product expression for silver dichromate, $Ag_2Cr_2O_7$, is:
$Ag_2Cr_2O_7$(s) $\rightleftharpoons$ $2Ag^+$(aq) + $Cr_2O_7^{2-}$ (aq) K_{sp} = [Ag^+]^2[$Cr_2O_7^{2-}$]
The solubility of $Ag_2Cr_2O_7$, converted from g/100 mL to M is:

$$Molar\ solubility = \left(\frac{8.3x10^{-3}\ g\ Ag_2Cr_2O_7}{100\,mL}\right)\left(\frac{1000\,mL}{L}\right)\left(\frac{mol\ Ag_2Cr_2O_7}{431.80\ g\ Ag_2Cr_2O_7}\right) = 1.9_{22} x10^{-4}\ M$$

Since 1 mole of $Ag_2Cr_2O_7$ dissociates to form 2 moles of Ag^+, the concentration of Ag^+ is 2(1.9_{22} x 10^{-4} M) = 3.8_{44} x 10^{-4} M. The concentration of $Cr_2O_7^{2-}$ is 1.9_{22} x 10^{-4} M because 1 mole of $Ag_2Cr_2O_7$ dissociates to form 1 mole of $Cr_2O_7^{2-}$.
K_{sp} = [Ag^+]^2[$Cr_2O_7^{2-}$] = (3.8_{44} x 10^{-4})^2(1.9_{22} x 10^{-4} M) = **2.8 x 10^{-11}**.

19.74 a) The solubility, S, in pure water equals $[Sr^{2+}]$ and $[CO_3^{2-}]$.

$K_{sp} = 5.4 \times 10^{-10} = [Sr^{2+}][CO_3^{2-}] = S^2$; $S =$ **2.3×10^{-5} M**

b) In 0.13 M $Sr(NO_3)_2$ the initial concentration of Sr^{2+} is 0.13 M.

Equilibrium $[Sr^{2+}] = 0.13 + S$ and equilibrium $[CO_3^{2-}] = S$ where S is the solubility of $SrCO_3$.

$K_{sp} = 5.4 \times 10^{-10} = [Sr^{2+}][CO_3^{2-}] = S(0.13 + S)$; $S =$ **4.2×10^{-9} M**

19.76 a) Write a reaction table that reflects an initial concentration of $Ca^{2+} = 0.060$ M. In this case, Ca^{2+} is the common ion.

Conc (M)	$Ca(IO_3)_2(s) \rightleftharpoons$	$Ca^{2+}(aq)$ +	$2IO_3^-(aq)$
initial	---	0.060	0
change	---	+S	+2S
equilibrium	---	0.060+S	2S

Assume that $0.060+S \approx 0.060$ because the amount of compound that dissolves will be negligible in comparison to 0.060 M.

From Table 19.2, $K_{sp}(Ca(IO_3)_2) = 7.1 \times 10^{-7}$.

$K_{sp} = [Ca^{2+}][IO_3^-]^2 = (0.060)(2S)^2 = 7.1 \times 10^{-7}$.

$S = 1.7_{20} \times 10^{-3}$

Check assumption: $1.7_{20} \times 10^{-3}$ M/0.060 M x 100% = 2.9% < 5%; assumption is good.

S represents both the molar solubility of Ca^{2+} and $Ca(IO_3)_2$, so the molar solubility of $Ca(IO_3)_2$ is **1.7×10^{-3} M**.

b) In this case the equilibrium concentration of Ca^{2+} is S, and the IO_3^- concentration is 0.060 + 2S. IO_3^- is the common ion in this problem.

Assume that $0.060 + 2S \approx 0.060$.

$K_{sp} = [Ca^{2+}][IO_3^-]^2 = (S)(0.060)^2 = 7.1 \times 10^{-7}$

$S = 1.9_{72} \times 10^{-4}$

Check assumption: $1.9_{72} \times 10^{-4}$ M/0.060 M x 100% = 0.3% < 5%; assumption is good.

S represents both the molar solubility of Ca^{2+} and $Ca(IO_3)_2$, so the molar solubility of $Ca(IO_3)_2$ is **2.0×10^{-4} M**.

19.78 The larger the K_{sp}, the larger the molar solubility if the number of ions are equal.

a) **$Mg(OH)_2$** with $K_{sp} = 6.3 \times 10^{-10}$ has higher molar solubility than $Ni(OH)_2$ with $K_{sp} = 6 \times 10^{-16}$.

b) **PbS** with $K_{sp} = 3 \times 10^{-25}$ has higher molar solubility than CuS with $K_{sp} = 8 \times 10^{-34}$.

c) **Ag_2SO_4** with $K_{sp} = 1.5 \times 10^{-5}$ has higher molar solubility than MgF_2 with $K_{sp} = 7.4 \times 10^{-9}$.

19.80 The larger the K_{sp}, the more water-soluble the compound if the number of ions are equal.

a) **$CaSO_4$** with $K_{sp} = 2.4 \times 10^{-5}$ is more water-soluble than $BaSO_4$ with $K_{sp} = 1.1 \times 10^{-10}$.

b) **$Mg_3(PO_4)_2$** with $K_{sp} = 5.2 \times 10^{-24}$ is more water soluble than $Ca_3(PO_4)_2$ with $K_{sp} = 1.2 \times 10^{-29}$.

c) **$PbSO_4$** with $K_{sp} = 1.6 \times 10^{-8}$ is more water soluble than AgCl with $K_{sp} = 1.8 \times 10^{-10}$.

19.82 a) $AgCl(s) \rightleftharpoons Ag^+(aq) + Cl^-(aq)$

Neither ion will act as an acid or base so **pH will not impact solubility**.

b) $SrCO_3(s) \rightleftharpoons Sr^{2+}(aq) + CO_3^{2-}(aq)$

The carbonate ion will act as a base:

$CO_3^{2-}(aq) + H_2O(l) \rightleftharpoons HCO_3^-(aq) + OH^-(aq)$

Changes in pH will change the $[CO_3^{2-}]$ so the **solubility of $SrCO_3$ will increase with decreasing pH.**

19.84 a) $Fe(OH)_2(s) \rightleftharpoons Fe^{2+}(aq) + 2OH^-(aq)$

The hydroxide ion is the anion of H_2O, a very weak acid, so it reacts with added H_3O^+:

$OH^-(aq) + H_3O^+(aq) \rightarrow 2H_2O(l)$

The added H_3O^+ consumes the OH^-, driving the equilibrium towards the right to dissolve more $Fe(OH)_2$. **Solubility increases with addition of H_3O^+ (decreasing pH).**

b) $CuS(s) + H_2O(l) \rightleftharpoons Cu^{2+}(aq) + HS^-(aq) + OH^-(aq)$

Both HS^- and OH^- are anions of weak acids, so both ions react with added H_3O^+. **Solubility increases with addition of H_3O^+ (decreasing pH).**

19.86 The ion-product expression for $Cu(OH)_2$ is $Q_{sp} = [Cu^{2+}][OH^-]^2$ and K_{sp} equals 2.2×10^{-20}. To decide if a precipitate will form, calculate Q_{sp} with the given quantities and compare it to K_{sp}.

$$[OH^-] = \left(\frac{0.075\,g\,KOH}{1.0\,L}\right)\left(\frac{1\,mol\,KOH}{56.11\,g\,KOH}\right)\left(\frac{1\,mol\,OH^-}{1\,mol\,KOH}\right) = 1.3_{37}\,x\,10^{-3}\ M$$

$[Cu^{2+}] = [Cu(NO_3)_2] = 1.0 \times 10^{-3}$ M

$Q_{sp} = (1.0 \times 10^{-3})(1.3_{37} \times 10^{-3})^2 = 1.8 \times 10^{-9}$

Q_{sp} is greater than K_{sp} ($1.8 \times 10^{-9} > 2.2 \times 10^{-20}$) so **$Cu(OH)_2$ will precipitate.**

19.88 The ion-product expression for $Ba(IO_3)_2$ is $Q_{sp} = [Ba^{2+}][IO_3^-]^2$ and K_{sp} equals 1.5×10^{-9}. To decide if a precipitate will form, calculate Q_{sp} with the given quantities and compare it to K_{sp}.

$$[Ba^{2+}] = \left(\frac{6.5\,mg\,BaCl_2}{500.\,mL\,sol'n}\right)\left(\frac{1000\,mL/L}{1000\,mg/g}\right)\left(\frac{1\,mol\,BaCl_2}{208.2\,g\,BaCl_2}\right)\left(\frac{1\,mol\,Ba^{2+}}{1\,mol\,BaCl_2}\right) = 6.2_{44}\,x10^{-5}\ M$$

$[IO_3^-] = [NaIO_3] = 0.033$ M

$Q_{sp} = [Ba^{2+}][IO_3^-]^2 = (6.2_{44} \times 10^{-5})(0.033)^2 = 6.8_{00} \times 10^{-8}$.

Since $Q_{sp} > K_{sp}$ ($6.8 \times 10^{-8} > 1.5 \times 10^{-9}$), **$Ba(IO_3)_2$ will precipitate.**

19.93 In the context of this equilibrium only, the increased solubility with added OH^- appears to be a violation of Le Châtelier's Principle. Before accepting this conclusion other possible equilibria must be considered. Lead is a metal ion and hydroxide ion is a ligand, so it is possible that a complex ion forms between the lead ion and hydroxide ion:

$$Pb^{2+}(aq) + nOH^-(aq) \rightleftharpoons Pb(OH)_n^{2-n}(aq)$$

This decreases the concentration of Pb^{2+} shifting the solubility equilibrium to the right to dissolve more PbS.

19.94 In many cases, a hydrated metal complex (e.g. $Hg(H_2O)_4^{2+}$) will exchange ligands when placed in a solution of another ligand (e.g. CN^-),

$Hg(H_2O)_4^{2+}(aq) + 4CN^-(aq) \rightleftharpoons Hg(CN)_4^{2-}(aq) + 4H_2O(l)$

Note that both sides of the equation have the same "overall" charge of –2. The chromium complex changes from +2 to –2 because water is a neutral *molecular* ligand whereas cyanide is an *ionic* ligand.

19.96 The two water ligands are replaced by two thiosulfate ion ligands. The +1 charge from the silver ion plus –4 charge from the two thiosulfate ions gives a net charge on the complex ion of –3.

$$Ag(H_2O)_2^+(aq) + 2S_2O_3^{2-}(aq) \rightleftharpoons Ag(S_2O_3)_2^{3-}(aq) + 2H_2O(l)$$

19.98 The reaction between SCN^- and Fe^{3+} produces the red complex $FeSCN^{2+}$. One can assume from the much larger concentration of SCN^- and large K_f that all of the Fe^{3+} ions react to form the complex. Calculate the initial concentrations of SCN^- and Fe^{3+} and write a reaction table in which x is the concentration of $FeSCN^{2+}$ formed.

$$[Fe^{3+}]_{init} = \frac{(0.0015\ M\ Fe^{3+})(0.50\ L)}{(0.50 + 0.50\ L)} = 7.5x10^{-4}\ M$$

$$[SCN^-]_{init} = \frac{(0.20\ M\ SCN^-)(0.50\ L)}{(0.50 + 0.50\ L)} = 0.10\ M$$

Conc(M)	$Fe^{3+}(aq)$ +	$SCN^-(aq)$	$\rightleftharpoons$	$FeSCN^{2+}$
Initial	7.5×10^{-4}	0.10		0
Change	-x	-x		+x
Equilibrium	$7.5x10^{-4} - x$	0.10 – x		x

It is reasonable to assume that x is much less than 0.10, so 0.10 – x ≈ 0.10. However, it is not reasonable to assume that 0.00075 – x ≈ 0.00075, because x may be significant in relation to such a small number.

$$K_f = \frac{[FeSCN^{2+}]}{[Fe^{3+}][SCN^-]} = \frac{x}{(7.5x10^{-4} - x)(0.10)} = 8.9x10^2$$

$$x = 6.6_{75}x10^{-2} - 89x$$

$$x = 7.4_{17}x10^{-4}$$

From the reaction table, $[Fe^{3+}]_{eq} = 7.5x10^{-4} - x$. Therefore, $[Fe^{3+}]_{eq} = 7.5x10^{-4} - 7.4_{17} \times 10^{-4}$ **= 1 x 10^{-5} M**.

19.100 The complex formation equilibrium is

$$Zn^{2+}(aq) + 4CN^-(aq) \rightleftharpoons Zn(CN)_4^{2-}(aq)$$

First calculate the initial concentration of Zn^{2+} and set up reaction table assuming that the reaction first goes to completion and then calculate back to find reactant concentrations.

$$[Zn^{2+}] = \left(\frac{0.82\ g\ ZnCl_2}{0.255\ L}\right)\left(\frac{1\ mol\ ZnCl_2}{136.29\ g\ ZnCl_2}\right) = 0.023_{59}\ M$$

Conc (M)	$Zn^{2+}(aq)$ +	$4CN^-(aq)$	$\rightleftharpoons$	$Zn(CN)_4^{2-}(aq)$
initial	0.023_{59}	0.150		0
change	$-0.023_{59}+x$	$-4(0.023_{59})$		$+0.023_{59}$
equilibrium	x	0.055_{64}		0.023_{59}

$$K_f = 4.2 \, x \, 10^{19} = \frac{[Zn(CN)_4^{\ 2-}]}{[Zn^{2+}][CN^-]^4} = \frac{0.023_{59}}{x(0.055_{64})^4}; \; x = 5.8_{60} \, x \, 10^{-17} \; M$$

$[Zn^{2+}]$ = **5.9 x 10^{-17} M**

$[Zn(CN)_4^{2-}]$ = **0.024 M**

$[CN^-]$ = **0.056 M**

19.102 Write the ion-product equilibrium reaction and the complex-ion equilibrium reaction. Sum the two reactions to yield an overall reaction; multiply the two constants to obtain $K_{overall}$. Write a reaction table where S = $[AgI]_{dissolved}$ = $[Ag(NH_3)_2^+]$.

Ion-product: $AgI(s) \rightleftharpoons Ag^+(aq) + I^-(aq)$

Complex-ion: $Ag^+(aq) + 2NH_3(aq) \rightleftharpoons Ag(NH_3)_2^+(aq)$

Overall: $AgI(s) + 2NH_3(aq) \rightleftharpoons Ag(NH_3)_2^+(aq) + I^-(aq)$

$K_{overall} = K_{sp} \times K_f = (8.3 \times 10^{-17})(1.7 \times 10^7) = 1.4_{11} \times 10^{-9}$

Reaction table:

Conc (M)	AgI(s) +	$2NH_3(aq)$	$\rightleftharpoons$	$Ag(NH_3)_2^+(aq)$ +	$I^-(aq)$
initial	------	2.5		0	0
change	------	-2S		+S	+S
equilibrium	------	2.5 – 2S		S	S

Assume that 2.5 – 2S ≈ 2.5 because $K_{overall}$ is so small.

$$K_{overall} = \frac{[Ag(NH_3)_2^{\ +}][I^-]}{[NH_3]^2} = \frac{S^2}{(2.5)^2} = 1.4_{11} x10^{-9} \Rightarrow S = [AgI] = \mathbf{9.4x10^{-5}\ M}$$

19.104 a) $Fe(OH)_3$ will precipitate first because its K_{sp} (1.6×10^{-39}) is smaller than the K_{sp} for $Cd(OH)_2$ at 8.1×10^{-15}.

b) The two ions are separated by adding just enough NaOH to precipitate the iron(III) hydroxide, but precipitating no more than 0.01% of the cadmium. The Fe^{3+} is found in the solid precipitate while the Cd^{2+} remains in the solution.

c) Precipitating 0.01% of the Cd^{2+} would not significantly change $[Cd^{2+}]$. Find the diluted $[Cd^{2+}]$ and calculate the $[OH^-]$ that would be present just before precipitation begins. This occurs when $Q_{sp} = K_{sp}$.

$[Cd^{2+}]$ = 0.25 M (125 mL/175 mL) = 0.17_{86} M

$K_{sp} = 8.1 \times 10^{-15} = [Cd^{2+}][OH^-]^2 = (0.17_{86})[OH^-]^2$

$[OH^-]$ = **2.1 x 10^{-7} M**

19.106 a) All ammonium salts are soluble, whereas some nickel salts are insoluble. Choose a reagent that forms an insoluble salt with Ni: examples include Na_2CO_3 and Na_3PO_4. Ni^{2+} will precipitate whereas NH_4^+ remains in solution. An alternative is to add a hydroxide salt. A precipitate of nickel hydroxide will form in the $NiCl_2$ solution in contrast to the release of ammonia gas (detect odor) from the reaction of the ammonium ion with the hydroxide in the NH_4Cl solution.

b) Both Ag^+ and Pb^{2+} form insoluble chlorides, however $PbCl_2$ is slightly soluble in water. Adding hot water dissolves the $PbCl_2$ but not the AgCl.

c) Addition of NaOH forms a red-brown Fe^{3+} precipitate ($Fe(OH)_3(s)$) and a white Al^{3+} precipitate ($Al(OH)_3(s)$). Further addition of OH^- forms the soluble Al^{3+} complex, $Al(OH)_4^-$, whereas the Fe^{3+} does not react further.

19.108 A formate buffer contains formate ($HCOO^-$) as the base and formic acid (HCOOH) as the acid. K_a for formic acid is 1.8×10^{-4}.

a) The buffer component ratio, $[HCOO^-]/[HCOOH]$, can be calculated from the equilibrium expression.

$$\frac{[HCOO^-]}{[HCOOH]} = \frac{K_a}{[H_3O^+]} = \frac{1.8 \times 10^{-4}}{1.8_{20} \times 10^{-4}} = \mathbf{0.99}$$

b) To prepare solutions set up equations for concentrations of formate and formic acid with x equal to the volume, in mL, of 1.0 M HCOOH added. The equations are based on the neutralization reaction between HCOOH and NaOH that produces $HCOO^-$.

$$HCOOH(aq) + NaOH(aq) \rightarrow HCOO^-(aq) + Na^+(aq) + H_2O(l)$$

$$[HCOO^-] = (1.0\ M\ NaOH)\left(\frac{600 - x\ mL\ 1.0\ M\ NaOH}{600\ mL\ sol'n}\right)\left(\frac{1\ mol\ HCOO^-}{1\ mol\ NaOH}\right)$$

$$[HCOOH] = (1.0\ M\ HCOOH)\left(\frac{x\ mL\ 1.0\ M\ HCOOH}{600\ mL\ sol'n}\right)$$

$$-(1.0\ M\ NaOH)\left(\frac{600 - x\ mL\ 1.0\ M\ NaOH}{600\ mL\ sol'n}\right)\left(\frac{1\ mol\ HCOO^-}{1\ mol\ NaOH}\right)$$

The component ratio equals 0.99 (from part a). Simplify the above equations and plug into ratio:

$$\frac{[HCOO^-]}{[HCOOH]} = \frac{\left(600 - x/600\ M\ HCOO^-\right)}{\left(x - (600 - x)/600\ M\ HCOOH\right)}$$

Solve for x:

$$0.99 = \frac{600 - x}{2x - 600};\ x = 400$$

Mixing **400 mL of 1.0 M HCOOH** and 600 – 400 = **200 mL of 1.0 M NaOH** gives a buffer of pH 3.74.

c) Final concentration of HCOOH from equation in part b:

$$[HCOOH] = (1.0\ M\ HCOOH)\left(\frac{400\ mL\ 1.0\ M\ HCOOH}{600\ mL\ sol'n}\right)$$

$$-(1.0\ M\ NaOH)\left(\frac{200\ mL\ 1.0\ M\ NaOH}{600\ mL\ sol'n}\right)\left(\frac{1\ mol\ HCOO^-}{1\ mol\ NaOH}\right) = \mathbf{0.33\ M}$$

19.110 The minimum urate ion concentration necessary to cause a deposit of sodium urate is determined by the K_{sp} for the salt. Convert solubility in g/100. mL to molar solubility and calculate K_{sp}. Substitute $[Na^+]$ and K_{sp} into the ion-product expression to find $[Ur^-]$.

Molar solubility of NaUr:

$$[NaUr] = \left(\frac{0.085\ g\ NaUr}{100.\ mL}\right)\left(\frac{1000\ mL}{1\ L}\right)\left(\frac{1\ mol\ NaUr}{190.10\ g\ NaUr}\right) = 4.4_{71} x 10^{-3}\ M = [Na^+] = [Ur^-]$$

$K_{sp} = [Na^+][Ur^-] = (4.4_{71} \times 10^{-3})(4.4_{71} \times 10^{-3}) = 1.9_{99} \times 10^{-5}$

When $[Na^+] = 0.15$ M, $[Ur^-] = K_{sp}/[Na^+] = (1.9_{99} \times 10^{-5})(0.15) = 1.3 \times 10^{-4}$.
The minimum urate ion concentration that will cause precipitation of sodium urate is **1.3 x 10⁻⁴ M**.

19.113 The buffer is made by starting with phosphoric acid and neutralizing some of the acid by adding sodium hydroxide:

$$H_3PO_4(aq) + OH^-(aq) \rightarrow H_2PO_4^-(aq) + H_2O(l)$$

Present initially is (0.50 L)(1.0 M H_3PO_4) = 0.50 mol H_3PO_4. Adding 0.80 mol NaOH converts all the phosphoric acid to dihydrogen phosphate ions (0.50 mol) and 0.30 mol NaOH are left. The remaining OH^- will react with the dihydrogen phosphate:

$$H_2PO_4^-(aq) + OH^-(aq) \rightarrow HPO_4^{2-}(aq) + H_2O(l)$$

The 0.50 mol $H_2PO_4^-$ reacts with the 0.30 mol OH^- to produce 0.30 mol HPO_4^{2-}. 0.20 mol $H_2PO_4^-$ will remain.

The pH is determined from the equilibrium involving the conjugate pair $H_2PO_4^-/HPO_4^{2-}$.

$$H_2PO_4^-(aq) + H_2O(l) \rightleftharpoons HPO_4^{2-}(aq) + H_3O^+(aq) \quad K_a = 6.3 \times 10^{-8}$$

$$pH = -\log(6.3x10^{-8}) + \log\left(\frac{0.30\,mol\,HPO_4^{2-}/0.50\,L}{0.20\,mol\,H_2PO_4^-/0.50\,L}\right) = \mathbf{7.38}$$

19.114 Step 1: Of the 5 cations listed, only Fe^{3+} forms a brown precipitate in the presence of $NH_3(aq)$. Al^{3+} is also insoluble in the presence of NH_3, but its precipitate is white. Because there is no white precipitate, Al^{3+} is not present.

Step 2: This step separates the brown precipitate from the solution. See step 5.

Step 3: The $Fe(OH)_3$ precipitate formed in step 1 is not amphoteric, so it does not react with OH^-. Since the solid remains, the addition of OH^- confirms this is $Fe(OH)_3$.

Step 4: The addition of NH_4Cl serves to add the acid NH_4^+ to neutralize excess OH^-. That reaction ($NH_4^+ + OH^- \rightarrow NH_3 + H_2O$) releases NH_3 to serve as a ligand to other metals. No mention is made of a blue solution, so Cu^{2+} is not present.

Step 5: The formation of white precipitate is likely AgCl. Potassium chloride, KCl, is soluble and Cu^{2+} forms a soluble green complex. Since no mention is made of a green solution, the presence of Cu^{2+} can be ruled out.

Step 6: To this point, the only unaccounted ion is K^+. Since potassium gives a distinct violet color in the flame test, and a pale blue color is observed, no K^+ is present.

Step 1 & 5 indicate that **Fe^{3+}** and **Ag^+** are present in the solution. The other steps rule out the presence of the other cations.

19.117 a) Use the equilibrium expression (or Henderson-Hasselbalch) and convert pH of 7.40 to $[H_3O^+]$.

$$\frac{[H_2CO_3]}{[HCO_3^-]} = \frac{[H_3O^+]}{K_a} = \frac{10^{-7.40}}{4.5x10^{-7}} = \mathbf{8.8x10^{-2}}$$

b) $$\frac{[H_2CO_3]}{[HCO_3^-]} = \frac{[H_3O^+]}{K_a} = \frac{10^{-7.20}}{4.5x10^{-7}} = \mathbf{0.14}$$

19.119 The buffer components will be TRIS, $(HOCH_2)_3CNH_2$, and its conjugate acid, $(HOCH_2)_3CNH_3^+$. The conjugate acid is formed from the reaction between TRIS and HCl. Since HCl is the limiting reactant in this reaction the concentration of conjugate acid will equal the starting concentration of HCl, 0.095 M. The concentration of TRIS is the initial concentration minus the amount reacted.

$$[TRIS] = \left(\frac{43.0\ g\ TRIS}{1.00\ L}\right)\left(\frac{1\ mol\ TRIS}{121.14\ g}\right) - 0.095\ M = \mathbf{0.260\ M}$$

$$pOH = pK_b + \log\frac{[acid]}{[base]} = 5.91 + \log\frac{0.095\ M}{0.260\ M} = 5.47$$

So, pH of the buffer is 14 - 5.47 = **8.53**

19.121 Zinc sulfide, ZnS, is much less soluble than manganese sulfide, MnS. Convert $ZnCl_2$ and $MnCl_2$ to ZnS and MnS by saturating the solution with H_2S; $[H_2S]_{sat'd}$ = 0.10 M. Adjust the pH so that the greatest amount of ZnS will precipitate and not exceed the solubility of MnS as determined by K_{sp}(MnS).

$K_{sp}(MnS) = [Mn^{2+}][HS^-][OH^-] = 3 \times 10^{-11}$

$[Mn^{2+}] = [MnCl_2] = 0.020$ M

$[HS^-]$ is calculated using the K_{a1} expression:

$$K_{a1} = \frac{[HS^-][H_3O^+]}{[H_2S]} = \frac{[HS^-][H_3O^+]}{0.10 - x} \approx \frac{[HS^-][H_3O^+]}{0.10} = 9x10^{-8}$$

Therefore, $[HS^-] = 9 \times 10^{-9}/[H_3O^+]$.

Substituting $[Mn^{2+}]$ and $[HS^-]$ into the K_{sp}(MnS) above gives:

$$K_{sp}(MnS) = (0.020)\left(\frac{9x10^{-9}}{[H_3O^+]}\right)[OH^-] = 3x10^{-11}$$

Substituting $K_w/[H_3O^+]$ for $[OH^-]$:

$$K_{sp}(MnS) = (0.020)\left(\frac{9x10^{-9}}{[H_3O^+]}\right)\left(\frac{K_W}{[H_3O^+]}\right) = 3x10^{-11}$$

$$[H_3O^+] = \sqrt{\frac{(0.020)(9x10^{-9})(1x10^{-14})}{(3x10^{-11})}} = 2._{45}\,x10^{-7}\ M$$

$$pH = -\log(2._{45}\,x10^{-7}) = \mathbf{6.6}$$

Maintain the pH below 6.6 to separate the ions as their sulfides.

19.124 An indicator changes color when the buffer component ratio of the two forms of the indicator changes from a value greater than 1 to a value less than one. The pH at which the ratio equals 1 is equal to pK_a. The midpoint in the pH range of the indicator is a good estimate of the pK_a of the indicator.

$pK_a = (3.4 + 4.8)/2 = 4.1$ $\qquad$ $K_a = 10^{-4.1} = \mathbf{8 \times 10^{-5}}$

19.127 a) Use a spreadsheet to quickly calculate ΔpH/ΔV and average volume for each data point. At the equivalence point, the pH changes drastically when only a small amount of base is added, therefore ΔpH/ΔV is at a maximum at the equivalence point.

V(mL)	pH	ΔpH/ΔV	$V_{average}$
0.00	1.00		
10.00	1.22	0.022	5.00
20.00	1.48	0.026	15.00
30.00	1.85	0.037	25.00
35.00	2.18	0.066	32.50
39.00	2.89	0.178	37.00
39.50	3.20	0.620	39.25
39.75	3.50	1.20	39.63
39.90	3.90	2.67	39.83
39.95	4.20	6.00	39.93
39.99	4.90	17.50	39.97
40.00	7.00	210	40.00
40.01	9.40	240	40.01
40.05	9.80	10.00	40.03
40.10	10.40	12.00	40.08
40.25	10.50	0.67	40.18
40.50	10.79	1.16	40.38
41.00	11.09	0.600	40.75
45.00	11.76	0.168	43.00
50.00	12.05	0.058	47.50
60.00	12.30	0.025	55.00
70.00	12.43	0.013	65.00
80.00	12.52	0.009	75.00

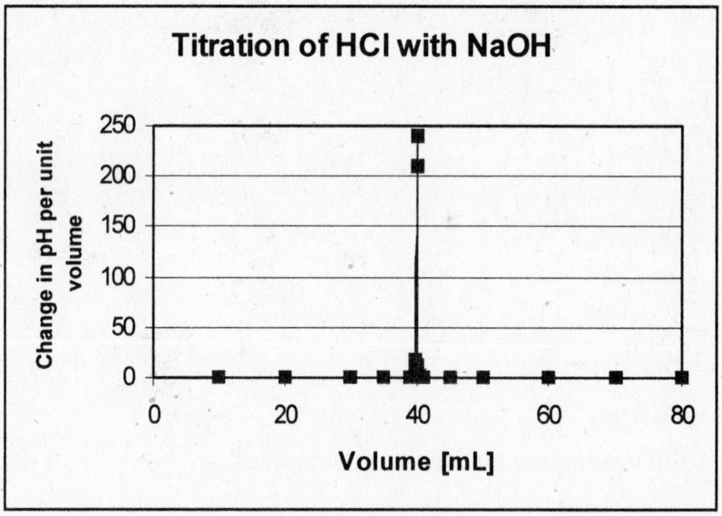

b) The maximum slope (equivalence point) occurs between V = 40.00 mL and V = 40.01 mL.

19.132 The equation that describes the behavior of a weak base in water is:

$B(aq) + H_2O(l) \rightleftharpoons BH^+(aq) + OH^-(aq)$

$$K_b = \frac{[BH^+][OH^-]}{[B]}$$

$$-\log K_b = -\log\left(\frac{[BH^+][OH^-]}{[B]}\right)$$

$$-\log K_b = -\log\left(\frac{[BH^+]}{[B]}\right) - \log[OH^-]$$

$$pK_b = -\log\left(\frac{[BH^+]}{[B]}\right) + pOH$$

$$pOH = pK_b + \log\left(\frac{[BH^+]}{[B]}\right)$$

19.136 To determine which species are present from a buffer system of a polyprotic acid, check the pK_a values for the one that is closest to the pH of the buffer. The two components involved in the equilibrium associated with this K_a are the principle species in the buffer. For carbonic acid pK_{a1} is closest to the pH of 7.4 so H_2CO_3 and HCO_3^- are the species present. For phosphoric acid pK_{a2} is closest to the pH so $H_2PO_4^-$ and HPO_4^{2-} are the principle species present.

$$\frac{[HPO_4^{2-}]}{[H_2PO_4^-]} = \frac{K_a}{[H_3O^+]} = \frac{2.3 \times 10^{-7}}{10^{-7.4}} = 5.8$$

The ratio of HPO_4^{2-} to $H_2PO_4^-$ is **5.8**.

19.138 Refer to Figure 19.4. Bromphenol blue is the best indicator as it is green in a fairly narrow range of 3.5 < pH < 4. The results from thymol blue (turns yellow at pH > 2.5) and methyl red (turns red at pH < 4.3) agree with the results obtained from the bromphenol blue. Therefore, a reasonable estimate for the rainwater pH is **3.5 to 4**.

19.141 a) To find the concentration of HCl after neutralizing the quinidine, calculate the concentration of quinidine and the amount of HCl required to neutralize it, remembering that the mole ratio for the neutralization is 2 mol HCl/1 mol quinidine.

$$mol\ quinidine = (33.85\ mg\ quinidine)\left(\frac{1\ g}{1000\ mg}\right)\left(\frac{1\ mol\ quinidine}{324.41\ g\ quinidine}\right) = 1.043_{43} \times 10^{-4}\ mol$$

$$mol\ HCl_{excess} = (6.55\ mL)\left(\frac{1\ L}{1000\ mL}\right)\left(\frac{0.150\ mol\ HCl}{L}\right)$$

$$-(1.043_{43} \times 10^{-4}\ mol\ quinidine)\left(\frac{2\ mol\ HCl}{1\ mol\ quinidine}\right) = 7.73_{81} \times 10^{-4}\ mol$$

$$V_{NaOH} = (7.73_{81} \times 10^{-4}\ mol\ HCl)\left(\frac{1\ mol\ NaOH}{1\ mol\ HCl}\right)\left(\frac{1\ L}{0.0133\ mol\ NaOH}\right)\left(\frac{1000\ mL}{L}\right) = \mathbf{58.2\ mL}$$

b) $$V_{NaOH} = (1.043_{43} \times 10^{-4}\ mol\ quin)\left(\frac{1\ mol\ NaOH}{1\ mol\ quin}\right)\left(\frac{1\ L}{0.0133\ mol\ NaOH}\right)\left(\frac{1000\ mL}{L}\right) = \mathbf{7.84\ mL}$$

c) When quinidine (Q) is first acidified, it has the general form QNH^+NH^+. At the first equivalence point, one of the acidified nitrogen atoms has completely reacted, leaving a single protonated form, $QNNH^+$. This form of quinidine can react with water as either an acid or a base so both must be considered. If the concentration of quinidine at the first equivalence point is greater than K_{b1} then the $[OH^-]$ at the first equivalence point can be estimated as $[OH^-] \approx \sqrt{(K_{b1} K_{b2})}$.

$$[OH^-] = \sqrt{K_{b1} K_{b2}} = \sqrt{(4.0 \times 10^{-6})(1.0 \times 10^{-10})} = 2.0 \times 10^{-8}\ M$$

pH = -log(1 x 10^{-14}/2.0 x 10^{-8}) = **6.30**

19.143 The Henderson-Hasselbalch equation demonstrates that pH changes when the ratio of acid to base in the buffer changes (pK_a is constant at a given temperature):

pH = pK_a + log([base]/[acid])

The pH of the A^-/HA buffer cannot be calculated because the identity of "A" and, thus, the value of pK_a are unknown. But the change in pH can be described:

ΔpH = log([base]/[acid])$_{final}$ - log([base]/[acid])$_{initial}$

In the given case log([base]/[acid])$_{initial}$ = 0 because [base] = [acid] so the change in pH is equal to the concentration ratio of base to acid after the addition of H_3O^+.

Consider the buffer prior to addition to the medium. When 0.0010 mol H_3O^+ is added to 1 L of the undiluted buffer, the $[A^-]$/[HA] ratio changes from 0.10/0.10 to (0.10 – 0.0010)/(0.10 + 0.0010). The change in pH is: ΔpH = log(0.099/0.101) = -0.009. If the undiluted buffer changes 0.009 pH units with addition of H_3O^+, how much can the buffer be diluted and still not change by 0.05 pH units (ΔpH < 0.05)?

Let x = fraction by which the buffer can be diluted. Assume 0.0010 mol H_3O^+ is added to 1L.

$$\log\frac{[base]}{[acid]} = \log\frac{(0.10x - 0.0010)}{(0.10x + 0.0010)} = 0.05$$

$$\frac{(0.10x - 0.0010)}{(0.10x + 0.0010)} = 10^{-0.05} = 0.89_{13}$$

$$0.10x - 0.0010 = 0.089_{13}x + 8.9_{13}x10^{-4}$$

$$0.010_{87}x = 0.0018_{91}$$

$$x = 0.17_4$$

The buffer concentration can be decreased by a factor of 0.17, or **170 mL** of buffer can be diluted to 1 L of medium. At least this amount should be used to adequately buffer the pH change.

19.146 To find the volume of rain, first convert the inches to yards and find the volume in yd^3. Then convert units to cm^3 and on to L.

$$(10.0\,acres)\left(\frac{4.840 \times 10^3\ yd^2}{acre}\right)\left(\frac{36\,in}{yd}\right)^2(1.00\,in\,rain) = 6.27_{26} \times 10^7\ in^3$$

$$(6.27_{26} \times 10^3\ in^3)\left(\frac{2.54\,cm}{in}\right)^3\left(\frac{1\,mL}{1\,cm^3}\right)\left(\frac{1\,L}{1000\,mL}\right) = 1.02_{79} \times 10^6\ L$$

a) At pH = 4.20, $[H_3O^+] = 10^{-4.20} = 6.3_{10} \times 10^{-5}$ M

mol H_3O^+ = (6.3_{10} x 10^{-5} M)(1.02_{79} x 10^6 L) = **65 mol**

b) $Volume_{lake} = (10.0\,acres)\left(\frac{4.840\,x\,10^3\,yd^2}{acre}\right)\left(\frac{3\,ft}{yd}\right)^2(10.0\,ft)\left(\frac{12\,in}{ft}\right)^3 = 2.25_{82}\,x\,10^{10}\,in^3$

$$Volume_{lake} = (2.25_{82}\,x\,10^{10}\,in^3)\left(\frac{2.54\,cm}{in}\right)^3\left(\frac{1\,mL}{1\,cm^3}\right)\left(\frac{1\,L}{1000\,mL}\right) = 1.23_{35}\,x\,10^8\,L$$

Total volume of lake after rain = 1.23_{35} x 10^8 L + 1.02_{79} x 10^6 L = 1.24_{38} x 10^8 L

$[H_3O^+]$ = $64._{86}$ mol H_3O^+/1.24_{38} x 10^8 L = 5.2_{15} x 10^{-7} M

pH = -log(5.2_{70} x 10^{-7}) = **6.28**

c) Each mol of H_3O^+ requires one mole of HCO_3^- for neutralization.

$$(64._{86}\,mol\,H_3O^+)\left(\frac{1\,mol\,HCO_3^-}{1\,mol\,H_3O^+}\right)\left(\frac{61.02\,g\,HCO_3^-}{mol\,HCO_3^-}\right) = \mathbf{4.0 x 10^3\,g}$$

19.148 Carbon dioxide dissolves in water to produce H_3O^+ ions:

$CO_2(g) \rightleftharpoons CO_2(aq)$

$CO_2(aq) + H_2O(l) \rightleftharpoons H_2CO_3(aq)$

$H_2CO_3(aq) \rightleftharpoons H_3O^+(aq) + HCO_3^-\,(aq)$

$$K_{a1} = \frac{[H_3O^+][HCO_3^-]}{[H_2CO_3]} = \frac{[H_3O^+][HCO_3^-]}{[CO_2]} = 4.5x10^{-7}$$

Let x = $[H_3O^+]$ = $[HCO_3^-]$

The molar concentration of CO_2, $[CO_2]$, depends on how much $CO_2(g)$ from the atmosphere can dissolve in pure water. At 25°C and 1 atm CO_2, 88 mL of CO_2 can dissolve in 100 mL of H_2O. The number of moles of CO_2 in 88 mL of CO_2 is:

$$n_{CO_2} = \frac{PV}{RT} = \frac{(1\,atm)(0.088\,L)}{(0.08206\,L \cdot atm/mol \cdot K)(298\,K)} = 3.5_{99}\,x10^{-3}\,mol$$

Since air is not pure CO_2, account for the volume fraction of air (0.033 L/100 L) when determining the molar concentration:

$$[CO_2] = \left(\frac{3.5_{99}\,x10^{-3}\,mol}{0.100\,L}\right)\left(\frac{0.033}{100}\right) = 1.1_{88}\,x10^{-5}\,M$$

$$\frac{x^2}{(1.1_{88}\,x10^{-5} - x)} = 4.5x10^{-7}$$

$$x^2 + 4.5x10^{-7}x - 5.3_{44}\,x10^{-12} = 0$$

$$x = 2.0_{98}\,x10^{-6} = [H_3O^+] \Rightarrow pH = \mathbf{5.68}$$

19.150 Initial concentrations of Pb^{2+} and $Ca(EDTA)^{2-}$ before reaction based on mixing 100. mL of 0.10 M $Na_2Ca(EDTA)$ with 1.5 L blood:

$$[Pb^{2+}] = \left(\frac{120\,\mu g\,Pb^{2+}}{0.100\,L}\right)\left(\frac{1.5\,L\,blood}{1.6\,L\,mixture}\right)\left(\frac{1\,g}{1x10^6\,\mu g}\right)\left(\frac{1\,mol\,Pb^{2+}}{207.2\,g\,Pb^{2+}}\right) = 5.4_{30}\,x\,10^{-6}\,M$$

$$[Ca(EDTA)^{2-}] = \frac{(0.10\,M\,Ca(EDTA)^{2-})(0.100\,L)}{1.6\,L} = 6.2_{50}\,x\,10^{-3}\,M$$

Set up a reaction table

Conc(M)	$[Ca(EDTA)]^{2-}(aq)$	+ $Pb^{2+}(aq)$ ⇌	$[Pb(EDTA)]^{2-}(aq)$	+ $Ca^{2+}(aq)$
initial	$6.2_{50} \times 10^{-3}$	$5.4_{30} \times 10^{-6}$	0	0
change	$-5.4_{30} \times 10^{-6}$	$-5.4_{30} \times 10^{-6}$	$+5.4_{30} \times 10^{-6}$	$+5.4_{30} \times 10^{-6}$
equilibrium	$6.2_{45} \times 10^{-3}$	x	$5.4_{30} \times 10^{-6}$	$5.4_{30} \times 10^{-6}$

$$K_c = 2.5 x 10^7 = \frac{[Pb(EDTA)^{2-}][Ca^{2+}]}{[Ca(EDTA)^{2-}][Pb^{2+}]} = \frac{(5.4_{30} x 10^{-6})^2}{(6.2_{45} x 10^{-3})(x)}$$

x = $[Pb^{2+}] = 1.8_{89} \times 10^{-16}$ M

Convert concentration from M to μg in 100 mL:

$$\left(\frac{1.8_{89} x 10^{-16} \ mol \ Pb^{2+}}{L}\right)\left(\frac{1 \ L}{1000 \ mL}\right)(100 \ mL)\left(\frac{207.2 \ g \ Pb^{2+}}{mol \ Pb^{2+}}\right)\left(\frac{1 x 10^6 \ \mu g}{g}\right) = 3.9 x 10^{-9} \ \mu g \ Pb^{2+}$$

The final concentration is **3.9 x 10^{-9} μg/100 mL**.

CHAPTER 20

THERMODYNAMICS: ENTROPY, FREE ENERGY, AND THE DIRECTION OF CHEMICAL REACTIONS

FOLLOW-UP PROBLEMS

20.1 a) **PCl_5(g)**. For substances with the same type of atoms and in the same physical state, entropy increases with increasing number of atoms per molecule because more types of molecular motion are available (Rule 5).

b) **$BaCl_2$(s)**. Entropy increases with increasing atomic size (Rule 5). The Ba^{2+} ion and Cl^- ion are larger than the Ca^{2+} ion and F^- ion, respectively.

c) **Br_2(g)**. Entropy increases from solid → liquid → gas.

20.2 Plan: Predict the sign of ΔS^0_{sys} by comparing the randomness of the products with the randomness of the reactants. Calculate ΔS^0_{sys} using Appendix B values.

Solution:

a) $2NaOH(s) + CO_2(g) \rightarrow Na_2CO_3(s) + H_2O(l)$

The ΔS^0_{sys} is predicted to decrease ($\Delta S^0_{sys} < 0$) because the more random, gaseous reactant is transformed into a more ordered, liquid product.

$\Delta S^0_{sys} = \Sigma(mS^0_{products}) - \Sigma(nS^0_{reactants})$

$\Delta S^0_{sys} = (S^0_{Na_2CO_3(s)} + S^0_{H_2O(l)}) - (2S^0_{NaOH(s)} + S^0_{CO_2(g)})$

ΔS^0_{sys} = {(1mol Na_2CO_3)(139 J/mol·K) + (1mol H_2O)(69.940 J/mol·K)} – {(2mol NaOH)(64.454 J/mol·K) + (1mol CO_2)(213.7 J/mol·K)} = **- 134 J/K**

b) $2Fe(s) + 3H_2O(g) \rightarrow Fe_2O_3(s) + 3H_2(g)$

The change in gaseous moles is zero, so the sign of ΔS^0_{sys} is difficult to predict. Iron(III) oxide has greater entropy than Fe because it is more complex, but this is offset by the greater molecular complexity of H_2O vs. H_2.

$\Delta S^0_{sys} = (S^0_{Fe_2O_3} + 3S^0_{H_2}) - (2S^0_{Fe} + 3S^0_{H_2O})$

ΔS^0_{sys} = {(1mol Fe_2O_3)(87.400 J/mol·K) + (3mol H_2)(130.6 J/mol·K)} – {(2mol Fe)(27.3 J/mol·K) + (3mol H_2O)(188.72 J/mol·K)} = **-141.6 J/K**

Check: $\Delta S^0_{sys} < 0$ as predicted for equation (a). The negative ΔS^0_{sys} for equation (b) reflects that greater entropy of H_2O vs. H_2 does outweigh the greater entropy of Fe_2O_3 vs. Fe.

20.3 Plan: Write the balanced equation for the reaction and calculate the ΔS^0_{sys} using Appendix B. Determine the ΔS^0_{surr} by first finding ΔH^0_{rxn}. Add ΔS^0_{surr} to ΔS^0_{sys} to verify that ΔS_{univ} is positive.

Solution:

$2FeO(s) + ½O_2(g) \rightarrow Fe_2O_3(s)$

ΔS^0_{sys} = {(1mol Fe_2O_3)(87.400 J/mol·K)} – {(2mol FeO)(60.75 J/mol·K) + (½ mol O_2)(205.0 J/mol·K)}

ΔS^0_{sys} = –136.6 J/K (entropy change is expected to be negative because Δn_{gas} = -1)

ΔH^0_{sys} = {(1mol Fe_2O_3)(-825.5 kJ/mol)} – {(2mol FeO)(-272.0 kJ/mol) + (½mol O_2)(0)}

ΔH^0_{sys} = –281.5 kJ

$\Delta S^0_{surr} = -(\Delta H^0_{rxn}/T) = -(-281.5 \text{ kJ}/298 \text{ K}) = 0.945 \text{ kJ/K} = 945 \text{ J/K}$

$\mathbf{\Delta S^0_{univ}} = \Delta S^0_{sys} + \Delta S^0_{surr} = (-136.6 \text{ J/K}) + (945 \text{ J/K}) = \mathbf{808 \text{ J/K}}$.

Because ΔS^0_{univ} is positive, the reaction is spontaneous at 298 K.

Check: This process is also known as rusting. Common sense tells us that rusting occurs spontaneously. Although the entropy change of the system is negative, the increase in entropy of the surroundings is large enough to offset ΔS^0_{sys}.

20.4 Plan: Calculate the ΔH^0_{rxn} using ΔH^0_f values from Appendix B. Calculate ΔS^0_{rxn} from tabulated S^0 values and apply Equation 20.6.

Solution:

$\Delta H^0_{rxn} = \{(2\text{mol } NO_2)(\Delta H^0_f(NO_2))\} - \{(2\text{mol } NO)(\Delta H^0_f(NO)) + (1\text{mol } O_2)(\Delta H^0_f(O_2))\}$

$\Delta H^0_{rxn} = \{(2\text{mol } NO_2)(33.2 \text{ kJ/mol})\} - \{(2\text{mol } NO)(90.29 \text{ kJ/mol}) + (1\text{mol } O_2)(0)\}$

$\Delta H^0_{rxn} = -114.2 \text{ kJ}$

$\Delta S^0_{rxn} = \{(2\text{mol } NO_2)(S^0_f(NO_2))\} - \{(2\text{mol } NO)(S^0_f(NO)) + (1\text{mol } O_2)(S^0_f(O_2))\}$

$\Delta S^0_{rxn} = \{(2\text{mol } NO_2)(239.9 \text{ J/mol·K})\} - \{(2\text{mol } NO)(210.65 \text{ J/mol·K}) + (1\text{mol } O_2)(205.0 \text{ J/mol·K})\} = -146.5 \text{ J/K}$

$\Delta G^0_{rxn} = \Delta H^0_{rxn} - T\Delta S^0_{rxn} = -114.2 \text{ kJ} - [(298 \text{ K})(-146.5 \text{ J/K})(1 \text{ kJ}/1000 \text{ J})] = \mathbf{-70.5 \text{ kJ}}$

Check: Verify that the equations contain the correct tabulated values and recalculate answer to double-check result.

20.5 Plan: Use ΔG^0_f values from Appendix B and apply Equation 20.7 to calculate ΔG^0_{rxn}.

Solution:

a) $\Delta G^0_{rxn} = \{(2\text{mol } NO_2)(\Delta G^0_f(NO_2))\} - \{(2\text{mol } NO)(\Delta G^0_f(NO)) + (1\text{mol } O_2)(\Delta G^0_f(O_2))\}$

$\Delta G^0_{rxn} = \{(2\text{mol } NO_2)(51 \text{ kJ/mol})\} - \{(2\text{mol } NO)(86.60 \text{ kJ/mol}) + (1\text{mol } O_2)(0))\}$

$\Delta G^0_{rxn} = \mathbf{-71 \text{ kJ}}$ (result agrees with answer in 20.4)

b) $\Delta G^0_{rxn} = \{(2\text{mol } CO)(\Delta G^0_f(CO))\} - \{(2\text{mol } C)(\Delta G^0_f(C)) + (1\text{mol } O_2)(\Delta G^0_f(O_2))\}$

$\Delta G^0_{rxn} = \{(2\text{mol } CO)(-137.2 \text{ kJ/mol})\} - \{(2\text{mol } C)(0) + (1\text{mol } O_2)(0)\}$

$\Delta G^0_{rxn} = \mathbf{-274.4 \text{ kJ}}$

Check: Verify that the equations contain the correct tabulated values and recalculate answer to double-check result.

20.6 Plan: Examine the equation $\Delta G^0 = \Delta H^0 - T\Delta S^0$ and determine which combination of enthalpy and entropy will describe the given reaction.

Solution: Two choices can already be eliminated:

1) When $\Delta H > 0$ (endothermic reaction) and $\Delta S < 0$ (entropy decreases), the reaction is always nonspontaneous, regardless of temperature, so this combination does not describe the reaction.

2) When $\Delta H < 0$ (exothermic reaction) and $\Delta S > 0$ (entropy increases), the reaction is always spontaneous, regardless of temperature, so this combination does not describe the reaction.

Two combinations remain: 3) $\Delta H^0 > 0$ and $\Delta S^0 > 0$, or 4) $\Delta H^0 < 0$ and $\Delta S^0 < 0$. If the reaction becomes spontaneous at –40°C, this means that ΔG^0 becomes negative at lower temperatures. Case 3) becomes spontaneous at higher temperatures, when the $-T\Delta S^0$ term is larger than the positive enthalpy term. By process of elimination, Case 4) describes the reaction.

Check: At a lower temperature, the negative ΔH^0 becomes larger than the positive $(-T\Delta S^0)$ value, so ΔG^0 becomes negative.

20.7 Plan: Write a balanced equation for the dissociation of hypobromous acid in water. The free energy of the reaction at standard state (part (a)) is calculated using Equation 20.10. The free energy of the reaction under non-standard state conditions is calculated using Equation 20.11.

Solution: $HBrO(aq) + H_2O(l) \rightleftharpoons BrO^-(aq) + H_3O^+(aq)$

a) $\Delta G^0 = -RT \ln K$

$$\Delta G^0 = -(8.314\ J/mol \cdot K)(1\ kJ/1000\ J)(298\ K)\ln(2.3x10^{-9}) = \mathbf{49\ kJ/mol}$$

b) $$Q = \frac{[BrO^-][H_3O^+]}{[HBrO]} = \frac{(0.10)(6.0x10^{-4})}{(0.20)} = 3.0x10^{-4}$$

$$\Delta G = \Delta G^0 + RT \ln Q$$

$$\Delta G = 49\ kJ/mol + (8.314\ J/mol \cdot K)(1\ kJ/1000\ J)(298\ K)\ln(3.0x10^{-4})$$

$$\Delta G = \mathbf{29\ kJ/mol}$$

Check: The value of K_a is very small, so it makes sense that ΔG^0 is a positive number (refer to Table 20.2). The natural log of a negative exponent gives a negative number ($\ln 3.0 \times 10^{-4}$), so the value of ΔG decreases with concentrations lower than the standard state 1M values.

END-OF-CHAPTER PROBLEMS

20.2 A spontaneous process occurs by itself (possibly requiring an initial input of energy) whereas a nonspontaneous process requires a continuous supply of energy to make it happen. It is possible to cause a nonspontaneous process to occur, but the process stops once the energy source is removed. A reaction that is found to be nonspontaneous under one set of conditions may be spontaneous under a different set of conditions (different temperature, different concentrations).

20.5 Vaporization is the change of a liquid substance to a gas so $\Delta S_{vaporization} = S_{gas} - S_{liquid}$. Fusion is the change of a solid substance into a liquid so $\Delta S_{fusion} = S_{liquid} - S_{solid}$. Vaporization involves a greater change in volume than fusion. Thus the transition from liquid to gas involves a greater entropy change than the transition from solid to liquid.

20.6 In an exothermic process, the *system* releases heat to its *surroundings*. The entropy of the surroundings increases because the temperature of the surroundings increases ($\Delta S_{surr} > 0$). In an endothermic process, the system absorbs heat from the surroundings and the surroundings become cooler. Thus, the entropy of the surroundings decreases ($\Delta S_{surr} < 0$). A chemical cold pack for injuries is an example of a spontaneous, endothermic chemical reaction.

20.8 a) **Spontaneous**. Evaporation occurs because a few of the liquid molecules have enough energy to break away from the intermolecular forces of the other liquid molecules and move spontaneously into the gas phase.

b) **Spontaneous**. A lion spontaneously chases an antelope without added force.

c) **Spontaneous**. An unstable substance decays spontaneously to a more stable substance.

20.10 a) **Spontaneous**. With a small amount of energy input, methane will continue to burn without additional energy (the reaction itself provides the necessary energy) until it is used up.

b) **Spontaneous**. The dissolved sugar molecules have more states they can occupy than the crystalline sugar, so the reaction proceeds in the direction of dissolution.

c) **Not spontaneous**. A cooked egg will not become raw again, no matter how long it sits or how many times it is mixed.

20.12 a) $\Delta S_{sys} > 0$. Melting is the change in state from solid to liquid. The solid state of a particular substance always has lower entropy than the same substance in the liquid state. Entropy increases during melting.

b) $\Delta S_{sys} < 0$. The entropy of most salt solutions is greater than the entropy of the solvent and solute separately. So entropy decreases as a salt precipitates.

c) $\Delta S_{sys} < 0$. Dew forms by the condensation of water vapor to liquid. Entropy of a substance in the gaseous state is greater than its entropy in the liquid state. Entropy decreases during condensation.

20.14 a) $\Delta S_{sys} > 0$. The process described is alcohol(l) $\rightarrow$ alcohol(g). The gas molecules have greater entropy than the liquid molecules.

b) $\Delta S_{sys} > 0$. The process described is a change from solid to gas, an increase in possible energy states for the system.

c) $\Delta S_{sys} > 0$. The perfume molecules have more possible locations in the larger volume of the room than inside the bottle. A system that has more possible arrangements has greater entropy.

20.16 a) $\Delta S_{sys} < 0$. Reaction involves a gaseous reactant and no gaseous products so entropy decreases. The number of particles also decreases indicating a decrease in entropy.

b) $\Delta S_{sys} < 0$. Gaseous reactants form solid product and number of particles decreases so entropy decreases.

c) $\Delta S_{sys} > 0$. When salts dissolve in water entropy generally increases.

20.18 a) $\Delta S_{sys} > 0$. The reaction produces gaseous CO_2 molecules that have greater entropy than the physical states of the reactants.

b) $\Delta S_{sys} < 0$. The reaction produces a net decrease in the number of gaseous molecules, so the system's entropy decreases.

c) $\Delta S_{sys} > 0$. The same reasoning in (a) applies here.

20.20 a) $\Delta S_{sys} > 0$. Decreasing the pressure increases the volume available to the gas molecules so entropy of the system increases.

b) $\Delta S_{sys} < 0$. Gaseous nitrogen molecules have greater entropy (more possible states) than dissolved nitrogen molecules.

c) $\Delta S_{sys} > 0$. Dissolved oxygen molecules have lower entropy than gaseous oxygen molecules.

20.22 a) **Butane** has the greater molar entropy because it has two additional C–H bonds that can vibrate and has greater rotational freedom around its bond. The presence of the double bond in 2-butene restricts rotation.

b) **Xe(g)** has the greater molar entropy because entropy increases with atomic size.

c) **CH_4(g)** has the greater molar entropy because gases in general have greater entropy than liquids.

20.24 a) Ethanol, **C_2H_5OH (l)**, is a more complex molecule than methanol, CH_3OH, and has the greater molar entropy.

b) When a salt dissolves the number of possible states for the ions increases. Thus, **$KClO_3$(aq)** has the greater molar entropy.

c) **K(s)** has greater molar entropy because K(s) has greater mass than Na(s).

20.26 a) diamond < graphite < charcoal. Diamond has a ordered, 3-dimensional crystalline shape, followed by graphite with an ordered 2-dimensional structure, followed by the amorphous structure of charcoal.

b) ice < liquid water < water vapor. Entropy increases as a substance changes from solid to liquid to gas.

c) O atoms < O_2 < O_3. Entropy increases with molecular complexity because there are more modes of movement (e.g. bond vibration) available to the complex molecules.

20.28 a) **ClO_4^-(aq) > ClO_3^-(aq) > ClO_2^-(aq).** The decreasing order of molar entropy follows the order of decreasing molecular complexity.

b) **NO_2(g) > NO(g) > N_2(g)**. N_2 has lower molar entropy than NO because N_2 consists of two of the same atoms while NO consists of two different atoms. NO_2 has greater molar entropy than NO because NO_2 consists of three atoms while NO consists of only two.

c) **Fe_3O_4(s) > Fe_2O_3(s) > Al_2O_3(s)**. Fe_3O_4 has greater molar entropy than Fe_2O_3 because Fe_3O_4 is more complex and more massive. Fe_2O_3 and Al_2O_3 contain the same number of atoms but Fe_2O_3 has greater molar entropy because iron atoms are more massive than aluminum atoms.

20.31 A system at equilibrium does not spontaneously produce more product or more reactant. For either reaction direction, entropy change of the system is exactly offset by the entropy change of the surroundings. Therefore, for system at equilibrium, $\Delta S_{univ} = \Delta S_{sys} + \Delta S_{surr} = 0$. However, for a system moving spontaneously toward equilibrium, $\Delta S_{univ} > 0$.

20.32 Since entropy is a state function the entropy changes can be found by summing the entropies of the products and subtracting the sum of the entropies of the reactants.

For the given reaction $\Delta S^0_{rxn} = 2S^0_{HClO(g)} - (S^0_{H_2O(g)} + S^0_{Cl_2O(g)})$. Rearranging this expression to solve for $S^0_{Cl_2O(g)}$ gives

$$S^0_{Cl_2O(g)} = 2S^0_{HClO(g)} - S^0_{H_2O(g)} - \Delta S^0_{rxn}$$

20.33 a) Prediction: $\Delta S^0 < 0$ because number of moles of (Δn) gas decreases.

ΔS^0 = {(1mol N_2O(g))(219.7 J/mol·K) + (1mol NO_2(g))(239.9 J/mol·K)} – {(3mol NO(g)) (210.65 J/mol·K)}

ΔS^0 = **-172.4 J/K**

b) Prediction: Sign difficult to predict because $\Delta n = 0$, but possibly $\Delta S^0 > 0$ because water vapor has greater complexity than H_2 gas.

ΔS^0 = {(2mol Fe(s))(27.3 J/mol·K) + (3mol H_2O(g))(188.72 J/mol·K)} – {(3mol H_2(g))(130.6 J/mol·K) + (1mol Fe_2O_3(s))(87.400 J/mol·K)}

ΔS^0 = **141.6 J/K**

c) Prediction: $\Delta S^0_{sys} < 0$ because $\Delta n = -5$.

ΔS^0 = {(1mol P_4O_{10}(s))(229 J/mol·K)} – {(1mol P_4(s))(41.1 J/mol·K) + (5mol O_2(g))(205.0 J/mol·K)}

ΔS^0 = **-837 J/K**

20.35 The balanced combustion reaction is

$$2C_2H_6(g) + 7O_2(g) \rightarrow 4CO_2(g) + 6H_2O(g)$$

ΔS^0_{rxn} = {4mol(S^0(CO_2)) + 6mol(S^0(H_2O))} – {2mol(S^0(C_2H_6) + 7mol(S^0(CO_2))}

= {4mol(213.7 J/mol K) + 6mol(188.72 J/mol K)} – {2mol(229.5 J/mol K) + 7mol(205.0 J/mol K)}

= **93.1 J/K**

The entropy value is not per mole of C_2H_6 but per 2 moles. Divide the calculated value by 2 to obtain entropy per mole of C_2H_6.

The positive sign of ΔS is expected because there is a net increase in the number of gas molecules from 9 moles as reactants to 10 moles as products.

20.37 The balanced chemical equation for the described reaction is:

$2NO(g) + 5H_2(g) \rightarrow 2NH_3(g) + 2H_2O(g)$

Because the number of moles of gas decreases, i.e. $\Delta n = 4 - 7 = -3$, the entropy is expected to decrease ($\Delta S^0_{sys} < 0$).

ΔS^0 = {(2mol NH_3)(193 J/mol·K)+(2mol H_2O)(188.72 J/mol·K)} – {(2mol NO)(210.65 J/mol·K)+(5mol H_2)(130.6 J/mol·K)}

ΔS^0 = **-311 J/K**

The calculated entropy matches the prediction.

20.39 The reaction for forming Cu_2O from copper metal and oxygen gas is

$$4Cu(s) + O_2(g) \rightarrow 2Cu_2O(s)$$

ΔS^0_{rxn} = {2mol(S^0(Cu_2O))} – {4mol(S^0(Cu)) + 1mol(S^0(O_2))}

= {2mol(93.1 J/mol K)} - {4mol(33.1 J/mol K) + 1mol(205.0 J/mol K)}

= -151.2 J/K to form two moles of Cu_2O

To form 1 mole of Cu_2O the entropy change is **–75.6 J/K**.

20.41 One mole of methanol is formed from its elements in their standard states according to the following equation: $C(g) + 2H_2(g) + ½O_2(g) \rightarrow CH_3OH(l)$

ΔS^0 = (S^0(CH_3OH)) – {S^0(C(graphite)) + 2(S^0(H_2)) + ½(S^0(O_2)}

ΔS^0 = {(1mol CH_3OH)(127 J/mol·K)} – {(1mol C)(5.686 J/mol·K) + (2mol H_2)(130.6 J/mol·K) + (½mol O_2)(205.0 J/mol·K)}

ΔS^0 = **-242 J/K**

20.44 Complete combustion of a hydrocarbon includes oxygen as a reactant and carbon dioxide and water as the products.

$$2C_2H_2(g) + 5O_2(g) \rightarrow 4CO_2(g) + 2H_2O(g)$$

ΔS^0_{rxn} = {4mol($S^0(CO_2)$) + 2mol($S^0(H_2O)$)} – {2mol($S^0(C_2H_2)$) + 5mol($S^0(O_2)$)}
= {4mol(213.7 J/mol·K) + 2mol(188.72 J/mol·K)} – {2mol(200.85 J/mol·K) + 5mol(205.0 J/mol·K)}
= **-194.4$_6$ J/K**

The entropy change for the combustion of 1 mole C_2H_2 equals ½ (-194.4$_6$ J/K) or **-97.2 J/ K.**

20.46 A spontaneous process has $\Delta S_{univ} > 0$. Since the Kelvin temperature is always positive, ΔG_{sys} must be negative ($\Delta G_{sys} < 0$) for a spontaneous process.

20.48 The reaction is endothermic ($\Delta H^0_{rxn} > 0$) and requires a lot of heat from its surroundings to be spontaneous. The removal of heat from the surroundings results in $\Delta S^0_{surr} < 0$. The only way an endothermic reaction can proceed spontaneously is if $\Delta S^0_{sys} >> 0$, effectively offsetting the decrease in surroundings entropy. In summary, the **$\Delta H^0_{rxn} > 0$ and $\Delta S^0_{sys} > 0$** for this reaction.

Alternatively, reaction spontaneity can be predicted by a single variable, ΔG. For a spontaneous process, $\Delta G^0 = \Delta H^0 - T\Delta S^0 < 0$. At high T, a positive ΔS^0_{sys} maintains a negative sign on ΔG^0.

An example of this type of process is the melting of a crystalline solid. This is a highly endothermic process, and the conversion of solid to liquid results in an increase in entropy.

20.49 For a given substance, the entropy changes greatly from one phase to another, e.g. from liquid to gas. However, the entropy changes little within a particular phase. As long as the substance does not change phase, the value of ΔS^0 is relatively unaffected by temperature.

20.50 The ΔG^0_{rxn} can be calculated from the individual ΔG^0_f's of the reactants and products found in Appendix B.

a) $\Delta G^0_{rxn} = \Sigma\{m\Delta G^0_f(\text{products})\} - \Sigma\{n\Delta G^0_f(\text{reactants})\}$

Both Mg(s) and O_2(g) are the standard states forms of their respective elements, so their ΔG^0_f's are zero.

ΔG^0_{rxn} = {(2mol MgO)(-569.0 kJ/mol)} – {(2mol Mg)(0) + (1mol O_2)(0)} = **-1138 kJ**

b) ΔG^0_{rxn} = {(2mol CO_2)(-394.4 kJ/mol) + (4mol H_2O)(-228.60 kJ/mol)} – {(2mol CH_3OH)(-161.9 kJ/mol) + 0)}

ΔG^0_{rxn} = **-1379.4 kJ**

c) ΔG^0_{rxn} = {(1mol $BaCO_3$)(-1139 kJ/mol)} – {(1mol BaO)(-520.4 kJ/mol) + (1mol CO_2)(-394.4 kJ/mol)}

ΔG^0_{rxn} = **-224 kJ**

20.52 a) ΔH^0_{rxn} = {2mol(ΔH_f^0(MgO))} – {2mol(ΔH_f^0(Mg)) + 1mol($\Delta H_f^0(O_2)$)}
= {2mol(-601.2 kJ/mol)} – {2mol (0) + 1mol(0)}
= -1202.$_4$ kJ

ΔS^0_{rxn} = {2mol(S^0(MgO))} – {2mol(S^0(Mg)) + 1mol($S^0(O_2)$)}
= {2mol(26.9 J/mol·K)} – {2mol(32.69 J/mol·K) + 1mol(205.0 J/mol·K)}
= -216.5$_8$ J/K

$\Delta G^0_{rxn} = \Delta H^0_{rxn} - T\Delta S^0_{rxn}$
= -1202.$_4$ kJ – {(298 K)(-216.5$_8$ J/K)(1 kJ/1000 J)} = **-1138 kJ**

b) $\Delta H^0_{rxn} = \{2mol(\Delta H_f^0(CO_2)) + 4mol(\Delta H_f^0(H_2O(g))\} - \{2mol(\Delta H_f^0(CH_3OH)) + 3mol(\Delta H_f^0(O_2))\}$

$= \{2mol(-393.5\ kJ/mol) + 4mol(-241.826\ kJ/mol)\} - \{2mol(-201.2\ kJ/mol) + 3mol(0)\}$

$= -1351.9\ kJ$

$\Delta S^0_{rxn} = \{2mol(S^0(CO_2)) + 4mol(S^0(H_2O(g))\} - \{2mol(S^0(CH_3OH)) + 3mol(S^0(O_2))\}$

$= \{2mol(213.7\ J/mol{\cdot}K) + 4mol(188.72\ J/mol{\cdot}K)\} - \{2mol(238\ J/mol{\cdot}K) + 3mol(205.0\ J/mol{\cdot}K)\}$

$= 91._{28}\ J/K$

$\Delta G^0_{rxn} = \Delta H^0_{rxn} - T\Delta S^0_{rxn}$

$= -1351.9\ kJ - \{(298\ K)(91._{28}\ J/K)(1\ kJ/1000\ J)\}$

$= \mathbf{-1379\ kJ}$

c) $\Delta H^0_{rxn} = \{1mol(\Delta H_f^0(BaCO_3(s))\} - \{1mol(\Delta H_f^0(BaO)) + 1mol(\Delta H_f^0(CO_2))\}$

$= \{1mol(-1219\ kJ/mol)\} - \{1mol(-548.1\ kJ/mol) + 1mol(-393.5\ kJ/mol)\}$

$= -277._4\ kJ$

$\Delta S^0_{rxn} = \{1mol(S^0(BaCO_3(s))\} - \{1mol(S^0(BaO)) + 1mol(S^0(CO_2))\}$

$= \{1mol(112\ J/mol{\cdot}K)\} - \{1mol(72.07\ J/mol{\cdot}K) + 1mol(213.7\ J/mol{\cdot}K)\}$

$= -173._{77}\ J/K$

$\Delta G^0_{rxn} = \Delta H^0_{rxn} - T\Delta S^0_{rxn}$

$= -277._4\ kJ - \{(298\ K)(-173._{77}\ J/K)(1\ kJ/1000\ J)\}$

$= \mathbf{-226\ kJ}$

20.54 a) Entropy decreases ($\Delta S^0 < 0$) because the number of moles of gas decreases from reactants (1½ moles) to products (1 mole). The oxidation (combustion) of CO requires initial energy input to start the reaction, but then releases energy (exothermic, $\Delta H^0 < 0$) that is typical of all combustion reactions.

b) Method 1: Calculate ΔG^0_{rxn} from ΔG^0_f's of products and reactants

$\Delta G^0_{rxn} = \{(1mol\ CO_2)(\Delta G^0_f(CO_2)\} - \{(1mol\ CO)(\Delta G^0_f(CO)) + (½\ mol)(\Delta G^0_f(O_2))\}$

$\Delta G^0_{rxn} = \{(1mol\ CO_2)(-394.4\ kJ/mol)\} - \{(1mol\ CO)(-137.2\ kJ/mol) + 0\} = \mathbf{-257.2\ kJ}$

Method 2: Calculate ΔG^0_{rxn} from ΔH^0 and ΔS^0 at 298 K (the zero superscript indicates a reaction at standard state, given in the appendix at 25°C)

$\Delta H^0_{rxn} = \{(1mol\ CO_2)(\Delta H^0_f(CO_2)\} - \{(1mol\ CO)(\Delta H^0_f(CO)) + (½\ mol)(\Delta H^0_f(O_2))\}$

$\Delta H^0_{rxn} = \{(1mol\ CO_2)(-393.5\ kJ/mol)\} - \{(1mol\ CO)(-110.5\ kJ/mol) + 0\} = \underline{-283.0\ kJ}$

$\Delta S^0_{rxn} = \{(1mol\ CO_2)(213.7\ J/mol{\cdot}K)\} - \{(1mol\ CO)(197.5\ J/mol{\cdot}K) + (½\ mol\ O_2)(205.0\ J/mol{\cdot}K)\} = \underline{-86.3\ J/K}$

$\Delta G^0_{rxn} = \Delta H^0_{rxn} - T\Delta S^0_{rxn} = (-283.0\ kJ) - (298\ K)(-86.3\ J/K)(1\ kJ/1000\ J) = \mathbf{-257.2\ kJ}$

20.56 Reaction is $Xe(g) + 3F_2(g) \rightarrow XeF_6(g)$

a) The change in entropy can be found from the relationship $\Delta G^0_{rxn} = \Delta H^0_{rxn} - T\Delta S^0_{rxn}$.

$\Delta S^0 = (\Delta H^0 - \Delta G^0)/T$

$= \{(-402\ kJ/mol) - (-280\ kJ/mol)\}/298\ K = \mathbf{-0.409\ kJ/mol{\cdot}K}$

b) The change in entropy at 500K is calculated from $\Delta G^0_{rxn} = \Delta H^0_{rxn} - T\Delta S^0_{rxn}$ with T at 500 K.

$\Delta G^0 = \Delta H^0 - T\Delta S^0 = (-402\ kJ/mol) - \{(500\ K)(-0.409\ kJ/mol{\cdot}K)\} = \mathbf{-198\ kJ/mol}$

20.58 a) ΔH^0_{rxn} = {(1mol CO)(ΔH^0_f(CO)) + (2mol H_2)(ΔH^0_f(H_2))} – {(1mol CH_3OH)(ΔH^0_f(CH_3OH))}
ΔH^0_{rxn} = {(1mol CO)(-110.5 kJ/mol) + (2mol H_2)(0)} – {(1mol CH_3OH)(-201.2 kJ/mol)}
ΔH^0_{rxn} = **90.7 kJ**
ΔS^0_{rxn} = {(1mol CO)(S^0_f(CO)) + (2mol H_2)(S^0_f(H_2))} – {(1mol CH_3OH)(S^0_f(CH_3OH))}
ΔS^0_{rxn} = {(1mol CO)(197.5 J/mol·K) + (2mol H_2)(130.6 J/mol·K)} – {(1mol CH_3OH)(238 J/mol·K)} = **221 J/K**

b) T_1 = 38 + 273 = 311 K ΔG^0 = 90.7 kJ – (311 K)(221 J/K)(1 kJ/1000 J) = **22.0 kJ**
T_2 = 138 + 273 = 411 K ΔG^0 = 90.7 kJ – (411 K)(221 J/K)(1 kJ/1000 J) = **-0.1 kJ**
T_3 = 238 + 273 = 511 K ΔG^0 = 90.7 kJ – (511 K)(221 J/K)(1 kJ/1000 J) = **-22.2 kJ**

c) The reaction is nonspontaneous at 38°C, near equilibrium at 138°C and spontaneous at 238°C.

20.60 At the normal boiling point, defined as the temperature at which the vapor pressure of the liquid equals 1 atm, the phase change from liquid to gas is at equilibrium. For a system at equilibrium the change in Gibbs free energy is zero. Since the gas is at 1 atm and the liquid assumed to be pure, the system is at standard state and ΔG^0 = 0. The temperature at which this occurs can be found from $\Delta G^0_{rxn} = 0 = \Delta H^0_{rxn} - T\Delta S^0_{rxn}$.
ΔH^0 = 1mol(ΔH_f^0(Br_2(g))) – 1mol(ΔH_f^0(Br_2(l))) = 30.91 kJ – 0 = 30.91 kJ
ΔS^0 = 1mol(S^0(Br_2(g))) – 1mol(S^0(Br_2(l))) = 245.38 J/K – 152.23 J/K = 93.15 J/K
$T_{bpt} = \Delta H^0/\Delta S^0$ = (30910 J)/(93.15 J/K) = **331.8 K**

20.62 a) The reaction for this process is H_2(g) + ½O_2(g) → H_2O(g). The coefficients are written this way (instead of 2H_2(g) + O_2(g) → 2H_2O(g)) because the problem specifies thermodynamic values "per (1) mol H_2", not per 2 mol H_2.
ΔH^0_{rxn} = {(1mol H_2O)(ΔH^0_f(H_2O))} – {(1mol H_2)(ΔH^0_f(H_2)) + (½mol O_2)(ΔH^0_f(O_2))}
ΔH^0_{rxn} = {(1mol H_2O)(-241.826 kJ/mol)} – {(1mol H_2)(0) + (½mol O_2)(0)}
ΔH^0_{rxn} = **-241.826 kJ**
ΔS^0_{rxn} = {(1mol H_2O)(S^0_f(H_2O))} – {(1mol H_2)(S^0_f(H_2)) + (½mol O_2)(S^0_f(O_2))}
ΔS^0_{rxn} = {(1mol H_2O)(188.72 J/mol·K)} – {(1mol H_2)(130.6 J/mol·K) + (½mol O_2)(205.0 J/mol·K)} = **-44.4 J/K**
ΔG^0_{rxn} = {(1mol H_2O)(ΔG^0_f(H_2O))} – {(1mol H_2)(ΔG^0_f(H_2)) + (½mol O_2)(ΔG^0_f(O_2))}
ΔG^0_{rxn} = {(1mol H_2O)(-228.60 kJ/mol)} – {(1mol H_2)(0) + (½mol O_2)(0)}
ΔG^0_{rxn} = **-228.60 kJ**

b) Because $\Delta H < 0$ and $\Delta S < 0$, the reaction will become nonspontaneous at higher temperatures because the positive (-TΔS) term becomes larger than the negative ΔH term.

c) The reaction becomes spontaneous at or below

$$T = \frac{\Delta H^0}{\Delta S^0} = \frac{-241.826\, kJ(1000\, J/kJ)}{-44.4\, J/K} = \mathbf{5.45x10^3\ K}.$$

20.64 a) An equilibrium constant that is much less than 1 indicates that very little product is made to reach equilbrium. The reaction thus is not spontaneous in the forward direction and ΔG^0 is a relatively large positive value.

b) A large negative ΔG^0 indicates that the reaction goes almost to completion. At equilibrium much more product is present than reactant so K >> 1. Q depends on initial conditions, not equilibrium conditions, so its value cannot be predicted from ΔG^0.

20.67 The standard free energy change, ΔG^0, occurs when all components of the system are in their standard states (do not confuse this with ΔG_f^0, the standard free energy of formation). Standard state is defined as 1 atm for gases, 1 M for solutes, and pure solids and liquids. Standard state does not specify a temperature because standard state can occur at any temperature. $\Delta G^0 = \Delta G$ when all concentrations equal 1 M and all partial pressures equal 1 atm. This occurs because the value of Q = 1 and lnQ = 0 in Equation 20.11.

20.68 For each reaction first find ΔG^0 then calculate K from $\Delta G^0 = -RT\ln K$.

a) ΔG^0 = 1mol(ΔG_f^0(NO_2(g)) – {1mol(ΔG_f^0(NO(g))) + ½ mol(ΔG_f^0 (O_2(g))}
= 51 kJ – 86.60 kJ – 0
= -35.$_{60}$ kJ

$$\ln K = \frac{(-35._{60}\ kJ)(1000\ J/kJ)}{-(8.314\ J/mol\cdot K)(298\ K)} = 14._{37}$$

$K = e^{14.37}$ = **1.7x10^6**

b) ΔG^0 = 1mol(ΔG_f^0 (H_2(g))) + 1mol(ΔG_f^0 (Cl_2(g))) - 2mol(ΔG_f^0 (HCl(g)))
= 0 + 0 – 2mol(-95.30 kJ/mol)
= 190.60 kJ

$$\ln K = \frac{(190.60\ kJ)(1000\ J/kJ)}{-(8.314\ J/mol\cdot K)(298\ K)} = -76.9_{30}$$

$K = e^{-76.930}$ = **3.89x10^{-34}**

c) ΔG^0 = 2mol(ΔG_f^0 (CO(g))) – {2mol(ΔG_f^0 (C(graphite))) + 1mol(ΔG_f^0 (O_2(g)))}
= 2mol(-137.2 kJ/mol) – 0 – 0
= -274.4 kJ

$$\ln K = \frac{(-274.4\ kJ)(1000\ J/kJ)}{-(8.314\ J/mol\cdot K)(298\ K)} = 110._{75}$$

$K = e^{110.75}$ = **1.26x10^{48}**

Note: You may get a different answer depending on how you rounded in earlier calculations.

20.70 The equilibrium constant, K, is related to ΔG^0 through the equation $K = e^{-(\Delta G^0/RT)}$. Calculate ΔG_{rxn}^0 using the ΔG_f^0 values from Appendix B.

a) ΔG_{rxn}^0 = {(2mol)(ΔG_f^0(H_2O)) + (2mol)(ΔG_f^0(SO_2))}–{(2mol)(ΔG_f^0(H_2S)) + (3mol)(ΔG_f^0(O_2))}
ΔG_{rxn}^0 = {(2mol)(-228.60 kJ/mol) + (2mol)(-300.2 kJ/mol)}–{(2mol)(-33 kJ/mol) + (3mol)(0)}
ΔG_{rxn}^0 = -991.$_6$ kJ (3 sig figs)

$$\frac{\Delta G}{RT} = \frac{-991.6\ kJ(1000\ J/kJ)}{(8.314\ J/mol\cdot K)(298\ K)} = -400._{23}$$

$K = e^{-(-400._{23})}$ = **6.57x10^{173}**

Comment: Depending on how you round ΔG_{rxn}^0, the value for K can vary by a factor of 2 or 3 because the inverse natural log varies greatly with small changes in ΔG_{rxn}^0. Your calculator might register "error" when trying to calculate e^{400} because it can't calculate exponents greater than 99. In this case, calculate $e^{200}e^{200} = (7.22_{60} \times 10^{86})^2 = (7.22_{60})^2 \times 10^{86\times2} = 52.2 \times 10^{172} = 5.22 \times 10^{173}$, which equals the first answer with rounding errors.

b) $\Delta G_{rxn}^0 = \{(1mol)(\Delta G_f^0(H_2O)) + (1mol)(\Delta G_f^0(SO_3))\} - \{(1mol)(\Delta G_f^0(H_2SO_4))\}$

$\Delta G_{rxn}^0 = \{(1mol)(-237.192\ kJ/mol) + (1mol)(-371\ kJ/mol)\} - \{(1mol)(-690.059\ kJ/mol)\}$

$\Delta G_{rxn}^0 = 81._{867}$ kJ (2 sig figs)

$$\frac{\Delta G}{RT} = \frac{81._{867}\ kJ(1000\ J/kJ)}{(8.314\ J/mol\cdot K)(298\ K)} = 33._{04}$$

$$K = e^{-(33._{04})} = \mathbf{5x10^{-15}}$$

c) $\Delta G_{rxn}^0 = \{(1mol)(\Delta G_f^0(NaCN)) + (1mol)(\Delta G_f^0(H_2O))\} - \{(1mol)(\Delta G_f^0(HCN)) + (1mol)(\Delta G_f^0(NaOH))\}$

NaCN(aq) is not listed in Appendix B, but Na^+(aq) and CN^-(aq) are listed. Since NaCN(aq) is equivalent to Na^+(aq) + CN^-(aq), the free energies of the cation and anion can be summed: $\Delta G_f^0(NaCN) = \Delta G_f^0(Na^+) + \Delta G_f^0(CN^-) = -261.87 + 166 = -95._{87}$ kJ/mol. The same holds true for NaOH(aq): $\Delta G_f^0(NaOH) = \Delta G_f^0(Na^+) + \Delta G_f^0(OH^-) = -261.87 + (-157.30) = -419.17$ kJ/mol

$\Delta G_{rxn}^0 = \{(1mol)(-95._{87}\ kJ/mol) + (1mol)(-237.192\ kJ/mol)\} - \{(1mol)(112\ kJ/mol) + (1mol)(-419.17\ kJ/mol)\}$

$\Delta G_{rxn}^0 = -25._{892}$ kJ

$$\frac{\Delta G}{RT} = \frac{-25._{892}\ kJ(1000\ J/kJ)}{(8.314\ J/mol\cdot K)(298\ K)} = -10._{45}$$

$$K = e^{-(-10._{45})} = \mathbf{3.5x10^{4}}$$

20.72 The solubility reaction for Ag_2S is

$$Ag_2S(s) + H_2O(l) \rightleftharpoons 2Ag^+(aq) + HS^-(aq) + OH^-(aq)$$

$$\Delta G^0 = \{2mol(\Delta G_{f\ Ag^+(aq)}^0) + 1mol(\Delta G_{f\ HS^-(aq)}^0) + 1mol(\Delta G_{f\ OH^-(aq)}^0)\} - \{1mol(\Delta G_{f\ Ag_2S(s)}^0) + 1mol(\Delta G_{f\ H_2O(l)}^0)\}$$

$$= \{(2mol)(77.111\ kJ/mol) + (1mol)(12.6\ kJ/mol) + (1mol)(-157.30\ kJ/mol)\} - \{(1mol)(-40.3\ kJ/mol) + (1mol)(-237.192\ kJ/mol)$$

$$= 287._{01}\ kJ$$

$$\ln K = \frac{(287._{01}\ kJ)(1000\ J/kJ)}{-(8.314\ J/mol\cdot K)(298\ K)} = -115._{84}$$

$$K = e^{-115._{84}} = \mathbf{5.0x10^{-51}}$$

20.74 Calculate ΔG_{rxn}^0, recognizing that I_2(s), not I_2(g), is the standard state for iodine. Solve for K_p using Equation 20.10.

$\Delta G_{rxn}^0 = \{(2mol)(\Delta G_f^0(ICl))\} - \{(1mol)(\Delta G_f^0(I_2)) + (1mol)(\Delta G_f^0(Cl_2))\}$

$\Delta G_{rxn}^0 = \{(2mol)(-6.075\ kJ/mol)\} - \{(1mol)(19.38\ kJ/mol) + (1mol)(0)\}$

$\Delta G_{rxn}^0 = -31.53$ kJ

$$K_p = e^{-(\Delta G/RT)}$$

$$\frac{\Delta G}{RT} = \frac{-31.53\,kJ(1000\,J/kJ)}{(8.314\,J/mol\cdot K)(298\,K)} = -12.73$$

$$K_p = e^{-(-12.73)} = \mathbf{3.4x10^5}$$

This calculation gives K_p and not K_c for a gas phase reaction because standard states for gases is defined as 1 atm. Using the ΔG^0 value based on partial pressures instead of concentrations of gases gives K_p instead of K_c.

20.76 ΔG^0 = -RTlnK = -(8.314 J/mol·K)(298 K)ln(1.7 x 10^{-5}) = 2.7 x 10^4 J/mol or **27 kJ/mol**.

The large positive ΔG^0 indicates that it would not be possible to prepare a solution with the concentrations of lead and chloride ions at the standard state concentration of 1M.

20.78 a) $\Delta G^0 = -RT\ln K = -(8.314\,J/mol\cdot K)(298\,K)\ln(9.1x10^{-6}) = -2.88x10^4\ J/mol = \mathbf{28.8\,kJ/mol}$

b) Since ΔG_{rxn}^0 is positive, the reaction direction as written is nonspontaneous. The reverse direction, formation of reactants, is spontaneous so the direction proceeds to the left.

c) Calculate the value for Q and then use to find ΔG.

$$Q = \frac{[Fe^{2+}]^2[Hg^{2+}]^2}{[Fe^{3+}]^2[Hg_2^{2+}]} = \frac{(0.010)^2(0.025)^2}{(0.20)^2(0.010)} = 1.5_{625}x10^{-4}$$

$$\Delta G_{298} = \Delta G^0 + RT\ln Q$$

$$\Delta G_{298} = 28.8\,kJ/mol + (8.314x10^{-3}\,kJ/mol\cdot K)(298\,K)\ln(1.5_{625}x10^{-4})$$

$$\Delta G_{298} = 28.8\,kJ/mol + (-21.7)$$

$$\Delta G_{298} = \mathbf{7.1 kJ/mol}$$

Because ΔG_{298} > 0 and Q > K, the reaction proceeds to left to equilibrium.

20.80 Formation of O_3 from O_2 : $3O_2(g) \rightleftharpoons 2O_3(g)$ or per mole of ozone: $3/2O_2(g) \rightleftharpoons O_3(g)$

a) To decide when production of ozone is favored both the signs of ΔH_f^0 and ΔS_f^0 for ozone are needed. From Appendix B: ΔH_f^0 = 143 kJ/mol and ΔS_f^0 = 238.82 J/mol·K – 3/2(205.0 J/mol·K) = -68.7 J/mol·K. The positive sign for ΔH_f^0 and the negative sign for ΔS_f^0 indicates the formation of ozone is favored at **no temperature**.

b) At 298 K, ΔG^0 can be most easily calculated from ΔG_f^0 values.

From ΔG_f^0: ΔG^0 = 2(163 kJ/mol O_3) – 0 = **326 kJ** for the formation of 2 moles of O_3.

c) $Q = \frac{[O_3]^2}{[O_2]^3} = \frac{(5x10^{-7})^2}{(0.21)^3} = 2.6_{99}\ x\,10^{-11}$

$$\Delta G = \Delta G^0 + RT\ln Q = 326\,kJ + ((0.008314\,kJ/mol\cdot K)(298\,K)(\ln(2.6_{99}\ x\,10^{-11}))) = \mathbf{266\,kJ}$$

20.83

	ΔS_{rxn}	ΔH_{rxn}	ΔG_{rxn}	Comment
(a)	+	–	–	**Spontaneous**
(b)	**(+)**	0	–	Spontaneous
(c)	–	+	**(+)**	Not spontaneous
(d)	0	**(–)**	–	Spontaneous
(e)	**(–)**	0	+	**Not spontaneous**
(f)	+	+	**(–)**	$T\Delta S > \Delta H$

a) The reaction is always spontaneous when $\Delta G_{rxn} < 0$, so there is no need to look at the other values other than to check the answer.
b) Because $\Delta G_{rxn} = \Delta H - T\Delta S = -T\Delta S$, ΔS must be positive for ΔG_{rxn} to be negative.
c) The reaction is always nonspontaneous when $\Delta G_{rxn} > 0$, so there is no need to look at the other values other than to check the answer.
d) Because $\Delta G_{rxn} = \Delta H - T\Delta S = \Delta H$, ΔH must be negative for ΔG_{rxn} to be negative.
e) This is the opposite of b).
f) Because $T\Delta S > \Delta H$, the subtraction of a larger positive term causes ΔG_{rxn} to be negative.

20.87 a) For the reaction $Q = \frac{[Hb \cdot CO][O_2]}{[Hb \cdot O_2][CO]}$ and since $[O_2] = [CO]$ then Q equals the ratio $[Hb{\cdot}CO]/[Hb{\cdot}O_2]$. This ratio can be calculated from $\Delta G^0 = -RT\ln K$

$\Delta G^0 = -RT\ln[Hb{\cdot}CO]/[Hb{\cdot}O_2]$

$$\frac{[Hb \cdot CO]}{[Hb \cdot O_2]} = e^{-\Delta G^0/RT} = e^{-(-14000\,J/(8.314\,J/mol\cdot K)(310.K))} = \mathbf{2.3x10^2}$$

b) By increasing the concentration of oxygen the equilibrium can be shifted in the direction of $Hb{\cdot}O_2$. Administer air with a higher concentration of O_2 to counteract the CO poisoning.

20.90 Sum the two reactions to yield an overall reaction. Because the process occurs at 85°C (358 K), the ΔG_{rxn}^0 cannot be calculated simply from the ΔG_f^0's which are tabulated at 298K. Instead, calculate ΔH_{rxn}^o and ΔS_{rxn}^o and use Equation 20.6.

$UO_2(s) + 4HF(g) \rightarrow$ ~~$UF_4(s)$~~ $+ 2H_2O(g)$

~~$UF_4(s)$~~ $+ F_2(g) \rightarrow UF_6(s)$

$UO_2(s) + 4HF(g) + F_2(g) \rightarrow UF_6(g) + 2H_2O(g)$

$\Delta H_{rxn}^o = \{(1mol)(\Delta H_f^o(UF_6)) + (2mol)(\Delta H_f^o(H_2O))\} - \{(1mol)(\Delta H_f^o(UO_2)) + (4mol)(\Delta H_f^o(HF)) + (1mol)(\Delta H_f^o(F_2))\}$

$\Delta H_{rxn}^o = \{(1mol)(-2197\ kJ/mol)) + (2mol)(-241.826\ kJ/mol)\} - \{(1mol)(-1085\ kJ/mol) + (4mol)(-273\ kJ/mol) + (1mol)(0)\}$

$\Delta H_{rxn}^o = -503._{65}\ kJ$

$\Delta S_{rxn}^o = \{(1mol)(S_f^o(UF_6)) + (2mol)(S_f^o(H_2O))\} - \{(1mol)(S_f^o(UO_2)) + (4mol)(S_f^o(HF)) + (1mol)(S_f^o(F_2))\}$

ΔS_{rxn}° = {(1mol)(225 J/mol·K) + (2mol)(188.72 J/mol·K)} – {(1mol)(77.0 J/mol·K) + (4mol)(173.67 J/mol·K) + (1mol)(202.7 J/mol·K)}

ΔS_{rxn}° = -371.$_{94}$ J/K

$\Delta G_{358}^{0} = \Delta H_{rxn}^{\circ} - T\Delta S_{rxn}^{\circ}$ = (-503.$_{65}$ kJ) – {(358 K)(-371.$_{94}$ J/K)(kJ/1000 J)} = **-370 kJ**

20.94 The reaction for the hydrogenation of ethene is

$$C_2H_4(g) + H_2(g) \rightarrow C_2H_6(g)$$

Calculate enthalpy, entropy and free energy from values in Appendix B, which are given at 25°C.

ΔH^{0}_{rxn} = 1mol($\Delta H_f^{0}(C_2H_6)$) – {1mol($\Delta H_f^{0}(C_2H_4)$) + 1mol($\Delta H_f^{0}(H_2)$)}
= (-84.667 kJ) – (52.47 + 0 kJ)
= **-137.14 kJ**

ΔS^{0}_{rxn} = 1mol($S^{0}(C_2H_6)$) – {1mol($S^{0}(C_2H_4)$) + 1mol($S^{0}(H_2)$)}
= (229.5 J/K) – {(219.22 J/K) + (130.6 J/K)}
= **-120.3 J/K**

ΔG^{0}_{rxn} = 1mol($\Delta G_f^{0}(C_2H_6)$) – {1mol($\Delta G_f^{0}(C_2H_4)$) + 1mol($\Delta G_f^{0}(H_2)$)}
= (-32.89 kJ) – (68.36 kJ + 0 kJ)
= **-101.25 kJ**

20.96 a) The chemical equation for this process is $3C(s) + 2Fe_2O_3(s) \rightarrow 3CO_2(g) + 4Fe(s)$

ΔH_{rxn}° = {(3mol)($\Delta H_f^{\circ}(CO_2)$) + (4mol)($\Delta H_f^{\circ}(Fe)$)} – {(3mol)($\Delta H_f^{\circ}(C)$) + (2mol)($\Delta H_f^{\circ}(Fe_2O_3)$)}

ΔH_{rxn}° = {(3mol)(-393.5 kJ/mol) + (4mol)(0)} – {(3mol)(0) + (2mol)(-825.5 kJ/mol)}

ΔH_{rxn}° = 470.5 kJ

ΔS_{rxn}° = {(3mol)($S_f^{\circ}(CO_2)$) + (4mol)($S_f^{\circ}(Fe)$)} – {(3mol)($S_f^{\circ}(C)$) + (2mol)($S_f^{\circ}(Fe_2O_3)$)}

ΔS_{rxn}° = {(3mol)(213.7 J/mol·K) + (4mol)(27.3 J/mol·K)} – {(3mol)(5.686 J/mol·K) + (2mol)(87.400 J/mol·K)}

ΔS_{rxn}° = 558.4 J/K

b) The reaction will be spontaneous at higher temperatures, where the $-T\Delta S$ term will be larger in magnitude than ΔH.

c) $\Delta G_{298}^{0} = \Delta H_{rxn}^{\circ} - T\Delta S_{rxn}^{\circ}$ = 470.5 kJ – (298 K)(0.5584 kJ/K) = 304.1 kJ

Because ΔG is positive, the reaction is not spontaneous.

d) The temperature at which the reaction becomes spontaneous is found by calculating $T = \Delta H/\Delta S$ = 470.5 kJ/(0.5584 kJ/K) = **842.6 K.**

20.101 Oxidation of iron: $4Fe(s) + 3O_2(g) \rightarrow 2Fe_2O_3(s)$

The change in Gibbs free energy for this reaction will be twice ΔG_f^{0} of Fe_2O_3 since the reaction forms the oxide from its elements and is reported per mole.

ΔG^{0} = 2mol(-743.6 kJ/mol) = -1487.2 kJ

Oxidation of aluminum: $4Al(s) + 3O_2(g) \rightarrow 2Al_2O_3(s)$

The change in Gibbs free energy for this reaction will be twice ΔG_f^{0} of Al_2O_3 since the reaction forms the oxide from its elements and is reported per mole.

ΔG^{0} = 2mol(-1582 kJ/mol) = -3164 kJ

With $\Delta G^{0} < 0$, both reactions are spontaneous at 25°C.

CHAPTER 21

ELECTROCHEMISTRY: CHEMICAL CHANGE AND ELECTRICAL WORK

FOLLOW-UP PROBLEMS

21.1 Plan: Follow the steps for balancing a redox reaction in acidic solution:

1. Divide into half-reactions
2. For each half- reaction balance a) atoms other than O and H, b) O atoms with H_2O, c) H atoms with H^+ and d) charge with e^-.
3. Multiply each half-reaction by an integer that will make the number of electrons lost equal to the number of electrons gained.
4. Add half-reactions and cancel substances appearing as both reactants and products.

Then add another step for basic solution:

5. Add hydroxide ions to neutralize H^+. Cancel water.

Solution:

1. Divide into half-reactions: group the reactants and products with similar atoms.

$MnO_4^-(aq) \rightarrow MnO_4^{2-}(aq)$

$I^-(aq) \rightarrow IO_3^-(aq)$

2. For each half- reaction balance

a) atoms other than O and H

Mn and I are balanced so no changes needed

b) O atoms with H_2O,

$MnO_4^-(aq) \rightarrow MnO_4^{2-}(aq)$ O already balanced

$I^-(aq) \rightarrow IO_3^-(aq)$ to balance 3 oxygens must be added as reactant, so add 3 H_2O to add the 3 oxygens:

$I^-(aq) + 3H_2O(l) \rightarrow IO_3^-(aq)$

c) H atoms with H^+

$MnO_4^-(aq) \rightarrow MnO_4^{2-}(aq)$ H already balanced

$I^-(aq) + 3H_2O(l) \rightarrow IO_3^-(aq)$ to balance 6 hydrogens must be added as product, so add $6H^+$ to products:

$I^-(aq) + 3H_2O(l) \rightarrow IO_3^-(aq) + 6H^+(aq)$

d) charge with e^-

$MnO_4^-(aq) \rightarrow MnO_4^{2-}(aq)$ total charge of reactants is –1 and of products is –2, so must add $1e^-$ to reactants to balance charge:

$MnO_4^-(aq) + e^- \rightarrow MnO_4^{2-}(aq)$

$I^-(aq) + 3H_2O(l) \rightarrow IO_3^-(aq) + 6H^+(aq)$ total charge is –1 for reactants and +5 for products, so must add $6e^-$ as product:

$I^-(aq) + 3H_2O(l) \rightarrow IO_3^-(aq) + 6H^+(aq) + 6e^-$

3. Multiply each half-reaction by an integer that will make the number of electrons lost equal to the number of electrons gained. One electron is gained and 6 are lost so reduction must be multiplied by 6 for the number of electrons to be equal.

$6\{MnO_4^-(aq) + e^- \rightarrow MnO_4^{2-}(aq)\}$

$I^-(aq) + 3H_2O(l) \rightarrow IO_3^-(aq) + 6H^+(aq) + 6e^-$

4. Add half-reactions and cancel substances appearing as both reactants and products.

$6MnO_4^-(aq) + 6e^- \rightarrow 6MnO_4^{2-}(aq)\}$

$\underline{I^-(aq) + 3H_2O(l) \rightarrow IO_3^-(aq) + 6H^+(aq) + 6e^-}$

$6MnO_4^-(aq) + I^-(aq) + 3H_2O(l) \rightarrow 6MnO_4^{2-}(aq) + IO_3^-(aq) + 6H^+(aq)$

5. Add hydroxide ions to neutralize H^+. Cancel water. The $6H^+$ are neutralized by adding $6OH^-$. The same number of hydroxide ions must be added to the reactants to keep the balance of O and H atoms on both sides of the reaction.

$6MnO_4^-(aq) + I^-(aq) + 3H_2O(l) + 6OH^-(aq) \rightarrow 6MnO_4^{2-}(aq) + IO_3^-(aq) + 6H^+(aq) + 6OH^-(aq)$

The neutralization reaction produces water: $6\{H^+ + OH^- \rightarrow H_2O\}$.

$6MnO_4^-(aq) + I^-(aq) + 3H_2O(l) + 6OH^-(aq) \rightarrow 6MnO_4^{2-}(aq) + IO_3^-(aq) + 6H_2O(l)$

Cancel water:

$6MnO_4^-(aq) + I^-(aq) + \cancel{3H_2O(l)} + 6OH^-(aq) \rightarrow 6MnO_4^{2-}(aq) + IO_3^-(aq) + \cancel{6}H_2O(l)$

Balanced reaction including spectator ions is

$6KMnO_4(aq) + KI(aq) + 6KOH(aq) \rightarrow 6K_2MnO_4(aq) + KIO_3(aq) + 3H_2O(l)$

Check: Check balance of atoms and charge:

Reactants:	Products:
6 Mn atoms	6 Mn atoms
1 I atom	1 I atom
30 O atoms	30 O atoms
6 H atoms	6 atoms
-13 charge	-13 charge

21.2 Plan: Given the solution and electrode compositions the two half cells involve the transfer of electrons 1) between chromium in $Cr_2O_7^{2-}$ and Cr^{3+} and 2) between Sn and Sn^{2+}. The negative electrode is the anode so the tin half-cell is where oxidation occurs. The graphite electrode with the chromium ion/chromate solution is where reduction occurs.

In the cell diagram show the electrodes and the solutes involved in the half-reactions. Include the salt bridge and wire connection between electrodes. Set up the two half-reactions and balance. (Note that the $Cr^{3+}/Cr_2O_7^{2-}$ half-cell is in acidic solution.) Write the cell notation placing the anode half-cell first then the salt bridge then the cathode half-cell.

Solution: Cell diagram:

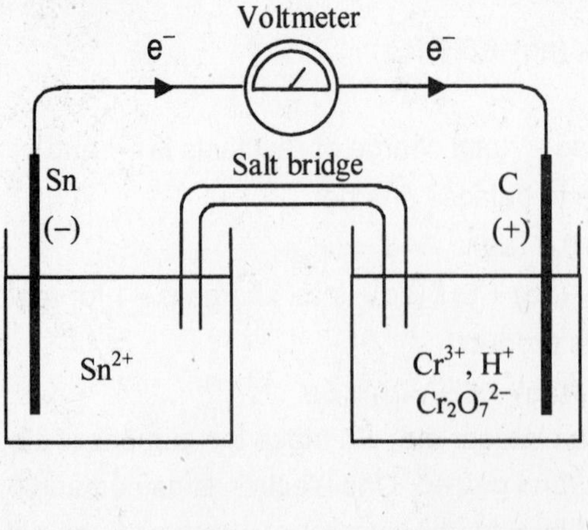

Balanced equations:

Anode is Sn/Sn^{2+} half-cell. Oxidation of Sn produces Sn^{3+}:

$$Sn(s) \rightarrow Sn^{2+}(aq)$$

All that needs to be balanced is charge:

$$Sn(s) \rightarrow Sn^{2+}(aq) + 2e^-$$

Cathode is the $Cr^{3+}/Cr_2O_7^{2-}$ half-cell. Check the oxidation number of chromium in each substance to determine which is reduced. Cr^{3+} oxidation number is +3 and chromium in $Cr_2O_7^{2-}$ has oxidation number +6. Going from +6 to +3 involves gain of electrons so $Cr_2O_7^{2-}$ is oxidized.

$$Cr_2O_7^{2-}(aq) \rightarrow Cr^{3+}(aq)$$

Balance Cr:

$$Cr_2O_7^{2-}(aq) \rightarrow 2Cr^{3+}(aq)$$

Balance O:

$$Cr_2O_7^{2-}(aq) \rightarrow 2Cr^{3+}(aq) + 7H_2O(l)$$

Balance H:

$$Cr_2O_7^{2-}(aq) + 14H^+(aq) \rightarrow 2Cr^{3+}(aq) + 7H_2O(l)$$

Balance charge:

$$Cr_2O_7^{2-}(aq) + 14H^+(aq) + 6e^- \rightarrow 2Cr^{3+}(aq) + 7H_2O(l)$$

Add two half-reactions multiplying the tin half-reaction by 3 to equalize the number of electrons transferred.

$$3Sn(s) + Cr_2O_7^{2-}(aq) + 14H^+(aq) \rightarrow 3Sn^{2+}(aq) + 2Cr^{3+}(aq) + 7H_2O(l)$$

Cell notation:

$$Sn(s) \mid Sn^{2+}(aq) \parallel H^+(aq), Cr_2O_7^{2-}(aq), Cr^{3+}(aq) \mid C(graphite)$$

21.3 <u>Plan</u>: Divide the reaction into half-reactions showing that Br_2 is reduced and V^{3+} is oxidized. Use Equation 21.3 to solve for E^0_{anode}.

<u>Solution</u>:

Half-reactions: cathode $Br_2(aq) + 2e^- \rightarrow 2Br^-(aq)$

anode $2V^{3+}(aq) + 2H_2O(l) \rightarrow 2VO^{2+}(aq) + 4H^+(aq) + 2e^-$

Equation 21.3: $E^0_{cell} = E^0_{cathode} - E^0_{anode}$

$E^0_{cathode}$ = 1.07 V from Sample Problem 21.3

E^0_{cell} = 1.39 V from problem

$E^0_{anode} = E^0_{cathode} - E^0_{cell}$ = 1.07 V – 1.39 V = **-0.32 V**

<u>Check</u>: Reverse the reaction and calculate E^0_{cell}, which should equal –1.39 V.

E^0_{cell} = -0.32 V – 1.07 V = -1.39 V

21.4 <u>Plan</u>: To determine if the reaction is spontaneous divide into half-reactions and calcualte E^0_{cell}. If E^0_{cell} is negative the reaction is not spontaneous, so reverse reaction to obtain the spontaneous reaction. Reducing strength increases with decreasing E^0.

<u>Solution</u>: Divide into half-reactions and balance:

$$Fe^{2+}(aq) + 2e^- \rightarrow Fe(s)$$

$$2Fe^{2+}(aq) \rightarrow 2Fe^{3+}(aq) + 2e^-$$

First half-reaction is reduction, so it's the cathode half-cell. Second half-reaction is oxidation, so it's the anode half-cell. Find half-reactions in Table 21.2.

$E^0_{cell} = E^0_{cathode} - E^0_{anode}$ = -0.44 V – 0.77 V = -1.21 V

The reaction is not spontaneous as written, so reverse the reaction:

$Fe(s) + 2Fe^{3+}(aq) \rightarrow 3Fe^{2+}(aq)$

E^0_{cell} is now +1.21 V so the reaction is spontaneous under standard state conditions.

When a substance acts as a reducing agent it is oxidized. Both Fe and Fe^{2+} can be oxidized, so they can act as reducing agents. But since Fe^{3+} cannot lose more electrons, it cannot act as a reducing agent. The stronger reducing agent between Fe and Fe^{2+} is the one with the smaller standard reduction potential. E^0 for Fe is –0.44 which is less than E^0 for Fe^{2+}, which is +0.77. So, Fe is a stronger reducing agent than Fe^{2+}. Ranking all three in order of decreasing reducing strength gives **Fe > Fe^{2+} > Fe^{3+}.**

21.5 Plan: Reaction is $Cd(s) + Cu^{2+}(aq) \rightarrow Cd^{2+}(aq) + Cu(s)$.

Given ΔG^0 both K and E^0_{cell} can be calculated using the relationships

$\Delta G^0 = -RT\ln K$ and $\Delta G^0 = -nFE^0_{cell}$.

Solution:

$$K = e^{-\Delta G^0/RT} = e^{-(-143000 J)/(8.314 J/mol\cdot K)(298 K)} = 1.17 x 10^{25}$$

$$E^0_{cell} = -\frac{\Delta G^0}{nF} = -\frac{-143000\ J}{(2\ mol\ e^-)(96485\ C/mol\ e^-)} = \mathbf{0.741 V}$$

Note that an alternative way to calculate E^0_{cell} is as equal to (RT/nF)lnK.

Check: Reverse the calculation of K:

$$\Delta G^0 = -RT\ln K = -(8.314\ J/mol\cdot K)(298\ K)\left(\ln(1.17\,x\,10^{25}\right) = -143\ kJ$$

The potential can easily be checked against the value calculated from Table 21.2.

$E^0_{cell} = E^0_{cathode} - E^0_{anode}$ = (0.34 V) – (-0.40 V) = 0.74 V

The potential value agrees with this literature value.

21.6 Plan: The problem is asking for concentration of iron ions when $E_{cell} - E^0_{cell}$ = 0.25 V.

Use the Nernst equation, $E_{cell} = E^0_{cell}$ – (RT/nF)lnQ to find $[Fe^{2+}]$.

Solution: Determining the cell reaction and E^0_{cell}:

$Fe(s) \rightarrow Fe^{2+}(aq) + 2e^-$	E^0 = -0.44 V
$Cu^{2+}(aq) + 2e^- \rightarrow Cu(s)$	E^0 = -0.34 V
$Fe(s) + Cu^{2+}(aq) \rightarrow Fe^{2+}(aq) + Cu(s)$	E^0_{cell} = -0.78 V

For the reaction $Q = [Fe^{2+}]/[Cu^{2+}]$ so Nernst equation is

$$E_{cell} = E^0_{cell} - \frac{RT}{nF}\ln\left(\frac{[Fe^{2+}]}{[Cu^{2+}]}\right)$$

Substituting in values from the problem:

$$0.25 V = -\frac{0.0592 V}{2}\log\left(\frac{[Fe^{2+}]}{0.30 M}\right)$$

$$\frac{[Fe^{2+}]}{0.30 M} = 10^{-8.4_{46}} = 3.5_{81} x 10^{-9};\ [Fe^{2+}] = \mathbf{1.1x10^{-9}\ M}$$

Check: When $[Fe^{2+}] = [Cu^{2+}]$ the log of $[Fe^{2+}]/[Cu^{2+}]$ equals 0 and $E_{cell} = E^0_{cell}$. A change to $[Fe^{2+}] = (0.1)[Cu^{2+}]$ gives $\log([Fe^{2+}]/[Cu^{2+}])$ = -1 and the potential changes by -(-1)(0.03 V) = 0.03 V. Use this to estimate the potential change that should occur if $[Fe^{2+}] = 1 \times 10^{-9}$ and $[Cu^{2+}]$ = 0.1 M. Estimating the ratio $[Fe^{2+}]/[Cu^{2+}] = 1 \times 10^{-9}/0.1 = 1 \times 10^{-8}$ gives log of the ratio equal to –8 to give a potential change of -(-8)(0.06 V/2) = 0.24 V which is almost the given difference in potential.

21.7 Plan: Half-cell B contains a higher concentration of gold ions so the ions will be reduced to decrease the concentration while in half-cell A with a lower $[Au^{3+}]$ gold metal will be oxidized to increase the concentration of gold ions. In the overall cell reaction the lower $[Au^{3+}]$ appears as a product and the higher $[Au^{3+}]$ appears as a reactant. This means that Q $= [Au^{3+}]_{lower}/[Au^{3+}]_{higher}$. Use the Nernst equation with $E^0_{cell} = 0$ to find E_{cell}. Oxidation of gold metal occurs at the anode, half-cell A, which is negative.

Solution:

$$E_{cell} = E^0_{cell} - \frac{RT}{nF}\ln Q = 0 - \frac{0.0592\,V}{3}\log\left(\frac{7.0 x 10^{-4}}{2.5 x 10^{-2}}\right) = \mathbf{0.031\,V}$$

Check: The potential of the cell can be estimated by realizing that each tenfold difference in concentration between the two half cells gives a 20 mV change in potential. The difference is more than tenfold but less than 100 times, so potential should fall between 20 and 40 mV.

21.8 Plan: In the electolysis the more easily reduced metal ion will be reduced at the cathode. Compare K^+ and Al^{3+} as to their location on the periodic table to find which has the higher ionization energy (IE increases up and across the table). Higher ionization energy of the metal means more easily reduced. At the anode the more easily oxidized nonmetal ion will be oxidized. For oxidation compare the electonegativity of the two nonmetals. The ion from the less electronegative element will be more easily oxidized. Note that the standard reduction potentials cannot be used in the case of molten salts because the E^0 values are based on aqueous solutions of ions, not liquid salts.

Solution: Al is above and to the right of K, so Al^{3+} is more easily reduced than K^+. The reaction at the cathode is

$Al^{3+}(l) + 3e^- \rightarrow Al(s)$

F is the most electronegative element so Br^- must be more easily oxidized than F^-. The reaction at the anode is

$2Br^-(l) \rightarrow Br_2(g) + 2e^-$

To add the two half-reaction the number of electrons must be equal:

$2\{Al^{3+}(l) + 3e^- \rightarrow Al(s)\}$

$3\{2Br^-(l) \rightarrow Br_2(g) + 2e^-\}$

$2Al^{3+}(l) + 6Br^-(l) \rightarrow 2Al(s) + 3Br_2(g)$

which is the overall reaction.

21.9 Plan: In aqueous $AuBr_3$ the species present are $Au^{3+}(aq)$, $Br^-(aq)$, H_2O, and very small amounts of H^+ and OH^-. The possible half-reactions are reduction of either Au^{3+} or H_2O and oxidation of either Br^- or H_2O. Whichever reduction and oxidation half-reactions are more spontaneous will take place with consideration of the overvoltage.

Solution: The two possible reductions are

$Au^{3+}(aq) + 3e^- \rightarrow Au(s)$ $E^0 = 1.50$ V

$2H_2O(l) + 2e^- \rightarrow H_2(g) + 2OH^-(aq)$ E = -0.42 V

The reduction of gold ions occurs because it has a higher reduction potential (more spontaneous) than reduction of water.

The two possible oxidations are

$2Br^-(aq) \rightarrow Br_2(l) + 2e^-$ $\quad -E^0 = -1.07$ V

$2H_2O(l) \rightarrow O_2(g) + 4H^+(aq) + 4e^-$ $\quad -E = -0.82$ V

Water appears to be more easily oxidized with the more positive potential, but the overvoltage for water makes the value closer to –1.2 V, so bromide ions are oxidized at the anode.

The two half-reactions that are predicted are

cathode: $Au^{3+}(aq) + 3e^- \rightarrow Au(s)$ $\quad E^0 = 1.50$ V

anode: $2Br^-(aq) \rightarrow Br_2(l) + 2e^-$ $\quad E^0 = 1.07$ V

21.10 Plan: Current is charge per time, so to find the time divide charge by current. To find the charge transferred first write the balanced half-reaction then calculate the charge from the grams of copper and moles of electrons transferred per molar mass of copper.

Solution:

$Cu^{2+}(aq) + 2e^- \rightarrow Cu(s)$ so 2 mol e^- per mole of Cu

$$Charge = (1.50\ g\ Cu)\left(\frac{1\ mol\ Cu}{63.55\ g\ Cu}\right)\left(\frac{2\ mol\ e^-}{1\ mol\ Cu}\right)\left(\frac{96485\ C}{1\ mol\ e^-}\right) = 4.55_{48}\ x\ 10^3\ C$$

$$Time = \frac{Charge}{Current} = \left(\frac{4.55_{48}\ x\ 10^3\ C}{4.75\ C/s}\right)\left(\frac{1\ min}{60\ s}\right) = \mathbf{16.0\ min}$$

Check: Check that units cancel.

(~~g Cu~~)(~~mol Cu/g Cu~~)(~~mol e~~$^-$/~~mol Cu~~)(C/~~mol e~~$^-$) = C

(~~C~~)(~~s~~/~~C~~)(min/~~s~~) = min

END-OF-CHAPTER PROBLEMS

21.1 Oxidation is the loss of electrons while reduction is the gain of electrons. In an oxidation-reduction reaction, electrons are transferred from the substance being oxidized to the substance being reduced. The oxidation number of the reactant being oxidized increases while the oxidation number of the reactant being reduced decreases.

21.3 No, one half-reaction cannot occur independent of the other because the electrons must be transferred from one substance to another. If one substance is losing electrons (oxidation half-reaction) another substance must be gaining those electrons (reduction half-reaction).

21.6 To remove protons from an equation, add an equal number of hydroxide ions to both sides to neutralize the H^+ and produce water: $H^+(aq) + OH^-(aq) \rightarrow H_2O(l)$.

21.8 Spontaneous reactions, $\Delta G_{sys} < 0$, take place in voltaic cells, which are also called galvanic cells. Nonspontaneous reactions take place in electrolytic cells and result in an increase in the free energy of the cell.

21.10 a) To decide which reactant is oxidized look at oxidation numbers. $\mathbf{Cl^-}$ is oxidized because its oxidation number increases from –1 to 0.

b) $\mathbf{MnO_4^-}$ is reduced because the oxidation number of Mn decreases from +7 to +2.

c) The oxidizing agent is the substance that causes the oxidation by accepting electrons. The oxidizing agent is the substance reduced in the reaction, so $\mathbf{MnO_4^-}$ is the oxidizing agent.

d) $\mathbf{Cl^-}$ is the reducing agent because it loses the electrons that are gained in the reduction.

e) **From $\mathbf{Cl^-}$**, which is losing electrons, **to $\mathbf{MnO_4^-}$**, which is gaining electrons.

f) $8H_2SO_4(aq) + 2KMnO_4(aq) + 10KCl(aq) \rightarrow 2MnSO_4(aq) + 5Cl_2(g) + 8H_2O(l) + 6K_2SO_4(aq)$

21.12 a) Divide into half-reactions:

$ClO_3^-(aq) \rightarrow Cl^-(aq)$

$I^-(aq) \rightarrow I_2(s)$

Balance elements other than O and H

$ClO_3^-(aq) \rightarrow Cl^-(aq)$ — chlorine is balanced

$2I^-(aq) \rightarrow I_2(s)$ — iodine now balanced

Balance O by adding H_2O

$ClO_3^-(aq) \rightarrow Cl^-(aq) + 3H_2O(l)$ — add 3 waters to add 3 O's to product

$2I^-(aq) \rightarrow I_2(s)$ — no change

Balance H by adding H^+

$ClO_3^-(aq) + 6H^+(aq) \rightarrow Cl^-(aq) + 3H_2O(l)$ — add 6 H^+ to reactants

$2I^-(aq) \rightarrow I_2(s)$ — no change

Balance charge by adding e^-

$ClO_3^-(aq) + 6H^+(aq) + 6e^- \rightarrow Cl^-(aq) + 3H_2O(l)$ — add 6 e^- to reactants

$2I^-(aq) \rightarrow I_2(s) + 2e^-$ — add 2 e^- to products

Multiply each half-reaction by an integer to equalize the number of electrons

$ClO_3^-(aq) + 6H^+(aq) + 6e^- \rightarrow Cl^-(aq) + 3H_2O(l)$ — multiply by 1 to give 6 e^-

$3\{2I^-(aq) \rightarrow I_2(s) + 2e^-\}$ — multiply by 3 to give 6 e^-

Add half-reactions to give balanced equation in acidic solution.

$\mathbf{ClO_3^-(aq) + 6H^+(aq) + 6I^-(aq) \rightarrow Cl^-(aq) + 3H_2O(l) + 3I_2(s)}$

Check balancing:

Reactants:		Products:	
	1 Cl		1 Cl
	3 O		3 O
	6 H		6 H
	6 I		6 I
	–1 charge		–1 charge

Oxidizing agent is $\mathbf{ClO_3^-}$ and reducing agent is $\mathbf{I^-}$.

b) Divide into half-reactions:

$MnO_4^-(aq) \rightarrow MnO_2(s)$

$SO_3^{2-}(aq) \rightarrow SO_4^{2-}(aq)$

Balance elements other than O and H

$MnO_4^-(aq) \rightarrow MnO_2(s)$ — Mn is balanced

$SO_3^{2-}(aq) \rightarrow SO_4^{2-}(aq)$ — S is balanced

Balance O by adding H_2O

$MnO_4^-(aq) \rightarrow MnO_2(s) + 2H_2O(l)$ — add 2 H_2O to products

$SO_3^{2-}(aq) + H_2O(l) \rightarrow SO_4^{2-}(aq)$ — add 1 H_2O to reactants

Balance H by adding H^+

$MnO_4^-(aq) + 4H^+(aq) \rightarrow MnO_2(s) + 2H_2O(l)$ add 4 H^+ to reactants

$SO_3^{2-}(aq) + H_2O(l) \rightarrow SO_4^{2-}(aq) + 2H^+(aq)$ add 2 H^+ to products

Balance charge by adding e^-

$MnO_4^-(aq) + 4H^+(aq) + 3e^- \rightarrow MnO_2(s) + 2H_2O(l)$ add 3 e^- to reactants

$SO_3^{2-}(aq) + H_2O(l) \rightarrow SO_4^{2-}(aq) + 2H^+(aq) + 2e^-$ add 2 e^- to products

Multiply each half-reaction by an integer to equalize the number of electrons

$2\{MnO_4^-(aq) + 4H^+(aq) + 3e^- \rightarrow MnO_2(s) + 2H_2O(l)\}$ multiply by 2 to give $6e^-$

$3\{SO_3^{2-}(aq) + H_2O(l) \rightarrow SO_4^{2-}(aq) + 2H^+(aq) + 2e^-\}$ multiply by 3 to give $6e^-$

Add half-reactions and cancel substances that appear as both reactants and products

$2MnO_4^-(aq) + \cancel{8}H^+(aq) + 3SO_3^{2-}(aq) + \cancel{3H_2O(l)} \rightarrow 2MnO_2(s) + \cancel{4}H_2O(l) + 3SO_4^{2-}(aq) + \cancel{6H^+(aq)}$

The balanced equation in acidic solution is:

$2MnO_4^-(aq) + 2H^+(aq) + 3SO_3^{2-}(aq) \rightarrow 2MnO_2(s) + H_2O(l) + 3SO_4^{2-}(aq)$

To change to basic solution, add OH^- to both sides of equation to neutralize H^+.

$2MnO_4^-(aq) + 2H^+(aq) + 2OH^-(aq) + 3SO_3^{2-}(aq) \rightarrow 2MnO_2(s) + H_2O(l) + 3SO_4^{2-}(aq) + 2OH^-(aq)$

Balanced equation in basic solution:

$\mathbf{2MnO_4^-(aq) + H_2O(l) + 3SO_3^{2-}(aq) \rightarrow 2MnO_2(s) + 3SO_4^{2-}(aq) + 2OH^-(aq)}$

Check balancing:

Reactants:	2Mn	Products:	2Mn
	18 O		18 O
	2 H		2 H
	3 S		3 S
	-8 charge		-8 charge

Oxidizing agent is MnO_4^- and reducing agent is SO_3^{2-}.

c) Divide into half-reactions:

$MnO_4^-(aq) \rightarrow Mn^{2+}(aq)$

$H_2O_2(aq) \rightarrow O_2(g)$

Balance elements other than O and H – Mn is balanced

Balance O by adding H_2O

$MnO_4^-(aq) \rightarrow Mn^{2+}(aq) + 4H_2O(l)$ add 4 H_2O to products

Balance H by adding H^+

$MnO_4^-(aq) + 8H^+(aq) \rightarrow Mn^{2+}(aq) + 4H_2O(l)$ add 8 H^+ to reactants

$H_2O_2(aq) \rightarrow O_2(g) + 2H^+(aq)$ add 2 H^+ to products

Balance charge by adding e^-

$MnO_4^-(aq) + 8H^+(aq) + 5e^- \rightarrow Mn^{2+}(aq) + 4H_2O(l)$ add 5 e^- to reactants

$H_2O_2(aq) \rightarrow O_2(g) + 2H^+(aq) + 2e^-$ add 2 e^- to products

Multiply each half-reaction by an integer to equalize the number of electrons

$2\{MnO_4^-(aq) + 8H^+(aq) + 5e^- \rightarrow Mn^{2+}(aq) + 4H_2O(l)\}$ multiply by 2 to give 10 e^-

$5\{H_2O_2(aq) \rightarrow O_2(g) + 2H^+(aq) + 2e^-\}$ multiply by 5 to give 10 e^-

Add half-reactions and cancel substances that appear as both reactants and products

$2MnO_4^-(aq) + \cancel{16}H^+(aq) + 5H_2O_2(aq) \rightarrow 2Mn^{2+}(aq) + 8H_2O(l) + 5O_2(g) + \cancel{10H^+(aq)}$

The balanced equation in acidic solution

$\mathbf{2MnO_4^-(aq) + 6H^+(aq) + 5H_2O_2(aq) \rightarrow 2Mn^{2+}(aq) + 8H_2O(l) + 5O_2(g)}$

Check balancing:

Reactants:	2Mn	Products:	2Mn
	18 O		18 O
	16 H		16 H
	+4 charge		+4 charge

Oxidizing agent is MnO_4^- and reducing agent is H_2O_2.

21.14 a) Balance the reduction half-reaction:

$Cr_2O_7^{2-}$(aq) → $2Cr^{3+}$(aq) balance Cr

$Cr_2O_7^{2-}$(aq) → $2Cr^{3+}$(aq) + $7H_2O$(l) balance O

$Cr_2O_7^{2-}$(aq) + $14H^+$(aq) → $2Cr^{3+}$(aq) + $7H_2O$(l) balance H

$Cr_2O_7^{2-}$(aq) + $14H^+$(aq) + $6e^-$ → $2Cr^{3+}$(aq) + $7H_2O$(l) balance charge

Balance the oxidation half-reaction:

Zn(s) → Zn^{2+}(aq) + $2e^-$ balance charge

Add the two half-reactions multiplying the oxidation half-reaction by 3 to equalize the electrons.

$Cr_2O_7^{2-}$(aq) + $14H^+$(aq) + 3Zn(s) → $2Cr^{3+}$(aq) + $7H_2O$(l) + $3Zn^{2+}$(aq)

Oxidizing agent is $Cr_2O_7^{2-}$ and reducing agent is Zn

b) Balance the reduction half-reaction:

MnO_4^-(aq) → MnO_2(s) + $2H_2O$(l) balance O

MnO_4^-(aq) + $4H^+$(aq) → MnO_2(s) + $2H_2O$(l) balance H

MnO_4^-(aq) + $4H^+$(aq) + $3e^-$ → MnO_2(s) + $2H_2O$(l) balance charge

Balance the oxidation half-reaction

$Fe(OH)_2$(s) + H_2O(l) → $Fe(OH)_3$(s) balance O

$Fe(OH)_2$(s) + H_2O(l) → $Fe(OH)_3$(s) + H^+(aq) balance H

$Fe(OH)_2$(s) + H_2O(l) → $Fe(OH)_3$(s) + H^+(aq) + e^- balance e^-

Add half-reactions after multiplying oxidation half-reaction by 3.

MnO_4^-(aq) + ~~4~~H^+(aq) + $3Fe(OH)_2$(s) + ~~3~~H_2O(l) → MnO_2(s) + ~~$2H_2O$(l)~~ + $3Fe(OH)_3$(s) + ~~$3H^+$(aq)~~

Add OH^- to both sides to neutralize the H^+ and convert H^+ + OH^- → H_2O

MnO_4^-(aq) + $3Fe(OH)_2$(s) + $2H_2O$(l) → MnO_2(s) + $3Fe(OH)_3$(s) + OH^-(aq)

Oxidizing agent is MnO_4^- and reducing agent is $Fe(OH)_2$.

c) Balance the reduction half-reaction:

NO_3^-(aq) → N_2(g) balance Cr

NO_3^-(aq) → N_2(g) + $3H_2O$(l) balance O

NO_3^-(aq) + $6H^+$(aq) → N_2(g) + $3H_2O$(l) balance H

NO_3^-(aq) + $6H^+$(aq) + $5e^-$ → N_2(g) + $3H_2O$(l) balance charge

Balance the oxidation half-reaction:

Zn(s) → Zn^{2+}(aq) + $2e^-$ balance charge

Add the half-reactions after multiplying the reduction half-reaction by 2 and the oxidation half-reaction by 5.

$2NO_3^-$(aq) + $12H^+$(aq) + 5Zn(s) → N_2(g) + $6H_2O$(l) + $5Zn^{2+}$(aq)

Oxidizing agent is NO_3^- and reducing agent is Zn.

21.16 a) Balance the reduction half-reaction:

$NO_3^-(aq) \rightarrow NO(g) + 2H_2O(l)$ balance O

$NO_3^-(aq) + 4H^+(aq) \rightarrow NO(g) + 2H_2O(l)$ balance H

$NO_3^-(aq) + 4H^+(aq) + 3e^- \rightarrow NO(g) + 2H_2O(l)$ balance charge

Balance oxidation half-reaction:

$4Sb(s) \rightarrow Sb_4O_6(s)$ balance Sb

$4Sb(s) + 6H_2O(l) \rightarrow Sb_4O_6(s)$ balance O

$4Sb(s) + 6H_2O(l) \rightarrow Sb_4O_6(s) + 12H^+(aq)$ balance H

$4Sb(s) + 6H_2O(l) \rightarrow Sb_4O_6(s) + 12H^+(aq) + 12e^-$ balance charge

Multiply reduction half-reaction by 4 and add half-reactions. Cancel common reactants and products.

$4NO_3^-(aq) + \cancel{16}H^+(aq) + 4Sb(s) + \cancel{6H_2O}(l) \rightarrow 4NO(g) + \cancel{8}H_2O(l) + Sb_4O_6(s) + \cancel{12H^+(aq)}$

Balanced equation in acidic solution:

$$\mathbf{4NO_3^-(aq) + 4H^+(aq) + 4Sb(s) \rightarrow 4NO(g) + 2H_2O(l) + Sb_4O_6(s)}$$

Oxidizing agent is NO_3^- and reducing agent is Sb.

b) Balance reduction half-reaction:

$BiO_3^-(aq) \rightarrow Bi^{3+}(aq) + 3H_2O(l)$ balance O

$BiO_3^-(aq) + 6H^+(aq) \rightarrow Bi^{3+}(aq) + 3H_2O(l)$ balance H

$BiO_3^-(aq) + 6H^+(aq) + 2e^- \rightarrow Bi^{3+}(aq) + 3H_2O(l)$ balance charge

Balance oxidation half-reaction:

$Mn^{2+}(aq) + 4H_2O(l) \rightarrow MnO_4^-(aq)$ balance O

$Mn^{2+}(aq) + 4H_2O(l) \rightarrow MnO_4^-(aq) + 8H^+(aq)$ balance H

$Mn^{2+}(aq) + 4H_2O(l) \rightarrow MnO_4^-(aq) + 8H^+(aq) + 5e^-$ balance H

Multiply reduction half-reaction by 2 and oxidation half-reaction by 2 to transfer 10 e^- in the overall reaction. Cancel H_2O and H^+ in reactants and products.

$5BiO_3^-(aq) + \cancel{30}H^+(aq) + 2Mn^{2+}(aq) + \cancel{8H_2O}(l) \rightarrow 5Bi^{3+}(aq) + \cancel{15}H_2O(l) + 2MnO_4^-(aq) + \cancel{16H^+(aq)}$

Balanced reaction in acidic solution:

$$\mathbf{5BiO_3^-(aq) + 14H^+(aq) + 2Mn^{2+}(aq) \rightarrow 5Bi^{3+}(aq) + 7H_2O(l) + 2MnO_4^-(aq)}$$

BiO_3^- is the oxidizing agent and Mn^{2+} is the reducing agent.

c) Balance the reduction half-reaction:

$Pb(OH)_3^-(aq) \rightarrow Pb(s) + 3H_2O(l)$ balance O

$Pb(OH)_3^-(aq) + 3H^+(aq) \rightarrow Pb(s) + 3H_2O(l)$ balance H

$Pb(OH)_3^-(aq) + 3H^+(aq) + 2e^- \rightarrow Pb(s) + 3H_2O(l)$ balance charge

Balance the oxidation half-reaction

$Fe(OH)_2(s) + H_2O(l) \rightarrow Fe(OH)_3(s)$ balance O

$Fe(OH)_2(s) + H_2O(l) \rightarrow Fe(OH)_3(s) + H^+(aq)$ balance H

$Fe(OH)_2(s) + H_2O(l) \rightarrow Fe(OH)_3(s) + H^+(aq) + e^-$ balance charge

Multiply oxidation half-reaction by 2 and add two half-reactions. Cancel H_2O and H^+.

$Pb(OH)_3^-(aq) + \cancel{3}H^+(aq) + 2Fe(OH)_2(s) + \cancel{2H_2O}(l) \rightarrow Pb(s) + \cancel{3}H_2O(l) + 2Fe(OH)_3(s) + \cancel{2H^+(aq)}$

Add OH^- to both sides to neutralize H^+.

$Pb(OH)_3^-(aq) + \cancel{H^+(aq) + OH^-(aq)} + 2Fe(OH)_2(s) \rightarrow Pb(s) + \cancel{H_2O(l)} + 2Fe(OH)_3(s) + OH^-(aq)$

Balanced reaction in basic solution:

$Pb(OH)_3^-(aq) + 2Fe(OH)_2(s) \rightarrow Pb(s) + 2Fe(OH)_3(s) + OH^-(aq)$

$Pb(OH)_3^-$ is the oxidizing agent and $Fe(OH)_2$ is the reducing agent.

21.18 a) Balance reduction half-reaction:

$MnO_4^-(aq) \rightarrow Mn^{2+}(aq) + 4H_2O(l)$	balance O
$MnO_4^-(aq) + 8H^+(aq) \rightarrow Mn^{2+}(aq) + 4H_2O(l)$	balance H
$MnO_4^-(aq) + 8H^+(aq) + 5e^- \rightarrow Mn^{2+}(aq) + 4H_2O(l)$	balance charge

Balance oxidation half-reaction:

$As_4O_6(s) \rightarrow 4AsO_4^{3-}(aq)$	balance As
$As_4O_6(s) + 10H_2O(l) \rightarrow 4AsO_4^{3-}(aq)$	balance O
$As_4O_6(s) + 10H_2O(l) \rightarrow 4AsO_4^{3-}(aq) + 20H^+(aq)$	balance H
$As_4O_6(s) + 10H_2O(l) \rightarrow 4AsO_4^{3-}(aq) + 20H^+(aq) + 8e^-$	balance charge

Multiply reduction half-reaction by 8 and oxidation half-reaction by 5 to transfer 40 e^- in overall reaction. Add the half-reactions and cancel H_2O and H^+.

$5As_4O_6(s) + 8MnO_4^-(aq) +$ ~~$64H^+$~~$(aq) +$ ~~50~~$H_2O(l) \rightarrow 20AsO_4^{3-}(aq) + 8Mn^{2+}(aq) +$ ~~$32H_2O$~~(l) $+$ ~~100~~$H^+(aq)$

Balanced reaction in acidic solution:

$5As_4O_6(s) + 8MnO_4^-(aq) + 18H_2O(l) \rightarrow 20AsO_4^{3-}(aq) + 8Mn^{2+}(aq) + 36H^+(aq)$

Oxidizing agent: MnO_4^-. Reducing agent: As_4O_6.

b) The reaction gives only one reactant, P_4. Since both products contain phosphorus divide the half-reactions so each inlcude P_4 as the reactant.

Balance reduction half-reaction:

$P_4(s) \rightarrow 4PH_3(g)$	balance P
$P_4(s) + 12H^+(aq) \rightarrow 4PH_3(g)$	balance H
$P_4(s) + 12H^+(aq) + 12e^- \rightarrow 4PH_3(g)$	balance charge

Balance oxidation half-reaction:

$P_4(s) \rightarrow 4HPO_3^{2-}(aq)$	balance P
$P_4(s) + 12H_2O(l) \rightarrow 4HPO_3^{2-}(aq)$	balance O
$P_4(s) + 12H_2O(l) \rightarrow 4HPO_3^{2-}(aq) + 20H^+(aq)$	balance H
$P_4(s) + 12H_2O(l) \rightarrow 4HPO_3^{2-}(aq) + 20H^+(aq) + 12e^-$	balance charge

Add two half-reactions and cancel H^+.

$2P_4(s) +$ ~~$12H^+$~~$(aq) + 12H_2O(l) \rightarrow 4HPO_3^{2-}(aq) + 4PH_3(g) +$ ~~20~~$H^+(aq)$

Balanced reaction in acidic solution:

$2P_4(s) + 12H_2O(l) \rightarrow 4HPO_3^{2-}(aq) + 4PH_3(g) + 8H^+(aq)$ or

$P_4(s) + 6H_2O(l) \rightarrow 2HPO_3^{2-}(aq) + 2PH_3(g) + 4H^+(aq)$

P_4 is both the oxidizing agent and reducing agent.

c) Balance the reduction half-reaction:

$MnO_4^-(aq) \rightarrow MnO_2(s) + 2H_2O(l)$	balance O
$MnO_4^-(aq) + 4H^+(aq) \rightarrow MnO_2(s) + 2H_2O(l)$	balance H
$MnO_4^-(aq) + 4H^+(aq) + 3e^- \rightarrow MnO_2(s) + 2H_2O(l)$	balance charge

Balance oxidation half-reaction:

$CN^-(aq) + H_2O(l) \rightarrow CNO^-(aq)$	balance O
$CN^-(aq) + H_2O(l) \rightarrow CNO^-(aq) + 2H^+(aq)$	balance H

$CN^-(aq) + H_2O(l) \rightarrow CNO^-(aq) + 2H^+(aq) + 2e^-$ balance charge

Multiply the oxidation half-reaction by 3 and reduction half-reaction by 2 to transfer 6 e^- in overall reaction. Add two half-reactions. Cancel H_2O and H^+.

$2MnO_4^-(aq) + 3CN^-(aq) + \cancel{8}H^+(aq) + \cancel{3H_2O}(l) \rightarrow 2MnO_2(s) + 3CNO^-(aq) + \cancel{6H^+}(aq) + \cancel{4}H_2O(l)$

Add 2 OH^- to both sides to neutralize H^+ and form H_2O.

$2MnO_4^-(aq) + 3CN^-(aq) + \cancel{2}H_2O(l) \rightarrow 2MnO_2(s) + 3CNO^-(aq) + \cancel{H_2O(l)} + 2OH^-(aq)$

Balanced reaction in basic solution:

$2MnO_4^-(aq) + 3CN^-(aq) + H_2O(l) \rightarrow 2MnO_2(s) + 3CNO^-(aq) + 2OH^-(aq)$

Oxidizing agent: MnO_4^-. Reducing agent: CN^-.

21.21 a) Balance reduction half-reaction:

$NO_3^-(aq) \rightarrow NO_2(g) + H_2O(l)$ balance O

$NO_3^-(aq) + 2H^+(aq) \rightarrow NO_2(g) + H_2O(l)$ balance H

$NO_3^-(aq) + 2H^+(aq) + e^- \rightarrow NO_2(g) + H_2O(l)$ balance charge

Balance oxidation half-reaction:

$Au(s) + 4Cl^-(aq) \rightarrow AuCl_4^-(aq)$ balance Cl

$Au(s) + 4Cl^-(aq) \rightarrow AuCl_4^-(aq) + 3e^-$ balance charge

Multiply reduction half-reaction by 3 and add half-reactions.

$Au(s) + 3NO_3^-(aq) + 4Cl^-(aq) + 6H^+(aq) \rightarrow AuCl_4^-(aq) + 3NO_2(g) + 3H_2O(l)$

b) Oxidizing agent: **NO_3^-**. Reducing agent: **Au**.

c) The HCl provides chloride ions that combine with the unstable gold ion to form the stable ion, $AuCl_4^-$.

21.22 a) **A** is the anode because by convention the anode is shown on the left.

b) **E** is the cathode because by convention the cathode is shown on the right.

c) **C** is the salt bridge providing electrical connection between the two solutions.

d) **A** is the anode so oxidation takes place there. Oxidation is the loss of electrons meaning that electrons are leaving the anode.

e) **E** is assigned a positive charge because it is the cathode.

f) **E** gains mass because the reduction of the metal ion produced the metal.

21.25 An active electrode is a reactant or product in the cell reaction whereas an inactive electrode is neither a reactant nor a product. An inactive electrode is used to conduct electricity when the half-cell reaction does not include a metal. Platinum and graphite are commonly used as inactive electrodes.

21.26 a) The metal **A** is being oxidized to form the metal cation. To form positive ions an atom must always lose electrons, so this half-reaction is always an oxidation.

b) The metal ion **B** is gaining electrons to form the metal B, so it is displaced.

c) The anode is the electrode at which oxidation takes place, so metal **A** is used as the anode.

d) Acid oxidizes metal B and metal B oxidizes metal A, so acid will oxidize metal A and **bubbles will form** when metal A is placed in acid. The same answer results if strength of reducing agents is considered. The fact that metal A is a better reducing agent than metal B indicates that if metal B reduces acid then metal A will also reduce acid.

21.27 a) If the zinc electrode is negative, oxidation takes place at the zinc electrode:

$Zn(s) \rightarrow Zn^{2+}(aq) + 2e^-$.

Reduction half-reaction: $Sn^{2+}(aq) + 2e^- \rightarrow Sn(s)$

Overall reaction: $Zn(s) + Sn^{2+}(aq) \rightarrow Zn^{2+}(aq) + Sn(s)$

b)

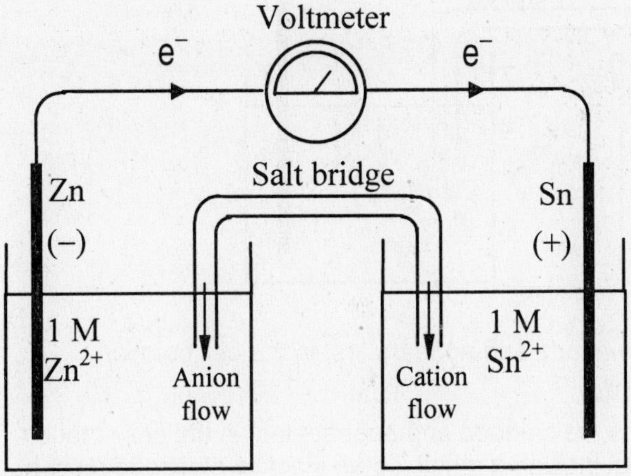

21.29 a) Electrons flow from the anode to the cathode, so **from the iron half-cell to the nickel half-cell**, left to right in the figure. By convention the anode appears on the left and the cathode on the left.

b) Oxidation occurs at the anode which is the electrode in the **iron** half-cell.

c) Electrons enter the reduction half-cell, the **nickel** half-cell in this example.

d) Electrons are consumed in the reduction half-reaction. Reduction takes place at the cathode, **nickel** electrode.

e) The anode is assigned a negative charge, so the **iron** electrode is negatively charged.

f) Metal is oxidized in the oxidation half-cell, so the **iron** electrode will decrease in mass.

g) The solution must contain nickel ions, so any nickel salt can be added. **1 M $NiSO_4$** is one choice.

h) KNO_3 is commonly used in salt bridges, the ions being **K^+ and NO_3^-**. Other salts are also acceptable answers.

i) **Neither** electrode could be replaced by an inactive electrode since both the oxidation and reduction half-reactions include the metal as either a reactant or a product.

j) Anions will move toward the half-cell in which positive ions are being produced. The oxidation half-cell produces Fe^{2+}, so salt bridge anions move **from right** (nickel half-cell) **to left** (iron half-cell).

k) Oxidation half-reaction: $Fe(s) \rightarrow Fe^{2+}(aq) + 2e^-$

Reduction half-reaction: $Ni^{2+}(aq) + 2e^- \rightarrow Ni(s)$

Overall cell reaction: $Fe(s) + Ni^{2+}(aq) \rightarrow Fe^{2+}(aq) + Ni(s)$

21.31 a) The cathode is assigned a positive charge, so the iron electrode is the cathode.

Reduction half-reaction: $Fe^{2+}(aq) + 2e^- \rightarrow Fe(s)$

Oxidation half-reaction: $Mn(s) \rightarrow Mn^{2+}(aq) + 2e^-$

Overall cell reaction: $Fe^{2+}(aq) + Mn(s) \rightarrow Fe(s) + Mn^{2+}(aq)$

b)

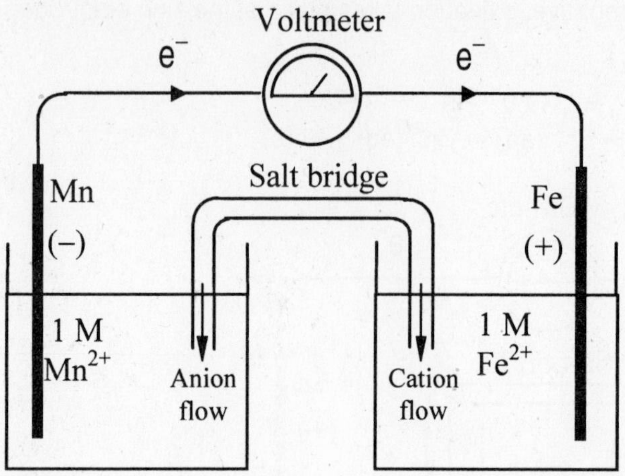

21.33 a) Al is oxidized, so it is the anode and appears first in the cell notation:

$Al(s)|Al^{3+}(aq)||Cr^{3+}(aq)|Cr(s)$

b) Cu^{2+} is reduced, so Cu is the cathode and appears last in the cell notation. The oxidation of SO_2 does not include a metal, so an inactive electrode must be used. Hydrogen ion must be included in the oxidation half cell.

$Pt|SO_2(g)|SO_4^{2-}(aq), H^+(aq)||Cu^{2+}(aq)|Cu(s)$

21.36 A negative E^0_{cell} indicates that the cell reaction is not spontaneous, $\Delta G^0 > 0$. The reverse reaction is spontaneous with $E^0_{cell} > 0$.

21.37 Similar to other state functions the sign of E^0 changes when a reaction is reversed. Unlike ΔG^0, ΔH^0 and S^0, E^0 is an intensive property, the ratio of energy to charge. When the coefficients in a reaction are multiplied by a factor, the values of ΔG^0, ΔH^0 and S^0 are multiplied by the same factor. But E^0 does not change because both the energy and charge are multiplied by the factor and their ratio remains unchanged.

21.38 a) Divide the balanced equation into reduction and oxidation half-reactions and add electrons. Add water and hydroxide ion to the half-reaction that includes oxygen.

Oxidation: $Se^{2-}(aq) \rightarrow Se(s) + 2e^-$

Reduction: $2SO_3^{2-}(aq) + 3H_2O(l) + 4e^- \rightarrow S_2O_3^{2-}(aq) + 6OH^-(aq)$

b) $E^0_{anode} = E^0_{cathode} - E^0_{cell} = -0.57\ V - 0.35\ V = \mathbf{-0.92\ V}$

21.40 a) The greater the reduction potential the greater the strength as an oxidizing agent. In order of decreasing strength: $\mathbf{Br_2 > Fe^{3+} > Cu^{2+}}$.

b) In order of increasing strength as oxidizing agent: $\mathbf{Ca^{2+} < Ag^+ < Cr_2O_7^{2-}}$.

21.42 a)

Oxidation:	$Co(s) \rightarrow Co^{2+}(aq) + 2e^-$	$-E^0 =$	0.28 V
Reduction:	$2H^+(aq) + 2e^- \rightarrow H_2(g)$	$E^0 =$	0.00 V
Overall reaction:	$Co(s) + 2H^+(aq) \rightarrow Co^{2+}(aq) + H_2(g)$	$E^0_{cell} =$	**0.28 V**

Reaction is **spontaneous** under standard state conditions because E^0_{cell} is positive.

b) Oxidation: $2\{Mn^{2+}(aq) + 4H_2O(l) \rightarrow MnO_4^-(aq) + 8H^+(aq) + 5e^-\}$ $-E^0 = -1.51$ V

Reduction: $5\{Br_2(l) + 2e^- \rightarrow 2Br^-(aq)\}$ $E^0 = +1.07$ V

Overall: $2Mn^{2+}(aq) + 5Br_2(l) + 8H_2O(l) \rightarrow 2MnO_4^-(aq) + 10Br^-(aq) + 16H^+(aq)$

$E^0_{cell} =$ **-0.44 V**

Reaction is **not spontaneous** under standard state conditions with $E^0_{cell} < 0$.

c) Oxidation: $Hg_2^{2+}(aq) \rightarrow 2Hg^{2+}(aq) + 2e^-$ $-E^0 = -0.92$ V

Reduction: $Hg_2^{2+}(aq) + 2e^- \rightarrow 2Hg(l)$ $E^0 = +0.85$ V

Overall: $2Hg_2^{2+}(aq) \rightarrow 2Hg^{2+}(aq) + 2Hg(l)$ $E^0_{cell} =$ **-0.07 V**

or $Hg_2^{2+}(aq) \rightarrow Hg^{2+}(aq) + Hg(l)$

Negative E^0_{cell} indicates reaction is **not spontaneous** under standard state conditions.

21.44 a) Oxidation: $Ag(s) \rightarrow Ag^+(aq) + e^-$ $-E^0 = -0.80$ V

Reduction: $Cu^{2+}(aq) + 2e^- \rightarrow Cu(s)$ $E^0 = +0.34$ V

Overall: $2Ag(s) + Cu^{2+}(aq) \rightarrow 2Ag^+(aq) + Cu(s)$ $E^0_{cell} =$ **-0.46 V**

Reaction **not spontaneous**.

b) Oxidation: $Cd(s) \rightarrow Cd^{2+}(aq) + 2e^-$ $-E^0 = +0.40$ V

Reduction: $Cr_2O_7^{2-}(aq) + 14H^+(aq) + 6e^- \rightarrow 2Cr^{3+}(aq) + 7H_2O(l)$ $E^0 = +1.33$ V

Overall: $Cr_2O_7^{2-}(aq) + 3Cd(s) + 14H^+(aq) \rightarrow 2Cr^{3+}(aq) + 3Cd^{2+}(aq) + 7H_2O(l)$

Reaction is **spontaneous**. $E^0_{cell} =$ **$+1.73$ V**

c) Oxidation: $Pb(s) \rightarrow Pb^{2+}(aq) + 2e^-$ $-E^0 = +0.13$ V

Reduction: $Ni^{2+}(aq) + 2e^- \rightarrow Ni(s)$ $E^0 = -0.25$ V

Overall: $Pb(s) + Ni^{2+}(aq) \rightarrow Pb^{2+}(aq) + Ni(s)$ $E^0_{cell} =$ **-0.12 V**

Reaction is **not spontaneous**.

21.46 Adding (1) and (2) to give a spontaneous reaction involves converting (1) to oxidation:

$3N_2O_4(g) + 2Al(s) \rightarrow 6NO_2^-(aq) + 2Al^{3+}(aq)$ $E^0_{cell} = 0.867 \text{ V} - (-1.66 \text{ V}) =$ **2.53 V**

Reverse (1) then add (1) and (3) to give a spontaneous reaction.

$2Al(s) + 3SO_4^{2-}(aq) + 3H_2O(l) \rightarrow 2Al^{3+}(aq) + 3SO_3^{2-}(aq) + 6OH^-(aq)$

$E^0_{cell} = 0.93 \text{ V} - (-1.66 \text{ V}) =$ **2.59 V**

Reverse (2) then add (2) and (3) for the spontaneous reaction:

$SO_4^{2-}(aq) + 2NO_2^-(aq) + H_2O(l) \rightarrow SO_3^{2-}(aq) + N_2O_4(g) + 2OH^-(aq)$

$E^0_{cell} = 0.93 \text{ V} - 0.867 \text{ V} =$ **0.06 V**

Rank oxidizing agents (substance being reduced) in order of increasing strength:

$Al^{3+} < N_2O_4 < SO_4^{2-}$

Rank reducing agents (substance being oxidized) in order of increasing strength:

$SO_3^{2-} < NO_2^- < Al$

21.48 Spontaneous reaction results when (2) is reversed and added to (1):

$2HClO(aq) + Pt(s) + 2H^+(aq) \rightarrow Cl_2(g) + Pt^{2+}(aq) + 2H_2O(l)$

$E^0_{cell} = 1.63 \text{ V} - 1.20 \text{ V} =$ **0.43 V**

Spontaneous reaction results when (3) is reversed and added to (1):

$2HClO(aq) + Pb(s) + SO_4^{2-}(aq) + 2H^+(aq) \rightarrow Cl_2(g) + PbSO_4(s) + 2H_2O(l)$

$E^0_{cell} = 1.63 \text{ V} - (-0.31 \text{ V}) = 1.94 \text{ V}$

Spontaneous reaction results when (3) is reversed and added to (2):

$Pt^{2+}(aq) + Pb(s) + SO_4^{2-}(aq) \rightarrow Pt(s) + PbSO_4(s)$

E^0_{cell} = 1.20 V – (-0.31 V) = **1.51 V**

Order of increasing strength as oxidzing agent: **$PbSO_4 < Pt^{2+} <$ HClO**

Order of increasing strength as reducing agent: **$Cl_2 <$ Pt < Pb + SO_4^{2-}**

21.50 Metal A + Metal B salt $\rightarrow$ solid colored product on metal A

Conclusion: Product is solid metal B. A is better reducing agent than B.

Metal B + acid $\rightarrow$ gas bubbles

Conclusion: Product is H_2 gas produced as result of reduction of H^+. B is better reducing agent than acid.

Metal A + Metal C salt $\rightarrow$ no reaction

Conclusion: C must be a better reducing agent than A.

Since C is a better reducing agent than A which is a better reducing agent than B and B reduces acid, then **C would also reduce acid to form H_2 bubbles**.

Order of strength of reducing agents: **C > A > B**.

21.53 At the negative (anode) electrode oxidation occurs so the overall cell reaction is

$A(s) + B^+(aq) \rightarrow A^+(aq) + B(s)$ with $Q = [A^+]/[B^+]$.

a) The reaction proceeds to the right because with $E_{cell} > 0$ (voltaic cell) the spontaneous reaction occurs. As the cell operates **$[A^+]$ increases and $[B^+]$ decreases**.

b) **E_{cell} decreases** because the cell reaction takes place to approach equilibrium, $E_{cell} = 0$.

c) E_{cell} and E^0_{cell} are related by the Nernst equation: $E_{cell} = E^0_{cell} - (RT/nF)\ln([A^+]/[B^+])$.
$E_{cell} = E^0_{cell}$ when $(RT/nF)\ln([A^+]/[B^+]) = 0$. This occurs when $\ln([A^+]/[B^+]) = 0$. $e^0 = 1$, so **$[A^+]$ must equal $[B^+]$** for E_{cell} to equal E^0_{cell}.

d) **Yes.** It is possible for E_{cell} to be less than E^0_{cell} when **$[A^+] > [B^+]$.**

21.55 In a concentration cell the overall reaction takes place to decrease the concentration of the more concentrated electrolyte. The more concentrated electrolyte is reduced, so it is in the **cathode** compartment.

21.56 The equilibrium constant can be found by combining two equations: $\Delta G^0 = -nFE^0$ and $\Delta G^0 = -RT\ln K$.

a) $E^0_{cell} = E^0_{cathode} - E^0_{anode}$ = 0.80 V — (-0.25 V) = 1.05 V; 2 electrons are transferred.

$$\ln K = \frac{nFE^0}{RT} = \frac{(2\ mol\ e^-)(96485\ C/mol\ e^-)(1.05\ J/C)}{(8.314\ J/mol \cdot K)(298\ K)} = 81.7_{81};\quad K = \mathbf{3x10^{35}}$$

b) E^0_{cell} = -0.74 V – (-0.44 V) = -0.30 V; 6 electrons are transferred.

$$\ln K = \frac{nFE^0}{RT} = \frac{(6\ mol\ e^-)(96485\ C/mol\ e^-)(-0.30\ J/C)}{(8.314\ J/mol \cdot K)(298\ K)} = -70._{10};\quad K = \mathbf{4x10^{-31}}$$

21.58 a) E^0_{cell} = -1.18 V – 0.80 V = -1.98 V; 2 electrons transferred.

$$\ln K = \frac{nFE^0}{RT} = \frac{(2\ mol\ e^-)(96485\ C/mol\ e^-)(-1.98\ J/C)}{(8.314\ J/mol \cdot K)(298\ K)} = -154._{22};\quad K = \mathbf{1x10^{-67}}$$

b) E^0_{cell} = 1.36 V – 1.07 V = 0.29 V

$$\ln K = \frac{nFE^0}{RT} = \frac{(2\,mol\,e^-)(96485\,C/mol\,e^-)(0.29\,J/C)}{(8.314\,J/mol\cdot K)(298\,K)} = 22._{59}\,;\quad K = \mathbf{6x10^9}$$

21.60 a) $\Delta G^0 = -nFE^0$ = -(2 mol e^-)(96485 C/mol e^-)(1.05 J/C) = -2.03 x 10^5 J or **-203 kJ**

Alternate calculation from K in problem 21.56:

ΔG^0 = -RTlnK = -(8.314 J/mol·K)(298 K)ln(3 x 10^{35}) = -2.02 x 10^5 J

b) $\Delta G^0 = -nFE^0$ = -(6 mol e^-)(96485 C/mol e^-)(-0.30 J/C) = 1.73 x 10^5 J or **173 kJ**

21.62 a) $\Delta G^0 = -nFE^0$ = -(2 mol e^-)(96485 C/mol e^-)(-1.98 J/C) = 3.82 x 10^5 J or **382 kJ**

b) $\Delta G^0 = -nFE^0$ = -(2 mol e^-)(96485 C/mol e^-)(0.29 J/C) = -5.59 x 10^4 J or **-55.9 kJ**

21.64 Find ΔG^0 from the fact that it equals –RTlnK. Then use ΔG^0 value to find E^0_{cell}.

ΔG^0 = -RTlnK = -(8.314 J/mol·K)(298 K)ln(5.0 x 10^3) = -2.1 x 10^4 J or **–21 kJ**

$E^0 = -\Delta G^0/nF$ = -(-2.1_{10} x 10^4 J)/(1 mol e^-)(96485 C/mol e^-) = 0.22 J/C = **0.22 V**

21.66 ΔG^0 = -RTlnK = -(8.314 J/mol·K)(298 K)ln(75) = -1.1 x 10^4 J or **–11 kJ**

$E^0 = -\Delta G^0/nF$ = -(-1.0_{70} x 10^4 J)/(2 mol e^-)(96485 C/mol e^-) = **0.055 V**

21.68 E^0_{cell} = 0.34 V – 0.00 V = 0.34 V

$$E_{cell} = E^0_{cell} - \frac{0.0592\,V}{n}\log\left(\frac{[H^+]^2}{[Cu^{2+}]}\right)$$

$$\log\left(\frac{1}{[Cu^{2+}]}\right) = -\frac{(0.25\,V - 0.34\,V)(2)}{0.0592\,V} = 3._{04}\,;\quad [Cu^{2+}] = \mathbf{9x10^{-4}\,M}$$

21.70 Spontaneous reaction is $Ni^{2+}(aq) + Co(s) \rightarrow Ni(s) + Co^{2+}(aq)$ with E^0_{cell} = 0.25 V – (-0.28 V) = 0.03 V.

a) Use the Nernst equation: $E_{cell} = E^0_{cell}$ – (0.0592 V/n)logQ.

E_{cell} = 0.03 V – (0.0592 V/2)log([Co^{2+}]/[Ni^{2+}])

= 0.03 V – (0.0592 V/2)log(0.20 M/0.80 M)

= 0.04$_{78}$ V = **0.05 V**

b) For the [Co^{2+}] to increase from 0.20 M to 0.45 M, a change of 0.25 M, the [Ni^{2+}] must decrease by the same amount, from 0.80 M to 0.55 M.

E_{cell} = 0.03 V – (0.0592 V/2)log(0.45 M/0.55 M)

= 0.03$_{26}$ V = **0.03 V**

c) From part b) notice that an increase in [Co^{2+}] leads to a decrease in cell potential. So, the concentration of cobalt ion must increase further to bring the potential down to 0.025 V. Thus the new concentrations will be [Co^{2+}] = 0.20 M + x and [Ni^{2+}] = 0.80 M – x.

0.025 V = 0.03 V – (0.0592 V/2)log[(0.20 + x)/(0.80 - x)]

x = 0.39$_{60}$

[Ni^{2+}] = 0.80 – 0.39$_{60}$ = **0.40 M**

d) At equilibrium E_{cell} = 0. To decrease the cell potential to 0, [Co^{2+}] increases and [Ni^{2+}] decreases.

0.00 V = 0.03 V – (0.0592 V/2)log[(0.20 + x)/(0.80 - x)]

x = 0.71_{12}

$[Co^{2+}]$ = 0.20 + 0.71 = **0.91 M**; $[Ni^{2+}]$ = 0.80 – 0.71 = **0.09 M**

21.72 The overall cell reaction proceeds to increase the 0.10 M H^+ concentration and decrease the 2.0 M H^+ concentration. So electrode A for the lower concentration cell is the anode.

Q for the cell equals $\dfrac{[H^+]^2_{anode}\,P_{H_2(cathode)}}{[H^+]^2_{cathode}\,P_{H_2(anode)}} = \dfrac{(0.10)^2(0.50)}{(2.0)^2(0.90)} = 0.0013_{89}$

E_{cell} = 0.00 V – (0.0592 V/2)log(0.0013_{89}) = **0.085 V**

21.74 Electrons flow from the anode, where oxidation occurs, to the cathode, where reduction occurs. Whether in a battery or not the flow of electrons is always from the anode to the cathode.

21.76 A D-sized battery is much larger than an AAA-sized battery, so the D-sized battery contains a greater amount of the cell components. The potential, however, is an intensive property and does not depend on the amount of the cell components. (Note that amount is different than concentration). The total amount of charge a battery can produce does depend on the amount of cell components so the D-sized battery produces more charge than the AAA-sized battery.

21.78 The teflon spacers keep the two metals separated so the copper cannot conduct electrons that would promote the corrosion of the iron skeleton. Oxidation of the iron by oxygen causes rust to form and the metal to corrode.

21.81 Sacrificial anodes are metals with E^0 less than that for iron, -0.44 V, so they are more easily oxidized than iron.

a) E^0(aluminum) = -1.66. Yes, except aluminum is known to resist corrosion because once it is covered by a coating of its oxide, no more aluminum corrodes. Therefore it would not be a good choice.

b) E^0(magnesium) = -2.37 V. Yes, magnesium is appropriate to act as a sacrificial anode.

c) E^0(sodium) = -2.71 V. Yes, except sodium reacts with water, so it would not be a good choice.

d) E^0(lead) = -0.13 V. No, lead is not appropriate to act as a sacrificial anode.

e) E^0(nickel) = -0.25 V. No, nickel is not appropriate to act as a sacrificial anode.

f) E^0(zinc) = -0.76 V. Yes, zinc is appropriate to act as a sacrificial anode.

g) E^0(chromium) = -0.74 V. Yes, chromium is appropriate to act as a sacrificial anode.

21.83 E^0_{cell} = -0.40 V – (-0.74 V) = 0.34 V

To reverse the reaction requires 0.34 V with the cell at standard state. A 1.5 V supplies more than enough potential, so the cadmium metal is oxidized to Cd^{2+} and chromium plates out.

21.85 The oxidation number of nitrogen in nitrate, NO_3^-, is +5 and cannot be oxidized further since nitrogen has only 5 electrons in its outer level. In nitrite, NO_2^-, on the other hand, the oxidation number of nitrogen is +3 so it can be oxidized to the +5 state.

21.87 a) At the anode bromide ions are oxidized to form bromine (**Br_2(g)**).

b) At the cathode sodium ions are reduced to form sodium metal (**Na(l)**).

21.89 Either iodide ions or fluoride ions can be oxidized at the anode. The ion that more easily loses an electron will form. Since I is less electronegative than F, I^- will more easily lose its electron and be oxidized at the anode. The product at the anode is **I_2** gas. The iodine is a gas because the temperature is high to melt the salts.

Either potassium or magnesium ions can be reduced at the cathode. Magnesium has greater ionization energy than potassium because magnesium is located up and to the right of potassium on the periodic table. The greater ionization energy means that magnesium ions will more readily add an electron (be reduced) than potassium ions. The product at the cathode is liquid **magnesium**.

21.91 **Bromine** gas forms at the anode because the electronegativity of bromine is less than that of chlorine. **Calcium** metal forms at the cathode because its ionization energy is greater than that of sodium (refer to Figure 8.15).

21.93 **Copper and bromine** can be prepared by electrolysis of their aqueous salts because their half-cell potentials are more positive than the potential for the electrolysis of water with overvoltage, ~1 V for reduction of water and ~1.4 V for the oxidation of water.

21.95 **Iodine, zinc and silver** can be prepared by electrolysis of their salt solutions because water is less readily oxidized than iodide ions and less readily reduced than zinc ions and silver ions.

21.97 a) Possible oxidations:

$2H_2O(l) \rightarrow O_2(g) + 4H^+(aq) + 4e^-$ $\quad -E \approx -1.4$ V with overvoltage

$2F^- \rightarrow F_2(g) + 2e^-$ $\quad -E^0 = -2.87$ V

Water is oxidized to produce oxygen gas (**O_2**) and hydronium ions (**H_3O^+**) at the anode.

Possible reductions:

$2H_2O(l) + 2e^- \rightarrow H_2(g) + 2OH^-(aq)$ $\quad E \approx -1$ V with overvoltage

$Li^+(aq) + e^- \rightarrow Li(s)$ $\quad E^0 = -3.05$ V

Water is reduced at the cathode to produce **H_2** gas and **OH^-**.

b) Possible oxidations:

$2H_2O(l) \rightarrow O_2(g) + 4H^+(aq) + 4e^-$ $\quad -E \approx -1.4$ V with overvoltage

Water is oxidized to produce oxygen gas (**O_2**) and hydronium ions (**H_3O^+**) at the anode.

Possible reductions:

$2H_2O(l) + 2e^- \rightarrow H_2(g) + 2OH^-(aq)$ $\quad E \approx -1$ V with overvoltage

$Sn^{2+}(aq) + 2e^- \rightarrow Sn(s)$ $\quad E^0 = -0.14$ V

$SO_4^{2-}(aq) + 4H^+(aq) + 2e^- \rightarrow SO_2(g) + 2H_2O(l)$ $\quad E \approx -0.63$ V

The potential for sulfate reduction is estimated from the Nernst equation using standard state concentrations and pressures for all reactants and products except H^+, which in pure water is 1×10^{-7} M.

$$E = 0.20V - \left(\frac{0.0592\,V}{2}\right)\log\left(\frac{1}{\left(1 x 10^{-7}\right)^4}\right) = -\,0.63\,V$$

The most easily reduced ion is Sn^{2+} so **tin metal** is formed at the cathode.

21.99 a) Possible oxidations:

$2H_2O(l) \rightarrow O_2(g) + 4H^+(aq) + 4e^-$ -E ≈ -1.4 V with overvoltage

Water is oxidized to produce **oxygen gas and hydronium ions** at the anode.

Possible reductions:

$2H_2O(l) + 2e^- \rightarrow H_2(g) + 2OH^-(aq)$ E ≈ -1 V with overvoltage

$Cr^{3+}(aq) + 3e^- \rightarrow Cr(s)$ $E^0 = -0.74$ V

$NO_3^-(aq) + 4H^+(aq) + 3e^- \rightarrow NO(g) + 2H_2O(l)$ E ≈ +0.13 V

The potential for nitrate reduction is estimated from the Nernst equation using standard state concentrations and pressures for all reactants and products except H^+, which in pure water is 1 x 10^{-7} M.

$$E = 0.96V - \left(\frac{0.0592\,V}{2}\right)\log\left(\frac{1}{\left(1x10^{-7}\right)^4}\right) = +0.13\,V$$

The most easily reduced ion is NO_3^- so **NO gas** is formed at the cathode.

b) Possible oxidations:

$2H_2O(l) \rightarrow O_2(g) + 4H^+(aq) + 4e^-$ -E ≈ -1.4 V with overvoltage

$2Cl^-(aq) \rightarrow Cl_2(g) + 2e^-$ $-E^0 = -1.36$ V

Chloride ions are oxidized to produce **chlorine gas** at the anode.

Possible reductions:

$2H_2O(l) + 2e^- \rightarrow H_2(g) + 2OH^-(aq)$ E ≈ -1 V with overvoltage

$Mn^{2+}(aq) + 2e^- \rightarrow Mn(s)$ $E^0 = -1.18$ V

Water is more readily reduced than manganese ions so **hydrogen gas and hydroxide ions** form at the cathode.

21.101 a) $(35.6\,g\,Mg)\left(\frac{1\,mol\,Mg}{24.31\,g\,Mg}\right)\left(\frac{2\,mol\,e^-}{1\,mol\,Mg}\right) = \mathbf{2.93\,mol\,e^-}$

b) $(2.92_{88}$ mol $e^-)(96485$ C/mol $e^-)$ = **2.83 x 10⁵ C**

c) $\frac{2.82_{59}\,x\,10^5\,C}{(2.50\,h)(3600\,s/h)} = \mathbf{31.4\,A}$

21.103 In the reduction of radium ions, Ra^{2+}, to radium metal two electrons are transferred.

$$(215\,C)\left(\frac{1\,mol\,e^-}{96485\,C}\right)\left(\frac{1\,mol\,Ra}{2\,mol\,e^-}\right)\left(\frac{226\,g\,Ra}{mol\,Ra}\right) = \mathbf{0.252\,g}$$

21.105 $\frac{(85.5\,g\,Zn)(1\,mol\,Zn/65.39\,g\,Zn)(2\,mol\,e^-/mol\,Zn)(96485\,C/mol\,e^-)}{23.0\,C/s} = \mathbf{1.10x10^4\,s}$

21.107 a) The sodium sulfate makes the water conductive so the current will flow through the water to complete the circuit, increasing the rate of electrolysis. Pure water, which contains very low (10^{-7} M) concentrations of H^+ and OH^-, does not conduct electricity.

b) Water is more readily reduced than sodium ions or sulfate ions. Water is the only substance present that can be oxidized.

21.109 $(0.755\ A)\left(\frac{1\ C/s}{1\ A}\right)\left(\frac{3600\ s}{1\ h}\right)\left(\frac{24\ h}{1\ day}\right)(2.00\ days)\left(\frac{1\ mol\ e^-}{96485\ C}\right)\left(\frac{1\ mol\ Zn}{2\ mol\ e^-}\right)\left(\frac{65.39\ g\ Zn}{mol\ Zn}\right) = \mathbf{44.2\ g\ Zn}$

21.111 a) First find the moles of hydrogen gas.

$$n = \frac{PV}{RT} = \frac{(10.0\ atm)(2.5\ x\ 10^6\ L)}{(0.08206\ atm \cdot L/mol \cdot K)(298\ K)} = 1.0_{22}\ x\ 10^6\ mol$$

Then find the coulombs knowing that there are two electrons transferred per mol of H_2.

$$(1.0_{22}\ x\ 10^6\ mol\ H_2)\left(\frac{2\ mol\ e^-}{1\ mol\ H_2}\right)\left(\frac{96485\ C}{mol\ e^-}\right) = 1.9_{73}\ x 10^{11}\ C = \mathbf{2.0\ x\ 10^{11}\ C}$$

b) Remember that 1 V equals 1 J/C.

$(1.24\ J/C)(1.9_{73}\ x\ 10^{11}\ C) = 2.4_{45}\ x\ 10^{11}\ J = \mathbf{2.4\ x\ 10^{11}\ J}$

c) $(2.4_{45}\ x\ 10^{11}\ J)\left(\frac{1\ kJ}{1000\ J}\right)\left(\frac{1\ kg\ oil}{4.0\ x\ 10^4\ kJ}\right) = \mathbf{6.1\ x\ 10^3\ kg}$

21.114 From the current 70.0% of the moles of product will be copper and 30.0% zinc.

$$mass\ \%\ copper = \frac{(70.0\ mol\ Cu)(63.55\ g/mol\ Cu)}{(70.0\ mol\ Cu)(63.55\ g/mol\ Cu) + (30.0\ mol\ Zn)(65.39\ g/mol\ Zn)}\ x\ 100 = \mathbf{69.4\%}$$

21.116 a) Volume of gold plated = $(2\pi)(2.50\ cm)^2(0.20\ mm)(1\ cm/10\ mm) = 0.78_{54}\ cm^3$

mass of gold = $(0.78_{54}\ cm^3)(19.3\ g/cm^3) = 15._{16}\ g$

$$(15._{16}\ g\ Au)\left(\frac{1\ mol\ Au}{197.0\ g\ Au}\right)\left(\frac{3\ mol\ e^-}{mol\ Au}\right)\left(\frac{96486\ C}{mol\ e^-}\right)\left(\frac{1\ s}{0.010\ C}\right)\left(\frac{1\ h}{3600\ s}\right)\left(\frac{1\ day}{24\ h}\right) = \mathbf{26\ days}$$

b) $(15._{16}\ g\ Au)\left(\frac{1\ troy\ ounce}{31.10\ g}\right)\left(\frac{\$320}{1\ troy\ ounce\ Au}\right) = \mathbf{\$160}$

21.119 For the hydrolysis of water the two half-reactions are

anode (+): $2H_2O(l) \rightarrow O_2(g) + 4H^+(aq) + 4e^-$

cathode (-): $2H_2O(l) + 2e^- \rightarrow H_2(g) + 2OH^-(aq)$

At the anode lead, H^+ is produced, the electrode is positive, and the filter paper remains colorless. At the cathode lead, OH^- is produced, the electrode is negative, and the filter paper turns pink.

21.122 a) E^0 for standard hygrogen electrode is 0.00V, for Pb/Pb^{2+} -0.13 V and for Cu/Cu^{2+} 0.34.

Cell with SHE and Pb/Pb^{2+}: E^0_{cell} = **0.13 V**

Cell with SHE and Cu/Cu^{2+}: E^0_{cell} = **0.34 V**

b) The anode (negative electrode) in cell with SHE and Pb/Pb^{2+} is **Pb**.

The anode in cell with SHE and Cu/Cu^{2+} is **platinum** in the SHE.

c) The precipitation of PbS decreases $[Pb^{2+}]$. Use Nernst equation to see how this impacts potential. Cell reaction is

$$Pb(s) + 2H^+(aq) \rightarrow Pb^{2+}(aq) + H_2(g)$$

$E_{cell} = E^0_{cell} - (0.0592\ V/2)\log([Pb^{2+}]P_{H2}/[H^+]^2)$

Decreasing the concentration of lead ions gives a negative value for the term $(0.0592\ V/2)\log([Pb^{2+}]P_{H2}/[H^+]^2)$. When this negative value is subtracted from E^0_{cell} cell potential **increases**.

d) Cell reaction: $Cu^{2+}(aq) + H_2(g) \rightarrow Cu(s) + 2H^+(aq)$

$E_{cell} = 0.34\ V - (0.0592\ V/2)\log([H^+]^2/[Cu^{2+}]P_{H2})$

$= 0.34\ V - (0.0592\ V/2)\log\{1/(1 \times 10^{-16})\}$

$= \mathbf{-0.13\ V}$

21.125 The three steps equivalent to the overall reaction $M^+(aq) + e^- \rightarrow M(s)$ are

1) $M^+(aq) \rightarrow M^+(g)$ Energy is $-\Delta H_{hydration}$
2) $M^+(g) + e^- \rightarrow M(g)$ Energy is $-IE$ or $-\Delta H_{ionization}$
3) $M(g) \rightarrow M(s)$ Energy is $-\Delta H_{atomization}$

The energy for step 3 is similar for all three elements, so the difference in the energy for the overall reaction depends on the values for $-\Delta H_{hydration}$ and $-IE$. The lithium ion has a much more negative hydration energy than Na^+ and K^+ because it is a smaller ion with large charge density that holds the water molecules more tightly. The amount of energy required to remove the waters surrounding the lithium ion offsets the lower ionization energy to make the overall energy for the reduction of lithium larger than expected.

21.127 The potentials in Table 21.2 give half-reactions that include ions in aqueous solutions. The presence of water must be considered in any cell. The substances with large positive E^0 values react to oxidize water, while those with large negative E^0 values react to reduce water. Thus, a cell containing a half-cell with large positive E^0 and one with large negative E^0 would not produce 6V, but the potential for the electrolysis of water, ~1.2 V. To obtain a battery with 6 V using aqueous solutions requires putting several cells of lower voltage in series.

21.129 a) Aluminum half-reaction: $Al^{3+}(aq) + 3e^- \rightarrow Al(s)$, so n = 3. Remember that 1 A = 1 C/s.

$$time = (1000\ kg\ Al)\left(\frac{1000\ g}{1\ kg}\right)\left(\frac{1\ mol\ Al}{26.98\ g}\right)\left(\frac{3\ mol\ e^-}{1\ mol\ Al}\right)\left(\frac{96485\ C}{1\ mol\ e^-}\right)\left(\frac{1\ s}{100{,}000\ C}\right) = \mathbf{1.07 \times 10^5\ s}$$

b) Multiply the time by the current and voltage remembering that 1 A = 1 C/s and 1 V = 1 J/C. Change units of J to kW·h.

$$(1.07_{29} \times 10^5\ s)\left(\frac{100{,}000\ C}{s}\right)\left(\frac{5.0\ J}{C}\right)\left(\frac{1\ kJ}{1000\ J}\right)\left(\frac{1\ kW \cdot h}{3.6 \times 10^3\ kJ}\right) = \mathbf{1.5 \times 10^4\ kW \cdot h}$$

c) From part b the 1.5×10^4 kW·h calculated is per 1000 kg of aluminum. Use the ration of kW·h to mass to find kW·h/lb and then use efficiency and cost per kW·h to find cost per pound.

$$\left(\frac{1.4_{90} \times 10^4\ kW \cdot h}{(1000\ kg\ Al)(0.90)}\right)\left(\frac{1\ kg}{1000\ g}\right)\left(\frac{454\ g}{1\ lb}\right)\left(\frac{0.90\ cents}{kW \cdot h}\right) = \mathbf{6.8\ cents / lb\ Al}$$

21.130 a) Electrons flow **from magnesium to the iron pipe** since magnesium is more easily oxidized than iron.

b) Current is charge per time. Charge can be calculated from the mass of magnesium by converting it to moles of magnesium and multiplying by 2 moles of electrons produced for each mole of magnesium and by Faradays constant to convert the moles of electrons to coulombs of charge. For units of amps time must be in seconds, so convert the 8.5 years to seconds.

$$\frac{(12\ kg\ Mg)\left(\frac{1000\ g}{1\ kg}\right)\left(\frac{1\ mol\ Mg}{24.31\ g\ Mg}\right)\left(\frac{2\ mol\ e^-}{1\ mol\ Mg}\right)\left(\frac{96485\ C}{1\ mol\ e^-}\right)}{(8.5\ yr)\left(\frac{365\ days}{1\ yr}\right)\left(\frac{24\ h}{1\ day}\right)\left(\frac{3600\ s}{h}\right)} = \mathbf{0.36\ A}$$

21.131 Statement: metal D + hot water → reaction.

Conclusion: D reduces water.

Statement: D + E salt → no reaction.

Conclusion: D does not reduce E salt, so E reduces D salt. E is better reducing agent than D.

Statement: D + F salt → reaction

Conclusion: D reduces F salt. D is better reducing agent than F.

If E metal and F salt were mixed the salt would be reduced producing F metal because E has the greatest reducing strength of the three metals (E is stronger than D and D is stronger than F). The ranking of increasing reducing strength is **F < D < E**.

21.134 a) Cell I: Oxidation number (O.N.) of H from 0 to +1, so 1 electron lost from each of 4 hydrogens for a total of 4 electrons. Oxygen O.N. goes from 0 to –2 indicating that 2 electrons are gained by each of the two oxygens for a total of 4 electrons. **Four electrons** are transferred in the reaction. ΔG can be calculated from potential:

$$\Delta G = -nFE = -(4\ mol\ e^-)\left(\frac{96485\ C}{1\ mol\ e^-}\right)(1.23\ V)\left(\frac{1\ J/C}{1\ V}\right)\left(\frac{1\ kJ}{1000\ J}\right) = \mathbf{-475\ kJ}$$

Cell II: In Pb(s) → $PbSO_4$ O.N. of Pb goes from 0 to +2 and in PbO_2 → $PbSO_4$ O.N. goes from +4 to +2. **Two electrons** are transferred the reaction.

$$\Delta G = -nFE = -(2\ mol\ e^-)\left(\frac{96485\ C}{1\ mol\ e^-}\right)(2.04\ V)\left(\frac{1\ J/C}{1\ V}\right)\left(\frac{1\ kJ}{1000\ J}\right) = \mathbf{-394\ kJ}$$

Cell III: O.N. of each of two Na atoms changes from 0 to +1 and O.N. of Fe changes from +2 to 0. **Two electrons** are transferred.

$$\Delta G = -nFE = -(2\ mol\ e^-)\left(\frac{96485\ C}{1\ mol\ e^-}\right)(2.35\ V)\left(\frac{1\ J/C}{1\ V}\right)\left(\frac{1\ kJ}{1000\ J}\right) = \mathbf{-453\ kJ}$$

b) Cell I: $mass\ of\ reactants = (2\ mol\ H_2)\left(\frac{2.016\ g\ H_2}{1\ mol\ H_2}\right) + (1\ mol\ O_2)\left(\frac{32.00\ g\ O_2}{1\ mol\ O_2}\right) = 36.03\ g$

$$\frac{w_{max}}{reactant\ mass} = \frac{-475\ kJ}{36.03\ g} = \mathbf{-13.2\ kJ/g}$$

Cell II:

$$mass\ of\ reactants = (1\ mol\ Pb)\left(\frac{207.2\ g\ Pb}{1\ mol\ Pb}\right) + (1\ mol\ PbO_2)\left(\frac{239.2\ g\ PbO_2}{1\ mol\ PbO_2}\right)$$

$$+(2\ mol\ H_2SO_4)\left(\frac{98.09\ g\ H_2SO_4}{1\ mol\ H_2SO_4}\right) = 642.6\ g$$

$$\frac{w_{max}}{reactant\ mass} = \frac{-394\ kJ}{642.6\ g} = -\mathbf{0.613 kJ / g}$$

Cell III:

$$mass\ of\ reactants = (2\ mol\ Na)\left(\frac{22.99\ g\ Na}{1\ mol\ Na}\right) + (1\ mol\ FeCl_2)\left(\frac{126.75\ g\ FeCl_2}{1\ mol\ FeCl_2}\right) = 172.73\ g$$

$$\frac{w_{max}}{reactant\ mass} = \frac{-453\ kJ}{172.73\ g} = -\mathbf{2.62 kJ / g}$$

Cell I has the highest (most energy released per gram) ratio because the reactants have very low mass while Cell II has the lowest ratio because the reactants are very massive.

21.135 The current traveling through both cells is the same, so the amount of silver is proportional to the amount of zinc based on their reduction half-reactions:

$Zn(s) \rightarrow Zn^{2+}(aq) + 2e^-$ and $Ag^+(aq) + e^- \rightarrow Ag(s)$

$$(1.2\ g\ Zn)\left(\frac{1\ mol\ Zn}{65.39\ g\ Zn}\right)\left(\frac{2\ mol\ e^-}{1\ mol\ Zn}\right)\left(\frac{1\ mol\ Ag}{1\ mol\ e^-}\right)\left(\frac{107.9\ g\ Ag}{1\ mol\ Ag}\right) = \mathbf{4.0\ g\ Ag}$$

21.138 a) Use the current and time to calculate total charge. Remember that the unit 1 A equals 1C/s, so the time must be converted to seconds.

$$(0.15\ A)\left(\frac{1 C / s}{1 A}\right)(54.0\ h)\left(\frac{3600\ s}{1 h}\right) = 2.9_{16}\ x\ 10^4\ C$$

The number of electrons transferred to form copper is calculated by dividing total charge by the charge per mole of electrons. Each mole of copper deposited requires 2 moles of electrons, so divide moles of electrons by 2 to get moles of copper. Then convert to grams of copper.

$$(2.9_{16}\ x\ 10^4\ C)\left(\frac{1\ mol\ e^-}{96485 C}\right)\left(\frac{1\ mol\ Cu}{2\ mol\ e^-}\right)\left(\frac{63.55\ g}{1\ mol\ Cu}\right) = \mathbf{9.6\ g\ Cu}$$

b) The initial concentration of Cu^{2+} is 1M and initial volume is 245 mL. Use this to calculate the initial moles of copper ions then subtract the number of moles of copper ions converted to copper metal.

$$(1\ M\ Cu^{2+})(0.245\ L) - (9.6_{03}\ g\ Cu)\left(\frac{1\ mol\ Cu}{63.55\ g\ Cu}\right) = 0.093_{89}\ mol\ Cu^{2+}$$

$[Cu^{2+}]$ remaining = 0.093_{89} mol Cu^{2+}/0.245 L = **0.38 M**

21.140 Examine each reaction to determine which reactant is the oxidizing agent by which reactant gains electrons in the reaction.

From reaction between $U^{3+} + Cr^{3+} \rightarrow Cr^{2+} + U^{4+}$, find that Cr^{3+} oxidizes U^{3+}.

From reaction between $Fe + Sn^{2+} \rightarrow Sn + Fe^{2+}$, find that Sn^{2+} oxidizes Fe.

From the fact that no reaction occurs between Fe and U^{4+}, find that Fe^{2+} oxidizes U^{3+}.

From reaction between $Cr^{3+} + Fe \rightarrow Cr^{2+} + Fe^{2+}$, find that Cr^{3+} oxidizes Fe.

From reaction between $Cr^{2+} + Sn^{2+} \rightarrow Sn + Cr^{3+}$, find that Sn^{2+} oxidizes Cr^{2+}.

Notice that nothing oxidzes Sn so Sn^{2+} must be the strongest oxidizing agent. Both Cr^{3+} and Fe^{2+} oxidize U^{3+} so U^{4+} must be the weakest oxidizing agent. Cr^{3+} oxidizes iron so Cr^{3+} is a stronger oxidizing agent than Fe^{2+}.

The half-reactions in order from strongest to weakest oxidizing agent:

$Sn^{2+}(aq) + 2e^- \rightarrow Sn(s)$

$Cr^{3+}(aq) + e^- \rightarrow Cr^{2+}(aq)$

$Fe^{2+}(aq) + 2e^- \rightarrow Fe(s)$

$U^{4+}(aq) + e^- \rightarrow U^{3+}(aq)$

21.141 a) Anode is where oxidation takes place. Reaction is $Cu(s) \rightarrow Cu^{2+}(aq) + 2e^-$.

b) Reduction takes place at the cathode: $Cu^{2+}(aq) + 2e^- \rightarrow Cu(s)$

c) Refer to problem 21.192 to find conversion factors, 1 metric ton = 1000 kg and 1 kW·h = 3.6 x 10^3 kJ. The reduction of copper has a half-cell potential of 0.34 V. From mass of copper find moles of electrons. Use moles of electrons to convert to charge in coulombs. Dividing the potential (0.34 V = 0.34 J/C) by the charge gives J, which is then converted to kW·h of electricity.

$$(1\,metric\,ton)\left(\frac{1000\,kg}{1\,metric\,ton}\right)\left(\frac{1000\,g}{1\,kg}\right)\left(\frac{1\,mol\,Cu}{63.55\,g\,Cu}\right)\left(\frac{2\,mol\,e^-}{1\,mol\,Cu}\right)\left(\frac{96485\,C}{1\,mol\,e^-}\right) = 3.036\,x\,10^9\,C$$

$$(0.34\,V)\left(\frac{1\,J/C}{1\,V}\right)(3.036\,x\,10^9\,C)\left(\frac{1\,kJ}{1000\,J}\right)\left(\frac{1\,kW\cdot h}{3.6\,x\,10^3\,kJ}\right) = \mathbf{2.9\,x\,10^2\,kW\cdot h}$$

d) Metals in the same group, Ag and Au are likely impurities along with Ni, Fe, Pd and Pt.

e) The potential applied is not sufficient to oxidize the metals with more positive E^0 (Au, Ag, Pd, Pt), so they remain in the impure "anode mud". Nickel and iron will be oxidized from the anode, but Ni^{2+} and Fe^{2+} will not be reduced at the cathode because their reduction potential is more negative than that of Cu^{2+}. Thus only Cu^{2+} ions are reduced at the cathode.

21.144 a) The calomel half-cell is the anode and the silver half-cell the cathode. The overall reaction is

$$2Ag^+(aq) + 2Hg(l) + 2Cl^-(aq) \rightarrow 2Ag(s) + Hg_2Cl_2(s)$$

and the cell potential is 0.80 V – 0.24 V = 0.56 V with n = 2.

Use the Nernst equation to find $[Ag^+]$ when E_{cell} = 0.60 V.

$$E_{cell} = 0.060\,V = 0.56\,V - \frac{0.0592\,V}{2}\log\frac{1}{[Ag^+]^2[Cl^-]^2}$$

The problem suggests assuming that $[Cl^-]$ is constant. Assume it is 1 M.

$$\log\left(\frac{1}{[Ag^+]^2}\right) = (0.060\,V - 0.56\,V)\left(-\frac{2}{0.0592\,V}\right) = 16._{89}$$

$\log[Ag^+]^2 = -16._{89}$ {math note: log(1/x) = -log(x)}

$[Ag^+]^2 = 10^{-16.89}$

$[Ag^+]^2 = 1.2_{88} \times 10^{-17}$

$[Ag^+] =$ **3.6×10^{-9} M**

Two significant figures comes from subtracting 0.56 from 0.060 that gives 0.50 with only two significant figures.

b) Again use the Nernst equation and assume $[Cl^-]$ = 1 M.

$$E_{cell} = 0.57\,V = 0.56\,V - \frac{0.0592\,V}{2}\log\frac{1}{[Ag^+]^2[Cl^-]^2}$$

$$\log\left(\frac{1}{[Ag^+]^2}\right) = -\log[Ag^+]^2 = (0.57\,V - 0.56\,V)\left(-\frac{2}{0.0592\,V}\right) = -0.3_{38}$$

$2\log[Ag^+] = 0.3_{38}$ {math note: $\log(x^n) = n\log(x)$}

$\log[Ag^+] = 0.1_{69}$

$[Ag^+] = 10^{+0.169} = 1.4_8$ M or **1.5 M**

21.146 a) Nonstandard cell: $E_{waste} = E^o_{cell} - (0.0592\ V)\log[Ag^+]_{waste}$

Standard cell: $E_{standard} = E^o_{cell} - (0.0592\ V)\log[Ag^+]_{standard}$

b) To find $[Ag^+]_{waste}$: $E_{waste} - E_{standard} = (-0.0592\ V)\log([Ag^+]_{waste}/[Ag^+]_{standard})$

$$[Ag^+]_{waste} = [Ag^+]_{standard}\left(10^{E_{standard} - E_{waste}/0.0592}\right)$$

c) Convert M to ng/L for both $[Ag^+]_{waste}$ and $[Ag^+]_{standard}$:

$$E_{waste} - E_{standard} = (-0.0592\,V)\log\left(\frac{\left(\frac{mol\ Ag^+{}_{waste}}{L}\right)\left(\frac{107.9\ g\ Ag}{1\,mol\ Ag}\right)\left(\frac{1\,x\,10^9\ ng}{1\,g}\right)}{\left(\frac{mol\ Ag^+{}_{standard}}{L}\right)\left(\frac{107.9\ g\ Ag}{1\,mol\ Ag}\right)\left(\frac{1\,x\,10^9\ ng}{1\,g}\right)}\right)$$

$$= (-0.0592\,V)\log\left(\frac{ng\ Ag^+{}_{waste}/L}{ng\ Ag^+{}_{standard}/L}\right)$$

$$ng\ Ag^+_{waste}/L = \left(ng\ Ag^+_{standard}/L\right)\left(10^{E_{standard} - E_{waste}/0.0592}\right)$$

d) $$0.003\,V = (-0.0592\,V)\log\left(\frac{ng\ Ag^+{}_{waste}/L}{1000\,ng\ Ag^+/L}\right)$$

$$\left(\frac{ng\ Ag^+{}_{waste}/L}{1000\,ng\ Ag^+/L}\right) = 10^{-(0.003/0.0592)} = 0.8_{90}$$

$ng\ Ag^+{}_{waste}/L$ = **9 x 10^2 ng/L**

e) Temperature is included in the RT/nF term which equals 0.0592 V/n at 25°C. To account for different temperatures insert the RT/nF term in place of 0.0592 V/n.

$E_{waste} - E_{standard} = (-RT_{waste}/nF)\log[Ag^+]_{waste} + (RT_{standard}/nF)\log[Ag^+]_{standard}$

$$\log[Ag^+]_{waste} = \frac{nF(E_{standard} - E_{waste})}{RT_{waste}} + \frac{T_{standard}\left(\log[Ag^+]_{standard}\right)}{T_{waste}}$$

$$[Ag^+]_{waste} = 10^{\left(nF(E_{standard} - E_{waste})/R + T_{standard}\log[Ag^+]_{standard}\right)/T_{waste}}$$

CHAPTER 22

THE ELEMENTS IN NATURE AND INDUSTRY

END-OF-CHAPTER PROBLEMS

22.2 Refer to Figure 22.4 to determine how metals most commonly occur in the crust. Iron forms iron(III) oxide, Fe_2O_3 (commonly known as hematite). Calcium forms calcium carbonate, $CaCO_3$, (commonly known as limestone). Sodium is commonly found in sodium chloride (halite), NaCl. Zinc is commonly found in zinc sulfide (sphalerite), ZnS.

22.3 a) Differentiation refers to the processes involved in the formation of the earth into regions (core, mantle and crust) of differing composition. Substances separated according to their densities with the more dense material in the core and the less dense in the crust.

b) The four most abundant elements are oxygen, silicon, aluminum and iron in order of decreasing abundance.

c) Oxygen is the most abundant element in the crust and mantle but is not found in the core. Silicon, aluminum, calcium, sodium, potassium and magnesium are also present in the crust and mantle but not in the core.

22.7 Plants produced O_2, slowly increasing the oxygen concentration in the atmosphere and creating an oxidative environment for metals. Evidence of Fe(II) deposits pre-dating Fe(III) deposits suggests this hypothesis is true. The decay of plant material and incorporation into the crust increased the concentration of carbon in the crust and created large fossil fuel deposits.

22.9 Fixation refers to the process of converting a nutrient in the gaseous phase into another, more readily usable form. Examples are the fixation of carbon by plants in the form of carbon dioxide and of nitrogen by bacteria in the form of nitrogen gas. Fixation of carbon dioxide gas by plants converts the CO_2 into carbohydrates during photosynthesis. Fixation of nitrogen gas by nitrogen-fixing bacteria involves the conversion of N_2 to ammonia and ammonium ions.

22.12 Atmospheric nitrogen is utilized by three fixation pathways: atmospheric, industrial, and biological. Atmospheric fixation requires high-temperature reactions (e.g. lightning) to convert N_2 into NO and other oxidized species. Industrial fixation involves mainly the formation of ammonia, NH_3, from N_2 and H_2. Biological fixation occurs in nitrogen-fixing bacteria that live in the roots of legumes.

According to Figure 22.6, "human activity" is referred to as "industrial fixation". This portion of the atmospheric nitrogen fixation cycle accounts for $36\text{x}10^6$ tons ÷ ($36+10+140+20 \times 10^6$ tons), or about 17%.

22.14 a) No gaseous phosphorus compounds are involved in the phosphorus cycle, so the atmosphere is not included in the phosphorus cycle.

b) Two roles of organisms in phosphorus cycle: 1) Plants excrete acid from their roots to convert PO_4^{3-} ions into more soluble $H_2PO_4^-$ ions which the plant can absorb. 2) Through excretion and decay after death, organisms return soluble phosphate compounds to the cycle.

22.17 a) First determine the amount of F present in 100. kg of fluorapatite, using the molar mass of $Ca_5(PO_4)_3F$ (504.31 g/mol). Take 15% of this amount, convert mol F to mol SiF_4, and use the ideal gas law to determine volume of SiF_4 gas.

$$amount\ F(mol) = (100.\ kg\ Ca_5(PO_4)_3F)\left(\frac{1000\ g}{kg}\right)\left(\frac{1\ mol\ Ca_5(PO_4)_3F}{504.31\ g}\right)\ x$$

$$\left(\frac{1\ mol\ F}{1\ mol\ Ca_5(PO_4)_3F}\right) = 198._{29}\ mol\ F$$

amount of F that forms SiF_4 = (198.$_{29}$ mol F)(0.15) = 29.$_{74}$ mol F

$$mol\ SiF_4 = (29._{74}\ mol\ F)\left(\frac{1\ mol\ SiF_4}{4\ mol\ F}\right) = 7.4_{36}\ mol\ SiF_4$$

$$V = \frac{nRT}{P} = \frac{(7.4_{36}\ mol)(0.08206\ atm \cdot L/mol \cdot K)(1450+273)K}{1.00\ atm} = \mathbf{1.1x10^3\ L}$$

b) Assume that all of the fluorine in sodium hexafluorosilicate is available as F^- when redissolved in drinking water. First calculate the moles of Na_2SiF_6 that form from 7.4$_{36}$ mol of SiF_4 according to the reaction stoichiometry. Then calculate the mass of F^- available from the Na_2SiF_6 that is produced in the reaction.

$$mol\ Na_2SiF_6 = (7.4_{36}\ mol\ SiF_4)\left(\frac{1\ mol\ Na_2SiF_6}{2\ mol\ SiF_4}\right) = 3.7_{18}\ mol\ Na_2SiF_6$$

$$mass\ F^- = (3.7_{18}\ mol\ Na_2SiF_6)\left(\frac{6\ mol\ F^-}{1\ mol\ Na_2SiF_6}\right)\left(\frac{19.00\ g\ F^-}{1\ mol\ F^-}\right) = 42_{3.9}\ g\ F^-$$

The definition of ppm ("parts per million") states that 1 ppm F^- = (1 g F^-)/(10^6 g H_2O). If 1 g F^- will fluoridate 10^6 g H_2O to a level of 1 ppm, how many grams (converted to mL using density, then converted from mL to L to m^3) of water can 424 g F^- fluoridate to the 1 ppm level? Necessary conversion factors: 1 m^3 = 1000 L, , d_{H2O} = 1.00 g/mL.

$$V(m^3) = (42_{3.9}\ g\ F^-)\left(\frac{10^6\ g\ H_2O}{1\ g\ F^-}\right)\left(\frac{mL\ H_2O}{1.00\ g\ H_2O}\right)\left(\frac{1\ L}{1000\ mL}\right)\left(\frac{1\ m^3}{1000\ L}\right) = 420\ m^3 = \mathbf{4.2x10^2\ m^3}$$

22.18 a) The iron ions form an insoluble salt, $Fe_3(PO_4)_2$ that decreases the yield of phosphorus. This salt is of limited value.

b) Each mole of $Ca_3(PO_4)_2$ contains 2 moles of P atoms and produces, at 100% yield, 0.5 mol of P_4. Use conversion factor 1 metric ton (T) = 1 x 10^3 kg.

$$(50\ T\ ore)\left(\frac{0.98\ T\ Ca_3(PO_4)_2}{1\ T\ ore}\right)\left(\frac{1 x 10^3\ kg}{1\ T}\right)\left(\frac{1 x 10^3\ mol\ Ca_3(PO_4)_2}{310.18\ kg\ Ca_3(PO_4)_2}\right) = 1.5_{80}\ x\ 10^5\ mol\ Ca_3(PO_4)_2$$

$$(1.5_{80}\ x\ 10^5\ mol\ Ca_3(PO_4)_2)\left(\frac{0.5\ mol\ P_4}{1\ mol\ Ca_3(PO_4)_2}\right)\left(\frac{123.88\ kg\ P_4}{1 x 10^3\ mol\ P_4}\right) = 9.7_{85}\ x\ 10^3\ kg\ P_4$$

$$(9.7_{85}\ x\ 10^3\ kg\ P_4)(0.90\ yield)\left(\frac{1\ T}{1 x 10^3\ kg}\right) = \mathbf{8.8\ T\ P_4}$$

22.20 a) Roasting involves heating the mineral in air (O_2) at high temperatures to convert the mineral to the oxide.

b) Smelting is the reduction of the metal oxide to the free metal using heat and a reducing agent like coke.

c) Flotation is a separation process in which the ore is removed from the gangue by exploiting the different abilities of the two to interact with detergent. The gangue sinks to the bottom and the lighter ore-detergent mix is skimmed off the top.

d) Refining is the final step in the purification process to yield the pure metal with no impurities.

22.25 a) Slag is a waste product of iron metallurgy formed by the reaction

$$CaO(s) + SiO_2(s) \rightarrow CaSiO_3(l)$$

In other words, slag is a byproduct of steel-making and contains the impurity SiO_2.

b) Pig iron, used to make cast iron products, is the impure product of iron metallurgy (containing 3-4% C) that is purified to steel.

c) Steel refers to the products of iron metallurgy, specifically alloys of iron containing small amounts of other elements including 1 – 1.5% carbon.

d) Basic-oxygen process refers to the process used to purify pig iron to form steel. The pig iron is melted and oxygen gas under high pressure is through the liquid metal. The oxygen oxidizes impurities to their oxides which then react with calcium oxide to form a liquid that is decanted. Molten steel is left after the basic-oxygen process.

22.27 Iron and nickel are more easily oxidized than copper, so they are separated from the copper in roasting step and conversion to slag. In the electrorefining process, all three metals are oxidized into solution, but only Cu^{2+} ions are reduced at the cathode to form Cu(s).

22.30 Hess's Law is used to evaluate the thermodynamic properties of each step in a process.

22.31 a) Aqueous salt solutions are mixtures of ions and water. When two half-reactions are possible at an electrode, the one with the more positive electrode potential occurs (Section 21.7). In this case, the two half-reactions are

$M^+ + e^- \rightarrow M^0$ $\quad E_{red}^0 = -3.05$ V, -2.93 V and -2.71 V for Li^+, K^+ and Na^+, respectively

$2H_2O + 2e^- \rightarrow H_2 + 2OH^-$ $\quad E_{red}^0 = -0.42$ V, with overvoltage ~ -1 V

In all of these cases, it is energetically more favorable to reduce H_2O to H_2 than to reduce M^+ to M.

b) The question asks if Ca could chemically reduce RbX, i.e. convert Rb^+ to Rb^0. In order for this to occur, Ca^0 loses electrons ($Ca^0 \rightarrow Ca^{2+} + 2e^-$) and Rb^+ gains an electron ($2Rb^+ + 2e^- \rightarrow 2Rb^0$). The reaction is written as follows:

$2RbX + Ca \rightarrow CaX_2 + 2Rb$

where $\Delta H = IE_1(Ca) + IE_2(Ca) - 2\,IE_1(Rb) = 590 + 1145 - 2(403) = +929$ kJ/mol.

Recall that Ca^0 acts as a *reducing agent* for the Rb^+ ion because it *oxidizes*. The energy required to remove an electron is the ionization energy. It requires more energy to ionize calcium's electrons, so it seems unlikely that Ca^0 could reduce Rb^+. Based on IE's and a positive ΔH for the forward reaction, it seems more reasonable that Rb^0 would reduce Ca^{2+}.

c) If the reaction is carried out at a temperature greater than 688°C (the boiling point of rubidium), the product mixture will contain gaseous Rb. This can be removed from the reaction vessel, causing a shift in equilibrium to form more Rb product. If the reaction is carried out between 688°C and 1484°C (b.p. for Ca), than Ca remains in the molten phase and remains separated from gaseous Rb.

d) The reaction of calcium with molten CsX is written as follows:

$2CsX + Ca \rightarrow CaX_2 + 2Cs$

where $\Delta H = IE_1(Ca) + IE_2(Ca) - 2\ IE_1(Cs) = 590 + 1145 - 2(376) = +983$ kJ/mol. This reaction is more unfavorable than for Rb, but Cs has a lower boiling point of 671°C. If the reaction is carried out between 671°C and 1484°C, then calcium can be used to separate gaseous Cs from molten CsX.

22.32 a) For each mole of Na metal produced 0.5 mole of Cl_2 is produced. Calculate amount of chlorine gas from stoichiometry, then use ideal gas law to find volume of chlorine gas.

$$(30.0\ kg\ Na)\left(\frac{1x10^3\ mol\ Na}{22.99\ kg\ Na}\right)\left(\frac{1\ mol\ Cl_2}{2\ mol\ Na}\right) = 652._{46}\ mol\ Cl_2$$

$$V = \frac{nRT}{P} = \frac{(652._{46}\ mol\ Cl_2)(0.08206\ atm \cdot L / mol \cdot K)(853\ K)}{1.0\ atm} = \mathbf{4.6x10^4\ L}$$

b) Two moles of electrons are passed through the cell for each mole of Cl_2 produced.

$$(652._{46}\ mol\ Cl_2)\left(\frac{2\ mol\ e^-}{1\ mol\ Cl_2}\right)\left(\frac{96485\ C}{1\ mol\ e^-}\right) = 1.25_{91}x10^8\ C = \mathbf{1.26x10^8\ C}$$

c) Current is charge per time with the amp unit equal to C/s.

$(1.25_{91} \times 10^8\ C) \div (75.0\ C/s) = 1.68 \times 10^6$ seconds or **466 hours**

22.35 a) Mg^{2+} is more difficult to reduce than H_2O, so $H_2(g)$ would be produced instead of Mg metal. $Cl_2(g)$ forms at the anode due to overvoltage (see problem 22.31a and Section 21.7).

b) The ΔH_f^0 for $MgCl_2(s)$ is –641.6 kJ/mol: $Mg(s) + Cl_2(g) \rightarrow MgCl_2(s)$ + heat

High temperature favors the reverse reaction (which is endothermic), so high temperature favor the formation of magnesium metal and chlorine gas.

22.37 a) Sulfur dioxide is the reducing agent and is oxidized to the +6 state (as sulfate ion, SO_4^{2-}).

b) The sulfate ion formed reacts as a base in the presence of acid to form the hydrogen sulfate ion.

$SO_4^{2-}(aq) + H^+(aq) \rightarrow HSO_4^-(aq)$

c) Skeleton redox equation: $SO_2(g) + H_2SeO_3(aq) \rightarrow Se(s) + HSO_4^-(aq)$

Reduction half-reaction:	$H_2SeO_3(aq) \rightarrow Se(s)$
balance O and H	$H_2SeO_3(aq) + 4H^+(aq) \rightarrow Se(s) + 3H_2O(l)$
balance charge	$H_2SeO_3(aq) + 4H^+(aq) + 4e^- \rightarrow Se(s) + 3H_2O(l)$
Oxidation half-reaction:	$SO_2(g) \rightarrow HSO_4^-(aq)$
balance O and H	$SO_2(g) + 2H_2O(l) \rightarrow HSO_4^-(aq) + 3H^+(aq)$
balance charge	$SO_2(g) + 2H_2O(l) \rightarrow HSO_4^-(aq) + 3H^+(aq) + 2e^-$

Multiply oxidation half-reaction by 2 and add the two half-reactions.

$H_2SeO_3(aq) + 2SO_2(g) + H_2O(l) \rightarrow Se(s) + 2HSO_4^-(aq) + 2H^+(aq)$

22.42 a) The oxidation state of copper in Cu_2S is +1 (sulfur is –2).
The oxidation state of copper in Cu_2O is +1 (oxygen is –2).
The oxidation state of copper in Cu is 0 (ox state is always 0 in the elemental form).

b) The reducing agent is the species that is *oxidized*. Sulfur changes oxidation state from –2 in Cu_2S to +4 in SO_2. Therefore, Cu_2S is the reducing agent, leaving Cu_2O as the oxidizing agent.

22.44 a) Use the surface area and thickness of the film to calculate its volume. Multiply the volume by the density to find mass of the aluminum oxide. Convert mass to moles aluminum oxide. The oxidation of aluminum involves the loss of 3 electrons for each aluminum atom or 6 electrons for each Al_2O_3 compound formed. From this find the number of electrons produced and multiply by Faraday's constant to find coulombs.

$$\left(2.3\,m^2\right)\left(20.\,x\,10^{-6}\,m\right)\left(\frac{100\,cm}{1\,m}\right)^3\left(\frac{3.97\,g\,Al_2O_3}{cm^3}\right)\left(\frac{1\,mol\,Al_2O_3}{101.96\,g\,Al_2O_3}\right)=1.7_{91}\,mol\,Al_2O_3$$

$$\left(1.7_{91}\,mol\,Al_2O_3\right)\left(\frac{6\,mol\,e^-}{1\,mol\,Al_2O_3}\right)\left(\frac{96485\,C}{mol\,e^-}\right)=\mathbf{1.0\,x\,10^6\,C}$$

b) Current in amps can be found by dividing charge in coulombs by time in seconds.

$$\frac{1.0_{37}\,x\,10^6\,C}{\left(15\,min\right)\left(\frac{60\,s}{min}\right)}=\mathbf{1.2\,x\,10^3\,A}$$

22.47 Free energy, ΔG, is the measure of the ability of a reaction to proceed. The direct reduction of ZnS follows the reaction:

$2ZnS(s) + C(graphite) \rightarrow 2Zn(s) + CS_2(g)$

ΔG_{rxn}^0 = {(2mol)(0) + (1mol)(66.9 kJ/mol)} – {(2mol)(-198 kJ/mol) + (1mol)(0)}

ΔG_{rxn}^0 = +463 kJ

Since ΔG_{rxn}^0 is positive, this reaction is not spontaneous. The stepwise reduction involves the conversion of ZnS to ZnO, followed by the final reduction step:

$2ZnO(s) + C(s) \rightarrow 2Zn(s) + CO_2(g)$

ΔG_{rxn}^0 = {(2mol)(0) + (1mol)(-394.4 kJ/mol)} – {(2mol)(-318.2 kJ/mol) + (1mol)(0)}

ΔG_{rxn}^0 = +242.0 kJ

This reaction is also not spontaneous, but the oxide reaction is less unfavorable.

22.48 The formation of sulfur trioxide is very slow at ordinary temperatures. Increasing the temperature can speed up the reaction, but the reaction is exothermic so increasing the temperature decreases the yield. Recall that yield, or the extent to which a reaction proceeds, is controlled by the thermodynamics of the reaction. Adding a catalyst increases the rate of the formation reaction but does not impact the thermodynamics, so a lower temperature can be used to enhance the yield. Catalysts are used in such a reaction to allow control of both rate and yield of the reaction.

22.51 a) The chlor-alkali process yields Cl_2, H_2 and NaOH.

b) The mercury-cell process yields higher purity NaOH, but produces Hg-polluted waters that are discharged into the environment.

22.52 a) The balanced equation for the oxidation of SO_2 to SO_3: $2SO_2(g) + O_2(g) \rightarrow 2SO_3(g)$

ΔG^0 = (2mol SO_3)(ΔG_f^0 of SO_3) – {(2mol SO_2)(ΔG_f^0 of SO_2) + (1mol O_2)(ΔG_f^0 of O_2)}

= (2 mol SO_3)(– 371 kJ/mol SO_3) - (2 mol SO_2)(-300.2 kJ/mol SO_2) - (1 mol O_2)(0)

= **-142 kJ**

The reaction is spontaneous at 25°C.

b) The rate of the reaction is very slow at 25°C, so the reaction does not produce significant amounts of SO_3 at room temperature.

c) Calculate standard state free energy at 500.°C (773 K) using values for ΔH^0 and ΔS^0 at 25°C with the assumption that these values are constant with temperature.

ΔH^0 = (2 mol SO_3)(ΔH_f^0 of SO_3) - (2 mol SO_2)(ΔH_f^0 of SO_2) - (1 mol O_2)(ΔH_f^0 of O_2)

= (2 mol SO_3)(– 396 kJ/mol SO_3) - (2 mol SO_2)(-296.8 kJ/mol SO_2) - (1 mol O_2)(0)

= -198 kJ

ΔS^0 = (2 mol SO_3)(S^0 of SO_3) - (2 mol SO_2)(S^0 of SO_2) - (1 mol O_2)(S^0 of O_2)

= (2 mol SO_3)(256.66 J/mol·K) - (2 mol SO_2)(248.1 J/mol·K) - (1 mol O_2)(205.0 J/mol·K)

= -187.9 J/K or –0.1879 kJ/K

$\Delta G_{500} = \Delta H^0 - T\Delta S^0$ = -198 kJ – (773 K)(-0.1879 kJ/K) = **-53 kJ**

The reaction is spontaneous at 500°C since free energy is negative.

d) The equilibrium constant at 500°C is smaller than the equilibrium constant at 25°C because the free energy at 500°C indicates the reaction does not go as far to completion as it does at 25°C.

Equilibrium constants can be calculated from $\Delta G^0 = -RT\ln K$.

At 25°C lnK = -(-142 kJ)/{(8.314 x 10^{-3} kJ/mol·K)(298 K)} = 57.3; K = $e^{57.3}$ = 7.8 x 10^{24}

At 500°C lnK = -(-53 kJ)/{(8.314 x 10^{-3} kJ/mol·K)(773 K)} = 8.25; K = $e^{8.25}$ = 3.8 x 10^3

The equilibrium constant decreases by a factor of 10^{20} with an increase in temperature from 25°C to 500°C.

e) Temperature below which the reaction is spontaneous at standard state can be calculated by setting ΔG^0 equal to zero and using the enthalpy and entropy values at 25°C to calculate the temperature.

$$T = \frac{\Delta H^0}{\Delta S^0} = \frac{-198\ kJ}{-0.1879\ kJ/K} = \mathbf{1.05x10^3\ K}$$

22.53 This problem deals with the stoichiometry of electrolysis (see Section 21.7). The balanced oxidation half reaction for the chlor-alkali process is given in Section 22.5:

$2Cl^-(aq) \rightarrow Cl_2(g) + 2e^-$

Follow the steps outlined in Figure 21.20, starting with the current given. Use the Faraday constant, *F*, (1 *F* = 9.65 x 10^4 C/mol e^-) and the fact that 1 mol of Cl_2 produces 2 mol e^- or 2 *F*, to convert coulombs to moles of Cl_2.

$$mol\ Cl_2 = \left(3x10^4\ \frac{C}{s}\right)(8\ hr)\left(3600\frac{s}{hr}\right)\left(\frac{1\ F}{9.65x10^4\ C}\right)\left(\frac{1\ mol\ Cl_2}{2\ F}\right) = 4._{48}\ x10^3\ mol$$

Convert moles of Cl_2 to lbs using the molar mass and the english-metric conversion factor.

$$lbs\ Cl_2 = (4._{48}\ x10^3\ mol)\left(\frac{70.90\ g\ Cl_2}{mol\ Cl_2}\right)\left(\frac{2.205\ lb}{1000\ g}\right) = \mathbf{7x10^2\ lb\ Cl_2}$$

22.57 a) Because P_4O_{10} is a drying agent, water is incorporated in the formation of phosphoric acid, H_3PO_4.

$P_4O_{10}(s) + 6H_2O(l) \rightarrow 4H_3PO_4(l)$

b) First calculate the concentration (M) of the resulting phosphoric acid solution.

$$M\ H_3PO_4 = \left(\frac{5.0\ g\ P_4O_{10}}{0.500\ L}\right)\left(\frac{1\ mol\ P_4O_{10}}{283.88\ g\ P_4O_{10}}\right)\left(\frac{4\ mol\ H_3PO_4}{1\ mol\ P_4O_{10}}\right) = 0.14_{09}\ mol\ H_3PO_4\ /\ L$$

Phosphoric acid is a weak acid and only partly dissociates to form H_3O^+ in water, based on its K_a. Since $K_{a1} >> K_{a2} >> K_{a3}$, assume that the H_3O^+ contributed by the second and third dissociation is negligible in comparison to the H_3O^+ contributed by the first dissociation.

$$K_a = 7.2x10^{-3} = \frac{[H_3O^+][H_2PO_4^-]}{[H_3PO_4]} = \frac{x^2}{0.14_{09} - x}$$

$$x^2 + 7.2x10^{-3}x - 1.0_{145} = 0$$

$$x = \frac{-(7.2x10^{-3}) \pm \sqrt{(7.2x10^{-3})^2 - 4(1)(-1.0_{145})}}{2} = \frac{-(7.2x10^{-3}) \pm 6.4_{11}x10^{-2}}{2} = 2.8_{45}x10^{-2}$$

Therefore, $[H_3O^+] = 2.8_{45}x10^{-2}$ M and pH = $-\log(2.8_{45}x10^{-2})$ = **1.55.**

22.59 The reaction taking place in the blast furnace to produce iron is

$Fe_2O_3(s) + 3CO(g) \rightarrow 2Fe(s) + 3CO_2(g)$

a) Three moles of carbon dioxide are produced for every 2 moles of Fe.

$$(8400.\,T\ Fe)\left(\frac{10^3\ kg}{1\,T}\right)\left(\frac{1x10^3\ mol\ Fe}{55.85\ kg\ Fe}\right)\left(\frac{3\ mol\ CO_2}{2\ mol\ Fe}\right)\left(\frac{44.01\ kg\ CO_2}{1x10^3\ mol\ CO_2}\right)\left(\frac{1\,t}{10^3\ kg}\right) = \mathbf{9.929x10^3\ t\ CO_2}$$

b) Combustion reaction for octane:

$2C_8H_{18}(l) + 25O_2 \rightarrow 16CO_2(g) + 18H_2O(l)$

$$(1.0x10^6\ autos)\left(\frac{5.0\ gal\ C_8H_{18}}{auto}\right)\left(\frac{3785\ mL}{1\ gal}\right)\left(\frac{0.74\ g\ C_8H_{18}}{mL}\right)\left(\frac{1\ mol\ C_8H_{18}}{114.22\ g\ C_8H_{18}}\right)$$

$$\left(\frac{16\ mol\ CO_2}{2\ mol\ C_8H_{18}}\right)\left(\frac{44.01\ kg\ CO_2}{1x10^3\ mol\ CO_2}\right)\left(\frac{1\,t}{10^3\ kg}\right) = \mathbf{4.3x10^4\ t\ CO_2}$$

The automobiles produce about five times as much carbon dioxide per day.

22.61 a) Step 2 of "Isolation and Uses of Magnesium" describes this reaction. Sufficient OH^- must be added to precipitate $Mg(OH)_2$. The necessary amount of OH^- can be determined from the K_{sp} of $Mg(OH)_2$.

$K_{sp} = 6.3 \times 10^{-10} = [Mg^{2+}][OH^-]^2$ where $[Mg^{2+}] = 0.051$ M.

$[OH^-] = \{(6.3 \times 10^{-10})/(0.051)\}^{1/2} = 1.1 \times 10^{-4}$ M.

Therefore, if $[OH^-] > 1.1 \times 10^{-4}$ M (pH > 10.04), then $Mg(OH)_2$ will precipitate.

b) The amount of OH^- that a saturated solution of $Ca(OH)_2$ provides is dependent on its K_{sp}.

$K_{sp} = 6.5 \times 10^{-6} = [Ca^{2+}][OH^-]^2 = (x)(2x)^2 = 4x^3$

$x = 1.1_{76}$ M; $[OH^-] = 2x = 2.3_{51} \times 10^{-2}$ M

Substitute this amount into the K_{sp} expression for $Mg(OH)_2$ to determine how much Mg^{2+} reacts with this amount of OH^-.

$K_{sp} = 6.3 \times 10^{-10} = [Mg^{2+}][OH^-]^2 = [Mg^{2+}](2.3_{51} \times 10^{-2})^2$

$[Mg^{2+}] = 1.1_{39} \times 10^{-6}$ M

This concentration is the amount of the original 0.051 M Mg^{2+} that was not precipitated. The percent precipitated is the difference between the remaining $[Mg^{2+}]$ and the initial $[Mg^{2+}]$ divided by the initial concentration.

% Mg^{2+} precipitated = {(0.051 M - $1.1_{39} \times 10^{-6}$ M)/(0.051)}100 = **99.998 %**

22.66 a) Balance the two steps and then add them.

(1) $16H_2S(g) + 16O_2(g) \rightarrow S_8(g) +$ ~~$8SO_2(g)$~~ $+ 16H_2O(g)$

(2) $16H_2S(g) +$ ~~$8SO_2(g)$~~ $\rightarrow 3S_8(g) + 16H_2O(g)$

$32H_2S(g) + 16O_2(g) \rightarrow 4S_8(g) + 32H_2O(g)$

The coefficients in the overall equation can be reduced since all are divisible by 4.

$8H_2S(g) + 4O_2(g) \rightarrow S_8(g) + 8H_2O(g)$

b) Replace oxygen with chlorine:

$8H_2S(g) + 8Cl_2(g) \rightarrow S_8(g) + 16HCl(g)$

Calculate ΔG^0 to determine if this reaction is thermodynamically possible.

ΔG^0 = (1 mol S_8)(49.1 kJ/mol) + (16 mol HCl)(-95.30 kJ/mol)
– (8 mol H_2S)(-33 kJ/mol) – (8 mol Cl_2)(0)
= -1212 kJ

Reaction is spontaneous.

c) Oxygen is readily available from the air, so chlorine is a more expensive reactant. In addition, water is less corrosive than HCl as a product.

22.69 a) CO has 10 valence electrons in its Lewis dot structure. Recall that in drawing the Lewis structure, first a bond is drawn between the two atoms and then the electrons are distributed around the surrounding atoms (most electronegative first):

Next satisfy the octet of each atom and minimize formal charges (FC = val e^- – (unshared e^- + ½ shared e^-): 

FC(C) = 4 – (2 + ½(6)) = -1   FC(O) = 6 – (2 + ½(6)) = +1

b) The C in CO bonds to the metal because its negative formal charge makes it the likely spot for electron donation.

22.71 a) Decide what is reduced and what is oxidized of the two ions present in molten NaOH. Then write half-reactions and balance them.

At the cathode sodium ions are reduced: $Na^+(l) + e^- \rightarrow Na(l)$

At the anode hydroxide ions are oxidized: $4OH^-(l) \rightarrow O_2(g) + 2H_2O(g) + 4e^-$

b) The overall cell reaction is $4Na^+(l) + 4OH^-(l) \rightarrow 4Na(l) + O_2(g) + 2H_2O(g)$ and the reaction between sodium metal and water is $2Na(l) + 2H_2O(g) \rightarrow 2NaOH(l) + H_2(g)$.

For each mole of water produced 2 moles of sodium are produced, but for each mole of water that reacts only 1 mole of sodium reacts. So, the maximum amount of sodium that reacts with water is half of the sodium produced. In the balanced cell reaction 4 moles of electrons are transferred for 4 moles of sodium.

$$\frac{4\,mol\ Na\ produced - 2\,mol\ Na\ reacted}{4\,mol\ e-} = \mathbf{0.5\ mol\ Na / mol\ e^-}$$

22.73 a) Nitric oxide destroys ozone by forming nitrogen dioxide and oxygen.

$NO(g) + O_3(g) \rightleftharpoons NO_2(g) + O_2(g)$

b) The order translates to the exponent in the rate law, and the rate is dependent on the concentration of the reactants.

$Rate_f = k_f[NO][O_3]$

$Rate_r = k_r[NO_2][O_2]$

c) Use $\Delta G = \Delta H - T\Delta S$ because the reaction does not take place at 298K.

ΔH_{rxn}^0 = {(1mol NO_2)(33.2 kJ/mol) + (1mol O_2)(0)} – {(1mol NO)(90.29 kJ/mol) + (1mol O_3)(143 kJ/mol)}

$\Delta H_{rxn}^0 = -200._{09}$ kJ

ΔS_{rxn}^0 = {(1mol NO_2)(239.9 J/mol·K) + (1mol O_2)(205.0 J/mol·K)} – {(1mol NO)(210.65 J/mol·K) + (1mol O_3)(238.82 J/mol·K)}

$\Delta S_{rxn}^0 = -4.5_7$ J/K

$\Delta G_{280}^0 = \Delta H_{rxn}^0 - T\,\Delta S_{rxn}^0 = (-200._{09}\ kJ) - (280.\ K)(-4.5_7\ J/K)(kJ/1000\ J) = \mathbf{-199\ kJ}$.

d) Assume that the reaction reaches equilibrium at 280. K. At equilibrium, the forward rate equals the reverse rate and their ratio equals the equilibrium constant (Equation 17.2): $k_f/k_r = K_{eq}$. Both ΔG and K_{eq} express the extent to which a reaction proceeds and are related by the equation $\Delta G = -RT\ \ln K_{eq}$.

$$K_{eq} = e^{-(\Delta G/RT)} = e^{-(-199x10^3\,J/(8.314\,J/mol\cdot K)(280.K))} = 1.3x10^{37}$$

Therefore, the ratio of rate constants, k_f/k_r, is **1.3 x 10^{37}** and the forward rate is much faster than the reverse rate.

22.74 a) The reaction for the fixation of carbon dioxide includes CO_2 and H_2O as reactants and $(CH_2O)_n$ and O_2 as products. The balanced equation is

$nCO_2(g) + nH_2O(l) \rightarrow (CH_2O)_n(s) + nO_2(g)$

b) The moles of carbon dioxide fixed equal the moles of O_2 produced.

$$\left(\frac{45\,g\,CO_2}{day}\right)\left(\frac{1\,mol\,CO_2}{44.01\,g\,CO_2}\right)\left(\frac{n\,mol\,O_2}{n\,mol\,CO_2}\right) = 1.0_{22}\ mol\ O_2\ /\ day$$

To find the volume of oxygen use the ideal gas equation. Temperature must be converted to K.

$$T(K) = \left(\frac{5}{9}\right)(T(F) - 32) + 273 = \left(\frac{5}{9}\right)(80. - 32) + 273 = 299._{67}\ K$$

$$V = \frac{(1.0_{22}\ mol\ O_2)(0.08206\ atm\cdot L/mol\cdot K)(299._{67}\ K)}{1.0\ atm} = \mathbf{25\ L}$$

c) The moles of air containing 1.0_{22} mol of CO_2 is

$(1.0_{22}$ mol $CO_2) \div (0.033$ mol $CO_2/100$ mol air$) = 3.0_{96} \times 10^3$ mol air

$$V = \frac{(3.0_{97}\, x\, 10^3\; mol)(0.08206\; atm \cdot L / mol \cdot K)(299._{67}\; K)}{1.0\; atm} = \mathbf{7.6\, x\, 10^4\ L\ air}$$

22.77 The balanced chemical equation for this reaction is:

$6HF(g) + Al(OH)_3(s) + 3NaOH(aq) \rightarrow Na_3AlF_6(aq) + 6H_2O(l)$

Determine which one of the three reactants is the limiting reactant.

1) The amount of $Al(OH)_3$ requires conversion of kg to g, and then determination of moles of cryolite formed.

$$mol\ Na_3AlF_6 = (353x10^3\ g\ Al(OH)_3)\left(\frac{1\,mol\ Al(OH)_3}{78.00\ g\ Al(OH)_3}\right)\left(\frac{1\,mol\ Na_3AlF_6}{1\,mol\ Al(OH)_3}\right) = 4.52_{56}x10^3\ mol$$

2) The amount of NaOH requires conversion from amount of solution to g, CO then determination of moles of cryolite formed.

$$mass\ NaOH = (1.10\,m^3\ sol'n)\left(\frac{10^3\ L}{1\,m^3}\right)\left(\frac{10^3\ mL}{1\,L}\right)\left(\frac{50.0\ g\ NaOH}{100.0\ g\ sol'n}\right)\left(\frac{1.53\ g\ sol'n}{mL\ sol'n}\right) = 8.41_5x10^5\ g$$

$$(8.41_5x10^5\ g\ NaOH)\left(\frac{1\,mol\ NaOH}{40.00\ g\ NaOH}\right)\left(\frac{1\,mol\ Na_3AlF_6}{3\,mol\ NaOH}\right) = 2.80_{49}x10^5\ mol\ Na_3AlF_6$$

3) The amount of HF requires conversion from volume to moles using the ideal gas law, and then determination of moles of cryolite formed.

$$n_{HF} = \frac{PV}{RT} = \frac{\left(315\,kPa\left(\frac{1\,atm}{101.325\,kPa}\right)\right)\left(225\,m^3\left(\frac{1000\,L}{m^3}\right)\right)}{\left(0.08206\frac{atm \cdot L}{mol \cdot K}\right)(89.5 + 273.15)K} = 2.35_{05}x10^4\ mol$$

$$(2.35_{05}x10^4\ mol\ HF)\left(\frac{1\,mol\ Na_3AlF_6}{6\,mol\ HF}\right) = 3.91_{75}x10^3\ mol\ Na_3AlF_6$$

Because the 225 m^3 of HF produces the smallest amount of Na_3AlF_6, it is the limiting reactant. Convert moles Na_3AlF_6 to grams Na_3AlF_6 using the molar mass, convert to kg, and multiply by 96.6% to determine the final yield.

$$mass\ cryolite = (3.91_{75}x10^3\ mol\ Na_3AlF_6)\left(\frac{209.95\ g\ Na_3AlF_6}{1\,mol\ Na_3AlF_6}\right)\left(\frac{kg}{10^3\ g}\right)(0.966) = \mathbf{795\ kg\ Na_3AlF_6}$$

22.78 a) The rate of effusion is inversely proportional to the square root of the molar mass of the molecule.

$$\frac{rate_{H_2}}{rate_{D_2}} = \sqrt{\frac{M_{H_2}}{M_{D_2}}} = \sqrt{\frac{4.02}{2.02}} = 1.41$$

The time for D_2 to effuse is 1.41 times greater than that for H_2.

$time_{D2}$ = (1.41)(14.5 min) = **20.4 min**

b) Set x equal to the number of effusion steps. The ratio of mol H_2 to mol D_2 is 99:1. Set up the equation to solve for x: $99/1 = (1.41)^x$.

When solving for an exponent take the log of both sides.

$\log(99) = \log(1.41)^x$

Remember that $\log(a^b) = b\log(a)$
$\log(99) = x\log(1.41)$
$x = 13.4$
To separate H_2 and D_2 to 99% purity requires 14 effusion steps.

22.80 a) Use the stoichiometric relationships found in the balanced chemical equation to find mass of Al_2O_3. Assume that 1 ton Al is an exact number.

$$mass\ Al_2O_3 = (1\ ton\ Al)\left(\frac{1\ mol\ Al}{26.98\ g\ Al}\right)\left(\frac{2\ mol\ Al_2O_3}{4\ mol\ Al}\right)\left(\frac{101.96\ g\ Al_2O_3}{1\ mol\ Al_2O_3}\right) = 1.889_{55}\ tons\ Al_2O_3$$

Therefore, **1.890 tons of Al_2O_3** are consumed in the production of 1 ton of pure Al. A conversion of (10^6 g Al)/(1 ton Al) can be inserted into the equation above, but it is canceled when Al_2O_3 is converted back from grams to tons.

b) Use a ratio of 3 mol C: 4 mol Al to find mass of graphite consumed.

$$mass\ C = (1\ ton\ Al)\left(\frac{1\ mol\ Al}{26.98\ g\ Al}\right)\left(\frac{3\ mol\ C}{4\ mol\ Al}\right)\left(\frac{12.01\ g\ C}{1\ mol\ C}\right) = 0.3338_{58}\ tons\ C$$

Therefore, **0.3339 tons of C** are consumed in the production of 1 ton of pure Al, assuming 100% efficiency.

c) The percent yield with respect to Al_2O_3 is **100%** because the actual plant requirement of 1.89 tons Al_2O_3 equals the theoretical amount calculated in part (a).

d) The amount of graphite used in reality to produce 1 ton of Al is greater than the amount calculated in (b). In other words, a 100% efficient reaction takes only 0.3339 tons of graphite to produce a ton of Al, whereas real production requires more graphite and is less than 100% efficient. Calculate the efficiency using a simple ratio:
$(0.45)(x) = (0.3339)(100\%)$
$x =$ **74 %**

e) For every 4 moles of Al produced, 3 moles of CO_2 are produced.

$$moles\ CO_2 = (1\ ton\ Al)\left(\frac{10^6\ g}{1\ ton}\right)\left(\frac{1\ mol\ Al}{26.98\ g\ Al}\right)\left(\frac{3\ mol\ CO_2}{4\ mol\ Al}\right) = 2.779_{84}x10^4\ mol\ CO_2$$

Assume that the operating condition of 1 atm is exact. Use the ideal gas law to calculate volume, given moles, temperature and pressure.

$$V(m^3) = \frac{nRT}{P} = \frac{(2.779_{84}x10^4\ mol)(0.08206)(960.+273)K\left(\frac{1\ m^3}{10^3\ L}\right)}{1\ atm} = \mathbf{2.813x10^3\ m^3\ CO_2}$$

22.85 Acid rain increases the leaching of PO_4^{3-} into the groundwater, due to the protonation of PO_4^{3-} to form HPO_4^{2-} and $H_2PO_4^-$. As shown in calculations a) & b), solubility increases from 6.4×10^{-7} M (in pure water) to 1.1×10^{-2} M (in acidic rainwater).

a) Solubility of a salt can be calculated from its K_{sp}. From Table 19.2, K_{sp} for $Ca_3(PO_4)_2$ is 1.2×10^{-29}. Set s = solubility, so $[Ca^{2+}] = 3s$ and $[PO_4^{3-}] = 2s$.
$K_{sp} = [Ca^{2+}]^3[PO_4^{3-}]^2 = (3s)^3(2s)^2 = 108s^5 = 1.2 \times 10^{-29}$.
$s =$ **6.4×10^{-7} M** $Ca_3(PO_4)_2$

b) Phosphate is derived from a weak acid, so the pH of the solution impacts exactly where the acid-base equilibrium lies. PO_4^{3-} can gain one H^+ to form HPO_4^{2-}. Gaining another H^+ gives $H_2PO_4^-$ and a last H^+ added gives H_3PO_4. The K_a values for phosphoric acid are $K_{a1} = 7.2 \times 10^{-3}$, $K_{a2} = 6.3 \times 10^{-8}$, $K_{a3} = 4.2 \times 10^{-13}$. To find the ratios of the various forms of phosphate, use the equilibrium expressions and the concentration of H^+: $[H^+] = 10^{-4.5} = 3.1_{62} \times 10^{-5}$ M.

$$\frac{[H_2PO_4^-]}{[H_3PO_4]} = \frac{K_{a1}}{[H^+]} = \frac{7.2 \times 10^{-3}}{3.1_{62} \times 10^{-5}} = 2.3 \times 10^2$$

$$\frac{[HPO_4^{2-}]}{[H_2PO_4^-]} = \frac{K_{a2}}{[H^+]} = \frac{6.3 \times 10^{-8}}{3.1_{62} \times 10^{-5}} = 2.0 \times 10^{-3}$$

$$\frac{[PO_4^{3-}]}{[HPO_4^{2-}]} = \frac{K_{a3}}{[H^+]} = \frac{4.2 \times 10^{-13}}{3.1_{62} \times 10^{-5}} = 1.3 \times 10^{-8}$$

From the ratios the concentration of $H_2PO_4^-$ is at least 100 times more than the concentration of any other species, so assume that dihydrogen phosphate is the dominant species and find the value for $[PO_4^{3-}]$.

$$[PO_4^{3-}] = (1.3 \times 10^{-8})[HPO_4^{2-}] = (1.3 \times 10^{-8})(2.0 \times 10^{-3})[H_2PO_4^-].$$

Substituting this into the K_{sp} expression gives

$K_{sp} = [Ca^{2+}]^3\{(2.6 \times 10^{-11})[H_2PO_4^-]\}^2$

Rearranging gives

$K_{sp}/(2.6 \times 10^{-11})^2 = [Ca^{2+}]^3[H_2PO_4^-]^2$

The concentration of calcium ions is still represented as 3s and the concentration of dihydrogen phosphate ion as 2s, since each $H_2PO_4^-$ comes from one PO_4^{3-}.

$K_{sp}/(2.6 \times 10^{-11})^2 = (3s)^3(2s)^2$

$(1.2 \times 10^{-29})/(2.6 \times 10^{-11})^2 = 108s^5$

s = **1.1×10^{-2} M**

CHAPTER 23

THE TRANSITION ELEMENTS AND THEIR COORDINATION COMPOUNDS

FOLLOW-UP PROBLEMS

23.1 Plan: Locate the element on the periodic table and use its position and atomic number to write a partial electron configuration. Add or subtract electrons to obtain the configuration for the ion. Partial electron configurations do not include the noble gas configuration and any filled f sublevels, but do include outer level electrons (n-level) and the n-1 level d orbitals.

Solution:

a) Ag is in the fifth row of the periodic table, so n = 5, and it is in group 1B(11). The partial electron configuration of the element is $5s^1 4d^{10}$. For the ion, Ag^+, remove one electron form the 5s orbital. The partial electron configuration of Ag^+ is $\mathbf{4d^{10}}$.

b) Cd is in the fifth row and group 2B(12), so the partial electron configuration of the element is $5s^2 4d^{10}$. Remove two electrons from the 5s orbital for the ion. The partial electron configuration of Cd^{2+} is $\mathbf{4d^{10}}$.

c) Ir is in the sixth row and group 8B(9). Partial electron configuration of the elements is $6s^2 5d^7$. Remove two electrons from 6s orbital and one from 5d for ion Ir^{3+}. The partial electron configuration of Ir^{3+} is $\mathbf{5d^6}$.

Check: Total the electrons to make sure they agree with configuration.

a) Ag^+ should have 47 – 1 = 46e^-. Configuration has 36e^- (from Kr configuration) plus the 10 electrons in $4d^{10}$. This also totals 46e^-.

b) Cd^{2+} should have 48 – 2 = 46e^-. Configuration is the same as Ag^+, so 46e^- as well.

c) Ir^{3+} should have 77 – 3 = 74e^-. Total in configuration is 54e^- (noble gas configuration) plus 14e^- (in f-sublevel) plus 6e^- (in partial configuration) which equals 74e^-.

23.2 Plan: Find Er on the periodic table and write the electron configuration for Er^{3+}. Place the outer level electrons in an electron diagram making sure to follow the aufbau principle and Hund's rule. Count the number of unpaired electrons.

Solution: Er is a lanthanide element in row 6 and with atomic number 68. The electron configuration of the element is $[Xe]6s^2 5d^1 4f^{11}$. Remove three electrons – 2 from 6s and 1 from 5d for the Er^{3+} ion, $\mathbf{[Xe]4f^{11}}$.

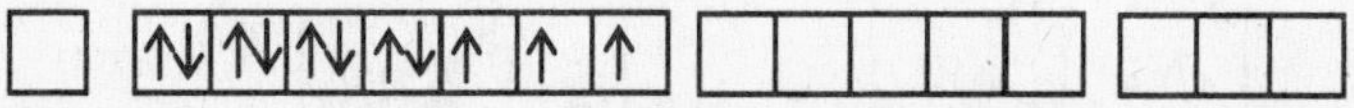

6s 4f 5d 6p

The number of unpaired electrons is **three**, all in the 4f sublevel.

Check: Total number of electrons for Er^{3+} is 68 – 3 = 65e^-. The configuration has 54 ([Xe] configuration) plus 11 in 4f to equal 65e^-.

23.3 Plan: Name the compound by following rules outlined in section 23.4. Use Table 23.8 for names of ligands and Table 23.9 for names of metal ions. Write the formula from the name by breaking down the name into pieces that follow naming rules.

Solution:

a) The compound consists of the cation $[Cr(H_2O)_5Br]^{2-}$ and anion Cl^-. The metal ion in the cation is chromium and its charge is found from the charges on the ligands and the total charge on the ion.

Total charge = (charge on Cr) + 5(charge on H_2O) + (charge on Br^-)

-2 = (charge on Cr) + 5(0) + (-1)

charge on Cr = +3

The name of the cation will end with chromium(III) to indicate the oxidation state of the chromium ion since there is more than one possible oxidation state for chromium. The water ligand is named aqua (Table 23.8). There are five water ligands, so pentaaqua describes the $(H_2O)_5$ ligands. The bromide ligand is named bromo. The ligands are named in alphabetical order, so aqua comes before bromo. The name of the cation is pentaaquabromochromium(III). Add the chloride for the anion to complete the name: **pentaaquabromochromium(III) chloride**.

b) The compound consists of a cation, barium ion, and the anion hexacyanocobaltate(III). The formula of the anion consists of six (from hexa) cyanide (from cyano) ligands and cobalt (from cobaltate) in the +3 oxidation state (from (III)). Putting the formula together gives $[Co(CN)_6]^{n+}$. To find the charge on the complex ion calculate from charges on ligand and metal ion:

total charge = 6(charge on CN^-) + (charge on cobalt ion)

= 6(-1) + (+3)

= -3

The formula of complex ion is $[Co(CN)_6]^{3+}$. Combining this with the cation, Ba^{2+}, gives the formula for the compound: $\mathbf{Ba_3[Co(CN)_6]_2}$. The three barium ions give a +6 charge and two anions give a –6 charge, so the net result is a neutral salt.

23.4 Plan: The given complex ion has a coordination number of 6 (*en* is bidentate), so it will have an octahedral arrangement of ligands. Stereoisomers can be either geometric or optical. For geometric isomers check if the ligands can be arranged either next to (*cis*) or opposite (*trans*) each other. Then check if the mirror image of any of the geometric isomers is not superimposable. If mirror image is not superimposable, the structure is an optical isomer.

Solution: The complex ion contains two NH_3 ligands and two chloride ligands, both of which can be arranged as either the *cis* or *trans* geometry. The possible combinations are 1) *cis*-NH_3 and *cis*-Cl, 2) *trans*-NH_3 and *cis*-Cl, and 3) *cis*-NH_3 and *trans*-Cl. Both NH_3 and Cl trans is not possible because the two bonds to ethylenediamine can be arranged only in this *cis*-position, which leaves only one set of *trans* positions.

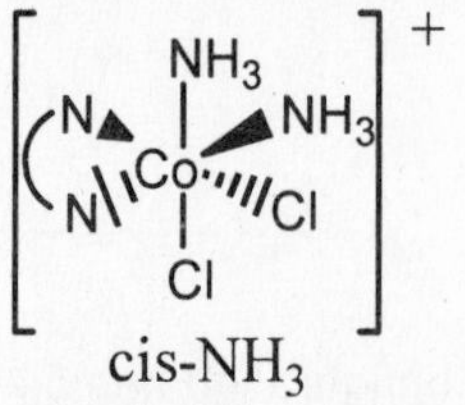

cis-NH_3
cis-Cl

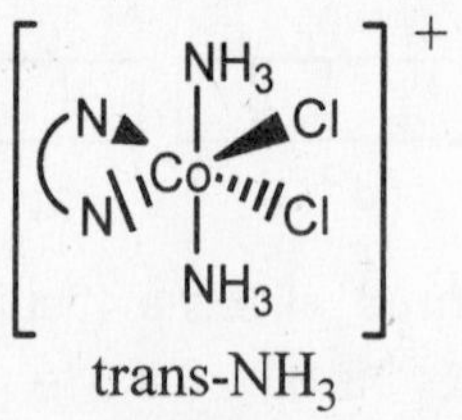

trans-NH_3
cis-Cl

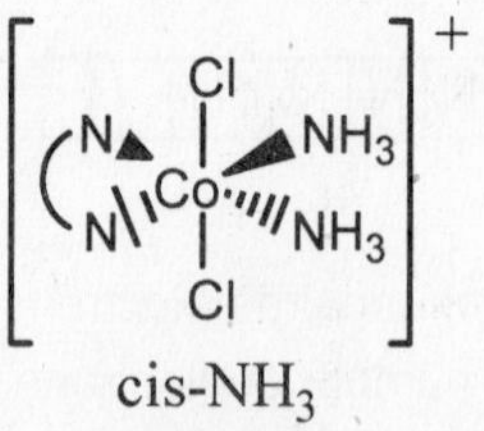

cis-NH_3
trans-Cl

The mirror images of the second two structures are superimposable since two of the ligands, either ammonia or chloride ion, are arranged in the trans position. When both types of ligands are in the cis arrangement the mirror image is not superimposable:

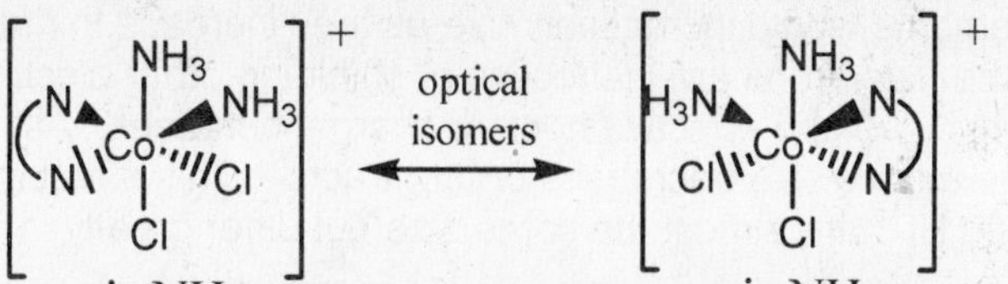

cis-NH_3 cis-Cl cis-NH_3 cis-Cl

There are four stereoisomers of $[Co(NH_3)_2(en)Cl_2]^+$.

23.5 Plan: Compare the two ions for oxidation state of the metal ion and the relative ability of ligands to split the d-orbital energies.

Solution: The oxidation number of vanadium in both ions is the same, so compare the two ligands. Ammonia is a stronger field ligand than water, so the complex ion **$[V(NH_3)_6]^{3+}$** absorbs visible light of higher energy than $[V(H_2O)_6]^{3+}$ absorbs.

23.6 Plan: Determine the charge on the manganese ion in the complex ion and the number of electrons in its d-orbitals. Since it is an octahedral ligand, the d-orbitals split into three lower energy orbitals and two higher energy orbitals. Check the spectrochemical series for whether CN^- is a strong or weak field ligand. If it is a strong field ligand fill the three lower energy orbitals before placing any electrons in the higher energy d-orbitals. If it is a weak field ligand place one electron in each of the five d orbitals, low energy before high energy, before pairing any electrons. After filling orbitals, count the number of unpaired electrons. The complex ion is low-spin if the ligand is a strong field ligand and high-spin if the ligand is weak field.

Solution: Figure the charge on manganese:

total charge = (charge on Mn) + 6(charge on CN^-)

charge on Mn = -3 – {6(-1)} = +3

Electron configuration of Mn^{3+} is $[Ar]3d^4$.

The ligand is CN^-, which is a strong field ligand, so the four d electrons will fill the lower energy d orbitals before any are placed in the higher energy d orbitals:

higher energy d-orbitals

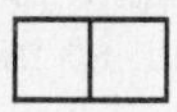

lower energy d-orbitals

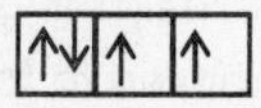

Two electrons are unpaired in $[Mn(CN)_6]^{3-}$. The complex is **low spin** since the cyanide ligand is a strong field ligand.

END-OF-CHAPTER PROBLEMS

23.2 a) All transition elements in Period 5 will have a "base" configuration of $[Ar]5s^2$, and will differ in the number of d electrons (x) that the configuration contains. Therefore, the general electron configuration is $\mathbf{1s^22s^22p^63s^23p^64s^23d^{10}4p^65s^24d^x}$.

b) A general electron configuration for Period 6 transition elements includes f sublevel electrons, which are lower in energy than the d sublevel. The configuration is $\mathbf{1s^22s^22p^63s^23p^64s^23d^{10}4p^65s^24d^{10}5p^66s^24f^{14}5d^x}$.

23.4 The maximum number of unpaired d electrons is **five** since there are five d orbitals. An example of an atom with five unpaired d electrons is Mn with electron configuration $[Ar]4s^23d^5$. An ion with five unpaired electrons is Mn^{2+} with electron configuration $[Ar]3d^5$.

23.6 a) One would expect that the elements would increase in size as they increase in mass from Period 5 to 6. Because the f electrons are ineffective at shielding other electrons from the attractive pull of the nucleus, the effective nuclear charge increases. As effective charge increases, the atomic size decreases or "contracts". This effect is significant enough that Zr^{4+} and Hf^{4+} are almost the same size but differ greatly in atomic mass.

b) The size increases from Period 4 to 5, but stays fairly constant from Period 5 to 6.

c) Atomic mass increases significantly from Period 5 to 6, but atomic radius (and thus volume) hardly increases, so Period 6 elements are very dense.

23.9 a) A paramagnetic substance is attracted to a magnetic field while a diamagnetic substance is slightly repelled by one.

b) Ions of transition elements often have unfilled d-orbitals that lead to unpaired electrons and paramagnetism. Ions of main group elements usually have a noble gas configuration with no partially filled levels. When orbitals are filled, electrons are paired and the ion is diamagnetic.

c) The d orbitals in the transition element ions are not filled which allows for promotion of an electron from a lower energy d orbital to a higher energy d orbital. The energy required for this promotion is relatively small and falls in the visible wavelength range. All orbitals are filled in a main-group element ion, so enough energy would have to be added to promote an electron to a higher energy level, not just another orbital within the same energy level. This amount of energy is relatively large and outside the visible range of wavelengths.

23.10 a) V: $1s^22s^22p^63s^23p^64s^23d^3$ Check: $2+2+6+2+6+2+3=23e^-$

b) Y: $1s^22s^22p^63s^23p^64s^23d^{10}4p^65s^24d^1$ Check: $2+2+6+2+6+2+10+6+2+1=39e^-$

c) Hg: $[Xe]6s^24f^{14}5d^{10}$ Check: $54+2+14+10=80e^-$

23.12 a) Osmium is in row 6 and group 8B(8). Electron configuration of Os is $\mathbf{[Xe]6s^24f^{14}5d^6}$.

b) Cobalt is in row 4 and group 8B(9). Electron configuration of Co is $\mathbf{[Ar]4s^23d^7}$.

c) Silver is in row 5 and group 1B(11). Electron configuration of Ag is $\mathbf{[Kr]5s^14d^{10}}$. Note that the filled d orbital is the preferred arrangement, so the configuration is not $5s^24d^9$.

23.14 Transition metals lose their s orbital electrons first in forming cations.

a) The two 4s electrons and 1 3d electron are removed to form Sc^{3+} with electron configuration [Ar] or $\mathbf{1s^22s^22p^63s^23p^6}$. There are **no unpaired electrons**.

b) The single 4s electron and one 3d electron are removed to form Cu^{2+} with electron configuration $\mathbf{[Ar]3d^9}$. There is **one unpaired electron**.

c) The two 4s electrons and 1 3d electron are removed to form Fe^{3+} with electron configuration $\mathbf{[Ar]3d^5}$. There are **five unpaired electrons**.

d) The two 5s electrons and 1 4d electron are removed to form Nb^{3+} with electron configuration $\mathbf{[Kr]4d^2}$. There are **two unpaired electrons**.

23.16 For groups 3B(3) to 7B(7) the highest oxidation state is equal to the group number.

a) Tantalum, Ta, is in group 5B(5), so highest oxidation state is **+5**.

b) Zirconium, Zr, is in group 4B(4), so highest oxidation state is **+4**.

c) Manganese, Mn, is in group 7B(7), so highest oxidation state is **+7**.

23.18 The elements in Group 6B(6) exhibit an oxidation state of +6. These elements include **Cr, Mo, and W**. Sg (Seaborgium) is also in Group 6B(6), but its lifetime is so short that chemical properties, like oxidation states within compounds, are impossible to measure.

23.20 Transition elements in their lower oxidation states act more like metals. The oxidation state of chromium in CrF_2 is +2 and in CrF_6 is +6 (use –1 oxidation state of fluorine to find oxidation state of Cr). **CrF_2** exhibits greater metallic behavior than CrF_6 because the chromium is in a lower oxidation state in CrF_2 than in CrF_6.

23.22 Table 23.3 lists the standard electrode potential for Cr^{2+} reduction (E^0 = -0.91 V). The CRC Handbook of Chemistry and Physics lists an electrode potential for Mo^{3+} reduction (E^0 = -0.200 V). The two equations can be reversed to yield oxidation equations:

$Cr(s) \rightarrow Cr^{2+}(aq) + 2e^-$ $\quad E^0 = 0.91V$

$Mo(s) \rightarrow Mo^{3+}(aq) + 3e^-$ $\quad E^0 = 0.200V$

The more positive the E^0 value, the more the reaction (as written) tends to occur. The reduction potential for Mo is lower, so **it is more difficult to oxidize Mo** than Cr. In addition, Mo's ionization energy is higher than that of Cr, so it is more difficult to remove electrons, i.e. oxidize, from Mo.

23.24 Oxides of transition metals become less basic (or more acidic) as oxidation state increases. The oxidation state of chromium in CrO_3 is +6 and in CrO is +2, based on the –2 oxidation state of oxygen. The oxide of the higher oxidation state, **CrO_3**, produces a more acidic solution.

23.28 a) The f-block contains 7 orbitals, so if one electron occupied each orbital a maximum of **7** electrons would be unpaired.

b) This corresponds to a half-filled f subshell.

23.30 a) Lanthanum is a transition element in row 6 with atomic number 57.

La: **$[Xe]6s^2 5d^1$**

b) Cerium is in the lanthanide series with atomic number 58.

Ce: $[Xe]6s^2 4f^1 5d^1$, so Ce^{3+}: **$[Xe]4f^1$**

c) Einsteinium is in the actinide series with atomic number 99.

Es: **$[Rn]7s^2 5f^{11}$**

d) Uranium is in the actinide series with atomic number 92.

U: $[Rn]7s^2 5f^3 6d^1$. Removing four electrons gives U^{4+} with configuration **$[Rn]5f^2$**.

23.32 a) Europium is in lanthanide series with atomic number 63. The configuration of Eu is $[Xe]6s^2 4f^7$. The stability of the half-filled f sublevel explains why the configuration is not $[Xe]6s^2 4f^6 5d^1$. The two 6s electrons are removed to form the Eu^{2+} ion, followed by electron removal in the f-block to form the other two ions:

Eu^{2+}: **$[Xe]4f^7$**

Eu^{3+}: **$[Xe]4f^6$**

Eu^{4+}: **$[Xe]4f^5$**

The stability of the half-filled f sublevel makes Eu^{2+} most stable.

b) Terbium is in the lanthanide series with atomic number 65. The configuration of Tb is $[Xe]6s^2 4f^9$. The two 6s electrons are removed to form the Tb^{2+} ion, followed by electron removal in the f-block to form the other two ions:

Tb^{2+}: **[Xe]4f^9**
Tb^{3+}: **[Xe]4f^8**
Tb^{4+}: **[Xe]4f^7**
Tb would demonstrate a +4 oxidation state because it has the half-filled sublevel.

23.34 The lanthanide element **gadolinium, Gd**, has electron configuration [Xe]6s^{2}4f^{7}5d^1 with **eight** unpaired electrons. The ion Gd^{3+} has **seven** unpaired electrons: [Xe]4f^7.

23.36 Valence-state electronegativity refers to the effective electronegativity of an element in a given oxidation state. The electronegativity of a metal in a higher oxidation state is higher than that of its parent element (oxidation state = 0) because the positive charge of the nucleus attracts the electrons more strongly. This higher electronegativity strengthens the M–O bond and makes the oxide more acidic because the O becomes a better electron pair acceptor. Thus, MnO (O.N. +2) is basic and Mn_2O_7 (O.N. +7) is acidic.

23.40 Mercury demonstrates an unusual physical property in that it exists in the liquid state at room temperature. There are two reasons for the weaker forces between mercury atoms. First, in its crystal structure each mercury atom is surrounded by 6 nearest neighbors instead of the usual 12. This decreases the forces holding mercury atoms, so they can move more freely than other metal atoms. Second, the bonding between metal atoms is metallic bonding, which involves only the 6s electrons in mercury compared to the s and d electrons as well in other metals.

Mercury occurs in the +1 oxidation state, unusual for its group, because two mercury +1 ions form a dimer, Hg_2^{2+} by sharing their unpaired 6s electrons in a covalent bond.

23.42 A reaction is favorable if the E^0 for the reaction as written is positive. Since Cr^{2+}(aq) is *prepared*, it must appear on the right side of the equation. Cr(s) and Cr^{3+}(aq) must appear as reactants. The middle equation will result in the cancellation of one of the needed reactants, so it is not included in the reaction summation. Reverse the third equation and multiply the first equation by 2 so that the electron transfer is balanced.

Reduction: $2(Cr^{3+}(aq) + e^- \rightarrow Cr^{2+})$ $E^0 = -0.41$ V (changing coefficients does not change E^0)

Oxidation: $Cr(s) \rightarrow Cr^{2+}(aq) + 2e^-$ $E^0 = +0.91$ V

Overall: $2Cr^{3+}(aq) + Cr(s) \rightarrow 3Cr^{2+}$ $E^0 = +0.50$ V

Since $E^0 > 0$, the reaction is favorable. So, **yes**, Cr^{2+} can be prepared from Cr and Cr^{3+}.

23.44 Check reduction standard potentials in Appendix D.

The standard potential for the reduction of MnO_4^- to Mn^{2+} is +1.51 V and the standard potential for the oxidation of Mn^{2+} to MnO_2 is –1.23 V. The reaction between MnO_4^- and Mn^{2+} to produce MnO_2 is +0.28 V, which indicates a spontaneous reaction:

$2\ MnO_4^-(aq) + 3Mn^{2+}(aq) + 2H_2O(l) \rightarrow 5MnO_2(s) + 4H^+(aq)$

23.47 The coordination number indicates the number of ligand atoms bonded to the central metal ion. The oxidation number represents the number of electrons lost to form the ion. The coordination number is unrelated to the oxidation number.

23.49 Coordination number of two indicates **linear** geometry.

Coordination number of four indicates either **tetrahedral or square planar** geometry.

Coordination number of six indicates **octahedral** geometry.

23.52 The complex ion has a negative charge.

23.55 a) Oxidation state of nickel is found from total charge on ion (+2 because two Cl^- charges equals –2) and charge on ligands:

charge on nickel = +2 – 6(0 charge on water) = +2

Name nickel as nickel(II) to indicate oxidation state. Ligands are six (hexa-) waters (aqua). Put together with chloride anions to give **hexaaquanickel(II) chloride**.

b) Cation is $[Cr(en)_3]^{n+}$ and anion is ClO_4^-, perchlorate ion (see chapter 2 for naming polyatomic ions). The charge on the cation is +3 to make a neutral salt in combination with 3 perchlorate ions. The ligand is ethylenediamine, which has 0 charge. The charge of the cation equals the charge on chromium ion, so chromium(III) is included in the name. The three ethylenediamine ligands, abbreviated en, are indicated by the prefix tris because the name of the ligand includes a numerical indicator, di-. The complete name is **tris(ethylenediamine)chromium(III) perchlorate**.

c) Cation is K^+ and anion is $[Mn(CN)_6]^{4-}$, the charge of 4- deduced from the four potassium ions in the formula. Oxidation state of Mn is -4 – {6(-1)} = +2. Name of CN^- ligand is cyano and six ligands is represented by the prefix hexa. Name of manganese in anion is manganate(II). The full name of compound is **potassium hexacyanomanganate(II).**

23.57 The charge of the central metal atom was determined in 23.55 because the roman numeral indicating oxidation state is part of the name. The coordination number, or number of ligand atoms bonded to the metal ion, is found by examining the bonded entities inside the square brackets to determine if they are unidentate, bidentate or polydentate.

a) The roman numeral "II" indicates a **+2** oxidation state. There are 6 water molecules bonded to Ni and each ligand is unidentate, so the coordination number is **6**.

b) The roman numeral "III" indicates a **+3** oxidation state. There are 3 ethylenediamine molecules bonded to Cr, but each ethylenediamine molecule contains two donor N atoms (bidentate). Therefore the coordination number is **6**.

c) The roman numeral "II" indicates a **+2** oxidation state. There are 6 unidentate cyano molecules bonded to Mn, so the coordination number is **6**.

23.59 a) Cation is K^+, potassium.

Anion is $[Ag(CN)_2]^-$ with name dicyanoargentate ion for two chloride ligands and the name of silver in anions, argentate(I).

Complete name is **potassium dicyanoargentate(I)**.

b) Cation is Na^+, sodium.

Anion is $[CdCl_4]^{2-}$ with name tetrachlorocadmate(II) ion with four chloride ligands, the oxidation state of cadmium shown as +2 and the name of cadmium in an anion as cadmate.

Complete name is **sodium tetrachlorocadmate(II).**

c) Cation is $[Co(NH_3)_4(H_2O)Br]^{2+}$ with the name tetraamineaquabromocobalt(III) ion with four ammonia ligands (tetraammine), one water ligand (aqua) and one bromide ligand (bromo). Oxidation state of cobalt is +3 indicated by (III) following cobalt in the name.

Anion is Br^-, bromide.

Complete name is **tetraammineaquabromocobalt(III) bromide**.

23.61 a) The counter ion is K^+, so the complex ion is $[Ag(CN)_2]^-$. Each cyano ligand has a –1 charge, so silver has a **+1** charge: +1 + 2(–1) = –1. Each cyano ligand is unidentate, so the coordination number is **2**.

b) The counter ion is Na^+, so the complex ion is $[CdCl_4]^{2-}$. Each chloride ligand has a –1 charge, so Cd has a **+2** charge: +2 + 4(–1) = –2. Each chloride ligand is unidentate, so the coordination number is **4**.

c) The counter ion is Br^-, so the complex ion is $[Co(NH_3)_4(H_2O)Br]^{2+}$. Both the ammine and aqua ligands are neutral. The bromide ligand has a –1 charge, so Co has a **+3** charge: +3 + 4(0) + 0 + (–1) = +2. Each ligand is unidentate, so the coordination number is **6**.

23.63 a) Cation is tetramminezinc ion. The tetraammine indicate four NH_3 ligands. Zinc has an oxidation state of +2, so the charge on the cation is +2.

Anion is SO_4^{2-}. Only one sulfate is needed to make a neutral salt.

The formula of the compound is $\mathbf{[Zn(NH_3)_4]SO_4}$.

b) Cation is pentaamminechlorochromium(III) ion. The ligands are 5 NH_3 from pentaammine, and one chloride form chloro. The chromium ion has a charge of +3, so the complex ion has a charge equal to +3 from chromium plus 0 from ammonnia plus –1 from chloride for a total of +2.

Anion is chloride, Cl^-. Two chloride ions are needed to make a neutral salt.

The formula of compound is $\mathbf{[Cr(NH_3)_5Cl]Cl_2}$.

c) Anion is bis(thiosulfato)argentate(I). Argentate(I) indicates silver in the +1 oxidation state. bis(thiosulfato) indicates 2 thiosulfate ligands, $S_2O_3^{2-}$. Total charge on anion is +1 plus 2(-2) to equal –3.

Cation is sodium, Na^+. Three sodium ions are needed to make a neutral salt.

The formula of compound is $\mathbf{Na_3[Ag(S_2O_3)_2]}$.

23.65 Coordination compounds act like electrolytes, i.e. they dissolve in water to yield charged species. However, the complex ion itself does not dissociate. The "number of individual ions per formula unit" refers to the number of ions that would form per coordination compound upon dissolution in water.

a) The counter ion is SO_4^{2-}, so the complex ion is $[Zn(NH_3)_4]^{2+}$. Each ammine ligand is unidentate, so the coordination number is **4**. Each molecule dissolves in water to form one SO_4^{2-} ion and one $[Zn(NH_3)_4]^{2+}$ ion, so **2 ions** form per formula unit.

b) The counter ion is Cl^-, so the complex ion is $[Cr(NH_3)_5Cl]^{2+}$. Each ligand is unidentate so the coordination number is **6**. Each molecule dissolves in water to form two Cl^- ions and one $[Cr(NH_3)_5Cl]^{2+}$ ion, so **3 ions** form per formula unit.

c) The counter ion is Na^+, so the complex ion is $[Ag(S_2O_3)_2]^{3-}$. Assuming that the thiosulfate ligand is unidentate, the coordination number is **2**. Each molecule dissolves in water to form 3 Na^+ ions and one $[Ag(S_2O_3)_2]^{3-}$ ion, so **4 ions** form per formula unit.

23.67 a) The cation is hexaaquachromium(III) with formula $[Cr(H_2O)_6]^{3+}$. Total charge of ion equals charge on chromium because water is neutral ligand. Six water ligands are indicated by *hexa* prefix to aqua.

The anion is SO_4^{2-}. To make neutral salt requires 3 sulfate ions for 2 cations. The compound formula is $\mathbf{[Cr(H_2O)_6]_2(SO_4)_3}$.

b) The anion is tetrabromoferrate(III) with formula $[FeBr_4]^-$. Total charge on ion equals +3 charge on iron plus 4 times –1 charge on each bromide ligand.

The cation is barium, Ba^{2+}. Two anions are needed for each barium ion to make a neutral salt. The compound formula is $\mathbf{Ba[FeBr_4]_2}$.

c) The anion is bis(ethylenediamine)platinum(II) ion. Charge on platinum is +2 and this equals the total charge on the complex ion because ethylenediamine is a neutral ligand. The *bis* preceding ethylenediamine indicates two ethylenediamine ligands, which are represented by the abbreviation "en".

The cation is $[Pt(en)_2]^{2+}$. The anion is CO_3^{2-}. One carbonate combines with one cation to produce a neutral salt. The compound formula is **$[Pt(en)_2]CO_3$**.

23.69 a) The counter ion is SO_4^{2-}, so the complex ion is $[Cr(H_2O)_6]^{3+}$. Each aqua ligand is unidentate, so the coordination number is **6**. Each molecule dissolves in water to form three SO_4^{2-} ions and two $[Cr(H_2O)_6]^{3+}$ ions, so **5 ions** form per formula unit.

b) The counter ion is Ba^{2+}, so the complex ion is $[FeBr_4]^-$. Each bromo ligand is unidentate, so the coordination number is **4**. Each molecule dissolves in water to form one Ba^{2+} ion and 2 $[FeBr_4]^-$ ions, so **3 ions** form per formula unit.

c) The counter ion is CO_3^{2-}, so the complex ion is $[Pt(en)_2]^{2+}$. Each ethylenediamine ligand is bidentate, so the coordination number is **4**. Each molecule dissolves in water to form one CO_3^{2-} ion and one $[Pt(en)_2]^{2+}$ ion, so **2 ions** forms per formula unit.

23.71 Ligands that form linkage isomers have two different possible donor atoms.

a) The nitrite ions **form linkage isomers** because it can bind to the metal ion through either the nitrogen or one of the oxygen atoms – both have a lone pair of electrons. Resonance Lewis structures are

b) Sulfur dioxide molecules **form linkage isomers** because both the sulfur and oxygen atoms can bind metal ions. Resonance Lewis structures are

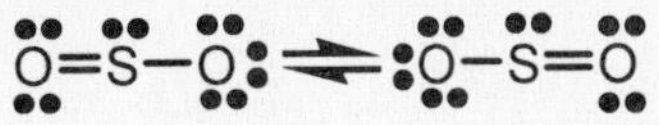

c) Nitrate ions have three oxygen atoms, all with a lone pair that can bond to the metal ion, but all of the oxygen atoms are equivalent, so there are **no linkage isomers**.

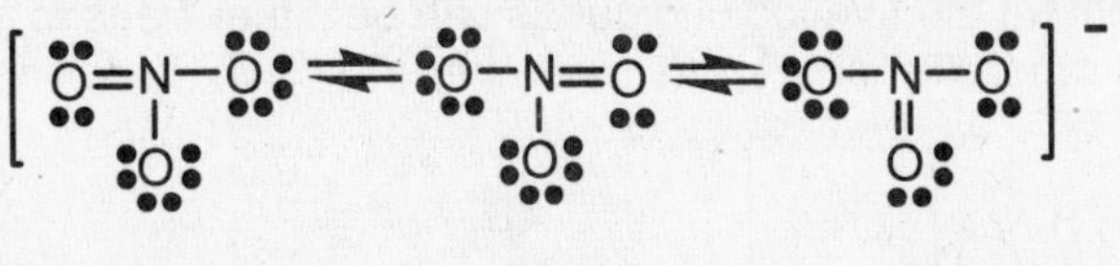

23.73 a) Platinum ion, Pt^{2+}, is a d^8 ion so ligand arrangement is square planar. A *cis* and *trans* geometric isomer exist for this complex ion:

cis isomer

trans isomer

No optical isomers exist because the mirror images of both compounds are superimposable on the original molecules. In general, a square planar molecule is superimposable on its mirror image.

b) A *cis* and *trans* geometric isomer exist for this complex ion:

cis isomer

trans isomer

c) Three geometric isomers exist for this molecule, although they are not named *cis* or *trans* because all ligands are different. A different naming system is used to indicate the relation of one ligand to another.

23.75 Types of isomers for coordination compounds are coordination isomers with different arrangements of ligands and counterions, linkage isomers with different donor atoms from the same ligand bound to the metal ion, geometric isomers with differences in ligand arrangement relative to other ligands., and optical isomers with mirror images that are not superimposable.

a) Platinum ion, Pt^{2+}, is a d^8 ion so ligand arrangement is square planar. The ligands are 2 Cl^- and 2 Br^- so the arrangement can be either both ligands *trans* or both ligands *cis* to form geometric isomers.

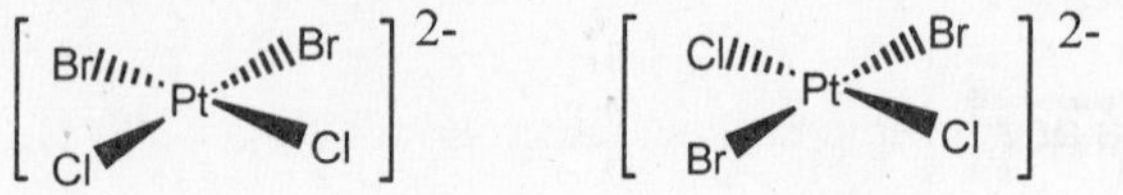

cis trans

b) The complex ion can form linkage isomers with the NO_2 ligand.

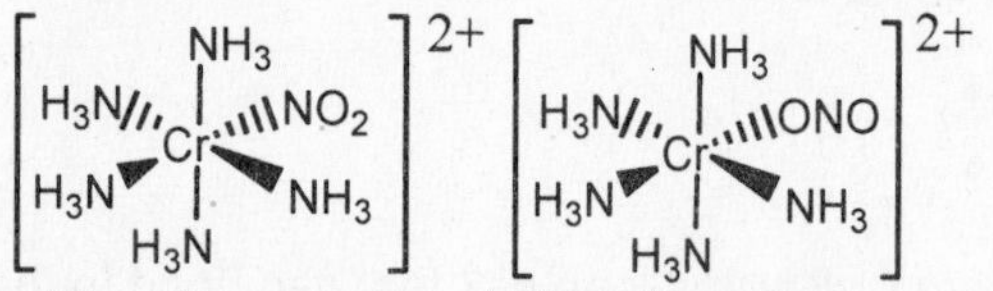

N as donor atom O as donor atom

c) In the octahedral arrangement the two iodide ligands can be either *trans* to each other, 180° apart, or *cis* to each other, 90° apart.

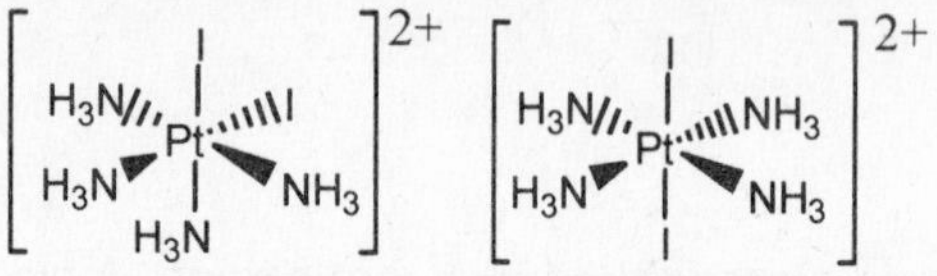

cis trans

23.77 The traditional formula does not correctly indicate that at least some of the chloride ions serve as counter ions. The NH_3 molecules serve as ligands, so n could equal 6 to satisfy the coordination number requirement. This complex has the formula $[Cr(NH_3)_6]Cl_3$. This is a correct formula because the name "chromium(III)" means that the chromium has a +3 charge, and this balances with the –3 charge provided by the three chloride counterions. However, when this compound dissociates in water, it produces 4 ions ($[Cr(NH_3)_6]^{3+}$ and 3 Cl^-) whereas NaCl only produces two ions. Other possible compounds are n=5, $CrCl_3 \cdot 5NH_3$, with formula $[Cr(NH_3)_5Cl]Cl_2$; n=4, $CrCl_3 \cdot 4NH_3$, with formula $[Cr(NH_3)_4Cl_2]Cl$; n=3, $CrCl_3 \cdot 3NH_3$, with formula $[Cr(NH_3)_3Cl_3]$. The compound **$[Cr(NH_3)_4Cl_2]Cl$** has a coordination number equal to 6 and produces two ions when dissociated in water, so it has an electrical conductivity similar to an equimolar solution of NaCl.

23.79 First find the charge on the palladium ion, then arrange ligands and counter ions to form complex.

a) charge on palladium = $-\{(+1 \text{ on } K^+) + (0 \text{ on } NH_3) + 3(-1 \text{ on } Cl^-)\} = +2$

Palladium(II) forms four-coordinate complexes. The four ligands in the formula are one NH_3 and three Cl^- ions. Formula of complex ion is $[Pd(NH_3)Cl_3]^-$. Combined with potassium cation the compound formula is **$K[Pd(NH_3)Cl_3]$.**

b) charge on palladium = –{2(–1 on Cl^-) + 2(0 on NH_3)} = +2

Palladium(II) forms four-coordinate complexes. The four ligands are 2 chloride ions and 2 ammonia molecules. Formula is **$[PdCl_2(NH_3)_2]$**.

c) charge on palladium = –{2(+1 on K^+) + 6(–1 on Cl^-)} = +4

Palladium(IV) forms six-coordinate complexes. The six ligands are the 6 chloride ions. Formula is **$K_2[PdCl_6]$.**

d) charge on palladium = –{4(0 on NH_3) + 4(–1 on Cl^-)} = +4

Palladium(IV) forms six-coordinate complexes. The ammonia molecules have to be ligands. The other two ligand bonds are formed with two of the chloride ions. The remaining two chloride ions are the counter ions. Formula is **$[Pd(NH_3)_4Cl_2]Cl_2$**.

23.81 a) Refer to Figure 23.14. Four empty orbitals of equal energy are "created" to receive the donated electron pairs from four ligands. The four orbitals are hybridized from an s, two p, and one d orbital from the previous *n* level to form 4 **dsp^2** orbitals.

b) Refer to Figure 23.15. One s and three p orbitals become four **sp^3** hybrid orbitals.

23.84 According to Table 23.11 absorption of **orange or yellow** light gives a blue solution.

23.85 a) The crystal field splitting energy is the energy difference between the two sets of d orbitals that result from the bonding of ligands to a central transition metal atom.

b) In an octahedral field of ligands, the ligands approach along the x, y, and z axes. The d_{x2-y2} and d_{z2} orbitals are located along the x, y, and z axes, so ligand interaction is "head on" and higher in energy than the other orbital-ligand interactions. The other orbital-ligand interactions occur "off-axis" because the d_{xy}, d_{yz}, and d_{xz} orbitals are located between the x, y, and z axes.

c) In a tetrahedral field of ligands, the ligands do not approach along the x, y, and z axes. Therefore, the ligand interaction is "head-on" for the d_{xy}, d_{yz}, and d_{xz} orbitals and "off-axis" for the d_{x2-y2} and d_{z2} orbitals. The crystal field splitting is reversed, and the d_{xy}, d_{yz}, and d_{xz} orbitals are higher in energy than the d_{x2-y2} and d_{z2} orbitals.

23.88 If Δ is greater than $E_{pairing}$ electrons will preferentially pair spins in the lower energy d orbitals before adding as unpaired electrons to the higher energy d orbitals. If Δ is less than $E_{pairing}$ electrons will preferentially add as unpaired electrons to the higher d orbitals before pairing spins in the lower energy d orbitals. The first case gives a complex that is low-spin and less paramagnetic than the high-spin complex formed in the latter case.

23.90 To determine the number of d electrons in a central metal ion, first write the electron configuration for the metal atom. Examine the formula of the complex to determine the charge on the central metal ion, and then write the ion's configuration.

a) Electron configuration of Ti: $[Ar]4s^23d^2$

Charge on Ti: Each chloride ligand has a –1 charge, so Ti has a +4 charge.

Electron configuration of Ti^{4+}: [Ar]

Ti^{4+} has **no d electrons**.

b) Electron configuration of Au: $[Xe]6s^14f^{14}5d^{10}$

Charge on Au: The complex ion has a –1 charge ($[AuCl_4]^-$) and each chloride ligand has a –1 charge, so Au has a +3 charge.

Electron configuration of Au^{3+}: $[Xe]4f^{14}5d^8$

Au^{3+} has **8 d electrons**.

c) Electron configuration of Rh: $[Kr]5s^24d^7$

Charge on Rh: Each chloride ligand has a –1 charge, so Rh has a +3 charge.

Electron configuration of Rh^{3+}: $[Kr]4d^6$

Rh^{3+} has **6 d electrons**.

23.92 a) charge on iridium = -{(+2 on Ca^{2+}) + 6(-1 on F^-)} = +4

Configuration of Ir^{4+} is $[Xe]4f^{14}5d^5$. **5** d electrons in Ir^{4+}.

b) charge on mercury = -{4(-1 on I^-)} –2 = +2

Configuration of Hg^{2+} is $[Xe]4f^{14}5d^{10}$. **10** d electrons in Hg^{2+}.

c) charge on cobalt = -(-4 on EDTA) – 2 = +2

Configuration of Co^{2+} is $[Ar]3d^7$. **7** d electrons in Co^{2+}.

23.94

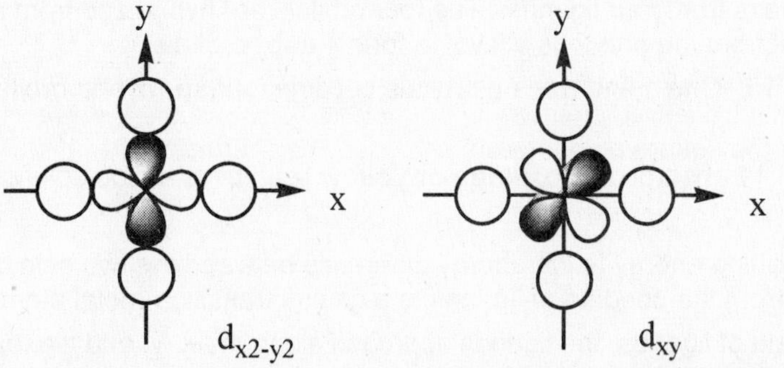

In an octahedral field of ligands, the ligands approach along the x, y, and z axes. The d_{x2-y2} orbital is located *along* the x and y axes, so ligand interaction is “head on”. The d_{xy} orbital is offset from the x and y axes by 90°, so ligand interaction is not “head on”. The “head on” interaction of the d_{x2-y2} orbital results in its higher energy.

23.96 a) Electron configuration of Ti^{3+} is $[Ar]3d^1$. With only one electron in the d orbitals the titanium(III) ion **cannot form** high and low spin complexes – all complexes will contain one unpaired electron and have the same spin.

b) Co^{2+} will form high and low spin complexes with 7 electrons in the d orbital.

c) Fe^{2+} will form high and low spin complexes with 6 electrons in the d orbital.

d) Electron configuration of Cu^{2+} is $[Ar]3d^9$ so in complexes with both strong and weak field ligands one electron will be unpaired and spin in both types of complexes is identical. Cu^{2+} **cannot form** high and low spin complexes

23.98 To draw the orbital-energy splitting diagram, first determine the number of d electrons in the transition metal ion. Examine the formula of the complex ion to determine the electron configuration of the metal ion, remembering that the ns electrons are lost first. Determine the coordination number from the number of ligands, recognizing that 6 ligands result in an octahedral arrangement and 4 ligands result in a tetrahedral or square planar arrangement. Refer to Figures 23.23, 23.25A and 23.25B for the energy difference of the d orbitals in each geometry. Weak-field ligands give the maximum number of unpaired electrons (high spin) while strong-field ligands lead to electron pairing (low spin).

a) Electron configuration of Cr: $[Ar]4s^13d^5$

Charge on Cr: The aqua ligands are neutral, so the charge on Cr is +3.

Electron configuration of Cr^{3+}: $[Ar]3d^3$

Six ligands indicate an octahedral arrangement. Using Hund's rule, fill the lower energy t_{2g} orbitals first, filling empty orbitals before pairing electrons within an orbital.

e_g ☐☐

t_{2g} [↑|↑|↑]

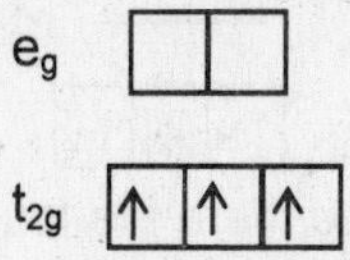

All lower energy orbitals are half-filled and H_2O is a weak-field ligand, so it is energetically unfavorable for any of the electrons to pair up. Therefore, the correct orbital-energy splitting diagram shows three unpaired electrons in the t_{2g} orbitals.

b) Electron configuration of Cu: $[Ar]4s^1 3d^{10}$

Charge on Cu: The aqua ligands are neutral, so Cu has a +2 charge.

Electron configuration of Cu^{2+}: $[Ar]3d^9$

Four ligands and a d^9 configuration indicate a square planar geometry (only filled d sublevel ions exhibit tetrahedral geometry). Figure 23.25B shows the d orbital-energy splitting diagram for the square planar geometry. Use Hund's rule to fill in the 9 d electrons.

d_{x2-y2} [↑]

d_{xy} [↑↓]

d_{z2} [↑↓]

d_{xz} [↑↓] [↑↓] d_{yz}

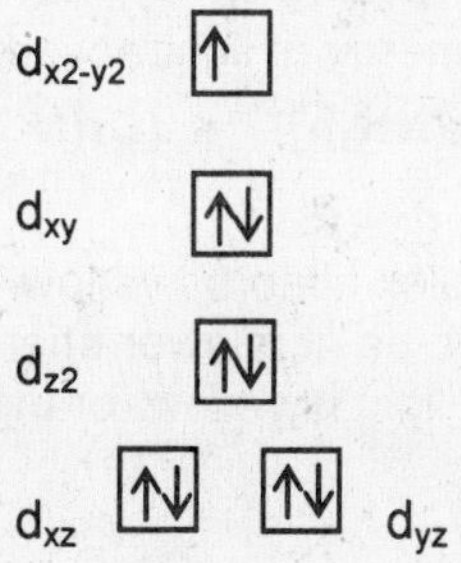

c) Electron configuration of Fe: $[Ar]4s^2 3d^6$

Charge on Fe: Each fluoride ligand has a –1 charge, so Fe has a +3 charge to make the overall complex charge equal to –3.

Electron configuration of Fe^{3+}: $[Ar]3d^5$

Six ligands indicate an octahedral arrangement. Use Hund's rule to fill the orbitals.

e_g [↑|↑]

t_{2g} [↑|↑|↑]

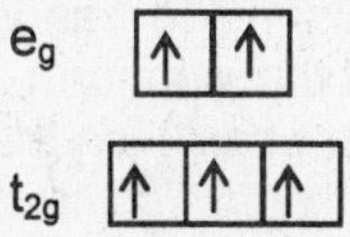

F^- is a weak field ligand, so the splitting energy, Δ, is not large enough to overcome the resistance to electron pairing. The electrons remain unpaired, and the complex is called *high-spin*.

23.100 a) Complex is octahedral so splitting is d_{xy}, d_{yz}, d_{xz} at lower energy than d_{z2} and d_{x2-y2}.

Mo^{3+} has 3 d electrons.

e_g ☐☐

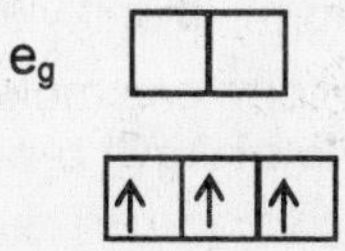

t_{2g}

b) Complex is octahedral so splitting is d_{xy}, d_{yz}, d_{xz} at lower energy than d_{z2} and d_{x2-y2}. Ni^{2+} has 8 d electrons.

e_g [↑][↑]

t_{2g} [↑↓][↑↓][↑↓]

c) Complex is square planar because there are four ligands and nickel has 8 d electrons. Complex is low spin because CN^- is a strong field ligand.

d_{x2-y2} [ ]

d_{xy} [↑↓]

d_{z2} [↑↓]

d_{xz} [↑↓] [↑↓] d_{yz}

23.102 Figure 23.22 describes the spectrum of splitting energy, Δ. NO_2^- is a stronger ligand than NH_3, which is stronger than H_2O. The energy of light absorbed increases as Δ increases.

$$[Cr(H_2O)_6]^{3+} < [Cr(NH_3)_6]^{3+} < [Cr(NO_2)_6]^{3-}$$

23.104 From Table 23.11 find that a violet complex absorbs yellow-green light. The light absorbed by a complex with a weaker ligand would be at a lower energy and higher wavelength. Light of lower energy than yellow-green light is yellow, orange or red light. The color observed would be **blue or green**.

23.107 The aqua ligand is weaker than the ammine ligand. The weaker ligand results in a lower splitting energy and absorbs a lower energy of visible light. The green hexaaqua complex appears green because it absorbs red light (opposite side of the Figure 23.16 color wheel). The hexaammine complex appears violet because it absorbs yellow light, which is higher in energy (shorter λ) than red light.

23.111 Electron configuration of Hg^+ is $[Xe]6s^14f^{14}5d^{10}$ and for Cu^+ is $[Ar]3d^{10}$. In the mercury(I) ion there is one electron in the 6s orbital that can form a covalent bond with the electron in the 6s orbital of another Hg^+ ion. In the copper(I) ion there are no electrons in the s orbital to bond with another copper(I) ion.

23.113 a) The coordination number of cobalt is **6**. The two Cl^- ligands are unidentate and the two ethylenediamine ligands are bidentate, so a total of 6 ligand atoms are connected to the central metal ion.

b) The counter ion is Cl^-, so the complex ion is $[Co(en)_2Cl_2]^+$. Each chloride ligand has a −1 charge and each en ligand is neutral, so cobalt has a **+3** charge: +3 + 2(0) + 2(−1) = +1.

c) One mole of complex dissolves in water to yield one mole of $[Co(en)_2Cl_2]^+$ ions and one mole of Cl^- ions. Therefore each formula unit yields **2** individual ions.

d) One mole of complex dissolves to form one mole of Cl^- ions, which reacts with the Ag^+ ion (from $AgNO_3$) to form **one mole of AgCl precipitate**.

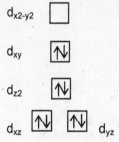

23.115 Assume a 100-g sample and convert the mass of each element to moles using the element's molar mass. Divide each mole amount by the smallest mole amount to determine whole number ratios of the elements to determine the empirical formula.

$$moles\ Pt = (38.8\ g\ Pt)\left(\frac{1\ mol\ Pt}{195.1\ g\ Pt}\right) = 0.199 \div 0.199 = 1$$

$$moles\ Cl = (14.1\ g\ Cl)\left(\frac{1\ mol\ Cl}{35.45\ g\ Cl}\right) = 0.398 \div 0.199 \approx 2$$

$$moles\ C = (28.7\ g\ C)\left(\frac{1\ mol\ C}{12.01\ g\ C}\right) = 2.39 \div 0.199 \approx 12$$

$$moles\ P = (12.4\ g\ P)\left(\frac{1\ mol\ P}{30.97\ g\ P}\right) = 0.400 \div 0.199 \approx 2$$

$$moles\ H = (6.02\ g\ H)\left(\frac{1\ mol\ H}{1.008\ g\ H}\right) = 5.97 \div 0.199 \approx 30$$

The empirical formula for the compound is $PtCl_2C_{12}P_2H_{30}$. Each triethylphosphine ligand, $P(C_2H_5)_3$ accounts for one phosphorous atom, 6 carbon atoms, and 15 hydrogen atoms. According to the empirical formula, there are two triethylphosphine ligands: 2 $P(C_2H_5)_3$ "equals" $C_{12}P_2H_{30}$. The compound must have two $P(C_2H_5)_3$ ligands and two chloro ligands per Pt ion. The formula is $[Pt[P(C_2H_5)_3]_2Cl_2]$. The central Pt ion has 4 ligands and is square planar, existing as either a *cis* or *trans* compound.

cis-dichlorobis(triethylphosphine)platinum(II)

trans-dichlorobis(triethylphosphine)platinum(II)

23.118 a) A plane that includes both ammonia ligands and the zinc ion is a plane of symmetry. The complex **does not have optical isomers** because it has a plane of symmetry.

b) The Pt^{2+} ion is d^8 so the complex is square planar. Any square planar complex has a plane of symmetry, so the complex **does not have optical isomers**.

c) The trans octahedral complex has the two chloride ions opposite each other. A plane of symmetry can be passed through the two chlorides and platinum ion so the trans-complex **does not have optical isomers**.

d) **No optical isomers** for same reason as in part c.

e) The cis-isomer does not have a plane of symmetry so it **does have optical isomers** as shown below.

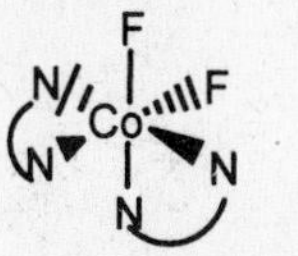

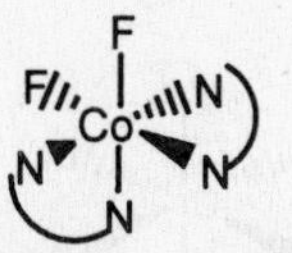

23.121 a) The first reaction shows no change in the number of particles. In the second reaction the number of reactant particles is less than the number of product particles. A decrease in the number of particles means a decrease in entropy while no change in number of particles indicates little change in entropy. Based on entropy change only, the first reaction is favored in comparison to the second reaction.

b) The ethylenediamine complex is more stable with respect to ligand exchange with water because the entropy change is unfavorable.

23.123 The compound has four ligands and a square planar geometry. The compound exhibits both geometric and linkage isomerism. The thiocyanate (SCN^-) ligand forms linkage isomers because it can bind to the metal ion through either the nitrogen ($SCN:^- \rightarrow$, isothiocyanato) or the sulfur ($NCS:^- \rightarrow$, thiocyanato) atoms.

There are three *cis* isomers:

cis-diamminebis(thiocyanato)platinum(II)

cis-diammineisothiocyanatothiocyanatoplatinum(II)

cis-diamminebis(isothiocyanato)platinum(II)

There are three *trans* isomers:

trans-diamminebis(thiocyanato)platinum(II)

trans-diammineisothiocyanatothiocyanatoplatinum(II)

trans-diamminebis(isothiocyanato)platinum(II)

23.125 $[Co(NH_3)_4(H_2O)Cl]^{2+}$

Name is **tetraammineaquachlorocobalt(III).**

Isomers are **geometric**:

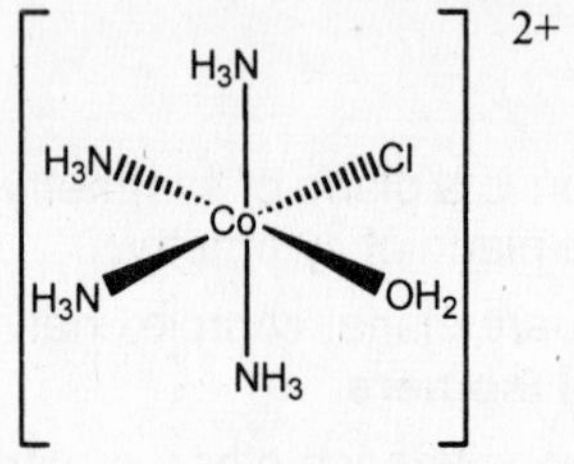

cis

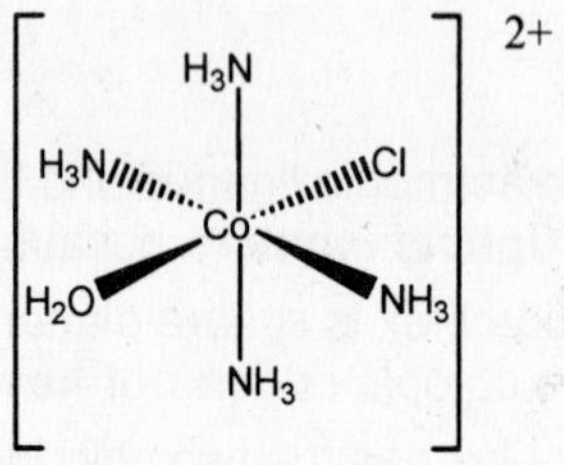

trans

$[Cr(H_2O)_3Br_2Cl]$

Name is **triaquadibromochlorochromium(III).**

Isomers are **geometric**:

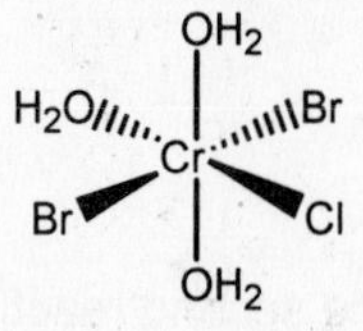

bromines are trans

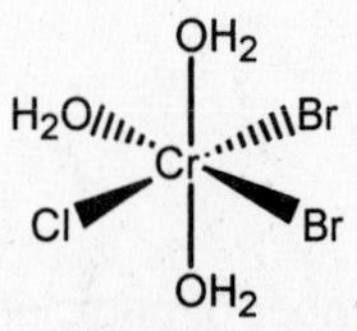

bromines are cis
waters are cis & trans

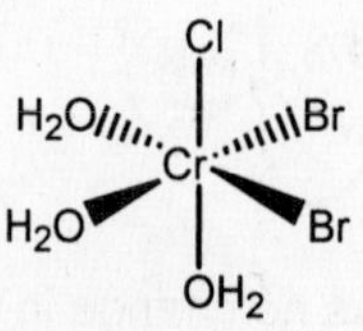

bromines and waters are cis

$[Cr(NH_3)_2(H_2O)_2Br_2]^{2+}$

Name is **diamminediaquadibromochromium(III).**

Isomers are **geometric and optical**:

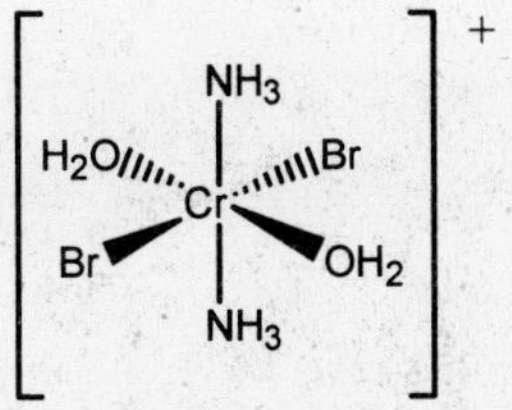

all ligands are trans

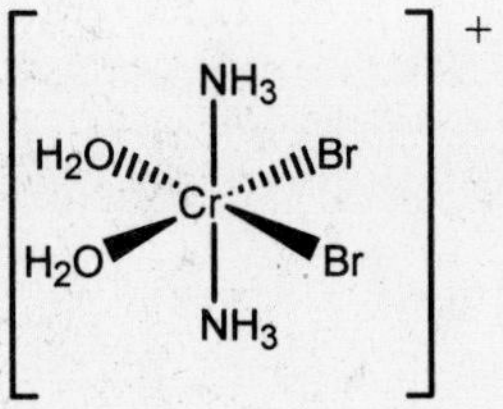

only NH_3 is trans

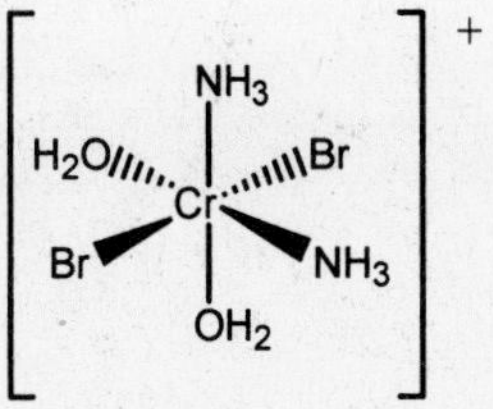

only Br is trans

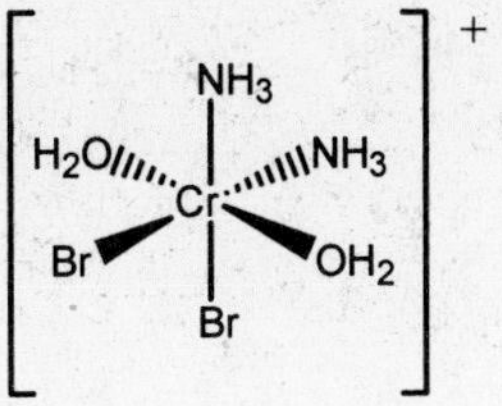

only H_2O is trans

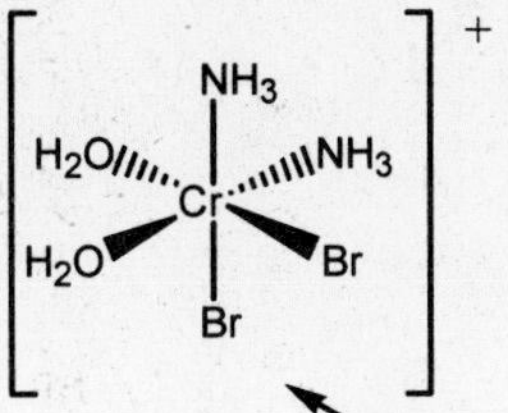

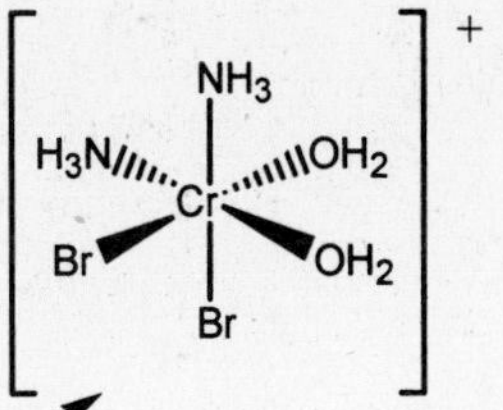

optical isomers
all 3 ligands are cis

23.128 a) The oxidation state of Mn in MnO_4^{2-} is **+6**, in MnO_4^- is **+7**, and in MnO_2 is **+4**.

b) The two half-reactions are

reduction:	$MnO_4^{2-}(aq) \rightarrow MnO_2(s)$
balancing gives:	$MnO_4^{2-}(aq) + 4H^+(aq) + 2e^- \rightarrow MnO_2(s) + 2H_2O(l)$
oxidation:	$MnO_4^{2-}(aq) \rightarrow MnO_4^-(aq) + e^-$

Multiply oxidation half-reaction by 2 and add to get

$\mathbf{3MnO_4^{2-}(aq) + 4H^+(aq) \rightarrow 2MnO_4^-(aq) + MnO_2(s) + 2H_2O(l)}$

CHAPTER 24

NUCLEAR REACTIONS AND THEIR APPLICATIONS

FOLLOW-UP PROBLEMS

24.1 Plan: Write a skeleton equation that shows an unknown nuclide, ${}^{A}_{Z}X$, undergoing β decay, ${}^{0}_{-1}\beta$, to form cesium-133, ${}^{133}_{55}Cs$. Conserve mass number and atomic number by ensuring the superscripts and subscripts equal one another on both sides of the equation. Determine the identity of X by using the periodic table to identify the element with atomic number equal to Z.

Solution: The unknown nuclide yields cesium-133 and a β particle:

$${}^{A}_{Z}X \rightarrow {}^{133}_{55}Cs + {}^{0}_{-1}\beta$$

To conserve atomic number, Z must equal 54. Element is xenon.

To conserve mass number, A must equal 133.

The identity of ${}^{A}_{Z}X$ is ${}^{133}_{54}Xe$.

The balanced equation is: ${}^{133}_{54}Xe \rightarrow {}^{133}_{55}Cs + {}^{0}_{-1}\beta$

Check: A = 133 = 133 + 0 and Z = 54 = 55+ (-1), so mass is conserved.

24.2 Plan: Nuclear stability is found in nuclides with an N/Z ratio that falls within the band of stability in Figure 24.2. Nuclides with an even N and Z, especially those nuclides that have magic numbers, are exceptionally stable. Examine the two nuclides to see which of these criteria can explain the difference in stability.

Solution: Phosphorus-31 has 16 neutrons and 15 protons, with an N/Z ratio of 1.07. Phosphorus-30 has 15 neutrons and 15 protons, with an N/Z ratio of 1.00. An N/Z ratio of 1.00 when Z < 20 could indicate a stable nuclide, but Figure 24.2 indicates a departure from the 1:1 line, even with smaller nuclides. **^{31}P** has an even N and a slightly higher N/Z ratio, accounting for its stability over ^{30}P.

24.3 Plan: Examine the N/Z ratio and determine which mode of decay will yield a product nucleus that is closer to the band.

Solution:

a) Iron-61 has an N/Z ratio of (61-26)/26 = 1.35, which is too high for this region of the band. Iron-61 will undergo **β decay**.

$${}^{61}_{26}Fe \rightarrow {}^{61}_{27}Co + {}^{0}_{-1}\beta \qquad \text{new N/Z} = (61\text{-}27)/27 = 1.26$$

b) Americium-241 has Z > 83, so it undergoes **α decay**.

$${}^{241}_{95}Am \rightarrow {}^{237}_{93}Np + {}^{4}_{2}He$$

24.4 Plan: Use the half-life of ^{24}Na to find *k*. Substitute the value of *k*, initial activity (A_0), and time of decay (4 days) into the integrated first-order rate equation to solve for activity at a later time (A_t).

Solution: $k = \ln 2/t_{1/2} = \ln 2/15\text{ hr} = 0.046_{21}\text{ hr}^{-1}$.

$\ln A_0 - \ln A_t = kt$

$\ln A_t = -kt + \ln A_0 = -(0.046_{21}\text{ hr}^{-1})(4\text{ days})(24\text{ hr/day}) + \ln(2.5 \times 10^9) = 17._{20}$

$\ln A_t = 17._{20}$

A_t = **3.0 x 10^7 d/s**

Check: In 4 days, a little more than 6 half-lives (6.4) have passed. A quick calculation on the calculator shows that 2.5 x 10^9 ÷ 2 = 1.25 x 10^9 (first half-life). If the "=" key is pressed again, the calculator divides the display number by 2 to show 6.25 x 10^8 (second half-life). Press the "=" key four more times to calculate the third through sixth half-lives to get 3.9 x 10^7 d/s. This number is slightly higher than the calculated answer because 6 half-lives have been calculated, not 6.4 half-lives.

24.5 Plan: The wood from the case came from a living organism, so A_0 equals 15.3 d/min·g. Substitute the current activity of the case (A_t), A_0 and the $t_{1/2}$ of carbon into the first- order rate expression and solve for t.

Solution: $t = \frac{1}{k}\ln\frac{A_0}{A_t} = \frac{1}{\ln 2 / t_{1/2}}\ln\frac{A_0}{A_t} = \frac{1}{\ln 2 / 5730\,yr}\ln\frac{15.3\,d/\min\cdot g}{9.41\,d/\min\cdot g} = 4018\,yr = \mathbf{4.02x10^3\ yr}$

Check: Since the activity of the bones is a little more than half of the modern activity, the age should be a little less than the half-life of carbon.

24.6 Plan: Uranium-235 has 92 protons and 143 neutrons in its nucleus. Calculate the mass defect (Δm) in one ^{235}U atom, convert to MeV and divide by 235 to obtain binding energy/nucleon.

Solution:

Δm = mass ^{235}U atom – [(92 x mass H atom) + (143 x mass neutron)]

Δm = 235.043924 amu – [(92 x 1.007825 amu) + (143 x 1.008665)]

Δm = | -1.915071 amu | = 1.915071 amu

$$\frac{binding\ energy}{nucleon} = \frac{(1.915071\,amu)(931.5\ MeV/amu)}{235\,nucleons} = \mathbf{7.591 MeV / nucleon}$$

The BE/nucleon of ^{12}C is 7.860 MeV/nucleon. The energy per nucleon holding the ^{235}U nucleus together is less than that for ^{12}C (7.591 < 7.860), so ^{235}U is less stable than ^{12}C.

Check: The answer is consistent with the expectation that uranium is less stable than carbon.

END-OF-CHAPTER PROBLEMS

24.1 a) Chemical reactions are accompanied by relatively small changes in energy while nuclear reactions are accompanied by relatively large changes in energy.

b) Increasing temperature increases the rate of a chemical reaction but has no effect on a nuclear reaction.

c) Both chemical and nuclear reaction rates increase with higher reactant concentrations.

d) If the reactant is limiting in a chemical reaction, then more reactant produces more product and the yield increases in a chemical reaction. The presence of more radioactive reagent results in more decay product, so a higher reactant concentration increases the yield in a nuclear reaction.

24.2 a) The percentage of sulfur atoms that are sulfur-32 is 95.02%, the same as the relative abundance of ^{32}S.

b) The atomic mass is larger than the isotopic mass of ^{32}S. Sulfur-32 is the lightest isotope, as stated in the problem, so the other 5% of sulfur atoms are heavier than 31.972070 amu. The average mass of all the sulfur atoms will therefore be greater than the mass of a sulfur-32 atom.

24.4 Refer to Table 24.2. Radioactive decay that produces a different element requires a change in *atomic number* (Z, number of protons). The *Table of Nuclides* in the CRC lists nuclides and their modes of decay.

a) Alpha decay produces an atom of a different element, i.e. a daughter with two less protons and two less neutrons.

Ex: $^{226}_{90}Th \rightarrow ^{222}_{88}Ra + ^{4}_{2}He$ (Z decreases by 2, N decreases by 2)

b) Beta decay produces an atom of a different element, i.e. a daughter with one more proton and one less neutron. A neutron is converted to a proton and β particle in this type of decay.

Ex: $^{78}_{33}As \rightarrow ^{78}_{34}Se + ^{0}_{-1}\beta$ (Z increases by 1, N decreases by 1)

c) Gamma decay does not produce an atom of a different element and Z and N remain unchanged.

d) Positron emission produces an atom of a different element, i.e. a daughter with one less proton and one more neutron. A proton is converted into a neutron and positron in this type of decay.

Ex: $^{90}_{43}Tc \rightarrow ^{90}_{42}Mo + ^{0}_{1}\beta$ (Z decreases by 1, N increases by 1)

e) Electron capture produces an atom of a different element, i.e. a daughter with one less proton and one more neutron. The net result of electron capture is the same as positron emission, but the two processes are different.

Ex: $^{128}_{56}Ba + ^{0}_{-1}e \rightarrow ^{128}_{55}Cs + h\nu$ (Z decreases by 1, N increases by 1)

24.6 A neutron-rich nuclide decays to convert neutrons to protons while a neutron-poor nuclide decays to convert protons to neutrons. The conversion of neutrons to protons occurs by beta decay:

$$^{1}_{0}n \rightarrow ^{1}_{1}p + ^{0}_{-1}\beta$$

The conversion of protons to neutrons occurs by either positron decay:

$$^{1}_{1}p \rightarrow ^{1}_{0}n + ^{0}_{1}\beta$$

or electron capture:

$$^{1}_{1}p + ^{0}_{-1}e \rightarrow ^{1}_{0}n$$

24.8 a) $^{234}_{92}U \rightarrow ^{230}_{90}Th + ^{4}_{2}He$

b) $^{232}_{93}Np + ^{0}_{-1}e \rightarrow ^{232}_{92}U$

c) $^{12}_{7}N \rightarrow ^{12}_{6}C + ^{0}_{1}\beta$

24.10 a) Process converts a neutron to proton so mass number is the same, but atomic number increases by one.

$$^{27}_{12}Mg \rightarrow ^{27}_{13}Al + ^{0}_{-1}\beta$$

b) Neutron emission decreases mass number, but not atomic number.

$$^{9}_{3}Li \rightarrow ^{8}_{3}Li + ^{1}_{0}n$$

c) The electron captured by the nucleus combines with a proton to form a neutron, so mass number is constant, but atomic number decreases by one.

$$^{103}_{46}Pd + ^{0}_{-1}e \rightarrow ^{103}_{45}Rh$$

24.12 a) In other words, an unknown nuclide decays to give Ti-48 and a positron.

$$^{A}_{Z}X \rightarrow ^{48}_{22}Ti + ^{0}_{1}\beta \text{ where } ^{A}_{Z}X = ^{48}_{23}V$$

b) In other words, an unknown nuclide captures an electron to form Ag-107.

$$^{A}_{Z}X + ^{0}_{-1}e \rightarrow ^{107}_{47}Ag \text{ where } ^{A}_{Z}X = ^{107}_{48}Cd$$

c) In other words, an unknown nuclide decays to give Po-206 and an alpha particle.

$$^{A}_{Z}X \rightarrow ^{206}_{84}Po + ^{4}_{2}He \text{ where } ^{A}_{Z}X = ^{210}_{86}Rn$$

24.14 a) $^{186}_{78}Pt + ^{0}_{-1}e \rightarrow ^{186}_{77}Ir$

b) $^{225}_{89}Ac \rightarrow ^{221}_{87}Fr + ^{4}_{2}He$

c) $^{129}_{52}Te \rightarrow ^{129}_{53}I + ^{0}_{-1}\beta$

24.16 The nuclide in (a) appears stable because its N and Z value are both magic numbers, but its N/Z ratio (1.50) is too high; nuclide (a) is unstable. The nuclide in (b) might look unstable because it Z value is an odd number, but its N/Z ratio (1.19) is in the band of stability, so nuclide (b) is stable. Nuclide (c) is unstable because its N/Z ratio (2.00) is too high.

24.18 For stable nuclides of elements with atomic number greater than 20, the ratio of number of neutrons to number of protons (N/Z) is greater than one. In addition, the ratio increases gradually as atomic number increases.

a) For the element iodine Z = 53. For iodine-127, N = 127 – 53 = 74. The N/Z ratio for ^{127}I is 74/53 = 1.4. Of the examples of stable nuclides given in the book ^{107}Ag has the closest atomic number to iodine. The N/Z ratio for ^{107}Ag is 1.3. Thus, it is likely that iodine with 6 additional protons is **stable** with an N/Z ratio of 1.4.

b) Tin is element number 50 (Z = 50). From part a) the stable nuclides of tin would have N/Z ratios approximately 1.3 to 1.4. N/Z ratio for ^{106}Sn is (106-50)/50 = 1.1. The nuclide ^{106}Sn is **unstable** with an N/Z ratio too small to be stable.

c) For ^{68}As, Z = 33 and N = 68-33 = 35 and N/Z = 1.1. Ratio is possible, but the nuclide is most likely **unstable** because there are an odd number of both protons and neutrons.

24.20 a) Nuclides with Z > 83 decay through **α decay**.

b) The N/Z ratio for Cr-48 is (48 – 24)/24 = 1.00. This number is below the band of stability because N is too low and Z is too high. To become more stable, the nucleus decays by converting a proton to a neutron, which is **positron decay**. Alternatively, a nucleus can capture an electron and convert a proton into a neutron through **electron capture**.

c) The N/Z ratio for Mn-50 is (50 – 25)/25 = 1.00. This number is also below the band of stability, so the nuclide undergoes **positron decay** or **electron capture**.

24.22 a) For carbon-15 N/Z = 9/6, so the nuclide is neutron-rich. To decrease the number of neutrons and increase the number of protons carbon-15 decays by **beta decay**.

b) The N/Z ratio for ^{120}Xe is 66/54 = 1.2. Around atomic number 50 the ratio for stable nuclides is larger than 1.2, so ^{120}Xe is proton-rich. To decrease the number of protons and increase the number of neutrons the xenon-120 nucleus either undergoes **positron emission** or **electron capture**.

c) Thorium-224 has an N/Z ratio of 134/90 = 1.5. All nuclides of elements above atomic number 83 are unstable and decay to decrease the number of both protons and neutrons. **Alpha decay** by thorium-224 is the most likely mode of decay.

24.24 Stability results from a favorable N/Z ratio, even numbers of N and/or Z, and the occurrence of magic numbers. The N/Z ratio of Cr-52 is 1.17, within the band of stability. The fact that Z is even doesn't account for the variation in stability because all isotopes of chromium have the same Z. However, Cr-52 has 28 neutrons, so N is both an even number and a magic number for this isotope only.

24.28 The equation for the nuclear reaction is

$$^{235}_{92}U \rightarrow {}^{207}_{82}Pb + \_\_{}^{0}_{-1}\beta + \_\_{}^{4}_{2}He$$

To determine the coefficients notice that the beta particles will not impact the mass number. Subtracting the mass number for lead from the mass number for uranium will give the total mass number for the alpha particles released, 235 – 207 = 28. Each alpha particle is a helium nucleus with mass number 4. The number of helium atoms is determined by dividing the total mass number change by 4, 28 ÷ 4 = 7 helium atoms or 7 alpha particles. The equation is now

$$^{235}_{92}U \rightarrow {}^{207}_{82}Pb + \_\_{}^{0}_{-1}e + 7\,{}^{4}_{2}He$$

To find the number of beta particles released examine the difference in number of protons (atomic number) between the reactant and products. Uranium, the reactant, has 92 protons. Atomic number in the products, lead atom and 7 helium nuclei, total 96. To balance the atomic numbers, four electrons (beta particles) must be emitted to give the total atomic number for the products as 96 – 4 = 92, the same as the reactant.

In summary, **7 alpha particles and 4 beta particles** are emitted in the decay of uranium-235 to lead-207.

24.31 No, it is not valid to conclude that $t_{1/2}$ equals 1 minute because the number of nuclei is so small (6 nuclei). Decay rate is an average rate and is only meaningful when the sample is macroscopic and contains a large number of nuclei, as in the second case. Because the second sample contains 6 x 10^{12} nuclei, the conclusion that $t_{1/2}$ = 1 minute is valid.

24.33 Specific activity of a radioactive sample is its decay rate per gram. Calculate the specific activity from the number of particles emitted per second and the mass of the sample.

$$\frac{decay\,rate}{mass} = \left(\frac{1.66\,x\,10^{6}\ dps}{1.55\,mg}\right)\left(\frac{1000\,mg}{1\,g}\right)\left(\frac{1\,Ci}{3.70\,x\,10^{10}\ dps}\right) = \mathbf{2.89x10^{-2}\ Ci/g}$$

24.35 A becquerel is a disintegration per second (d/s). A disintegration can be any of the decay particles mentioned in Section 24.1 (α, β, positron, etc.). Therefore, the emission of one α particle equals one disintegration.

$$specific\ activity\,(Bq/g) = \frac{(7.4x10^{4}\ d/\min)(1\min/60\sec)}{8.58\ \mu g(1\,g/10^{6}\ \mu g)} = 1.4x10^{8}\ d/s\cdot g = \mathbf{1.4x10^{8}\ Bq/g}$$

24.37 Decay constant is the rate constant for the first order reaction:

$$rate = -\frac{\Delta N}{\Delta t} = k[N]$$

$$k = \left(-\frac{1\,atom\,disintegrates}{1\,h}\right)\left(\frac{1}{1\,x\,10^{12}\,atoms}\right) = \mathbf{1x10^{-12}\,h^{-1}}$$

24.39 The rate constant, *k*, relates the number of radioactive nuclei to their decay rate through the equation A = *k*N. The number of radioactive nuclei is calculated by converting moles to atoms using Avogadro's number. The decay rate is 1.39 x 10^5 d/yr or more simply, 1.39 x 10^5 yr^{-1} (the disintegrations are assumed).

$$k = \frac{A}{N} = \frac{1.39x10^5\,yr^{-1}}{1.00x10^{-12}\,mol(6.022x10^{23}\,atoms/mol)} = \mathbf{2.31x10^{-7}\,yr^{-1}}$$

24.41 Radioactive decay is a first-order process so the integrated rate law is lnN_0 – lnN_t = kt. To calculate the fraction of bismuth-212 remaining after 3.75 x 10^3 h, first find the value of k from the half-life then calculate the fraction remaining with A_0 set to 1.

$$k = \frac{\ln 2}{t_{1/2}} = \frac{\ln 2}{1.01\,yr} = 0.686_{28}\,yr^{-1}$$

$$\ln N_t = \ln(1) - (0.686_{28}\,yr^{-1})(3.75\,x\,10^3\,h)\left(\frac{1\,day}{24\,h}\right)\left(\frac{1\,yr}{365\,days}\right) = -0.293_{78}$$

The fraction remaining equals $e^{-0.29378}$ = 0.745_{44}. Multiplying this fraction by the initial mass, 2.00 mg, gives **1.49 mg** of bismuth-212 remaining after 3.75 x 10^2 h.

24.43 Lead-206 is a stable daughter of ^{238}U. Since all of the ^{206}Pb came from ^{238}U, the starting amount of ^{238}U was (270 μmol + 110 μmol) = 380 μmol = N_0. The amount of ^{238}U at time t (current) is 270 μmol = N_t. Substitute these values into the first-order rate expression, substitute ln2/$t_{1/2}$ for *k*, and solve for t.

$$\ln\frac{N_t}{N_0} = -kt$$

$$\ln\frac{N_t}{N_0} = -\left(\frac{\ln 2}{t_{1/2}}\right)t$$

$$\ln\frac{(270\,\mu mol\,^{238}U)}{(270+110)\,\mu mol\,^{238}U} = -\left(\frac{\ln 2}{4.5x10^9\,yr}\right)t$$

$$-0.34_{17} = -(1.5_{40}x10^{-10}\,yr^{-1})t$$

t = **2.2x10⁹ yr**

24.45 Use the conversion factor, 1 Ci = 3.70 x 10^{10} disintegrations per second.

$$(1.0\,qt\,milk)\left(\frac{1\,L}{1.057\,qt}\right)\left(\frac{1000\,mL}{1\,L}\right) = 946\,mL$$

$$(946\,mL)\left(\frac{6\,x\,10^{-11}\ mCi}{1\,mL\ milk}\right)\left(\frac{1\,Ci}{1000\,mCi}\right)\left(\frac{3.70\,x\,10^{10}\ d/s}{1\,Ci}\right)\left(\frac{60\,s}{1\,min}\right)=\mathbf{1x10^2\ d/min}$$

24.47 Both N_t and N_0 are given: The number of nuclei present currently, N_t, is found from the moles of ^{232}Th. Each fission track represents one nucleus that disintegrated, so the number of nuclei disintegrated is added to the number of nuclei currently present to determine the initial number of nuclei, N_0. The rate constant, *k*, is calculated from the half-life. All values are substituted into the first-order decay equation to find t.

N_t: $$(2.15x10^{-15}\ mol\ ^{232}Th)\left(\frac{6.022x10^{23}\ nuclei\ ^{232}Th}{mol\ ^{232}Th}\right)=1.2_{646}x10^{9}\ nuclei\ ^{232}Th$$

N_0: $$9.5x10^{4}\ tracks\left(\frac{1\,nuclei}{1\,track}\right)+1.2_{646}x10^{9}\ nuclei=1.2_{647}x10^{9}\ nuclei$$

k: $$k=\ln 2/t_{1/2}=\ln 2/1.4x10^{10}\ yr=4.9_{511}x10^{-11}\ yr^{-1}$$

$$\ln\frac{N_t}{N_0}=-kt\ \Rightarrow\ t=-\frac{\ln\frac{N_t}{N_0}}{k}=-\frac{\ln\frac{1.2_{646}x10^{9}}{1.2_{647}x10^{9}}}{4.9_{511}x10^{-11}\ yr^{-1}}=\mathbf{1.6x10^{6}\ yr}$$

24.50 Both gamma radiation and neutron beams have no charge so neither is deflected by electric or magnetic fields. Neutron beams differ from gamma radiation in that a neutron has mass approximately equal to that of a proton. Researchers observed that a neutron beam could induce the emission of protons from a substance. Gamma rays do not cause such emissions.

24.52 Protons are repelled from the target nuclei due to the interaction of like, positive charges. Higher energy is required to overcome the repulsion.

24.53 a) Alpha particle is reactant with ^{10}B and a neutron is one product. The mass number for reactants is 10 + 4 = 14. So the missing product must have mass number 14 – 1 = 13. The total atomic number for reactants is 5 + 2 = 7, so the atomic number for the missing product is 7.

^{10}B (α,n) ^{13}N

$$^{10}_{5}B+\,^{4}_{2}He\rightarrow\,^{1}_{0}n+\,^{13}_{7}N$$

b) For reactants mass number is 28 + 2 = 30 and atomic number is 14 + 1 = 15. The given product has mass number 29 and atomic number 15, so the missing product particle has mass number 1 and atomic number 0. The particle is thus a neutron.

^{28}Si (d,n) ^{29}P

$$^{28}_{14}Si+\,^{2}_{1}H\rightarrow\,^{1}_{0}n+\,^{29}_{15}P$$

c) For products mass number is 2 + 244 = 246 and atomic number is 98. The given reactant particle is an alpha particle with mass number 4 and atomic number 2. The missing reactant must have mass number of 246 – 4 = 242 and atomic number 98 – 2 = 96. Element 96 is Cm.

^{242}Cm (α, 2n) ^{244}Cf

$$^{242}_{96}Cm+\,^{4}_{2}He\rightarrow 2\,^{1}_{0}n+\,^{244}_{98}Cf$$

24.58 Ionizing radiation is more dangerous to children because their rapidly dividing cells are more susceptible to radiation than an adult's slower dividing cells.

24.60 a) The rad is the amount of radiation energy absorbed per body mass in kg.

$$\left(\frac{3.3 x 10^{-7}\ J}{135 lb}\right)\left(\frac{2.205 lb}{1 kg}\right)\left(\frac{1 rad}{1 x 10^{-2}\ J/kg}\right) = \mathbf{5.4 x 10^{-7}\ rad}$$

b) Conversion factor is 1 rad = 0.01 Gy

(5.4 x 10^{-7} rad)(0.01 Gy/1 rad) = **5.4 x 10^{-9} Gy**

24.62 a) Convert the given information to units of J/kg.

$$gray\,(J/kg) = \frac{(6.0x10^{5}\ \beta)(8.74x10^{-14}\ J/\beta)}{70.\ kg} = \mathbf{7.5x10^{-10}\ Gy}$$

b) Convert grays to rads and multiply rads by RBE to find rems. Convert rems to mrems.

rem = rads x RBE = (7.5 x 10^{-10} Gy)(1 rad/0.01 Gy)(1.0) = 7.5 x 10^{-8} rem

mrem = (7.5 x 10^{-8} rem)(1000 mrem/rem) = **7.5 x 10^{-5} mrem**.

c) Sv = (7.5 x 10^{-8} rem)(0.01 Sv/rem) = **7.5 x 10^{-10} Sv**.

24.65 Use the time and disintegrations per second (Bq) to find the number of ^{60}Co atoms that disintegrate which equals the number of β particles emitted.

(27.0 min)(60 s/min)(475 Bq)(1 d·s^{-1}/1 Bq) = 7.69$_{50}$ x 10^{5} disintegrations

The dose in rads is calculated as energy absorbed per body mass.

(7.69$_{50}$ x 10^{5} β particles/1.588 g)(5.05 x 10^{-14} J/particle)(1000 g/kg) = 2.45 x 10^{-5} J/kg

Since 1 rad = 0.01J/kg, the dose is **2.45 x 10^{-3} rad**.

24.67 NAA does not destroy the sample like other chemical analyses do. Neutrons bombard a non-radioactive sample, "activating" or energizing individual atoms within the sample to create radioisotopes. The radioisotopes decay back to their original state (thus the sample is not destroyed) by emitting radiation that is different for each isotope.

24.70 The oxygen in formaldehyde comes from **methanol** because the oxygen isotope in the methanol reactant ends up in the formaldehyde product. The oxygen isotope in the chromic acid reactant appears in the water product, not the formaldehyde product. The isotope traces the oxygen in methanol to the oxygen in formaldehyde.

24.73 Energy is released when a nuclide forms from nucleons. This is the reverse process shown in Section 24.6, i.e. nucleons → nucleus + energy. The nuclear binding energy is the amount of energy holding the nucleus together. Energy is absorbed to break the nucleons apart and is released when nucleons "come together".

24.75 a) $$(0.01861\ MeV)\left(\frac{1\,eV}{1 x 10^{-6}\ MeV}\right) = \mathbf{1.861x10^{4}\ eV}$$

b) $$(1.861 x 10^{4}\ eV)\left(\frac{1.602 x 10^{-19}\ J}{1\,eV}\right) = \mathbf{2.981x10^{-15}\ J}$$

24.77 Convert moles of ^{239}Pu to atoms of ^{239}Pu using Avogadro's number. Multiply number of atoms by energy/nucleus and convert MeV to J using the conversion factor 1 MeV = 1.602 x 10^{-13} J.

$$energy\,(J) = \left(1.0\,mol\ ^{239}Pu\right)\left(\frac{6.022x10^{23}\ atoms}{mol}\right)\left(\frac{5.243\ MeV}{atom}\right)\left(\frac{1.602x10^{-13}\ J}{MeV}\right) = \mathbf{5.1x10^{11}\ J}$$

One mol of ^{239}Pu, or ~240 g ^{239}Pu releases about 100 million kilojoules!

24.79 The mass defect is calculated from the mass of 8 protons and 8 neutrons vs. the mass of the oxygen nuclide.

Δm = mass ^{16}O atom – [(8 x mass H atom) + (8 x mass neutron)]

Δm = 15.994915 amu – [(8 x 1.007825 amu) + (8 x 1.008665 amu)] = -0.137005 amu

Δm = |-0.137005 amu| = 0.137005 amu

a) $\Delta E = \left(\frac{0.137005\,amu}{16\,nucleons}\right)\left(\frac{931.5\ MeV}{1\,amu}\right) = \mathbf{7.976\ MeV\ /\ nucleon}$

b) $\Delta E = \left(\frac{0.137005\,amu}{^{16}O\ atom}\right)\left(\frac{931.5\ MeV}{1\,amu}\right) = \mathbf{127.6\ MeV\ /\ atom}$

c) $\Delta E = mc^2 = (0.137005\,g\,/\,mol)\left(\frac{1\,kg}{1000\,g}\right)(2.9979\,x\,10^8\ m\,/\,s)^2 = 1.2313\,x\,10^{13}\ J\,/\,mol$

In kJ/mol the binding energy is **1.2313 x 10^{10} kJ/mol**

An alternate method for calculating BE per mole is found in 24.81c).

24.81 a) Calculate Δm, convert mass deficit to MeV, and divide by 59 nucleons.

Δm = mass ^{59}Co atom – [(27 x mass H atom) + (32 x mass neutron)]

Δm = 58.933198 amu – [(27 x 1.007825 amu) + (32 x 1.008665 amu)]

Δm = |-0.555357 amu| = 0.555357 amu

$$\frac{binding\ energy}{nucleon} = \frac{(0.555357\ amu)(931.5\ MeV\ /\ amu)}{59\ nucleons} = \mathbf{8.768\ MeV\ /\ nucleon}$$

b) The Δm calculated in (a) is the mass deficit per atom. This is converted to binding energy per atom using the 913.5 MeV/amu conversion factor.

$$\frac{binding\ energy}{atom} = (0.555357\ amu\ /\ atom)(931.5\ MeV\ /\ amu) = \mathbf{517.3\ MeV\ /\ atom}$$

c) Convert atoms to moles using Avogadro's number, and convert MeV to kJ.

$$\frac{binding\ energy}{mol} = (0.555357\ amu\ /\ atom)(931.5\ MeV\ /\ amu)$$

$$x(1.602x10^{-13}\ J\ /\ MeV)(6.022x10^{23}\ atoms\ /\ mol)\left(\frac{1kJ}{1000J}\right) = \mathbf{4.991x10^{10}\ kJ\ /\ mol}$$

An alternate method for calculating BE per mole is found in 24.79c).

24.85 In both radioactive decay and fission radioactive particles are emitted, but the process leading to the emission is different in the two cases. Radioactive decay is a spontaneous process in which unstable nuclei emit radioactive particles and energy. Fission occurs as the result of high-energy bombardment of nuclei with small particles that cause the formation of smaller nuclides, radioactive particles and energy.

In a chain reaction all fission events are not the same. The collision between the small particle emitted in the fission and the large nucleus can lead to splitting the large nuclei in a number of ways to produce several different products.

24.88 The water serves to moderate the energy of the neutrons so that they are better able to cause a fission reaction. Heavy water (${}^{2}_{1}H_2O$ or D_2O) is a better moderator because it does not absorb neutrons as well as light water (${}^{1}_{1}H_2O$) does, so more neutrons are available to initiate the fission process. However, D_2O does not occur naturally in great abundance so production of D_2O adds to the cost of a heavy water reactor. In addition, if heavy water does absorb a neutron, it becomes *tritiated*, i.e. it contains the isotope tritium, ${}^{3}_{1}H$, which is radioactive.

24.91 In the stellar nucleogenesis model, the lighter nuclides (up to ${}^{24}Mg$) are produced in a "helium burning" stage where continual heavier nuclei are formed by the fusion of helium nuclei which have a mass of 4 amu.

24.96 a) First determine the amount of activity released by the ${}^{239}Pu$ for the duration spent in the body (16 hr) using the relationship A = *k*N. The rate constant is derived from the half-life and N is calculated using the molar mass and Avogadro's number.

$$k = \frac{\ln 2}{t_{1/2}} = \left(\frac{\ln 2}{2.41x10^4\ yr}\right)\left(\frac{yr}{365.25\ day}\right)\left(\frac{1\ day}{24\ hr}\right)\left(\frac{hr}{3600\ s}\right) = 9.11_{39}x10^{-13}\ s^{-1}$$

$$N = \left(1.00\mu g\ {}^{239}Pu\right)\left(\frac{1\ g}{10^6\ \mu g}\right)\left(\frac{mol\ {}^{239}Pu}{239\ g\ {}^{239}Pu}\right)\left(\frac{6.022x10^{23}\ atoms\ {}^{239}Pu}{mol\ {}^{239}Pu}\right) = 2.51_{97}x10^{15}\ atoms$$

$$A = kN = \left(9.11_{39}x10^{-13}\ s^{-1}\right)\left(2.51_{97}x10^{15}\right) = 2.29_{64}x10^3\ d/s$$

Each disintegration releases 5.15 MeV, so d/s can be converted to MeV. Convert MeV to J (using 1.602 x 10^{-13} J = 1 MeV) and J to rad (using 0.01 J/kg = 1 rad).

$$E(rad) = \left(\frac{2.29_{64}x10^3\ d/s}{85\ kg}\right)\left(\frac{5.15\ MeV}{d}\right)\left(\frac{1.602x10^{-13}\ J}{MeV}\right)\left(\frac{1\ rad}{0.01\ J/kg}\right) = 2.22_{89}x10^{-9}\ rad/s$$

Over 16 hours, the exposure is

$$E(rad) = \left(2.22_{89}x10^{-9}\ rad/s\right)\left(16\ hr\right)\left(3600\ s/hr\right) = \mathbf{1.28x10^{-4}\ rad}$$

b) Since 0.01 Gy=1 rad, the worker receives (1.28x10^{-4} rad)(0.01 Gy/rad) = **1.28 x 10^{-6} Gy**.

24.98 Find the rate constant from the rate of decay and the initial number of atoms. Use rate constant to calculate half-life.

Initial number of atoms:

$$\left(5.4\ \mu g\ {}^{226}RaCl_2\right)\left(\frac{1\ g}{1x10^6\ \mu g}\right)\left(\frac{1\ mol\ {}^{226}RaCl_2}{297\ g\ {}^{226}RaCl_2}\right)\left(\frac{1\ mol\ {}^{226}Ra}{1\ mol\ {}^{226}RaCl_2}\right)\left(\frac{6.022\ x\ 10^{23}\ atoms}{mol}\right)$$

$$= 1.0_{95}\ x\ 10^{16}\ atoms\ of\ {}^{226}Ra$$

$$k = \frac{1.5 x 10^5\ Bq}{1.0_{95}\ x\ 10^{16}\ {}^{226}Ra\ atoms}\left(\frac{1\ d\ s^{-1}}{1\ Bq}\right) = 1.3_{70}\ x\ 10^{-11}\ s^{-1}$$

$t_{½} = \ln 2/k = 0.693/1.3_{70} \times 10^{-11}\ s^{-1}$ = **5.1×10^{10} s**

24.101 Determine how many grams of AgCl are dissolved in 1 mL of solution. The activity of the radioactive Ag^+ indicates how much AgCl dissolved, given a starting sample with a specific activity (175 nCi/g).

$$Conc\,(g/mL) = \left(\frac{1.25x10^{-2}\ Bq\ Ag^+}{mL}\right)\left(\frac{1\,Ci}{3.70x10^{10}\ Bq}\right)\left(\frac{10^9\ nCi}{1\,Ci}\right)\left(\frac{g\ AgCl}{175\,nCi}\right) = 1.93_{05}x10^{-6}\ g\ AgCl/mL$$

Convert g/mL to mol/L (molar solubility) using the molar mass of AgCl.

$M(mol/L) = \left(1.93_{05}x10^{-6}\ g\ AgCl/mL\right)\left(1\,mol\ AgCl/143.35\,g\right)\left(1000\,mL/L\right)$ = **1.35×10^{-5} M**

24.103 $k = \ln 2/(7.0 \times 10^8\ yr) = 9.9_{02} \times 10^{-10}\ yr^{-1}$

$\ln(N_t/N_0) = -kt = -(9.9_{02} \times 10^{-10}\ yr^{-1})(2.8 \times 10^9\ yr) = -2.7_{73}$

$N_t/N_0 = e^{-2.773}$ = **0.063**

24.106 a) Find the rate constant, *k*, using any two data pairs. Calculate $t_{1/2}$ using *k*.

$$\ln\frac{N_t}{N_0} = -kt$$

$$\ln\frac{2000}{5000} = -k(8\,hr) \Rightarrow k = 0.114_{54}\ hr^{-1}\ \ (assume\ time\ is\ exact\ and\ \gamma\ rate\ is\ known\ to\ 3\,SF's)$$

$t_{1/2} = \ln 2/k = 0.693/0.114\ hr^{-1}$ = **6.05 hr** ($t_{1/2}$ will vary slightly depending on which data pair is chosen)

b) The percentage of isotope *remaining* is the fraction remaining after 2.0 hr (A_t where t = 2.0 hr) divided by the initial amount (A_0), i.e. fraction remaining is A_t/A_0. Solve the first order rate expression for A_t/A_0, and then subtract from 100% to get fraction *lost*.

$$\ln\frac{N_t}{N_0} = -\left(0.114_{54}\ hr^{-1}\right)\left(2.0\,hr\right) = -0.23$$

$$\frac{N_t}{N_0} = 0.79$$

The fraction lost is 1.00 – 0.79 = 0.21, so **21%** of the isotope is lost upon preparation.

24.108 a) After 10 years, $fraction\ remaining = \left(\frac{1}{2}\right)^{\left(10.0\,yr/5730\,yr\right)}$ = **0.999**

b) After 10.0×10^3 years, $fraction\ remaining = \left(\frac{1}{2}\right)^{\left(10.0x10^3\,yr/5730\,yr\right)}$ = **0.298**

c) After 10.0×10^4 years, $fraction\ remaining = \left(\frac{1}{2}\right)^{\left(10.0x10^4\,yr/5730\,yr\right)}$ = **5.58×10^{-6}**

d) After 10 years, so little of the carbon-14 in the sample has decayed that is difficult to distinguish it from an undecayed sample. After 10.0 x 10^4 years (over 17 half lives), so little activity remains, making it difficult to measure reliably. In both cases, the relative error in determining age will be high. In sample (b), a significant portion of the carbon-14 has decayed (0.702), and a significant portion still remains (0.298). Therefore the error in determining age will be relatively small.

24.111 a) β decay by vanadium-52 produces **chromium-52**.

$${}^{51}_{23}V + {}^{1}_{0}n \rightarrow {}^{52}_{23}V \rightarrow {}^{52}_{24}Cr + {}^{0}_{-1}\beta$$

$${}^{51}_{23}V\,(n,\beta)\,{}^{52}_{24}Cr$$

b) Positron emission by copper-64 produces **nickel-64**.

$${}^{63}_{29}Cu + {}^{1}_{0}n \rightarrow {}^{64}_{29}Cu \rightarrow {}^{64}_{28}Ni + {}^{0}_{1}\beta$$

$${}^{63}_{29}Cu\left(n,\beta^{+}\right){}^{64}_{28}Ni$$

c) β decay by aluminum-28 produces **silicon-28**.

$${}^{27}_{13}Al + {}^{1}_{0}n \rightarrow {}^{28}_{13}Al \rightarrow {}^{28}_{14}Si + {}^{0}_{-1}\beta$$

$${}^{27}_{13}Al\,(n,\beta)\,{}^{28}_{14}Si$$

24.114 The *production rate* of radon gas (volume/hour) which is also the *decay rate* of ^{226}Ra. The decay rate, or activity, is proportional to the number of radioactive nuclei decaying, or the number of atoms in 1.000 g of ^{226}Ra, using the relationship A = *k*N. Calculate the number of atoms in the sample, and find *k* from the half-life. Convert the activity in units of nuclei/time (also disintegrations per unit time) to volume/time using the ideal gas law.

$$N = \left(1.000\,g\,{}^{226}Ra\right)\left(\frac{1\,mol\,{}^{226}Ra}{226.025402\,g\,{}^{226}Ra}\right)\left(\frac{6.022x10^{23}\,atoms}{1\,mol\,{}^{226}Ra}\right) = 2.664_{30}x10^{21}\,atoms\,{}^{226}Ra$$

$$k = \frac{\ln 2}{t_{1/2}} = \frac{\ln 2}{1599\,yr} = 4.334_{88}x10^{-4}\,yr^{-1}$$

$$A = kN = \left(4.334_{88}x10^{-4}\,yr^{-1}\right)\left(2.664_{30}x10^{21}\,atoms\,{}^{226}Ra\right)\left(\frac{1\,yr}{8766\,h}\right) = 1.317_{52}x10^{14}\,d/h$$

This result means that 1.317$_{52}$ x 10^{14} ^{226}Ra nuclei are decaying into ^{222}Rn nuclei every hour. Convert atoms of ^{222}Rn into volume of gas using the ideal gas law.

$$n = \left(1.317_{52}x10^{14}\,{}^{222}Rn\,atoms\right)\left(1\,mol/6.022x10^{23}\,atoms\right) = 2.187_{84}x10^{-10}\,mol$$

$$V = \frac{nRT}{P} = \frac{(2.187_{84}x10^{-10}\,mol)(0.08206\,L\cdot atm/mol\cdot K)(273.15\,K)}{1\,atm} = 4.904x10^{-9}\,L$$

Therefore, radon gas is produced at a rate of **4.904x10^{-9} L/h**. Note: Activity could have been calculated as decay in moles/time, removing Avogadro's number as a multiplication and division factor in the calculation.

24.117 The change from 8.0×10^4 β particles per month to 1.0×10^4 β particles per month requires 3 half- lives:

$$8.0 \times 10^4 \xrightarrow{1} 4.0 \times 10^4 \xrightarrow{2} 2.0 \times 10^4 \xrightarrow{3} 1.0 \times 10^4$$

The time equals 3 x $t_{½}$ = **87 yr.**

24.119 a) Convert pCi to Bq using the conversion factor 1 Ci = 3.70×10^{10} Bq and 10^{12} pCi = 1 Ci.

$$activity\,(Bq) = (4.0\ pCi)\left(\frac{1\,Ci}{10^{12}\ pCi}\right)\left(\frac{3.70x10^{10}\ Bq}{1\,Ci}\right) = 0.14_8\ Bq$$

The safe level is **0.15 Bq/L.**

b) Use the first-order rate expression to find the activity at the later time (t = 8.5 days).

$$\ln\frac{A_t}{A_0} = -kt$$

$$\ln\frac{A_t}{43.5\ pCi\ /\ L} = -\left(\frac{\ln 2}{3.82\ days}\right)(8.5\,days)$$

$$\frac{A_t}{43.5\ pCi\ /\ L} = 0.21_{39}$$

$$A_t = 9.3_{04}\ pCi\ /\ L$$

Using the conversion scheme in (a), 9.3_{04} pCi/L converts to **0.34 Bq/L.**

c) The desired activity is 0.15 Bq/L, however the room air currently contains 0.34 Bq/L. Rearrange the first-order rate expression to solve for time.

$$t = \frac{1}{k}\ln\left(\frac{A_0}{A_t}\right) = \frac{1}{\ln 2\ /\ 3.82\ days}\ln\left(\frac{0.34\ Bq\ /\ L}{0.15\ Bq\ /\ L}\right) = \mathbf{4.7\ days}$$

It takes 4.7 days more to reach the recommended EPA level. A total of 13.2 (4.7+ 8.5) days is required to reach recommended levels when the room was initially measured at 43.5 pCi/L.

24.122 The energy per time is calculated from the disintegrations per time (1.0 mCi) and the energy per disintegration (5.59 MeV).

$$(1.0\,mCi)\left(\frac{3.70\,x\,10^7\ dps}{1\,mCi}\right)\left(\frac{5.59\ MeV}{disintegration}\right)\left(\frac{1.602\,x\,10^{-13}\ J}{1\ MeV}\right) = 3.13_{34}\ x\,10^{-5}\ J\ /\ s$$

Convert the mass of the child to kg:

(48 lb)(1 kg/2.205 lb) = $21._{77}$ kg

Time for 1.0 mrad to be absorbed:

$$(1.0\,mrad)\left(\frac{1\,rad}{1000\,mrad}\right)\left(\frac{1\,x\,10^{-2}\ J\ /\ kg}{1\,rad}\right)\left(\frac{21._{77}\ kg}{3.31_{34}\ x\,10^{-5}\ J\ /\ s}\right) = \mathbf{6.6\ s}$$

24.125 Because the 1941 wine has a little over twice as much tritium in it, just over one half-life has passed between the two wines. Therefore the older wine was produced before 1929 (1941 – 12.26) but not much earlier than that. To find the number of years back in time, use the first order rate expression, where A_0 = 2.32A, A_t = A and t = years transpired between the manufacture date and 1941.

$$\ln\left(\frac{A_0}{A_t}\right) = kt$$

$$\ln\left(\frac{2.32A}{A}\right) = \left(\frac{\ln 2}{12.26\ yr}\right)t$$

$t = 14.89\ yr$

The wine was produced in (1941 – 15) = **1926**.

24.129 a) When 1.00 kg of antimatter annihilates 1.00 kg of matter the change in mass is $\Delta m = 0 - 2.00$ kg = -2.00 kg. The energy released is calculated from $\Delta E = \Delta mc^2$.

$\Delta E = (-2.00\ kg)(2.9979 \times 10^8\ m/s)^2 = -1.80 \times 10^{17}$ J or **1.80×10^{14} kJ is released**.

b) Assuming that four hydrogen atoms fuse to form the two protons and two neutrons in one helium atom, the energy released can be calculated from the binding energy of helium-4. Figure 24.12 shows a binding energy of 7.0 MeV per nucleon for ^{4}He.

(7.0 MeV/nucleon)(1.602×10^{-13} J/MeV) = $1.1_{21} \times 10^{-12}$ J per H atom fused.

Total energy released in each collision

= ($1.1_{21} \times 10^{-12}$ J/atom)(1.00×10^5 atoms) = $1.1_{21} \times 10^{-7}$ J/collision

1.00 kg of antihydrogen is 992 mol antihydrogen using atomic mass of antihydrogen equal to that for hydrogen. That means that in 1.00 kg there are

(6.022×10^{23} atoms/mol)(992 mol) = $5.97_{38} \times 10^{26}$ atoms

Each collision involves one antihydrogen atom so the energy released per kg antihydrogen is

($5.97_{38} \times 10^{26}$ atoms)($1.1_{21} \times 10^{-7}$ J/atom) = 6.7×10^{19} J or **6.7×10^{16} kJ**

c) From the above calculations the procedure in part b) with excess hydrogen produces more energy per kg of antihydrogen.

24.135 Einstein's equation is $E = mc^2$, which is modified to $E = \Delta mc^2$ to reflect a mass deficit. The speed of light, c, is 2.99792×10^8 m/s. The mass of exactly one amu is 1.66054×10^{-27} kg (inside back cover). When the quantities are multiplied together, the unit will be kg·m^2/s^2, which is also the unit of joules. Convert J to MeV using the conversion factor 1.602×10^{-13} J = 1 MeV.

$$E = \Delta mc^2 = \left(1.66054x10^{-27}\ kg\right)\left(2.99792x10^8\ m/s\right)^2\left(\frac{1\ J}{kg \cdot m^2/s^2}\right)\left(\frac{1\ MeV}{1.602x10^{-13}\ J}\right) = \mathbf{9.316x10^2\ MeV}$$

Notes

Notes

Notes

Notes

Notes

Notes

Notes

Notes

Notes

Notes

Notes

Notes

Notes

Notes